P9-AFZ-964

Graduate Texts in Contemporary Physics

Springer
New York
Berlin
Heidelberg
Barcelona
Budapest
Hong Kong
London
Milan
Paris
Santa Clara
Singapore
Tokyo

Graduate Texts in Contemporary Physics

R.N. Mohapatra: **Unification and Supersymmetry: The Frontiers of Quark-Lepton Physics, 2nd Edition**

R.E. Prange and S.M. Girvin (eds.): **The Quantum Hall Effect**

M. Kaku: **Introduction to Superstrings**

J.W. Lynn (ed.): **High-Temperature Superconductivity**

H.V. Klapdor (ed.): **Neutrinos**

J.H. Hinken: **Superconductor Electronics: Fundamentals and Microwave Applications**

M. Kaku: **Strings, Conformal Fields, and Topology: An Introduction**

A. Auerbach: **Interacting Electrons and Quantum Magnetism**

Yu.M. Ivanchenko and A.A. Lisyansky: **Physics of Critical Fluctuations**

P. Di Francesco, P. Mathieu, and D. Sénéchal: **Conformal Field Theories**

B. Felsager: **Geometry, Particles, and Fields**

Bjørn Felsager

Geometry, Particles, and Fields

With 219 Illustrations

Bjørn Felsager
Mathematics Department, Odense University
The Niels Bohr Institute, The University of Copenhagen
Haslev Gynasium & HF
Skolegade 31
DK-4690 Haslev
Denmark

Library of Congress Cataloging-in-Publication Data
Felsager, Bjørn.
Geometry, particles, and fields / Bjørn Felsager.
p. cm. — (Graduate texts in contemporary physics)
Includes index.
ISBN 0-387-98267-1 (hardcover : alk. paper)
1. Field theory (Physics) 2. Geometry, Differential.
3. Particles (Nuclear physics) I. Title. II. Series.
QC793.3.F5F44 1997
530.14—dc21 97-15252

Printed on acid-free paper.

Production managed by Anthony Guardiola; manufacturing supervised by Jacqui Ashri.
Typeset by Bartlett Press, Marietta, GA.
Printed and bound by Edwards Brothers, Inc., Ann Arbor, MI.
Printed in the United States of America.

9 8 7 6 5 4 3 2 1

ISBN 0-387-98267-1 Springer-Verlag New York Berlin Heidelberg SPIN 10629814

Preface

The present book is an attempt to present modern field theory in an elementary way. It is written mainly for students, and for this reason it presupposes little knowledge in advance except for a standard course in calculus (on the level of multiple integrals) and a standard course in classical physics (including classical mechanics, special relativity, and electrodynamics). The main emphasis is placed on the presentation of the central concepts, not on mathematical rigor. Hopefully this textbook will prove useful to high-energy physicists who want to get acquainted with the basic concepts of differential geometry. Mathematicians may also have fun reading about the application of central concepts from differential geometry in theoretical physics.

To set the stage I have in the first part included a self-contained introduction to field theory leading up to recent important concepts like solitons and instantons. There are two main themes in part one: On the one hand, I discuss the structure of gauge theory, exemplified by ordinary electromagnetism. This include a derivation of the Bohm–Aharonov effect and the flux quantization of magnetic vortices in a superconductor. On the other hand, I discuss the structure of nonlinear field theory, exemplified by the ϕ^4-model and the sine-Gordon model in $(1 + 1)$-dimensional space–time. This includes the construction of a topological charge, the particle interpretation of the kink-solution; and finally, the relevance of the kink-solution for the tunnel effect in quantum mechanics is pointed out. Although the present text deals mainly with the classical aspects of field theory, I have also touched the quantum-mechanical aspects using path-integral techniques.

In part two I have included a self-contained introduction to differential geometry. The main emphasis is placed on the so-called exterior calculus of differential forms, which permits on the one hand the construction of various differential operators—the exterior derivative, the codifferential, and the Laplacian—and on the other hand the construction of a covariant integral. But I also investigate metrics and various related concepts, especially Christoffel fields, geodesics, and conformal mappings.

Apart from an introduction to the basic concepts in differential geometry, the second part contains a number of illustrative applications. The Lagrangian formalism is put on covariant (i.e., geometrical) form. A detailed discussion of magnetic monopoles, including the Lagrangian formalism for monopoles and the quantization of magnetic charges, continues the investigation of gauge theories initiated in part one. Further examples of nonlinear models are presented: The Heisenberg ferromagnet, the exceptional ϕ^4-model and the abelian Higgs's model (including a discussion of the Nambu strings and their relationship with the Nielsen-Olesen vortices). Finally, symmetry transformations and their associated conservation laws (i.e., Noether's theorem) are investigated in great detail.

Acknowledgments

A project like this would never have been completed were it not for the moral and financial support of a great number of persons and institutions.

From the mathematics department, Odense University, I would especially like to thank my scientific advisor Erik Kjær Pedersen and Hans Jørgen Munkholm, who have followed the project through all its various stages. I am also grateful to Ole Hjort Rasmussen for many stimulating discussions about geometry.

From the Niels Bohr Institute I would like to thank my scientific advisor Poul Olesen (who originally suggested to me to take a closer look at the geometrical and topological structure of gauge theories and who encouraged me to give the lectures upon which this book is based). I am also grateful for moral support from Torben Huus.

Helge Kastrup Olesen looked over a preliminary version of the manuscript and taught me a lot about English grammar. Carsten Claussen has been of invaluable help to me. He has read the whole manuscript in several versions and has suggested innumerable improvements.

I would also like to thank the secretaries, Lisbeth Larsen at Odense University and Vera Rothenberg at the Niels Bohr Institute, who with great patience and professional skill typed major parts of the manuscript.

Finally, I would like to thank Odense Universitets publikationskonto for financial support for the printing of the manuscript, and Lørup and Holck's fonde (at the Niels Bohr Institute) for a generous donation for typing assistance.

In regard to the present corrected reprint of the first edition, I would like to thank the staff at Springer-Verlag for their generous support and cooperation. Especially I would like to thank Physics Editor Tom von Foerster for his never ending faith in the project (which eventually set the wheels in motion) and Senior Production Editor Anthony Guardiola for his excellent and competent steering through the intricate process of recreating the manuscript.

Haslev, Denmark Bjørn Felsager

Contents

PART I

Basic Properties of Particles and Fields

General References to Part I

R. P. Feynman, R. B. Leighton, and M. Sands, *The Feynman Lectures on Physics*, McGraw-Hill, New York (1965).

P. A. M. Dirac, *The Principles of Quantum Mechanics*, second edition, Oxford (1935).

L. I. Schiff, *Quantum Mechanics*, McGraw-Hill Kogakusha (1968).

de Gennes, *Superconductivity of Metals and Alloys*, W. A. Benjamin, New York (1966).

A. O. Barut, *Electrodynamics and Classical Theory of Fields and Particles*, Macmillan, New York (1964).

S. Coleman, "Classical Lumps and Their Quantum Descendants" in Erice Lectures, *New Phenomena in Subnuclear Physics*, Plenum Press, New York and London (1975).

S. Coleman, "The Uses of Instantons," in Erice Lectures, *The Whys of Subnuclear Physics*, Plenum Press, New York and London (1977).

R. P. Feynman and A. R. Hibbs, *Quantum Mechanics and Path Integrals*, McGraw-Hill, New York (1964).

L. S. Schulman, *Techniques and Applications of Path Integration*, Wiley, New York, (1981).

B. S. Deaver and W. M. Fairbanks, "Experimental evidence for quantized flux in superconductivity cylinders," *Phys. Rev. Lett.* **7** (1961) 43.

D. Bohm and Y. Aharonov, "Significance of electromagnetic potentials in the quantum theory," *Phys. Ref.* **115** (1959) 485.

D. Bohm and Y. Aharonov, "Further considerations on electromagnetic potentials in the quantum theory," *Phys. Rev.* **123** (1961) 1511.

G. Möllenstedt and W. Bayh, "The continuous variation of the phase of electron waves in field free space by means of the magnetic vector potential of a solenoid," *Phys. Blätt.* **18** (1962) 299.

Abrikosov, *Soviet Phys. Jetp.* **5** (1957) 1174.

Perring and Skyrme, "A model unified field equation, *Nucl. Phys.* **31** (1962) 550.

J. H. Van Vleck, *Proc. Nat. Akad. Sci.* **14** (178) 1928.

E. Gildener and A. Patrascioiu, *Phys. Rev.* **D16**, 423.

CHAPTER 1

Electromagnetism

§ 1.1 The Electromagnetic Field

The fundamental quantities of the electromagnetic field are the *field strengths*

$$\mathbf{E}(t, \mathbf{x}) \quad \text{and} \quad \mathbf{B}(t, \mathbf{x}).$$

The electric field strength is an ordinary vector field; i.e., to each point in space we have attached a vector. This vector may depend on time, too. The magnetic field strength is a pseudovector field; i.e., to each point in space we have attached a *pseudovector*, which may depend on time, too.

Although we are going to discuss mathematical concepts in greater detail later on, let us clarify the situation a little. An ordinary vector (like $\mathbf{E}$) is nothing but a directed line segment PQ connecting two points in Euclidean space. It is defined before we introduce a coordinate system. (See Figure 1.1.)

But a pseudovector can be specified only if we also specify an orientation. Consider two ordered sets of linearly independent vectors $(\mathbf{U}_1, \mathbf{U}_2, \mathbf{U}_3)$ and $(\mathbf{V}_1, \mathbf{V}_2, \mathbf{V}_3)$. We can decompose one set in terms of the other set in the

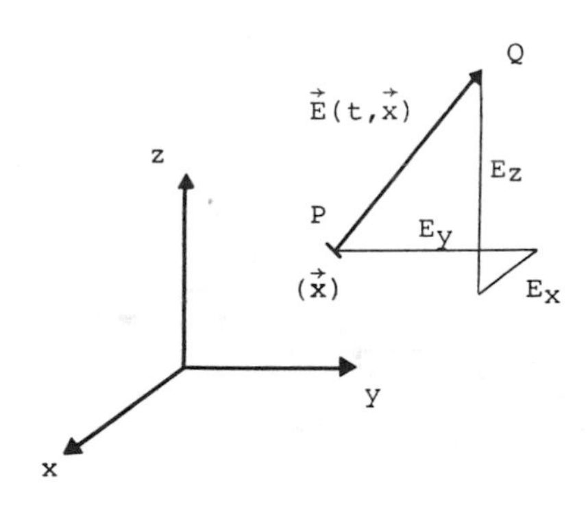

Figure 1.1.

following way:

$$\mathbf{V}_j = \mathbf{U}_i A^i_j.$$

We say that the two sets are equivalent if $\det(A^i_j) > 0$. In this way all possible ordered sets $(\mathbf{U}, \mathbf{V}, \mathbf{W})$ are divided into two classes. We arbitrarily choose one of these classes to represent *positive* oriented sets and the other class to represent *negative* oriented sets.

Once we have chosen a specific orientation, the pseudovector is represented by an ordinary vector $\mathbf{B} = (B_x, B_y, B_z)$, but if we exchange the orientation, then the pseudovector is represented by the opposite vector: $\mathbf{B}' = -\mathbf{B} = (-B_x, -B_y, -B_z)$.

The best known example of a pseudovector is the *cross product* of two ordinary vectors: $\mathbf{U} \times \mathbf{V}$. It is sometimes denoted by $\mathbf{U} \wedge \mathbf{V}$ and referred to as the *wedge product*, but we shall avoid this notation, which we reserve for differential forms. Once we have chosen a specific orientation, then $\mathbf{U} \times \mathbf{V}$ is specified by the following requirements:

1. The length of $\mathbf{U} \times \mathbf{V}$ is equal to the area of the parallelogram spanned by $\mathbf{U}$ and $\mathbf{V}$.
2. $\mathbf{U} \times \mathbf{V}$ is perpendicular to the parallelogram spanned by $\mathbf{U}$ and $\mathbf{V}$.
3. $(\mathbf{U}, \mathbf{V}, \mathbf{U} \times \mathbf{V})$ generates a positive orientation.

(See Figure 1.2).

Back to business! The field strengths $\mathbf{E}$ and $\mathbf{B}$ can be measured using the interaction between charged particles and an electromagnetic field. If a particle carries a charge q, it will experience a force $\mathbf{F}$ that depends on both the particle's position $\mathbf{x}$ and velocity $\mathbf{v}$:

$$\mathbf{F} = q(\mathbf{E} + \mathbf{v} \times \mathbf{B}). \tag{1.1}$$

(Observe that $\mathbf{v} \times \mathbf{B}$ is an ordinary vector because $\mathbf{B}$ is a pseudovector!) This force is known as the *Lorentz force*. Using the principles of special relativity, this

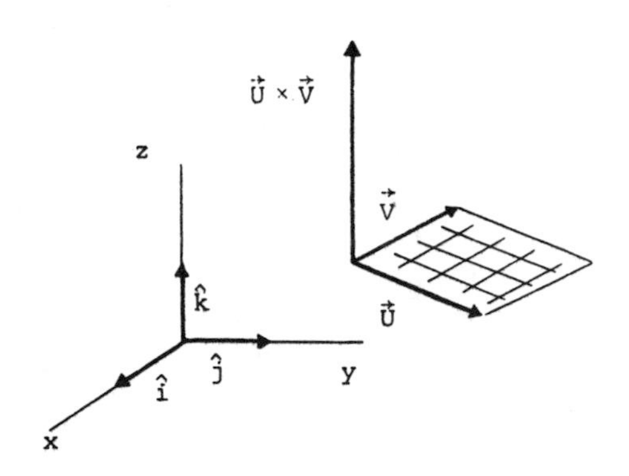

Figure 1.2.

means that the particle has the following equation of motion:

$$\frac{d\mathbf{p}}{dt} = q(\mathbf{E} + \mathbf{v} \times \mathbf{B}), \quad \mathbf{p} = \frac{m\mathbf{v}}{\sqrt{1 - \frac{v^2}{c^2}}} \quad (c = \text{velocity of light}). \tag{1.2}$$

This relation is of extreme importance, and actually it serves to define the electromagnetic field. If the charge is small, we may neglect the influence of the particle on electromagnetic field. Using a swarm of test particles moving through the electromagnetic field, we may analyze their motion and hence determine the field strengths: $\mathbf{E}(t, \mathbf{x})$ and $\mathbf{B}(t, \mathbf{x})$.

The field strengths themselves evolve in space and time, and hence they, too, obey some equations of motion, the *Maxwell equations*:

$$\nabla \cdot \mathbf{B} = 0 \quad \text{(no magnetic poles)} \tag{1.3}$$

$$\frac{\partial \mathbf{B}}{\partial t} + \nabla \times \mathbf{E} = \mathbf{0} \quad \text{(Faraday's Law)} \tag{1.4}$$

$$\nabla \cdot \mathbf{E} = \frac{\rho}{\epsilon_0} \quad \text{(Gauss's Law)} \tag{1.5}$$

$$\frac{\partial \mathbf{E}}{\partial t} - c^2 \nabla \times \mathbf{B} = -\frac{\mathbf{j}}{\epsilon_0} \quad \text{(Ampère's Law)} \tag{1.6}$$

Here ρ is the charge density, $\mathbf{j}$ the current, and ϵ_0 the permittivity of empty space. Furthermore, we have introduced the vector operator $\nabla = \left(\frac{\partial}{\partial x}, \frac{\partial}{\partial y}, \frac{\partial}{\partial z} \right)$.

In the following table we have collected some useful formulas:

$\phi(t, \mathbf{x})$ is a scalar field; $\nabla\phi(t, \mathbf{x})$ is a vector field; the *gradient*.

$\mathbf{E}(t, \mathbf{x})$ is a vector field; $\nabla \cdot \mathbf{E}(t, \mathbf{x})$ is a scalar field; the *divergence*.

$\mathbf{B}(t, \mathbf{x})$ is a pseudovector field; $\nabla \times \mathbf{B}(t, \mathbf{x})$ is a vector field; the *curl*.

$$\Delta = \frac{\partial^2}{\partial x^2} + \frac{\partial^2}{\partial y^2} + \frac{\partial^2}{\partial z^2} \quad : \text{the } \textit{Laplacian}$$

$$\mathbf{A} \times \mathbf{B} = -\mathbf{B} \times \mathbf{A} \tag{1.7}$$

$$\mathbf{A} \cdot (\mathbf{B} \times \mathbf{C}) = (\mathbf{A} \times \mathbf{B}) \cdot \mathbf{C} \tag{1.8}$$

$$\mathbf{A} \times (\mathbf{B} \times \mathbf{C}) = \mathbf{B}(\mathbf{A} \cdot \mathbf{C}) - (\mathbf{A} \cdot \mathbf{B})\mathbf{C} \tag{1.9}$$

$$(\mathbf{A} \times \mathbf{B}) \times \mathbf{C} = (\mathbf{A} \cdot \mathbf{C})\mathbf{B} - \mathbf{A}(\mathbf{B} \cdot \mathbf{C}) \tag{1.10}$$

$$\nabla \cdot (\nabla\phi) = \Delta\phi \tag{1.11}$$

$$\nabla \times (\nabla\phi) = \mathbf{0} \tag{1.12}$$

$$\nabla \cdot (\nabla \times \mathbf{B}) = 0 \tag{1.13}$$

$$\nabla \times (\nabla \times \mathbf{B}) = \nabla(\nabla \cdot \mathbf{B}) - \Delta \mathbf{B} \tag{1.14}$$

Let us deduce some of the important consequences of Maxwell's equations. If we take the divergence of (1.6), we get

$$\nabla \cdot \left(\frac{\partial \mathbf{E}}{\partial t}\right) - c^2 \nabla \cdot (\nabla \times \mathbf{B}) = -\frac{1}{\epsilon_0}\, \nabla \cdot \mathbf{j} \Rightarrow$$
$$\frac{\partial}{\partial t}\,(\nabla \cdot \mathbf{E}) - 0 = -\frac{1}{\epsilon_0}\, \nabla \cdot \mathbf{j} \Rightarrow \frac{1}{\epsilon_0}\, \frac{\partial \rho}{\partial t} = -\frac{1}{\epsilon_0}\, \nabla \cdot \mathbf{j},$$

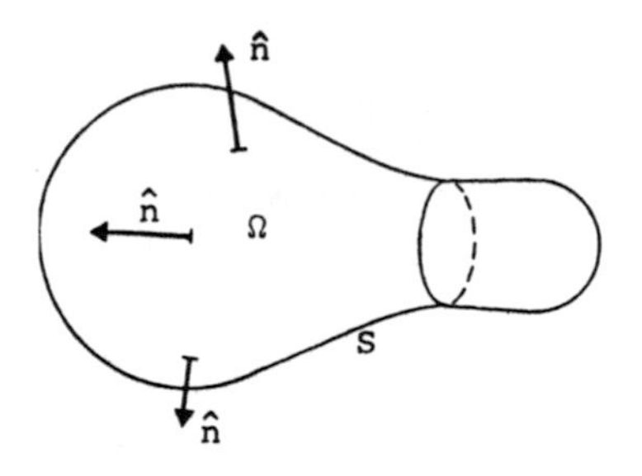

Gauss' theorem:

$$\int_\Omega \vec{\nabla} \cdot \vec{E}\, dV = \int_S \vec{E} \cdot \hat{n}\, dA \tag{1.15}$$

Ω : inside region bounded by S
S : Closed surface
n̂ : Unit normal pointing outwards

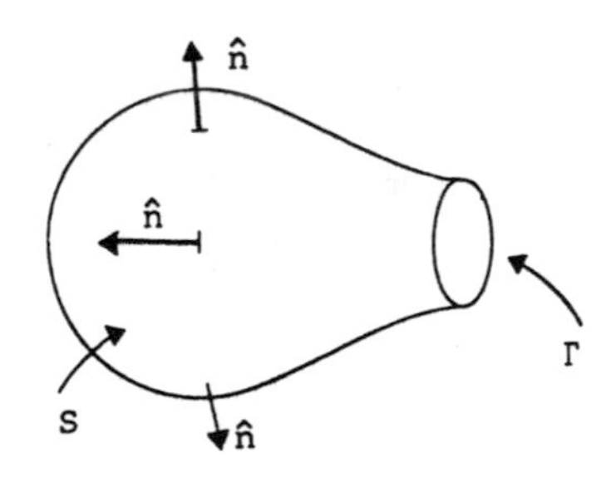

Stokes' theorem:

$$\int_S (\vec{\nabla} \times \vec{B}) \cdot \hat{n}\, dA = \oint_\Gamma \vec{B} \cdot d\vec{r} \tag{1.16}$$

S : Surface bounded by Γ
Γ : Closed loop
n̂ : Unit normal pointing outwards

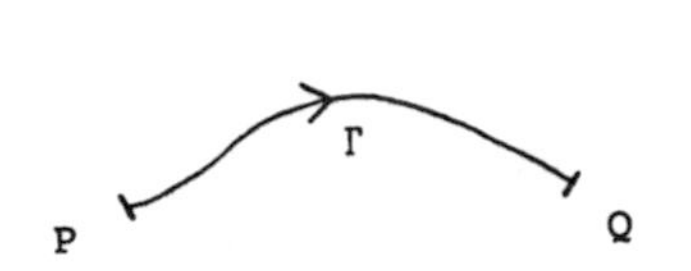

Theorem of line-integrals:

$$\int_\Gamma \vec{\nabla}\phi \cdot d\vec{r} = \phi(Q) - \phi(P)$$

Γ : Smooth curve with end-points P and Q.

FIGURE 1.3. Figure 1.3 illustrates three important theorems: (a) Gauss's theorem; (b) Stokes's theorem; (c) Theorem of line integrals.

where we have used (1.13) and (1.5). Omitting the trivial common factor, we have thus shown that

$$\frac{\partial \rho}{\partial t} + \nabla \cdot \mathbf{j} = 0. \tag{1.17}$$

This equation is known as the *equation of continuity*. It has the important consequence of *charge conservation*. To see this, consider the total charge of the system

$$Q = \int \rho d^3x,$$

where we have integrated over all space at a specific time. We assume, of course, that the system is confined in space, so that the charge density and all currents vanish sufficiently far away. Hence, all our integrals are well-defined. Now formally, Q might depend on time because the charge density $\rho(t, \mathbf{x})$ may very well depend on time! But due to the equation of continuity, we get

$$\frac{dQ}{dt} = \int \frac{\partial \rho}{\partial t} d^3x = -\int \nabla \cdot \mathbf{j} d^3x.$$

Using Gauss's theorem (1.15), this can be rearranged as

$$\frac{dQ}{dt} = -\int_{\substack{\text{surface at} \\ \text{infinity}}} (\mathbf{j} \cdot \hat{n}) dA = 0$$

because the system is confined, so that $\mathbf{j}$ vanishes at infinity. This is our first example of a *conservation law*, and we emphasize that it depends strongly on the equations of motion, i.e., the Maxwell equations. Later on, we shall discuss examples of conserved quantities that are conserved independently of the equations of motion. They will be conserved for topological reasons rather than for dynamical reasons.

§ 1.2 The Introduction of Gauge Potentials in Electromagnetism

Now take the Maxwell equation (1.3). If we try the following assumption,

$$\mathbf{B} = \nabla \times \mathbf{A}, \tag{1.18}$$

where $\mathbf{A}$ is a vector field, we see that the Maxwell equation (1.3) is automatically satisfied:

$$\nabla \cdot \mathbf{B} = \nabla \cdot (\nabla \times \mathbf{A}) = 0$$

due to (1.13). Hence, we might look for solutions to Maxwell's equations of the form (1.18). However, it is not at all evident that all solutions should automatically be of this form. To make this more precise, we consider a specific region in space Ω and a pseudovector field $\mathbf{B}$ defined on Ω in such a way that

$$\nabla \cdot \mathbf{B} = 0$$

throughout the whole region Ω. Then it is in general *not* possible to find a vector field $\mathbf{A}$ with the property (1.18). Actually, the existence of $\mathbf{A}$ depends very strongly on the topological properties of the space region Ω. Only for very simple space regions (e.g., the interior of a cube) is it always possible to find a suitable $\mathbf{A}$. But let us neglect these difficulties for a moment and just look for solutions to the Maxwell equations of the form (1.18).

Substituting this assumption into (1.4), we get

$$\frac{\partial}{\partial t}(\nabla \times \mathbf{A}) + \nabla \times \mathbf{E} = \mathbf{0},$$

which implies that

$$\nabla \times \left(\frac{\partial \mathbf{A}}{\partial t} + \mathbf{E}\right) = \mathbf{0}.$$

Using the same trick as before, we try the assumption

$$\mathbf{E} + \frac{\partial \mathbf{A}}{\partial t} = -\nabla\phi,$$

where ϕ is a scalar field. (Observe the sign!) This obviously solves (1.4), because

$$\nabla \times \left(\frac{\partial \mathbf{A}}{\partial t} + \mathbf{E}\right) = -\nabla \times (\nabla\phi) = \mathbf{0}$$

due to (1.12). Hence, if we look for solutions to the Maxwell equations of the form

$$\mathbf{B} = \nabla \times \mathbf{A}, \qquad \mathbf{E} = -\nabla\phi - \frac{\partial \mathbf{A}}{\partial t}, \tag{1.19}$$

then we have automatically solved the first two of the Maxwell equations: (1.3) and (1.4) The new fields ϕ and $\mathbf{A}$ are called *gauge potentials*, and from now on we will be very much concerned with their properties!

First, let us substitute (1.19) into the two remaining Maxwell equations:

$$\frac{\rho}{\epsilon_0} = \nabla \cdot \mathbf{E} = \nabla \cdot \left(-\nabla\phi - \frac{\partial \mathbf{A}}{\partial t}\right) = -\nabla\phi - \frac{\partial}{\partial t}\nabla \cdot \mathbf{A}$$

and

$$-\frac{\mathbf{j}}{\epsilon_0} = \frac{\partial \mathbf{E}}{\partial t} - c^2\nabla \times \mathbf{B} = \frac{\partial}{\partial t}\left(-\nabla\phi - \frac{\partial \mathbf{A}}{\partial t}\right) - c^2\nabla \times (\nabla \times \mathbf{A})$$

$$= -\nabla\frac{\partial\phi}{\partial t} - \frac{\partial^2 \mathbf{A}}{\partial t^2} - c^2\nabla(\nabla \cdot \mathbf{A}) + c^2\nabla\mathbf{A}.$$

Rearranging these two equations, we finally get the *equations of motion for the gauge potentials*:

$$\left(\Delta\phi - \frac{1}{c^2}\frac{\partial^2\phi}{\partial t^2}\right) = -\frac{\rho}{\epsilon_0} - \frac{\partial}{\partial t}\left[\nabla \cdot \mathbf{A} + \frac{1}{c^2}\frac{\partial\phi}{\partial t}\right], \tag{1.20a}$$

$$\left(\Delta\mathbf{A} - \frac{1}{c^2}\frac{\partial^2\mathbf{A}}{\partial t^2}\right) = -\frac{1}{\epsilon_0 c^2}\mathbf{j} + \nabla\left[\nabla \cdot \mathbf{A} + \frac{1}{c^2}\frac{\partial\phi}{\partial t}\right]. \tag{1.20b}$$

Observe that we have artificially introduced $\frac{1}{c^2}\frac{\partial^2\phi}{\partial t^2}$ on both sides of (1.20a) to make it look like (1.20b). This may at first sight not seem very impressive. Although we have reduced the number of equations from four to two, they are still complicated. In particular, they are still mixed in ϕ and $\mathbf{A}$. But now remember that we are actually not looking for the gauge potentials ϕ and $\mathbf{A}$ but for the field strengths $\mathbf{E}$ and $\mathbf{B}$, because they are the only ones that can be measured!

Now, suppose we have found a solution to the Maxwell equations (1.3)–(1.6) represented by the gauge potentials ϕ_0 and $\mathbf{A}_0$:

$$\mathbf{B}_0 = \nabla \times \mathbf{A}_0, \qquad \mathbf{E}_0 = -\nabla\phi_0 - \frac{\partial \mathbf{A}_0}{\partial t}\,.$$

Then we can immediately find other gauge potentials representing the *same field strengths* $\mathbf{B}_0$ and $\mathbf{E}_0$. To see this, let $\chi(t, \mathbf{x})$ be an arbitrary scalar field. Then

$$\phi = \phi_0 - \frac{\partial}{\partial t}\chi, \qquad \mathbf{A} = \mathbf{A}_0 + \nabla\chi$$

will do the job:

$$\nabla \times \mathbf{A} = \nabla \times \mathbf{A}_0 + \nabla \times (\nabla\chi) = \nabla \times \mathbf{A}_0 = \mathbf{B}_0$$

by (1.12), and

$$-\nabla\phi - \frac{\partial \mathbf{A}}{\partial t} = -\nabla\phi_0 + \nabla\frac{\partial\chi}{\partial t} - \frac{\partial \mathbf{A}_0}{\partial t} - \frac{\partial}{\partial t}\nabla\chi = -\nabla\phi_0 - \frac{\partial \mathbf{A}_0}{\partial t} = \mathbf{E}_0.$$

Furthermore, the new gauge potentials ϕ and $\mathbf{A}$ will solve exactly the same equations of motion as ϕ_0 and $\mathbf{A}_0$. By substituting ϕ and $\mathbf{A}$ into (1.20a) and (1.20b), we easily find that the terms containing χ drop out. By representing the electromagnetic field through the gauge potentials ϕ and $\mathbf{A}$, we therefore have discovered a strange symmetry:

The field strengths and the equations of motion are unchanged under the transformations

$$\phi \to \phi - \frac{\partial\chi}{\partial t}; \qquad \mathbf{A} \to \mathbf{A} + \nabla\chi. \tag{1.21}$$

Such a transformation is called *gauge transformation*, and since physics is unchanged under this transformation, we speak of *gauge symmetry*. Hence, we see that a physical system, like the electromagnetic field, is described not only by a single gauge potential $(\phi, \mathbf{A})$, but by a whole family of gauge potentials differing only by a gauge transformation. Consequently, we may parametrize the solution in the following way:

$$(\phi_\chi, \mathbf{A}_\chi) = \left(\phi_0 - \frac{\partial\chi}{\partial t}\,, \mathbf{A}_0 + \nabla\chi\right).$$

By picking a special member of this family, we say that we have chosen a specific *gauge*. This freedom of choosing a gauge is very important.

Let us now return to the equations of motion (1.20a) and (1.20b). If $(\phi_0, \mathbf{A}_0)$ denotes a specific solution, we may then choose χ such that $(\phi, \mathbf{A})$ satisfies

$$\nabla \cdot \mathbf{A} + \frac{1}{c^2}\frac{\partial \phi}{\partial t} = 0. \tag{1.22}$$

To see this, we substitute

$$\phi = \phi_0 - \frac{\partial \chi}{\partial t}, \qquad \mathbf{A} = \mathbf{A}_0 + \nabla \chi.$$

We then get

$$\nabla \cdot \mathbf{A}_0 + \nabla \cdot (\nabla \chi) + \frac{1}{c^2}\frac{\partial \phi_0}{\partial t} - \frac{\partial^2 \chi}{\partial t^2} = 0;$$

i.e.,

$$\Delta \chi - \frac{1}{c^2}\frac{\partial^2 \chi}{\partial t^2} = -\left[\nabla \cdot \mathbf{A}_0 + \frac{1}{c^2}\frac{\partial \phi_0}{\partial t}\right].$$

This is a wave equation where we know the source term:

$$-\left[\nabla \cdot \mathbf{A}_0 + \frac{1}{c^2}\frac{\partial \phi_0}{\partial t}\right].$$

We can solve it, and hence we have obtained the appropriate *gauge condition* (1.22). Observe that the solution is not unique: We may always add a solution to the homogeneous wave equation

$$\Delta \chi - \frac{1}{c^2}\frac{\partial^2 \chi}{\partial t^2} = 0. \tag{1.23}$$

Hence, we are still allowed to make gauge transformations $\phi \to \phi - \frac{\partial \chi}{\partial t}$, $\mathbf{A} \to \mathbf{A} + \nabla \chi$ without spoiling condition (1.22), provided that χ satisfies (1.23)! The gauge condition (1.22) is called the *Lorenz*[1] *condition*, and we say that we work in the *Lorenz gauge*. In this gauge the equations of motion simplify considerably, because the equations for ϕ and $\mathbf{A}$ decouple:

$$\left(\Delta - \frac{1}{c^2}\frac{\partial^2}{\partial t^2}\right)\phi = -\frac{\rho}{\epsilon_0}, \qquad \left(\Delta - \frac{1}{c^2}\frac{\partial^2}{\partial t^2}\right)\mathbf{A} = -\frac{1}{\epsilon_0 c^2}\mathbf{j} \tag{1.24}$$

These equations are beautiful wave equations, and they can be solved by standard techniques. In passing, we observe that the Lorenz condition (1.22)

$$\nabla \cdot \mathbf{A} + \frac{1}{c^2}\frac{\partial \phi}{\partial t} = 0$$

has exactly the same form as the equation of continuity (1.17)

$$\nabla \cdot \mathbf{j} + \frac{\partial \rho}{\partial t} = 0.$$

[1] L. V. Lorenz (1829–1891). Danish physicist who, independently of Maxwell, constructed a theory of light as electromagnetic waves.

In gauge theories it is a standard "trick" to let the gauge condition resemble the characteristic equation of the source.

§ 1.3 Magnetic Flux

To get some experience with gauge potentials, we will deduce a formula for magnetic flux. Let us consider a surface S bounded by the closed loop Γ. (See Figure 1.4.) We are interested in the magnetic flux passing through the surface at a specific time $t = t_0$. By definition, the flux is given by the formula

$$\Phi = \int_S \mathbf{B} \cdot \hat{n}\, dA.$$

Substituting from equation (1.18), we get

$$\Phi = \int_S (\nabla \times \mathbf{A}) \cdot \hat{n}\, dA,$$

which by Stokes's theorem (1.16) may be rewritten as

$$\Phi = \oint_\Gamma \mathbf{A} \cdot d\mathbf{r}. \tag{1.25}$$

Hence, we can express the magnetic flux in terms of $\mathbf{A}$. Observe that we have put *no* restriction on the electromagnetic field. It need not be static or anything else. This formula has important consequences. Consider a closed surface S. We assume that the electromagnetic field is smooth in a neighborhood of S, but it may be singular at a finite number of points inside S. (The electromagnetic field created by an electron is singular at the actual position of the electron.) Let us now calculate the magnetic flux through S:

$$\Phi = \int_S \mathbf{B} \cdot \hat{n}\, dA = \oint_\Gamma \mathbf{A} \cdot d\mathbf{r}.$$

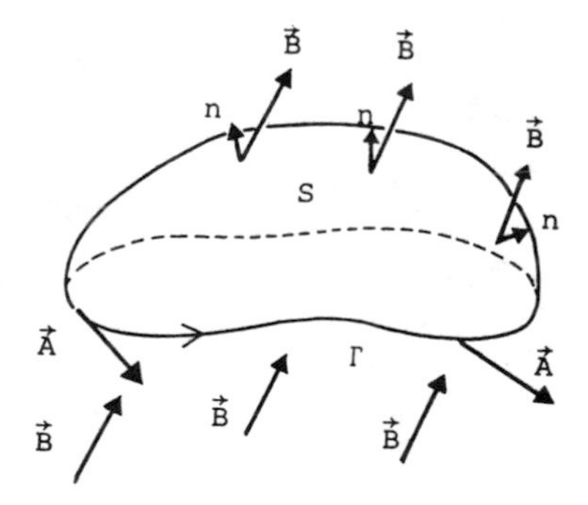

Figure 1.4.

But since S is closed, Γ is empty. Consequently, the last integral vanishes, and $\Phi = 0$.

You may feel that I am cheating you! But let me try to convince you in another way: Choose a closed loop Γ lying on S that divides S into two regions: S_1 and S_2. (See Figure 1.5.) The corresponding fluxes through S_1 and S_2 will be denoted by Φ_1 and Φ_2. We then have

$$\Phi_1 = \int_{S_1} \mathbf{B} \cdot \hat{n}\, dA = \oint_\Gamma \mathbf{A} \cdot d\mathbf{r}, \qquad \Phi_2 = \int_{S_2} \mathbf{B} \cdot \hat{n}\, dA = -\oint_\Gamma \mathbf{A} \cdot d\mathbf{r},$$

because this time we integrate the other way around Γ. Adding the two fluxes, we get

$$\Phi = \Phi_1 + \Phi_2 = 0$$

as postulated. So we have shown:

Theorem 1.1. *If the electromagnetic field is generated from a gauge potential* $(\phi, \mathbf{A})$, *then the magnetic flux through any closed surface automatically vanishes!*

To understand this result, let us consider the static Coulomb field created by a single electron. Let S be the surface of a ball with radius r and center at the position of the electron. (See Figure 1.6.) Then the normal component of the electric field $\mathbf{E}$ is easily found:

$$\mathbf{E} \cdot \hat{n} = \frac{q}{4\pi\epsilon_0} \cdot \frac{1}{r^2}.$$

It is constant on the surface S, and the electric flux is then easy to calculate:

$$\int_S \mathbf{E} \cdot \hat{n}\, dA = \frac{q}{4\pi\epsilon_0} \cdot \frac{1}{r^2} \int_S dA = \frac{q}{4\pi\epsilon_0} \frac{1}{r^2} 4\pi r^2 = \frac{q}{\epsilon_0}.$$

All these computations were made just to remind you that the *electric flux through*

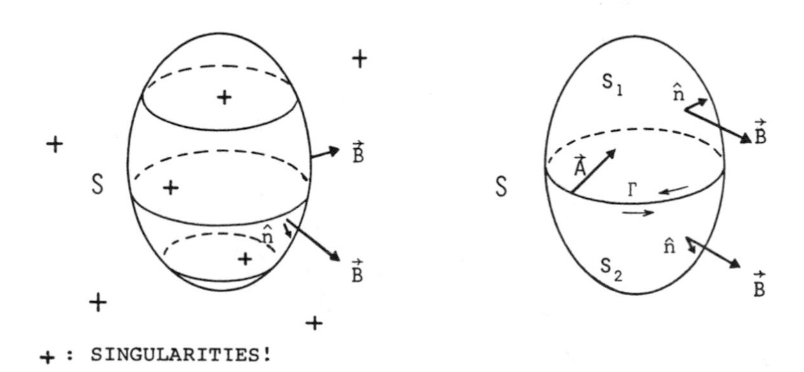

Figure 1.5.

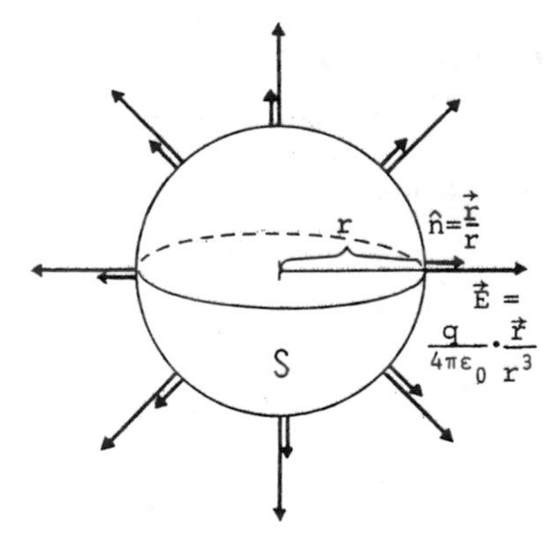

Figure 1.6.

a closed surface is proportional to the charge contained inside it. Therefore, the above result about the magnetic flux simply excludes the possibility of magnetic charges, i.e., magnetic monopoles. Of course, this is not surprising, because we start with the Maxwell equations

$$\nabla \cdot \mathbf{B} = 0, \qquad \nabla \cdot \mathbf{E} = \frac{\rho}{\epsilon_0}, \qquad \text{etc.}$$

where we have included an electric charge density but not a magnetic charge density.

We are now prepared to consider a space region Ω with a magnetic field that cannot be derived from a gauge potential **A**. (See Figure 1.7.) Consider

a. The electromagnetic field created by a resting electron.
b. The electromagnetic field created by a hypothetical magnetic monopole.

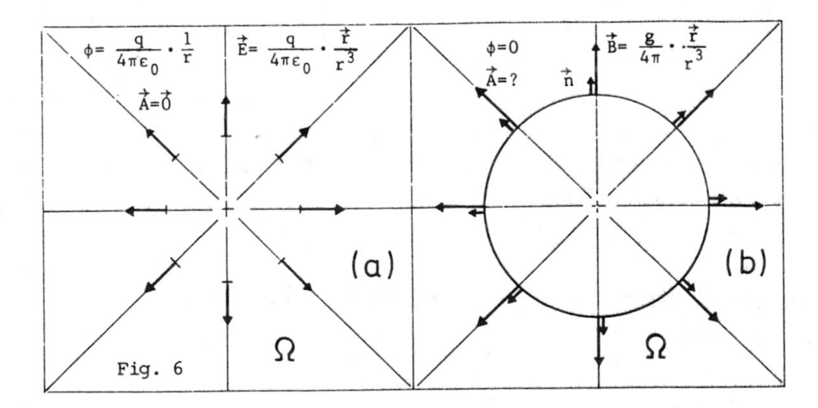

Figure 1.7.

In both cases the fields are singular at the origin $\mathbf{0}$, so we exclude that point. The remaining part of space is denoted by Ω:

$$\Omega = \mathbb{R}^3 \setminus \{\mathbf{0}\}.$$

In Ω the electromagnetic field solves the following Maxwell equations:

$$\nabla \cdot \mathbf{B} = 0, \qquad \nabla \times \mathbf{B} = \mathbf{0},$$
$$\nabla \cdot \mathbf{E} = 0, \qquad \nabla \times \mathbf{E} = \mathbf{0}.$$

This can be shown by a direct computation, and we emphasize that they will hold for both cases (a) and (b). Observe that in Ω there are neither electric charges or magnetic charges. Furthermore, the Coulomb field and the monopole field do *not* differ by their field equations. But they differ when we try to represent them by gauge potentials!

The Coulomb field may be represented by the gauge potential

$$\phi = \frac{q}{4\pi\epsilon_0} \cdot \frac{1}{r}, \qquad \mathbf{A} = \mathbf{0}$$

because this immediately leads to

$$\mathbf{B} = \nabla \times \mathbf{A} = \mathbf{0} \quad \text{and} \quad \mathbf{E} = -\nabla\phi - \frac{\partial \mathbf{A}}{\partial t} = \frac{q}{4\pi\epsilon_0} \frac{\mathbf{r}}{r^3}.$$

But the monopole field cannot be represented by a gauge potential $(\phi, \mathbf{A})$! This is because the magnetic flux through the unit sphere S is different from zero:

$$\Phi = \int_S \mathbf{B} \cdot \hat{n}\, dA = \frac{g}{4\pi} \cdot 4\pi = g \neq 0.$$

We immediately see that this magnetic field contradicts Theorem 1.1. Thus, we have shown that even if

$$\nabla \cdot \mathbf{B} = 0$$

in a space region Ω, we cannot in general conclude that $\mathbf{B}$ may be written in the form (1.18)

$$\mathbf{B} = \nabla \times \mathbf{A}.$$

We may summarize the preceding discussion in the following way: The concept of magnetic monopoles and the concept of gauge potentials seem to be in conflict with each other! Later on, we shall spend much time constructing a theory where magnetic monopoles and gauge potentials live in peaceful coexistence.

§ 1.4 Illustrative Example: The Gauge Potential of a Solenoid

Up to this point we have discussed gauge potentials in very general terms only, so let us pause to compute a gauge potential in a special situation. Consider a long

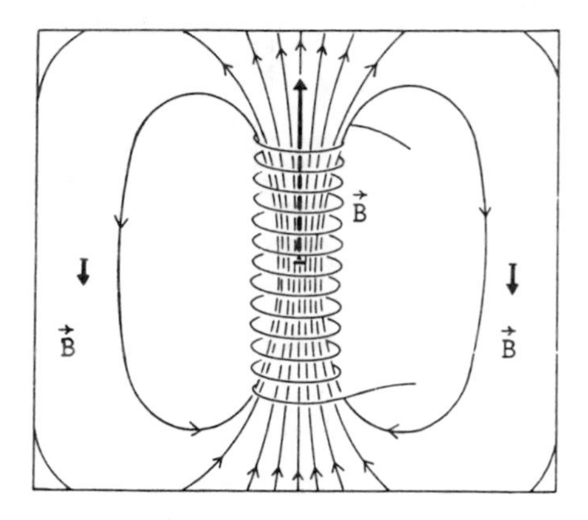

Figure 1.8.

coil of wire wound in a tight helix. (See Figure 1.8.) A current through the wire will produce a very strong magnetic field inside the solenoid, and if the solenoid is very long compared to its diameter, the magnetic field outside the solenoid will be negligible. Let us consider the ideal case of an infinitely long solenoid. Then we can safely put $\mathbf{B} = \mathbf{0}$ outside it. Inside the solenoid the magnetic field $\mathbf{B}$ will everywhere be pointing in the same direction. Now let us try to compute the gauge potential $\mathbf{A}$, which is the only gauge potential we need to be concerned about.

Inside the solenoid we have a constant magnetic field:

$$\mathbf{B} = (0, 0, B_0).$$

This may be represented by

$$\mathbf{A} = \frac{B_0}{2}(-y, x, 0) = \frac{1}{2}\mathbf{B} \times \mathbf{r},$$

as can be easily verified. This choice of gauge has nice properties:

1. The magnitude of the gauge field is proportional to the distance ρ from the z-axis.
2. It is always perpendicular to the z-axis and the radial vector $\boldsymbol{\rho}$.

Outside the solenoid the magnetic field vanishes: $\mathbf{B} = \mathbf{0}$. Hence, it would be tempting to put $\mathbf{A} = \mathbf{0}$, but this is wrong! If we consider a closed loop Γ as shown in Figure 1.9, then the magnetic flux passing through Γ is

$$\Phi = \int_S \mathbf{B} \cdot \hat{n}\, dA = B_0 \pi a^2,$$

where a is the radius of the helix, but it can also be expressed as the line integral

$$\Phi = \oint \mathbf{A} \cdot d\mathbf{r}, \tag{1.25}$$

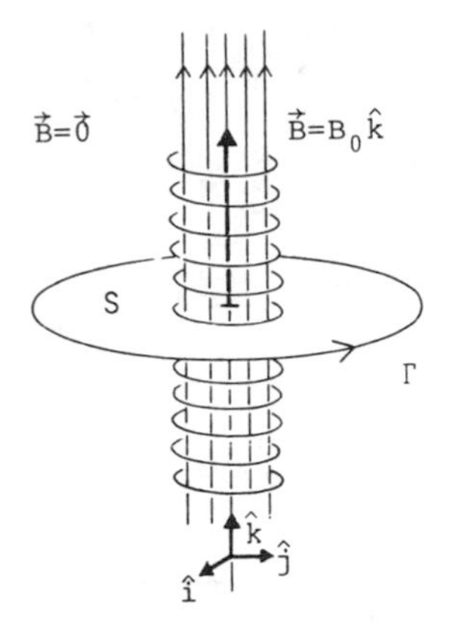

Figure 1.9.

and this is certainly inconsistent with the choice $\mathbf{A} = \mathbf{0}$! What can we do then? First, we observe that the gauge potential inside the solenoid is circulating around the z-axis, but then it is tempting to let the gauge potential outside the solenoid do the same;

$$\mathbf{A} \propto (-y, x, 0).$$

Second, we know that the line integral should have the constant value $B_0 \pi a^2$. But if we choose Γ as a circle with radius ρ_0, then it is clear that the line integral is proportional to the magnitude of $\mathbf{A}$ and proportional to the radius ρ_0. Hence, $|\mathbf{A}|$ must vary inversely proportional to ρ_0:

$$|\mathbf{A}| \propto \frac{1}{\sqrt{x^2 + y^2}}.$$

Combining these two observations, we look for $\mathbf{A}$ of the form

$$\mathbf{A} = k \left(-\frac{y}{x^2 + y^2}, \frac{x}{x^2 + y^2}, 0 \right).$$

The first thing we should check is this: Does its curl really vanish? This is easily verified. The second thing we should check is whether it produces the correct magnetic flux or not. The tangential component of $\mathbf{A}$ is $k \cdot \frac{1}{\rho_0}$, and so we get

$$\Phi = \oint_\Gamma \mathbf{A} \cdot d\mathbf{r} = k \cdot \frac{1}{\rho_0} \cdot 2\pi \rho_0 = 2\pi k.$$

(See Figure 1.10.) Hence, we can determine the constant k, because

$$2\pi k = B_0 \pi a^2; \quad \text{i.e., } k = \frac{1}{2} B_0 a^2.$$

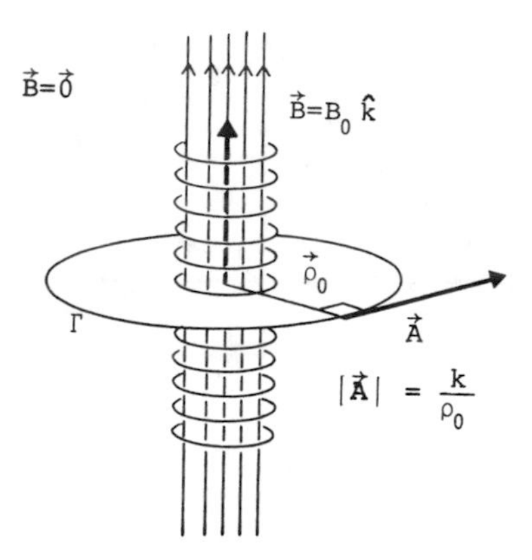

Figure 1.10.

Thus, we have completely solved our problem:

$$\mathbf{A}_{\text{inside}} = \frac{B_0}{2}\,(-y, x, 0);$$

$$\mathbf{A}_{\text{outside}} = \frac{B_0}{2}\cdot\frac{a^2}{x^2+y^2}\,(-y, x, 0).$$

Observe that $\mathbf{A}_{\text{inside}}$ and $\mathbf{A}_{\text{outside}}$ match continuously on the boundary of the solenoid.

It might seem puzzling that we can have a nonvanishing gauge potential **A** in a space region where the magnetic field vanishes! At first it might seem that it occurred only because we worked in a "bad" gauge, so maybe if we used a gauge transformation, we could "gauge away" **A**? But we have already killed this hope because we have seen that $\mathbf{A} = \mathbf{0}$ is incompatible with the magnetic flux passing through the solenoid! Hence, we cannot gauge away **A** throughout the exterior of the solenoid.

However, it is possible to gauge away the gauge potential almost everywhere! To see this, we consider the function

$$\phi(x, y) = \arctan\frac{y}{x}\,.$$

Observe that ϕ simply produces the polar angle and that ϕ is smooth except at the negative x-axis, where it makes a jump from $+\pi$ to $-\pi$. (See Figure 1.11.) First we compute the derivatives of ϕ:

$$\frac{\partial\phi}{\partial x} = \frac{\partial}{\partial x}\arctan\frac{y}{x} = \frac{-\frac{y}{x^2}}{1+\left(\frac{y}{x}\right)^2} = \frac{-y}{x^2+y^2}\,,$$

$$\frac{\partial\phi}{\partial y} = \frac{\partial}{\partial y}\arctan\frac{y}{x} = \frac{\frac{1}{x}}{1+\left(\frac{y}{x}\right)^2} = \frac{x}{x^2+y^2}\,.$$

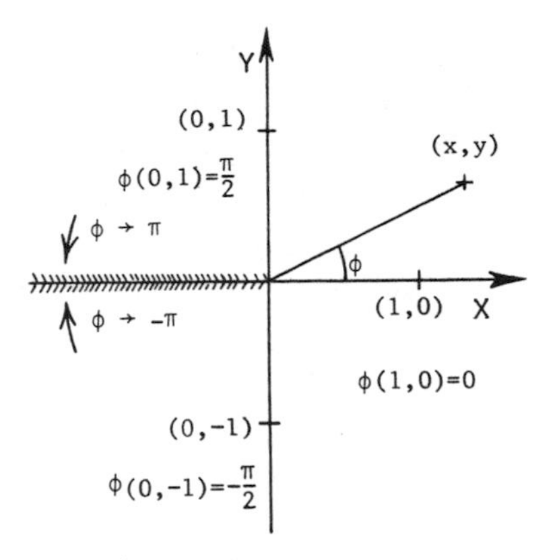

Figure 1.11.

They look very much like A_x and A_y!

Now consider the gauge transformation with

$$\chi(t, x, y, z) = -\frac{B_0 a^2}{2}\,\phi(x, y).$$

This produces the equivalent gauge potential

$$\mathbf{A}' = \mathbf{A} + \nabla\chi = \frac{B_0 a^2}{2}\begin{bmatrix} -\dfrac{y}{x^2+y^2} \\ \dfrac{x}{x^2+y^2} \\ 0 \end{bmatrix} - \frac{B_0 a^2}{2}\begin{bmatrix} -\dfrac{y}{x^2+y^2} \\ \dfrac{x}{x^2+y^2} \\ 0 \end{bmatrix} = \begin{bmatrix} 0 \\ 0 \\ 0 \end{bmatrix}.$$

Hence, we have managed to gauge away $\mathbf{A}$ except for the half plane $y = 0, x < 0$, and χ is singular. Since there is symmetry about the z-axis, we may ignore the z-coordinate for a moment. The problem is then really two-dimensional, and we see that everything is all right except on the negative x-axis. Since χ is discontinuous along the negative x-axis, we conclude that $\nabla\chi$ becomes singular—like the δ-function—on the negative x-axis. (See Figure 1.12.) If we insist on performing this *singular gauge transformation*, i.e., insist on using a *singular gauge*, we see that we have concentrated a singularity in the gauge potential along the negative x-axis. Such a string, on which the gauge potential is singular, is called a *Dirac string*. We will discuss such strings in greater detail later on.

Now, what about the fact that

$$\Phi = \oint \mathbf{A} \cdot d\mathbf{r} = B_0 \pi a^2?$$

Does it hold in the singular gauge, too? Yes! When we integrate, we necessarily pass the negative x-axis where $\mathbf{A}'$ has a δ-like singularity, and this gives the correct contribution to the line integral. (See Figure 1.13.)

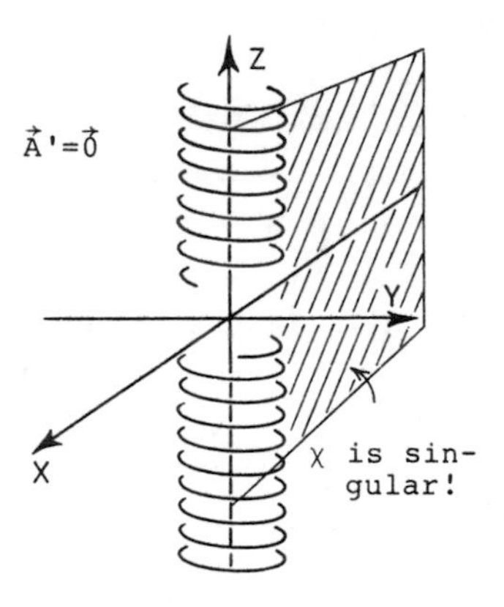

Figure 1.12.

We may clarify these remarks by recalling the definition of a δ-function. It is defined by the property that

$$\int_{-\infty}^{+\infty} \delta(x) f(x) dx = f(0)$$

for any smooth function that vanishes at infinity. Now consider the Heaviside function $\theta(x)$ defined by

$$\theta(x) = \begin{cases} 1 & \text{for } x \geq 0, \\ 0 & \text{for } x < 0. \end{cases}$$

This function makes a jump of height 1 at $x = 0$, and we are going to show that

$$\delta(x) = \frac{d\theta}{dx} .$$

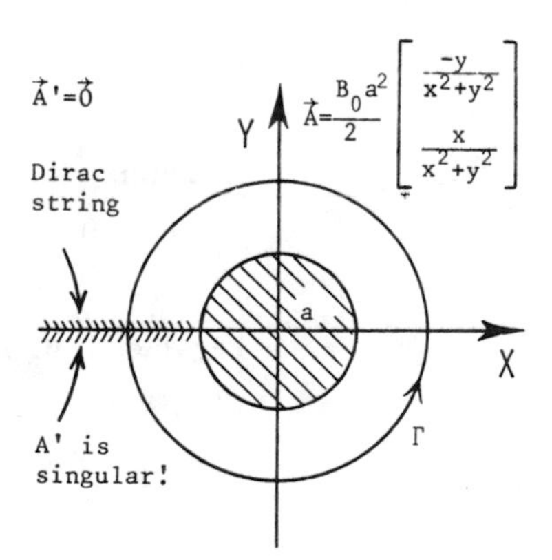

Figure 1.13.

In fact, we immediately find that

$$\int_{-\infty}^{+\infty} \frac{d\theta}{dx} f(x)dx = [\theta(x)f(x)]_{-\infty}^{+\infty} - \int_{-\infty}^{+\infty} \theta(x)f'(x)dx$$

$$= -\int_{0}^{\infty} f'(x)dx = -[f(x)]_0^{\infty} = f(0),$$

where we have integrated by parts and extensively used that f vanishes at infinity. Hence, we have demonstrated in full detail that if a function makes a jump, then the derivatives have δ-like singularities.

Worked Exercise 1.4.1
Problem: Prove the following formula:

$$\int_{-\infty}^{+\infty} f(x)\delta(g(x))dx = \frac{f(x_0)}{|g'(x_0)|} \; ; \qquad g(x_0) = 0,$$

where g is a monotonic function!

§ 1.5 Relativistic Formulation of the Theory of Electromagnetism

Up to those point we have carefully separated space and time, and we have not used Lorentz invariance. However, several times it will be more convenient to translate the results into a Lorentz invariant form. We assume that the reader is familiar with the basic principles of special relativity, but to fix notation we have collected some useful formulas:

In relativistic formulas we put $c = \hbar = \epsilon_0 = 1$ (natural units)

Indices:

Space–time indices: $\alpha, \beta, \gamma, \ldots, \mu, \nu = 0, 1, 2, 3$

Space indices: $i, j, k, l = 1, 2, 3$

Other indices: $a, b, c, d = 1, 2, \ldots, n$

Space–time diagram:

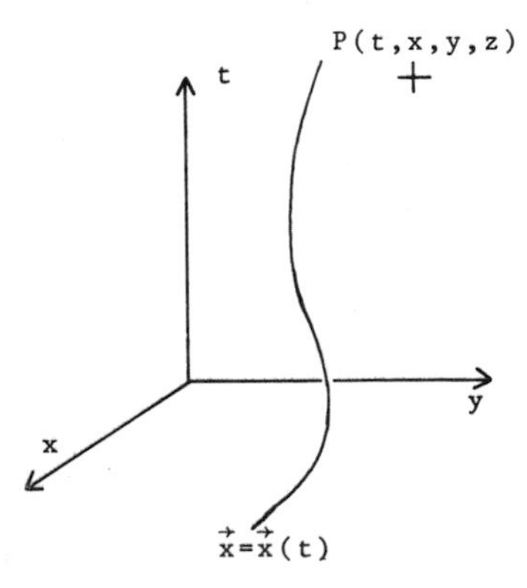

Coordinates:

$x^\alpha : \alpha = 0, 1, 2, 3$

$x^0 = t \quad x^1 = x \quad x^2 = y \quad x^3 = z$

Parametrization of world line:

$x^\alpha = x^\alpha(\tau)$

τ : Proper time

$dt = \gamma \, d\tau$

$$\gamma = \frac{1}{\sqrt{1 - v^2}}$$

Metric

$$\eta_{\alpha\beta} : \alpha \left\{ \begin{array}{c|cccc} & & \overbrace{\quad\quad\quad\quad\quad}^{\beta} & & \\ & 0 & 1 & 2 & 3 \\ 0 & -1 & 0 & 0 & 0 \\ 1 & 0 & +1 & 0 & 0 \\ 2 & 0 & 0 & +1 & 0 \\ 3 & 0 & 0 & 0 & +1 \end{array} \right.$$

Four vector.

$k^\alpha = (k^0, k^i) = (\omega, \mathbf{k})$

$k_\alpha = \eta_{\alpha\beta} k^\beta = (-\omega, \mathbf{k})$

ω : Scalar part

$\mathbf{k}$: Vector part

$k_\alpha k^\alpha = -\omega^2 + \mathbf{k}^2$

Tensor:

$$S_{\alpha\beta} = \begin{bmatrix} S_{00} & S_{0j} \\ S_{i0} & S_{ij} \end{bmatrix} \qquad S^\alpha{}_\beta = \begin{bmatrix} -S_{00} & -S_{0j} \\ S_{i0} & S_{ij} \end{bmatrix} \qquad S^{\alpha\beta} = \begin{bmatrix} S_{00} & -S_{0j} \\ -S_{i0} & S_{ij} \end{bmatrix}$$

Velocity:

$$\mathbf{v} = \left(\frac{dx}{dt}, \frac{dy}{dt}, \frac{dz}{dt} \right)$$

$$U^\alpha = \frac{dx^\alpha}{d\tau} = (\gamma, \gamma\mathbf{v}) \qquad U^\alpha U_\alpha = -1 \quad \text{(four-velocity)}$$

U^α is a time-like unit vector, pointing towards the future

Energy, momentum:

$$E = m\gamma, \quad \mathbf{p} = m\gamma\mathbf{v}$$

$$P^\alpha = m \frac{dx^\alpha}{d\tau} = (E, \mathbf{p}) \quad \text{(four-momentum)}$$

$$-P_\alpha P^\alpha = m^2 = \text{square of rest mass}$$

Charge density, current:

$$\rho, \mathbf{j} = \rho\mathbf{v}$$

$$J^\alpha = (\rho, \mathbf{j}) \quad \text{(four-current)}$$

Gauge potentials:

$$\phi, \mathbf{A}$$

$$A^\alpha = (\phi, \mathbf{A}) \qquad A_\alpha = (-\phi, \mathbf{A})$$

Derivatives:

$$\frac{\partial}{\partial t}, \nabla$$

$$\partial_\alpha = \frac{\partial}{\partial x^\alpha} = \left(\frac{\partial}{\partial t}, \nabla \right)$$

$$\partial_\alpha \partial^\alpha = -\frac{\partial^2}{\partial t^2} + \Delta = \Box : \quad \text{d'Alembertian}$$

Field strengths:

$$\mathbf{E}, \mathbf{B}$$

$$F_{\alpha\beta} = \begin{bmatrix} 0 & -E_x & -E_y & -E_z \\ E_x & 0 & B_z & -B_y \\ E_y & -B_z & 0 & B_x \\ E_z & B_y & -B_x & 0 \end{bmatrix} \quad \text{(field tensor)}$$

Lorentz force:

$$\mathbf{F} = \frac{d\mathbf{p}}{dt} = q(\mathbf{E} + \mathbf{v} \times \mathbf{B})$$

$$F^\alpha = \frac{dP^\alpha}{d\tau} = qF^\alpha_{\ \beta}U^\beta \tag{1.26}$$

Maxwell equations:

$$\nabla \cdot \mathbf{B} = 0 \qquad \frac{\partial \mathbf{B}}{\partial t} + \nabla \times \mathbf{E} = \mathbf{0}$$

$$\nabla \cdot \mathbf{E} = \frac{\rho}{\epsilon_0} \qquad \frac{\partial \mathbf{E}}{\partial t} - c^2 \nabla \times \mathbf{B} = -\frac{\mathbf{j}}{\epsilon_0}$$

$$\partial_\alpha F_{\beta\gamma} + \partial_\beta F_{\gamma\alpha} + \partial_\gamma F_{\alpha\beta} = 0 \tag{1.27}$$

$$\partial_\beta F^{\alpha\beta} = J^\alpha \tag{1.28}$$

Equation of continuity:

$$\frac{\partial \rho}{\partial t} + \nabla \cdot \mathbf{j} = 0$$

$$\partial_\alpha J^\alpha = 0 \tag{1.29}$$

Gauge potentials:

$$\mathbf{B} = \nabla \times \mathbf{A};\ \mathbf{E} = -\nabla\phi - \frac{\partial \mathbf{A}}{\partial t}$$

$$F_{\alpha\beta} = \partial_\alpha A_\beta - \partial_\beta A_\alpha \tag{1.30}$$

Gauge transformation:

$$\phi \to \phi - \frac{\partial \chi}{\partial t}$$

$$\mathbf{A} \to \mathbf{A} + \nabla\chi$$

$$A_\alpha \to A_\alpha + \partial_\alpha \chi \tag{1.31}$$

Lorenz gauge:

$$\nabla \cdot \mathbf{A} + \frac{1}{c^2}\frac{\partial \phi}{\partial t} = 0$$

$$\partial_\alpha A^\alpha = 0 \tag{1.32}$$

Equation of motion for the gauge potentials:

$$\Delta\phi - \frac{1}{c^2}\frac{\partial^2\phi}{\partial t^2} = -\frac{\rho}{\epsilon_0} - \frac{\partial}{\partial t}\left(\nabla\cdot\mathbf{A} + \frac{1}{c^2}\frac{\partial\phi}{\partial t}\right)$$

$$\Delta\mathbf{A} - \frac{1}{c^2}\frac{\partial^2\mathbf{A}}{\partial t^2} = -\frac{1}{\epsilon_0 c^2}\,\mathbf{j} + \nabla\left(\nabla\cdot\mathbf{A} + \frac{1}{c^2}\frac{\partial\phi}{\partial t}\right)$$

$$\partial_\alpha(\partial_\beta A^\beta) - (\partial_\beta\partial^\beta)A_\alpha = J_\alpha \tag{1.33}$$

Observe that the gauge potentials ϕ and $\mathbf{A}$ have been collected into a single four-vector field $A^\alpha = (\phi, \mathbf{A})$. This four-dimensional gauge potential will be referred to as the *Maxwell field.*

Worked Exercise 1.5.1
Problem: Use the relativistic formulation to reexamine the introduction of potentials, the equation of continuity, and the equation of motion for the gauge potential.

§ 1.6 The Energy–Momentum Tensor

For a more exciting topic, we will discuss the energy–momentum tensor of the Maxwell field. As preparation, we consider a system of N particles with positions $\mathbf{x}_n(t)$ and charges q_n.

The charge density ρ is defined as

$$\rho(t, \mathbf{x}) = \sum_n q_n \delta^3(\mathbf{x} - \mathbf{x}_n(t)),$$

and the current $\mathbf{j}$ is defined by

$$\mathbf{j}(t, \mathbf{x}) = \sum_n q_n \delta^3(\mathbf{x} - \mathbf{x}_n(t)) \cdot \frac{d\mathbf{x}_n}{dt}\ .$$

We may collect ρ and $\mathbf{j}$ into the four-vector

$$J^\alpha = \sum_n q_n \delta^3(\mathbf{x} - \mathbf{x}_n(t))\,\frac{dx_n^\alpha}{dt}\ .$$

(For $\alpha = 0$ we remember that $x_n^0 = t$.) This is *not* manifestly a Lorentz invariant, but we may rearrange it.

To prove the Lorentz invariance of J^α, we use the invariant parametrization of the particle trajectories $x_n^\alpha = x_n^\alpha(\tau)$, where τ is the proper time. Using Exercise 1.4.1, we rewrite the expression for J^α:

$$\begin{aligned}
J^\alpha &= \sum_n q_n \delta^3(\mathbf{x} - \mathbf{x}_n(t)) \frac{dx_n^\alpha}{dt} \\
&= \sum_n q_n \frac{\delta^3(x^i - x_n^i(\tau)) \frac{dx_n^\alpha}{d\tau}\Big|_{t=x_n^0(\tau)}}{\frac{dt}{d\tau}\Big|_{t=x_n^0(\tau)}} \\
&= \int_{\tau=-\infty}^{+\infty} \sum_n q_n \delta^3(x^i - x_n^i(\tau)) \frac{dx_n^\alpha}{d\tau} \delta(t - x_n^0(\tau)) d\tau \\
&= \int \sum_n q_n \delta^4(x^\beta - x_n^\beta(\tau)) \frac{dx_n^\alpha}{d\tau} d\tau.
\end{aligned} \tag{1.34}$$

The last expression is obviously Lorentz invariant. Observe that we may deduce the equation of continuity directly from the definition:

$$\begin{aligned}
\frac{\partial J^0}{dx^0} = \frac{\partial \rho}{\partial t} &= \sum q_n \frac{\partial}{\partial t} \delta^3(\mathbf{x} - \mathbf{x}_n(t)) \\
&= \sum_n q_n \left[\frac{\partial}{\partial x_n^i} \delta^3(\mathbf{x} - \mathbf{x}_n(t) \right] \frac{dx_n^i}{dt} \\
&= -\sum_n q_n \left[\frac{\partial}{\partial x^i} \delta^3(\mathbf{x} - \mathbf{x}_n(t)) \right] \frac{dx_n^i}{dt} \\
&= -\frac{\partial}{\partial x^i} \sum_n q_n \delta^3(\mathbf{x} - \mathbf{x}_n(t)) \frac{dx_n^i}{dt} = -\frac{\partial}{\partial x^i} J^i.
\end{aligned}$$

Hence, we have obtained the conservation of charge without using the Maxwell equations, which of course is not surprising in this very simple model where we have a finite, but fixed, number of particles with fixed charges.

This was entertainment. Now we must get down to business. Each of the particles carries a four-momentum given by $P_n^\alpha(t) = (E_n(t), \mathbf{p}_n(t))$. Hence, the *density of energy and momentum* is given by

$$T_{\text{MECH}}^{\alpha 0}(t, \mathbf{x}) = \sum_n P_n^\alpha(t) \delta^3(\mathbf{x} - \mathbf{x}_n(t)),$$

and the *current of energy–momentum* is given by

$$T_{\text{MECH}}^{\alpha j}(t, \mathbf{x}) = \sum_n P_n^\alpha(t) \delta^3(\mathbf{x} - \mathbf{x}_n(t)) \frac{dx_n^j}{dt}.$$

We may unite them into a single formula:

$$T_{\text{MECH}}^{\alpha\beta}(t, \mathbf{x}) = \sum_n P_n^\alpha(t) \delta^3(\mathbf{x} - \mathbf{x}_n(t)) \frac{dx_n^\beta}{dt}. \tag{1.35}$$

The important thing is to recognize that this is a tensor. To see this, we rewrite the expression for $T^{\alpha\beta}_{\text{MECH}}$:

$$T^{\alpha\beta}_{\text{MECH}}(t, \mathbf{x}) = \sum_n \frac{P^\alpha_n(\tau) \frac{dx_n}{d\tau}\Big|_{t=x^0_n} \delta^3(x^i_0 - x^i_n(\tau))}{\frac{dt}{d\tau}\Big|_{t=x^0_n(\tau)}}$$

$$= \int_{-\infty}^{+\infty} \sum_n P^\alpha_n(\tau) \frac{dx^\beta_n}{d\tau} \delta^3(x^i - x^i_n(\tau))\delta(t - x^0_n(\tau))d\tau;$$

i.e.,

$$T^{\alpha\beta}_{\text{MECH}} = \sum_n \int_{-\infty}^{+\infty} P^\alpha_n(\tau) \frac{dx^\beta_n}{d\tau} \delta^4(x^\mu - x^\mu_n(\tau))d\tau, \tag{1.36}$$

which immediately shows that it is a tensor. The components of this so-called *energy–momentum tensor* have the following meaning:

$$T^{\alpha\beta} = \begin{bmatrix} T^{00} & T^{0j} \\ T^{i0} & T^{ij} \end{bmatrix} = \begin{bmatrix} \text{energy density} : \epsilon & \text{energy current} : \mathbf{S} \\ \text{momentum density} : \mathbf{g} & \text{momentum current} \\ & \text{“stress-tensor”} \end{bmatrix}. \tag{1.37}$$

Now we want the total energy to be conserved. Hence, we expect the energy density and energy current to satisfy an equation of continuity:

$$\frac{\partial \epsilon}{\partial t} + \nabla \cdot \mathbf{S} = 0 \Leftrightarrow \partial_0 T^{00} + \partial_j T^{0j} = 0 \Leftrightarrow \partial_\beta T^{0\beta} = 0.$$

In the same way, the total momentum should be conserved. Hence, the density of the x-component and the x-component of the current should satisfy an equation of continuity:

$$\partial_0 T^{10} + \partial_j T^{1j} = 0 \Leftrightarrow \partial_\beta T^{1\beta} = 0,$$

etc.! Consequently, the conservation of the total energy and momentum can be expressed by an equation of continuity, satisfied by the energy–momentum tensor:

$$\partial_\beta T^{\alpha\beta} = 0. \tag{1.38}$$

Returning to the system of N particles, we get

$$\partial_\beta T^{\alpha\beta}_{\text{MECH}} = \partial_\beta \left[\sum_n P^\alpha_n(t) \frac{dx^\beta_n}{dt} \delta^3(\mathbf{x} - \mathbf{x}_n(t)) \right]$$

$$= \frac{\partial}{\partial t} \left[\sum_n P^\alpha_n(t) \delta^3(\mathbf{x} - \mathbf{x}_n(t)) \right]$$

$$+ \frac{\partial}{\partial x^i} \left[\sum_n P^\alpha_n \frac{dx^i_n}{dt} \delta^3(\mathbf{x} - \mathbf{x}_n(t)) \right]$$

$$= \sum_n \frac{dP_n^\alpha}{dt}\,\delta^3(\mathbf{x} - \mathbf{x}_n(t))$$
$$+ \sum_n P_n^\alpha(t)\left[\frac{\partial}{\partial x_n^i}\,\delta^3(x^j - x_n^j(t))\right]\frac{dx_n^i}{dt}$$
$$+ \sum_n P_n^\alpha(t)\frac{dx_n^i}{dt}\left[\frac{\partial}{\partial x^i}\,\delta^3(x^j - x_n^j(t))\right].$$

The last two terms cancel each other, and we have shown that

$$\partial_\beta T^{\alpha\beta}_{\text{MECH}} = \sum_n \frac{dP_n^\alpha}{dt}\,\delta^3(\mathbf{x} - \mathbf{x}_n(t)) = \sum_n \int \frac{dP_n^\alpha}{d\tau}\,\delta^4(x - x_n(\tau))d\tau. \quad (1.39)$$

If the particles are *free*, i.e., they are not charged and they do not experience any forces, etc., then $P_n^\alpha(t)$ are constants, and we get immediately

$$\partial_\beta T^{\alpha\beta}_{\text{MECH}} = 0 \qquad \text{(for a system of free particles!)}.$$

However, if they are charged, they will create an electromagnetic field and then experience forces. In this case, we get

$$\partial_\beta T^{\alpha\beta}_{\text{MECH}} \neq 0,$$

and of course, this is not surprising, because we have only counted the *mechanical* part of energy and momentum and not the part of it stored in the electromagnetic field! Nothing can prevent mechanical energy from being converted to field energy and vice versa!

Now, introducing the energy–momentum tensor of the electromagnetic field $T^{\alpha\beta}_{\text{EL}}$, the total energy–momentum tensor may be decomposed as

$$T^{\alpha\beta} = T^{\alpha\beta}_{\text{MECH}} + T^{\alpha\beta}_{\text{EL}},$$

and the conservation of energy and momentum requires

$$0 = \partial_\beta T^{\alpha\beta} = \partial_\beta T^{\alpha\beta}_{\text{MECH}} + \partial_\beta T^{\alpha\beta}_{\text{EL}} \Rightarrow \partial_\beta T^{\alpha\beta}_{\text{EL}} = -\partial_\beta T^{\alpha\beta}_{\text{MECH}}.$$

Since we know $\partial_\beta T^{\alpha\beta}_{\text{MECH}}$, we may use this to determine $T^{\alpha\beta}_{\text{EL}}$!

$$\partial_\beta T^{\alpha\beta}_{\text{EL}} = -\sum_n \frac{dP_n^\alpha}{dt}\,\delta^3(\mathbf{x} - \mathbf{x}_n(t)) = -\sum_n \frac{dP_n^\alpha}{d\tau}\frac{d\tau}{dt}\,\delta^3(\mathbf{x} - \mathbf{x}_n(t))$$
$$= -\sum_n q_n F^\alpha{}_\beta \frac{dx_n^\beta}{d\tau}\frac{d\tau}{dt}\,\delta^3(\mathbf{x} - \mathbf{x}_n(t)) \quad \text{(due to the Lorentz force)}$$
$$= -F^\alpha{}_\beta \sum_n q_n \frac{dx_n^\beta}{dt}\,\delta^3(\mathbf{x} - \mathbf{x}_n(t)) = -F^\alpha{}_\beta J^\beta.$$

Thus we have shown that

$$\partial_\beta T^{\alpha\beta}_{\text{EL}} = -F^\alpha{}_\beta J^\beta. \quad (1.40)$$

We must rearrange the term on the right-hand side.

Worked Exercise 1.6.1
Problem: Show that Maxwell's equations imply the formula

$$F^{\alpha}_{\ \beta} J^{\beta} = \partial_{\beta} \left[F^{\alpha}_{\ \gamma} F^{\gamma\beta} + \frac{1}{4}\, \eta^{\alpha\beta} F^{\gamma\delta} F_{\gamma\delta} \right].$$

From this exercise one can immediately read off

$$T^{\alpha\beta}_{\text{EL}} = -F^{\alpha}_{\ \gamma} F^{\gamma\beta} - \frac{1}{4}\, \eta^{\alpha\beta} F^{\gamma\delta} F_{\gamma\delta}. \tag{1.41}$$

This was a very long computation! Let us summarize the results: We have introduced two energy–momentum tensors: $T^{\alpha\beta}_{\text{MECH}}$ and $T^{\alpha\beta}_{\text{EL}}$. They obey equations (1.39) and (1.40),

$$\partial_{\beta} T^{\alpha\beta}_{\text{MECH}} = \sum_{n} \frac{dP^{\alpha}_{n}}{dt}\, \delta^{3}(\mathbf{x} - \mathbf{x}_{n}(t)), \qquad \partial_{\beta} T^{\alpha\beta}_{\text{EL}} = -F^{\alpha}_{\ \beta} J^{\beta}.$$

For a system of *free* particles this immediately implies

$$\partial_{\beta} T^{\alpha\beta}_{\text{MECH}} = 0,$$

because $P^{\alpha}_{n}(t)$ are constants. In a similar way, a *free* electromagnetic field obeys

$$\partial_{\beta} T^{\alpha\beta}_{\text{EL}} = 0,$$

because $J^{\beta} = 0$ for a free field. Finally, if we have a system of charged particles interacting with the electromagnetic field, we get

$$\partial_{\beta} T^{\alpha\beta}_{\text{MECH}} = -\partial_{\beta} T^{\alpha\beta}_{\text{EL}},$$

because the total energy–momentum tensor $T^{\alpha\beta} = T^{\alpha\beta}_{\text{MECH}} + T^{\alpha\beta}_{\text{EL}}$ obeys

$$\partial_{\beta} T^{\alpha\beta} = 0.$$

Exercise 1.6.2
Problem: Show that $T^{\alpha\beta}_{\text{MECH}}$ and $T^{\alpha\beta}_{\text{EL}}$ are *symmetric* tensors, i.e., $T^{\alpha\beta} = T^{\beta\alpha}$.

When we know the energy–momentum tensor of the electromagnetic field, we can in particular find the energy density and the momentum density:

$$\begin{aligned} \epsilon(t, \mathbf{x}) &= T^{00} = -F^{0}_{\ \gamma} F^{\gamma 0} - \frac{1}{4}\, (F^{\gamma\delta} F_{\gamma\delta}) \\ &= \mathbf{E}^{2} + \frac{1}{4}\, (2\mathbf{B}^{2} - 2\mathbf{E}^{2}) = \frac{1}{2}\, (\mathbf{E}^{2} + \mathbf{B}^{2}), \\ \mathbf{g}(t, \mathbf{x}) &= T^{i0} = -F^{i}_{\ \gamma} F^{\gamma 0} = -F^{i}_{\ j} F^{j0} = \mathbf{E} \times \mathbf{B}; \end{aligned}$$

i.e.,

$$\epsilon(t, \mathbf{x}) = \frac{\epsilon_0}{2}\, (\mathbf{E}^{2} + c^{2}\mathbf{B}^{2}) \qquad \mathbf{g}(t, \mathbf{x}) = \epsilon_0 (\mathbf{E} \times \mathbf{B}), \tag{1.42}$$

where we have reintroduced ϵ_0 and c.

We will finish with some remarks about energy–momentum tensors in general: Let us consider the conservation laws. We have seen that $\partial_{\alpha} J^{\alpha} = 0$ implies conservation of charge and that $\partial_\beta T^{\alpha\beta} = 0$ implies conservation of energy and momentum. Actually, they imply something more:

Theorem 1.2 (Abraham's theorem).

a. *If $J^\alpha(x)$ is a field of four-vectors, then $\partial_\alpha J^\alpha = 0$ implies that $Q = \int_{x^0=t^0} J^0 d^3x$ is a Lorentz-scalar; i.e., it is independent of the observer.*

b. *If $T^{\alpha\beta}(x)$ is a field of Lorentz tensors, then $\partial_\beta T^{\alpha\beta} = 0$ implies that $P^\alpha = \int T^{\alpha 0} d^3x$ transform as the components of an ordinary four-vector.*

This will be proved in Section 11.1.

Observe that in the cases we have been discussing, $T^{\alpha\beta}$ is *symmetric* (see Exercise 1.6.2).

This has an important consequence. Consider

$$M^{\alpha\beta\gamma}(x) = x^\alpha T^{\beta\gamma} - x^\beta T^{\alpha\gamma}.$$

(Observe that $M^{\alpha\beta\gamma}$ is not a tensor because x^α is the coordinate of a point and not the component of a vector!) Taking the divergence, we find

$$\partial_\gamma M^{\alpha\beta\gamma} = \delta^\alpha_\gamma T^{\beta\gamma} + x^\alpha(\partial_\gamma T^{\beta\gamma}) - \delta^\beta_\gamma T^{\alpha\gamma} - x^\beta(\partial_\gamma T^{\alpha\gamma}) = T^{\beta\alpha} - T^{\alpha\beta}.$$

Hence, the symmetry of $T^{\alpha\beta}$ implies that $J^{\alpha\beta}(t^0) = \int M^{\alpha\beta 0} d^3\mathbf{x}$ is conserved! Now $J^{\alpha\beta}$ is the four-dimensional angular momentum! For instance,

$$\begin{aligned}\frac{1}{2}\epsilon_{ijk} J^{jk} &= \int \frac{1}{2}\epsilon_{ijk}(x^j T^{k0} - x^k T^{j0}) d^3\mathbf{x} = \int \epsilon_{ijk} x^j T^{k0} d^3\mathbf{x} \\ &= \int (\mathbf{r} \times \mathbf{g})_i d^3\mathbf{x},\end{aligned}$$

where $\mathbf{g}$ is the momentum density. Since we expect this to be conserved for a closed system, *we must demand that the energy–momentum tensor be symmetric.*

Solutions to Worked Exercies

Solution to 1.4.1

Suppose $g(x)$ is a monotonic function, and consider

$$\int_{x=-\infty}^{+\infty} f(x)\delta(g(x))dx = \int_{x=-\infty}^{+\infty} \frac{f(x)}{g'(x)}\delta(g(x))dg(x)$$

$$= \begin{cases} \displaystyle\int_{u=-\infty}^{+\infty} \frac{f(g^{-1}(u))}{g'(g^{-1}(u))}\delta(u)du = \frac{f(g^{-1}(0))}{g'(g^{-1}(0))} = \frac{f(x_0)}{g'(x_0)}, \\ \qquad \text{where } g(x_0) = 0 \text{ and } g \text{ is increasing;} \\ \displaystyle -\int_{u=-\infty}^{u=+\infty} \frac{f(g^{-1}(u))}{g'(g^{-1}(u))}\delta(u)du = -\frac{f(g^{-1}(0))}{g'(g^{-1}(0))} = \frac{f(x_0)}{-g'(x_0)}, \\ \qquad \text{where } g(x_0) = 0 \text{ and } g \text{ is decreasing.} \end{cases}$$

If g is increasing, then $g'(x_0) > 0$, and if g is decreasing, then $g'(x_0) < 0$. Hence, we may collect the above in the following formula:

$$\int_{x=-\infty}^{+\infty} f(x)\delta(g(x))dx = \frac{f(x_0)}{|g'(x_0)|}; \qquad g(x_0) = 0.$$

Solution to 1.5.1

Some time ago we solved the Maxwell equations

$$\nabla \cdot \mathbf{B} = 0, \qquad \frac{\partial \mathbf{B}}{\partial t} + \nabla \times \mathbf{E} = \mathbf{0}$$

with the assumption, that $\mathbf{B}$ and $\mathbf{E}$ could be reexpressed through the potentials ϕ and $\mathbf{A}$ as follows

$$\mathbf{B} = \nabla \times \mathbf{A}, \qquad \mathbf{E} = -\nabla\phi - \frac{\partial \mathbf{A}}{\partial t}.$$

Now we have reformulated these Maxwell equations as

$$0 = \partial_\alpha F_{\beta\gamma} + \partial_\beta F_{\gamma\alpha} + \partial_\gamma F_{\alpha\beta},$$

and we solve them with the assumption, that $F_{\alpha\beta}$ can be written in the following form

$$F_{\alpha\beta} = \partial_\alpha A_\beta - \partial_\beta A_\alpha.$$

We can easily check this by explicit computation:

$$\begin{aligned}
\partial_\alpha F_{\beta\gamma} + \partial_\beta F_{\gamma\alpha} + \partial_\gamma F_{\alpha\beta} &= \partial_\alpha(\partial_\beta A_\gamma - \partial_\gamma A_\beta) + \partial_\beta(\partial_\gamma A_\alpha - \partial_\alpha A_\gamma) \\
&\quad + \partial_\gamma(\partial_\alpha A_\beta - \partial_\beta A_\alpha) \\
&= (\partial_\alpha\partial_\beta - \partial_\beta\partial_\alpha)A_\gamma + (\partial_\gamma\partial_\alpha - \partial_\alpha\partial_\gamma)A_\beta \\
&\quad + (\partial_\beta\partial_\gamma - \partial_\gamma\partial_\beta)A_\alpha = 0
\end{aligned}$$

due to the fact that partial derivatives commute ($\partial_\alpha\partial_\beta = \partial_\beta\partial_\alpha$). From the two remaining Maxwell equations

$$\nabla \cdot \mathbf{E} = \frac{\rho}{\epsilon_0}, \qquad \frac{\partial \mathbf{E}}{\partial t} - c^2(\nabla \times \mathbf{B}) = -\frac{\mathbf{j}}{\epsilon_0}$$

we deduced the equation of continuity

$$\frac{\partial \rho}{\partial t} + \nabla \cdot \mathbf{j} = 0.$$

Now we have combined these Maxwell equations into a single equation:

$$\partial_\beta F^{\alpha\beta} = J^\alpha.$$

Taking the divergence, we get

$$\partial_\alpha \partial_\beta F^{\alpha\beta} = \partial_\alpha J^\alpha,$$

but $\partial_\alpha \partial_\beta F^{\alpha\beta} = 0$ because $\partial_\alpha \partial_\beta$ is symmetric in $\alpha\beta$, while $F^{\alpha\beta}$ is antisymmetric in $\alpha\beta$. Hence, we recover the equation of continuity

$$\partial_\alpha J^\alpha = 0.$$

Finally, we may deduce the equations of motion for the Maxwell field A_α. Substituting $F_{\alpha\beta} = \partial_\alpha A_\beta - \partial_\beta A_\alpha$ in the Maxwell equation $\partial_\beta F^{\alpha\beta} = J^\alpha$, we get

$$\begin{aligned} \partial_\beta(\partial^\alpha A^\beta - \partial^\beta A^\alpha) = J^\alpha &\Leftrightarrow \partial^\alpha(\partial_\beta A^\beta) - (\partial_\beta \partial^\beta) A^\alpha = J^\alpha \\ &\Rightarrow \partial^\alpha(\partial_\beta A^\beta) - \Box A^\alpha = J^\alpha. \end{aligned}$$

If we choose the Lorentz gauge, where $\partial_\alpha A^\alpha = 0$, these equations simplify to

$$\Box A^\alpha = -J^\alpha.$$

Solution to 1.6.1

Using (1.28) we get

$$\underbrace{F^\alpha_{\ \beta} J^\beta = F^\alpha_{\ \beta} \partial_\gamma F^{\beta\gamma}}_{(1.28)} = \partial_\gamma [F^\alpha_{\ \beta} F^{\beta\gamma}] - [\partial_\gamma F^\alpha_{\ \beta}] F^{\beta\gamma}.$$

The last term is a mess, and we must rearrange it separately:

$$\begin{aligned}
& \text{Index Gymnastics!} \\
[\partial_\gamma F^\alpha_{\ \beta}] F^{\beta\gamma} &= [\partial^\gamma F^{\alpha\beta}] F_{\beta\gamma} \\
&= \frac{1}{2} (\partial^\gamma F^{\alpha\beta}) F_{\beta\gamma} + \frac{1}{2} (\partial^\gamma F^{\alpha\beta}) F_{\beta\gamma} \quad [\beta \to \gamma \text{ and } \gamma \to \beta] \\
&= \frac{1}{2} (\partial^\gamma F^{\alpha\beta}) F_{\beta\gamma} \to \frac{1}{2} (\partial^\beta F^{\alpha\gamma}) F_{\gamma\beta} \\
& \qquad [F^{\alpha\gamma} = -F^{\gamma\alpha} \text{ and } F_{\gamma\beta} = -F_{\beta\gamma}] \\
&= \frac{1}{2} (\partial^\gamma F^{\alpha\beta}) F_{\beta\gamma} + \frac{1}{2} (\partial^\beta F^{\gamma\alpha}) F_{\beta\gamma} \\
&= \frac{1}{2} [\partial^\gamma F^{\alpha\beta} + \partial^\beta F^{\gamma\alpha}] F_{\beta\gamma} = -\frac{1}{2} [\partial^\alpha F^{\beta\gamma}] F_{\beta\gamma} \quad \text{(by 1.27)} \\
&= -\frac{1}{4} \partial^\alpha [F^{\beta\gamma} F_{\beta\gamma}] = -\frac{1}{4} \partial_\gamma (\eta^{\alpha\gamma} F^{\beta\delta} F_{\beta\delta}). \\
& \qquad (\text{since } \partial^\alpha = \eta^{\alpha\gamma} \partial_\gamma)
\end{aligned}$$

This was the complicated part of the computation. Now we are almost through:

$$\partial_\beta T^{\alpha\beta}_{\mathrm{EL}} = -F^\alpha_{\ \beta} J^\beta = -\partial_\gamma [F^\alpha_{\ \beta} F^{\beta\gamma}] - \frac{1}{4}\,\partial_\gamma [\eta^{\alpha\gamma} F^{\beta\delta} F_{\beta\delta}]$$

$$= -\partial_\gamma \left[F^\alpha_{\ \beta} F^{\beta\gamma} + \frac{1}{4}\,\eta^{\alpha\gamma} F^{\beta\delta} F_{\beta\delta} \right].$$

In the last expression we interchange the dummy indices β and γ and get

$$\partial_\beta T^{\alpha\beta}_{\mathrm{EL}} = -\partial_\beta \left[F^\alpha_{\ \gamma} F^{\gamma\beta} + \frac{1}{4}\,\eta^{\alpha\beta} F^{\gamma\delta} F_{\gamma\delta} \right].$$

CHAPTER 2

Interaction of Fields and Particles

§ 2.1 Introduction

As we have seen, it is possible to introduce a gauge potential $A_0^\alpha = (\phi_0, \mathbf{A}_0)$ describing an electromagnetic field, and this potential has the peculiar property that the gauge transformed potential (1.31)

$$A^\alpha = A_0^\alpha + \partial^\alpha \chi$$

describes the same electromagnetic field and satisfies the same equation of motion.

Now, a *classical* particle interacts with an electromagnetic field in the following way: The field produces a force, and this force depends only on the values of the field strengths at the momentary position of the particle. What is going on elsewhere is completely irrelevant.

Hence if there is a space–time region where **E** and **B** vanish, a classical charged particle will experience nothing. But we have seen that even if there are no field strengths in a space–time region Ω, there may be a nontrivial gauge potential **A** in

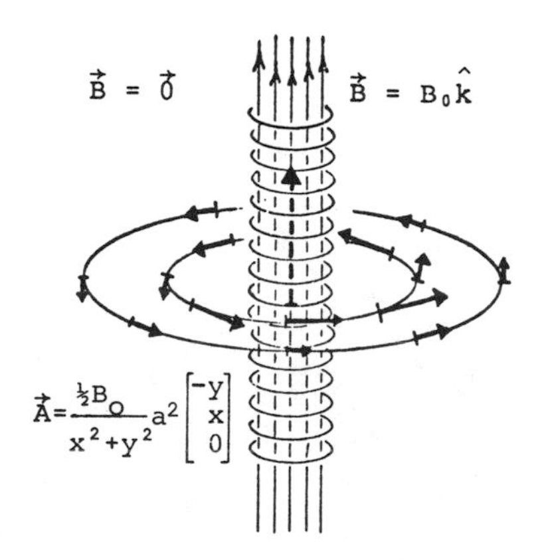

Figure 2.1.

this region. Here nontriviality means that the gauge potential cannot be completely gauged away. For instance, we have seen that outside a solenoid there is no magnetic field, but there is a nontrivial gauge potential

$$\mathbf{A} = \frac{B_0}{2}\frac{a^2}{x^2+y^2}\begin{bmatrix} -y \\ x \\ 0 \end{bmatrix}.$$

Thus one might ask, Do there exist physically measurable effects associated with the motion of a charged particle completely outside the solenoid? In other words, can we distinguish between the situation where there is no current in the solenoid and the situation where there is just by letting a charged particle go around the solenoid?

Clearly, if there is such an effect, it cannot be a *classical* one, because a classical charged particle will experience no forces in either situation. However, there could be a quantum-mechanical effect!

To study this possibility we are faced with the problem of how to quantize the physical systems above. If the fields are strong, so that the particle can be regarded as a test particle, then we are allowed to keep the fields on a classical level. We treat them as *external fields*. However, the motion of the particle should be quantized. Hence, instead of characterizing the particle by its classical trajectory $\mathbf{x} = \mathbf{x}(t)$, we introduce a Schrödinger wave function $\psi(\mathbf{r}, t)$ with the following well-known interpretation: At time t the absolute square $|\psi(\mathbf{r}, t)|^2$ is the probability density for finding the particle at position $\mathbf{r}$. Instead of the usual Newtonian equation of motion

$$m\ddot{\mathbf{x}} = -\nabla V(\mathbf{x}) \qquad \text{where the dots represent time derivatives,} \tag{2.1}$$

the dynamical evolution of the quantum-mechanical system is governed by the *Schrödinger equation* ($\hbar = \frac{h}{2\pi}$; h = Planck's constant)

$$i\hbar = \frac{\partial\psi}{\partial t} = \left(-\frac{\hbar^2}{2m}\Delta + V(\mathbf{r})\right)\psi(\mathbf{r}, t). \tag{2.2}$$

Although the Schrödinger equation when there are electromagnetic fields present is probably familiar, it will be rewarding to examine the quantization procedure carefully.

§ 2.2 The Lagrangian Formalism for Particles: The Nonrelativistic Case

The first step in quantizing a system consists in casting the problem into the *Lagrangian* form. Classically, it is a compact way of deriving the equations of motion. Consider for simplicity a one-dimensional motion of a particle, say along the x-axis. We want to consider those motions of the particle where it moves from the space–time point $A(t_1, x_1)$ to the space–time point $B(t_2, x_2)$. The question is, How can we find the classical path leading from A to B?

First, we will introduce a function $L(x, \dot{x})$ of two variables. It is called the *Lagrangian*, and the explicit form will be derived later. To each path leading from A to B we now associate the number

$$S = \int_{t_1}^{t_2} L(x, \dot{x}) dt \tag{2.3}$$

called the *action* associated with that path. (See Figure 2.2.) We want to choose the path $x = x(t)$ in such a way that the action has an *extremum*. Suppose $x = x_0(t)$ produces such an extremum, and consider another path

$$x = x_0(t) + \epsilon y(t); \qquad y(t_1) = y(t_2) = 0.$$

We will allow ϵ to vary, but we will keep y fixed for a moment. We know that the action

$$S(\epsilon) = \int_{t_1}^{t_2} L(x_0(t) + \epsilon \cdot y(t), \dot{x}_0 + \epsilon \cdot \dot{y}) dt,$$

which may be considered a function of ϵ, has an extremum for $\epsilon = 0$. Consequently,

$$\begin{aligned} 0 = S'(0) &= \int_{t_1}^{t_2} \left[\frac{\partial L}{\partial x} y(t) + \frac{\partial L}{\partial(\dot{x})} \dot{y}(t) \right] dt \\ &= \left[\frac{\partial L}{\partial(\dot{x})} y(t) \right]_{t_1}^{t_2} + \int_{t_1}^{t_2} \left[\frac{\partial L}{\partial x} y(t) - \frac{d}{dt} \frac{\partial L}{\partial(\dot{x})} y(t) \right] dt \\ &= \int_{t_1}^{t_2} \left[\frac{\partial L}{\partial x} - \frac{d}{dt} \frac{\partial L}{\partial(\dot{x})} \right] y(t) dt, \end{aligned}$$

where we have used the boundary conditions $0 = y(t_1) = y(t_2)$. However, $y(t)$ was an arbitrarily chosen function. Therefore, the above identity is consistent only if $x = x_0(t)$ satisfies the following differential equation, known as the *Euler-*

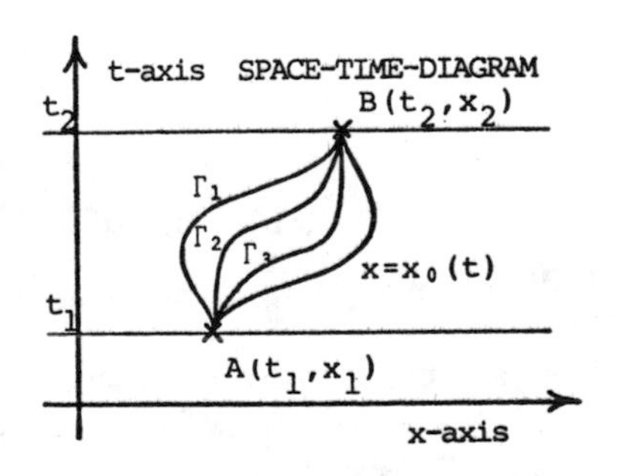

Figure 2.2.

Lagrange equation:

$$\frac{\partial L}{\partial x} - \frac{d}{dt}\frac{\partial L}{\partial(\dot{x})} = 0. \tag{2.4}$$

Illustrative example: One-dimensional motion in a potential $V(x)$. Let us try to determine the Lagrangian in such a way that the Euler–Lagrange equation (2.4)

$$\frac{\partial L}{\partial x} = \frac{d}{dt}\frac{\partial L}{\partial \dot{x}}$$

reproduces Newton's equation of motion (2.1)

$$-\frac{\partial V}{\partial x} = m\ddot{x} = \frac{d}{dt}(m\dot{x}).$$

We immediately read off the following relationships:

$$\frac{\partial L}{\partial x} = -\frac{\partial V}{\partial x} \Rightarrow L(x, \dot{x}) = (\text{terms involving } \dot{x}) - V(x),$$

$$\frac{\partial L}{\partial x} = m\dot{x} \Rightarrow L(x, \dot{x}) = \frac{1}{2}m(\dot{x})^2 + (\text{terms involving } x).$$

That is,

$$L(x, \dot{x}) = \frac{1}{2}m(\dot{x})^2 - V(x) + \text{constant}.$$

Consequently, we have determined the following suitable Lagrangian:

$$L(x, \dot{x}) = \frac{1}{2}m(\dot{x})^2 - V(x). \tag{2.5}$$

The above scheme is easily generalized to more complicated systems. Suppose we have a system characterized by the following set of *generalized coordinates* $q^1, \ldots, q^n$. If we have two particles, for instance, then q^1, q^2, q^3 could be the Cartesian coordinates of the center of mass, and q^4, q^5, q^6 could be the polar coordinates of the vector connecting particle 1 and particle 2. (See Figure 2.3.) These generalized coordinates take their values in a subset Ω of the n-dimensional space $\mathbb{R}^n$. We call Ω the *configuration space*. The dynamical evolution of the system is described by a curve $q^i = q^i(t)$ in the configuration space. We now introduce a Lagrangian $L(q^i, \dot{q}^i, t)$, which may depend explicitly on time, too.

Next we consider paths connecting the point $A(t_1, q_1^1, \ldots, q^n)$ with the point $B(t_2, q_2^1, \ldots, q_2^n)$. With each of these paths we associate the action

$$S = \int_{t_1}^{t_2} L(q^i, \dot{q}^i, t)dt,$$

and we want to determine a path $q^i = q_0^i(t)$ extremizing the action. Again we suppose that $q^i = q_0^i(t)$ actually extremizes it and consider another nearby path:

$$q^i = q_0^i(t) + \epsilon^i y(t), \qquad y(t_1) = y(t_2) = 0.$$

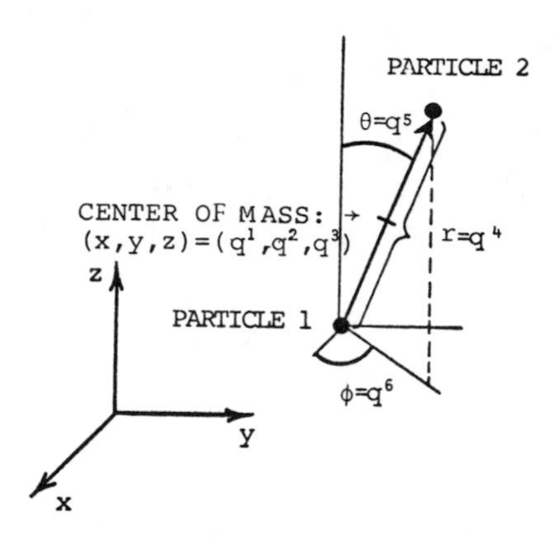

Figure 2.3.

We will allow ϵ^i to vary, but we keep $y(t)$ fixed for a moment. Consider the action

$$S(\epsilon^1, \ldots, \epsilon^n) = \int_{t_1}^{t_2} L(q_0^i + \epsilon^i y, \dot{q}_0^i + \epsilon^i \dot{y}, t)dt$$

as a function of $\epsilon^1, \ldots, \epsilon^n$. It assumes an extremal value for $(\epsilon^1, \ldots, \epsilon^n) = (0, \ldots, 0)$. Therefore,

$$\begin{aligned} 0 = \frac{\partial S}{\partial \epsilon}\Big|_{(0,\ldots,0)} &= \int_{t_1}^{t_2} \left[\frac{\partial L}{\partial q^i}\, y + \frac{\partial L}{\partial \dot{q}^i}\, \dot{y} \right] dt \\ &= \int_{t_1}^{t_2} \left[\frac{\partial L}{\partial q^i} - \frac{d}{dt}\frac{\partial L}{\partial \dot{q}^i} \right] y(t)\, dt. \end{aligned}$$

But this is consistent only if

$$\frac{\partial L}{\partial q^i} - \frac{d}{dt}\frac{\partial L}{\partial \dot{q}^i} = 0, \qquad i = 1, \ldots, n. \tag{2.6}$$

Thus each of the components satisfies the appropriate Euler–Lagrange equation.

Let us return to the case where a single particle moves in three-dimensional space. If it moves under the influence of a potential $V(\mathbf{x})$ then we can immediately extend the above analysis. It corresponds to a Lagrangian

$$L(\mathbf{x}, \mathbf{v}, t) = \frac{1}{2}\, m\mathbf{v}^2 - V(\mathbf{x}), \tag{2.7}$$

which reproduces the well-known Newtonian equation of motion

$$m\,\frac{d^2\mathbf{x}}{dt^2} = -\nabla V(\mathbf{x}). \tag{2.8}$$

But we are really interested in the motion of a charged particle in an electromagnetic field. This corresponds to the equation of motion

$$m \frac{d^2\mathbf{x}}{dt^2} = q(\mathbf{E} + \mathbf{v} \times \mathbf{B}), \tag{2.9}$$

and this equation cannot immediately be reproduced by a Lagrangian of the form (2.7) for two reasons:

1. The force may contain an explicit time dependence because the field strengths may depend explicitly on time. (Consider for instance the electric field $\mathbf{E}$ in a capacitor joined to a circuit with an oscillating current.)
2. The force is velocity-dependent due to the $\mathbf{v} \times \mathbf{B}$ term.

Hence we can try replace the simple-minded potential $V(\mathbf{x})$ by a generalized potential $U(\mathbf{x}, \mathbf{v}, t)$ that may depend explicitly on position, velocity, and time. So we look for a Lagrangian of the form

$$L(\mathbf{x}, \mathbf{v}, t) = \frac{1}{2} m\mathbf{v}^2 - U(\mathbf{x}, \mathbf{v}, t). \tag{2.10}$$

Substituting this Lagrangian into the Euler–Lagrange equations (2.6), we get

$$0 = \frac{\partial L}{\partial \mathbf{x}} - \frac{d}{dt}\frac{\partial L}{\partial \mathbf{v}} = -\frac{\partial U}{\partial \mathbf{x}} - \frac{d}{dt}(m\mathbf{v}) + \frac{d}{dt}\frac{\partial U}{\partial \mathbf{v}};$$

i.e.,

$$m \frac{d^2x}{dt^2} = -\frac{\partial U}{\partial \mathbf{x}} + \frac{d}{dt}\frac{\partial U}{\partial \mathbf{v}}.$$

To determine the generalized potential, we must rearrange the expression for the Lorentz force (2.9)

$$m \frac{d^2\mathbf{x}}{dt^2} = q(\mathbf{E} + \mathbf{v} \times \mathbf{B}).$$

First, we introduce the gauge potentials $(\phi, \mathbf{A})$ through the well-known relations (1.19)

$$\mathbf{B} = \nabla \times \mathbf{A}; \qquad \mathbf{E} = -\nabla\phi - \frac{\partial \mathbf{A}}{\partial t},$$

giving

$$m \frac{d^2\mathbf{x}}{dt^2} = q\left(-\nabla\phi - \frac{\partial \mathbf{A}}{\partial t} + \mathbf{v} \times (\nabla \times \mathbf{A})\right).$$

Clearly, this is a mess, and we will have to rearrange the $\mathbf{v} \times (\nabla \times \mathbf{A})$ term. Using the standard rules of vector analysis, we get

$$\begin{aligned}\mathbf{v} \times (\nabla \times \mathbf{A}) &= \nabla(\mathbf{v} \cdot \mathbf{A}) - (\mathbf{v} \cdot \nabla)\mathbf{A} \\ &= \frac{\partial}{\partial \mathbf{x}}(\mathbf{v} \cdot \mathbf{A}) - v \cdot \frac{\partial \mathbf{A}}{\partial \mathbf{x}} = \frac{\partial}{\partial \mathbf{x}}(\mathbf{v} \cdot \mathbf{A}) - \frac{\partial \mathbf{A}}{\partial \mathbf{x}}\frac{d\mathbf{x}}{dt}.\end{aligned}$$

Substituting this into the equation of motion, we now obtain

$$\begin{aligned} m\,\frac{d^2\mathbf{x}}{dt^2} &= q\left(-\frac{\partial\phi}{\partial\mathbf{x}} - \frac{\partial\mathbf{A}}{\partial t} + \frac{\partial}{\partial\mathbf{x}}\,(\mathbf{v}\cdot\mathbf{A}) - \frac{\partial\mathbf{A}}{\partial\mathbf{x}}\frac{d\mathbf{x}}{dt}\right) \\ &= q\left(-\frac{\partial}{\partial\mathbf{x}}\,[\phi - \mathbf{v}\cdot\mathbf{A}] - \frac{\partial\mathbf{A}}{\partial t} - \frac{\partial\mathbf{A}}{\partial\mathbf{x}}\cdot\frac{d\mathbf{x}}{dt}\right) \\ &= q\left(-\frac{\partial}{\partial\mathbf{x}}\,[\phi - \mathbf{v}\cdot\mathbf{A}] - \frac{\partial\mathbf{A}}{\partial t}\right), \end{aligned}$$

and we are almost through. Performing the trivial substitution

$$\mathbf{A} = \frac{\partial}{\partial\mathbf{v}}\,(\mathbf{A}\cdot\mathbf{v}) = \frac{\partial}{\partial\mathbf{v}}\,(\mathbf{A}\cdot\mathbf{v} - \phi),$$

which is permitted because ϕ does *not* depend on $\mathbf{v}$, we get

$$m\,\frac{d^2\mathbf{x}}{dt^2} = q\left(-\frac{\partial}{\partial\mathbf{x}}\,[\phi - \mathbf{v}\cdot\mathbf{A}] + \frac{d}{dt}\frac{\partial}{\partial\mathbf{v}}\,[\phi - \mathbf{v}\cdot\mathbf{A}]\right),$$

and from this formula we can immediately read off the generalized potential

$$U(\mathbf{x}, \mathbf{v}, t) = q[\phi - \mathbf{v}\cdot\mathbf{A}]. \tag{2.11}$$

Thus we have found the total Lagrangian:

$$L(\mathbf{x}, \mathbf{v}, t) = \frac{1}{2}\,m\mathbf{v}^2 - q\phi(t, \mathbf{r}) + q\mathbf{A}(t, \mathbf{r})\cdot\mathbf{v}. \tag{2.12}$$

We may think of the above result in the following way: If we have a system without electromagnetic fields, then to any path joining $A(t_1, \mathbf{x}_1)$ and $B(t_2, \mathbf{x}_2)$ there corresponds an action

$$S_0 = \int_{t_1}^{t_2}\left(\frac{1}{2}\,m\mathbf{v}^2 - V(\mathbf{x})\right)dt.$$

If we switch on an external electromagnetic field $(\phi, \mathbf{A})$, then we have to add another piece of action, the *interaction* term:

$$S_I = \int_{t_1}^{t_2}[-q\phi + q\mathbf{A}\cdot\mathbf{v}]dt.$$

The total action now becomes: $S = S_0 + S_I$. We may rewrite the interaction term in a more elegant manner. Let us denote the path joining the space–time points $A(t_1, \mathbf{x}_1)$ and $B(t_2, \mathbf{x}_2)$ by Γ. (See Figure 2.4.) We then obtain

$$\begin{aligned} S_I &= -q\int_{t_1}^{t_2}\phi\,dt + q\int_{t_1}^{t_2}\mathbf{A}\cdot\frac{d\mathbf{r}}{dt}\,dt = q\left[\int_{t_1}^{t_2}-\phi\,dt + \int_{t_1}^{t_2}\mathbf{A}\cdot d\mathbf{r}\right] \\ &= q\int_A^B[A_0dx^0 + A_idx^i]; \end{aligned}$$

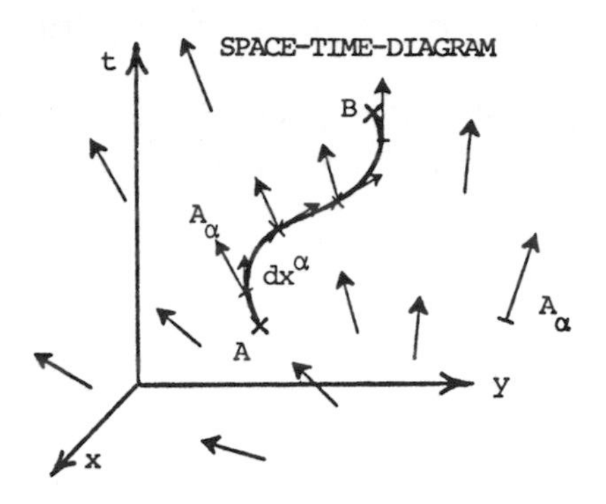

Figure 2.4.

i.e.,

$$S_I = q \int_\Gamma A_\alpha dx^\alpha. \tag{2.13}$$

In the Lagrangian formalism this formula replaces the Lorentz force! Observe that in this formalism the particle interacts directly with the gauge potential and not with the field strengths. Now, remember the gauge symmetry: We are allowed to replace A_α by $A_\alpha + \partial_\alpha \chi$ without changing any physics. This is obvious if we look at the equation of motion

$$m \frac{d^2\mathbf{x}}{dt^2} = q(\mathbf{E} + \mathbf{v} \times \mathbf{B}),$$

because the field strengths are gauge invariant! But does there exist a quick way to see it in the Lagrangian formalism? What we want to show is that the form of the Lagrangian immediately implies that the motion of a particle is unaffected by a gauge transformation. It should be possible to see this without explicitly computing the equation of motion. The particle is following a path Γ_0 that extremizes the full action $S = S_0 + S_I$. Performing a gauge transformation, we get a new interaction term:

$$\begin{aligned} S_I' &= q \int_\Gamma A_\alpha' dx^\alpha = q \int_\Gamma (A_\alpha + \partial_\alpha \chi) dx^\alpha \\ &= q \int A_\alpha dx^\alpha + q \int (\partial_\alpha \chi) dx^\alpha = S + q[\chi(B) - \chi(A)]. \end{aligned} \tag{2.14}$$

But the last term does not at all depend on the path Γ! For any path, the action will therefore be shifted by the same amount. But then Γ_0 is still an extremal path, and we have explicitly shown the gauge invariance.

§ 2.3 Basic Principles of Quantum Mechanics

We are now prepared for the quantization procedure! There are actually two alternative ways of quantizing a system:

1. The path integral technique (Feynman's procedure).
2. The canonical quantization.

The first is the procedure that serves our purpose best, so we will start with that one following Feynman [1964]. The fundamental concept in quantum mechanics is the *probability amplitude*. Let us assume that we have an electron gun A emitting electrons, a screen B with two holes, and a screen C with a detector that can measure the arrival of electrons. (See Figure 2.5.)

An *event* is the registration of an electron in the detector at a specific position x. To each event we associate a probability amplitude $\phi(x)$, which is a complex number, and the probability $P(x)$ for the corresponding event is found by squaring the amplitude:

$$P(x) = |\phi(x)|^2. \tag{2.15}$$

Furthermore, the electron may have passed through either of the holes 1 and 2. Hence we have two *alternative ways* in which the event can occur. Each of these is characterized by a probability amplitude, $\phi_1(x)$ and $\phi_2(x)$. The total amplitude is found by adding the two amplitudes:

$$\phi(x) = \phi_1(x) + \phi_2(x). \tag{2.16}$$

In applying this principle it is important that we have no way of verifying experimentally which of the alternatives actually occurred. If we perform the experiment in such a way that we can decide which of the alternative trajectories the electron followed, then interference is lost, and we see no wave pattern! (See Figure 2.6.)

Now consider an experiment with polarized light. We have a beam of photons that has passed through a polarizer 0. In the direction of the beam we have put two more polarizers, 1 and 2. (See Figure 2.7.)

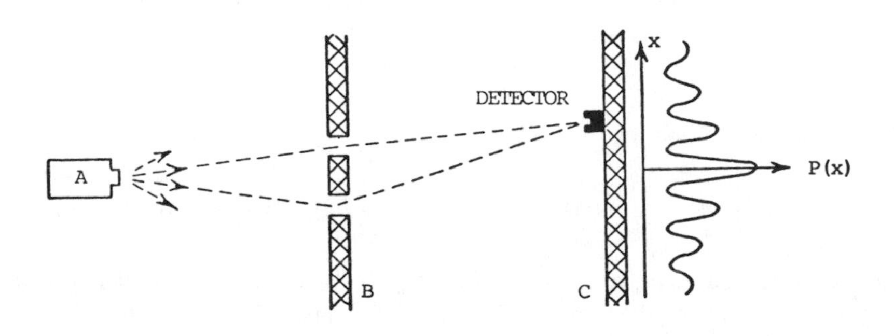

Figure 2.5.

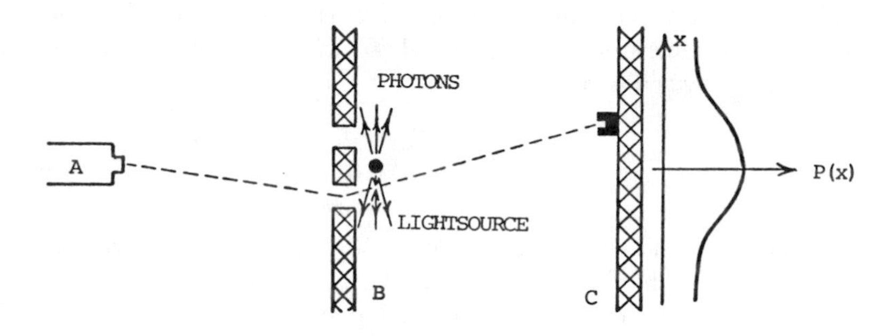

Figure 2.6.

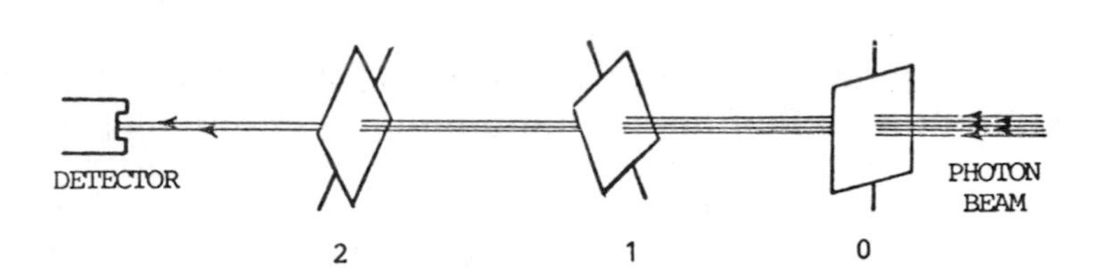

Figure 2.7.

For a given photon, there is a probability amplitude ϕ_{10} that it will pass through polarizer 1, and once it has passed through that polarizer, there is a probability amplitude ϕ_{21} that it will pass through polarizer 2. We are interested in the situation where a given photon actually hits the detector after passing polarizer 2. This amplitude ϕ_{20} is found by multiplying the amplitudes corresponding to the *individual steps*:

$$\phi_{20} = \phi_{21}\phi_{10}. \tag{2.17}$$

Observe that this time there are no alternative ways to arrive at the detector. There is only one way, which consists of two steps. First, the photon has to pass through polarizer 1, and then it has to pass through polarizer 2.

We have now stated all the fundamental laws of probability amplitudes. They are basic laws; i.e., they cannot be derived from some underlying principles. We simply have to accept them as a starting point for the quantum-mechanical description of a system. For later reference we have collected them in the following scheme:

The Basic Principles of Quantum Mechanics:

1. To each event is associated a probability amplitude ϕ. The probability of the event is found by squaring the amplitude:

$$P = |\phi|^2. \tag{2.15}$$

2. If an event may classically occur in several alternative ways $i = 1, \ldots, n$ each of which is characterized by the amplitude ϕ_i, then the total amplitude of the event is found by addition:

$$\phi = \sum_{i=1}^{n} \phi_i. \tag{2.16}$$

3. If an event occurs in a way that can be decomposed into several individual steps $j = 1, \ldots, m$ each of which is characterized by the amplitude ϕ_j, then the total amplitude of the event is found by multiplication:

$$\phi = \prod_{j=1}^{m} \phi_j. \tag{2.17}$$

Now we are prepared for the discussion of a system consisting of a single particle. Consider two fixed times t_1 and t_2. At time t_1 we have a probability amplitude $\psi(\mathbf{r}, t_1)$ of finding the particle at $\mathbf{r}$. At time t_2 we have another probability amplitude $\psi(\mathbf{r}, t_2)$ of finding the particle at $\mathbf{r}$. The amplitude $\psi(\mathbf{r}, t)$ is of course nothing but the Schrödinger wave function. The dynamical problem we must solve is this: How does $\psi(\mathbf{r}, t)$ develop over time? We will solve it by finding a formula that connects $\psi(\mathbf{r}, t)$ at times $t = t_1$ and $t = t_2$.

Consider $\psi(\mathbf{r}_2, t_2)$. It is the probability amplitude of finding the particle at $\mathbf{r}_2$ at time t_2. *Now let $K(\mathbf{r}_2; t_2 \mid \mathbf{r}_1; t_1)$ denote the probability amplitude for the event that a particle emitted at $\mathbf{r}_1$ at time t_1 will be observed at $\mathbf{r}_2$ at time t_2.* Then we can argue in the following way. A particle that arrives at $\mathbf{r}_2$ at time t_2 must have been somewhere at time t_1! Hence there are several alternative paths the particle could have followed, each characterized by the position at time t_1. Each of these possibilities is composed of two individual steps:

1. The particle was at the point $\mathbf{r}_1$ at time t_1 (amplitude $= \psi(\mathbf{r}_1, t)$).
2. The particle was emitted from $\mathbf{r}_1$ at time t_1 and observed at $\mathbf{r}_2$ at time t_2 (amplitude $= K(\mathbf{r}_2; t_2 \mid \mathbf{r}_1; t_1)$).

According to (2.17), each possibility is then characterized by the amplitude

$$K(\mathbf{r}_2; t_2 \mid \mathbf{r}_1; t_1)\psi(\mathbf{r}_1, t_1).$$

To find the total amplitude $\psi(\mathbf{r}_2, t_2)$ for arriving at $\mathbf{r}_2$ at time t_2, we must sum up all these amplitudes (cf. (2.16)); i.e., we get

$$\psi(\mathbf{r}_2; t_2) = \int K(\mathbf{r}_2; t_2 \mid \mathbf{r}_1; t_1)\psi(\mathbf{r}_1; t_1) d\mathbf{r}_1. \tag{2.18}$$

This is the basic dynamical equation of the theory. Although it is an integral equation, we shall see later on that it is completely equivalent to the *Schrödinger equation* (2.2).

§ 2.4 Path Integrals: The Feynman Propagator

Consider now the amplitude $K(\mathbf{r}_2;\, t_2 \mid \mathbf{r}_1;\, t_1)$. It is called the *Feynman propagator*, and if we can determine that, we will control the dynamical evolution of the Schrödinger wave function! Now, $K(\mathbf{r}_2;\, t_2 \mid \mathbf{r}_1;\, t_1)$ is the amplitude for the event that a particle that was emitted at $\mathbf{r}_1$ at time t_1 will be observed at $\mathbf{r}_2$ at time t_2. To come from $(\mathbf{r}_1, t_1)$ to $(\mathbf{r}_2, t_2)$, a classical particle must have followed some path Γ. (See Figure 2.8.) Consequently, the classical particle could have moved in several ways, each corresponding to a path leading from $A(\mathbf{r}_1, t_1)$ to $B(\mathbf{r}_2, t_2)$. Let $\phi_\Gamma(B \mid A)$ denote the amplitude that a particle emitted at A and observed at B actually moved along the path Γ. Then we get

$$K(B \mid A) = \int \phi_\Gamma(B \mid A) D[\Gamma], \tag{2.19}$$

where we sum over all paths Γ leading from A to B.

Obviously, this sum is very complicated, since there is an infinity of paths! It is called a *path integral*. The precise definition of such a sum is very complicated, and as we are interested only in the basic physical ideas, we will continue writing the path integral in the above naive form!

We have reduced our problem to that of finding $\phi_\Gamma(B \mid A)$. Now, this cannot be deduced from basic principles, so we must simply postulate the value of it. This is where we make contact with the Lagrangian formulation of classical mechanics.

To each path Γ we have associated a classical action

$$S(\Gamma) = \int_{t_1}^{t_2} L\left(\mathbf{x}, \frac{d\mathbf{x}}{dt}, t\right) dt.$$

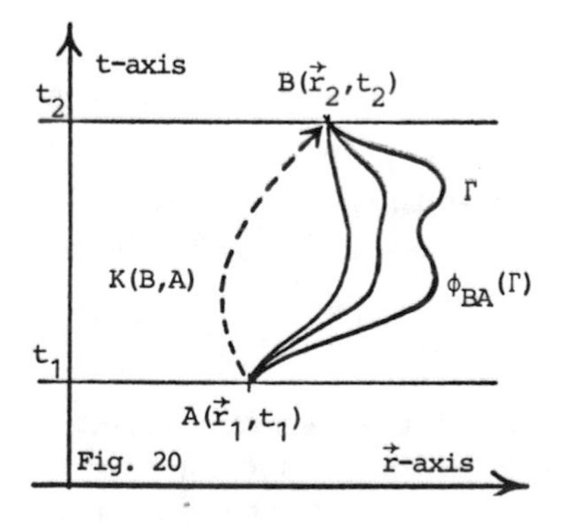

Figure 2.8.

Following a suggestion from Dirac, Feynman now used the following expression (compare Dirac [1935], §33, pp. 125–126):

$$\phi_\Gamma(B \mid A) = \exp\left[\frac{i}{\hbar}\, S(\Gamma)\right]. \tag{2.20}$$

Since each path represents a possible history and each history contributes a phase factor, i.e., a complex number of modulus 1, this equation is often referred to as *Feynman's principle of the democratic equality of all histories!*

This principle leads to the following formula for the propagator:

$$K(\mathbf{r}_2; t_2 \mid \mathbf{r}_1; t_1) = \int_{\mathbf{x}(t_1)=\mathbf{r}_1}^{\mathbf{x}(t_2)=\mathbf{r}_2} \exp\left\{\frac{i}{\hbar}\int_{t_1}^{t_2} L\left(\mathbf{x}, \frac{d\mathbf{x}}{dt}, t\right) dt\right\} D[\mathbf{x}(t)], \tag{2.21}$$

where we sum over all paths $\mathbf{x} = \mathbf{x}(t)$ leading from $(\mathbf{r}_1, t_1)$ to $(\mathbf{r}_2, t_2)$.

Now, this is a very complicated formula, and one might wonder what happened to classical physics? In classical physics we are used to considering only a single path, the path that extremizes the action. In the above formula all paths are on the same footing. There is no special reference to the classical path. But how can the classical path then play such a major role in our everyday experiences?

To understand this, we must first try to characterize what we expect to be a typical classical problem and what we expect to be a typical quantum-mechanical problem.

First, one might be tempted to say that a typical classical problem is one where the involved actions are very great compared to $\hbar$, and a typical quantum mechanical problem is one where the involved actions are comparable to $\hbar$. However, this is not a good formulation, because the action is not a definite number. It is only determined up to a constant, and hence the action for single path has no absolute meaning. But if we take two paths Γ_1 and Γ_2, then the difference $\Delta S = S(\Gamma_2) - S(\Gamma_1)$ has a unique meaning. Therefore, we are allowed to say, *A typical classical problem is a problem in which small changes in the path can produce a great change ΔS compared to $\hbar$, and a typical quantum-mechanical problem is one in which even great changes in the path only produce a change ΔS comparable to $\hbar$.* (See Figure 2.9.)

Suppose we consider a classical situation and pick an arbitrary path Γ_1. It contributes to the propagator with a phase factor $\exp\left[\frac{i}{\hbar}\, S(\Gamma_1)\right]$. But very close to the path Γ_1 there will be another path Γ_2 that differs in the action by $\pi\hbar$, but then the corresponding amplitudes will cancel each other, because

$$\exp\left[\frac{i}{\hbar}\, S(\Gamma_2)\right] = -\exp\left[\frac{i}{\hbar}\, S(\Gamma_1)\right].$$

Thus they do not contribute to the general sum, due to the destructive interference. Although this is the general situation, it is not always true. Let Γ_{cl} be the classical path. Then $S(\Gamma_{\text{cl}})$ is extremal, and all the nearby paths will have almost the same action. Hence we will have constructive interference from all the paths close to the classical one! Since the main contribution to the propagator comes

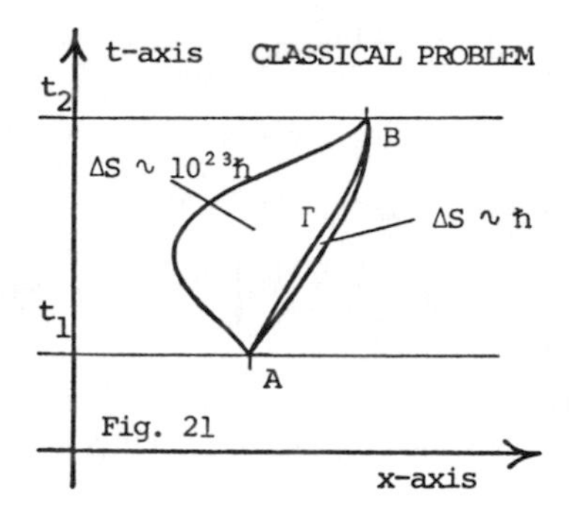

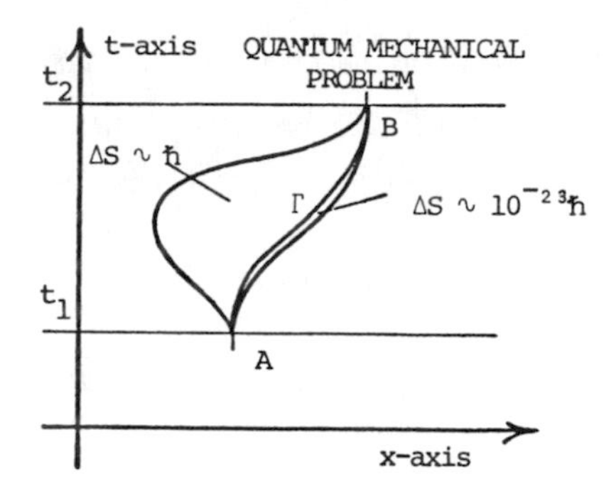

Figure 2.9.

from paths near the classical path, and since they all have approximately the same action as the classical path, we can put as a first approximation

$$K(\mathbf{x}_2; t_2 \mid \mathbf{x}_1; t_1) \sim \exp\left[\frac{i}{\hbar}\int_{t_1}^{t_2} L\left(\mathbf{x}_{\mathrm{cl}}, \frac{d\mathbf{x}_{\mathrm{cl}}}{dt}, t\right) dt\right]. \tag{2.22}$$

This is known as the *classical approximation*. (See Figure 2.10.)

As we have seen, the main contribution to the propagation comes from a "strip" around the classical path where the action varies slowly and where the changes in the action are smaller than $\pi\hbar$. For a typical classical problem, this strip is very "thin," but for a typical quantum-mechanical problem, the strip is very "broad." Consequently, the classical path loses its significance in typical quantum-mechanical situations, say an electron orbiting around a nucleus. The path of the electron is "smeared" out.

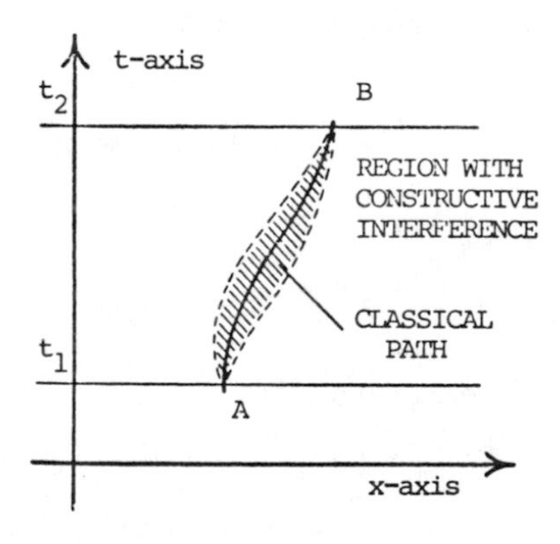

Figure 2.10.

Worked Exercise 2.4.1
Problem: Consider the quadratic Lagrangian

$$L(x, \dot{x}, t) = a(t)x^2 + b(t)\dot{x}^2 + c(t)x\dot{x} + d(t)x + e(t)\dot{x} + f(t).$$

Show starting from (2.21) that the Feynman propagator is given by the formula

$$K(x_2; t_2|x_1; t_1) = A(t_2, t_1) \exp\left[\frac{i}{\hbar}\int_{t_1}^{t_2} L(x_{\text{cl}}, \dot{x}_{\text{cl}}, t)dt\right]. \tag{2.23}$$

Assume now that the coefficients $a(t)$, $b(t)$, and $c(t)$ do not depend on time; i.e., they are constants. Show that the Feynman propagator then is given by the formula

$$K(x_2; t_2|x_1; t_1) = A(t_2 - t_1) \exp\left[\frac{i}{\hbar}\int_{t_1}^{t_2} L(x_{\text{cl}}, \dot{x}_{\text{cl}}, t)dt\right]. \tag{2.24}$$

In both cases $A(\cdot)$ denotes an unknown function, and x_{cl} denotes the classical path.

The above quadratic Lagrangian includes, for instance, the following cases:

a. A free particle: $L = \frac{1}{2}m\dot{x}^2$.
b. The harmonic oscillator: $L = \frac{1}{2}m\dot{x}^2 - \frac{1}{2}m\omega^2x^2$.
c. A forced oscillator: $L = \frac{1}{2}m\dot{x}^2 - \frac{1}{2}m\omega^2x^2 - \lambda(t)x$.

Before we use formula (2.24), let us point out another feature of the propagator. By definition, $K(x_2; t_2 \mid x_1; t_1)$ is the probability amplitude for finding the particle at the point x_2 at time t_2 when we know that it was emitted at x_1 at time t_1. Let us keep x_1 and t_1 fixed for the moment and regard $K(x_2; t_2 \mid x_1; t_1)$ as a function of x_2 and t_2:

$$K_{(x_1,t_1)}(x, t) = K(x; t \mid x_1; t_1).$$

Then $K_{(x_1,t_1)}(x, t)$ is simply a Schrödinger wave function, since it denotes the amplitude for finding the particle at x! But it is a Schrödinger wave function corresponding to a very special situation, because at time t_1 we know exactly where the particle is: It is at $x = x_1$. The amplitude is *not* smeared out at $t = t_1$! This can also be seen from the integral equation (2.18) for an arbitrary Schrödinger wave function

$$\psi(x_2, t_2) = \int K(x_2; t_2 \mid x_1; t_1)\psi(x_1, t_1)dx_1.$$

If we put $t_2 = t_1$, we simply get

$$\psi(x_2, t_1) = \int K(x_2; t_1 \mid x_1; t_1)\psi(x_1, t_1)dx_1.$$

But then we can immediately read off that

$$K(x_2; t_1 \mid x_1; t_1) = \delta(x_2 - x_1).$$

Hence $K_{(x_1,t_1)}(x, t)$ reduces to a δ-function at time t_1;

$$K_{(x_1,t_1)}(x, t) = \delta(x - x_1).$$

Since $K_{(x_1,t_1)}(x,t)$ itself is a Schrödinger wave function, it must furthermore satisfy the integral equation (2.18). We have thus deduced the following important group property of the propagator:

$$K(x_3; t_3 \mid x_1; t_1) = \int K(x_3; t_3 \mid x_2; t_2) K(x_2; t_2 \mid x_2; t_1) dx_2. \tag{2.25}$$

§ 2.5 Illustrative Example: The Free Particle Propagator

We now are in a position where we can calculate the propagator for a free particle. Here $L(x; \dot{x}) = \frac{1}{2} m\dot{x}^2$, and we can use Exercise 2.4.1 about quadratic Lagrangians:

$$\begin{aligned} K(x_2, t_2 \mid x_1, t_1) &= A(t_2 - t_1) \exp\left[\frac{i}{\hbar}\int_{t_1}^{t_2} \frac{m}{2}\, \dot{x}_{\text{cl}}^2 dt\right] \\ &= A(t_2 - t_1) \exp\left[\frac{i}{\hbar}\frac{m}{2}\frac{(x_2 - x_1)^2}{t_2 - t_1}\right]. \end{aligned} \tag{2.26}$$

(The classical path is a straight line; i.e.,

$$\dot{x}_{\text{cl}} = \frac{x_2 - x_1}{t_2 - t_1}$$

is a constant!) To finish, we must determine the function A. Substituting (2.26) into the integral equation (2.25), we get

$$\begin{aligned} &A(t_3 - t_1) \exp \frac{i}{\hbar}\frac{m}{2}\frac{(x_3 - x_1)^2}{t_3 - t_1} \\ &\quad = A(t_3 - t_2) A(t_2 - t_1) \\ &\quad \cdot \int_{x_2=-\infty}^{x_2=+\infty} \exp\left[\frac{i}{\hbar}\frac{m}{2}\left(\frac{(x_3 - x_2)^2}{t_3 - t_2} + \frac{(x_2 - x_1)^2}{t_2 - t_1}\right)\right] dx_2. \end{aligned}$$

Using the algebraic formula

$$\begin{aligned} &\frac{(x_3 - x_2)^2}{t_3 - t_2} + \frac{(x_2 - x_1)^2}{t_2 - t_1} \\ &= \frac{t_3 - t_1}{(t_3 - t_2)(t_2 - t_1)} \cdot \left[x_2 - \frac{x_3(t_2 - t_1) + x_1(t_3 - t_2)}{t_3 - t_1}\right]^2 + \frac{(x_3 - x_1)^2}{t_3 - t_1}, \end{aligned}$$

we may rewrite the last integral:

$$\begin{aligned} &\int_{x_2=-\infty}^{x_2=+\infty} \exp\left[\frac{i}{\hbar}\frac{m}{2}\left(\frac{(x_3 - x_2)^2}{t_3 - t_2} + \frac{(x_2 - x_1)^2}{t_2 - t_1}\right)\right] dx_2 \\ &\quad = \exp \frac{i}{\hbar}\frac{m}{2}\frac{(x_3 - x_1)^2}{t_3 - t_1} \int_{u=-\infty}^{u=+\infty} \exp\left[\frac{i}{\hbar}\frac{m}{2}\frac{(t_3 - t_1)}{(t_3 - t_2)(t_2 - t_1)} u^2\right] du, \end{aligned}$$

where we have introduced

$$u = x_2 - \frac{x_3(t_2 - t_1) + x_1(t_3 - t_2)}{t_3 - t_1}$$

to get rid of a lot of junk. The total integral formula is thus reduced to

$$\frac{A(t_3 - t_1)}{A(t_3 - t_2)A(t_2 - t_1)} \exp\left[\frac{i}{\hbar}\frac{m}{2}\frac{(x_3 - x_1)^2}{t_3 - t_1}\right]$$

$$= A(t_3 - t_2)A(t_2 - t_1) \exp\left[\frac{i}{\hbar}\frac{m}{2}\frac{(x_3 - x_1)^2}{t_3 - t_1}\right]$$

$$\cdot \int_{u=-\infty}^{u=+\infty} \exp\left[\frac{i}{\hbar}\frac{m}{2}\frac{(t_3 - t_1)}{(t_3 - t_2)(t_2 - t_1)}u^2\right] du.$$

Notice that x_2 has completely disappeared! The integral does not depend on x_3 and x_1, so both sides have the same dependence on x_3 and x_1. Everything fits beautifully. Using the formula

$$\int_{x=-\infty}^{x=+\infty} \exp\left[-ax^2\right] dx = \sqrt{\frac{\pi}{a}}, \tag{2.27}$$

which may be used for a complex a too, we finally get

$$\frac{A(t_3 - t_1)}{A(t_3 - t_2)A(t_2 - t_1)} = \sqrt{\frac{2\pi i\hbar}{m}\frac{(t_3 - t_2)(t_2 - t_1)}{(t_3 - t_1)}}$$

$$= \frac{\sqrt{\frac{m}{2\pi i\hbar(t_3 - t_1)}}}{\sqrt{\frac{m}{2\pi i\hbar(t_3 - t_2)}}\sqrt{\frac{m}{2\pi i\hbar(t_2 - t_1)}}}.$$

From this we see that up to a trivial phase factor $\exp(i\alpha t)$, which we normalize to 1, the function $A(t)$ is given by

$$A(t) = \sqrt{\frac{m}{2\pi i\hbar t}}.$$

So now we have computed in full detail the *free-particle propagator*

$$K(x_2; t_2 \mid x_1; t_1) = \sqrt{\frac{m}{2\pi i\hbar(t_2 - t_1)}} \exp\left[\frac{i}{\hbar}\frac{m}{2}\frac{(x_2 - x_1)^2}{t_2 - t_1}\right]. \tag{2.28}$$

Exercise 2.5.1
Problem: Show by an explicit calculation that the propagator of a free particle reduces to a δ-function in the limit as $t \to 0^+$.

Let us try to become familiar with the free-particle propagator. We know that $K(x; t \mid 0; 0)$ is the Schrödinger wave function of a free particle emitted from $x = 0$ at the time $t = 0$:

$$\psi(x, t) = K(x, t \mid 0, 0) = \sqrt{\frac{m}{2\pi i\hbar t}} \exp\left[\frac{i}{\hbar}\frac{m}{2}\frac{x^2}{t}\right]. \tag{2.29}$$

Let us focus our attention on a specific point (x_0, t_0). If the free particle is observed at $x = x_0$ at time t_0, we would say that it classically had momentum

$$p_0 = mv_0 = m\,\frac{x_0}{t_0}$$

and energy

$$E_0 = \frac{1}{2}\,mv_0^2 = \frac{1}{2}\,m\,\frac{x_0^2}{t_0^2}\,.$$

If we investigate the variation of the *phase* $\frac{m}{2\hbar}\,\frac{x^2}{t}$ in the vicinity of $(x_0; t_0)$, we find

$$\begin{aligned}
\psi(x, t) &= \sqrt{\frac{m}{2\pi i\hbar t}}\,\exp\left[\frac{i}{\hbar}\,\frac{m}{2}\,\frac{x^2}{t}\right] \\
&\approx \sqrt{\frac{m}{2\pi i\hbar t}}\,\exp\left\{\frac{im}{2\hbar}\left[\frac{x_0^2}{t_0} + \frac{\partial}{\partial x}\Big|_{x_0,t_0}\left(\frac{x^2}{t}\right)\cdot(x - x_0)\right.\right. \\
&\qquad \left.\left. + \frac{\partial}{\partial t}\Big|_{x_0,t_0}\left(\frac{x^2}{t}\right)\cdot(t - t_0) + \cdots\right]\right\} \\
&= \sqrt{\frac{m}{2\pi i\hbar t}}\,\exp\left[\frac{i}{\hbar}\left[m\,\frac{x_0}{t_0}\,x - \frac{1}{2}\,m\,\frac{x_0^2}{t_0^2}\,t\right]\right].
\end{aligned}$$

Thus, very close to (x_0, t_0), the wave function varies like

$$\psi(x, t) \approx \sqrt{\frac{m}{2\pi i\hbar t}}\,\exp\left[\frac{i}{\hbar}\,[p_0 x - E_0 t]\right] \qquad \text{for } (x, t) \approx (x_0, t_0). \quad (2.30)$$

This is closely related to the *Einstein–de Broglie rule*, according to which a particle with momentum p and energy E is associated with a wave function with wave length $\lambda = \frac{h}{p}$ and frequency $\nu = E/h$:

$$\exp\left[i\left(\frac{2\pi x}{\lambda} - \frac{\nu t}{2\pi}\right)\right] = \exp\left[\frac{i}{\hbar}\,(px - Et)\right].$$

While $\psi(x, t)$ represents a particle that is located at $x = 0$ at the time $t = 0$, we would also like to have a Schrödinger wave function representing a free particle with specific energy and momentum. At time $t = 0$ we suggest the following amplitude:

$$\psi(x, 0) = \exp\left[\frac{i}{\hbar}\,px\right],$$

where p is a constant. At a later time t we can compute $\psi(x, t)$ using the integral equation (2.18).

$$\psi(x, t) = \int_{x_1=-\infty}^{x_1=+\infty} \sqrt{\frac{m}{2\pi i\hbar t}}\,\exp\left[\frac{i}{\hbar}\,\frac{m}{2}\,\frac{(x - x_1)^2}{5}\right]\exp\left[\frac{i}{\hbar}\,px_1\right]dx_1.$$

Using the algebraic identity

$$px_1 + \frac{m}{2}\frac{(x-x_1)^2}{t} = \frac{m}{2t}\left[x_1 - \left(x - \frac{pt}{m}\right)\right]^2 + px - \frac{p^2}{2m}t,$$

we get

$$\psi(x,t) = \exp\left\{\frac{i}{\hbar}\left[px - \frac{p^2}{2m}t\right]\right\}\sqrt{\frac{m}{2\pi i\hbar t}}$$
$$\int_{x_1=-\infty}^{x_1=+\infty} \exp\left\{\frac{i}{\hbar}\frac{m}{2t}\left[x_1 - \left(x - \frac{pt}{m}\right)\right]^2\right\} dx_1 .$$

Using the substitution $u = x_1 - \left(x - \frac{pt}{m}\right)$, this may be reduced to

$$\begin{aligned}\psi(x,t) &= \exp\left\{\frac{i}{\hbar}\left[px - \frac{p^2}{2m}t\right]\right\}\sqrt{\frac{m}{2\pi i\hbar t}}\int_{u=-\infty}^{u=+\infty}\exp\left[\frac{i}{\hbar}\frac{m}{2}u^2\right]du \\ &= \exp\left\{\frac{i}{\hbar}\left[px - \frac{p^2}{2m}t\right]\right\}.\end{aligned}$$

Thus the free particle is represented by a solution of the form

$$\exp\left\{\frac{i}{\hbar}\left[px - \frac{p^2}{2m}t\right]\right\},$$

which according to Einstein–de Broglie's rule must be interpreted as a particle with momentum p and energy

$$E = \frac{p^2}{2m}.$$

Let us summarize: We have played a little with the free propagator and discovered two exact wave functions:

$$\psi(x,t) = \sqrt{\frac{m}{2\pi i\hbar t}}\exp\left[\frac{i}{\hbar}\cdot\frac{m}{2}\cdot\frac{x^2}{t}\right], \tag{2.31}$$

which represents a free particle, strictly located at $x = 0$ at time $t = 0$, and

$$\psi(x,t) = \exp\left[\frac{i}{\hbar}\left[px - \frac{p^2}{2m}t\right]\right], \tag{2.32}$$

which represents a free particle with a specific momentum p and specific energy

$$E = \frac{p^2}{2m}.$$

§ 2.6 The Bohm–Aharonov Effect: The Lorentz Force

Having described the basic principles of the quantization procedure, we will return to our original problem: The motion of a charged particle in an external electromagnetic field. Consider a path Γ connecting $A(x_1, t_1)$ and $B(x_2, t_2)$. Before the external field is switched on, this path contributes the phase factor

$$\phi_\Gamma^0(B \mid A) = \exp\left[\frac{i}{\hbar}\int_{t_1}^{t_2}\frac{1}{2}m\mathbf{v}^2 dt\right]$$

to the propagator, but after we have switched on the electromagnetic field, we must add to the action the interaction term (2.13)

$$S_I = q\int A_\alpha dx^\alpha .$$

Consequently, the phase factor is changed into

$$\begin{aligned}\phi_\Gamma(B \mid A) &= \exp\left[\frac{i}{\hbar}\left[S_0(\Gamma) + S_I(\Gamma)\right]\right]\\ &= \exp\left[\frac{i}{\hbar}S_0(\Gamma)\right]\exp\left[\frac{i}{\hbar}S_I(\Gamma)\right] = \phi_\Gamma^0(B \mid A)\phi_\Gamma^I(B \mid A).\end{aligned}$$

So we see that the external field produces a change obtained simply by multiplying the original amplitude by the *gauge phase factor*

$$\phi_\Gamma^I(B \mid A) = \exp\left[\frac{i}{\hbar}q\int A_\alpha dx^\alpha\right]. \tag{2.33}$$

This is the quantum-mechanical law, which replaces the Lorentz force!

Now we are ready to give a *qualitative* discussion of the slit experiment. We first look at a situation where there is no external field. (See Figure 2.11.)

If the electron gun emits electrons with a characteristic momentum p, then the electron wave function has de Broglie wavelength λ, where $\lambda = h/p$. Constructive interferences are expected when the difference between the two path lengths is an

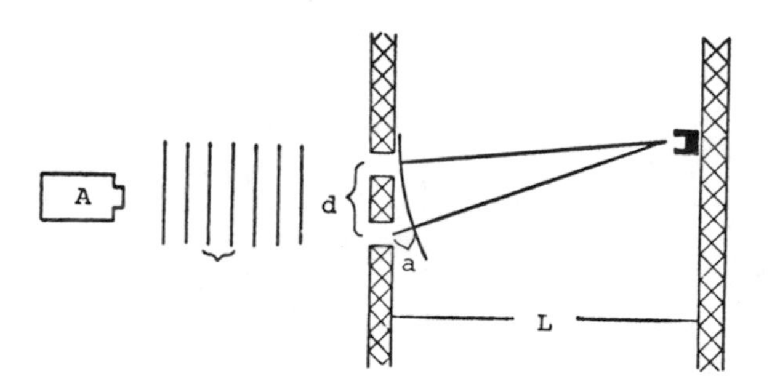

Figure 2.11.

integer multiple of λ. If the difference between the path lengths is a, then the phase difference in the electron wave functions is

$$\Delta^0 = \frac{a}{\lambda} \cdot 2\pi.$$

But if we assume that $L \gg \lambda$, where L is the distance between slit and screen and d is the distance between the two holes, then simple geometric considerations show that

$$\tan\theta = \frac{x}{L} \quad \text{and} \quad \tan\theta = \frac{a/2}{d/2} = \frac{a}{d};$$

i.e.,

$$\frac{x}{L} = \frac{a}{d}.$$

(See Figure 2.12.) Consequently, we may replace a by x, thus obtaining for the phase difference at the point x on the screen

$$\Delta^0(x) = \frac{xd}{L\lambda} 2\pi. \tag{2.34}$$

If x is zero, then $\Delta^0 = 0$, and we have constructive interference. When $x = \frac{\lambda L}{2d}$, then $\Delta^0 = \pi$, and we have destructive interference, etc.! Thus we have obtained a qualitative understanding of the wave pattern on the screen!

Now suppose there is a pure static magnetic field between the slit and the screen. We know that this will produce a change of the phase factor associated with a "classical" electron arriving at the point x. (See Figure 2.13.) (Strictly speaking we should use the classical trajectories, which are circles. If the field is not too strong and the momenta of the electrons are sufficiently high, then the radii of the circles will be great compared with L, and we can safely replace the arcs of the circles with straight lines.)

Now, the main contribution to the phase comes from the paths Γ_1 and Γ_2 shown in Figure 2.13. Let us look at Γ_1. Before the external field is switched on, a "classical"

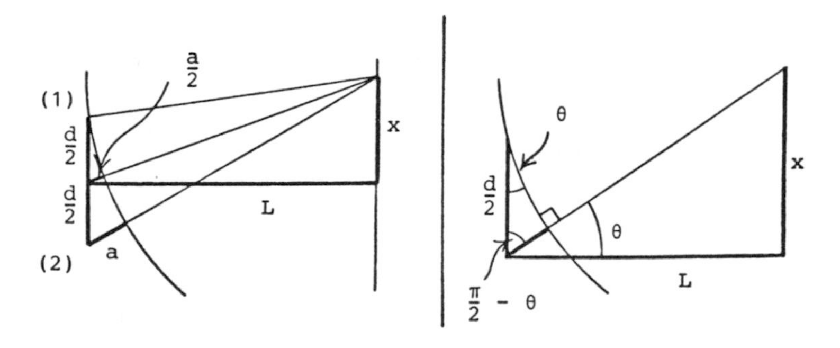

Figure 2.12.

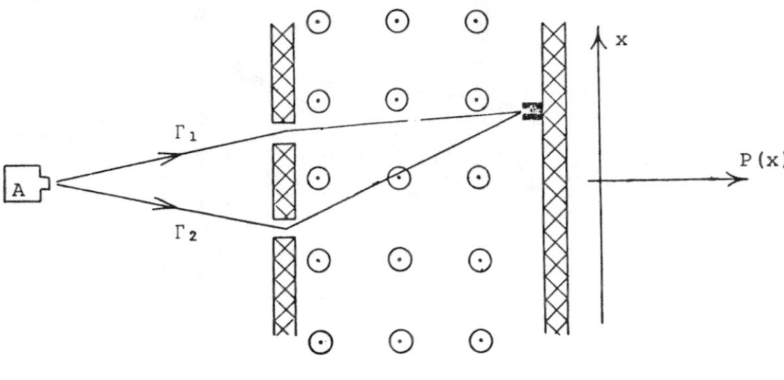

Figure 2.13.

electron moving along Γ_1 will be associated with a phase factor $\exp[i\theta_1]$, i.e., a phase θ_1.

But according to (2.33) switching on the external field will change this phase into

$$\theta_1 - \frac{q}{\hbar}\int_{\Gamma_1}\phi\, dt + \frac{q}{\hbar}\int_{\Gamma_1}\mathbf{A}\cdot d\mathbf{r} = \theta_1 + \frac{q}{\hbar}\int_{\Gamma_1}\mathbf{A}\cdot d\mathbf{r}.$$

(Observe that $\phi = 0$, as there is no electric field.) In a similar way, a "classical" electron moving along Γ_2 will now be associated with the modified phase

$$\theta_2 + \frac{q}{\hbar}\int_{\Gamma_2}\mathbf{A}\cdot d\mathbf{r}.$$

Therefore, the phase difference between the electron wave functions is given by

$$\Delta = \theta_2 - \theta_1 + \frac{q}{\hbar}\left(\int_{\Gamma_2} - \int_{\Gamma_1}\right)\mathbf{A}\cdot d\mathbf{r} = \Delta^0 + \frac{q}{\hbar}\oint\mathbf{A}\cdot d\mathbf{r}. \tag{2.35}$$

Here Δ^0 is the phase difference when there is *no* external field present, and it was calculated in (2.34). The loop integral is found by integrating forwards along the path Γ_2 and backwards along Γ_1. But since we are in a static situation, $\oint \mathbf{A}\cdot d\mathbf{r}$ is nothing but the magnetic flux enclosed by the $\Gamma_2 - \Gamma_1$ loop! This is the formula we are looking for.

We will "use" a very special kind of magnetic field. Under certain circumstances, iron crystals can grow in the form of very long, microscopically thin filaments called *whiskers*. These whiskers can be magnetized, and they then act like very tiny solenoids! Hence we have a purely static magnetic field concentrated in the whiskers (the field outside is extremely small and we may neglect it), but even though there is no magnetic field outside the whiskers, there is a nontrivial gauge potential $\mathbf{A}$. This is exactly what we are after. (See Figure 2.14.)

Now we can show that there is a measurable effect when we let electrons move outside the whiskers. Phase factors corresponding to paths lying on one side of

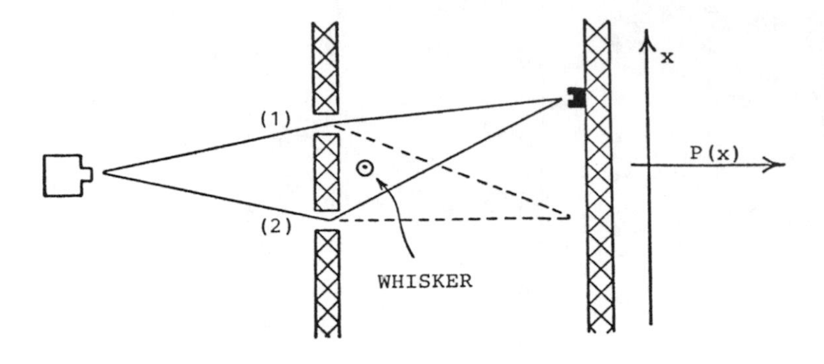

Figure 2.14.

the whiskers interfere with phase factors corresponding to paths lying on the other side of the whiskers, and at the screen, the whisker (i.e., the gauge potential) has therefore produced the following change of phase:

$$\frac{q}{\hbar}\oint \mathbf{A}\cdot d\mathbf{r} = \frac{q}{\hbar}\,\Phi. \tag{2.36}$$

Here Φ is the magnetic flux through the whisker. Observe that the phase difference is independent of the position x on the screen. Hence the total wave pattern on the screen will be shifted a constant amount! This effect is called the *Bohm–Aharonov effect*. (See Figure 2.15.)

Classically, we can only find out whether or not there is a whisker if we hit it directly. But quantum-mechanically, we can detect the whisker because the gauge potential changes the phases of electrons passing by. This remarkable effect was predicted by Bohm and Aharonov (Bohm–Aharonov (1959, 1961)) and later verified experimentally (Möllenstedt and Bayh (1962)).

One might still feel uncomfortable about the entire analysis. We have studied the quantum-mechanical interaction of charged particles with the Maxwell field and

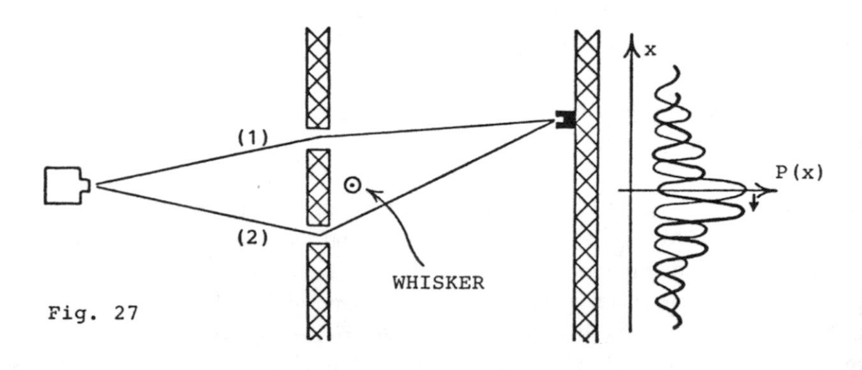

Figure 2.15.

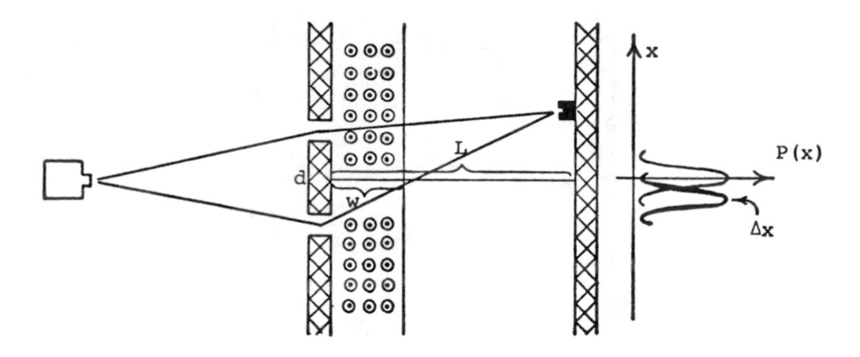

Figure 2.16.

seen that it produces strange effects. We will now derive something well known. We will show that this theory reproduces the Lorentz force, too. For simplicity, we consider only a very simple situation.

We consider the slit experiment once more, but this time we put a homogeneous magnetic field behind the slit. This homogeneous, static field is concentrated in a region of width W, where $W \ll L$. (See Figure 2.16.) If we had no magnetic field, there would be a phase difference (2.34)

$$\Delta^0 = \frac{xd}{L\lambda} 2\pi$$

for electrons arriving at the screen. But the presence of a magnetic field will produce a phase shift proportional to the magnetic flux. In our case, the magnetic flux is approximately BWd. Therefore, the actual phase difference is given by

$$\Delta = \frac{xd}{L\lambda} 2\pi + \frac{q}{\hbar} BWd.$$

The point of maximum intensity will be determined by

$$0 = \frac{xd}{L\lambda} 2\pi + \frac{q}{\hbar} BWd; \qquad \text{i.e., } x = -\frac{BW\lambda Lq}{2\pi\hbar}.$$

In the classical limit we would interpret this in the following way: We would see electrons traveling through the magnetic field experience a shift in their direction of movement. Consequently, they must have experienced a force. (See Figure 2.17.) As the electrons have the velocity

$$v = \frac{p}{m},$$

they experience the magnetic field during the time

$$\Delta t = \frac{W}{v}.$$

If t denotes the time it takes to travel from the slit to the screen, then t is given by

$$t = \frac{L}{v}.$$

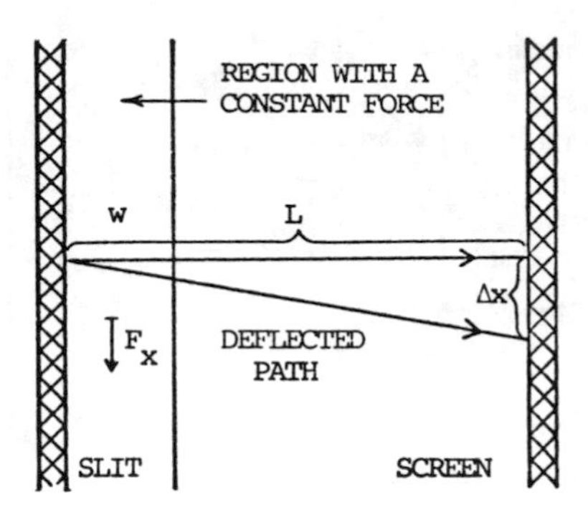

Figure 2.17.

As the electrons hit the slit at the point

$$x = -\frac{BW\lambda Lq}{2\pi\hbar},$$

the passage through the magnetic field has produced a velocity in the x-direction given by

$$\Delta v_x = \frac{x}{t} = -BWq \cdot \frac{\lambda p}{mh}.$$

Consequently, we would say classically that the electrons experienced a force in the x-direction given by

$$F_x = \frac{\Delta p_x}{\Delta t} = -Bvq \cdot \frac{\lambda p}{h}.$$

Using Einstein–de Broglie's rule

$$\lambda = \frac{h}{p},$$

we finally get

$$F_x = -qvB = q(\mathbf{v} \times \mathbf{B})_x,$$

and this is exactly the Lorentz force!

§ 2.7 Gauge Transformation of the Schrödinger Wave Function

Let us now return to the general theory! We have seen that we can represent the electromagnetic field by the gauge potential A_α and have also discovered the gauge symmetry: Physics is unchanged if we replace the gauge potential A_α with the new potential

$$A'_\alpha = A_\alpha + \partial_\alpha \chi.$$

Now, how do these ideas fit together with our quantum-mechanical description of a particle interacting with the Maxwell field A_α?

Consider a specific path Γ_1 in space–time connecting the space–time points A and B. We have seen that the external field A_α produces a change in the original amplitude corresponding to multiplication by the gauge phase factor (2.33)

$$\phi_\Gamma(B \mid A) = \exp\left[\frac{i}{\hbar}\, S^I(\Gamma)\right] = \exp\left[\frac{iq}{\hbar} \int_\Gamma A_\alpha dx^\alpha\right],$$

but the interaction term S^I is *not* gauge invariant! If we perform a gauge transformation, it is changed according to the rule (2.14)

$$S^I(\Gamma) \to S'^I(\Gamma) = S^I(\Gamma) + q[\chi(B) - \chi(A)].$$

Consequently, the gauge phase factor $\phi_\Gamma(B \mid A)$ is *not* gauge invariant, but transforms according to the rule

$$\phi_\Gamma(B \mid A) \to \phi'_\Gamma(B \mid A) = \exp\left[\frac{iq}{\hbar}\, \chi(B)\right] \phi_\Gamma(B \mid A) \exp\left[-\frac{iq}{\hbar}\, \chi(A)\right]. \tag{2.37}$$

But this has consequences for the propagator! We know that the propagator is given by the path-integral formula

$$K(B \mid A) = \int \phi_\Gamma(B \mid A) D[\Gamma].$$

We can decompose the phase factor $\phi_\Gamma(B \mid A)$ into two parts. One, $\phi^0_\Gamma(B \mid A)$, that describes the system in the absence of gauge potentials, and one, $\phi^I_\Gamma(B \mid A)$, that describes the interaction with the gauge potential. Consequently, we get

$$K(B \mid A) = \int \phi^0_\Gamma(B \mid A)\phi^I_\Gamma(B \mid A) D[\Gamma].$$

Performing the gauge transformation, we then get the new propagator

$$\begin{aligned} K'(B \mid A) &= \int \phi^0_\Gamma(B \mid A)\phi'^I_\Gamma(B \mid A) D[\Gamma] \\ &= \int \phi^0_\Gamma(B \mid A) \exp\left[\frac{iq}{\hbar}\, \chi(B)\right] \phi^I_\Gamma(B \mid A) \exp\left[-\frac{iq}{\hbar}\, \chi(A)\right]. \end{aligned}$$

But the phase factors

$$\exp\left[\frac{iq}{\hbar}\, \chi(B)\right]; \qquad \exp\left[-\frac{iq}{\hbar}\, \chi(A)\right]$$

do not depend on the path Γ, so we can pull them outside the path integral:

$$\begin{aligned} K'(B \mid A) = \exp\left[\frac{iq}{\hbar}\, \chi(B)\right] &\left\{\int \phi^0_\Gamma(B \mid A)\phi^I_\Gamma(B \mid A) D[\Gamma]\right\} \\ &\cdot \exp\left[-\frac{iq}{\hbar}\, \chi(A)\right]. \end{aligned}$$

Thus we have shown that

$$
\begin{aligned}
K(B \mid A) &\to K'(B \mid A) \\
&= \exp\left[\frac{iq}{\hbar}\,\chi(B)\right] K(B \mid A) \exp\left[-\frac{iq}{\hbar}\,\chi(A)\right],
\end{aligned} \tag{2.38}
$$

so that the propagator of a charged particle transforms in exactly the same way as the gauge phase factor!

This has consequences for the Schrödinger wave function! The Schrödinger wave function $\psi(\mathbf{r}, t)$ satisfies the integral equation (2.18)

$$
\psi(\mathbf{r}_2; t_2) = \int K(\mathbf{r}_2; t_2 \mid \mathbf{r}_1; t_1)\psi(\mathbf{r}_1; t_1)d\mathbf{r}_1,
$$

and since we have seen that the propagator is changed under a gauge transformation, we conclude that the Schrödinger wave function must change too:

$$
\psi(\mathbf{r}, t) \to \psi'(\mathbf{r}, t).
$$

To find the gauge transformed wave function ψ', we observe that it must satisfy the integral equation

$$
\begin{aligned}
\psi'(\mathbf{r}_2; t_2) &= \int K'(\mathbf{r}_2; t_2 \mid \mathbf{r}_1; t_1)\psi'(\mathbf{r}_1; t_1)d\mathbf{r}_1 \\
&= \int \exp\left[\frac{iq}{\hbar}\,\chi(\mathbf{r}_2; t_2)\right] K(\mathbf{r}_2; t_2 \mid \mathbf{r}_1; t_1) \\
&\quad \cdot \exp\left[-\frac{iq}{\hbar}\,\chi(\mathbf{r}_1, t_1)\right] \psi'(\mathbf{r}_1; t_1)d\mathbf{r}_1.
\end{aligned}
$$

The first phase factor $\exp\left[\frac{iq}{\hbar}\,\chi(\mathbf{r}_2, t_2)\right]$ does not depend on the integration variable $\mathbf{r}_1$, so we may pull it outside the integral, thereby obtaining

$$
\begin{aligned}
&\exp\left[-\frac{iq}{\hbar}\,\chi(\mathbf{r}_2; t_2)\right] \psi'(\mathbf{r}_2; t_2) \\
&\quad = \int K(\mathbf{r}_2, t_2 \mid \mathbf{r}_1, t_1) \exp\left[-\frac{iq}{\hbar}\,\chi(\mathbf{r}_1; t_1)\right] \psi'(\mathbf{r}_1; t_1)d\mathbf{r}_1.
\end{aligned}
$$

Comparing this with the integral equation for ψ, we immediately obtain

$$
\exp\left[-\frac{iq}{\hbar}\,\chi(\mathbf{r}, t)\right] \psi'(\mathbf{r}, t) = \psi(\mathbf{r}, t);
$$

i.e.,

$$
\psi(\mathbf{r}, t) \to \psi'(\mathbf{r}, t) = \exp\left[\frac{iq}{\hbar}\,\chi(\mathbf{r}, t)\right] \psi(\mathbf{r}, t). \tag{2.39}
$$

This is really beautiful! Remember that the only physically measurable quantity associated with the wave function ψ is the square of the norm, $|\psi(\mathbf{r}, t)|^2$; but this is *completely unchanged* by the gauge transformation. Hence we see that quantum-mechanically too, all gauge potentials differing by a gauge transformation represent the same physical state.

We also see that when a charged particle interacts with an electromagnetic field, then the phase of the wave function $\psi = |\psi| \exp[i\phi]$ can be chosen completely arbitrarily at any space–time point!

§ 2.8 Quantum Mechanics of a Charged Particle as a Gauge Theory

We have found a number of formulas that describe the effect of performing gauge transformations. These formulas may be cast into a more transparent form if we introduce some fancy language.

Observe that the wave function $\psi(\mathbf{r}, t)$ is changed simply by multiplying it by a phase factor

$$\psi(\mathbf{r}, t) \rightarrow \exp\left[\frac{iq}{\hbar}\chi(\mathbf{r}, t)\right]\psi(\mathbf{r}, t).$$

If we perform two gauge transformations

$$\begin{aligned}\psi &\overset{1}{\rightarrow} \exp\left[\frac{iq}{\hbar}\chi_1\right]\psi^2 \\ &\overset{2}{\rightarrow} \exp\left[\frac{iq}{\hbar}\chi_2\right]\left[\exp\left[\frac{iq}{\hbar}\chi_1\right]\right]\psi = \exp\left[\frac{iq}{\hbar}(\chi_1+\chi_2)\right]\psi,\end{aligned}$$

then this is equivalent to a single gauge transformation with the combined factor

$$\exp\left[\frac{iq}{\hbar}\chi_2\right]\cdot\exp\left[\frac{iq}{\hbar}\chi_1\right] = \exp\left[\frac{iq}{\hbar}(\chi_2+\chi_1)\right].$$

Thus the phase factors constitute a *group*, the group of all complex numbers of modulus 1. This group is denoted by $U(1)$, and it is called the *unitary group in one dimension*. (Remember that a unitary matrix is a complex matrix with the property

$$AA^* = A^*A = I,$$

where A^* is the conjugate transpose of A. The unitary 2×2 matrices constitute the group $U(2)$; the unitary 3×3 matrices constitute the group $U(3)$; etc. Now, a 1×1 matrix is nothing but an ordinary complex number z, and the Hermitian conjugate is just the ordinary conjugate $z = x + iy \rightarrow \bar{z} = x - iy$. Hence $U(1)$ consists of all complex numbers with the property $z\bar{z} = 1$; i.e., it consists exactly of the phase factors.)

We have now attached a group G to our gauge transformations. It is called the *gauge group*, and in this case $G = U(1)$. The group $U(1)$ is an abelian group; i.e.,

$$g_1 \cdot g_2 = g_2 \cdot g_1 \qquad \text{for all } g_1, g_2 \in U(1)$$

(contrary to $U(2)$!), and therefore electromagnetism is called an *abelian gauge theory*. If we write $X = (\mathbf{r}, t)$ for abbreviation, then to *every space–time point X we have attached a group element*

$$g(X) = \exp\left[\frac{iq}{\hbar}\,\chi(X)\right]. \tag{2.40}$$

This group element is responsible for the gauge transformation. First, we consider the Schrödinger wave function $\psi(X)$. It transforms according to the rule

$$\psi(X) \to \psi'(X) = g(X)\psi(X). \tag{2.41}$$

A quantity transforming in this way is called a *gauge vector*.

Exercise 2.8.1
Problem: Prove that

$$\partial_\alpha \chi = -\frac{iq}{\hbar}\,(\partial_\alpha g)g^{-1},$$

where g is given by (2.40).

Then there is the gauge potential $A_\alpha(X)$. According to (1.31) and the above exercise, the gauge potential transforms according to the rule

$$A_\alpha(X) \to A'_\alpha(X) = A_\alpha(X) - i\,\frac{\hbar}{q}\,[\partial_\alpha g(X)]g^{-1}(X). \tag{2.42}$$

We shall explore the geometrical significance of this strange formula later on.

Finally, we have the field strengths $F_{\alpha\beta}(X)$, which transform according to the rule

$$F_{\alpha\beta}(X) \to F'_{\alpha\beta}(X) = F_{\alpha\beta}(X), \tag{2.43}$$

and a quantity with this property is called a *gauge scalar*.

As a consequence of the gauge symmetry of our theory we immediately see that any physically observable quantity must be a *gauge scalar*. Let us play a little with gauge scalars! First, we consider the Schrödinger wave function $\psi(X)$. As

$$\psi(X) \to g(X)\psi(X) \quad \text{and} \quad \bar{\psi}(X) \to \bar{g}(X)\bar{\psi}(X) = g^{-1}(X)\psi(X),$$

we see that $\psi\bar{\psi}$ is a gauge scalar. Actually, it is observable. It is the *probability density*.

Then consider the Maxwell field $A_\alpha(X)$. Here we can form the gauge scalar

$$F_{\alpha\beta}(X) = \partial_\alpha A_\beta(X) - \partial_\beta A_\alpha(X).$$

But this is not the only one. Consider a closed loop Γ in space–time. Then the integral

$$\oint_\Gamma A_\alpha dx^\alpha$$

is a gauge scalar. To see this, we perform a gauge transformation:

$$\oint_\Gamma A_\alpha dx^\alpha \to \int_\Gamma A'_\alpha dx^\alpha = \oint_\Gamma A_\alpha dx^\alpha + \oint_\Gamma \partial_\alpha \chi dx^\alpha = \oint_\Gamma A_\alpha dx^\alpha.$$

(Here $\partial_\alpha \chi$ does not contribute to the integral because we consider a *closed* loop.) Is the above loop integral observable? Well, consider two space–time points A and B connected by the paths Γ_1 and Γ_2. (See Figure 2.18.) Then an external field will produce a phase shift between electrons moving from A to B along Γ_1 or along Γ_2. This phase shift is given by

$$\exp\left[\frac{iq}{\hbar}\left(\int_{\Gamma_2} A_\alpha dx^\alpha - \int_{\Gamma_1} A_\alpha dx^\alpha\right)\right] = \exp\left[\frac{iq}{\hbar}\oint_\Gamma A_\alpha dx^\alpha\right],$$

where $\Gamma = \Gamma_2 - \Gamma_1$. This phase shift is in principle measurable, being nothing but the Bohm–Aharonov effect! Thus we conclude that

$$\exp\left[\frac{iq}{\hbar}\oint A_\alpha dx^\alpha\right]$$

is in principle measurable quantum-mechanically.

Worked Exercise 2.8.2.
Problem: Consider the parallelogram shown in Figure 2.19. Show that the field strength $F_{\alpha\beta}$ is related to the loop integral $\oint A_\alpha dx^\alpha$ through the following formula:

$$F_{\alpha\beta} X \Delta x^\beta \delta x^\alpha = \lim_{\epsilon\to 0}\frac{1}{\epsilon^2}\oint_{\Gamma_\epsilon} A_\alpha dx^\alpha.$$

Finally, we look at the Schrödinger wave function $\psi(X)$ and the Maxwell field $A_\alpha(X)$ at the same time. What can we make out of them? Consider the derivative $\partial_\alpha\psi$ of ψ. This is *not* a gauge vector, as

$$\partial_\alpha\psi \to \partial_\alpha\psi' = \partial_\alpha[g(X)\psi(X)] = g(X)\partial_\alpha\psi(X) + [\partial_\alpha g(X)]\psi(X).$$

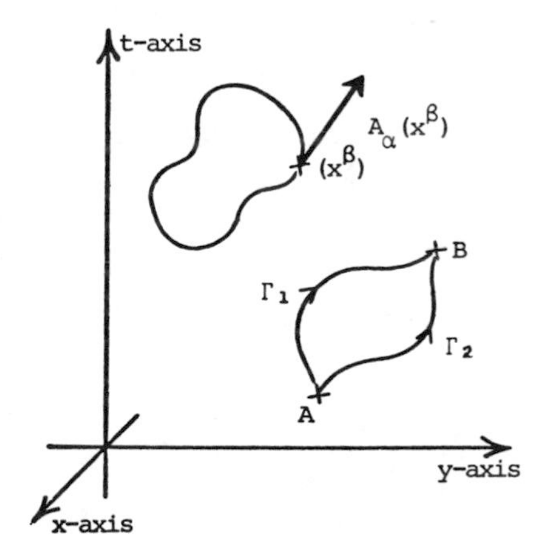

Figure 2.18.

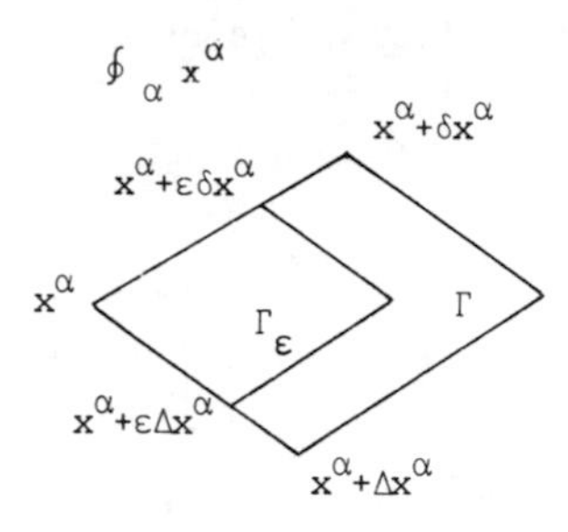

Figure 2.19.

The last term spoils the game, but containing $\partial_\alpha g(X)$, it reminds us of A_α. Consider $A_\alpha \psi$. It transforms in the following way:

$$A_\alpha \psi \to A'_\alpha \psi' = \left[A_\alpha - i\, \frac{\hbar}{q} (\partial_\alpha g) g^{-1} \right] g\psi$$
$$= g(X) A_\alpha(X)\psi(X) - i\, \frac{\hbar}{q} [\partial_\alpha g(X)]\psi(X).$$

Hence the same spoiling term occurs! Consequently,

$$\partial_\alpha \psi - i\, \frac{q}{\hbar}\, A_\alpha \psi = \left(\partial_\alpha - i\, \frac{q}{\hbar}\, A_\alpha \right) \psi$$

is a *gauge vector*; i.e., it has the correct transformation properties:

$$\left(\partial_\alpha - i\, \frac{q}{\hbar}\, A_\alpha \right) \psi \to \left(\partial_\alpha - i\, \frac{q}{\hbar}\, A'_\alpha \right) \psi' = g(X) \left(\partial_\alpha - \frac{iq}{\hbar}\, A_\alpha \right) \psi. \quad (2.44)$$

The differential operator

$$D_\alpha = \partial_\alpha - \frac{iq}{\hbar}\, A_\alpha$$

is called the *gauge covariant derivative*, and it plays a very important role in constructing gauge-invariant theories!

The discovery of the gauge covariant derivative allows us to write down nontrivial *gauge scalars* based upon $\psi(X)$ and $A_\alpha(X)$; e.g.,

$$\bar{\psi} D_\alpha \psi, \qquad (\overline{D_\alpha \psi})(D_\beta \psi), \qquad F_{\alpha\beta} (\overline{D^\alpha \psi})(D^\beta \psi),$$

or written out explicitly,

$$\bar{\psi}(X) \left[\partial_\alpha \psi(X) - i\, \frac{q}{\hbar}\, A_\alpha(X)\psi(X) \right],$$

$$\left[\partial_\alpha \bar{\psi} + \frac{iq}{\hbar}\, A_\alpha \bar{\psi} \right] \left[\partial_\beta \psi - \frac{iq}{\hbar}\, A_\beta \psi \right],$$

Table 2.1. The Basic Ingredients of a Gauge Theory:

A gauge theory consists of a gauge group G and a gauge potential.
Example: Electromagnetism is a gauge theory based upon the gauge group $U(1)$ and the Maxwell field $A_\alpha(x)$.
To perform a gauge transformation one must attach a group element $g(X)$ *to each space–time point* X.

(2.45)	*gauge covariant derivative*	$D_\alpha = \partial_\alpha - \frac{iq}{\hbar} A_\alpha$	
(2.46)	*gauge vectors*	$\psi(X)$ $D_\alpha \psi(X)$	$\psi'(X) = g(X)\psi(X)$ $D'\psi'(X) = g(X) D_\alpha \psi(X)$
(2.47)	*gauge scalars*	$F_{\alpha\beta} = \partial_\alpha A_\beta - \partial_\beta A_\alpha$ $\oint A_\alpha dx^\alpha$	$F'_{\alpha\beta} = F_{\alpha\beta}$ $\oint A'_\alpha dx^\alpha = \oint A_\alpha dx^\alpha$
(2.48)	*gauge potential*	$A_\alpha(X)$	$A'_\alpha(X) = A_\alpha(X) - \frac{i\hbar}{q}[\partial_\alpha g]g^{-1}$
(2.49)	*gauge phase factor*	$\phi_\Gamma(B \mid A)$ $= \exp\left[\frac{iq}{\hbar}\int_\Gamma A_\alpha dx^\alpha\right]$	$\phi'_\Gamma(B \mid A)$ $= g(B)\phi_\Gamma(B \mid A) g^{-1}(A)$

etc.

Exercise 2.8.3
Problem: Let $\psi(X)$ be a gauge vector. Show that

$$[D_\mu; D_\nu]\psi = -\frac{iq}{\hbar} F_{\mu\nu}\psi.$$

So we have walked right into our first gauge theory! We summarize the most important points in Table 2.1:

§ 2.9 The Schrödinger Equation in the Path Integral Formalism

At this point we will get in contact with the conventional approach to *quantum mechanics* based upon the *Schrödinger equation*. In studying dynamical problems we have been used in the propagator, which governs the dynamical evolution through the integral equation (2.18)

$$\psi(\mathbf{r}_2; t_2) = \int K(\mathbf{r}_2; t_2 \mid \mathbf{r}_1; t_1)\psi(\mathbf{r}_1; t_1)d\mathbf{r}_1.$$

We now want to show that this integral equation actually is equivalent to the Schrödinger equation. For simplicity we will only consider the one-dimensional motion of a particle in a potential $V(x)$. In this case the Lagrangian is given by (2.5),

$$L(x, \dot{x}, t) = \frac{1}{2} m\dot{x}^2 - V(x),$$

and the propagator by (2.21)

$$K(x; t_2 \mid y; t_1) = \int_{x(t_1)=x_1}^{x(t_2)=x_2} \exp\left\{\frac{i}{\hbar}\int_{t_1}^{t_2}\left[\frac{m}{2}\dot{x}^2 - V(x)\right]dt\right\} D[x(t)].$$

To convert the integral equation (2.18) into a differential equation, we put $t_1 = t$ and $t_2 = t + \Delta t$:

$$\psi(x; t + \Delta t) = \int K(x; t + \Delta t \mid y; t)\psi(y; t)dy.$$

Since Δt is going to be very small, we expect the phase factor to be approximately equal to

$$\exp\left\{\frac{i}{\hbar}\left[\frac{m}{2}\frac{(x-y)^2}{(\Delta t)} - V(x)\right]\Delta t\right\}.$$

Of course, this is not true for a very "wild" path, but the wild paths kill each other due to the rapid oscillations. (See Figure 2.20.) Consequently, the "infinitesimal" propagator is given by

$$K(x; t + \Delta t \mid y; t) \approx A \exp\left\{\frac{i}{\hbar}\left[\frac{m}{2}\frac{(x-y)^2}{\Delta t} - V(x)\Delta t\right]\right\}$$

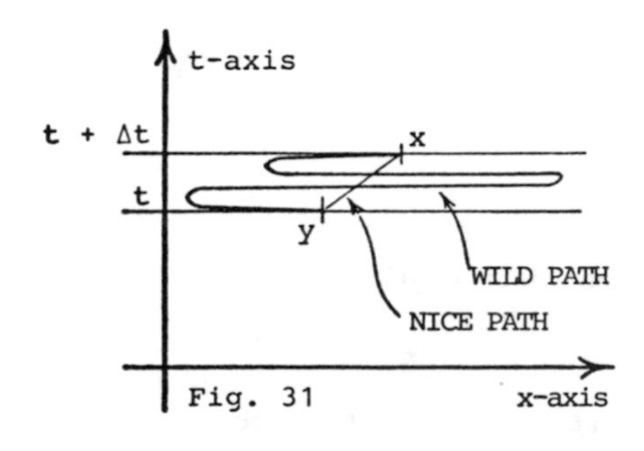

Figure 2.20.

in the limit of very small Δt. A is a normalization constant that arises from summing up all the amplitudes in the path integral, and we will determine its value in a moment. Inserting this approximation into the integral equation, we get

$$\psi(x, t + \Delta t) \approx A \int_{-\infty}^{+\infty} \exp\left\{\frac{i}{\hbar}\left[\frac{m}{2}\frac{(y-x)^2}{\Delta t} - V(x)\Delta t\right]\right\}\psi(y, t)dt$$

$$= A \exp\left[-\frac{i}{\hbar}V(x)\Delta t\right]\int_{-\infty}^{+\infty} \exp\left[\frac{i}{\hbar}\frac{m}{2}\frac{\eta^2}{\Delta t}\right]\psi(x+\eta, t)d\eta, \tag{2.50}$$

where we have put $y = x + \eta$. Observe that the integral is wildly oscillating for large values of η. Hence the main contribution comes from values of η where η^2 is comparable to Δt. We may therefore expand $\psi(x + \eta, t)$ to second order in η (second-order corrections in η correspond to first-order corrections in Δt because of the phase factor

$$\exp\left[\frac{i}{\hbar}\frac{m}{2}\frac{\eta^2}{\Delta t}\right].$$

Doing this we get

$$\psi(x+\eta, t) \approx \psi(x, t) + \eta\frac{\partial\psi}{\partial x} + \frac{1}{2}\eta^2\frac{\partial^2\psi}{\partial x^2}.$$

Inserting this into (2.50), we now obtain

$$\psi(x, t+\Delta t) \approx A \exp\left[-\frac{i}{\hbar}V(x)\Delta t\right]$$

$$\int_{-\infty}^{+\infty} \exp\left[\frac{i}{\hbar}\frac{m}{2}\frac{\eta^2}{\Delta t}\right]\left\{\psi(x, t) + \eta\frac{\partial\psi}{\partial x} + \frac{1}{2}\eta^2\frac{\partial^2\psi}{\partial x^2}\right\}d\eta.$$

The term $\eta\,\frac{\partial\psi}{\partial x}$ does not contribute to the integral because it is an odd function of η! Thus we have shown

$$\psi(x, t+\Delta t) \approx A \exp\left\{-\frac{1}{\hbar}V(x)\Delta t\right\}$$

$$\cdot\left[\psi(x, t)\int_{-\infty}^{+\infty} \exp\left[\frac{im}{2\hbar}\frac{\eta^2}{\Delta t}\right]d\eta + \frac{1}{2}\frac{\partial^2\psi}{\partial x^2}\int_{-\infty}^{+\infty}\eta^2\exp\left[\frac{im}{2\hbar}\frac{\eta^2}{\Delta t}\right]d\eta\right].$$

Here the first integral is a Gaussian integral of the standard type, cf. (2.27), while the second is obtained from the formula

$$\int_{-\infty}^{+\infty} x^2\exp[-ax^2]dx = \frac{1}{2a}\sqrt{\frac{\pi}{a}},$$

which follows from (2.27) when we differentiate with respect to a. In this way we finally obtain

$$\begin{aligned}\psi(x, t+\Delta t) \approx A\sqrt{\frac{2\pi i\hbar\Delta t}{m}} \exp\left[-\frac{1}{\hbar} V(x)\Delta t\right]\psi(x,t) \\ + A\sqrt{\frac{2\pi i\hbar\Delta t}{m}}\frac{i\hbar\Delta t}{2m} \exp\left[-\frac{i}{\hbar} V(x)\Delta t\right]\frac{\partial^2\psi}{\partial x^2}.\end{aligned} \tag{2.51}$$

First, we observe that this fixes the value of A because when $\Delta t \to 0$, the left-hand side approaches $\psi(x, t)$. But this is consistent only if

$$A = \sqrt{\frac{m}{2\pi i\hbar\Delta t}}. \tag{2.52}$$

This should be compared with the free-particle propagator (2.28). In fact, the "infinitesimal propagator" can now be decomposed as follows:

$$\begin{aligned}&K(x; t+\Delta t \mid y; t) \\ &\approx \sqrt{\frac{m}{2\pi i\hbar\Delta t}} \exp\left[\frac{i}{\hbar}\frac{m}{2}\frac{(x-y)^2}{\Delta t}\right]\cdot\exp\left[-\frac{i}{\hbar}V(x)\Delta t\right] \\ &= K^0(x; t+\Delta \mid y; t)\phi^I(x; t+\Delta t \mid y; t),\end{aligned} \tag{2.53}$$

where $K^0(B \mid A)$ is the free-particle propagator and $\phi^I_\Gamma(B \mid A)$ is the phase factor corresponding to the interaction, cf. (2.33),

$$\phi^I_\Gamma(B \mid A) = \exp\left[-\frac{i}{\hbar}\int_\Gamma V(x)dt\right].$$

Inserting (2.52) into (2.51), we now get

$$\psi(x, t+\Delta t) \approx \exp\left[-\frac{i}{\hbar}V(x)\Delta t\right]\psi(x,t) + \frac{i\hbar\Delta t}{2m}\cdot\frac{\partial^2\psi}{\partial x^2};$$

i.e.,

$$\frac{\psi(x, t+\Delta t) - \psi(x,t)}{\Delta t} \approx \frac{i\hbar}{2m}\cdot\frac{\partial^2\psi}{\partial x^2} + \frac{\exp[-(i/\hbar)V(x)\Delta t] - 1}{\Delta t}\psi(x,t),$$

which in the limit $\Delta t \to 0$ reduces to

$$\frac{\partial\psi}{\partial t} = \frac{i\hbar}{2m}\frac{\partial^2\psi}{\partial x^2} - \frac{i}{\hbar}V(x)\psi(x,t).$$

Finally, this may be rearranged as

$$i\hbar\frac{\partial\psi}{\partial t} = -\frac{\hbar^2}{2m}\cdot\frac{\partial^2\psi}{\partial x^2} + V(x)\psi(x,t). \tag{2.54}$$

So we have captured the Schrödinger equation! By using exactly the same procedure, we can find the Schrödinger equation for a charged particle in an external field. However, it becomes technically more complicated, partly because it is now a three-dimensional motion and partly because the generalized potential

$$U\left(\mathbf{r}, \frac{d\mathbf{r}}{dt}, t\right) = q\phi(\mathbf{r}, t) - \mathbf{A} \cdot \frac{d\mathbf{r}}{dt} \tag{2.11}$$

has a more complicated structure. We will leave this as an exercise: (Details can be found in Schulman (1981)).

Exercise 2.9.1
Problem: Consider a three-dimensional motion of a particle in an ordinary potential $V(\mathbf{x})$. Use the path integral formalism to derive the Schrödinger equation

$$i\hbar \frac{\partial \psi}{\partial t} = -\frac{\hbar^2}{2m} \partial_i \partial^i \psi + V(\mathbf{x})\psi.$$

Exercise 2.9.2
Problem: Consider the three-dimensional motion of a particle in the generalized potential

$$U\left(\mathbf{r}, \frac{d\mathbf{r}}{dt}, t\right) = q\phi(\mathbf{r}, t) - \mathbf{A}(\mathbf{r}, t) \cdot \frac{d\mathbf{r}}{dt}.$$

Use the path integral formalism to derive the Schrödinger equation

$$i\hbar \frac{\partial \psi}{\partial t} = -\frac{\hbar^2}{2m} \left(\partial_i - i\frac{q}{\hbar} A_i\right)\left(\partial^i - i\frac{q}{\hbar} A^i\right) + q\phi\psi.$$

§ 2.10 The Hamiltonian Formalism

The other method available to quantize a system is the *canonical quantization*. Here the starting point is the *Hamiltonian* formalism.

Consider a Lagrangian $L(q^i, \dot{q}^i, t)$, $i = 1, \ldots, n$. To each of the generalized coordinates q^i we associate a *conjugate momentum* p_i defined by

$$p_i = \frac{\partial L}{\partial \dot{q}^i}; \qquad i = 1, \ldots, n. \tag{2.55}$$

Observe that p_i is a function of $q^i, \dot{q}^i, t$:

$$p_i = p_i(q^i, \dot{q}^i, t), \qquad i = 1, \ldots, n.$$

Sometimes we may invert these equations and express $\dot{q}^i$ as a function of q^i, p_i, t. If this is the case, we say that the system under consideration is a *canonical system*.

For a canonical system we then define the *Hamiltonian*

$$H(q^i, p_i, t) = p_i \cdot \dot{q}^i(q^j, p_j, t) - L(q^i, \dot{q}^i(p_j, q_j, t), t). \tag{2.56}$$

We shall often abbreviate this as $H = p_i\dot{q}^i - L$, but we should always remember that $\dot{q}^i$ should be regarded as a function of the canonical variables p_i, q^i. Observe that we are using the Einstein summation convention! An expression like $p_i\dot{q}^i$ implies a summation over $i = 1, \ldots, n$.

The Hamiltonian is very useful. We may use it to cast the equations of motion into a very beautiful form. Consider the partial derivatives

$$\frac{\partial H}{\partial p_i} \quad \text{and} \quad \frac{\partial H}{\partial q^i}.$$

Let us work them out:

$$\frac{\partial H}{\partial p_i} = \frac{\partial}{\partial p_i}\left[p_j \dot{q}^j - L\right] = \delta^i_j \dot{q}^j + p_j \frac{\partial \dot{q}^j}{\partial p_i} - \frac{\partial L}{\partial \dot{q}^j}\frac{\partial \dot{q}^j}{\partial p_i} = \dot{q}^i,$$

by the definition (2.55) of the momentum conjugate to q^j. Next we have

$$\frac{\partial H}{\partial q^i} = \frac{\partial}{\partial q^i}\left[p_j \dot{q}^j - L\right] = p_j \frac{\partial \dot{q}^j}{\partial q^i} - \frac{\partial L}{\partial q^i} - \frac{\partial L}{\partial \dot{q}^j}\frac{\partial \dot{q}^j}{\partial q^i} = -\frac{\partial L}{\partial q^i}.$$

From the Euler–Lagrange equation (2.6) we get

$$\frac{\partial L}{\partial q^i} = \frac{d}{dt}\frac{\partial L}{\partial \dot{q}^i} = \frac{d}{dt} p_i = \dot{p}_i,$$

leading to

$$\frac{\partial H}{\partial p_i} = \dot{q}^i \quad \text{and} \quad \frac{\partial H}{\partial q^i} = -\dot{p}_i. \tag{2.57}$$

These equations are referred to as *Hamilton's equations*, and they are completely equivalent to the Euler–Lagrange equations. To check these ideas in a trivial example, we consider a one-dimensional motion in a potential $V(x)$. From the Lagrangian (2.5)

$$L(x, \dot{x}) = \frac{1}{2} m\dot{x}^2 - V(x),$$

we see that

$$p = \frac{\partial L}{\partial \dot{x}} = m\dot{x},$$

so that the momentum conjugate to x in this simple case coincides with the usual kinematic momentum. The system is canonical, and we have

$$\dot{x} = \frac{1}{m} p.$$

Hence we can find the Hamiltonian

$$H(p, x) = p\dot{x} - L = \frac{p^2}{2m} + V(x), \tag{2.58}$$

which reduces to the usual formula for the mechanical energy. Finally, we get Hamilton's equation

$$\dot{x} = \frac{\partial H}{\partial p} = \frac{1}{m} p,$$

$$\dot{p} = -\frac{\partial H}{\partial x} = -V'(x),$$

which are obviously equivalent to Newton's equations of motion. For a time-independent system we may identify H with the total energy of the system. To establish this, we consider a system characterized by a Lagrangian that does not depend explicit on time:

$$L = L(q_i, \dot{q}_i).$$

Then the Hamiltonian, too, does not depend explicitly on time:

$$H = p_i \dot{q}^i(q^j, p_j) - L(q^i; \dot{q}^i(q^j, p_j)).$$

Consider a trajectory, $q^i = q^i(t)$ that satisfies the equations of motion. Then H is constant along this trajectory:

$$\frac{dH}{dt} = \frac{dH}{dp_i}\frac{dp_i}{dt} + \frac{\partial H}{\partial q^i}\frac{dq^i}{dt} = \dot{q}^i \dot{p}_i - \dot{p}_i \dot{q}^i = 0,$$

according to Hamilton's equations! Hence H *is a constant of motion* along the classical path. That this constant of motion is actually the total energy follows from the illustrative example above, cf. (2.58).

Let $A(q^i; p_i)$ and $B(q^i; p_i)$ be functions of the generalized coordinates $q^1, \ldots, q^n$ and their conjugate momenta $p_1, \ldots, p_n$. We define the *Poisson bracket* of A and B as follows:

$$\{A; B\} = \frac{\partial A}{\partial q^i}\frac{\partial B}{\partial p_i} - \frac{\partial B}{\partial q^i}\frac{\partial A}{\partial p_i}, \tag{2.59}$$

where a summation over i is implied as usual.

Exercise 2.10.1
Problem: Show that the Poisson brackets satisfy the following simple rules:

1. $\{A; B\} = -\{B; A\}$ (Skew symmetry).
2. $\{A; c\} = 0$ whenever c is a number, i.e., a constant function of q^i and p_i.
3. $\left\{\begin{matrix} \{A_1 + A_2; B\} = \{A_1, ; B\} + \{A_2; B\} \\ \{\lambda A; B\} = \lambda\{A; B\} \end{matrix}\right\}$ Linearity.
4. $\{A_1 A_2; B\} = \{A_1; B\}A_2 + A_1\{A_2; B\}$.
5. $\{A; \{B; C\}\} + \{B; \{C; A\}\} + \{C; \{A; B\}\} = 0$ (Jacobi identity).

Exercise 2.10.2
Problem: Consider the ith coordinate function

$$q^i(q^1, \ldots, q^n; p_1, \ldots, p_n) = q^i.$$

Similarly, we can consider the ith component of the conjugate momentum as a special function:

$$p_i(q^1, \ldots, q^n; p_1, \ldots, p_n) = p_i.$$

Show that q^i and p_i satisfy the rule

$$\{q^i; q^j\} = \{p_i; p_j\} = 0 \quad \text{and} \quad \{q^i; p_j\} = \delta^i_j. \tag{2.60}$$

Exercise 2.10.3
Problem: Let $F(q^1, \ldots, q^n; p_1, \ldots, p_n, t)$ be a function of the generalized coordinates, their conjugate momenta, and the time t. Then q^i and p_i themselves depend on

time, and their time dependence is governed by Hamilton's equations (2.57). Show that the total time derivative of F along a trajectory in phase space is given by

$$\frac{dF}{dt} = \frac{\partial F}{\partial t} + \{H; F\}. \tag{2.61}$$

(Note that if F does not depend explicitly on time, then F is a constant of motion if and only if the Poisson bracket $\{H, F\}$ vanishes.)

Exercise 2.10.4
Problem: Let $F(q^1, \ldots, q^n)$ be an analytic function. Show that

1. $\{p_i, F(q^1, \ldots, q^n)\} = -\frac{\partial F}{\partial q_i}(q^1, \ldots, q^n)$.
2. $\{q^i, F(q^1, \ldots, q^n)\} = q^i F(q^1, \ldots, q^n)$.

(Hint: expand $F(q^1, \ldots, q^n)$ in a power series and show that $\{p_i, q_i^n\} = -nq_i^{n-i}$, where we exceptionally do not imply a summation over i!)

§ 2.11 Canonical Quantization and the Schrödinger Equation

Now, the canonical quantization runs as follows: The generalized coordinates $(q^1, \ldots, q^n)$ specify a position in configuration space. Quantum-mechanically, we characterize the system by its Schrödinger wave function $\psi(q^1, \ldots, q^n; t)$, which gives us the probability amplitude for finding the system at the point $(q^1, \ldots, q^n)$ as a function of time t. The dynamical evolution of the wave function is governed by the Schrödinger equation, which is constructed in the following way. Consider a Hamiltonian $E = H(q^i, p_i)$. On the left-hand side substitute $E \to i\hbar\frac{\partial}{\partial t}$, and on the right-hand side substitute $p_i \to -i\hbar\frac{\partial}{\partial q^i}$. Then both sides are converted to differential equations, and the Schrödinger equation reads

$$i\hbar\frac{\partial}{\partial t}\psi = H\left(q^i, -i\hbar\frac{\partial}{\partial q^i}, t\right)\psi(q^i, t). \tag{2.62}$$

To check these ideas in a trivial example, we consider a one-dimensional motion in a potential $V(x)$. We have previously found the Hamiltonian: $H(x, p) = p^2/2m + V(x)$. Thus we get the ordinary Schrödinger equation

$$i\hbar\frac{\partial\psi}{\partial t} = -\frac{\hbar^2}{2m}\left[\frac{\partial^2}{\partial x^2} + V(x)\right]\psi(x, t). \tag{2.54}$$

Let us attack a more interesting problem. Suppose a particle is moving in an external electromagnetic field, and let us try to quantize this problem in the canonical way: From

$$L(\mathbf{x}, \mathbf{v}, t) = \frac{1}{2}m\mathbf{v}^2 - q\phi(\mathbf{x}, t) + q\mathbf{A}(x, t)\cdot\mathbf{v} \tag{2.12}$$

we see that the canonical momentum is given by

$$\mathbf{p} = \frac{\partial L}{\partial \mathbf{v}} = m\mathbf{v} + q\mathbf{A},$$

and hence these equations do not just reproduce the kinematic momenta! This system is canonical, too, and

$$\mathbf{v} = \frac{1}{m}\,\mathbf{p} - \frac{q}{m}\,\mathbf{A}.$$

Thus, we can construct a Hamiltonian

$$H(\mathbf{x}, \mathbf{p}, t) = \mathbf{p} \cdot \mathbf{v} - L = \frac{1}{2m}\,(\mathbf{p} - q\mathbf{A})^2 + q\phi.$$

We may rearrange this slightly:

$$E - q\phi = \frac{1}{2m}\,(\mathbf{p} - q\mathbf{A})^2. \tag{2.63}$$

Here we see something interesting: To pass from a *free* particle, where $E = \frac{p^2}{2m}$, to a particle in an external electromagnetic field, we have to make the following substitutions:

$$\begin{aligned} E &\to E - q\phi, \\ \mathbf{p} &\to \mathbf{p} - q\mathbf{A}, \end{aligned} \tag{2.64}$$

and this simple rule is known as the *rule of minimal coupling*. It is of very great importance, and we shall investigate it more closely later on.

For the moment, we want to pass to the Schrödinger equation

$$\left[i\hbar\,\frac{\partial}{\partial t} - q\phi\right]\psi(\mathbf{r}, t) = \frac{[-\hbar\nabla - q\mathbf{A}]^2}{2m}\,\psi(\mathbf{r}, t). \tag{2.65}$$

Observe that the Schrödinger wave function interacts directly with the gauge potential $(\phi, \mathbf{A})$ and not with the field strengths! It should also be observed that the equation is *gauge covariant*. To see this, we rewrite it slightly:

$$i\hbar\left[\frac{\partial}{\partial t} + i\,\frac{q}{\hbar}\,\phi\right]\psi = \frac{-\hbar^2\left[\nabla - \frac{iq}{\hbar}\mathbf{A}\right]^2}{2m}\,\psi.$$

Hence to pass from a free particle to a particle in an external field, we make the following substitutions:

$$\frac{\partial}{\partial t} \to \frac{\partial}{\partial t} + i\,\frac{q}{\hbar}\,\phi, \qquad \nabla \to \nabla - \frac{iq}{\hbar}\,\mathbf{A}. \tag{2.66}$$

These substitutions can be written more compactly as

$$\partial_\alpha \to \partial_\alpha - i\,\frac{q}{\hbar}\,A_\alpha = D_\alpha. \tag{2.67}$$

Consequently, we have simply *converted the ordinary derivatives to gauge covariant derivatives*! This is also referred to as *the rule of minimal coupling*. Using this, we can rewrite the Schrödinger equation as

$$i\hbar D_0\psi = \frac{\hbar^2}{2m}\,D_i D^i\psi. \tag{2.68}$$

Here we explicitly see that it is a gauge covariant equation, since both sides transform as gauge vectors. Thus, if they are equal to each other in one particular gauge, then they are identical in all gauges!

Consider a Hamiltonian operator

$$\hat{H} = \hat{H}\left(q^i, -i\hbar\frac{\partial}{\partial q^i}\right).$$

If we introduce the following inner product for wave functions,

$$\langle \psi \mid \phi \rangle = \int \bar{\psi}(x)\phi(x)dx,$$

then it can be shown under very general assumptions that the operator $\hat{H}$ is *Hermitian*; i.e.,

$$\langle \hat{H}\psi \mid \phi \rangle = \langle \psi \mid \hat{H}\phi \rangle,$$

where ψ, ϕ are arbitrary wave functions.

Worked Exercise 2.11.1
Problem: Let $\hat{H}$ be a Hermitian operator. Show that the *eigenvalues* are real (i.e., if $\hat{H}\psi = E\psi$, then E is a real number) and that the *eigenfunctions* belonging to two different eigenvalues are orthogonal (i.e., if $\hat{H}\psi_1 = E_1\psi$ and $\hat{H}\psi_2 = E_2\psi_2$ where $E_1 \neq E_2$, then $\langle \psi_1 \mid \psi_2 \rangle = 0$).

Furthermore, we may generally assume that the set of eigenfunctions $\{\psi_n(x)\}$ is *complete*, i.e., that we may expand an arbitrary wave function as a superposition of eigenfunctions:

$$\psi(x) = \sum_n a_n \psi_n(x).$$

Here the coefficient a_n is determined by the relation

$$a_m = \langle \psi_m \mid \psi \rangle,$$

provided that the set of eigenfunctions is normalized; i.e.,

$$\langle \psi_m \mid \psi_n \rangle = \delta_{mn}.$$

Worked Exercise 2.11.2
Problem: Let $\{\psi_n(x)\}$ be a complete orthonormal set of eigenfunctions of the Hamiltonian operator $\hat{H}$. Show that

$$\sum_n \overline{\psi_n(x_1)}\psi_n(x_2) = \delta(x_1 - x_2). \tag{2.69}$$

Consider an eigenfunction of $\hat{H}$, i.e., a wave function $\psi_n(x)$ with the property

$$\hat{H}\psi_n(x) = E_n \cdot \psi_n(x).$$

This eigenfunction can immediately be extended to a solution of the Schrödinger equation

$$\psi_n(x, t) = \psi_n(x)\exp\left[-\frac{i}{\hbar}E_n t\right].$$

In accordance with the Einstein–de Broglie rule, this is interpreted as the state of a particle with energy E_n. Observe that the probability distribution is time-independent:

$$|\psi_n(x, t)|^2 = |\psi_n(x)|^2,$$

and we therefore say that the wave function represents a *stationary state*.

Let $\psi(x)$ be an arbitrary wave function. Then we can decompose it as a superposition of eigenfunctions:

$$\psi(x) = \sum_n a_n \psi_n(x).$$

But the Schrödinger equation is linear, so we can immediately extend this to the solution

$$\psi(x, t) = \sum_n a_n \psi_n(x) \cdot \exp\left[-\frac{i}{\hbar} E_n t\right],$$

which reduces to $\psi(x)$ for $t = 0$. Once we know a complete set of eigenfunctions, we therefore control the dynamical evolution of Schrödinger wave functions.

This suggests that the Feynman propagator itself can be expressed through a complete set of eigenfunctions.

Worked Exercise 2.11.3
Problem: Let $\{\psi(x)\}$ be a complete orthonormal set of eigenfunctions to the Hermitian operator $\hat{H}$. Show that the Feynman propagator can be expanded as

$$K(x_2; t_2 \mid x_1; t_1) = \sum_n \overline{\psi_n(x_1)} \psi_n(x_2) \exp\left[-\frac{i}{\hbar} E_n(t_2 - t_1)\right]. \tag{2.70}$$

Let us make a final comment about the canonical quantization. In the preceding discussion the Schrödinger wave function has played a central role. However, it is possible to avoid it. The physical quantities are then represented by Hermitian operators that do not necessarily operate on the space of Schrödinger wave functions. These operators cannot be chosen arbitrarily:

In the classical context, the physical quantities are represented as functions $A(q^i; p_i)$ of the generalized coordinates and their conjugate momenta. *In the quantum context, these functions are replaced by Hermitian operators in such a way that their Poisson bracket is replaced by the commutator.*

$$\{A; B\} \to \frac{1}{i\hbar} [\hat{A}, \hat{B}]. \tag{2.71}$$

Exercise 2.11.4
Problem: Show that the commutator $[\hat{A}; \hat{B}] = \hat{A}\hat{B} - \hat{B}\hat{A}$ satisfies the rules corresponding to Exercise 2.10.1.

In particular, the generalized coordinates q^i and their canonical momenta p_i must be replaced by Hermitian operators satisfying the so-called *Heisenberg commutation rules*:

$$[\hat{q}^i; \hat{q}^j] = [\hat{p}_i; \hat{p}_j] = 0; \qquad [\hat{q}^i; \hat{p}_j] = i\hbar\delta^i_j. \tag{2.72}$$

Compare (2.60).

Exercise 2.11.5
Problem: Show that if we replace the generalized coordinate q^i by the multiplication operator $\hat{q}^i = q^i$ and the canonical momentum p_i by the differential operator

$$\hat{p}_i = -i\hbar \frac{\partial}{\partial q^i} ,$$

then $\hat{q}^i$ and $\hat{p}_i$ satisfy the Heisenberg commutation rules.

Exercise 2.11.6
Problem: Show that if $\hat{q}^i$ and $\hat{p}_i$ satisfy Heisenberg's commutation rules, then their commutator with other operators satisfies rules corresponding to Exercise 2.10.4.

§ 2.12 Illustrative Example: Superconductors and Flux Quantization

We finish this chapter by reviewing another experiment that shows how gauge potentials may produce unexpected quantum-mechanical effects! Consider a suitable piece of metal, say aluminum. If we cool it down, it becomes superconducting, and interesting things happen. (See Feynman (1964), de Gennes (1966)). Suppose we originally had an external magnetic field. This would penetrate into the metal when the temperature is high, but it turns out that when the temperature falls below a certain critical temperature T_c, then the external field is expelled. This is the famous Meissner effect. (See Figure 2.21.)

It turns out that currents are produced in the outer layers of the metal, and these currents prevent the magnetic field from penetrating into the metal.

This is an interesting situation. We have a magnetic field **B** that stays entirely outside the lump of metal, but there is a gauge potential **A** too, and it may very well penetrate into the metal! We now change the experiment a little. We take a ring whose width is great compared to the penetration depth of the magnetic field. We place it in an external magnetic field **B** at room temperature. (See Fig-

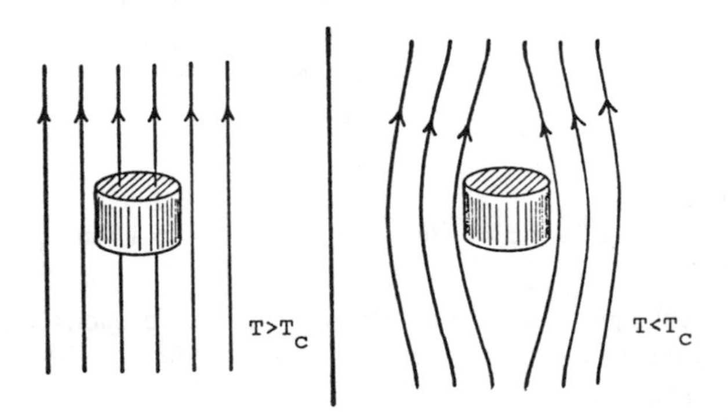

Figure 2.21.

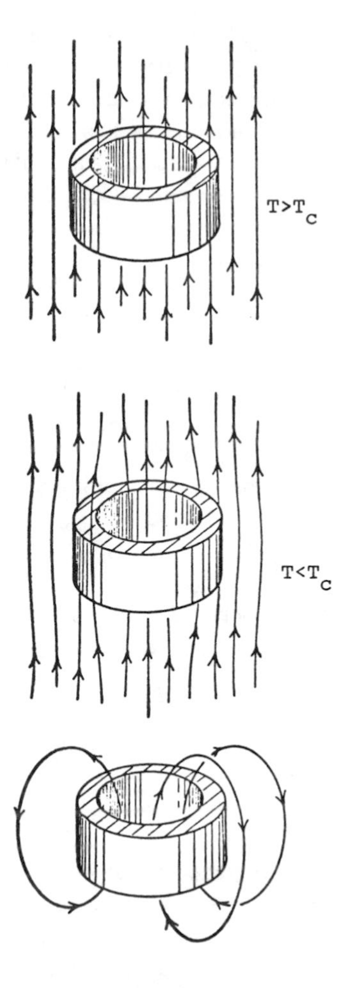

Figure 2.22.

ure 2.22.) What happens? The magnetic field is spread throughout the whole space, and in particular, it penetrates into the metal.

Then we cool down the ring, and when we pass below the critical temperature T_c, the Meissner effect occurs. The magnetic field is expelled due to surface currents in the superconducting ring. Hence some of the field lines pass outside the ring, and some of them pass through the hole in the ring. But now we remove the external field. However, this does not mean that the magnetic field disappears completely. Part of the field passing through the hole is trapped by the surface currents. So the superconducting ring now acts much like a solenoid! Observe that inside the metal there is no magnetic field **B**, only a gauge potential **A**. Since we have trapped a magnetic field **B** that passes through the ring, there is a magnetic flux Φ through the ring. It is this flux that we are going to examine!

Let us try to get a qualitative understanding of the situation. According to BCS-theory, the electrons in the metal will form pairs in the superconducting state. They are called *Cooper pairs*. Now, the electrons are fermions, but the Cooper pairs act like bosons! This has the important consequence that the Pauli principle no longer applies. Two electrons cannot occupy the same state, but there is nothing to prevent two Cooper pairs from occupying the same state! Actually, bosons have a strong tendency to occupy the same state.

Suppose $\psi(\mathbf{r}, t)$ denotes the Schrödinger wave function for a Cooper pair. We may actually assume that there is a macroscopic number of Cooper pairs all described by the same wave function $\psi(\mathbf{r}, t)$. This has important consequences. The absolute square $|\psi(\mathbf{r}, t)|^2$ is the probability of finding a specific Cooper pair at the point $\mathbf{r}$. If N is the total number of Cooper pairs, then

$$N|\psi(\mathbf{r}, t)|^2$$

is simply the *density* of Cooper pairs at the point $\mathbf{r}$. If, furthermore, $q = -2e$ is the charge of a Cooper pair, we conclude that

$$\rho(\mathbf{r}, t) = qN|\psi(\mathbf{r}, t)|^2$$

is the *charge density* of the Cooper pairs. Since a solution to the Schrödinger equation is only determined up to a constant, we may redefine it:

$$\psi(\mathbf{r}, t) \to \sqrt{|q|N}\, \psi(\mathbf{r}, t) = \psi'(\mathbf{r}, t).$$

So now we have a macroscopic number of Cooper pairs described by a single wave function,

$$\psi'(\mathbf{r}, t),$$

and $-|\psi'(\mathbf{r}, t)|^2$ simply denotes the charge density at $\mathbf{r}$. Observe that although $|\psi|^2$ in general only has a statistical meaning, $|\psi'|^2$ will in this case denote a macroscopic physical quantity! In what follows we will drop the prime and simply denote the wave function by $\psi(\mathbf{r}, t)$. We may decompose it in the following way:

$$\psi(\mathbf{r}, t) = \sqrt{|\rho(\mathbf{r}, t)|}\, \exp[i\phi(\mathbf{r}, t)]. \tag{2.73}$$

Here $\rho(\mathbf{r}, t)$ is the charge density, which has a direct physical meaning, and $\phi(\mathbf{r}, t)$ is the phase, which has no physical meaning, as we can change it by performing a gauge transformation!

In the superconducting state, the number of Cooper pairs is conserved, and so is the charge. If we denote the Cooper current by $\mathbf{j}$, we therefore conclude that ρ and $\mathbf{j}$ obey an equation of continuity (1.17):

$$\frac{\partial \rho}{\partial t} + \nabla \cdot \mathbf{j} = 0.$$

We can use this to find an expression for the Cooper current. $\rho = -\psi\bar{\psi}$ implies that

$$\frac{\partial \rho}{\partial t} = -\frac{\partial \psi}{\partial t}\, \bar{\psi} - \psi\, \frac{\partial \bar{\psi}}{\partial t}.$$

Using the Schrödinger equation, we can rearrange this as

$$\frac{\partial \rho}{\partial t} = \frac{i\hbar}{2m}\left\{\psi\left[\nabla + \frac{iq}{\hbar}\mathbf{A}\right]^2\bar{\psi} - \bar{\psi}\left[\nabla - \frac{iq}{\hbar}\mathbf{A}\right]^2\psi\right\}$$
$$= \frac{i\hbar}{2m}\nabla\left\{\psi\nabla\bar{\psi} - \bar{\psi}\nabla\psi + \frac{2iq}{\hbar}\mathbf{A}(\psi\bar{\psi})\right\},$$

which may be rewritten as

$$\frac{\partial \rho}{\partial t} = \frac{i\hbar}{2m}\nabla\left\{\psi\left(\nabla + \frac{iq}{\hbar}\mathbf{A}\right)\bar{\psi} - \bar{\psi}\left(\nabla - \frac{iq}{\hbar}\mathbf{A}\right)\psi\right\}.$$

From this expression we immediately read off the Cooper current:

$$j_i = \frac{i\hbar}{2m}\left[\bar{\psi}D_i\psi - \psi\overline{D_i\psi}\right]. \tag{2.74}$$

This shows us that the Cooper current is a gauge scalar, as it ought to be, and that it is a real quantity. In the following we will always assume that the density ρ of Cooper pairs is constant throughout our piece of metal! This is not strictly correct at the boundary, where it falls rapidly to zero. Except for that, it is a very reasonable assumption that the Cooper pairs do not "crowd" together but are evenly spaced.

Exercise 2.12.1
Problem: Show that the expression for the Cooper current can be rearranged as

$$\mathbf{j} = \frac{\rho\hbar}{m}\left[\nabla\phi - \frac{q\mathbf{A}}{\hbar}\right], \tag{2.75}$$

provided that ρ is constant.

Using Exercise 2.12.1 we can now derive the Meissner effect. Since there is no external electric field involved, the Maxwell equations (1.3) and (1.6) reduce to

$$\nabla\cdot\mathbf{B} = 0; \qquad \nabla\times\mathbf{B} = \frac{\mathbf{j}}{\epsilon_0 c^2}. \tag{2.76}$$

But from (2.75) we then get

$$\nabla\times(\nabla\times\mathbf{B}) = \frac{1}{\epsilon_0 c^2}\nabla\times\mathbf{j} = -\frac{\rho q}{\epsilon_0 m c^2}\nabla\times\mathbf{A} = -\frac{\rho q}{\epsilon_0 m c^2}\mathbf{B},$$

where we have used (1.12) to get rid of the term $\nabla\times(\nabla\phi)$. Furthermore, using (1.14) and (1.3), the left-hand side can be rearranged as

$$\nabla\times(\nabla\times\mathbf{B}) = \nabla(\nabla\cdot\mathbf{B}) - \Delta\mathbf{B} = -\Delta\mathbf{B}.$$

Hence we end up with the simple equation (the London equation)

$$\Delta\mathbf{B} = \frac{1}{\lambda_L^2}\mathbf{B}, \tag{2.77}$$

where

$$\lambda_L = \sqrt{\frac{\rho q}{\epsilon_0 m c^2}} \tag{2.78}$$

is the so-called London length. That equation (2.77) indeed implies the Meissner effect is left as an exercise; see below.

Exercise 2.12.2
Problem: Consider a semi-infinite superconductor occupying the half-space: $z > 0$. Let us apply a constant field $\mathbf{B} = B_0(1, 0, 0)$ parallel to the surface. Show that inside the superconductor the solution to (2.78) is given by

$$\mathbf{B}(z) = B_0 \left(\exp\left[-\frac{z}{\lambda_L} \right], 0, 0 \right), \qquad z > 0,$$

so that the magnetic field vanishes exponentially for $z > \lambda_L$. (λ_L is typically a few hundred Ångstrom.)

Observe that (2.76) not only implies $\mathbf{B}$ vanishes inside the superconductor; the same thing holds for the Cooper current $\mathbf{j}$.

Consider now a superconducting ring as shown in Figure 2.23. Inside the ring at the curve Γ the Cooper current vanishes, as does the magnetic field! Using (2.75) we therefore get

$$0 = \frac{\rho\hbar}{m} \left(\nabla\phi - \frac{q}{\hbar} \mathbf{A} \right)$$

i.e., $\nabla\phi = \frac{q}{\hbar} \mathbf{A}$. But now we can compute the magnetic flux inside the ring, because

$$\Phi = \oint_\Gamma \mathbf{A} \cdot d\mathbf{r} = \frac{\hbar}{q} \oint_\Gamma \nabla\phi \cdot d\mathbf{r}. \tag{2.79}$$

It might be tempting to say that this is equal to zero! But let us look a little more closely at the phase ϕ. The wave function

$$\psi(\mathbf{r}, t) = \sqrt{|\rho(\mathbf{r}, t)|} \exp[i\phi(\mathbf{r}, t)]$$

is only nontrivial in the ring, and all we can demand is that it be single-valued. But then nothing can prevent ϕ from making a jump of $2\pi n$. If that is the case, $\nabla\phi$ will contain a δ-like singularity, and the integral need not vanish! (Compare the discussion in Section 1.4.) To compute the line integral, we assume that $\nabla\phi$ makes the jump at the point B. If B^+ and B^- are points extremely close to B on

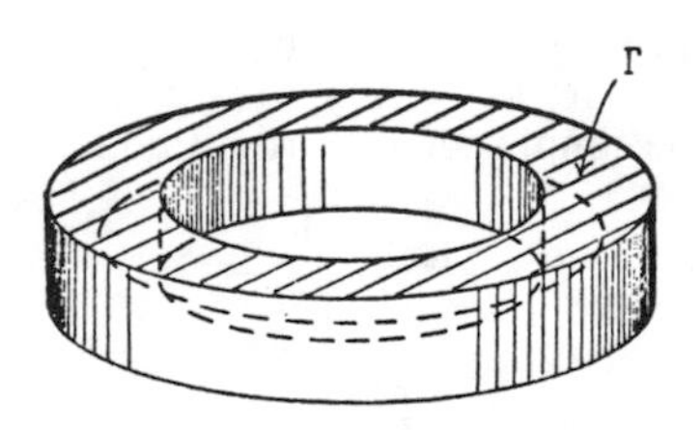

Figure 2.23.

each side of B, we get

$$\Phi = \oint \mathbf{A} \cdot d\mathbf{r} \approx \frac{\hbar}{q} \int_{B^-}^{B^+} \nabla\phi \cdot d\mathbf{r}$$
$$= \frac{\hbar}{q} [\phi(B^+) - \phi(B^-)] \to \frac{\hbar}{q} 2\pi n \qquad \text{as } B^+, B^- \to B.$$

Consequently, we get

$$\Phi = n\Phi_0, \qquad \text{where } \Phi_0 = \frac{2\pi\hbar}{|q|}. \tag{2.80}$$

So the trapped flux is quantized! This remarkable effect has been established experimentally (Deaver and Fairbanks (1961)). The quantity Φ_0 is the fundamental flux quantum. We have already seen that a vanishing magnetic field inside a superconductor implies a vanishing Cooper current. In the next exercise we will show that the opposite holds too: If the Cooper current vanishes, then so does the magnetic field.

Exercise 2.12.3
Problem: (a) Show that the expression (2.74) for the Cooper current can be rearranged as

$$j_i = \frac{i\hbar}{m} \bar{\psi} D_i \psi, \tag{2.81}$$

provided that the Cooper density ρ is constant. (b) Use this expression and Exercise 2.8.3 to prove directly that a vanishing Cooper current implies a vanishing magnetic field.

In the previous discussion we have assumed that the superconducting state of a metal depends only on the temperature. Actually, it also depends on the strength of an external magnetic field. Let us discuss this in some detail for a special kind of superconductor known as type II superconductor. Consider a superconduction cylinder. (See Figure 2.24.) Outside the cylinder we have a solenoid. Suppose a weak current flows in the coil. Then it will produce a magnetic field of strength B, which is expelled from the cylinder due to the Meissner effect. When we increase the current, B will reach a critical value B_{c_1}, where the superconducting state begins to break down. Thin vortices are formed where the normal state of

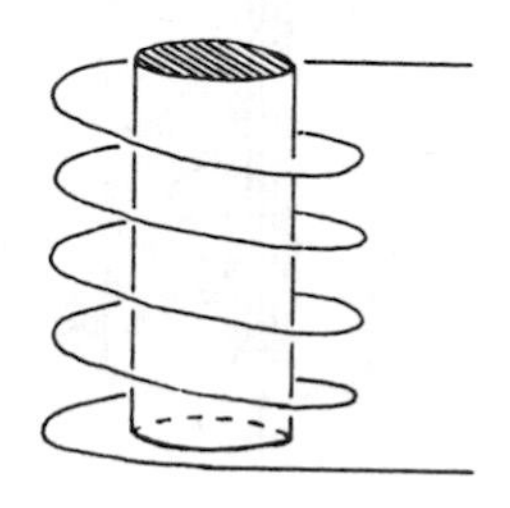

Figure 2.24.

the metal is reestablished. The magnetic field begins to penetrate into the metal through these vortices. As the magnetic field strength increases, more and more vortices are formed, and when we approach another critical field strength B_{c_2}, only small superconducting regions are still distributed throughout the cylinder. When we finally pass B_{c_2}, the superconducting state breaks down completely, and the cylinder is now back in its normal state. If we decrease the current again, then the same things happen in the reverse order.

Now let us concentrate on a single vortex. Let us enclose it by a great circle Γ as shown in Figure 2.25. Far away from the vortex, the magnetic field, and thus the Cooper current as well, vanishes. The magnetic flux through the vortex is given by

$$\Phi = \oint_\Gamma \mathbf{A} \cdot d\mathbf{r} = \frac{\hbar}{q} \oint_\Gamma \nabla\phi \cdot d\mathbf{r}$$

by the same arguments as for the superconducting ring. But then we see that the magnetic flux Φ through the vortex is necessarily *quantized*. The existence of quantized vortices was predicted by Abrikosov (1956), and they are therefore called *Abrikosov vortices*.

We have previously stated that charged particles may interact with the gauge potential A_μ in a space–time region Ω, even if the field strength $F_{\mu\nu}$ vanishes identically throughout this region. This interaction is a pure quantum-mechanical effect, the Bohm–Aharonov effect (see Section 2.6).

We may throw light on this using our results concerning the flux quantization. Consider two identical rings A and B. (See Figure 2.26.) In the following experiment the two rings are placed at room temperature. Inside ring B we also place a tiny solenoid. In this solenoid we have a current that produces exactly one-half of a flux quantum. Hence in the beginning of the experiment, the flux through A is 0, while the flux through B is $\frac{1}{2}\,\Phi_0$. Now we cool down the two rings, and they become superconducting. What happens? In the first ring nothing happens. But in the second ring the preceding analysis concerning flux quantization is clearly valid. The superconducting ring will allow only a quantized flux through the ring.

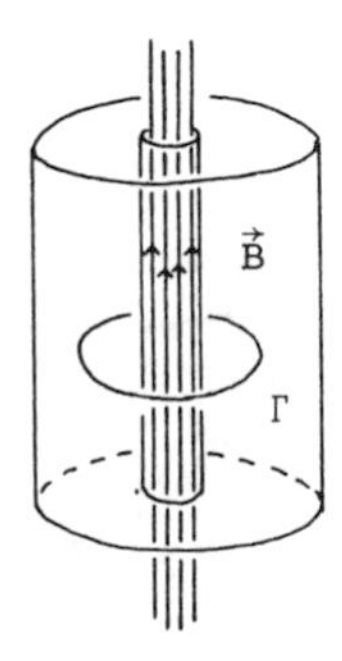

Figure 2.25.

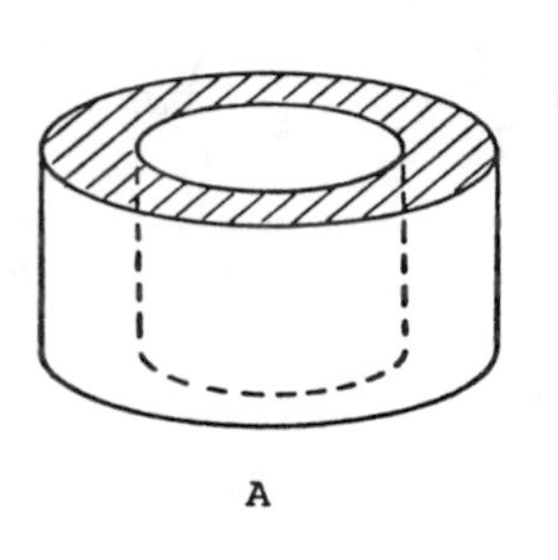

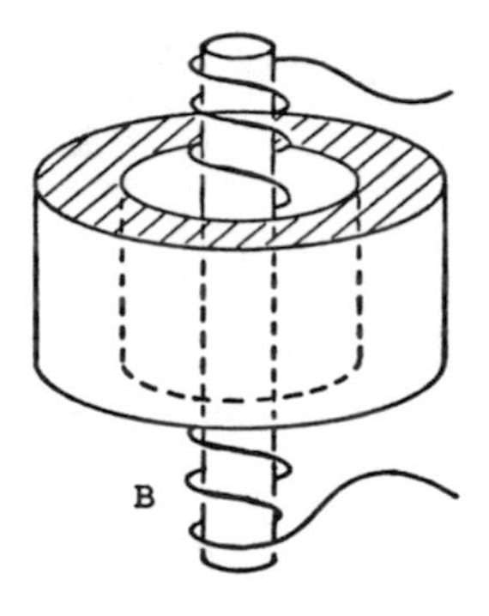

Figure 2.26.

Thus a Cooper current is produced in the outer layers, and this contributes to the flux, so that the total flux becomes 0 or Φ_0.

But what is the origin of this Cooper current? If the Cooper pair interacted only with the field strength, this would be a mystery, because there is only a nonvanishing field strength inside the solenoid. But as we have seen, there is a nontrivial gauge potential outside the solenoid, and this penetrates into the ring. This suggests that it is the interaction between the gauge potential and the Cooper pairs that is responsible for the Copper current.

This is also in accordance with the Schrödinger equation

$$i\hbar\left(\frac{\partial}{\partial t} + i\,\frac{q}{\hbar}\,\phi\right)\psi(\mathbf{r}, t) = -\frac{\hbar^2}{2m}\left(\nabla - i\,\frac{q}{\hbar}\,\mathbf{A}\right)^2\psi(\mathbf{r}, t), \tag{2.65}$$

which shows us that the Schrödinger wave function interacts directly with the gauge potential.

Solutions to Worked Exercises

Solution to 2.4.1

Now let $x = x_{\text{cl}}(t)$ be the classical path. Then we can write an arbitrary path in the following form:

$$x = x_{\text{cl}}(t) + y(t), \quad \text{where } y(t_1) = y(t_2) = 0.$$

Let us expand L using Taylor's theorem:

$$L(x_{\text{cl}} + y; \dot{x}_{\text{cl}} + \dot{y}; t) = L(x_{\text{cl}}; \dot{x}_{\text{cl}}; t) + \frac{\partial L}{\partial \dot{x}}\,y + \frac{\partial L}{\partial x}\,\dot{y}$$
$$+ \frac{1}{2}\left[\frac{\partial^2 L}{\partial x \partial x}\,y^2 + 2\,\frac{\partial^2 L}{\partial x \partial \dot{x}}\,y\dot{y} + \frac{\partial^2 L}{\partial \dot{x} \partial \dot{x}}\,\dot{y}^2\right].$$

This expansion is exact because L is quadratic! Then we can rewrite the action:

$$S = \int_{t_1}^{t_2} L(x_{\mathrm{cl}}, \dot{x}_{\mathrm{cl}}, t)dt + \int_{t_1}^{t_2} \left(\frac{\partial L}{\partial x}\, y + \frac{\partial L}{\partial \dot{x}}\, \dot{y} \right) dt$$
$$+ \frac{1}{2} \int_{t_1}^{t_2} \left[\frac{\partial^2 L}{\partial x \partial x}\, y^2 + 2\, \frac{\partial^2 L}{\partial x \partial \dot{x}}\, y\dot{y} + \frac{\partial^2 L}{\partial \dot{x} \partial \dot{x}}\, \dot{y}^2 \right] dt,$$

where the term in the middle vanishes, namely,

$$\int_{t_1}^{t_2} \left(\frac{\partial L}{\partial x}\, y + \frac{\partial L}{\partial \dot{x}}\, \dot{y} \right) dt = \int_{t_1}^{t_2} \left(\frac{\partial L}{\partial x} - \frac{d}{dt}\frac{\partial L}{\partial \dot{x}} \right) y\, dt = 0$$

because x_{cl} satisfies Euler's equation! The last term can easily be computed explicitly:

$$S = \int_{t_1}^{t_2} L(x_{\mathrm{cl}}, \dot{x}_{\mathrm{cl}}, t)dt + \int_{t_1}^{t_2} \left[a(t)y^2 + b(t)\dot{y}^2 + c(t)y\dot{y} \right] dt.$$

Substituting this into the path integral, we get

$$K(x_2; t_2 \mid x_1; t_1) = \exp\left\{ \frac{i}{\hbar}\, S[x_{\mathrm{cl}}] \right\}$$
$$\cdot \int_{y(t_1)=0}^{y(t_2)=0} \exp\left[\frac{i}{\hbar} \int_{t_1}^{t_2} (a(t)y^2 + b(t)\dot{y}^2 + c(t)y\dot{y})dt \right] D[y(t)].$$

But x_1 and x_2 are not at all involved in this last path integral, so it can depend only on t_1 and t_2! Consequently, we have shown that

$$K(x_2; t_2 \mid x_1; t_1) = A(t_1; t_2) \exp\left[\frac{i}{\hbar} \int_{t_1}^{t_2} L(x_{\mathrm{cl}}, \dot{x}_{\mathrm{cl}}, t)dt \right].$$

If furthermore, the coefficients $a(t)$, $b(t)$ and $c(t)$ do not depend on time, then we can easily show that

$$A(t_1; t_2) = A(0; t_2 - t_1).$$

Observe that

$$\int_{t_1}^{t_2} \left[ay^2(t) + b\dot{y}^2(t) + cy\dot{y} \right] dt = \int_{t_1+\Delta t}^{t_2+\Delta t} \left[ay^2(t - \Delta t) + \cdots \right] dt$$
$$= \int_{t_1+\Delta t}^{t_2+\Delta t} \left[ay'^2(t) + \cdots \right] dt,$$

where we have introduced $y'(t) = y(t - \Delta t)$. Thus

$$A(t_1; t_2) = \int_{y(t_1)=0}^{y(t_2)=0} \exp\left[\frac{i}{\hbar} \int_{t_1}^{t_2} ay^2(t) + \cdots dt \right] D[y(t)]$$
$$= \int_{y'(t_1+\Delta t)=0}^{y'(t_2+\Delta t)=0} \exp\left[\frac{i}{\hbar} \int_{t_1+\Delta t}^{t_2+\Delta t} ay'^2(t) + \cdots dt \right] D[y'(t)]$$
$$= A(t_1 + \Delta t; t_2 + \Delta t),$$

from which the desired result follows immediately when we put: $\Delta t = -t_1$.

Solution to 2.8.2

$$A_\alpha\left(x^\beta + \frac{\epsilon}{2}\,\delta x^\beta\right) = A_\alpha(x) + \frac{\epsilon}{2}\frac{\partial A_\alpha}{\partial x^\beta}\,\delta x^\beta + \ldots$$

$$A_\alpha\left(x^\beta + \frac{\epsilon}{2}\,\Delta x^\beta + \epsilon\delta x^\beta\right) = A_\alpha(x) + \frac{\epsilon}{2}\frac{\partial A_\alpha}{\partial x^\beta}\,\Delta x^\beta + \epsilon\,\frac{\partial A_\alpha}{\partial x^\beta}\,\delta x^\beta + \ldots$$

$$A_\alpha\left(x^\beta + \frac{\epsilon}{2}\,\Delta x^\beta\right) = A_\alpha(x) + \frac{\epsilon}{2}\frac{\partial A_\alpha}{\partial x^\beta}\,\Delta x^\beta + \ldots$$

$$A_\alpha\left(x^\beta + \epsilon\Delta x^\beta + \frac{\epsilon}{2}\,\delta x^\beta\right) = A_\alpha(x) + \epsilon\,\frac{\partial A_\alpha}{\partial x^\beta}\,\Delta x^\beta + \frac{\epsilon}{2}\frac{\partial A_\alpha}{\partial x^\beta}\,\delta x^\beta + \ldots$$

At the end of the calculation we are going to let $\epsilon \to 0$; hence we need only to compute the lowest-order terms. In this approximation we get

$$\begin{aligned}
\oint_{\Gamma_\epsilon} A_\alpha dx^\alpha &= \int_A^B A_\alpha dx^\alpha + \int_B^C A_\alpha dx^\alpha + \int_C^D A_\alpha dx^\alpha + \int_D^A A_\alpha dx^\alpha \\
&\approx \left[A_\alpha(x) + \frac{1}{2}\,\epsilon\,\frac{\partial A_\alpha}{\partial x^\beta}\,\Delta x^\beta\right]\epsilon\Delta x^\alpha \\
&\quad + \left[A_\alpha(x) + \epsilon\,\frac{\partial A_\alpha}{\partial x^\beta}\,\Delta x^\beta + \frac{\epsilon}{2}\frac{\partial A_\alpha}{\partial x^\beta}\,\delta x^\beta\right]\epsilon\delta x^\alpha \\
&\quad - \left[A_\alpha(x) + \epsilon\,\frac{\partial A_\alpha}{\partial x^\beta}\,\delta x^\beta + \frac{\epsilon}{2}\frac{\partial A_\alpha}{\partial x^\beta}\,\Delta x^\beta\right]\epsilon\Delta x^\alpha \\
&\quad - \left[A_\alpha(x) + \frac{1}{2}\,\epsilon\,\frac{\partial A_\alpha}{\partial x^\beta}\,\delta x^\beta\right]\epsilon\delta x^\alpha \\
&= \epsilon^2\left[\frac{\partial A_\alpha}{\partial x^\beta}\,\Delta x^\beta\delta x^\alpha - \frac{\partial A_\alpha}{\partial x^\beta}\,\delta x^\beta\Delta x^\alpha.\right]
\end{aligned}$$

Interchanging the dummy indices α and β in the last term, we finally get

$$\oint A_\alpha dx^\alpha = \epsilon^2\left[\partial_\alpha A_\beta - \partial_\beta A_\alpha\right]\Delta x^\beta\delta x^\alpha.$$

Taking the limit, we obtain the exact result

$$F_{\alpha\beta}(x)\cdot\Delta x^\beta\delta x^\alpha = \lim_{\epsilon\to 0}\frac{1}{\epsilon^2}\oint_{\Gamma_\epsilon} A_\alpha dx^\alpha.$$

Solution to 2.11.1

a. Let E be an eigenvalue and let ψ be the corresponding normalized eigenfunction

$$\bar{E} = \bar{E}\langle\psi \mid \psi\rangle = \langle E\psi \mid \psi\rangle = \langle \hat{H}\psi \mid \psi\rangle.$$

Using the Hermiticity of $\hat{H}$, this is rearranged as

$$= \langle\psi \mid \hat{H}\psi\rangle = \langle\psi \mid E\psi\rangle = \langle\psi \mid \psi\rangle E = E.$$

Thus $\bar{E} = E$, and therefore E is real.

b. Let E_1, E_2 be different eigenvalues and let ψ_1, ψ_2 be the corresponding eigenfunctions. Then we get

$$\begin{aligned} E_1\langle\psi_1 \mid \psi_2\rangle &= \langle E_1\psi_1 \mid \psi_2\rangle = \langle\hat{H}\psi_1 \mid \psi_2\rangle = \langle\psi_1 \mid \hat{H}\psi_2\rangle \\ &= \langle\psi_1 \mid E_2\psi_2\rangle = \langle\psi_1 \mid \psi_2\rangle E_2. \end{aligned}$$

As $E_1 \neq E_2$, this implies $\langle\psi_1 \mid \psi_2\rangle = 0$.

Solution to 2.11.2

Let $\phi(x)$ be an arbitrary wave function. We can decompose it into the complete set of eigenfunctions

$$\phi(x) = \sum_m a_m \psi_m(x).$$

Using this decomposition, we get

$$\begin{aligned} &\int \left(\sum_n \overline{\psi_n}(x_1)\psi_n(x_2) \right) \phi(x_1) dx_1 \\ &= \sum_n \psi_n(x_2) \sum_m a_m \int \overline{\psi_n}(x_1)\psi_m(x_1) dx_1 \\ &= \sum_n \psi_n(x_2) \sum_m a_m \delta_{mn} \\ &= \sum_n a_n \psi_n(x_2) = \phi(x_2). \end{aligned}$$

But this clearly shows that

$$\sum_n \psi_n(x_1)\psi_n(x_2) = \delta(x_1 - x_2).$$

Solution to 2.11.3

For fixed (x_1, t_1) we know that the Feynman propagator is a solution to the Schrödinger equation. Hence we may decompose it as

$$K(x_2; t_2 \mid x_1; t_1) = \sum_n a_n(x_1; t_1)\psi_n(x_2) \exp\left[-\frac{i}{\hbar} E_n t_2\right].$$

The coefficients can still depend on x_1 and t_1. For $t_2 = t_1$ we know that the Feynman propagator reduces to a δ-function. Thus we get

$$\delta(x_1 - x_2) = \sum_n a_n(x_1; t_1)\psi_n(x_2) \exp\left[-\frac{i}{\hbar} E_n t_1\right].$$

This forces us to put

$$a_n(x_1; t_1) = a_n(x_1) \exp\left[\frac{i}{\hbar} E_n t_1\right],$$

since the left-hand side is independent of t_1. The above formula then reduces to

$$\delta(x_1 - x_2) = \sum_n a_n(x_1)\psi_n(x_2).$$

But a comparison with Exercise 2.11.2 then gives us

$$a_n(x_1) = \overline{\psi_n}(x_1),$$

and we are through.

CHAPTER 3

Dynamics of Classical Fields

§ 3.1 Illustrative Example: The Lagrangian Formalism for a String

We have already discussed the Lagrangian formulation of the dynamics of a system with a finite number of degrees of freedom, say a finite number of particles moving in an external field. Now we want to include the dynamics of fields. We will start by considering a one-dimensional string. It has an important property: If the string is disturbed at one place, then this disturbance may propagate along the string. We can understand this in an intuitive way. The string consists of "atoms." Each atom interacts with its nearest neighbors. Hence, if one atom is disturbed, this disturbance has influence on its neighbor. But this disturbance of a neighbor has influence on the neighbor of the neighbor, etc.! In this way a traveling wave is created that propagates along the string!

In our model the string is composed of "atoms" that in the equilibrium state are evenly spaced throughout the x-axis. (See Figure 3.1.) The important assumption is that the "atoms" are coupled to each other through forces proportional to their relative displacements (Hooke forces).

We will enumerate the "atoms" with an integer n, so that the equilibrium position of the nth "atom" is $x_n = a_n$. If we set the "atoms" in motion, then the nth "atom" will be displaced an amount q_n from its equilibrium position.

When the string is put into vibration, its dynamical evolution is described by the functions

$$q_n = q_n(t), \qquad n = \ldots, -2, -1, 0, 1, 2, \ldots.$$

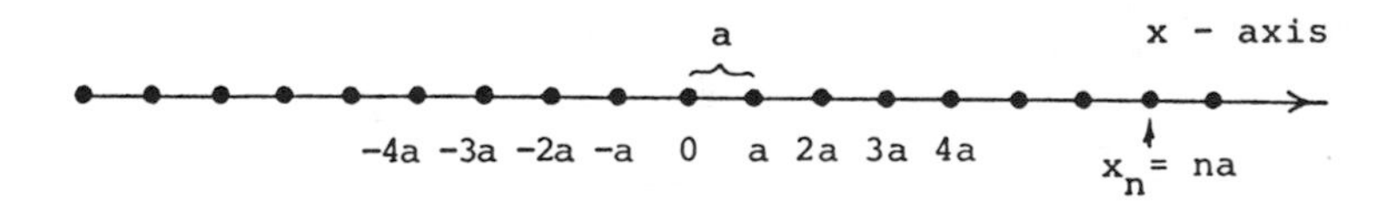

Figure 3.1.

The kinetic energy of the nth "atom" is

$$\frac{1}{2} m\dot{q}_n^2,$$

and the potential energy associated with the separation of atom n and $n+1$ is:

$$\frac{1}{2} mv^2[q_{n+1} - q_n]^2.$$

Hence the total *Lagrangian* for the system is

$$L = \sum_{n=-\infty}^{+\infty} \left(\frac{1}{2} m\dot{q}_n^2 - \frac{1}{2} mv^2[q_{n+1} - q_n]^2 \right). \tag{3.1}$$

Let us determine the equations of motion. Using (2.6) we get

$$m\ddot{q}_i = mv^2(q_{i+1} - 2q_i + q_{i-1}). \tag{3.2}$$

It is easy to check that the equations of motion actually allow wave solutions. If we put

$$q_n(t) = A\cos(kx_n - \omega t), \tag{3.3}$$

then this will represent a traveling wave. Inserting this into the equation of motion, we get

$$-m\omega^2 A\cos(kx_n - \omega t) = mv^2\cos(kx_n - \omega t)(2\cos(k \cdot a) - 1),$$

or

$$\omega^2 = 2v^2(1 - \cos(k \cdot a)). \tag{3.4}$$

Thus (3.3) is a solution to the equation of motion, provided that (3.4) is satisfied. Relation (3.4) is called a *dispersion relation*. Observe that for small k, i.e., in the long wave-length limit, we may expand the cosine, getting

$$\omega \approx \pm v\sqrt{2\left(1 - 1 + \frac{1}{2} k^2a^2\right)} = \pm va \cdot k, \tag{3.5}$$

which is a linear dispersion relation.

Now we want to investigate a *continuous* string, where there are "atoms" everywhere. We can do this by letting $a \to 0$ in our discrete model. We say that we pass to the *continuum limit*.

Now, instead of describing the displacements by the infinite set of numbers $q_j(t)$, we will represent them by a smooth function $q(x, t)$ giving the displacement of the "atom" with equilibrium position x. (See Figure 3.2.)

In the discrete model, the mass density is m/a. When we pass to the continuum limit, we suppose that it approaches a constant ρ, the mass density of the continuous string; i.e.,

$$\frac{m}{a} \to \rho \qquad \text{as } a \to 0.$$

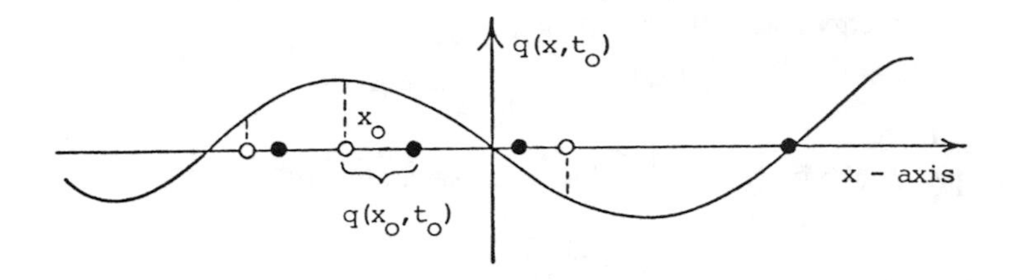

Figure 3.2.

Finally, it will be necessary to make an assumption about ν. In the discrete model, the velocity of a wave in the long wave-length limit is (compare (3.3) and (3.5))

$$\frac{\omega}{k} = \nu a.$$

We assume that it approaches a constant c, the velocity of a traveling wave in the continuous string:

$$\nu a \to c \qquad \text{as } a \to 0.$$

With these preliminaries we can investigate what happens to the Lagrangian in the continuum limit. Let us take a look at the kinetic energy:

$$\begin{aligned}\sum_{n=-\infty}^{+\infty} \frac{1}{2} m\dot{q}_n^2(t) &= \sum_{x_n=-\infty}^{x_n=+\infty} \frac{1}{2} \frac{m}{a} [\dot{q}(x_n, t)]^2 a \\ &= \sum_{x_n=-\infty}^{x_n=+\infty} \frac{1}{2} \frac{m}{a} [\dot{q}(x_n, t)]^2 \Delta x_n \\ &\to \int_{x=-\infty}^{x=+\infty} \frac{1}{2} \rho[\dot{q}(x, t)]^2 dx.\end{aligned}$$

We can treat the potential energy in a similar way. From the observation

$$q_{n+1}(t) - q_n(t) = q(x_{n+1}, t) - q(x_n, t) \approx a \frac{\partial q}{\partial x} (x_n, t)$$

we get

$$\begin{aligned}\sum_{n=-\infty}^{+\infty} \frac{1}{2} m\nu^2[q_{n+1}(t) - q_n(t)]^2 &\approx \sum_{x_n=-\infty}^{x_n=+\infty} \frac{1}{2} \frac{m}{a} (\nu a)^2 \left[\frac{\partial q}{\partial x} (x_n, t)\right]^2 a \\ &= \sum_{x_n=-\infty}^{x_n=+\infty} \frac{1}{2} \frac{m}{a} (\nu a)^2 \left[\frac{\partial q}{\partial x} (x_n, t)\right]^2 \Delta x_n \\ &\to \int_{x=-\infty}^{x=+\infty} \frac{1}{2} \rho c^2 \left[\frac{\partial q}{\partial x} (x, t)\right]^2 dx.\end{aligned}$$

Thus for the total Lagrangian we get

$$\sum_{n=-\infty}^{+\infty} \frac{1}{2} m\dot{q}_n^2 - \frac{1}{2} mv^2[q_{n+1} - q_n]^2 \\ \to \int_{x=-\infty}^{+\infty} \left[\frac{1}{2} \rho \left(\frac{\partial q}{\partial t} \right)^2 - \frac{1}{2} \rho c^2 \left(\frac{\partial q}{\partial x} \right)^2 \right] dx, \tag{3.6}$$

showing that in the continuum limit *the Lagrangian is expressed as an integral over space*. The integrand is called the *Lagrangian density*,

$$L\left(q, \frac{\partial q}{\partial t}, \frac{\partial q}{\partial x}\right) = \frac{1}{2} \rho \left(\frac{\partial q}{\partial t} \right)^2 - \frac{1}{2} \rho c^2 \left(\frac{\partial q}{\partial x} \right)^2. \tag{3.7}$$

Observe that it contains not only $\frac{\partial q}{\partial t}$ but also $\frac{\partial q}{\partial x}$! Where did the space derivative $\frac{\partial q}{\partial x}$ come from? It came from the term $(q_{n+1} - q_n)^2$ in the discrete model. Hence, it reflects the *property of local interactions*. Each point in space interacts with its nearest neighbors.

In a similar way we may analyze the equations of motion. In the discrete model we have

$$\ddot{q}_n(t) = v^2[q_{n+1} - 2q_n + q_{n-1}].$$

We may rearrange the term on the right side:

$$\begin{aligned} q_{n+1} - 2q_n + q_{n-1} &= [q(x_{n+1}, t) - q(x_n, t)] - [q(x_n, t) - q(x_n, t)] \\ &\approx a\left[\frac{\partial q}{\partial x} \left(x_n + \frac{q}{2}, t \right) - \frac{\partial q}{\partial x} \left(x_n - \frac{q}{2}, t \right) \right] \\ &\approx a^2 \left[\frac{\partial^2 q}{\partial x^2} (x_n, t) \right]. \end{aligned}$$

Thus we obtain

$$\ddot{q}(x_n, t) \approx v^2 a^2 \frac{\partial^2 q}{\partial x^2} (x_n, t) \to c^2 \frac{\partial^2 q}{\partial x^2}; \qquad \text{i.e.,} \quad \frac{1}{c^2} \frac{\partial^2 q}{\partial t^2} = \frac{\partial^2 q}{\partial x^2}. \tag{3.8}$$

As before, we may look for a solution representing a traveling wave:

$$q(x, t) = A \cos(kx - \omega t). \tag{3.9}$$

If we insert this into the equation of motion (3.8), we get

$$-\omega^2 A \cos(kx - \omega t) = -c^2 k^2 A \cos(kx - \omega t);$$

i.e.,

$$\omega = \pm ck. \tag{3.10}$$

Hence, (3.9) is a solution to the equation of motion, provided that ω satisfies the linear dispersion relation (3.10).

§ 3.2 The Lagrangian Formalism for Relativistic Fields

We should now be motivated for the abstract field theory. We start with a field $\phi(t, \mathbf{x})$ defined throughout space–time. The value of the field at a particular point $\mathbf{x}_0$, $\phi(t, \mathbf{x}_0)$, corresponds to the stretching $q(t, \mathbf{x}_0)$ in the preceding example. The dynamics of the field are governed by a Lagrangian $\mathcal{L}$, which by analogy with the preceding example we write as

$$\mathcal{L} = \int L d^3\mathbf{x}. \tag{3.11}$$

The Lagrangian density L depends not only on the time derivative, but on the space derivatives as well

$$L = L(\phi, \partial_\mu \phi).$$

The presence of space derivatives $\partial_i \phi$ reflects the principle of local interactions.

If we choose two times t_1 and t_2, we may specify the field at these times. Any smooth function $\phi(t, \mathbf{x})$ that satisfies the boundary conditions

$$\phi(t_1, \mathbf{x}) = \phi_1(\mathbf{x}) \quad \text{and} \quad \phi(t_2, \mathbf{x}) = \phi_2(\mathbf{x})$$

represents a possible history of the field. To each such history we associate the action

$$S = \int_{t_1}^{t_2} \mathcal{L}\, dt = \int_{t_1}^{t_2} \int L d^3\mathbf{x}\, dt;$$

i.e.,

$$S = \int_\Omega L d^4 x, \tag{3.12}$$

where Ω is the four-dimensional region between the hyperplanes $t = t_1$ and $t = t_2$. (See Figure 3.3.) As usual, we want to determine a history $\phi(t, \mathbf{x})$ that extremizes the action. This, of course, leads to the equation of motion for the field. Now suppose that $\phi_0(t, \mathbf{x})$ really extremizes the action. Consider another history,

$$\phi(t, \mathbf{x}) = \phi_0(t, \mathbf{x}) + \epsilon \eta(t, \mathbf{x}),$$

where $\eta(t, \mathbf{x})$ satisfies the boundary conditions $\eta(t_1, \mathbf{x}) = \eta(t_2, \mathbf{x}) = 0$.

Then the action

$$S(\epsilon) = \int_\Omega L(\phi_0 + \epsilon\eta, \partial_\mu \phi_0 + \epsilon \partial_\mu \eta) d^4 x$$

has an extremal value when $\epsilon = 0$. Consequently, we get

$$0 = \frac{dS}{d\epsilon}\Big|_{\epsilon=0} = \int_\Omega \left[\frac{\partial L}{\partial \phi}\Big|_{\epsilon=0} \cdot \eta + \frac{\partial L}{\partial(\partial_\mu \phi)}\Big|_{\epsilon=0} \cdot \partial_\mu \eta \right] d^4 x$$

$$= \int_\Omega \left[\frac{\partial L}{\partial \phi}\Big|_{\phi_0} - \partial_\mu \frac{\partial L}{\partial(\partial_\mu \phi)}\Big|_{\phi_0} \right] \eta d^4 x,$$

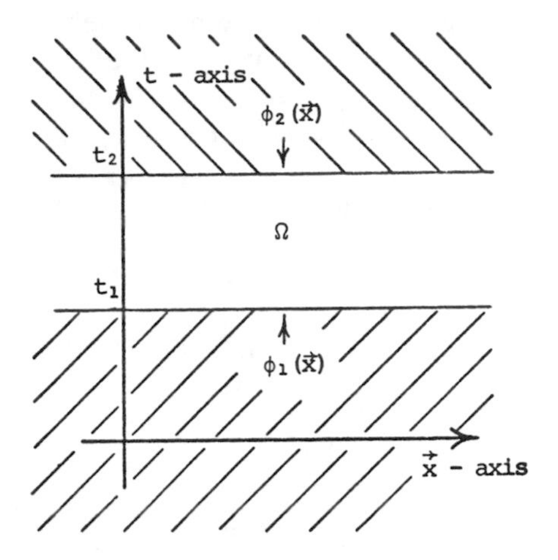

Figure 3.3.

where we have neglected the surface terms due to the boundary conditions on η. But $\eta(t, \mathbf{x})$ was arbitrarily chosen. Therefore, the above result is consistent only if ϕ_0 satisfies the differential equation

$$\frac{\partial L}{\partial \phi} - \partial_\mu \frac{\partial L}{\partial(\partial_\mu \phi)} = 0. \tag{3.13}$$

This generalizes the Euler–Lagrange equation for a system with a finite number of degrees of freedom. Observe that all the derivatives occur! This has an important consequence: *The equation of motion is Lorentz-invariant, provided that L is a Lorentz scalar.*

We would also like to discuss fields with several components $\phi_a, a = 1, \ldots, n$ (like the Maxwell field A_α). We leave the deduction of the equations of motion as an exercise.

Exercise 3.2.1
Problem: Suppose the field has several components: ϕ_a, $a = 1, \ldots, n$. Show that each of the components must satisfy the appropriate Euler–Lagrange equation

$$\frac{\partial L}{\partial \phi_a} - \partial_\mu \frac{\partial L}{\partial(\partial_\mu \phi_a)} = 0, \qquad a = 1, \ldots, n. \tag{3.14}$$

We may also discuss the energy–momentum corresponding to our field ϕ_a. A direct generalization of the Hamiltonian method suggests that the energy density is given by the *Hamiltonian density*

$$H = \frac{\partial L}{\partial(\partial_0 \phi_a)} \partial_0 \phi_a - L \tag{3.15}$$

(compare with (2.56)). We expect energy to propagate throughout space, so we must also have an energy current **s** associated with the field. Since the total energy

$$E = \int H d^3\mathbf{x}$$

should be conserved, we must demand that H and **s** satisfy an equation of continuity

$$\frac{\partial H}{\partial t} + \nabla \cdot \mathbf{s} = 0.$$

We can use this to determine an expression for the energy current **s**:

$$\begin{aligned}\frac{\partial H}{\partial t} &= \partial_0 \left[\frac{\partial L}{\partial(\partial_0 \phi_a)} \partial_0 \phi_a - L \right] \\ &= \partial_0 \frac{\partial L}{\partial(\partial_0 \phi_a)} \cdot \partial_0 \phi_a + \frac{\partial L}{\partial(\partial_0 \phi_a)} \partial_0 \partial_0 \phi_a \\ &\quad - \frac{\partial L}{\partial \phi_a} \partial_0 \phi_a - \frac{\partial L}{\partial(\partial_\mu \phi_a)} \cdot \partial_0 \partial_\mu \phi_a.\end{aligned}$$

Using the Euler–Lagrange equations (3.14), we get

$$\partial_0 \frac{\partial L}{\partial(\partial_0 \phi_a)} = \frac{\partial L}{\partial \phi_a} - \partial_i \frac{\partial L}{\partial(\partial_i \phi_a)}.$$

Therefore,

$$\frac{\partial H}{\partial t} = -\partial_i \frac{\partial L}{\partial(\partial_i \phi_a)} \cdot \partial_0 \phi_a - \frac{\partial L}{\partial(\partial_i \phi_a)} \partial_0 \partial_i \phi_a = -\partial_i \left[\frac{\partial L}{\partial(\partial_i \phi_a)} \partial_0 \phi_a \right].$$

From this we can immediately read off the *energy current*

$$s^i = \frac{\partial L}{\partial(\partial_i \phi_a)} \partial_0 \phi_a. \tag{3.16}$$

We know that the energy density corresponds to the T^{00}-component of the energy–momentum tensor and that the energy current s^i corresponds to the T^{0i} component (compare (1.37)), but then the above expressions for H and s^i suggest that we put

$$\overset{\circ}{T}{}^{\alpha\beta} = \frac{-\partial L}{\partial(\partial_\beta \phi_a)} \partial^\alpha \phi_a + \eta^{\alpha\beta} L. \tag{3.17}$$

If L is a Lorentz scalar, this is a tensor that reproduces H and **s**.

Exercise 3.2.2
Problem: Use the equations of motions to show that the above energy–momentum tensor is conserved:

$$\partial_\beta \overset{\circ}{T}{}^{\alpha\beta} = 0.$$

From Exercise 3.2.2 we learn that $\overset{\circ}{T}{}^{\alpha\beta}$ produces a conserved energy–momentum! So everything should be all right. $\overset{\circ}{T}{}^{\alpha\beta}$ is called the *canonical energy–momentum tensor*. However, there is one point where we must be careful: $\overset{\circ}{T}{}^{\alpha\beta}$ *need not* be symmetric, contrary to our previous demand (Section 1.6) based upon conservation of angular momentum. Consider the expression (3.17) for the canonical energy–momentum tensor:

$$\overset{\circ}{T}{}^{\alpha\beta} = \frac{-\partial L}{\partial(\partial_\beta \phi_a)} \partial^\alpha \phi_a + \eta^{\alpha\beta} L.$$

Obviously, the last term is symmetric, but the first term need not be so! What can we do if the canonical energy–momentum tensor is not symmetric? We must repair it! Suppose we can find a tensor $\Theta^{\alpha\beta}$ with the following properties:

1. $\overset{\circ}{T}{}^{\alpha\beta} + \Theta^{\alpha\beta}$ is symmetric.
2. $\partial_\beta \Theta^{\alpha\beta} = 0$. (3.18)
3. $\int \Theta^{\alpha 0} d^3\mathbf{x} = 0$.

Then $\overset{\circ}{T}{}^{\alpha\beta} + \Theta^{\alpha\beta}$ is a symmetric tensor that is conserved, and it reproduces the same energy–momentum as $\overset{\circ}{T}{}^{\alpha\beta}$! Hence, we might call $T^{\alpha\beta} = \overset{\circ}{T}{}^{\alpha\beta} + \Theta^{\alpha\beta}$ the *true energy–momentum tensor*. There exists a systematic method to construct $\Theta^{\alpha\beta}$ (the method of Rosenfeld and Belinfante; see for instance: Barut (1964), Ch. 3, Sec. 4). But the method is very complicated, and we will not have to use it. Later on (Ch. 11), we shall see how it can be constructed from a different point of view.

Okay, suppose we want to construct *free field theory*. What should we demand about L? That depends on what we expect of it! Although we will not quantize the fields, it might help to look at the following *very naive* discussion:

A particle of mass m, energy E, and momentum $\mathbf{p}$ is represented by the wave function

$$e^{i/\hbar(\mathbf{p}\cdot\mathbf{x}-Et)} = e^{ip_\mu x^\mu}$$

(where we have put $\hbar = 1$). When we quantize a field theory of a free field, we expect that the field represents particles (quanta). Therefore, we might look for solutions to the equations of motion of the form

$$\phi(x) = e^{ip_\mu x^\mu}, \qquad \text{where } p_\mu p^\mu = -m^2 \tag{3.19}$$

and we might interpret these solutions as the wave functions of the quanta of the field. For instance, if we quantize the Maxwell field, we expect it to represent massless particles = photons. (We might be worried about the fact that we allow *complex* solutions in a classical theory of fields. If we are very worried, we may look for solutions of the form $\cos(p_\mu x^\mu)$!)

Let us summarize our expectations: We want to construct a free field theory based upon a Lagrangian density with the following properties:

1. L is Lorentz scalar.
2. H is positive definite. (3.20)
3. The equations of motion allow solutions of the form $e^{ip_\mu x^\mu}$, where p_μ satisfies the dispersion relation $p_\mu p^\mu = -m^2$.

Property (1) guarantees that we are constructing a relativistic theory; (2) guarantees that the energy density is positive; and (3) guarantees that if we quantize the field, it will represent *free* particles with a mass m, where

$$-m^2 = p_\mu p^\mu .$$

§ 3.3 The Hamiltonian Formalism for Relativistic Fields

In analogy with the Hamiltonian formalism for a particle, we can now associate to each component of the field a *conjugate momentum* defined by

$$\pi^a = \frac{\partial L}{\partial(\partial_0 \phi_a)} . \tag{3.21}$$

Here π^a should be considered as a function of ϕ_a, $\partial_0\phi_a$, $\partial_i\phi_a$. If we can invert the relation and obtain $\partial_0\phi_a$ as a function of ϕ_a, π^a, $\partial_i\phi_a$, we say that the system is *canonical*.

Exercise 3.3.1
Problem: Consider a canonical system. Show that the Euler equations are equivalent to the following *Hamilton equations*:

$$\frac{\partial H}{\partial \pi^a} = \partial_0\phi_a , \qquad \frac{\partial H}{\partial \phi_a} = -\partial_0\pi^a + \partial_i \frac{\partial H}{\partial(\partial_i\phi_a)} . \tag{3.22}$$

Consider a suitable function space M. A *functional* F is a map, $F : M \to \mathbb{R}$ that maps a function into a real number. If M is the set of all smooth functions $f : [a, b] \to \mathbb{R}$, then we can, for instance, consider the following functionals:

$$F_1[f] = \int_a^b f(x)dx; \qquad F_2[f] = \int_a^b \frac{1}{2}(f'(x))^2 dx.$$

If the functional F is sufficiently nice, we can introduce a *functional derivative*. Consider first the case of an ordinary smooth function $f : \mathbb{R}^n \to \mathbb{R}$. Let x_0 be given. Then we can expand f in a neighborhood of x_0:

$$f(x_0 + y) = f(x_0) + \frac{\partial f}{\partial x^i} y^i + \cdots .$$

To formalize this we consider the map

$$y \to \frac{\partial f}{\partial \epsilon} (x_0 + \epsilon y)_{|\epsilon=0} .$$

This is a linear map. Thus, we can write it in the form

$$\frac{\partial f}{\partial \epsilon}_{|\epsilon=0} (x_0 + \epsilon y) = a_i(x_0)y^i .$$

Here the coefficients $a_i(x_0)$ depend on x_0, and actually, they are the partial derivatives of f:

$$a_i(x_0) = \frac{\partial f}{\partial x^i} (x_0).$$

This motivates the following definition: Consider a functional F. Let f_0 be given. Then we can expand F in a neighborhood of f_0:

$$F[f_0 + g] = F[f_0] + \int \frac{\delta F}{\delta f}\Big|_{f=f_0}(x)g(x)dx + \cdots .$$

To formalize this, we consider the map

$$g \to \frac{\partial F}{\partial \epsilon}\Big|_{\epsilon=0}[f_0 + \epsilon g].$$

This is a linear map. Thus, we can write it in the form

$$\frac{\partial}{\partial \epsilon}\Big|_{\epsilon=0} F[f_0 + \epsilon g] = \int k(x)g(x)dx.$$

The function $k(x)$ depends on f_0, and we define it to be the *functional derivative* of F at the function f_0:

$$\frac{\delta F}{\delta f}\Big|_{f=f_0} = k(x).$$

Note that when we perform the variation $f_0 \to f_0 + \epsilon g$, then we will always assume that g vanishes on the boundary.

Exercise 3.3.2
Problem: Consider the following functionals:

$$F_1[f] = \int_{t_1}^{t_2} \left[\frac{1}{2} m \left(\frac{df}{dt} \right)^2 - V(f) \right] dt,$$

$$F_2[f] = f(x_0), \qquad F_3[f] = f'(x_0).$$

Show that the functional derivatives are given by

$$\frac{\delta F_1}{\delta f} = -m \frac{d^2 f}{dt^2} - V'(f), \quad \frac{\delta F_2}{\delta f} = \delta(x - x_0), \quad \frac{\delta F_3}{\delta f} = -\delta'(x - x_0).$$

Exercise 3.3.3
Problem: (a) Show that the Euler–Lagrange equation can be written in the form

$$\frac{\delta S}{\delta \phi_a} = 0, \tag{3.23}$$

where

$$S[\phi_a] = \int L(\phi_a, \partial_\mu \phi_a) d^4x.$$

(b) Show that Hamilton's equations can be written in the form

$$\frac{\delta H}{\delta \phi_a} = -\frac{\partial \pi^a}{\partial t}, \qquad \frac{\delta H}{\delta \pi^a} = \frac{\partial \phi_a}{\partial t}, \tag{3.24}$$

where

$$H\left[\phi_a; \pi^a\right] = \int H(\phi_a; \partial_i \phi_a; \pi^a) d^3x.$$

Let F and G be two functionals of the field components and their conjugate momenta, i.e.,

$$F\left[\phi_a;\pi^a\right] = \int_{\mathbb{R}^3} F(\phi_a;\partial_i\phi_a;\pi^a)d^3\mathbf{x}$$

and a similar expression for G. We may then define their *Poisson bracket* in analogy with (2.59):

$$\{F;G\} = \int_{\mathbb{R}^3}\left(\frac{\delta F}{\delta\phi_a}\frac{\delta G}{\delta\pi^a} - \frac{\delta F}{\delta\pi^a}\frac{\delta G}{\delta\phi_a}\right)d^3\mathbf{x}. \tag{3.25}$$

Exercise 3.3.4
Problem: Show that the Poisson bracket of two functionals satisfies the properties listed in Exercise 2.10.1.

Exercise 3.3.5
Problem: Consider a field theory. For fixed $(t,\mathbf{x}_1)$ we can define a functional through the formula

$$\phi_a \to \phi_a(t,\mathbf{x}_1).$$

This functional, which depends on t and $\mathbf{x}_1$, will simply be denoted by $\phi_a(t,\mathbf{x}_1)$.
In a similar way we can define a functional

$$\pi^b \to \pi^b(t,\mathbf{x}_1),$$

which we simply denote by $\pi^b(t,\mathbf{x}_1)$. Show that the functionals $\phi_a(t,\mathbf{x}_1)$ and $\pi^b(t,\mathbf{x}_2)$ satisfy the rules

$$\{\phi_a(t,\mathbf{x}_1);\phi_b(t,\mathbf{x}_2)\} = \{\pi^a(t,\mathbf{x}_1);\pi^b(t,\mathbf{x}_2)\} = 0,$$
$$\{\phi_a(t,\mathbf{x}_1);\pi^b(t,\mathbf{x}_2)\} = \delta_a^b\delta^3(\mathbf{x}_1-\mathbf{x}_2).$$

Hint: Rearrange the functionals $\phi_a(t,\mathbf{x}_1)$ and $\pi^b(t,\mathbf{x}_2)$ in the form

$$\phi_a(t,\mathbf{x}_1) = \int \phi_a(t,\mathbf{x})\delta^3(\mathbf{x}-\mathbf{x}_1)d^3\mathbf{x},$$

$$\pi^b(t,\mathbf{x}_2) = \int \pi^b(t,\mathbf{x})\delta^3(\mathbf{x}-\mathbf{x}_2)d^3\mathbf{x},$$

and use this to show

$$\frac{\delta\phi_a(t,\mathbf{x}_1)}{\delta\phi_b} = \delta_a^b\delta^3(\mathbf{x}-\mathbf{x}_1), \qquad \frac{\delta\phi_a(t,\mathbf{x}_1)}{\delta\pi^b} = 0,$$

$$\frac{\delta\pi^b(t,\mathbf{x}_2)}{\delta\phi_a} = 0, \qquad \frac{\delta\pi^b(t,\mathbf{x}_2)}{\delta\pi^a} = \delta_a^b\delta^3(\mathbf{x}-\mathbf{x}_2).$$

Exercise 3.3.6
Problem: Consider a canonical system. Let F be a functional of the fields and their conjugate momenta. We will also allow F to depend explicitly on time:

$$F[\phi_a;\pi^a;t] = \int F(\phi_a;\partial_i\phi_a;\pi^a;t)d^3\mathbf{x}.$$

The field components ϕ_a and their conjugate momenta themselves evolve in time according to Hamilton's equation (3.20). Show that the total time derivative along a field history is given by

$$\frac{dF}{dt} = \frac{\partial F}{\partial t} + \{F; H\}.$$

Let us make a short comment on how to quantize a field theory. As in the elementary quantum mechanics we can use two different strategies:

a. *Path-integral formalism*: This is closely connected to the Lagrangian formalism. Consider a field configuration $\phi_a^{(1)}(\mathbf{x})$ at time t_1 and another field configuration $\phi_a^{(2)}(\mathbf{x})$ at time t_2. We are interested in the *transition amplitude* $\langle \phi_a^{(2)} \mid \phi_a^{(1)} \rangle$, i.e., the probability amplitude for finding the field in the state $\phi_a^{(2)}$ at time t_2 when we know that it was in the state $\phi_a^{(1)}$ at time t_1. To come from the configuration $\phi_a^{(1)}$ to $\phi_a^{(2)}$, the field must have developed according to some history $\phi_a(t, \mathbf{x})$ that interpolates between $\phi_a^{(1)}$ and $\phi_a^{(2)}$:

$$\phi_a(t_1, \mathbf{x}) = \phi_a^{(1)}(\mathbf{x}); \qquad \phi_a(t_2, \mathbf{x}) = \phi_a^{(2)}(\mathbf{x}).$$

To each such history we have associated the action

$$S[\phi_a] = \int_{t_1}^{t_2} \int_{\mathbb{R}^3} L(\phi_a, \partial_\mu \phi_a) d^3\mathbf{x}\, dt.$$

A straightforward generalization of Feynman's principle (2.21) then gives the result

$$\langle \phi_a^{(2)} \mid \phi_a^{(1)} \rangle = \int_{\phi_a(t_1,\mathbf{x})=\phi_a^{(1)}(\mathbf{x})}^{\phi_a(t_2,\mathbf{x})=\phi_a^{(2)}(\mathbf{x})} \exp\left\{\frac{i}{\hbar}\, S[\phi_a]\right\} D[\phi_a(t, \mathbf{x})], \tag{3.26}$$

where we sum over all histories interpolating between $\phi_a^{(1)}$ and $\phi_a^{(2)}$.

b. *Canonical quantization*: This is closely connected to the Hamiltonian formalism. Consider a canonical field theory. The physical quantities are represented as *functionals* $f[\phi_a, \pi^b]$ of the field components and their conjugate momenta. In the quantum context these functionals are replaced by Hermitian operators in such a way that their Poisson brackets are replaced by commutators

$$\{F, G\} \rightarrow \frac{1}{i\hbar}\, [\hat{F}, \hat{G}]. \tag{3.27}$$

As we have seen, the field components $\phi_a(t, \mathbf{x}_1)$ and their conjugate momenta $\pi^b(t, \mathbf{x}_2)$ can themselves be interpreted as functionals. According to Exercise 3.2.7 they must, therefore, be replaced by Hermitian operators satisfying the *Heisenberg Commutation Rules*:

$$[\hat{\phi}_a(t, \mathbf{x}_1); \hat{\phi}_b(t, \mathbf{x}_2)] = [\hat{\pi}^a(t, \mathbf{x}_1); \hat{\pi}^b(t, \mathbf{x}_2)] = 0,$$
$$[\hat{\phi}_a(t, \mathbf{x}_1); \hat{\pi}^b(t, \mathbf{x}_2)] = i\hbar\delta_a^b\delta^3(\mathbf{x}_2 - \mathbf{x}_1).$$

As an illustration of the preceding ideas we take a look at the following Lagrangian density:

$$L(\phi, \partial_\mu \phi) = -\frac{1}{2}\, (\partial_\mu \phi)(\partial^\mu \phi) - U(\phi). \tag{3.28}$$

This is obviously a Lorentz scalar, and if we write out the first term explicitly,

$$L(\phi, \partial_\mu \phi) = \frac{1}{2}\left[\left(\frac{\partial \phi}{\partial t}\right)^2 - \left(\frac{\partial \phi}{\partial x}\right)^2\right] - U(\phi),$$

we see that it is a direct generalization of the string Lagrangian (3.7). The last term can be interpreted as a potential energy density. It can easily be included in the string model, too, if we simply assume that each "atom" in the string has a potential energy $U(x)$ arising, e.g., from the gravitational potential. Notice, too, that the above Lagrangian has the same form as the nonrelativistic Lagrangian for a point particle (2.5).

With this choice of the Lagrangian, the conjugate momentum is given by

$$\pi = \frac{\partial \phi}{\partial t};$$

i.e., it coincides with the kinematic momentum. Furthermore, the Hamiltonian density reduces to

$$H = \pi \partial_0 \phi - L = \frac{1}{2}\left[\left(\frac{\partial \phi}{\partial t}\right)^2 + \left(\frac{\partial \phi}{\partial x}\right)^2\right] + U(\phi). \tag{3.29}$$

This is positive definite, as it ought to be, provided that the potential energy density is positive definite. In the following we will assume that the minimum of U is zero and that this minimum is attained only when ϕ vanishes identically.

Therefore, the Lagrangian satisfies the first two conditions in (3.30), which must be valid for any field theory if we are going to make sense out of it. The third condition is specifically related to free-field theories. In our case, the equation of motion is given by

$$\partial_\mu \partial^\mu \phi = U'(\phi); \qquad \text{i.e.,}\ \frac{\partial^2 \phi}{\partial t^2} - \frac{\partial^2 \phi}{\partial x^2} = -U'(\phi), \tag{3.30}$$

which should be compared with Newton's equation of motion. If we make a Taylor expansion of the potential energy density, using that $U(0)$ and $U'(0)$ vanish by assumption, the equation of motion reduces to

$$\partial_\mu \partial^\mu \phi = U''(0)\phi + \frac{1}{2}\, U'''(0)\phi^2 + \cdots.$$

This will allow solutions of the form $\phi(x) = \epsilon \exp[i p_\mu x^\mu]$, provided that the potential is purely quadratic. That follows immediately from the formulas

$$\begin{aligned}
\partial_\mu \partial^\mu \phi &= -p_\mu p^\mu \epsilon \exp[i p_\mu x^\mu], \\
U'(\phi) &= \left\{U''(0) + \frac{1}{2}\, U'''(0)\epsilon \exp[i p_\mu x^\mu] + \cdots\right\} \epsilon \exp[i p_\mu x^\mu].
\end{aligned}$$

In the case of a quadratic potential we get the dispersion relation

$$-p_\mu p^\mu = U''(0),$$

which shows that $U''(0)$ is the square of the mass of the particle in the theory. In the general case, $\epsilon \exp[ip_\mu x^\mu]$ will never be an exact solution, but it will be approximate solution, provided that ϵ is so small that we can neglect the higher-order terms. In the weak field limit we can therefore treat the field theory with a general potential as a free field theory, where the mass-square of the particle in the theory is still given by $U''(0)$. As the field quanta in the general case are no longer free, they must exert forces on each other. We therefore say that they are self-interacting.

Notice that the free field case, where U is quadratic, corresponds to a linear equation of motion, while the self-interacting case corresponds to a nonlinear equation of motion.

§ 3.4 The Klein–Gordon Field

Now we are in a position to construct our first explicit field theory. We start by constructing the equations of motions. Consider the energy–momentum relation $p_\mu p^\mu = -m^2$. If we perform the substitution $p_\mu = -i\partial_\mu$, this leads to the equation of motion

$$(\partial_\mu \partial^\mu - m^2)\phi(x) = 0. \tag{3.31}$$

This is known as the *Klein–Gordon equation*, and it is obviously Lorentz invariant. To check that it has the solutions we want, we observe that

$$(\partial_\mu \partial^\mu - m^2) \exp[ip_\mu x^\mu] = (-p_\mu p^\mu - m^2) \exp[ip_\mu x^\mu],$$

and this shows us that $\phi(x) = \exp[ip_\mu x^\mu]$ is a solution to the Klein–Gordon equation, provided that p_μ satisfies $p_\mu p^\mu = -m^2$.

Now we must find the Lagrangian density L. We know that the equation of motion

$$\partial_\mu \frac{\partial L}{\partial(\partial_\mu \phi)} = \frac{\partial L}{\partial \phi}$$

must reproduce

$$-\partial_\mu \partial^\mu \phi = -m^2 \phi.$$

From this we immediately read off

$$\begin{aligned} \frac{\partial L}{\partial(\partial_\mu \phi)} &= -\partial^\mu \phi \to L = -\frac{1}{2}\,(\partial_\mu \phi)(\partial^\mu \phi) + \text{terms involving } \phi, \\ \frac{\partial L}{\partial \phi} &= -m^2 \phi \to L = -\frac{1}{2}\,m^2 \phi^2 + \text{terms involving } \partial_\mu \phi. \end{aligned}$$

So we have reconstructed the Lagrangian density:

$$L(\phi, \partial_\mu \phi) = -\frac{1}{2}\,(\partial_\mu \phi)(\partial^\mu \phi) - \frac{1}{2}\,m^2 \phi^2. \tag{3.32}$$

Since ϕ is a Lorentz scalar, the Lagrangian density is a Lorentz scalar, too. Finally, we should check that the energy density is positive definite:

$$H = \frac{\partial L}{\partial(\partial_0\phi)}\,\partial_0\phi - L = -\frac{1}{2}\,(\partial^0\phi)(\partial_0\phi) + \frac{1}{2}\,(\partial_i\phi)(\partial^i\phi) + \frac{1}{2}\,m^2\phi^2;$$

i.e.,

$$H = \frac{1}{2}\left[\left(\frac{\partial\phi}{\partial t}\right)^2 + (\nabla\phi)^2 + m^2\phi^2\right]. \tag{3.33}$$

This is obviously positive definite, so everything is okay!

Let us look a little more closely at the Lagrangian density. It is quadratic in ϕ and $\partial_\mu\phi$, which is typical for a *free field* theory. Furthermore, it consists of two terms:

a. A term $-\frac{1}{2}\,(\partial_\mu\phi)(\partial^\mu\phi)$ that is quadratic in the derivatives and that acts like a *kinetic energy* term. Observe, however, that it includes the space derivatives, too:

$$-\frac{1}{2}\,(\partial_\mu\phi)(\partial^\mu\phi) = \frac{1}{2}\left(\frac{\partial\phi}{\partial t}\right)^2 - \frac{1}{2}\,(\nabla\phi)^2.$$

If we compare this with the model for the continuous string (Section 3.1), we see that $-\frac{1}{2}\,(\nabla\phi)^2$ should be counted as a potential energy term, due to the local interaction.

b. A term $-\frac{1}{2}\,m^2\phi^2$ that is quadratic in ϕ. It is called the *mass term*, because m gives the mass of the field quanta. Observe the sign! It will be crucial later on.

Exercise 3.4.1

Problem: Determine the conjugate momentum of the Klein–Gordon field and show that the theory of the Klein–Gordon field is a canonical field theory.

Determine the Hamiltonian functional $H[\phi, \pi]$ and verify by an explicit calculation that Hamilton's equations reproduce the Klein–Gordon equation.

At this point we have looked only for solutions of the form

$$\phi(x) = \exp[ip_\mu x^\mu]$$

that had an important significance on the quantum-mechanical level. We might look for other solutions. The simplest possible classical solution is a static spherically symmetric solution $\phi(t, \mathbf{x}) = \phi(r)$. Substituting this equation, the Klein–Gordon equation reduces to

$$m^2\phi(r) = \frac{1}{r}\,\frac{d^2}{dr^2}\,(r\phi); \qquad \text{i.e.,}\quad \frac{d^2}{dr^2}\,(r\phi) = m^2[r\phi].$$

From this we find the solution

$$\phi(t, \mathbf{x}) = \phi(r) = \frac{c}{r}\,e^{\pm mr}. \tag{3.34}$$

This solution is singular at the origin. Only the decreasing solution is physically acceptable, because the contribution to energy from infinity otherwise explodes.

How can we interpret the solution (3.34)? In electromagnetism we have a similar solution, the Coulomb solution. There we use the assumption, that ϕ and $\mathbf{A}$ can be written as follows $\phi(t, \mathbf{x}) = \phi(r), \mathbf{A}(t, \mathbf{x}) = \mathbf{0}$, where ϕ and $\mathbf{A}$ denote respectively the scalar potential and the vector potential. Using these equations, the Maxwell equations reduce to

$$0 = \frac{1}{r}\frac{d^2}{dr^2}[r\phi] \Rightarrow \phi(r) = a + \frac{b}{r} .$$

This correspond to a spherically symmetric electric field, the Coulomb field

$$\mathbf{E} = -\nabla\phi - \frac{\partial \mathbf{A}}{\partial t} = -\frac{b}{r^2}\frac{\mathbf{r}}{r} .$$

The field $\phi(r) = \frac{b}{r}$ is interpreted in the following way: At the point of singularity, $r = 0$, we have an electrically charged particle, say a proton, acting as a source for the electromagnetic field. The field itself is interpreted as a potential for the electromagnetic force $\mathbf{E} = -\nabla\phi$ that other charged particles will experience. On the quantum-mechanical level, two charged particles will interact by exchanging photons. (See Figure 3.4.) The photons, i.e., the quanta of the electromagnetic field, will transfer momentum, and when the momentum of a particle changes, it experiences a force.

Let us make the same interpretation of the static spherically symmetric solution of the Klein–Gordon equation. The electromagnetic field is responsible for the electromagnetic interaction between protons. Let us assume that the Klein–Gordon field is responsible for the strong interaction between protons. When we quantize the electromagnetic field, we get massless photons. In the same way, we assume that we get π-mesons when we quantize the Klein–Gordon field. Two protons (or a proton and a neutron) then interact strongly by exchanging π-mesons: (see Figure 3.5). This exchange produces the strong forces. The strong force is then derived from the potential

$$\phi(r) = \frac{c}{r} e^{-mr} .$$

In other words, at the position of the singularity, we have a particle, say a proton, acting as the source of the Klein–Gordon field. It produces the potential $\phi(r) =$

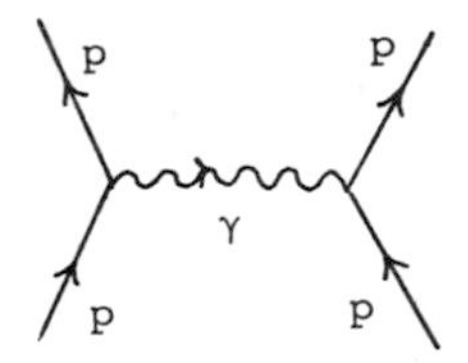

Figure 3.4.

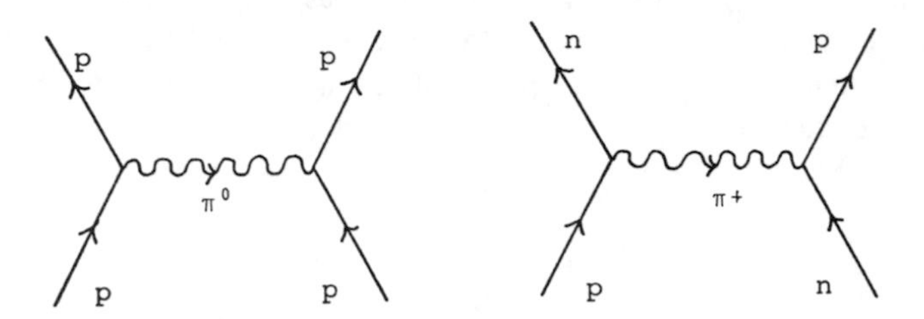

Figure 3.5.

$\frac{c}{r}\, e^{-mr}$ and hence the force that other strongly interacting particles will experience. The potential $\phi(r) = \frac{c}{r}\, e^{-mr}$ is called the *Yukawa potential.*

Contrary to the Coulomb potential, we see that the Yukawa potential is exponentially damped! Hence, it has only a *finite* range. Let us estimate this range. If y is a length measured in meters and x is the mass measured in MeV, then in our units, where $c = \hbar = 1$, it can be shown that

$$y \approx \frac{2 \cdot 10^{-13}}{x} .$$

The observed mass of the π-meson is $m = 140$ MeV, and hence the typical range of the force is

$$\ell = \frac{2 \cdot 10^{-13}}{140} \text{ meters} = 1.4 \cdot 10^{-15} \text{ meters.}$$

Now, that is exactly the typical distance between protons and neutrons in the atomic nucleus, and the Yukawa potential can therefore very well account for the strong force binding the nucleus together. It also explains why this force was never observed classically. In fact, the extremely short range means that the force operates on space–time regions so small that the quantum-mechanical effects dominate completely.

§ 3.5 The Maxwell Field

The next field we attack is the Maxwell field A_α. Here we know the equation of motion (1.33)

$$(\partial_\mu \partial^\mu) A^\nu - \partial^\nu (\partial_\mu A^\mu) = 0.$$

Can we find a simple Lagrangian density that leads to this equation? Since the equation of motion involves only derivatives of A_α, we expect that L involves only $\partial_\mu A_\nu$:

$$L = L(\partial_\mu A_\nu),$$

and we should determine it so that

$$\partial_\mu \frac{\partial L}{\partial(\partial_\mu A_\nu)} = \partial_\mu \partial^\mu A^\nu - \partial_\mu(\partial^\nu A^\mu) = \partial_\mu(\partial^\mu A^\nu - \partial^\nu A^\mu).$$

This is a mess, so we will use a trick. The equations of motion are gauge invariant. Now, the simplest way to construct a gauge-invariant theory is to use a gauge-invariant action. This guarantees that whenever A_μ extremizes the action, then the same thing will be true for $A_\mu + \partial_\mu \chi$. Note, however, that it is not the only possibility when we wish to construct a gauge-invariant theory (compare the discussion in Section 2.2), but let us try it!

The starting point is a gauge-invariant Lagrangian density L. Since L involves only derivatives of A_α we must try to construct a gauge-invariant quantity out of the derivatives. But this must be the field strength

$$F_{\alpha\beta} = \partial_\alpha A_\beta - \partial_\beta A_\alpha.$$

Thus we expect that L depends only on $F_{\alpha\beta}$:

$$L = L(F_{\alpha\beta}).$$

But the simplest quantity we can construct out of $F_{\alpha\beta}$ that is gauge and Lorentz invariant is the square

$$F_{\alpha\beta} F^{\alpha\beta}.$$

So we expect that

$$L = k F_{\alpha\beta} F^{\alpha\beta}, \tag{3.35}$$

where k is an arbitrary constant.

Worked Exercise 3.5.1
Problem: Show that

$$\frac{\partial(F_{\alpha\beta} F^{\alpha\beta})}{\partial(\partial_\mu A_\nu)} = 4F^{\mu\nu}.$$

Using Exercise 3.4.1, we see that the equation (3.35) leads to the correct equations of motion:

$$0 = \partial_\mu \frac{\partial L}{\partial(\partial_\mu A_\nu)} = 4k\partial_\mu F^{\mu\nu}.$$

To determine k, we investigate the energy–momentum tensor

$$\overset{\circ}{T}{}^{\alpha\beta} = \frac{-\partial L}{\partial(\partial_\beta A_\nu)} \partial^\alpha A_\nu + \eta^{\alpha\beta} L = -4k \left[F^{\beta\nu} \partial^\alpha A_\nu - \frac{1}{4} \eta^{\alpha\beta} F_{\gamma\delta} F^{\gamma\delta} \right].$$

But here something is wrong! It is not symmetric, and in fact, it is not gauge invariant, due to the term $\partial^\alpha A_\nu$. This suggests that we should make the replacement

$$\partial^\alpha A_\nu \to \partial^\alpha A_\nu - \partial_\nu A^\alpha = F^\alpha{}_\nu,$$

because the new term is gauge invariant. Therefore, we try to repair the canonical energy–momentum tensor $\overset{\circ}{T}{}^{\alpha\beta}$ with the correction term

$$\Theta^{\alpha\beta} = 4kF^{\beta\nu}\partial_\nu A^\alpha.$$

Is this legal? Remember that $\Theta^{\alpha\beta}$ should possess three characteristic properties (3.18). First, we observe that $\overset{\circ}{T}{}^{\alpha\beta} + \Theta^{\alpha\beta}$ is symmetric:

$$\overset{\circ}{T}{}^{\alpha\beta} + \Theta^{\alpha\beta} = -4k\left[F^{\beta\nu}F^{\alpha}_{\ \nu} - \frac{1}{4}\eta^{\alpha\beta}F_{\gamma\delta}F^{\gamma\delta}\right].$$

Second, we observe that $\Theta^{\alpha\beta}$ is conserved:

$$\begin{aligned}\partial_\beta\Theta^{\alpha\beta} &= 4k\partial_\beta\left[F^{\beta\nu}\partial_\nu A^\alpha\right]\\ &= 4k\left[\left(\partial_\beta F^{\beta\nu}\right)\partial_\nu A^\alpha + F^{\beta\nu}\partial_\beta\partial_\nu A^\alpha\right] = 0,\end{aligned}$$

because the field's equation of motion kills the first term, and the antisymmetry of the field strength kills the symmetric tensor $\partial_\beta\partial_\nu A^\alpha$!

Finally, we observe that $\Theta^{\alpha\beta}$ does not contribute to the energy–momentum of the Maxwell field:

$$\begin{aligned}\int\Theta^{\alpha 0}d^3\mathbf{x} &= 4k\int F^{0\nu}\partial_\nu A^\alpha d^3\mathbf{x} = 4k\int F^{0i}\partial_i A^\alpha d^3\mathbf{x}\\ &= -4k\int(\partial_i F^{0i})A^\alpha d^3\mathbf{x} = -4k\int(\partial_\mu F^{0\mu})A^\alpha d^3\mathbf{x} = 0,\end{aligned}$$

where we have neglected the surface terms because the field strength vanishes at infinity. The last integral vanishes due to the field's equation of motion.

So we have succeeded in repairing the canonical energy–momentum tensor. Comparing the energy–momentum tensor

$$\overset{\circ}{T}{}^{\alpha\beta} + \Theta^{\alpha\beta} = -4k\left[B^{\beta\nu}F^{\alpha}_{\ \nu} - \frac{1}{4}\eta^{\alpha\beta}F_{\gamma\delta}F^{\gamma\delta}\right]$$

with the true energy–momentum tensor for the Maxwell field (1.41)

$$T^{\alpha\beta} = -F^{\alpha}_{\ \gamma}F^{\gamma\beta} - \frac{1}{4}\eta^{\alpha\beta}F^{\gamma\delta}F_{\gamma\delta},$$

we finally obtain the result $k = -1/4$! Thus we have constructed the Lagrangian density for the Maxwell field:

$$L_{EM} = -\frac{1}{4}F_{\alpha\beta}F^{\alpha\beta}. \tag{3.36}$$

Exercise 3.5.2
Problem: Determine the conjugate momenta of the Maxwell field A_μ and show that the theory of the Maxwell field is *not* a canonical field theory.

As we have seen, L_{EM} is Lorentz and gauge invariant, and it produces a positive definite energy density

$$H = -F^0_{\ \gamma} F^{\gamma 0} - \frac{1}{4} F^{\gamma\delta} F_{\gamma\delta} = \frac{1}{2} (\mathbf{E}^2 + \mathbf{B}^2).$$

(Compare with (1.42).) It remains to investigate the equations of motion (1.33):

$$(\partial_\mu \partial^\mu) A_\nu - \partial_\nu (\partial^\mu A_\mu) = 0.$$

We look for solutions of the form

$$A_\mu = \epsilon_\mu \exp[i p_\mu x^\mu]. \tag{3.37}$$

Here ϵ_μ is a Lorentz vector, called the *polarization vector*. If we insert the wave function into the equations of motions, we get

$$0 = (-p_\mu p^\mu \epsilon_\nu + p_\nu (p^\mu \epsilon_\mu)) \exp[i p_\mu x^\mu].$$

Hence $A_\mu = \epsilon_\mu \exp[i p_\mu x^\mu]$ is a solution of the equations of motion, provided that

$$(p_\mu p^\mu) \epsilon_\nu = (p^\mu \epsilon_\mu) p_\nu.$$

This algebraic condition has two kinds of solutions:

a. $p_\mu p^\mu \neq 0$.
b. $p_\mu p^\mu = 0$.

Let us examine condition (a). If $p_\mu p^\mu \neq 0$, we conclude that

$$\epsilon_\nu = \frac{(p^\mu \epsilon_\mu)}{(p^\mu p_\mu)} p_\nu.$$

Consequently, ϵ_ν is proportional to p_ν, and we see that the equations of motion allow solutions of the form

$$A_\nu = \alpha p_\nu \exp[i p_\mu x^\mu]. \tag{3.38}$$

But these solutions are trivial. They are *not* observable, because we can gauge them away! Choosing $\chi = i\alpha \exp[i p_\mu x^\mu]$, we get

$$A'_\nu = A_\nu + \partial_\nu \chi = \alpha p_\nu \exp[i p_\mu x^\mu] - \alpha p_\nu \exp[i p_\mu x^\mu] = 0.$$

We therefore disregard them!

Then we are left with case (b). If $p_\mu p^\mu = 0$, we conclude that

$$0 = (p^\mu \epsilon_\mu) p_\nu,$$

but this is consistent only if

$$p^\mu \epsilon_\mu = 0.$$

Thus, we conclude that the equations of motions allow solutions of the form

$$A_\nu = \epsilon_\nu \exp[i p_\mu x^\mu], \tag{3.39}$$

with $p_\mu p^\mu = 0$ and $p_\mu \epsilon^\mu = 0$.

Sine $p_\mu p^\mu = 0$, *the photons are massless*! These solutions are nontrivial; i.e., we cannot completely gauge them away. However, it is not all the components ϵ_μ, $\mu = 0, 1, 2, 3$, that carry physical information. To see this, we make a gauge transformation choosing $\chi = i\alpha \exp[ip_\mu x^\mu]$. This leads to the transformed potential

$$A'_\nu = A_\nu + \partial_\nu \chi = \epsilon_\nu \exp[ip_\mu x^\mu] - \alpha p_\nu \exp[ip_\mu x^\mu] = (\epsilon_\nu - \alpha p_\nu) \exp[ip_\mu x^\mu].$$

So if we perform a gauge transformation, we can change the polarization vector according to the rule

$$\epsilon_\mu \to \epsilon'_\mu = \epsilon_\mu - \alpha p_\mu. \tag{3.40}$$

To investigate the consequences of this gauge freedom, we consider a wave traveling along the z-axis. Then the four-momentum is given by

$$p^\mu = (\omega, 0, 0, k),$$

where $\omega = -k$ because the photon is massless. Due to the condition $p^\mu \epsilon_\mu = 0$, we conclude that

$$\omega\epsilon_0 - \omega\epsilon_3 = 0; \qquad \text{i.e., } \epsilon_0 = \epsilon_3.$$

Thus, the wave function of the photon has the form

$$A_\mu = [\epsilon_3, \epsilon_1, \epsilon_2, \epsilon_3] \exp[i\omega(z - t)].$$

But we are still allowed to make gauge transformations:

$$\epsilon'_\mu = [\epsilon_3 - \alpha\omega, \epsilon_1, \epsilon_2, \epsilon_3 - \alpha\omega].$$

Hence, if we choose $\alpha = \epsilon_3/\omega$, we have gauged away ϵ_0 and ϵ_3 completely! If we split the polarization vector into its different components, then

ϵ_0 is called the scalar part;
ϵ_1, ϵ_2 is called the transverse part;
ϵ_3 is called the longitudinal part.

(See Figure 3.6.) The above result then shows that we can gauge away the scalar photons and the longitudinal photons, but we cannot change the transverse part. Consequently, all the physical information is contained in the transverse parts ϵ_1 and ϵ_2!

Exercise 3.4.3
Problem: Compute the complex field strengths corresponding to the above traveling wave and show explicitly that they depend only on ϵ_1 and ϵ_2.

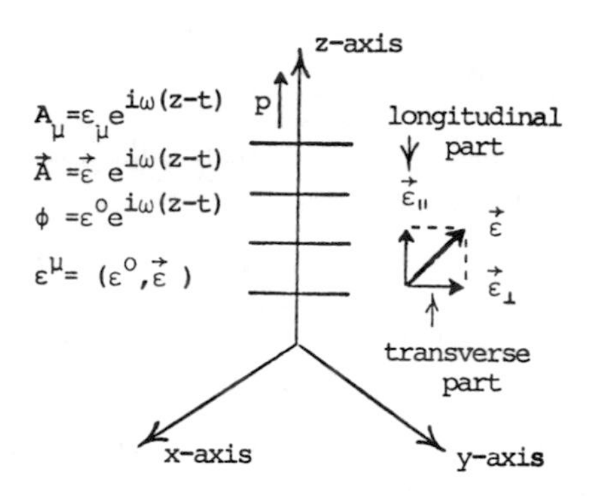

Figure 3.6.

§ 3.6 Spin of the Photon – Polarization of Electromagnetic Waves

Let us investigate the physical meaning of ϵ_1 and ϵ_2 a little more closely. Each polarization vector ϵ_μ may be decomposed in the following way:

$$[\epsilon_0, \epsilon_1, \epsilon_2, \epsilon_3] = \epsilon_0[1, 0, 0, 0] + \epsilon_+[0, 1, i, 0] + \epsilon_-[0, 1, -i, 0] + \epsilon_3[0, 0, 0, 1],$$

where

$$\epsilon_\pm = \frac{1}{2}(\epsilon_1 \mp i\epsilon_2).$$

We may decompose the wave function

$$A_\mu = \epsilon_\mu \exp[i\omega(z-t)]$$

in exactly the same way. We want to show that this decomposition is closely related to spin! To see this, we must investigate what happens if we perform a rotation about an axis, say the z-axis.

Now, a rotation is nothing but a special kind of Lorentz transformation

$$x'^\mu = \alpha^\mu{}_\nu x^\nu$$

represented by the matrix $(\alpha^\mu{}_\nu)$, where

$$\alpha^\mu{}_\nu = \begin{bmatrix} 1 & 0 & 0 & 0 \\ 0 & \cos\theta & \sin\theta & 0 \\ 0 & -\sin\theta & \cos\theta & 0 \\ 0 & 0 & 0 & 1 \end{bmatrix}.$$

Here θ is the angle of rotation. In the new coordinate system we have a new coordinate representation of the wave:

$$A'_\nu = A_\mu \breve{\alpha}^\mu{}_\nu = \epsilon_\mu \exp[-ip_\delta x^\delta]\breve{\alpha}^\mu{}_\nu = \epsilon'_\mu \exp[-ip_\delta x^\delta],$$

with

$$\epsilon'_\nu = \epsilon_\mu \breve{\alpha}^\mu{}_\nu$$

and $(\breve{\alpha}^\mu{}_\nu)$ the reciprocal matrix of $(\alpha^\mu{}_\nu)$. Hence, the rotation attacks only the polarization vector. We now get

$$[1, 0, 0, 0] \to [1, 0, 0, 0] \begin{bmatrix} 1 & 0 & 0 & 0 \\ 0 & \cos\theta & -\sin\theta & 0 \\ 0 & \sin\theta & \cos\theta & 0 \\ 0 & 0 & 0 & 1 \end{bmatrix} = [1, 0, 0, 0],$$

and similarly for $[0, 0, 0, 1]$. Finally, $[0, 1, \pm i, 0]$ transforms into

$$[0, 1, \pm i, 0] \begin{bmatrix} 1 & 0 & 0 & 0 \\ 0 & \cos\theta & -\sin\theta & 0 \\ 0 & \sin\theta & \cos\theta & 0 \\ 0 & 0 & 0 & 1 \end{bmatrix}$$
$$= [0, \cos\theta \pm i\sin\theta, -\sin\theta \pm i\cos\theta, 0] = \exp(\pm i\theta) \cdot [0, 1, \pm i, 0].$$

Thus, if we let R_θ denote the rotation operator, then

$$[1, 0, 0, 0]\exp[i\omega(z-t)] \quad \text{and} \quad [0, 0, 0, 1]\exp[i\omega(z-t)]$$

are eigenfunctions with eigenvalue 1, and

$$[0, 1, i, 0]\exp[i\omega(z-t)] \quad \text{and} \quad [0, 1, -i, 0]\exp[i\omega(z-t)]$$

are eigenfunctions with eigenvalues: $e^{\pm i\theta}$.

But the rotation operator is given by $R_\theta = e^{(i/\hbar)\theta I_z}$, where I_z is the operator of angular momentum around the z-axis. We then see that the wave function for a scalar or a longitudinal photon carries spin projection 0, while the wave functions for the transverse photons carry spin projection ± 1. We should, however, remember that only the transverse photons are observable! Thus, a photon is a spin 1 particle, and it has either spin projection 1 in the direction of the momentum or it has spin projection -1 in the direction of momentum, but it never has spin projection 0! We therefore say that the photon has *helicity* ± 1. (See Figure 3.7.)

The preceding discussion may seem to bear little resemblance to what we have previously learned about electromagnetism. Maybe the following remarks will clarify this. We have looked for solutions of the form (3.37)

$$A_\mu(x) = \epsilon_\mu \exp[ip_\mu x^\mu].$$

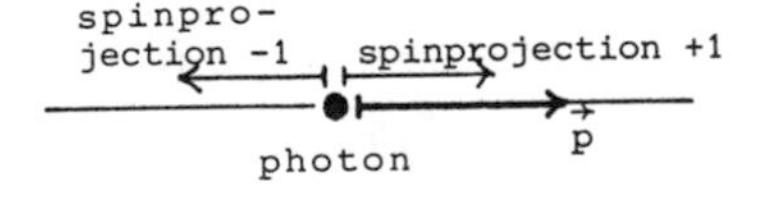

Figure 3.7.

The above solution is complex-valued, and hence it can have relevance only on the quantum-mechanical level. But clearly, the real and imaginary parts will solve the Maxwell equations too, and they are real solutions, so that they may have relevance on the classical level. We want to find out how we can interpret the above solution classically and quantum-mechanically.

Let us discuss the classical interpretation first. We know that Maxwell's equations allow plane waves as solutions,

$$A_\mu(x) = \epsilon_\mu \cos(k_\mu x^\mu), \qquad k_\mu k^\mu = 0.$$

If we introduce a coordinate system where the 3-dimensional wave vector $\mathbf{k}$ points along the z-axis, then this formula reduces to

$$A_\mu(x) = \epsilon_\mu \cos(\omega z - \omega t).$$

We may decompose it into a scalar part and a vector part:

$$A^\mu = (\phi, \mathbf{A}), \qquad \phi = \epsilon^0 \cos(\omega z - \omega t);$$
$$\epsilon^\mu = (\epsilon^0, \boldsymbol{\epsilon}), \qquad \mathbf{A} = \boldsymbol{\epsilon} \cos(\omega z - \omega t).$$

But as we have already seen, we can gauge ϵ^0 and ϵ^3 away!

If we work in the special gauge, where $\epsilon^0 = \epsilon^3 = 0$, the scalar potential drops out, and the vector potential is represented by

$$\mathbf{A} = \boldsymbol{\epsilon} \cos(\omega z - \omega t), \tag{3.41}$$

where $\boldsymbol{\epsilon} = [\epsilon^1, \epsilon^2, 0]$ is orthogonal to the wave vector $\mathbf{k}$. We can also calculate the field strengths. Observe that since $\mathbf{A}$ is time dependent, this wave will also represent an electric field:

$$\begin{aligned} \mathbf{B} &= \nabla \times \mathbf{A} = -(\boldsymbol{\epsilon} \times \mathbf{k}) \sin(\omega z - \omega t), \\ \mathbf{E} &= -\frac{\partial \mathbf{A}}{\partial t} = \boldsymbol{\epsilon}\omega \sin(\omega z - \omega t). \end{aligned} \tag{3.42}$$

Here the electric field is pointing along the polarization vector, and the magnetic field is orthogonal to both the wave vector and the polarization vector. (See Figure 3.8.) As $\mathbf{B}$ and $\mathbf{E}$ are pointing in *constant directions* in space, we speak about *linearly polarized* light.

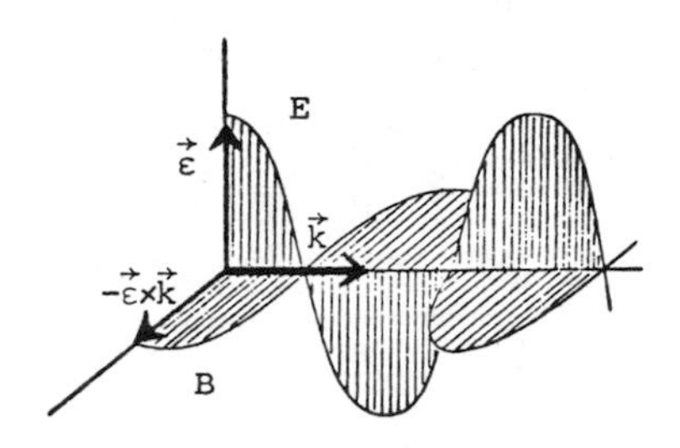

Figure 3.8.

The density of momentum $\mathbf{g}$ is pointing along Poynting's vector ($c = \epsilon_0 = 1$):

$$\mathbf{g} = \mathbf{E} \times \mathbf{B} = \mathbf{k}\omega \sin^2(\omega z - \omega t).$$

Hence, there is a simple connection between the momentum density $\mathbf{g}$ and the wave vector $\mathbf{k}$ that represents the momentum of a single photon ($\mathbf{p} = \hbar\mathbf{k}$).

Similarly, the energy density is given by ($c = \epsilon_0 = 1$)

$$\epsilon(\mathbf{x}, t) = \frac{1}{2}(\mathbf{E}^2 + \mathbf{B}^2) = \omega^2 \sin^2(\omega z - \omega t).$$

Let us consider a big box at a certain time, say $t = 0$. Assume that the box contains N photons. Each has an energy given by $\hbar\omega$ and a momentum given by $\hbar\mathbf{k}$, i.e., in relativistic units, where $\epsilon_0 = c = \hbar = 1$, we see that the total energy and momentum are given by

$$E_{\text{tot}} = N\omega, \qquad P_{\text{tot}} = N\mathbf{k}.$$

Now let us compare this with the similar classical calculation. Here the total energy is represented by

$$E_{\text{tot}} = \int \epsilon(\mathbf{x}, 0) d^3x = \omega^2 \int \sin^2 \omega z \, dx \, dy \, dz = \frac{\omega^2 V}{2},$$

$$\mathbf{P}_{\text{tot}} = \int \mathbf{g}(\mathbf{x}, t) d^3x = \omega\mathbf{k} \int \sin^2 \omega z \, dx \, dy \, dz = \frac{\omega\mathbf{k} V}{2},$$

where V is the volume of the box. Thus, we see that the classical and quantum-mechanical calculations are consistent, provided that we put

$$N = \frac{\omega V}{2}.$$

We have also found waves representing spin:

$$[0, 1, i, 0] \exp[i(\omega z - \omega t)].$$

We extract the real part and get the classical wave:

$$\phi = 0, \qquad \mathbf{A} = \begin{bmatrix} -\cos(\omega z - \omega t) \\ \sin(\omega z - \omega t) \\ 0 \end{bmatrix}, \tag{3.43}$$

which solves the Maxwell equations. In this case, the vector potential is circulating around the wave vector with a frequency ω. Let us compute the field strengths:

$$\begin{aligned} \mathbf{E} &= -\frac{\partial \mathbf{A}}{\partial t} = \omega \begin{bmatrix} -\sin(\omega z - \omega t) \\ \cos(\omega z - \omega t) \\ 0 \end{bmatrix}, \\ \mathbf{B} &= \nabla \times \mathbf{A} = \omega \begin{bmatrix} -\cos(\omega z - \omega t) \\ \sin(\omega z - \omega t) \\ 0 \end{bmatrix}. \end{aligned} \tag{3.44}$$

Clearly, they are rotating too, and we speak about *circularly polarized* light.

Then we turn to the *quantum-mechanical* interpretation of the complex wave solution:

$$A^{\mu}(x) = \epsilon^{\mu} \exp[i(\mathbf{kx} - \omega t)],$$

which we interpret as the quantum-mechanical wave function for a *photon* with momentum $\mathbf{p} = \hbar\mathbf{k}$ and energy $E = \hbar\omega$.

Again, we work in the special gauge where ϵ^0 and ϵ^3 disappear:

$$A^{\mu}(x) = \begin{bmatrix} 0 \\ \epsilon^1 \\ \epsilon^2 \\ 0 \end{bmatrix} \exp[i(\mathbf{kx} - \omega t)].$$

We may decompose this wave function along the x-axis and the y-axis:

$$\begin{aligned} A^{\mu}(x) &= \epsilon^1 \begin{bmatrix} 0 \\ 1 \\ 0 \\ 0 \end{bmatrix} \exp[i(\mathbf{kx} - \omega t)] + \epsilon^2 \begin{bmatrix} 0 \\ 0 \\ 1 \\ 0 \end{bmatrix} \exp[i(\mathbf{kx} - \omega t)] \\ &= \epsilon^1 A^{\mu}_{(1)}(x) + \epsilon^2 A^{\mu}_{(2)}(x). \end{aligned}$$

As the solutions are only determined up to a factor, we may normalize the solution so that

$$(\epsilon^1)^2 + (\epsilon^2)^2 = 1;$$

i.e., $\epsilon^1 = \cos\theta$ and $\epsilon^2 = \sin\theta$, where θ is the angle between ϵ and the x-axis. Since all states can be obtained as a linear combination of $A^{\mu}_{(1)}$ and $A^{\mu}_{(2)}$, these states represent a frame for the linearly polarized states. $A^{\mu}_{(1)}$ represents a photon polarized along the x-axis, and $A^{\mu}_{(2)}$ represents a photon polarized along the y-axis. In the general state, $A^{\mu}(x) = \cos\theta A^{\mu}_{(1)}(x) + \sin\theta A^{\mu}_{(2)}(x)$; $\cos\theta$ will give the probability amplitude for finding the photon polarized along the x-axis, and $\sin\theta$ will give the probability amplitude for finding the photon polarized along the y-axis.

An experiment may be performed in the following way: If we have a beam of photons, a light beam, that passes through a polarizer, then some of the photons will go through the polarizer, while some of them will be absorbed. After the passage, the light is *polarized*. All photons are now in the same state; they are polarized along the characteristic axis of the polarizer. Now, suppose we insert a second polarizer that is rotated an angle θ relative to the first polarizer. What is the intensity I of the beam after the passage of the second polarizer in terms of the intensity I_0 of the polarized beam?

Let us introduce a coordinate system where the x-axis points in the direction of the axis of the second polarizer. Then the polarized beam is characterized by a polarization vector

$$\epsilon = \begin{bmatrix} \cos\theta \\ \sin\theta \end{bmatrix},$$

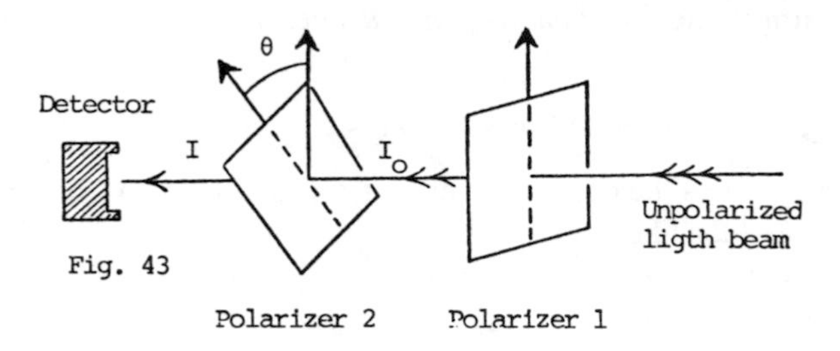

Figure 3.9.

and the wave is decomposed according to

$$A^{\mu} = \cos\theta A^{\mu}_{(1)} + \sin\theta A^{\mu}_{(2)}.$$

(See Figure 3.9.)

(Observe that we have a macroscopic number of photons all occupying the same state. This is possible because photons are *bosons*. Compare with the discussion of superconductivity.)

Consider a single photon. The probability amplitude that it is polarized along the x-axis is $\cos\theta$. Thus, there is a probability $\cos^2\theta$ that a given photon goes through the second polarizer!

But as there is a macroscopic number of photons in the same state, we may actually interpret $\cos^2\theta$ as a macroscopic physical quantity. If N_i is the total number of photons before passage through the second polarizer and N_f is the number of photons after passage through the second polarizer, then

$$N_f = N_i \cos^2\theta.$$

As the number of photons is proportional to the classical intensity of the light beam, we may rearrange this as

$$I_f = I_i \cos^2\theta, \tag{3.45}$$

and this is a formula that can be tested very simply in a classical experiment.

§ 3.7 The Massive Vector Field

We have seen that the Lagrangian density of the Maxwell field contains only a kinetic energy term and no mass term. Correspondingly, the photons are massless! Now, suppose we add a mass term to this Lagrangian density and consider the following field theory:

$$L = -\frac{1}{4} F_{\alpha\beta}F^{\alpha\beta} - \frac{1}{2} m^2 A_\alpha A^\alpha. \tag{3.46}$$

The first thing we observe is that we have spoiled the gauge invariance. Therefore, this field A_α is *not* a gauge field. So in what follows, A_α and $A_\alpha + \partial_\alpha\chi$

represent different physical situations. Let us find the equations of motion. Using (3.14), we get

$$m^2 A^\beta = \partial_\alpha F^{\alpha\beta}.$$

These equations imply that

$$m^2 \partial_\beta A^\beta = \partial_\beta \partial_\alpha F^{\alpha\beta} = 0.$$

Hence, the Lorentz condition $\partial_\alpha A^\alpha = 0$ is automatically fulfilled for the massive field. But then the equations of motion simplify considerably:

$$(\partial_\mu \partial^\mu - m^2) A^\alpha = 0 \quad \text{and} \quad \partial_\alpha A^\alpha = 0. \tag{3.47}$$

Each of the components A^α thus satisfies the Klein–Gordon equation, and the field A^α will clearly represent free particles with mass m. We call A^α a *massive vector field*. Observe, however, that the components A^α are not independent, as they are connected through the Lorenz condition. Therefore, only three of the components are independent. As we shall see in a moment, this means that the scalar part A_0 can be eliminated completely. The remaining three degrees of freedom correspond to spin. Using exactly the same method as we did in connection with the Maxwell field, we can show that the massive vector particle is a spin 1 particle. The projection of the spin onto the z-axis assumes the values -1, 0, $+1$, and this time the longitudinal part also contributes! The massive vector field is *not* a gauge field.

Worked Exercise 3.7.1
Problem: (a) Calculate the canonical energy–momentum tensor for the massive vector field and show that it is *not* symmetric. (b) Show that we can repair the canonical energy–momentum tensor by adding the following correction term:

$$\theta^{\alpha\beta} = -\partial_\rho (F^{\beta\rho} A^\alpha),$$

leading to the true energy–momentum tensor

$$T^{\alpha\beta} = \left\{ -F^\alpha{}_\rho F^{\rho\beta} - \frac{1}{4}\,\eta^{\alpha\beta} F^{\gamma\delta} F_{\gamma\delta} \right\} + m^2 \left\{ A^\alpha A^\beta - \frac{1}{2}\,\eta^{\alpha\beta} A^\gamma A_\gamma \right\}. \tag{3.48}$$

(c) Show that the energy density T^{00} corresponding to the true energy–momentum tensor is positive definite.

To conclude, we have presented *very naive* semiclassical arguments to motivate three kinds of fields, which are listed in Table 3.1.

We have seen that they represent spin 0 particles, massless spin 1 particles (photons), and massive spin 1 particles (vector particles).

§ 3.8 The Cauchy Problem

In what follows we will treat our fields as classical fields; i.e., we will admit only *real* solutions to the field equations.

Table 3.1.

Name	Lagrangian density	Equation of motion	
Klein–Gordon field	$-\frac{1}{2}(\partial_\mu\phi)(\partial^\mu\phi) - \frac{1}{2}m^2\phi^2$	$(\partial_\mu\partial^\mu - m^2)\phi = 0$	(3.49)
Maxwell field	$-\frac{1}{4}F_{\alpha\beta}F^{\alpha\beta}$	$(\partial_\mu\partial^\mu)A_\nu - \partial_\nu(\partial_\mu A^\mu) = 0$	(3.50)
Massive vector field	$-\frac{1}{4}F_{\alpha\beta}F^{\alpha\beta} - \frac{1}{2}m^2 A_\alpha A^\alpha$	$(\partial_\mu\partial^\mu - m^2)A_\alpha = 0$ $\partial_\alpha A^\alpha = 0$	(3.51)

The first thing we will study is the *Cauchy problem*. If we have a single particle moving in a one-dimensional space, it has the equation of motion

$$m\,\frac{d^2x}{dt^2} = -V'(x).$$

Now, fix a time t_1 and prescribe the value of x and the rate of change of x at this particular moment:

$$x(t_1) = x_1, \qquad \frac{dx}{dt}\Big|_{t_1} = v_1.$$

What can we say about the dynamical evolution of the system once we have prescribed these initial data? This is the Cauchy problem.

In this case there is a simple answer. The above differential equation is an ordinary second-order equation; hence, it has a unique solution once we have specified the value of x and the rate of change of x at a given moment. It is because of this that we say that x is a *dynamical quantity* and that the system in question is uniquely characterized by this single dynamical quantity.

Now let us look at the Klein–Gordon field. Again we fix at a time t_1, and we prescribe the value of the field and the rate of change at that moment:

$$\phi(t_1, \mathbf{x}) = \phi_1(\mathbf{x})$$
$$\frac{\partial\phi}{\partial t}(t_1, \mathbf{x}) = \psi_1(\mathbf{x}).$$

(See Figure 3.10.) What can we say about the dynamical evolution? Well, let us write out the equations of motion:

$$(\partial_\mu\partial^\mu - m^2)\phi = 0, \qquad \text{or} \qquad \frac{\partial^2\phi}{\partial t^2} = \Delta\phi - m^2\phi. \tag{3.49}$$

This is a second-order equation, and thus it has a unique solution once we have

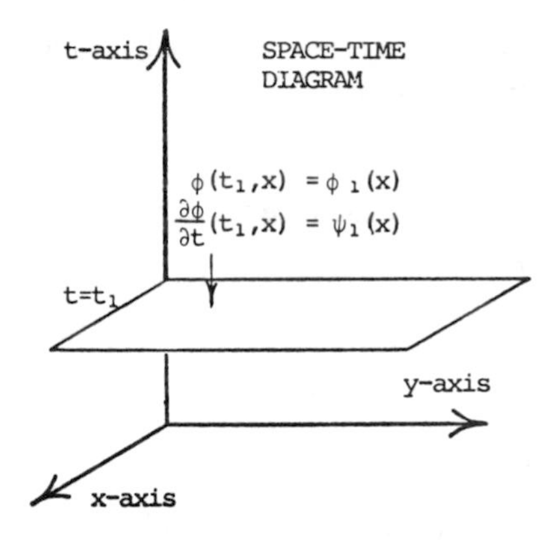

Figure 3.10.

prescribed ϕ and $\frac{\partial\phi}{\partial t}$ at a given moment. In fact, it is easy to see that we may construct the dynamical evolution directly by using an iteration procedure. We know the value of ϕ and $\frac{\partial\phi}{\partial t}$ at the time t_1. We may then find them at time $t = t_1 + \delta t$ using the formula

$$\phi(t_1 + \delta t, \mathbf{x}) \approx \phi(t_1, \mathbf{x}) + \delta t \, \frac{\partial\phi}{\partial t}(t_1, \mathbf{x}),$$

$$\begin{aligned}\frac{\partial\phi}{\partial t}(t_1 + \delta t, \mathbf{x}) &\approx \frac{\partial\phi}{\partial t}(t_1, \mathbf{x}) + \delta t \, \frac{\partial^2\phi}{\partial t^2}(t_1, \mathbf{x}) \\ &= \frac{\partial\phi}{\partial t}(t_1, \mathbf{x}) + \delta t[\Delta\phi(t_1, \mathbf{x}) - m^2\phi(t_1, \mathbf{x})].\end{aligned}$$

But when we know ϕ and $\frac{\partial\phi}{\partial t}$ at time $t = t_1 + \delta t$, we may repeat the procedure to find ϕ and $\frac{\partial\phi}{\partial t}$ at time $t = t_1 + 2\delta t$, etc. Thus we can reconstruct the dynamical evolution of the Klein–Gordon field. This shows that the field variable ϕ is a dynamical quantity completely characterizing the system.

Then we look at the Maxwell field A_α. We select a time $t = t_1$ and prescribe the values of A_α and $\frac{\partial A_\alpha}{\partial t}$ at this time. Now, what can we say about the dynamical evolution of the Maxwell field? *Nothing*! And here is the reason why: We know that if we find a solution A_α to the equations of motion, then the gauge transformation

$$A_\alpha \to A_\alpha + \partial_\alpha \chi$$

produces another solution. Thus, if we choose χ such that it is zero in a neighborhood of $t = t_1$, this will not disturb the boundary conditions, and we see that there are infinitely many solutions to the equations of motion all satisfying the boundary conditions.

On the other hand, we have seen that the equations of motion are second-order differential equations

$$(\partial_\mu \partial^\mu) A^\nu - \partial^\nu (\partial_\mu A^\mu) = 0,$$

so we might wonder what is wrong?

There is a subtle reason for this. Remember that the full equations for a Maxwell field with a source look as follows (1.33):

$$(\partial_\mu \partial^\mu) A^\nu - \partial^\nu (\partial_\mu A^\mu) = -J^\nu.$$

These equations automatically guarantee the conservation of charge:

$$-\partial_\nu J^\nu = (\partial_\mu \partial^\mu)(\partial_\nu A^\nu) - (\partial_\nu \partial^\nu)(\partial_\mu A^\mu) = 0.$$

But this conservation has a price. It means that the differential operator

$$(\partial_\mu \partial^\mu) A^\nu - \partial^\nu (\partial_\mu A^\mu)$$

obeys a special identity:

$$\partial_\nu \left[(\partial_\mu \partial^\mu) A^\nu - \partial^\nu (\partial_\mu A^\mu) \right] = 0. \tag{3.52}$$

Suppose we isolate ∂_0 on the left side:

$$\partial_0 \left[(\partial_\mu \partial^\mu) A^0 - \partial^0 (\partial_\mu A^\mu) \right] = -\partial_i \left[(\partial_\mu \partial^\mu) A^i - \partial^i (\partial_\mu A^\mu) \right].$$

This equation is interesting. On the right we have time derivatives of maximum order two, but then the same thing must be true for the left! Removing ∂_0, we see that

$$(\partial_\mu \partial^\mu) A^0 - \partial^0 (\partial_\mu A^\mu)$$

can only contain time derivatives of maximum order one! So if we consider the equation of motion for the scalar part A^0,

$$(\partial_\mu \partial^\mu) A^0 - \partial^0 (\partial_\mu A^\mu) = 0,$$

we observe that there is something wrong. It is not a second-order equation. By explicit computation we can easily find that it reduces to

$$\Delta \phi + \frac{\partial}{\partial t} (\nabla \cdot \mathbf{A}) = 0; \qquad \text{i.e., } \nabla \mathbf{E} = 0. \tag{3.53}$$

This is not an equation of motion at all! This shows us clearly that the scalar part ϕ is *not* a dynamical quantity. In fact, we cannot even prescribe ϕ and $\frac{\partial \phi}{\partial t}$ at the initial moment t_1, because once we have prescribed

$$\mathbf{A}(t_1, \mathbf{x}) \qquad \text{and} \qquad \frac{\partial \mathbf{A}}{\partial t} (t_1, \mathbf{x}),$$

then ϕ has to obey

$$\Delta \phi = -\frac{\partial}{\partial t} (\nabla \cdot \mathbf{A}).$$

For this reason we say that equation (3.53) is an *equation of constraint*. This leaves us with the three remaining field variables $\mathbf{A}$ as dynamical variables. We may rewrite their equations of motion as:

$$(\partial_\mu \partial^\mu) A^i - \partial^i (\partial_\mu A^\mu) = 0;$$

i.e.,

$$\frac{\partial^2 \mathbf{A}}{\partial t^2} = \Delta \mathbf{A} - \nabla \left[\nabla \cdot \mathbf{A} + \frac{\partial \phi}{\partial t} \right], \tag{3.54}$$

so the Maxwell equations reduce to one equation of constraint and three equations of motion. But we still have not solved our dynamical problems. We are still allowed to make gauge transformations, and thus the dynamical evolution is completely undetermined. To get rid of this we must choose a specific gauge. It is tempting to use the Lorenz gauge (1.22)

$$0 = \partial_\alpha A^\alpha = \frac{\partial \phi}{\partial t} + \nabla \cdot \mathbf{A},$$

not only because the Lorenz condition is Lorentz invariant, but also because the massive vector field automatically obeys this condition. Using the Lorenz condition, we can now eliminate $\frac{\partial \phi}{\partial t}$ too. That solves our dynamical problems.

The only true dynamical quantities are the three space components $\mathbf{A}$. They are governed by the equations of motion

$$\frac{\partial^2 \mathbf{A}}{\partial t^2} = \Delta \mathbf{A}. \tag{3.55}$$

The scalar part ϕ is eliminated by the equation of constraint and the Lorenz condition:

$$\Delta \phi = -\frac{\partial}{\partial t} (\nabla \cdot \mathbf{A}), \qquad \frac{\partial \phi}{\partial t} = -\nabla \cdot \mathbf{A}. \tag{3.56}$$

We should, however, be a little careful with the iteration procedure. There are no problems with the $\mathbf{A}$-variables. But consider the scalar part. It should satisfy (3.56) at all times if our method is to be consistent.

Worked Exercise 3.8.1
Problem: Use the iteration procedure and the equation of motion to show that the equation of constraint and the Lorenz condition are automatically preserved at all times.

From this exercise it follows that the method is in fact consistent. We might also be worried about the fact that we can still perform gauge transformations

$$A_\mu \to A_\mu + \partial_\mu \chi,$$

provided that χ satisfies (1.23)

$$\frac{\partial^2 \chi}{\partial t^2} = \Delta \chi.$$

Thus, we might think that we could repeat the argument from before to show that there are infinitely many solutions to the equations of motion all satisfying the boundary condition. But this is not so. The only solution to equation (1.23) that satisfies the initial data

$$\chi(\mathbf{x}, 0) = 0 \quad \text{and} \quad \frac{\partial \chi}{\partial t}(\mathbf{x}, 0) = 0$$

is the χ-function that vanishes identically!

In the case of the *massive vector field* we can show by a similar analysis that only the space components A^i are true dynamical quantities. This time the Lorenz condition is automatically satisfied, and therefore the scalar part A^0 is automatically eliminated.

Observe that when we constructed the Lagrangian density for the massive vector field

$$L = -\frac{1}{4} F_{\alpha\beta} F^{\alpha\beta} - \frac{1}{2} m^2 A_\alpha A^\alpha, \tag{3.46}$$

we directly generalized the Lagrangian density for the gauge potential. In particular, we kept the gauge-invariant kinetic energy term

$$L = -\frac{1}{4} (\partial_\alpha A_\beta - \partial_\beta A_\alpha)(\partial^\alpha A^\beta - \partial^\beta A^\alpha) - \frac{1}{2} m^2 A_\alpha A^\alpha.$$

But the massive vector field is *not* a gauge field, so we might wonder whether it would not have been much easier to start with the following Lagrangian density:

$$L' = -\frac{1}{2} (\partial_\alpha A_\beta)(\partial^\alpha A^\beta) - \frac{1}{2} m^2 A_\beta A^\beta. \tag{3.57}$$

This would be a direct generalization of the Klein–Gordon field. Computing the equations of motion, we easily find that

$$(\partial_\mu \partial^\mu - m^2) A^\beta = 0. \tag{3.58}$$

Consequently, each of the components satisfies the Klein–Gordon equation. Thus, in this case, all four components are true dynamical variables. But why did we not choose that Lagrangian density? Because as we have seen, the scalar part A^0 represents a spin 0 particle, while the vector part A^i represents a spin 1 particle. Hence, if we adopted the Lagrangian density L', we would construct a field theory that when quantized would represent a mixture of spin 0 and spin 1 particles! Furthermore, this would lead to troubles with the energy–momentum tensor. This is the content of the following exercise:

Exercise 3.8.2
Problem: Calculate the canonical energy–momentum tensor corresponding to the Lagrangian (3.57). Show that it is symmetric but that the energy density is indefinite.

§ 3.9 The Complex Klein–Gordon Field

Now suppose that we have two real classical fields ϕ_1 and ϕ_2. We will base their dynamics upon the Lagrangian density

$$L = -\frac{1}{2} (\partial_\mu \phi_1)(\partial^\mu \phi_1) - \frac{1}{2} (\partial_\mu \phi_2)(\partial^\mu \phi_2) - \frac{1}{2} m^2 \phi_1^2 - \frac{1}{2} m^2 \phi_2^2. \tag{3.59}$$

Clearly, there are no interactions between the two Klein–Gordon fields, as their equations of motion decouple:

$$(\partial_\mu \partial^\mu - m^2)\phi_1 = 0 \quad \text{and} \quad (\partial_\mu \partial^\mu - m^2)\phi_2 = 0. \tag{3.60}$$

Now we perform a very useful trick. We fuse the two real fields ϕ_1 and ϕ_2 into a single *complex* field:

$$\phi = \phi_1 + i\phi_2. \tag{3.61}$$

We can then rearrange the Lagrangian density as follows:

$$L = -\frac{1}{2}\,(\overline{\partial_\mu \phi})(\partial^\mu \phi) - \frac{1}{2}\,m^2 \bar{\phi}\phi. \tag{3.62}$$

This is only a fancy way of writing the same thing, but what about the equations of motion? To derive the Euler–Lagrange equations we should vary ϕ_1 and ϕ_2 independently, getting

$$\frac{\partial L}{\partial \phi_1} = \partial_\mu \frac{\partial L}{\partial(\partial_\mu \phi_1)} \quad \text{and} \quad \frac{\partial L}{\partial \phi_2} = \partial_\mu \frac{\partial L}{\partial(\partial_\mu \phi_2)}\,.$$

But due to the formulas

$$\phi = \phi_1 + i\phi_2, \qquad \bar{\phi} = \phi_1 - i\phi_2$$

and

$$\phi_1 = \frac{1}{2}\,(\phi + \bar{\phi}), \qquad \phi_2 = \frac{1}{2i}\,(\phi - \bar{\phi}),$$

these equations of motion are equivalent to the following equations of motion:

$$\frac{\partial L}{\partial \phi} = \partial_\mu \frac{\partial L}{\partial(\partial_\mu \phi)} \quad \text{and} \quad \frac{\partial}{\partial \bar{\phi}} = \partial_\mu \frac{\partial L}{\partial(\partial_\mu \bar{\phi})}\,, \tag{3.63}$$

where we *formally* treat ϕ and $\bar{\phi}$ as independent variables:

$$L = L(\phi, \bar{\phi}, \partial_\mu \phi, \partial_\mu \bar{\phi}).$$

For instance, we get in the above case (3.62)

$$0 = \frac{\partial L}{\partial \bar{\phi}} - \partial_\mu \frac{\partial L}{\partial(\partial_\mu \bar{\phi})} = -\frac{1}{2}\,m^2 \phi + \frac{1}{2}\,\partial_\mu \partial^\mu \phi,$$

or

$$0 = \frac{1}{2}\left[\partial_\mu \partial^\mu - m^2\right]\phi,$$

which, if we decompose ϕ, reproduces the Klein–Gordon equation for the real and imaginary parts. So the complex field obeys the Klein–Gordon equation too:

$$(\partial_\mu \partial^\mu - m^2)\phi = 0, \qquad \phi = \phi_1 + i\phi_2. \tag{3.64}$$

Now, returning to the Lagrangian density (3.26)

$$L = -\frac{1}{2}\,(\overline{\partial_\mu \phi})(\partial^\mu \phi) - \frac{1}{2}\,m^2 \bar{\phi}\phi,$$

we observe that it is invariant under the substitution

$$\phi(x) \to e^{i\alpha}\phi(x) \quad \text{and} \quad \bar{\phi}(x) \to e^{-i\alpha}\bar{\phi}(x). \tag{3.65}$$

Thus, we have discovered a symmetry. But when there is a symmetry, there should also be a conservation law. (This is known as *Noether's theorem*). Let us examine this a little more closely. Let Ω be an *arbitrary* space–time region, and consider the corresponding action

$$S = \int_{\Omega} L(\phi, \bar{\phi}, \partial_\mu \phi, \partial_\mu \bar{\phi}) d^4x. \tag{3.66}$$

Since L is invariant under the substitutions

$$\phi \to e^{i\alpha}\phi, \qquad \bar{\phi} \to e^{-i\alpha}\bar{\phi},$$

we conclude that

$$S(\alpha) = \int_{\Omega} L(e^{i\alpha}\phi, e^{-i\alpha}\bar{\phi}, e^{i\alpha}\partial_\mu\phi, e^{-i\alpha}\partial_\mu\bar{\phi}) d^4x$$

is a constant function of α. Hence, differentiating with respect to α we get

$$\begin{aligned} 0 &= S'(0) \\ &= \int_{\Omega} i \left[\left(\phi \frac{\partial L}{\partial \phi} + \partial_\mu \phi \frac{\partial L}{\partial(\partial_\mu \phi)} \right) - \left(\bar{\phi} \frac{\partial L}{\partial \bar{\phi}} + \partial_\mu \bar{\phi} \frac{\partial L}{\partial(\partial_\mu \bar{\phi})} \right) \right] d^4x. \end{aligned}$$

But since Ω was chosen arbitrarily, this is consistent only if

$$i \left[\left(\phi \frac{\partial L}{\partial \phi} + \partial_\mu \phi \frac{\partial L}{\partial(\partial_\mu \phi)} \right) - \left(\bar{\phi} \frac{\partial L}{\partial \bar{\phi}} + \partial_\mu \bar{\phi} \frac{\partial L}{\partial(\partial_\mu \bar{\phi})} \right) \right] = 0.$$

Using the equations of motion, we can rearrange this as

$$\begin{aligned} 0 &= i \left[\left(\phi \partial_\mu \frac{\partial L}{\partial(\partial_\mu \phi)} + \partial_\mu \phi \frac{\partial L}{\partial(\partial_\mu \phi)} \right) \right. \\ &\qquad \left. - \left(\bar{\phi} \partial_\mu \frac{\partial L}{\partial(\partial_\mu \bar{\phi})} + \partial_\mu \bar{\phi} \frac{\partial L}{\partial(\partial_\mu \bar{\phi})} \right) \right] \\ &= i \partial_\mu \left[\phi \frac{\partial L}{\partial(\partial_\mu \phi)} - \bar{\phi} \frac{\partial L}{\partial(\partial_\mu \bar{\phi})} \right]. \end{aligned}$$

Consequently, we conclude that because of the symmetry *and* the equations of motion, the following four-vector,

$$J^\mu = i \left[\phi \frac{\partial L}{\partial(\partial_\mu \phi)} - \bar{\phi} \frac{\partial L}{\partial(\partial_\mu \bar{\phi})} \right], \tag{3.67}$$

will obey the equation of continuity (1.29)

$$\partial_\mu J^\mu = 0.$$

For this reason we will call J^μ a *current*, although it need not have anything to do with electromagnetism! In the same spirit we will refer to

$$Q = \int_{t=t^0} J^0 d^2\mathbf{x}$$

as a *charge*, although it is not necessarily an electric charge!

A symmetry like $\phi(x) \to e^{i\alpha}\phi(x)$ is called an *internal symmetry*, in contrast to a *space–time symmetry*, where the Lagrangian density is invariant under a *coordinate transformation*

$$y^\mu = \alpha^\mu{}_\nu x^\nu.$$

But internal symmetries and space–time symmetries have one thing in common: They produce conservation laws! For instance, we have just seen that the symmetry $\phi(x) \to e^{i\alpha}\phi(x)$ leads to a *conserved current* J^μ!

Exercise 3.9.1
Notation: Let $W_\mu = A_\mu^{(1)} + iA_\mu^{(2)}$ be a complex vector field. Let $G_{\mu\nu}$ denote the corresponding complex field strengths

$$G_{\mu\nu} = \partial_\mu W_\nu - \partial_\nu W_\mu.$$

The Lagrangian for the massive vector field is immediately generalized to

$$L = -\frac{1}{4}\bar{G}_{\mu\nu}G^{\mu\nu} - \frac{1}{2}m^2\bar{W}_\mu W^\mu.$$

Problem: (a) Show that the Lagrangian is invariant under the internal symmetry

$$W_\mu \to e^{i\alpha}W_\mu, \qquad \bar{W}_\mu \to e^{-i\alpha}\bar{W}_\mu.$$

(b) Show that the corresponding conserved current is given by

$$J^\nu = \frac{i}{2}(W_\mu\bar{G}^{\mu\nu} - \bar{W}_\mu G^{\mu\nu}).$$

§ 3.10 The Theory of Electrically Charged Fields as a Gauge Theory

Let us now examine the complex Klein–Gordon field (3.62)

$$L = -\frac{1}{2}(\overline{\partial_\mu\phi})(\partial^\mu\phi) - \frac{1}{2}m^2\bar{\phi}\phi.$$

In this case the conserved current J^μ is

$$J^\mu = \frac{i}{2}(\phi\partial^\mu\bar{\phi} - \bar{\phi}\partial^\mu\phi). \tag{3.68}$$

Can we give a physical interpretation of this current? Suppose we wish to construct a theory of an *electrically charged field*, then we certainly expect that we could define a reasonable four-current J^μ, and this electric current should satisfy the equation of continuity (1.29)

$$\partial_\mu J^\mu = 0.$$

But then the complex Klein–Gordon field is an obvious candidate. However, the complex field $\phi = \phi_1 + i\phi_2$ actually consists of *two* classical fields ϕ_1 and ϕ_2. Hence, if we quantize it, we expect that it represents two kinds of particles!

Can we understand this in a simple way? Yes! In a relativistic theory we may have production of a particle–antiparticle pair. For instance, a photon *may* split into an electron–positron pair

$$\gamma \to e^- + e^+,$$

and if the particle has positive charge, then the antiparticle has negative charge. Thus, if we want charge conservation, the theory of a charged field must represent both of these particles, one with positive charge and one with negative charge!

This makes the complex Klein–Gordon field an even more obvious candidate. But if ϕ is going to be a charged field, it must interact with the Maxwell field. In fact, the current J^μ must act as a source for the Maxwell field. The question is then whether we can construct a suitable interaction term for the total system consisting of the charged field and the electromagnetic field

$$L = L_\phi + L_I + L_{A^\alpha} = -\frac{1}{2}\,(\overline{\partial_\mu\phi})(\partial^\mu\phi) - \frac{1}{2}\,m^2\bar{\phi}\phi + L_I - \frac{1}{4}\,F_{\alpha\beta}F^{\alpha\beta}.$$

To answer this we may take advantage of the *gauge symmetry*! We know that the theory has to be invariant under the gauge transformation (1.31)

$$A_\alpha \to A_\alpha + \partial_\alpha\chi,$$

but then we can use what we have learned studying quantum–mechanical systems.

The combined theory is gauge invariant provided that we exchange the usual derivatives ∂_α with the gauge covariant derivatives (2.44)

$$D_\alpha = \partial_\alpha - ieA_\alpha.$$

Therefore, we suggest the following total Lagrangian:

$$L = \frac{1}{2}\left[\overline{(\partial_\mu - ieA_\mu)\phi}\right]\left[(\partial^\mu - ieA^\mu)\phi\right] - \frac{1}{2}\,m^2\bar{\phi}\phi - \frac{1}{4}\,F_{\alpha\beta}F^{\alpha\beta}. \quad (3.69)$$

This is invariant under the combined gauge transformation

$$\phi(x) \to e^{ie\chi(x)}\cdot\phi(x) \quad \text{and} \quad A_\alpha(x) \to A_\alpha(x) + \partial_\alpha\chi. \quad (3.70)$$

Using this total Lagrangian, we have, in fact, gained something. In the original Lagrangian we were allowed only to make the transformation

(a) $$\phi(x) \to e^{i\alpha}\phi(x),$$

where α was a constant. But in the combined theory we are allowed to make the transformation

(b) $$\phi(x) \to e^{i\alpha(x)}\phi(x),$$

where $\alpha(x)$ is space–time dependent.

It is customary to refer to (a) as a gauge transformation of the first kind and to (b) as a gauge transformation of the second kind. Hence, incorporating the Maxwell field, we have extended the symmetry from a gauge symmetry of the first kind to a gauge symmetry of the second kind.

We have seen that there is a natural way to incorporate the interaction between the complex Klein–Gordon field and the Maxwell field. In what follows we shall refer to $\phi = \phi_1 + i\phi_2$ as a *charged* Klein–Gordon field, and we shall base the theory upon the Lagrangian density (3.69)

$$L = -\frac{1}{2}(\overline{D_\mu\phi})(D^\mu\phi) - \frac{1}{2}m^2\bar{\phi}\phi - \frac{1}{4}F_{\alpha\beta}F^{\alpha\beta}.$$

Now, what about the current? Since we have extended the Lagrangian, it is no longer necessarily equal to (3.68). It may contain terms involving A_μ:

$$J^\mu = \frac{i}{2}\left[\phi\partial^\mu\bar{\phi} - \bar{\phi}\partial^\mu\phi\right] + \text{terms involving } A_\mu.$$

In fact, the old expression must necessarily be wrong because it is not gauge invariant! This suggests that we simply exchange the usual derivatives ∂_μ with the gauge covariant derivatives D_μ, giving

$$J^\mu \stackrel{?}{=} \frac{i}{2}\left[\phi\overline{D^\mu\phi} - \bar{\phi}D^\mu\phi\right]. \tag{3.71}$$

This is a real, gauge invariant quantity, and in the absence of the Maxwell field, it reduces to the old expression (3.68). Is it conserved?

The easiest way to see this is to calculate the Maxwell equations. From

$$L = -\frac{1}{4}F_{\alpha\beta}F^{\alpha\beta} - \frac{1}{2}\left[\overline{(\partial_\mu - ieA_\mu)\phi}\right]\left[(\partial^\mu - ieA^\mu)\phi\right] - \frac{1}{2}m^2\bar{\phi}\phi$$

we get

$$\partial_\mu \frac{\partial L}{\partial(\partial_\mu A_\alpha)} = -[(\partial_\mu\partial^\mu)A^\alpha - \partial^\alpha(\partial_\mu A^\mu)] = -\partial_\mu F^{\mu\alpha}$$

and

$$\frac{\partial L}{\partial A_\alpha} = \frac{ie}{2}\left[\phi\overline{D^\alpha\phi} - \bar{\phi}D^\alpha\phi\right],$$

which immediately shows that

$$-\partial_\mu F^{\mu\alpha} = -\left[(\partial_\mu\partial^\mu)A^\alpha - \partial^\alpha(\partial_\mu A^\mu)\right] = \frac{ie}{2}\left[\phi\overline{D^\alpha\phi} - \bar{\phi}D^\alpha\phi\right].$$

Hence, we read off the conserved electromagnetic current

$$J^\alpha = \frac{ie}{2}\left[\phi\overline{D^\alpha\phi} - \bar{\phi}D^\alpha\phi\right], \tag{3.72}$$

which up to a constant factor reproduces (3.68).

Exercise 3.10.1
Problem: Consider the charged Klein–Gordon field. Show that the interaction term can be rearranged as

$$L_I = J^\alpha A_\alpha - e^2 A_\mu A^\mu \bar{\phi}\phi,$$

where J^α is the electromagnetic current (3.72).

Exercise 3.10.2
Problem: Consider a charged particle moving in an electromagnetic field. Show that the interaction term (2.14) can be rearranged as

$$S_I = \int J^\alpha A_\alpha d^4x, \tag{3.73}$$

where J^α is the electromagnetic current (1.34).

Exercise 3.10.3
Notation: Consider a charged massive vector field W_μ. Here the rule of minimal coupling corresponds to the substitution

$$G_{\mu\nu} \to \nabla_\mu W_\nu - \nabla_\nu W_\mu$$

with

$$\nabla_\mu W_\nu = \partial_\mu W_\nu - ieA_\mu W_\nu.$$

Problem: Show that the electromagnetic current is given by

$$J^\mu = \frac{ie}{2}\left[W_\mu(\overline{\nabla^\mu W^\nu - \nabla^\nu W^\mu}) - \bar{W}_\mu(\nabla^\mu W^\nu - \nabla^\nu W^\mu)\right].$$

§ 3.11 Charge Conservation as a Consequence of Gauge Symmetry

We conclude the discussion of charged fields with some general remarks. We saw above that it was easy to couple a *complex* field $\phi = \phi_1 + i\phi_2$ to the Maxwell field A_α. If ϕ was described by the free Lagrangian density

$$L_0 = L_0(\phi, \partial_\mu\phi, \bar{\phi}, \partial_\mu\bar{\phi}),$$

then the total Lagrangian density was obtained simply by replacing the ordinary derivatives with the gauge covariant derivatives and adding the usual piece for the Maxwell field:

$$L = L_0(\phi, D_\mu\phi, \bar{\phi}, \overline{D_\mu\phi}) - \frac{1}{4} F_{\alpha\beta}F^{\alpha\beta}.$$

This Lagrangian density is obviously gauge invariant. We can now give a general definition of the current associated with the charged field! Consider the action for the combined system

$$S = S_0 + S_I + S_{A_\alpha}.$$

We have decomposed it into the actions corresponding to the free fields ϕ and A_α, and the interaction term.

Consider the interaction term

$$S_I = S_I(\phi, A_\alpha).$$

If we produce a small change in the Maxwell field,

$$A_\alpha \to A_\alpha + \delta A_\alpha,$$

this will produce a small change in the interaction term,

$$S_I \to S_I + \delta S_I.$$

Here δS_I will depend linearly on δA_α, so we can write it as

$$\delta S_I = \int_\Omega J^\alpha(x)\delta A_\alpha(x) d^4x. \tag{3.74}$$

The coefficient J^α that measures the *response* of the interaction will be *defined* to be the current! Compare with Exercise 3.10.2 in the case of charged particles. We must show that this is in agreement with our previous ideas:

First, we observe that

$$S_I = \int_\Omega L_I(\phi, \partial_\mu\phi, A_\mu) d^4x, \tag{3.75}$$

where the interaction term does not depend on $\partial_\mu A_\nu$ since the presence of the Maxwell field comes from the gauge covariant derivatives. If we replace A_α by $A_\alpha + \epsilon\delta A_\alpha$ and consider the displaced action

$$S_I(\epsilon) = \int_\Omega L_I(\phi, \partial_\mu\phi, A_\mu + \epsilon\delta A_\mu) d^4x,$$

then

$$\delta S_I = \frac{dS_I}{d\epsilon}\Big|_{\epsilon=0} = \int_\Omega \frac{\partial L_I}{\partial A_\mu}\, \delta A_\mu d^4x,$$

from which we read off that

$$J^\alpha(x) = \frac{\partial L_I}{\partial A_\alpha}. \tag{3.76}$$

This is in agreement with the Maxwell equations

$$\partial_\mu \frac{\partial L}{\partial(\partial_\mu A_\alpha)} = \frac{\partial L}{\partial A_\alpha}; \qquad \text{i.e.,} \quad \frac{\partial L_I}{\partial A_\alpha} = -\partial_\mu F^{\mu\alpha},$$

since the interaction term is the only term that depends explicitly on A_α.

We can now give another argument for the conservation of electric charge. We will use only the part of the action containing ϕ:

$$S_\phi = S_0 + S_I = \int_\Omega L_0 d^4x + \int_\Omega L_I d^4x.$$

This is a gauge invariant action. Performing a gauge transformation

$$\phi(x) \to e^{ie[\epsilon\chi(x)]}\phi(x); \qquad A_\mu(x) \to A_\mu(x) + \epsilon\partial_\mu\chi(x)$$

we get

$$S(\epsilon) = \int_\Omega L_0(e^{ie[\epsilon\chi(x)]}\phi(x), \ldots) d^4x$$
$$+ \int_\Omega L_I(e^{ie[\epsilon\chi(x)]}\phi(x), \ldots, A_\mu(x) + \epsilon\partial_\mu\chi(x)) d^4x,$$

which is constant.

Hence, we conclude as usual that

$$0 = \frac{dS_\phi}{d\epsilon}\Big|_{\epsilon=0}.$$

Let us make some assumptions. First we assume that ϕ solves the equation of motion (3.63). Next we assume that $\chi(x)$ vanishes on the boundary Ω. This has the following consequences:

$$e^{ie[\epsilon\chi(x)]}\phi(x) = \phi(x) + \epsilon\psi(x),$$

where $\psi(x)$ is a smooth function vanishing on the boundary of Ω. (We are not interested in an explicit expression of $\psi(x)$.) Now, observe that since ϕ solves the equation of motion, it will *not* contribute to $\frac{dS_\phi}{d\epsilon}\big|_{\epsilon=0}$ when we actually perform the differentiation! Thus we get

$$0 = \frac{dS_\phi}{d\epsilon}\Big|_{\epsilon=0} = \int_\Omega \frac{\partial L_I}{\partial A_\mu}\partial_\mu\chi(x)d^4x = \int_\Omega J^\mu(x)\partial_\mu\chi(x)d^4x.$$

Performing a partial integration, we finally obtain

$$0 = -\int_\Omega (\partial_\mu J^\mu)\chi(x)d^4x.$$

But as $\chi(x)$ was arbitrarily chosen, we deduce that

$$\partial_\mu J^\mu = 0.$$

So we have captured our equation of continuity!

The above argument was very general and abstract! So maybe it is good to summarize the conclusions:

We are studying the interaction between a charged field $\phi = \phi_1 + i\phi_2$ and the Maxwell field A_α:

$$L = L_0(\phi, \partial_\mu\phi) + L_I(\phi, \partial_\mu\phi, A_\alpha) + L_{A^\alpha}(F_{\alpha\beta}).$$

From the total Lagrangian density we get the equations of motion

$$\partial_\alpha\partial^\alpha\phi = \cdots,$$

$$(\partial_\beta\partial^\alpha)A^\beta - (\partial_\beta\partial^\beta)A^\alpha = \cdots.$$

Usually we use Maxwell's equations to identify the current J^β. It is equal to the right-hand side of the Maxwell equation

$$(\partial_\beta\partial^\alpha)A^\beta - (\partial_\beta\partial^\beta)A^\alpha = J^\alpha[\phi; \partial_\mu\phi; A_\mu].$$

The conservation law then follows automatically from the form of the Maxwell equation. We do not use the equations of motion of the charged field at all!

The new argument shows that the dynamics of the Maxwell field are superfluous. We need only bother about the ϕ-field

$$L_\phi = L_0(\phi; \partial_\mu\phi) + L_I(\phi; \partial_\mu; A_\alpha).$$

From this Lagrangian density we get the equations of motion of the charged field (3.63). The current is then identified through he interaction term (3.76), and the conservation law $\partial_\beta J^\beta = 0$ follows automatically from the equations of motion of the charged field. *We do not use the dynamics of the Maxwell field at all!*

What we use, however, is the gauge symmetry, so again we see that *gauge invariance implies a conserved current.*

Worked Exercise 3.11.1
Problem: Consider the complex Klein–Gordon field ϕ coupled to the Maxwell field. Write down the equations of motion for ϕ and write down an expression for the current J^μ.

Show by explicit calculation that the equations of motion for ϕ imply that J^μ is conserved.

§ 3.12 The Equivalence of Real and Complex Field Theories

As we have seen, it is very easy to couple a complex field $\phi = \phi_1 + i\phi_2$ to the Maxwell field A_α. All we have to do is to exchange the ordinary derivatives ∂_α with the gauge covariant derivatives $D_\alpha = \partial_\alpha - ieA_\alpha$.

Of course, there is nothing mysterious about the fact that we use complex fields. It is nothing but a trick that makes life easier, and we could easily have avoided it.

To see this, let us return to the problem of constructing a charged field. The starting point is the free Lagrangian (3.59)

$$L_0 = -\frac{1}{2}(\partial_\mu\phi_1)(\partial^\mu\phi_1) - \frac{1}{2}(\partial_\mu\phi_2)(\partial^\mu\phi_2) - \frac{1}{2}m^2\phi_1^2 - \frac{1}{2}m^2\phi_2^2$$

(we must have two classical fields because the theory should incorporate both particles and antiparticles).

If we collect the two components into a single vector

$$\overset{=}{\phi}_| = \begin{bmatrix}\phi_1\\ \phi_2\end{bmatrix}, \tag{3.77}$$

we may rearrange the Lagrangian as follows:

$$L_0 = -\frac{1}{2}\left(\partial_\mu\overset{=}{\phi}_|\right)^+\left(\partial^\mu\overset{=}{\phi}_|\right) - \frac{1}{2}m^2\overset{=+}{\phi}_|\overset{=}{\phi}_|. \tag{3.78}$$

But then it is obvious that it possesses the symmetry

$$\begin{bmatrix}\phi_1\\ \phi_2\end{bmatrix} \to \begin{bmatrix}\cos\alpha & -\sin\alpha\\ \sin\alpha & \cos\alpha\end{bmatrix}\begin{bmatrix}\phi_1\\ \phi_2\end{bmatrix}, \tag{3.79}$$

where α is a constant. Of course, this is completely equivalent to formula (3.70),

$$\phi \to e^{i\alpha}\phi; \qquad \text{i.e., } \phi_1 + i\phi_2 \to (\cos\alpha + i\sin\alpha)(\phi_1 + i\phi_2).$$

In the same way, we can construct a gauge covariant derivative

$$\overline{\overline{D}}_\mu = \partial_\mu - eA_\mu \begin{bmatrix} 0 & -1 \\ 1 & 0 \end{bmatrix}. \tag{3.80}$$

Operating on the combined field, we find that

$$D_\mu \overline{\overline{\phi}}_| = \partial_\mu \begin{bmatrix} \phi_1 \\ \phi_2 \end{bmatrix} - eA_\mu \begin{bmatrix} \phi_2 \\ \phi_1 \end{bmatrix}, \tag{3.81}$$

which is completely equivalent to the formula

$$D_\mu \phi = (\partial_\mu - ieA_\mu)(\phi_1 + i\phi_2) = \partial_\mu(\phi_1 + i\phi_2) - eA_\mu(-\phi_2 + i\phi_1).$$

Hence, we can write down a gauge invariant Lagrangian density

$$L_\phi = -\frac{1}{2}(\overline{\overline{D}}_\mu \overline{\overline{\phi}}_|)^+ (\overline{\overline{D}}^\mu \overline{\overline{\phi}}_|) - \frac{1}{2} m^2 \overline{\overline{\phi}}_|^{\,+} \overline{\overline{\phi}}_|. \tag{3.82}$$

In this case "gauge invariance" means that it is invariant under the combined transformation

$$\begin{bmatrix} \phi_1 \\ \phi_2 \end{bmatrix} \to \begin{bmatrix} \cos(e\chi(x)) & -\sin(e\chi(x)) \\ \sin(e\chi(x)) & \cos(e\chi(x)) \end{bmatrix} \begin{bmatrix} \phi_1 \\ \phi_2 \end{bmatrix}. \tag{3.83}$$

$$A_\alpha(x) \to A_\alpha(x) + \partial_\alpha \chi(x)$$

Table 3.2 is a dictionary that allows us to pass from the "complex" formulation to the "real" formulation:

Concerning the gauge phase factor, you should observe that

$$\begin{bmatrix} 0 & -1 \\ 1 & 0 \end{bmatrix}^2 = -\begin{bmatrix} 1 & 0 \\ 0 & 1 \end{bmatrix}$$

From the formula

$$\exp[x\overline{\overline{A}}] = \sum_{n=0}^{\infty} \frac{(x\overline{\overline{A}})^n}{n!}$$

we therefore obtain

$$\begin{aligned} \exp\left\{x \begin{bmatrix} 0 & -1 \\ 1 & 0 \end{bmatrix}\right\} &= \left(1 - \frac{1}{2!}x^2 + \frac{1}{4!}x^4 - \cdots\right) \begin{bmatrix} 1 & 0 \\ 0 & 1 \end{bmatrix} \\ &\quad + \left(x - \frac{x^3}{3!} + \frac{x^5}{5!} - \cdots\right) \begin{bmatrix} 0 & -1 \\ 1 & 0 \end{bmatrix} \\ &= \cos x \begin{bmatrix} 1 & 0 \\ 0 & 1 \end{bmatrix} + \sin x \begin{bmatrix} 0 & -1 \\ 1 & 0 \end{bmatrix} \\ &= \begin{bmatrix} \cos x & -\sin x \\ \sin x & \cos x \end{bmatrix}. \end{aligned}$$

Thus the gauge phase factor lives in the gauge group, as any decent gauge phase factor ought to do! So we see that there is nothing sacred about complex numbers!

Table 3.2.

Gauge group	$SO(2)$	$U(1)$
Group element	$\begin{bmatrix} \cos\alpha & -\sin\alpha \\ \sin\alpha & \cos\alpha \end{bmatrix}$	$\exp[i\alpha]$
Gauge vector	$\overset{=}{\phi}_{\vert} = \begin{bmatrix} \phi_1 \\ \phi_2 \end{bmatrix}$	$\phi = \phi_1 + i\phi_2$
Gauge covariant derivative	$D_\mu = \partial_\mu - eA_\mu \begin{bmatrix} 0 & -1 \\ 1 & 0 \end{bmatrix}$	$D_\mu = \partial_\mu - ieA_\mu$
Gauge scalar	$\overset{=}{\phi}_{\vert}^{+} \overset{=}{\phi}_{\vert} = \phi_1^2 + \phi_2^2$	$\bar{\phi}\phi = (\phi_1 - i\phi_2)(\phi_1 + i\phi_2)$
Gauge phase factor	$\exp\left\{\begin{bmatrix} 0 & -1 \\ 1 & 0 \end{bmatrix} e \int A_\alpha dx^\alpha\right\}$	$\exp\left[ie \int_\Gamma A_\alpha dx^\alpha\right]$

Solutions to Worked Exercises

Solution to 3.5.1

$$\begin{aligned}\frac{\partial(F_{\alpha\beta}F^{\alpha\beta})}{\partial(\partial_\mu A_\nu)} &= \frac{\partial(\eta^{\alpha\gamma}\eta^{\beta\delta}F_{\alpha\beta}F_{\gamma\delta})}{\partial(\partial_\mu A_\nu)} \\ &= \eta^{\alpha\gamma}\eta^{\beta\delta}\left[\frac{\partial F_{\alpha\beta}}{\partial(\partial_\mu A_\nu)}F_{\gamma\delta} + F_{\alpha\beta}\frac{\partial F_{\gamma\delta}}{\partial(\partial_\mu A_\nu)}\right].\end{aligned}$$

Thus, we must first compute $\frac{\partial F_{\alpha\beta}}{\partial(\partial_\mu A_\nu)}$:

$$\frac{\partial F_{\alpha\beta}}{\partial(\partial_\mu A_\nu)} = \frac{\partial(\partial_\alpha A_\beta - \partial_\beta A_\alpha)}{\partial(\partial_\mu A_\nu)} = \delta^{\mu\nu}_{\alpha\beta} - \delta^{\mu\nu}_{\beta\alpha}.$$

Inserting this into the first equation, we get

$$\begin{aligned}\frac{\partial(F_{\alpha\beta}F^{\alpha\beta})}{\partial(\partial_\mu A_\nu)} &= \eta^{\alpha\gamma}\eta^{\beta\delta}\left[\left(\delta^{\mu\nu}_{\alpha\beta} - \delta^{\mu\nu}_{\beta\alpha}\right)F_{\gamma\delta} + F_{\alpha\beta}\left(\delta^{\mu\nu}_{\gamma\delta} - \delta^{\mu\nu}_{\delta\gamma}\right)\right] \\ &= \left(\delta^{\mu\nu}_{\alpha\beta} - \delta^{\mu\nu}_{\beta\alpha}\right)F^{\alpha\beta} + F^{\gamma\delta}\left(\delta^{\mu\nu}_{\gamma\delta} - \delta^{\mu\nu}_{\delta\gamma}\right) \\ &= F^{\mu\nu} - F^{\nu\mu} + F^{\mu\nu} - F^{\nu\mu} = 2(F^{\mu\nu} - F^{\nu\mu}) = 4F^{\mu\nu}\end{aligned}$$

(due to the antisymmetry of $F^{\mu\nu}$); so we have shown once and for all that

$$\frac{\partial(F_{\alpha\beta}F^{\alpha\beta})}{\partial(\partial_\mu A_\nu)} = 4F^{\mu\nu}.$$

Solution to 3.7.1

(a)

$$\begin{aligned}\overset{\circ}{T}{}^{\alpha\beta} &= \frac{-\partial L}{\partial(\partial_\beta A_\rho)}\,\partial^\alpha A_\rho + \eta^{\alpha\beta}L\\ &= F^{\beta\rho}\partial^\alpha A_\rho - \frac{1}{4}\,\eta^{\alpha\beta}F_{\gamma\delta}F^{\gamma\delta} - \frac{1}{2}\,\eta^{\alpha\beta}m^2 A_\gamma A^\gamma.\end{aligned}$$

Here the first term is not symmetric, while the last two terms are trivially symmetric.

(b) Observe first that using the equations of motion, we may rearrange the expression for $\theta^{\alpha\beta}$:

$$\theta^{\alpha\beta} = -(\partial_\rho F^{\beta\rho})A^\alpha - F^{\beta\rho}\partial_\rho A^\alpha = -m^2 A^\beta A^\alpha - F^{\beta\rho}\partial_\rho A^\alpha.$$

Here the first term is born symmetric, but its presence is very important for the verification of the various properties $\theta^{\alpha\beta}$ ought to have:

1. $\overset{\circ}{T}{}^{\alpha\beta} + \theta^{\alpha\beta} = \left[-F^{\alpha}_{\ \rho}F^{\rho\beta} - \frac{1}{4}\,\eta^{\alpha\beta}F_{\gamma\delta}F^{\gamma\delta}\right] + \left[m^2 A^\alpha A^\beta - \frac{1}{2}\,\eta^{\alpha\beta}m^2 A_\gamma A^\gamma\right].$ This is clearly symmetric. Observe that the first piece is identical to the energy–momentum tensor for the electromagnetic field
2. $\partial_\beta\theta^{\alpha\beta} = -\partial_\beta\partial_\rho(F^{\beta\rho}A^\alpha) = 0$, since $\partial_\beta\partial_\rho$ is symmetric in $\beta\rho$.
3. $\int\theta^{\alpha 0}d^3x = -\int\partial_\rho(F^{0\rho}A^\alpha)d^3x = -\int\partial_i(F^{0i}A^\alpha)d^3x$. But the last integral can be converted to a surface integral that vanishes automatically, provided that the fields A^α vanish sufficiently quickly

(c)

$$\begin{aligned}T^{00} &= \left[-F^{0}_{\ \rho}F^{\rho 0} - \frac{1}{4}\,F_{\gamma\delta}F^{\gamma\delta}\right] + \left[m^2 A^0 A^0 + \frac{1}{2}\,m^2 A_\gamma A^\gamma\right]\\ &= \frac{1}{2}\,(\mathbf{E}^2 + \mathbf{B}^2) + \frac{1}{2}\,m^2\left[(A^0)^2 + (A^1)^2 + (A^2)^2 + (A^3)^2\right]\end{aligned}$$

which is clearly positive definite.

Solution to 3.8.1

From the equation of constraint and the Lorenz condition we immediately find that

$$\frac{\partial^2\phi}{\partial t^2} = -\frac{\partial}{\partial t}\,(\nabla\cdot\mathbf{A}) = \Delta\phi.$$

If we prescribe $\mathbf{A}(t_1, \mathbf{x})$, and $\frac{\partial\mathbf{A}}{\partial t}\,(t_1, \mathbf{x})$, we can determine $\phi(t_1, \mathbf{x})$ and $\frac{\partial\phi}{\partial t}\,(t_1, \mathbf{x})$ from the equations

$$\Delta\phi(t_1, \mathbf{x}) = -\nabla\cdot\frac{\partial\mathbf{A}}{\partial t}\,(t_1, \mathbf{x}); \qquad \frac{\partial\phi}{\partial t}\,(t_1, \mathbf{x}) = -\nabla\cdot\mathbf{A}(t_1, \mathbf{x}),$$

and we can then use iteration to determine ϕ and $\frac{\partial \phi}{\partial t}$ at $t = t_1 + \delta t$:

$$\phi(t_1 + \delta t, \mathbf{x}) \approx \phi(t_1, \mathbf{x}) + \delta t \, \frac{\partial \phi}{\partial t} \, (t_1, \mathbf{x}),$$

$$\frac{\partial \phi}{\partial t} \, (t_1 + \delta t, \mathbf{x}) \approx \frac{\partial \phi}{\partial t} \, (t_1, \mathbf{x}) + \delta t \, \frac{\partial^2 \phi}{\partial t^2} \, (t_1, \mathbf{x}) = \frac{\partial \phi}{\partial t} \, (t_1, \mathbf{x}) + \delta t \cdot \Delta t(t_1, \mathbf{x}).$$

We must then show that the equations of constraint and the Lorenz condition are still valid at $t = t_1 + \delta t$.

Equation of constraint:

$$\Delta \phi(t_1 + \delta t, \mathbf{x}) = \Delta \phi(t_1, \mathbf{x}) + \delta t \cdot \Delta \, \frac{\partial \phi}{\partial t} \, (t_1, x).$$

Using the equation of constraint and the Lorenz condition at time t_1, this is rearranged as

$$= -\nabla \cdot \frac{\partial \mathbf{A}}{\partial t} \, (t_1, \mathbf{x}) + \delta t \cdot \Delta(-\nabla \cdot \mathbf{A}(t_1, \mathbf{x}))$$
$$= -\nabla \left[\frac{\partial \mathbf{A}}{\partial t} \, (t_1, \mathbf{x}) + \delta t \cdot \Delta \mathbf{A}(t_1, \mathbf{x}) \right].$$

Using the equation of motion, this is rearranged as

$$= -\nabla \left[\frac{\partial \mathbf{A}}{\partial t} \, (t_1, \mathbf{x}) + \delta t \cdot \frac{\partial^2 \mathbf{A}}{\partial t^2} \, (t_1, \mathbf{x}) \right] = -\nabla \cdot \frac{\partial \mathbf{A}}{\partial t} \, (t_1 + \delta t, \mathbf{x}).$$

Lorenz condition:

$$\frac{\partial \phi}{\partial t} \, (t_1 + \delta t, \mathbf{x}) = \frac{\partial \phi}{\partial t} \, (t_1, \mathbf{x}) + \delta t \cdot \Delta \phi(t_1, \mathbf{x}).$$

Using the equation of constraint and the Lorenz condition at time t_1, this is rearranged as

$$= -\nabla \cdot \mathbf{A}(t_1, \mathbf{x}) - \delta t \, \frac{\partial}{\partial t} \, (\nabla \cdot \mathbf{A}(t_1, \mathbf{x}))$$
$$= -\nabla \left[\mathbf{A}(t_1, \mathbf{x}) + \delta t \cdot \frac{\partial \mathbf{A}}{\partial t} \, (t_1, \mathbf{x}) \right] = -\nabla \cdot \mathbf{A}(t_1 + \delta t, \mathbf{x}).$$

Solution to 3.11.1

The starting point is the free Lagrangian

$$L_0 = -\frac{1}{2} \, (\overline{\partial_\mu \phi})(\partial^\mu \phi) - \frac{1}{2} \, m^2 \bar{\phi} \phi.$$

We make it gauge invariant by using the rule of minimal coupling:

$$
\begin{aligned}
L_\phi(\phi, \partial_\mu\phi, A_\alpha) &= L_0(\phi, D_\mu\phi) \\
&= \left[-\frac{1}{2}\,(\overline{\partial_\mu\phi})(\partial^\mu\phi) - \frac{1}{2}\,m^2\bar{\phi}\phi\right] \\
&\quad + \frac{ie}{2}\,A_\mu[\phi\partial^\mu\bar{\phi} - \bar{\phi}\partial^\mu\phi] - \frac{1}{2}\,e^2 A_\mu A^\mu\bar{\phi}\phi.
\end{aligned}
$$

From this we get the equations of motion (3.63)

$$
(\partial_\mu\partial^\mu)\phi = m^2\phi + ie[A_\mu\partial^\mu\phi + \partial_\mu(A^\mu\phi)] + e^2 A_\mu A^\mu\phi,
$$

and we can immediately read off the interaction term

$$
L_I = \frac{ie}{2}\,A_\mu[\phi\partial^\mu\bar{\phi} - \bar{\phi}\partial^\mu\phi] - \frac{1}{2}\,e^2 A_\mu A^\mu\bar{\phi}\phi.
$$

From this we get the current

$$
J^\mu = \frac{\partial L_I}{\partial A_\mu} = \frac{ie}{2}\,[\phi\partial^\mu\bar{\phi} - \bar{\phi}\partial^\mu\phi] - e^2 A_\mu\bar{\phi}\phi.
$$

Of course, this is identical with the previously found current (3.72)! Then we compute $\partial_\mu J^\mu$:

$$
\partial_\mu J^\mu = \frac{ie}{2}\,[\phi(\partial_\mu\partial^\mu)\bar{\phi} - \bar{\phi}(\partial_\mu\partial^\mu)\phi] - e^2\partial_\mu[A^\mu\bar{\phi}\phi].
$$

From the equations of motions we now get

$$
\begin{aligned}
&\frac{ie}{2}\,[\phi(\partial_\mu\partial^\mu)\bar{\phi} - \bar{\phi}(\partial_\mu\partial^\mu)\phi] \\
&\quad = \frac{ie}{2}\,[\phi(m^2\bar{\phi} - ie(A_\mu\partial^\mu\bar{\phi} + \partial_\mu(A^\mu\bar{\phi})) - e^2 A_\mu A^\mu\bar{\phi}) \\
&\qquad - \bar{\phi}(m^2\phi + ie(A_\mu\partial^\mu\phi + \partial_\mu(A^\mu\phi)) - e^2 A_\mu A^\mu\phi)] \\
&\quad = e^2\partial_\mu[A^\mu\bar{\phi}\phi].
\end{aligned}
$$

Inserting this into the equation above, we get

$$
\partial_\mu J^\mu = e^2\partial_\mu[A^\mu\bar{\phi}\phi] - e^2\partial_\mu[A^\mu\bar{\phi}\phi] = 0.
$$

Consequently, we have reproduced the equation of motion without using Maxwell's equations!

CHAPTER 4

Solitons

§ 4.1 Nonlinear Field Theories with a Degenerate Vacuum

In this chapter we will discuss in some detail a few important models in classical field theory that attracted great attention due to their so-called "topological" properties. To keep the discussion as simple as possible, we will only consider classical field theories in (1 + 1)-space–time dimensions. (Although some of the results have suitable generalizations to higher dimensions, these generalizations are by no means trivial.)

The theories we are going to consider will be nonlinear field theories; i.e., in the nonrelativistic case, with which we shall be mostly concerned, they are based upon a Lagrangian density of the form (3.28)

$$L = -\frac{1}{2}(\partial_\mu \phi)(\partial^\mu \phi) - U(\phi),$$

where the potential energy density U is no longer quadratic in ϕ (cf. the discussion in Section 3.3). The associated equation of motion is given by (3.30)

$$\partial_\mu \partial^\mu \phi = U'(\phi),$$

and to ensure a positive energy density, we shall as usual assume that the potential is positive definite:

$$U(\phi) \geq 0\,.$$

To obtain nontrivial results, we shall, however, furthermore assume that it has *more than one (global) minimum*.

Two examples are of particular interest:

ILLUSTRATIVE EXAMPLE 4.1 (The ϕ^4-model). In this model the potential energy density is given by the following fourth-order polynomial in ϕ:

$$U(\phi) = \frac{\lambda}{4}\left(\phi^2 - \frac{\mu^2}{\lambda}\right)^2 = \frac{\lambda}{4}\phi^4 - \frac{1}{2}\mu^2\phi^2 + \frac{\mu^4}{4\lambda}\,. \tag{4.1}$$

Notice the sign of the quadratic term, which is the opposite of the usual mass term (cf. the discussion in Section 3.4). The potential energy density has two distinct minima,

$$\phi_+ = \frac{\mu}{\sqrt{\lambda}} \quad \text{and} \quad \phi_- = -\frac{\mu}{\sqrt{\lambda}}\ ; \tag{4.2}$$

and has the characteristic shape indicated in Figure 4.1a (known as double well).

ILLUSTRATIVE EXAMPLE 4.2 (The sine-Gordon model). In this case the potential energy density is given by

$$U(\phi) = \frac{\mu^2}{\lambda^2}\,(1 - \cos\lambda\phi), \tag{4.3}$$

which clearly is periodic in ϕ. The potential energy density, which has the characteristic shape shown in Figure 4.1b, thus has an infinite series of minima:

$$\phi_n = n\,\frac{2\pi}{\lambda}\ ; \qquad n = \ldots, -2, -1, 0, 1, 2, \ldots . \tag{4.4}$$

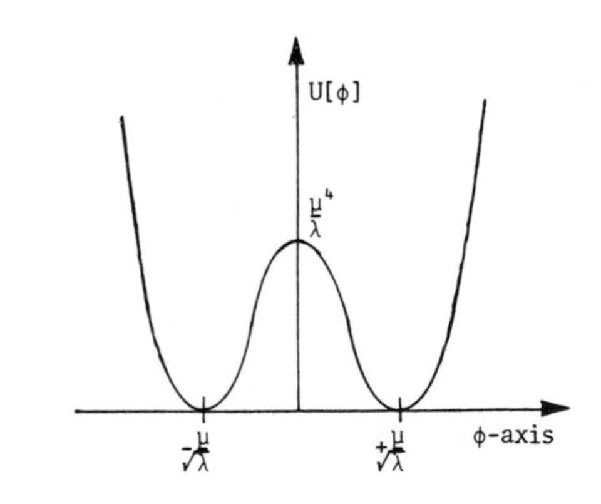

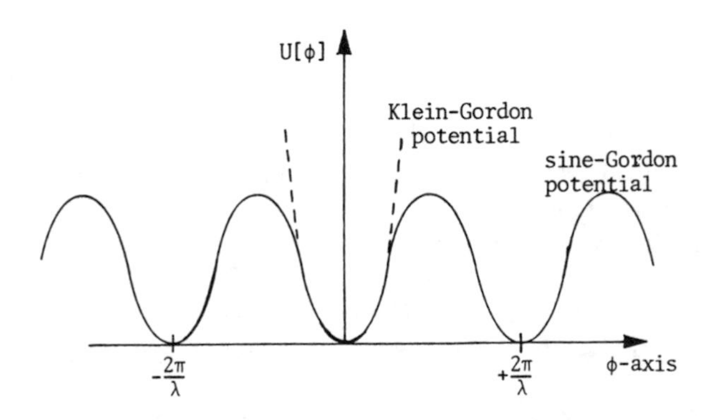

Figure 4.1.

Consider the equation of motion

$$\partial_\mu \partial^\mu \phi = \frac{\mu^2}{\lambda} \sin \lambda\phi. \tag{4.5}$$

In the weak field limit where $|\phi| \ll 1$, we can safely replace $\sin \lambda\phi$ with $\lambda\phi$, whereby we recover the Klein–Gordon equation (3.31)

$$\partial_\mu \partial^\mu \phi = \mu^2 \phi.$$

Now because the equation of motion reduces to the Klein–Gordon equation in the weak field limit and because the exact equation involves a sine function, it has become customary to refer to equation (4.5) as the *sine-Gordon equation*. So that is the origin of the funny name for our model.[1]

Remark. This model has a famous analogue in classical mechanics that allows one to "visualize" the basic properties of the model:

Imagine a fixed string on the x-axis. To this string we attach an infinite equidistant series of pendulums. (See Figure 4.2.) These pendulums are only allowed to move perpendicular to the string; i.e., they can rotate in the yz-plane. The position of a single pendulum at $x_n = a_n$ is then completely characterized by its angle $\phi(x_n, t)$ relative to the y-axis. (For convenience we imagine that the y-axis is pointing downwards.) Finally, we introduce a small coupling between neighboring pendulums, e.g., by connecting them with small strings. We can then write down the energy for this system. A single pendulum contributes with

a. its kinetic energy

$$\frac{1}{2} mr^2 \left[\frac{\partial \phi(x_n, t)}{\partial t} \right]^2,$$

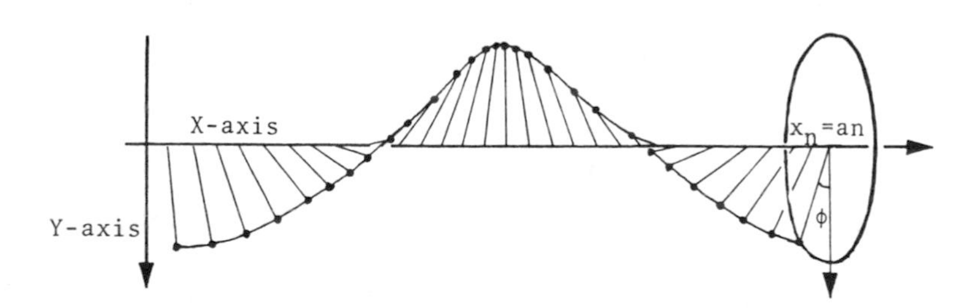

Figure 4.2.

[1] According to Coleman, the name was invented by Finkelstein. In Coleman [1975] there is a quotation from a letter from Finkelstein: "I am sorry that I ever called it the sine-Gordon equation. It was a private joke between me and Julio Rubinstein, and I never used it in print. By the time he used it as the title of a paper he had earned his Ph.D. and was beyond the reach of justice."

b. its interaction energy coming from its coupling to the neighboring pendulums

$$\frac{1}{2}k^2\left[\phi(x_{n+1},t)-\phi(x_n,t)\right]^2,$$

c. its potential energy due to the gravitational field

$$mgr[1-\cos\phi(x_n,t)].$$

The total energy is therefore given by

$$H=\sum_{n=-\infty}^{+\infty}\frac{1}{2}mr^2\left[\frac{\partial\phi}{\partial t}(x_n,t)\right]^2+\frac{1}{2}k^2\left[\phi(x_{n+1},t)-\phi(x_n,t)\right]^2$$
$$+\,mgr[1-\cos\phi(x_n,t)].$$

Furthermore, the dynamics are controlled by the Lagrangian

$$L=T-V=\sum\frac{1}{2}mr^2\left[\frac{\partial\phi}{\partial t}\right]^2-\frac{1}{2}k^2[\phi(x_{n+1},t)-\phi(x_n,t)]^2$$
$$-\,mgr[1-\cos\phi(x_n,t)].$$

(Compare with the discussion in Section 3.1.) We then pass to the continuum limit where $a\to 0$, while at the same time $m/a\to\rho$ and $ka\to\sigma$. In this way we obtain the following Lagrangian:

$$L=\int_{x=-\infty}^{+\infty}\left[\frac{1}{2}\rho r^2\left[\frac{\partial\phi}{\partial t}\right]^2-\frac{1}{2}\sigma\left[\frac{\partial\phi}{\partial x}\right]^2-\rho gr(1-\cos\phi)\right]dx.$$

Therefore, we see for a suitable choice of parameters the infinite system of pendulums is equivalent to the classical field theory in $(1+1)$ dimensions characterized by the potential energy density (4.3).

Notice that in both of the above examples the potential $U(\phi)$ possesses a discrete symmetry that transforms one minimum into another. In the ϕ^4-model it is the reflection

$$\phi\to-\phi,$$

while in the sine-Gordon model it is the translation

$$\phi\to\phi+\frac{2\pi}{\lambda}.$$

The existence of such a discrete symmetry is a typical feature for the kind of model we are going to examine.

We proceed to investigate various configurations in a nonlinear field theory. In a classical theory we will always restrict ourselves to *configurations with a finite total energy*. Since the total energy is given by

$$H=\int_{-\infty}^{\infty}\left[\frac{1}{2}\left(\frac{\partial\phi}{\partial t}\right)^2+\frac{1}{2}\left(\frac{\partial\phi}{\partial x}\right)^2+U(\phi)\right]dx, \tag{4.6}$$

the assumption of finite energy naturally leads to the boundary condition

$$\frac{\partial\phi}{\partial t} \to 0; \qquad \frac{\partial\phi}{\partial x} \to 0; \qquad U(\phi) \to 0 \qquad \text{as } |x| \to \infty. \tag{4.7}$$

At infinity the field is consequently static and approaches a constant value

$$\phi_{\pm\infty} = \lim_{x\to\pm\infty} \phi(x, t)$$

due to the first two conditions. The third condition states that the asymptotic value must in fact be one of the global minima for the potential; i.e.,

$$U(\phi_{\pm\infty}) = 0.$$

Among the configurations with finite energy we have especially the *vacuum configurations*. The classical vacuum is characterized by having the lowest possible energy; i.e., it is characterized by a vanishing energy density. It follows from (4.6) that a classical vacuum must satisfy the equations

$$\frac{\partial\phi}{\partial t} = \frac{\partial\phi}{\partial x} = 0 \quad \text{and} \quad U(\phi) = 0. \tag{4.8}$$

Thus a classical vacuum is represented by a constant field

$$\phi(x, t) = \phi_0,$$

where ϕ_0 is one of the global minima for the potential, i.e.,

$$U(\phi_0) = 0.$$

Notice that the assumption about several global minima for the potential implies that there are several classical vacua. A nontrivial nonlinear field theory is thus characterized by a degenerate classical vacuum!

Next we consider small fluctuations around a classical vacuum ϕ_0:

$$\phi(x, t) = \phi_0 + \eta(x, t).$$

In the weak field limit, where $|\eta(x, t)| \ll 1$, we can neglect the self-interaction of the field; i.e., we need only keep the quadratic terms in the Lagrangian density. In terms of the shifted field $\eta(x, t)$ we thus get

$$L(\eta, \partial_\mu\eta) \approx -\frac{1}{2}\,\partial_\mu\eta\partial^\mu\eta - \frac{1}{2}\,U''(\phi_0)\eta^2.$$

But this is exactly a free field Lagrangian; cf. (3.32). In the quantized version of the theory, small oscillations around a classical vacuum ϕ_0 consequently represent free particles with the mass–square $U''(\phi_0)$. If the fluctuations are large, we can of course no longer neglect the self-interaction; i.e., when the density of field quanta is high, the field quanta no longer act as free particles.

If the various classical vacua are all connected by a symmetry transformation, then $U''(\phi_0)$ will be the same for all the classical vacua; i.e., the field quanta associated with two different classical vacua will have the same mass. In the ϕ^4-model, e.g., the mass–square of the field quanta is given by

$$U''(\phi_\pm) = 2\mu^2, \tag{4.9}$$

while in the sine-Gordon model it is given by

$$U''(\phi_n) = \mu^2. \tag{4.10}$$

§ 4.2 Topological Charges

As we have noticed, finiteness of the energy implies that the asymptotic values

$$\phi_{\pm\infty} = \lim_{x \to \pm\infty} \phi(x, t)$$

are independent of time. These asymptotic values furthermore correspond to classical vacua.

Let E denote the space consisting of all smooth finite-energy configurations. As a consequence of the degeneracy of the classical vacuum, this space breaks up into different sectors characterized by the different possibilities for the asymptotic behavior. Consider, e.g., the ϕ^4-theory. Here the boundary conditions are given by

$$\lim_{x \to \pm\infty} \phi(x, t) = \pm \frac{\mu}{\sqrt{\lambda}} .$$

This gives four different possibilities for the asymptotic behavior:

I	$\phi_{-\infty} = -\frac{\mu}{\sqrt{\lambda}}$; $\phi_{+\infty} = -\frac{\mu}{\sqrt{\lambda}}$	II	$\phi_{-\infty} = -\frac{\mu}{\sqrt{\lambda}}$; $\phi_{+\infty} = +\frac{\mu}{\sqrt{\lambda}}$
III	$\phi_{-\infty} = +\frac{\mu}{\sqrt{\lambda}}$; $\phi_{+\infty} = -\frac{\mu}{\sqrt{\lambda}}$	IV	$\phi_{-\infty} = +\frac{\mu}{\sqrt{\lambda}}$; $\phi_{+\infty} = +\frac{\mu}{\sqrt{\lambda}}$

Thus E breaks up into four sectors, which we label E_{--}, E_{-+}, E_{+-}, and E_{++}. The vacuum solutions $\phi_\pm$ belong to the sectors E_{++} and E_{--}. These sectors are consequently referred to as *vacuum sectors*.

What can we say about the remaining nontrivial sectors? Consider a given smooth finite energy configuration $\tilde{\phi}$. A *smooth deformation* of $\tilde{\phi}$ is a one-parameter family of finite energy configurations $\phi_\lambda(x)$ such that $\phi_\lambda(x)$ depends smoothly upon λ and such that $\phi_\lambda(x)$ reduces to the given configuration $\tilde{\phi}(x)$ when $\lambda = 0$. Clearly, a smooth deformation cannot "break" the boundary conditions; i.e.,

$$\lim_{x \to \pm\infty} \phi_\lambda(x) = \phi_{\pm\infty}$$

is independent of λ. A smooth deformation thus necessarily stays within a single sector. We may think of $\phi_\lambda(x)$ as representing a *perturbation* of the given configuration $\tilde{\phi}$. It follows that a smooth perturbation of a classical vacuum necessarily stays within the corresponding vacuum sector. On the other hand, a configuration from a nontrivial sector cannot be obtained by perturbing a classical vacuum. The nontrivial sectors are therefore called *nonperturbative sectors*.

Notice that if the model possesses a discrete symmetry operation that relates the different classical vacua, then this symmetry operation will also relate the different sectors to each other; e.g., in the ϕ^4-model, the inversion

$$\phi \curvearrowright -\phi$$

will interchange the vacuum sectors; i.e., it will map E_{--} onto E_{++}. Similarly, it interchanges the nonperturbative sectors E_{-+} and E_{+-}. Thus the structure of the two vacuum sectors and similarly the structure of the two nonperturbative sectors are completely equivalent.

As another example we consider the sine-Gordon model. It possesses the translational symmetry

$$\phi \curvearrowright \phi + \frac{2\pi}{\lambda} p, \qquad p \text{ an integer.}$$

The different sectors in the sine-Gordon model can be labeled by a pair of integers (n, m) such that a field configuration belonging to $E_{n,m}$ satisfies the boundary conditions

$$\lim_{x\to-\infty} \phi(x, t) = \frac{2\pi}{\lambda} n; \qquad \lim_{x\to+\infty} \phi(x, t) = \frac{2\pi}{\lambda} m.$$

The translation operator then maps the sector $E_{n,m}$ onto the sector $E_{n+p,m+p}$. In the sine-Gordon model there is therefore a discrete series of qualitatively different types of sectors labeled by a single integer $q = m - n$ (where $q = 0$ corresponds to the vacuum sectors).

Next, we look for conserved quantities in our nonlinear models. Due to our boundary conditions, the asymptotic values

$$\lim_{x\to\pm\infty} \phi(x, t)$$

are conserved; i.e., they are independent of time. Let us especially focus upon their difference

$$Q = \phi_{+\infty} - \phi_{-\infty}. \tag{4.11}$$

This can be interpreted as a conserved charge with the corresponding charge density given by

$$\rho[\phi] = \frac{\partial\phi}{\partial x}\,; \tag{4.12}$$

i.e.,

$$Q = \int_{-\infty}^{+\infty} \frac{\partial\phi}{\partial x}\, dx = \phi(+\infty, t) - \phi(-\infty, t). \tag{4.13}$$

We will now compare this charge with the charge associated with the complex Klein–Gordon field, cf. the discussion in Section 3.9,

Exercise 4.2.1
Problem: Let $\phi = \phi^1 + i\phi^2$ be a complex Klein–Gordon field. Show that the charge corresponding to the conserved current (3.67) is given by

$$Q = \int_{\mathbb{R}^3} \epsilon_{ab}\phi^a \frac{\partial \phi^b}{\partial t}\, d^3x. \tag{4.14}$$

The main difference between the structures of (4.13) and (4.14) is that the ordinary charge (4.14) depends upon the time derivative of the field as well. Notice that the conservation of the ordinary charge presupposes that ϕ^a obeys the equations of motion: Differentiating (4.14) with respect to time, we get

$$\frac{dQ}{dt} = \int_{\mathbb{R}^3} \epsilon_{ab}\phi^a \frac{\partial^2 \phi^b}{\partial t^2}\, d^3x.$$

Using the Klein–Gordon equation (3.31)

$$\frac{\partial^2 \phi^b}{\partial t^2} = \Delta\phi^b - m^2\phi^b,$$

the above integral can be rearranged as follows:

$$\frac{dQ}{dt} = \int_{\mathbb{R}^3} \epsilon_{ab}\phi^a \Delta\phi^b d^3x = \int_{\mathbb{R}^3} \nabla\left(\epsilon_{ab}\phi^a \nabla\phi^b\right) d^3x.$$

But this can be converted into a surface integral at infinity. Provided that the gradients vanish sufficiently rapidly at infinity, the surface integral vanishes, and the charge is conserved. The conservation of an ordinary charge is thus dependent not only upon the boundary conditions at infinity, but also upon the dynamics of the field.

On the other hand, the conservation of the new charge (4.13) is independent of dynamics. It depends only upon the boundary conditions. For this reason it is called a *topological charge*, and the corresponding conservation law is called a *topological conservation law*.

Now, whenever we have a conserved charge, we expect it to be associated to a conserved current, i.e., a current that together with the charge density obeys the equation of continuity (1.17). In the present case, the topological charge density (4.12) is associated to the relativistic current

$$J^\mu = \epsilon^{\mu\nu}\partial_\nu\phi. \tag{4.15}$$

Notice that this current is trivially conserved:

$$\partial_\mu J^\mu = \epsilon^{\mu\nu}\partial_\mu\partial_\nu\phi = 0,$$

due to the symmetry of $\partial_\mu\partial_\nu$. I emphasize once again that the dynamics of the field have not been evoked in the above derivation of the equation of continuity. Since the current (4.15) is thus "automatically" conserved, it is called a *topological current*.

Let us investigate the topological charge (4.13) a little more closely. Notice first that it is conserved due to the requirement of finite energy, but that the structure of the potential energy density is irrelevant in this connection. Thus the topological charge (4.13) will be well-defined and conserved for *any* relativistic scalar field theory in $(1+1)$-dimensional space–time.

In an ordinary scalar field theory we will have a potential energy density $U[\phi]$ with a unique minimum at $\phi = 0$. But in that case the boundary conditions (4.7) reduce to

$$\lim_{x \to \pm\infty} \phi(x, t) = 0.$$

Thus the topological charge is completely trivial in an ordinary scalar field theory, since it automatically vanishes.

As another example we may consider a model where the potential energy density is absent; i.e., it is based upon the Lagrangian density

$$L = -\frac{1}{2}\, \partial_\mu \phi \partial^\mu \phi.$$

In that case the classical vacuum is characterized by ϕ being an arbitrary constant. Thus the topological charge can take any value.

Between these two extremes we have the nonlinear models in which we are particularly interested. In these models we have a potential energy density with a discrete set of global minima. The topological charge can therefore only take a discrete set of values. In the ϕ^4-model, e.g., the topological charge is of the form

$$Q = 0, \pm \frac{2\mu}{\sqrt{\lambda}}, \tag{4.16}$$

while in the sine-Gordon model it is given by

$$Q = \frac{2\pi}{\lambda} q, \qquad q \text{ an integer.} \tag{4.17}$$

Since the topological charges in this kind of model can take only a discrete set of values, we say that the topological charge is *quantized*. But notice that it is quantized already on a classical level! We speak about topological quantization.

Observe that the topological charge vanishes for a classical vacuum. *Nonperturbative configurations thus correspond to nontrivial values of the topological charge.*

We end this section with a remark about the sine-Gordon model. In its mechanical analogue, the classical vacuum corresponds to the configuration where all the "pendulums" are pointing downwards. In general, finite energy implies that sufficiently far away the "pendulums" are static and point downwards. When we go along the x-axis from $-\infty$ to $+\infty$, the pendulums will rotate around the x-axis, but they will necessarily rotate an *integral* number of times. Thus each finite energy configuration is characterized by a *winding number*. (See Figure 4.3.) Clearly, the winding number coincides, apart from a constant, with the topological charge in the sine-Gordon model.

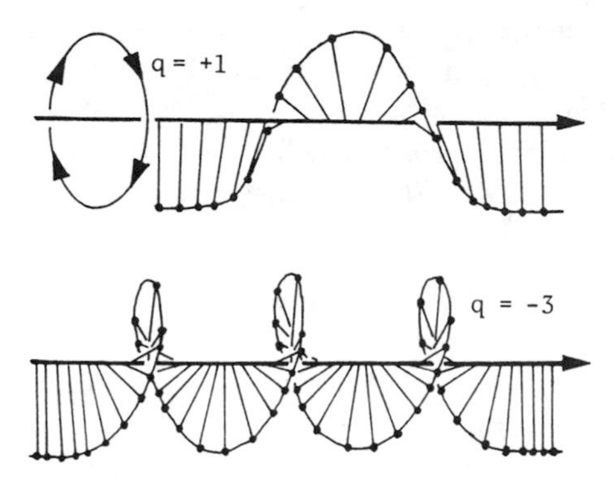

Figure 4.3.

§ 4.3 Solitary Waves

We proceed to look for wave-like solutions. In a free field theory we would naturally look for *plane wave* solutions, i.e., solutions of the form

$$\phi(x, t) = A \cos(kx - \omega t).$$

Although they have infinite energy themselves, we can construct superpositions of plane waves, which represent solutions with finite energy. (This corresponds simply to a Fourier analysis of such a solution.) In a nonlinear field theory such plane waves can never be exact solutions. Instead, we will look for solutions of the form

$$\phi(x, t) = f(x - vt). \tag{4.18}$$

Such a solution will be called a *traveling wave*. In particular, we will be interested in traveling waves with finite energy; i.e., if we introduce the variable, $\xi = x - vt$, then f must satisfy the boundary conditions

$$\lim_{\xi \to \pm\infty} f(\xi) = \phi_{\pm\infty}. \tag{4.19}$$

As f is essentially constant sufficiently far away, we say that such a wave is localized. A solution of the form (4.18) that satisfies the boundary conditions (4.19) is called a *solitary wave*.

Let us write out the equation of motion (3.30) explicitly:

$$-\frac{\partial^2 \phi}{\partial t^2} + \frac{\partial^2 \phi}{\partial x^2} = U'[\phi].$$

Inserting the equation (4.18), this reduces to the ordinary differential equation

$$(1 - v^2)\frac{d^2 f}{d\xi^2} = U'[f].$$

If we multiply this equation by $df/d\xi$ on both sides, it can be rearranged as

$$\frac{1}{2}(1-v^2)\frac{d}{d\xi}\left[\left(\frac{df}{d\xi}\right)^2\right] = U'[f]\frac{df}{d\xi} \,.$$

But that can be immediately integrated, whereby we obtain

$$\frac{1}{2}(1-v^2)\left(\frac{df}{d\xi}\right)^2 = U[f] + C.$$

The integration constant must, however, vanish, since both $\frac{df}{d\xi}$ and $U[f]$ vanish at infinity. Thus the equation of motion reduces to

$$\frac{df}{d\xi} = \pm\gamma\sqrt{2U[f]}, \qquad \text{with } \gamma = \frac{1}{\sqrt{1-v^2}} \,. \tag{4.20}$$

This first-order differential equation can be integrated, too:

$$\lambda(\xi - \xi_0) = \pm\int \frac{df}{\sqrt{2U[f]}} \,. \tag{4.21}$$

This formula gives ξ as a function of f; i.e., we have actually determined the inverse function. Notice that the boundary conditions are satisfied. Close to a zero, ϕ_0, of the potential, we can expand the potential as follows:

$$U[f] \approx \frac{1}{2}\,U''[\phi_0](f-\phi_0)^2.$$

Thereby the solution (4.21) reduces to

$$\gamma(\xi - \xi_0) \approx \pm\int \frac{df}{\sqrt{U''(\phi_0)}|f-\phi_0|} = \pm\frac{\ln|f-\phi_0|}{\sqrt{U''(\phi_0)}} \,.$$

Thus ξ goes to infinity when f approaches a classical vacuum. The asymptotic behavior of f is therefore given by

$$|f(\xi) - \phi_0| \approx \exp\left\{\pm\sqrt{U''[\phi_0]}\gamma(\xi - \xi_0)\right\} \qquad \text{as } |\xi| \to \infty. \tag{4.22}$$

Clearly, we get two types of solitary waves: The first type is *increasing* (corresponding to the plus sign in equation (4.20)). It interpolates between two neighboring classical vacua, and it is known as a *kink*. The other one is decreasing and similarly interpolates between two adjacent classical vacua. It is known as an *antikink*. (See Figure 4.4.)

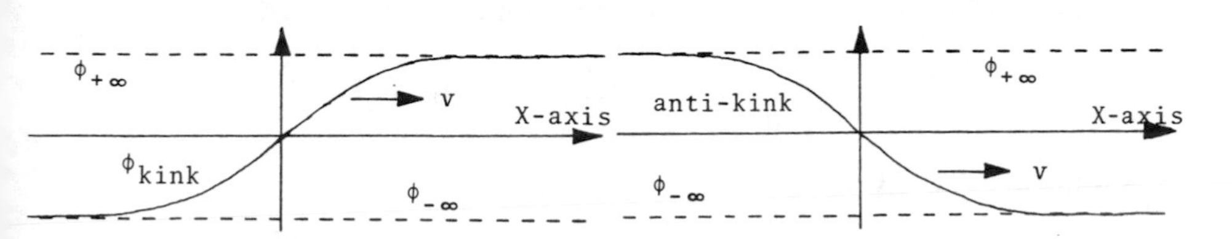

Figure 4.4.

Okay, let us look at some specific examples:
In the ϕ^4-model the kink is given by

$$\gamma(\xi - \xi_0) = \sqrt{\frac{2}{\lambda}} \int_0^f \frac{d\tilde{f}}{\frac{\mu^2}{\lambda} - \tilde{f}^2} = \frac{\sqrt{2}}{\mu} \operatorname{arctanh}\left[\frac{\sqrt{\lambda}}{\mu} f\right];$$

i.e.,

$$\phi_{\text{kink}}(x,t) = f(x - vt) = \frac{\mu}{\lambda} \tanh\left\{\frac{\mu}{\sqrt{2}} \gamma(x - vt - \xi_0)\right\}. \tag{4.23}$$

In the sine-Gordon model the kink is similarly given by

$$\begin{aligned}
\gamma(\xi - \xi_0) &= \frac{\lambda}{\sqrt{\mu}} \int_{\pi/\lambda}^f \frac{d\tilde{f}}{\sqrt{2(1 - \cos\lambda\tilde{f})}} \\
&= \frac{1}{\sqrt{\mu}} \int_{\frac{1}{2}\pi}^{\frac{1}{2}\lambda f} \frac{d\tilde{f}}{\sqrt{\frac{1}{2}(1 - \cos 2\tilde{f})}} \\
&= \frac{1}{\sqrt{\mu}} \int_{\frac{1}{2}\pi}^{\frac{1}{2}\lambda f} \frac{d\hat{f}}{\sin\hat{f}} = \ln\left(\tan\left(\frac{\lambda f}{4}\right)\right);
\end{aligned}$$

i.e.,

$$\phi_{\text{kink}}(x,t) = f(x - vt) = \frac{4}{\lambda} \arctan\left\{\exp[\mu\gamma(x - vt - \xi_0)]\right\}. \tag{4.24}$$

We can also easily determine the energy of the kink. Inserting the equation (4.18) into the energy functional (4.6), this reduces to

$$H[\phi_{\text{kink}}] = \int_{-\infty}^{+\infty} \left(\frac{1}{2}\frac{1}{\gamma^2}\left(\frac{df}{d\xi}\right)^2 + U[f]\right) d\xi.$$

Using the equation of motion (4.20), this reduces to

$$\begin{aligned}
H[\phi_{\text{kink}}] &= \int_{-\infty}^{+\infty} \left(\frac{df}{d\xi}\right)^2 d\xi \\
&= \gamma \int_{-\infty}^{+\infty} \frac{df}{d\xi} \sqrt{2U[f]}\, d\xi = \gamma \int_{\phi_{-\phi}}^{\phi_{+\infty}} \sqrt{2U[f]}\, df.
\end{aligned} \tag{4.25}$$

Thus the kink energy can be computed directly from the potential. We need not know the explicit form of the kink at all!

Once again, we look at some specific examples: In the ϕ^4-model we get

$$H[\phi_{\text{kink}}] = \gamma \int_{-\mu/\sqrt{\lambda}}^{+\mu/\sqrt{\lambda}} \sqrt{\frac{1}{2}\lambda} \left(\frac{\mu^2}{\lambda} - f^2\right) df = \gamma\, \frac{2\sqrt{2}}{3} \frac{\mu^3}{\lambda}, \tag{4.26}$$

while in the sine-Gordon model we obtain

$$H[\phi_{\text{kink}}] = \gamma \int_0^{2\pi/\lambda} \sqrt{2\frac{\mu^2}{\lambda^2}(1 - \cos\lambda f)}\, df = \gamma\, \frac{8\mu}{\lambda^2}. \tag{4.27}$$

Notice that the solitary waves constructed above interpolate between different vacua. Consequently, they belong to the nonperturbative sectors. Since they actually interpolate between *neighboring* vacua, they carry the smallest possible nontrivial topological charge. The corresponding sectors are called the *kink sectors*, respectively the *antikink sectors*.

§ 4.4 Ground States for the Nonperturbative Sectors

Now that we have gained some familiarity with solitary waves, we will rederive them from a somewhat different point of view. As we have seen, the space of all finite energy configurations E breaks up into disconnected sectors characterized by the different possibilities for the asymptotic behavior of the finite energy configurations. In each sector we will now look for a *ground state*, i.e., a configuration $\phi_0(x)$ that has the lowest possible static energy in that sector. Since the ground state minimizes the static energy

$$H_{\text{static}}[\phi] = \int \frac{1}{2}\left(\frac{d\phi}{dx}\right)^2 + U[\phi(x)]dx,$$

it must satisfy the associated Euler–Lagrange equation

$$\frac{d^2\phi_0}{dx^2} = U'[\phi_0(x)].$$

Notice that the ground state extends to a static solution $\phi(x,t) = \phi_0(x)$ of the full field equations (3.30). Actually, the following holds:

The ground state for a given sector corresponds to the solution of the field equations that has the lowest possible energy.

PROOF. To see this, let $\phi(x,t)$ be a nonstatic solution in the corresponding sector. Then there is a space–time point (x_0, t_0) such that

$$\frac{\partial\phi}{\partial t}(x_0, t_0) \neq 0.$$

Consider, then, another solution specified by the initial data

$$\psi(x, t_0) = \phi(x, t_0) \quad \text{and} \quad \frac{\partial\psi}{\partial t}(x, t_0) = 0.$$

It has the same asymptotic behavior at infinity as ϕ. Thus it belongs to the same sector. Furthermore, it has the same static energy as ϕ, but it misses the kinetic energy! Consequently, a nonstatic solution cannot minimize the energy in a given sector. □

Okay, let us look for possible ground states. In a vacuum sector the ground state is simply given by the corresponding classical vacuum. But what about the nonperturbative sectors? To look for possible ground states in these sectors, we will first examine the static solutions in our model (since a ground state is necessarily a static solution, although the converse need not be true). The field equation for a static solution reduces to

$$\frac{d^2\phi}{dx^2} = U'[\phi]. \tag{4.28}$$

This can be integrated once (using the same trick as before), whereby we get

$$\left(\frac{d\phi}{dx}\right)^2 = 2U[\phi].$$

Thus we get two first-order equations:

$$\frac{d\phi}{dx} = \sqrt{2U[\phi]}; \qquad \frac{d\phi}{dx} = -\sqrt{2U[\phi]}. \tag{4.29}$$

They have two kind of solutions:

a. On the one hand, there are *constant* solutions, corresponding to the zeros of U; i.e., the constant solutions reproduce the classical vacua.
b. On the other hand, there are *monotone* solutions, which interpolate between two adjacent vacua:

$$x - x_0 = \pm \int \frac{d\phi}{\sqrt{2U[\phi]}}. \tag{4.30}$$

(See Figure 4.5.) They correspond precisely to the static kink and the static antikink.

This suggests that the kink is the ground state in the kink sector!

Notice that there are *no* static solutions that pass through a classical vacuum. If the model possesses more than two classical vacua, there will consequently be sectors (consisting of configurations that interpolate between nonneighboring vacua) that have *no* ground state!

Observe that if $\phi(x,t)$ is any solution to the field equations, then so is the boosted configuration

$$\tilde{\phi}(x,t) = \phi[\gamma(x - vt); \gamma(t + vx)] \tag{4.31}$$

due to the Lorentz invariance. If we apply this to the static solution $\phi(x)$, we thus produce a solitary wave:

$$\phi(x,t) = \phi[\gamma(x - vt)].$$

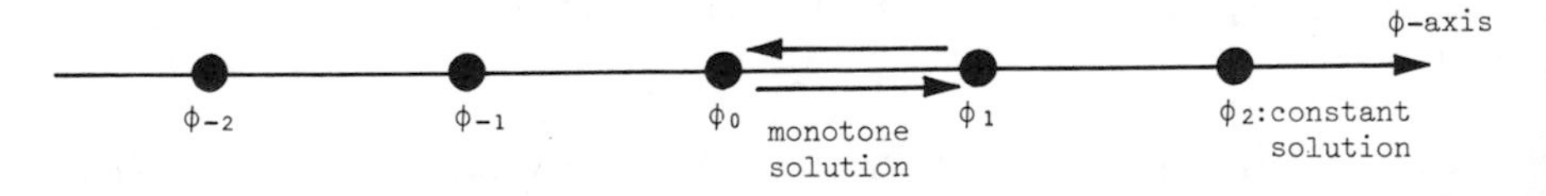

Figure 4.5.

We can therefore easily get back our solitary waves once we have determined the static solutions!

Next we want to tackle the problem of whether the static kink really *is* a ground state configuration for the kink sector. This will be shown using a beautiful trick going back to Bogomolny. He showed that in many interesting models the static energy can be decomposed as follows:

$$H_{\text{static}}[\phi] = \int_{-\infty}^{\infty}\left[P\left(\phi;\ \frac{d\phi}{dx}\right)\right]^2 dx + \{\text{topological term}\}. \tag{4.32}$$

Here $P\left(\phi,\ \frac{d\phi}{dx}\right)$ is a first-order differential operator acting on ϕ, while the topological term depends only upon the asymptotic behavior of the field at infinity; i.e., it is constant throughout each sector. Provided that the topological term is positive, we therefore get the following bound for the static energy:

$$H_{\text{static}} \geq \{\text{topological term}\}. \tag{4.33}$$

Furthermore, this bound is saturated only provided that ϕ solves the first-order differential equation

$$P\left(\phi;\ \frac{d\phi}{dx}\right) = 0. \tag{4.34}$$

Any solution to this first-order differential equation is thus a ground state. A decomposition of the type (4.32) is known as a *Bogomolny* decomposition. The associated first-order differential equation (4.34) is known as the *ground state equation*.

Let us try to apply this to our (1 + 1)-dimensional models. Guided by the kink equations (4.29), we try the following decomposition:

$$H_{\text{static}} = \frac{1}{2}\int_{-\infty}^{\infty}\left\{\frac{d\phi}{dx} \mp \sqrt{2U[\phi]}\right\}^2 dx + \{\text{remainder term}\}.$$

The remainder term is given by

$$\pm\int_{\infty}^{\infty}\frac{d\phi}{dx}\sqrt{2U[\phi]}\,dx = \pm\int_{\phi_{-\infty}}^{\phi_{+\infty}}\sqrt{2U[\phi]}\,d\phi.$$

Thus it *is* a topological term dependent only upon the asymptotic behavior at infinity! Let us specifically look at a model with a finite or infinite number of classical vacua,

$$\ldots, \phi_{-2}, \phi_{-1}, \phi_0, \phi_1, \phi_2, \ldots,$$

where the different classical vacua are related by a symmetry transformation. In the sector E_n consisting of all configurations that interpolate between ϕ_0 and ϕ_n, we can now apparently rearrange the above decomposition as follows:

$$H_{\text{static}}[\phi] = \frac{1}{2}\int_{\infty}^{\infty}\left\{\frac{d\phi}{dx} \mp \sqrt{2U[\phi]}\right\}^2 dx + n\int_{\phi_0}^{\phi_1}\sqrt{2U[\phi]}\,d\phi. \tag{4.35}$$

Thus we have shown that

a. *The energy in the sector E_n is bounded below by*

$$H_{\text{static}} \geq |n| \int_{\phi_0}^{\phi_1} \sqrt{2U[\phi]}\, d\phi. \tag{4.36}$$

b. *A configuration in the sector E_n is a ground state if and only if it satisfies the first-order differential equation* (4.29)

$$\frac{d\phi}{dx} = \pm\sqrt{2U[\phi]} \qquad \begin{cases} + : n \text{ positive}, \\ - : n \text{ negative}. \end{cases}$$

This shows especially that the kink *is* the ground state for the kink sector. It also shows that the energy of the static kink is given by

$$H[\phi_{\text{kink}}] = \int_{\phi_0}^{\phi_1} \sqrt{2U[\phi]}\, d\phi, \tag{4.37}$$

which of course is in agreement with our earlier result (4.25).

The Bogomolny decomposition can also be used to examine the higher-charged sectors. Consider as a specific example the sine-Gordon model. We know in advance that there are no exact ground states in the higher-charged sectors. We know also that the static energy is bounded below by

$$H_{\text{static}}[\phi] \geq |n| H[\phi_{\text{kink}}].$$

Now let $x_1, x_2, \ldots, x_n$ be widely separated consecutive points on the x-axis and furthermore, let $\phi_+(x)$ denote the static kink solution centered at $x = 0$ (i.e., we put $v = \xi_0 = 0$ in (4.24)). It satisfies the boundary conditions

$$\lim_{x \to -\infty} \phi_+(x) = 0; \qquad \lim_{x \to +\infty} \phi_+(x) = \frac{2\pi}{\lambda}.$$

Consider now the superposition

$$\tilde{\phi}_{[n:x_1,\ldots,x_n]}(x) = \phi_+(x - x_1) + \phi_+(x - x_2) + \cdots + \phi_+(x - x_n).$$

It clearly satisfies the boundary conditions

$$\lim_{x \to -\infty} \tilde{\phi}(x) = 0; \qquad \lim_{x \to +\infty} \tilde{\phi}(x) = n\,\frac{2\pi}{\lambda},$$

so that it belongs to the sector E_n. (See Figure 4.6.) Notice furthermore that the kink solution is strongly localized, since it approaches the classical vacuum exponentially; cf. (4.22). Outside a small region of width $1/\mu$, the kink solution essentially reduces to the classical vacuum. This has the following consequence for the above superposition:

a. Except when we are close to the centers $x_1, x_2, \ldots, x_n$, the configuration $\tilde{\phi}(x)$ is exponentially close to a classical vacuum.
b. Close to a center, x_k, it behaves like a kink solution. In particular, the difference

$$\frac{d\tilde{\phi}}{dx} - \sqrt{2U[\tilde{\phi}]}$$

is exponentially small.

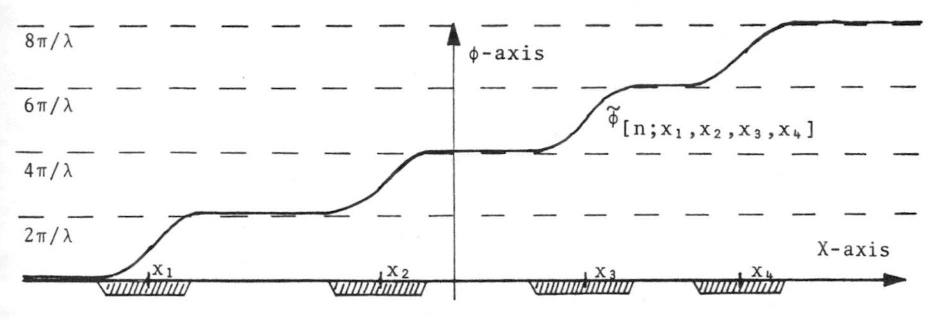

Figure 4.6.

As a consequence, the above configuration has a static energy given by

$$H_{\text{static}}[\tilde{\phi}] = nH[\phi_{\text{kink}}] + \{\text{exponentially small error}\}. \tag{4.38}$$

It is thus extremely close to the lower bound given by the Bogomolny decomposition. Furthermore, we clearly have

$$H_{\text{static}}[\tilde{\phi}] \to nH[\phi_{\text{kink}}] \qquad \text{as } (x_{k+1} - x_k) \to \infty, \qquad k = 1, \ldots, n-1.$$

Thus we can find *approximate ground states* consisting of widely separated kinks.

Apart from a configuration consisting of widely separated kinks, we may also consider strings of widely separated kinks *and* antikinks. Let us this time look at the ϕ^4-model, and let us even specialize to one of the vacuum sectors, say E_{--}. As before, $\phi_+(x)$ denotes the static kink centered at $x = 0$, while $\phi_-(x)$ denotes the static antikink centered at $x = 0$. We now consider the following configuration E_{--}:

$$\tilde{\phi}(x) = \phi_+(x - x_1) + \phi_-(x - x_2) + \cdots + \phi_+(x - x_{2n-1}) + \phi_-(x - x_{2n}).$$

Notice that this time we must let the kinks and antikinks alternate, and furthermore, since we are in a vacuum sector, there must be an even number of centers. (See Figure 4.7.) As before, we see that

$$\left(\frac{d\tilde{\phi}}{dx}\right)^2 = 2U[\tilde{\phi}] + \{\text{exponentially small error}\}.$$

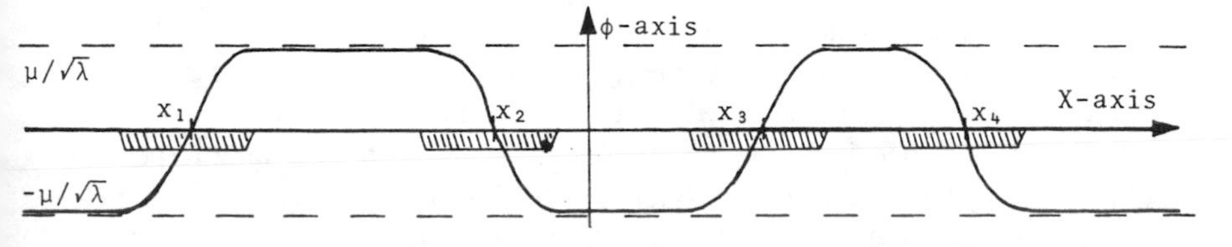

Figure 4.7.

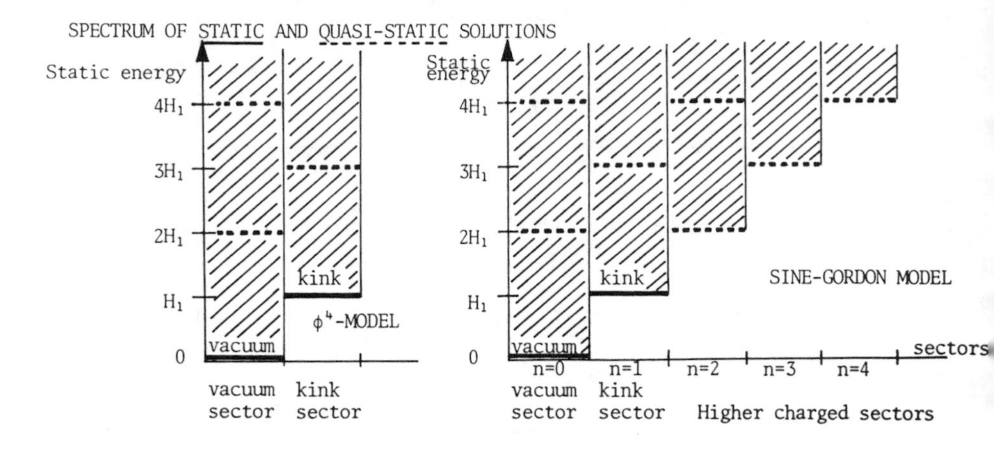

Figure 4.8.

When the kinks and antikinks are widely separated, the configuration $\tilde{\phi}(x)$ is thus almost a static solution. Since the static energy is thus not truly stationary under deformations, we call such a configuration a *quasi-stationary* configuration. Although a quasi-stationary configuration need not be an approximative ground state, it will almost be a local minimum for the static energy. This is clear, since any deformation of the quasi-stationary configuration (except for a pure displacement of the centers) will either destroy one of the classical vacua involved or it will destroy one of the kinks/antikinks involved. Thus the static energy will increase (or at least stay constant). (See Figure 4.8.)

§ 4.5 Solitons

Observe that asymptotically, the kink solution approach the Yukawa potential (3.34). But it is a solution of quite a different type! The Yukawa potential (3.34) is a singular solution to the Klein–Gordon equation, and (just like the Coulomb field) it signals the presence of an external point source, i.e., a foreign particle. But the kink is a smooth solution everywhere, and consequently, it is *not* associated with any external sources of the sine-Gordon field. Notice, furthermore, that they are localized solutions, where the energy is concentrated within a small region; i.e., they represent a small extended object. In accordance with this, kink solutions in nonlinear field theories are generally interpreted as particles—an interpolation that the case of the sine-Gordon model goes back to Perring and Skyrme (1962). Since such particle are associated with solitary wave solutions, they are referred to as *solitons*. We will now look at some of the properties of the kink solutions that justify this interpretation:

The kink solution $\phi_+(x - x_0)$ represents a soliton at rest centered at $x = x_0$. In the same way the boosted solution (4.31) represents a soliton moving with the velocity v!

We can also consider the energy and momentum of a single soliton. It is given by the total energy and momentum contained in the field, i.e., by

$$P^\mu = \int_{-\infty}^{+\infty} T^{\mu 0} dx,$$

where the energy–momentum tensor is given by

$$T^{\mu\nu} = \partial^\mu \phi \partial^\nu \phi + \eta^{\mu\nu} \left[\frac{1}{2} \partial_\alpha \phi \partial^\alpha \phi + U(\phi) \right].$$

We know that P^μ transforms as a Lorentz vector, and in this example it is furthermore reasonable to localize it, since the energy is concentrated in a very small region! (See Figure 4.9.) Thus we can attach the Lorentz vector to the center of the soliton. For a soliton at rest, the energy E is equal to the rest mass M

$$M = \int_{-\infty}^{+\infty} T^{00} dx = \int_{-\infty}^{+\infty} \left\{ \frac{1}{2} \left(\frac{\partial \phi}{\partial x} \right)^2 + U[x] \right\} dx,$$

and the momentum $\mathbf{p}$ vanishes because the solution is static. Due to the Lorentz covariance of P^μ, we therefore see that a soliton moving with velocity v has energy and momentum

$$P = M\gamma v; \qquad E = M\gamma \qquad \text{(in units where } c = 1). \tag{4.39}$$

The kink solution therefore represents a free particle with rest mass M, while the antikink solution represents the antiparticle corresponding to the soliton.

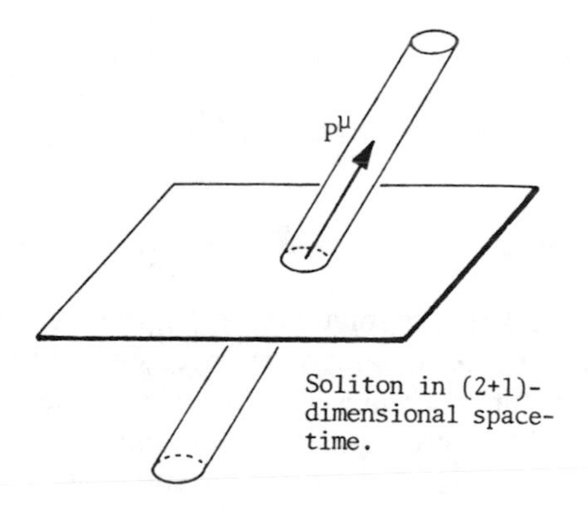

Figure 4.9.

We now specialize to the sine-Gordon model. In the higher-charged sectors, where the winding number is numerically greater than one, it is tempting to see whether we can find solutions that represent several solitons. There are no static solutions. When several solitons are present, they consequently no longer behave like free particles. This means that two solitons exert forces on each other. This phenomenon is closely associated to the nonlinearity of the sine-Gordon equation: The superposition of two kink solutions is no longer a solution. We have seen, however, that if we choose the superposition carefully, then it is "almost" a solution.

Worked Exercise 4.5.1
Problem: (a) Consider the *strict* superposition

$$\phi(x,t) = \phi_+[\gamma(x+vt+x_0)] + \phi_+[\gamma(x-vt-x_0)] - \frac{2\pi}{\gamma}\,.$$

Show that it satisfies

$$\tan\left[\frac{\lambda\phi}{4}\right] = \frac{\sinh(\mu\gamma x)}{\cosh[\mu\gamma(vt+x_0)]}\,.$$

(b) Show that it has the asymptotic expansion

$$\tan\left[\frac{\lambda\phi}{4}\right] \underset{t\to\infty}{\sim} e^{-\mu\gamma x_0}\frac{\sinh[\mu\gamma x]}{\cosh[\mu\gamma vt]}\,. \tag{4.40}$$

Based on computer simulations, Perring and Skyrme guessed the following analytic expression for an *exact* time-dependent solution (cf. the asymptotic expansion (4.40)!):

$$\tan\left[\frac{\lambda\phi_{ss}}{4}\right] = v\,\frac{\sinh[\mu\gamma x]}{\cosh[\mu\gamma vt]}\,. \tag{4.41}$$

Remark. It is definitely not "trivial" to verify that this is in fact a solution. In the end of this section we shall present a general method to solve the equations of motion whereby we can derive (4.41) "relatively" easy.

Exercise 4.5.2
Problem: Show that the above solution (4.41) can be expanded asymptotically as

$$\phi_{ss}(t,x) \underset{t\to-\infty}{\sim} \phi_+[\gamma(x+vt+x_0)] + \phi_+[\gamma(x-vt-x_0)] - \frac{2\pi}{\lambda}\,; \tag{4.22a}$$

$$\phi_{ss}(t,x) \underset{t\to+\infty}{\sim} \phi_+[\gamma(x+vt-x_0)] + \phi_+[\gamma(x-vt+x_0)] - \frac{2\pi}{\lambda}\,. \tag{4.22b}$$

According to Exercise 4.5.2, the solution (4.41) has a very simple interpretation: In the remote past it represents two widely separated solitons moving towards each other, and in the remote future it represents two widely separated solitons moving away from each other. Thus, it describes a *scattering process* between solitons! (See Figure 4.10.) Observe that the two solitons are identical, so it is impossible to say whether the two solitons move through each other or they are scattered elastically: All we can say is that there are two particles going into the collision

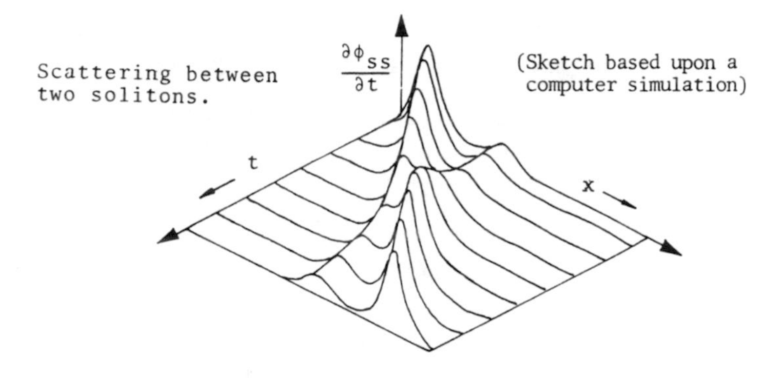

Figure 4.10.

center and two particles going out. (See Figure 4.11.) Each of these particles carries an energy and momentum P_1^μ and P_2^μ that can be calculated by integrating the energy–momentum tensor over the central region occupied by the particle (outside this region, the energy–momentum tensors die off exponentially, and we can safely neglect it). If we look back on Abraham's theorem (Section 1.6), it is now clear that we have decomposed the total energy momentum as

$$P^\mu = P_1^\mu + P_2^\mu,$$

where P_1^μ, P_2^μ themselves transform like Lorentz vectors. (Of course this decomposition works only in the remote past and the remote future, where the solitons are widely separated. During the collision it has no meaning to speak of individual solitons!) But the total energy–momentum stored in the field is conserved. Thus

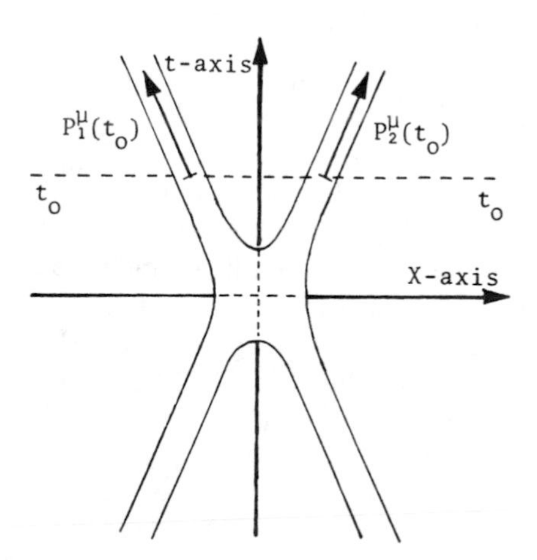

Figure 4.11.

we get

$$P_1^\mu(-\infty) + P_2^\mu(-\infty) = P_1^\mu(+\infty) + P_2^\mu(+\infty); \tag{4.43}$$

i.e., the total energy and momentum of the solitons are conserved during the scattering. We can also compute the energy of the 2-soliton solution: This is most easily done using the asymptotic expressions (4.42). Since they consist of two widely separated solitons moving with velocity v, we simply get

$$M[\phi] = 2M\gamma = \frac{2M}{\sqrt{1-v^2}}, \tag{4.44}$$

where M is the rest mass of a single soliton.

Perring and Skyrme found two other interesting analytic solutions. They can be obtained from (4.41) using various "analytic continuations." First, we apply the "symmetry-transformation" $(x, t) \curvearrowright (it, ix)$ and obtain the singular complex-valued solution

$$\tan\left[\frac{\lambda\phi(t,x)}{4}\right] = v\frac{\sinh\left[\mu\frac{i}{\sqrt{1-v^2}}t\right]}{\cosh\left[\mu\frac{v}{\sqrt{1-v^2}}x\right]}\left(= iv\frac{\sin\left[\mu\frac{1}{\sqrt{1-v^2}}t\right]}{\cos\left[\mu\frac{v}{\sqrt{1-v^2}}x\right]}\right).$$

Then we perform the substitution $v \curvearrowright \frac{1}{v}$ whereby we obtain the regular solution

$$\tan\left[\frac{\lambda\phi_{s\bar{s}}(t,x)}{4}\right] = \frac{1}{v}\frac{\sinh[\mu\gamma vt]}{\cosh[\mu\gamma x]}. \tag{4.45}$$

Exercise 4.5.3
Problem: (a) Show that (4.45) represents a scattering process between a soliton and an antisoliton by investigating the asymptotic form of the solution as $t \to -\infty$ and $t \to +\infty$.

(b) Show that its energy is the same as the 2-soliton solution

$$H[\phi_{s\bar{s}}] = 2M\gamma. \tag{4.46}$$

Performing the "analytic continuation" $v \to iv$, (4.45) can furthermore be transformed into the periodic solution

$$\tan\left[\frac{\lambda\phi_b}{4}\right] = \frac{1}{v}\frac{\sin[\mu\Gamma vt]}{\cosh[\mu\Gamma x]} \qquad \text{with } \Gamma = \frac{1}{\sqrt{1+v^2}}. \tag{4.47}$$

This solution is strictly localized at all times within the region $|x| < (\mu\Gamma)^{-1}$, since it falls off exponentially due to the denominator. It has winding number 0, so it belongs to the vacuum sector. It represents clearly a periodic oscillating configuration, called a *breather*. (See Figure 4.12.) The energy of the breather can also be obtained by "analytic continuation":

$$H[\phi_b] = 2M\Gamma = \frac{2M}{\sqrt{1+v^2}}. \tag{4.48}$$

Observe that it is less than the total energy of a widely separated soliton–antisoliton pair. The breather is interpreted as a *bound state* of a soliton and an antisoliton.

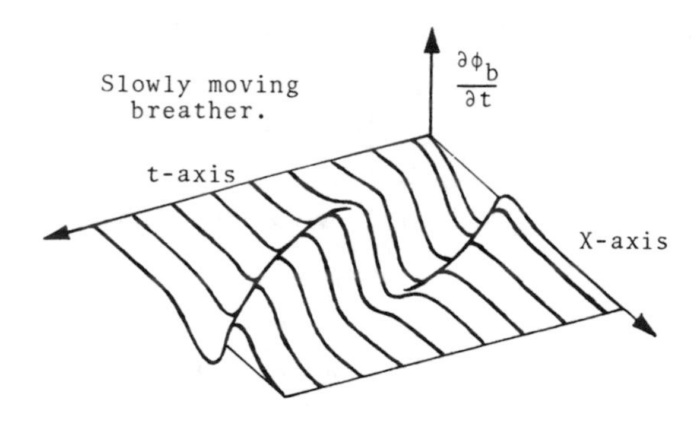

Figure 4.12.

For this reason the particle it represents is called a *bion*. (Bion is an abbreviation for bi-soliton bound state.)

When investigating the breather it is in fact more natural to introduce the cyclic frequency ω and the "amplitude" η:

$$\omega = \mu \Gamma v; \qquad \eta = \frac{1}{v}. \tag{4.49}$$

In terms of these variables the breather takes the form

$$\tan\left[\frac{\lambda \phi_b}{4}\right] = \eta \frac{\sin(\omega t)}{\cosh(\eta \omega x)} \qquad \text{with } \eta = \frac{\sqrt{\mu^2 - \omega^2}}{\omega}. \tag{4.50}$$

Notice that when $t = n\,\frac{\pi}{\lambda}$ (n an integer), the breather momentarily degenerates to a classical vacuum. At these times the energy of the breather is thus purely kinetic.

Exercise 4.5.4
Problem: Show, by explicit calculation, that the energy of the breather at time $t = 0$ is given by

$$E = 2M\left(1 - \frac{\omega^2}{\mu^2}\right)^{1/2}, \tag{4.51}$$

where M is the soliton mass; cf. the earlier obtained result (4.48).

Now suppose that

$$\omega \ll \mu.$$

Let us furthermore assume that $\sin[\omega t]$ is positive, say $0 < t < \frac{\pi}{\lambda}$. We then get the following asymptotic expansions:

$$\tan\left[\frac{\lambda \phi_b}{4}\right] \approx \begin{cases} \exp\left\{\mu\left(x + \mu^{-1}\ln\left[\frac{2\mu \sin \omega t}{\omega}\right]\right)\right\} & \text{when } x \ll 0, \\ \frac{\mu \sin \omega t}{2\omega} & \text{when } x \approx 0, \\ \exp\left\{\mu\left(x - \mu^{-1}\ln\left[\frac{2\mu \sin \omega t}{\omega}\right]\right)\right\} & \text{when } x \gg 0. \end{cases}$$

On the other hand, we get from (4.24) that a static kink, respectively antikink, is given by

$$\tan\left[\frac{\lambda\phi_+}{4}\right] = \exp[\pm\mu x]. \tag{4.52}$$

When t is not too close to 0 or $\frac{\pi}{\lambda}$, we know by assumption that

$$\frac{\mu \sin \omega t}{\omega}$$

is a large positive number. Consequently, we can interpret the asymptotic behavior as follows: We have

a. a soliton to the far left with the center

$$x_0(t) = -\mu^{-1} \ln\left[\frac{2\mu \sin \omega t}{\omega}\right],$$

b. an antisoliton to the far right with the center

$$x_0(t) = -\mu^{-1} \ln\left[\frac{2\mu \sin \omega t}{\omega}\right],$$

c. a classical vacuum $\left(\phi \approx \frac{2\pi}{\lambda}\right)$ in the central region between the soliton and the antisoliton.

This thus confirms our interpretation of the breather as a bound state of a soliton and an antisoliton.

The breather belongs to the vacuum sector. Thus it is *not* a nonperturbative configuration. Notice the following two extreme limits of the breather solution:

(a) When $\omega \to 0$, the mass of the breather approaches twice the soliton mass, and we further get

$$\tan\left[\frac{\lambda\phi_b}{4}\right] \to \frac{\mu t}{\cosh[\mu x]} \qquad \text{as } \omega \to 0.$$

But this is the same as the limit of the soliton–antisoliton solution (4.45) when we let $v \to 0$. In this limit, the soliton–antisoliton pair thus becomes unbound.

(b) When $\omega \to \mu$, the mass approaches 0, while $\phi_b \to 0$. In this limit the breather is thus just a small perturbation of the vacuum.

§ 4.6 The Bäcklund Transformation

The existence of exact multisoliton solutions is a peculiar property of the sine-Gordon model; e.g., there are no exact multisoliton solutions in the ϕ^4-model.

Interestingly enough, it turns out that there exists a systematic method for solving the sine-Gordon equation known as the inverse spectral transformation. Due to its complexity, we shall, however, confine ourselves to a discussion of the *Bäcklund transformation*, which is a related powerful technique that allows a systematic computation of all the multisoliton solutions.

To simplify the analysis, we introduce the so-called *light-cone coordinates*

$$x^+ = \frac{1}{2}(x+t) \qquad x^- = \frac{1}{2}(x-t). \tag{4.53}$$

The derivatives with respect to the light-cone coordinates are given by

$$\partial_+ = \frac{\partial}{\partial x} + \frac{\partial}{\partial t}; \qquad \partial_- = \frac{\partial}{\partial x} - \frac{\partial}{\partial t}.$$

In light-cone coordinates the sine-Gordon equation thus reduces to

$$\partial_+\partial_-\phi = \frac{\mu^2}{\lambda}\sin[\lambda\phi]. \tag{4.54}$$

For simplicity we put $\lambda = \mu = 1$ in the following. The crucial idea is to reduce the solution of this second-order differential equation to a pair of first-order differential equations; cf. the philosophy behind the Bogomolny decomposition. In the last century Bäcklund discovered the following remarkable pair of first-order differential equations:

$$\begin{aligned} \partial_+\left[\frac{\phi_1-\phi_0}{2}\right] &= a\sin\left[\frac{\phi_1+\phi_0}{2}\right]; \\ \partial_-\left[\frac{\phi_1+\phi_0}{2}\right] &= \frac{1}{a}\sin\left[\frac{\phi_1-\phi_0}{2}\right]. \end{aligned} \tag{4.55}$$

Differentiating the first Bäcklund equation with respect to x^-, we get

$$\partial_-\partial_+\left[\frac{\phi_1-\phi_0}{2}\right] = a\cos\left[\frac{\phi_1+\phi_0}{2}\right]\partial_-\left[\frac{\phi_1+\phi_0}{2}\right].$$

Using the second Bäcklund transformation, the right-hand side can be further reduced to

$$\partial_-\partial_+\left[\frac{\phi_1-\phi_0}{2}\right] = \cos\left[\frac{\phi_1+\phi_0}{2}\right]\sin\left[\frac{\phi_1-\phi_0}{2}\right]. \tag{4.56}$$

Similarly, we get from the second Bäcklund equation

$$\partial_+\partial_-\left[\frac{\phi_1+\phi_0}{2}\right] = \cos\left[\frac{\phi_1-\phi_0}{2}\right]\sin\left[\frac{\phi_1+\phi_0}{2}\right]. \tag{4.57}$$

Consequently, we get by adding and subtracting (4.56) and (4.57)

$$\partial_+\partial_-\phi_1 = \sin[\phi_1]; \qquad \partial_+\partial_-\phi_0 = \sin[\phi_0]. \tag{4.58}$$

We have therefore shown that *any two functions* ϕ_0, ϕ_1 *that satisfy the Bäcklund equations* (4.55) *must necessarily solve the sine-Gordon equation too*! In other words, the sine-Gordon equation is the *integrability condition* for the Bäcklund equations.

We can now reformulate the above observation in the following way: Suppose we have been given a solution ϕ_0 to the sine-Gordon equation. Then we can insert it into the Bäcklund equations and solve them with respect to ϕ_1. Since the new solution depends upon the old solution ϕ_0, we have in this way constructed an operator B_a that transforms a given solution of the sine-Gordon equation into another solution. The operator B_a is known as the *Bäcklund transformation* (with the scale parameter a).

As an example we apply the Bäcklund transformation to the classical vacuum. The Bäcklund transformed vacuum ϕ solves the first-order differential equation

$$\partial_+\phi = 2a \sin\frac{\phi}{2}\,; \qquad \partial_-\phi = \frac{2}{a}\sin\phi_2\,.$$

Introducing the new variables

$$\xi = ax^+ + \frac{1}{a}x^-\,; \qquad \eta = ax^+ - \frac{1}{a}x^-$$

they reduce to

$$\partial_\xi\phi = 2\sin\frac{\phi}{2}\,; \qquad \partial_\eta\phi = 0$$

with the obvious solution

$$\tan\left[\frac{\phi}{4}\right] = \exp[\xi - \xi_0] = \exp[\gamma(x - vt - \xi_0)] \quad \text{with } v = \left(\frac{1}{a} - a\right) \Big/ \left(\frac{1}{a} + a\right). \tag{4.59}$$

But that is precisely the kink solution (4.24)! By applying the Bäcklund transformation to the classical vacuum, we therefore create a single soliton.

In principle, we can now obtain the two-solution (4.41) by Bäcklund transforming the one-soliton solution, etc. Remarkably enough, it turns out, however, that further integration of the Bäcklund equations can be reduced to pure algebra. This important observation is due to Bianchi, who showed that successive Bäcklund transformations commute and that furthermore, the following nonlinear superposition principle holds for ϕ_0, $\phi_1 = B_{a_1}[\phi_0]$, $\phi_2 = B_{a_2}[\phi_0]$, and $\phi_3 = B_{a_2}[\phi_1] = B_{a_1}[\phi_2]$:

$$\tan\left[\frac{\phi_3 - \phi_0}{4}\right] = \frac{a_2 + a_1}{a_2 - a_1}\tan\left[\frac{\phi_2 - \phi_1}{4}\right]. \tag{4.60}$$

(See Figure 4.13.) (This is known as *Bianchi's permutability theorem*.) We leave the details of the proof as an exercise:

Worked Exercise 4.6.1
Introduction: Consider the following Bäcklund transforms of a given solution ϕ_0: $\phi_1 = B_{a_1}[\phi_0]$; $\phi_2 = B_{a_2}[\phi_0]$; $\phi_3' = B_{a_2}[\phi_1]$; $\phi_3'' = B_{a_1}[\phi_2]$. We are going to show that the integration constants leading from ϕ_1 to ϕ_3', respectively from ϕ_2 to ϕ_3'', can be chosen in such a way that ϕ_3' coincides with ϕ_3''. (See Figure 4.14.)

Problem: (a) Assume for a moment that the Bäcklund transformations B_{a_1} and B_{a_2} commute and denote the common value of ϕ_3' and ϕ_3'' by ϕ_3. Show that ϕ_3 has to

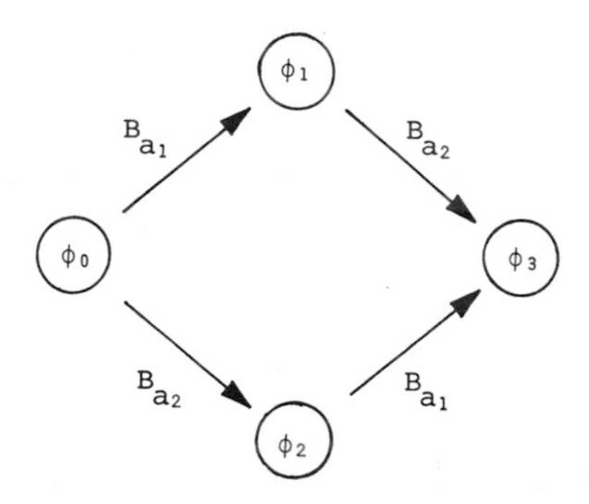

Figure 4.13.

satisfy the consistency relation

$$a_1 \sin A_+ = a_2 \sin A_- \quad \text{with} \quad A_\pm = \frac{1}{4}\,[(\phi_3 - \phi_0) \pm (\phi_2 - \phi_1)]. \tag{4.61}$$

(b) Return to the general situation where the Bäcklund transformations need not a priori commute. From ϕ_0, ϕ_1, and ϕ_2 we now construct a new function ϕ_3 by imposing the condition (4.61). Show that ϕ_3 *is* the Bäcklund transform of ϕ_1 with scale parameter a_2; i.e., show that ϕ_3 satisfies the Bäcklund equations

$$\partial_+ \left(\frac{\phi_3 - \phi_1}{2} \right) = a_2 \sin \left(\frac{\phi_3 + \phi_1}{2} \right);$$

$$\partial_- \left(\frac{\phi_3 + \phi_1}{2} \right) = \frac{1}{a_2} \sin \left(\frac{\phi_3 - \phi_1}{2} \right).$$

(Similarly, it follows that ϕ_3 is the Bäcklund transform of ϕ_2 with the scale parameter a_1. This shows precisely that the integration constants can be chosen such that ϕ_3' and ϕ_3'' coincide with the function ϕ_3 defined by (4.61).)

(c) Show that the consistency relation (4.61) is equivalent to the Bianchi identity (4.60).

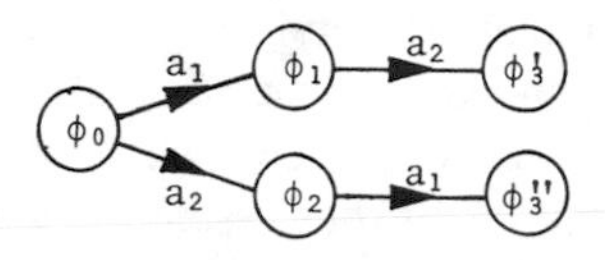

Figure 4.14.

In the Bianchi identity (4.60) we now substitute for ϕ_0 the classical vacuum and for ϕ_1 and ϕ_2 two kink solutions with different velocities. We then obtain

$$\begin{aligned}\tan\left[\frac{\phi}{4}\right] &= \frac{a_2+a_1}{a_2-a_1}\frac{\tan\left[\frac{\phi_2}{4}\right]-\tan\left[\frac{\phi_1}{4}\right]}{1+\tan\left[\frac{\phi_2}{4}\right]\tan\left[\frac{\phi_1}{4}\right]}\\ &= \frac{a_2+a_1}{a_2-a_1}\frac{\exp\left[\xi_2-\xi_0''\right]-\exp\left[\xi_1-\xi_0'\right]}{1+\exp\left[\xi_2+\xi_1-\xi_0''-\xi_0'\right]}\\ &= \frac{a_2+a_1}{a_2-a_1}\frac{\sinh\left[\frac{1}{2}(\xi_2-\xi_1-\xi_0''+\xi_0')\right]}{\cosh\left[\frac{1}{2}(\xi_2+\xi_1-\xi_0''-\xi_0')\right]}.\end{aligned} \tag{4.62}$$

For simplicity, we further put $a = a_1 = -\frac{1}{a_2}$ (i.e., $v_2 = -v_1$) and $\xi_0' = \xi_0'' = 0$. Then (4.62) precisely reduces to (4.41). This constitutes the promised derivation of the two-soliton solution.

Remark. If one has patience enough, one can work out 3-soliton solutions, 4-soliton solutions, etc. In fact, it is possible to write out the general N-soliton/antisoliton solution explicitly. It is given by the following simple expression:

$$\cos\phi = 1 - 2\partial_\mu\partial^\mu \ln\{\det[\overset{=}{M}]\}; \qquad M_{ij} = \frac{2}{a_i+a_j}\cosh\left[\frac{\theta_i+\theta_j}{2}\right]. \tag{4.63}$$

For soliton/antisoliton solutions we put

$$a_i^2 = \frac{1-v_i}{1+v_i}; \qquad \theta_i = \pm\gamma_i(x - v_i t - \xi_i); \qquad |v_i| < 1; \qquad v_i \neq v_j.$$

Bions are included by allowing the a_i's to be complex. In that case, they must occur in pairs of Hermitian conjugate numbers:

$$a_{i+1} = \bar{a}_i \qquad (\xi_{i+1} = \bar{\xi}_i).$$

The formula for θ_i is then modified to

$$\theta_i = \frac{1}{2}\left[\left(a_i + \frac{1}{a_i}\right)x + \left(a_i - \frac{1}{a_i}\right)t - \xi_i\right].$$

§ 4.7 Dynamical Stability of Solitons

Let us now turn to the question of the stability of solitons. Actually, this has already been solved, since we have shown that the static kink is the ground state for the kink sector, and as such it is clearly stable. However, the argument was based upon the Bogomolny decomposition. It is thus essentially a topological argument; i.e., we have shown that the soliton is *topologically stable*. It is not always possible to give topological arguments for the stability of a static solution. It will therefore be useful to rederive the stability of the soliton from a conventional dynamical point of view. By analyzing the dynamical stability we will furthermore be able to extract useful information about the particle spectrum in the model.

Okay, we want to investigate the stability of a static solution $\phi_0(x)$. This is done by studying small fluctuations around the configuration

$$\phi(x, t) \approx \phi_0(x) + \eta(x, t), \qquad |\eta(x, t)| \ll 1.$$

Inserting this into the equation of motion, we can expand the potential energy density to the lowest order:

$$U'[\phi(x, t)] \approx U'[\phi_0(x)] + U''[\phi_0(x)]\eta(x, t).$$

Thus the *linearized equation of fluctuations* becomes:

$$\partial_\mu \partial^\mu \eta = U''[\phi_0(x)]\eta(x, t). \tag{4.64}$$

This is not an ordinary Klein-Gordon equation because $U''[\phi_0(x)]$ is a function of x. Since it is not explicitly time-dependent, we can, however, decompose η into normal modes:

$$\eta(x, t) = \sum_k \exp[-i\omega_k t]\phi_k(x). \tag{4.65}$$

Inserting this into (4.64) we get

$$\sum_k \left(\omega_k^2 + \frac{\partial^2}{\partial x^2}\right) \phi_k(x) \exp[-i\omega_k t] = \sum_k U''[\phi_0(x)]\phi_k(x) \exp[i\omega_k t];$$

i.e., each mode is a solution to the eigenvalue problem:

$$\left\{-\frac{\partial^2}{\partial x^2} + U''[\phi_0(x)]\right\} \phi_k(x) = \omega_k^2 \phi_k(x). \tag{4.66}$$

Notice that this is just an ordinary Schrödinger equation for a "particle moving in the potential $W(x) = U''[\phi_0(x)]$." It can be solved (in principle), and we should especially look for the eigenvalues. Observe, in particular, that if there is a *negative* eigenvalue, then the corresponding cyclic frequency ω_k must be purely imaginary. Thus the "oscillation" $\exp[i\omega_k t]$ in fact grows exponentially in the future (or in the past). Clearly, this contradicts the stability, according to which a small fluctuation must stay small. Thus we have deduced the following criterion for dynamical stability:

A static solution $\phi_0(x)$ is dynamically stable, provided that the eigenvalue problem

$$\left\{-\frac{\partial^2}{\partial x^2} + U''[\phi_0(x)]\right\} \phi_k(x) = \omega_k^2 \phi_k(x) \tag{4.66}$$

has no negative eigenvalues.

We can also derive this criterion by looking at the static energy functional. A stable static solution must at least be a local minimum of the static energy functional. Consider now a small perturbation

$$\phi(x) = \phi_0(x) + \epsilon\psi(x).$$

Expanding the potential energy density to second order, we get

$$H[\phi] = H[\phi_0] + \epsilon \int \left[\frac{d\phi_0}{dx}\frac{d\psi}{dx} + U'[\phi_0]\psi \right] dx + \frac{1}{2}\epsilon^2 \int \left[\left(\frac{d\psi}{dx}\right)^2 + U''[\phi_0(x)]\psi^2(x) \right] dx.$$

In the last two terms we perform a partial integration whereby we get

$$H[\phi] = H[\phi_0] + \epsilon \int \left\{ -\frac{d^2\phi_0}{dx^2} + U'[\phi_0] \right\} \psi(x)dx + \frac{1}{2}\epsilon^2 \int \psi(x) \left\{ -\frac{d^2}{dx^2} + U''[\phi_0] \right\} \psi(x)dx.$$

But here the middle term vanishes due to the equation of motion; cf. (4.28). Thus the change in the static energy is of the second order in ϵ:

$$\delta H[\psi] = \frac{1}{2}\epsilon^2 \int \psi(x) \left\{ -\frac{d^2}{dx^2} + U''[\phi_0(x)] \right\} \psi(x)dx. \tag{4.67}$$

We must now demand that the quadratic form $\delta H[\psi]$ be positive (semi-) definite. If we solve the eigenvalue problem

$$\left\{ -\frac{d^2}{dx^2} + U''[\phi_0(x)] \right\} \phi_k(x) = \omega_k^2 \phi_k(x),$$

we can now decompose the fluctuation $\psi(x)$ into the eigenfunctions

$$\psi(x) = \sum a_k\phi_k(x).$$

Thus the quadratic form (4.67) reduces to

$$\delta H[\psi] = \frac{1}{2}\epsilon^2 \sum_k a_k^2\omega_k^2 . \tag{4.68}$$

Thus we see again that a negative eigenvalue would be catastrophic, since fluctuations along the corresponding mode $\phi_k(x)$ would lower the static energy.

Okay, having motivated the eigenvalue problem (4.66), we now proceed to solve it. Notice that the "potential" $W(x) = U''[\phi_0(x)]$ is strongly localized and satisfies the boundary conditions

$$\lim_{x\to\infty} W(x) = U''[\phi_{\pm\infty}]. \tag{4.69}$$

But here the right-hand side reproduces precisely the mass–square of the field quantum; cf. the discussion in Section 4.1. Making a trivial shift in the eigenvalue, we can further reduce the eigenvalue problem (4.66) to the case where the "potential" vanishes at infinity.

ILLUSTRATIVE EXAMPLE: THE SPECTRUM FOR THE SCHRÖDINGER OPERATOR.

Let us look at the following equivalent problem: *Find the eigenvalues for the second-order differential operator*

$$-\frac{d^2\phi}{dt^2} + W(t)\phi = \lambda\phi(t), \tag{4.70}$$

where $W(t)$ vanishes exponentially at infinity. (Notice that we have momentarily replaced the position variable x with the time variable t. This is because we can now appeal to our intuitive knowledge of classical mechanics. This is a standard strategy in physics: When we want to discuss the qualitative behavior of a solution to a differential equation, we try to find a mechanical analogue.) If we rearrange the above differential equation as follows,

$$\frac{d^2\phi}{dt^2} = -\lambda\phi(t) + W(t)\phi(t),$$

we recognize it as Newton's equation of motion for a particle of unit mass that is under the influence partly of a constant force $-\lambda$ and partly of a time-dependent force $W(t)$. Notice that the time-dependent force acts only for a very short time. For simplicity, we assume in the following that $W(t)$ is a negative function; i.e., it corresponds to an attractive force.

When discussing the particle motion, $x = \phi(t)$, we now distinguish between two cases (see Figure 4.15):

(a) *Positive eigenvalues*, $\lambda = \omega^2$ (*the continuous spectrum*). When λ is positive, the constant force is attractive; i.e., it drags the particle toward the center $x = 0$. In the remote past it will therefore oscillate:

$$x(t) \underset{t\to-\infty}{\approx} a_{-\infty}\cos\omega t + b_{-\infty}\sin\omega t.$$

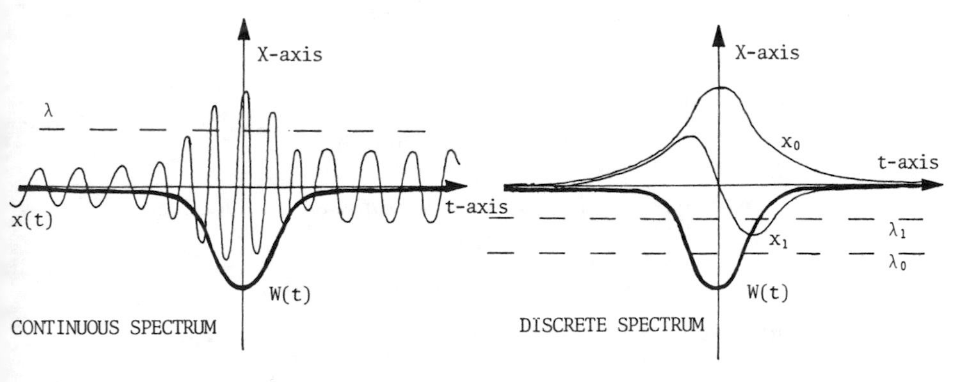

Figure 4.15.

For a short time it comes also under the influence of the time-dependent force $W(t)$, but it is then left again to the sole influence of the constant force; i.e., in the remote future it oscillates again:

$$x(t) \underset{t\to+\infty}{\approx} a_{+\infty}\cos\omega t + b_{+\infty}\sin\omega t.$$

(See Figure 4.15a.) Except for a change in amplitude and a phase shift, it thus has the same qualitative behavior in the past and the future.

(b) *Negative eigenvalues*, $\lambda = -\omega^2$ *(the discrete spectrum)*. In this case the constant force is repulsive. We shall investigate the motion of a particle that in the remote past is "almost" at rest at the origin; i.e., the particle motion is subject to the following asymptotic behavior:

$$\phi(t) \underset{t\to-\infty}{\approx} \exp[\omega t].$$

What will happen with this particle? We will be especially interested in the following question: Under what circumstances does it return to the origin in the remote future. (This corresponds to an eigenstate.)

Were it not for the attractive force $W(t)$, the particle would just fly off to infinity. Thus for λ less than the minimum of $W(t)$, the total force is always a repulsive force, and the particle flies off. When λ passes the minimum of $W(t)$, there will be a short time where the particle experiences an attractive force too. For a value λ_0 sufficiently high above the minimum, this attractive force will be just sufficient to decelerate the particle and send it back to the origin, so that when $W(t)$ dies off, the particle has just the reverse of the "escape velocity" (i.e., it approaches the origin exponentially); see Figure 4.15b. When λ is slightly higher than λ_0, the attractive force will become greater. Thus the velocity obtained will be too great, and the particle will pass through the origin and escape to infinity. For an even higher value λ_1, the attractive force will now be great enough to decelerate the particle, send it back through the origin, decelerate it again, and leave it with the reverse of the "escape velocity," so that once again it approaches the origin exponentially; see Figure 4.15b. And that is how the game proceeds!

Equipped with the experiences from the mechanical analogue, the following classical theorem should not come as a great surprise:

Theorem 4.1. *The spectrum of the second-order differential operator*

$$-\frac{d^2}{dx^2} + W(x) \qquad \textit{with} \lim_{x\to\pm\infty} W(x) = 0$$

consists of a continuous spectrum $\lambda = k^2$ *with positive eigenvalues extending from zero to infinity and a finite discrete spectrum*

$$-\infty < \lambda_0 < \lambda_1 < \lambda_2 < \cdots < \lambda_n < 0.$$

The eigenfunctions of the continuous spectrum can be characterized by the following asymptotic behavior ($\lambda = k^2$)*:*

$$\phi_k(x) \approx \begin{cases} \exp[ikx] & \textit{when } x \ll 0, \\ \alpha(k)\exp[ikx] + \beta(k)\exp[-ikx] & \textit{when } x \gg 0. \end{cases} \tag{4.71}$$

The eigenfunctions $\phi_\kappa(x)$ of the discrete spectrum can be characterized by the following asymptotic behavior ($\lambda_\kappa = -\kappa^2$):

$$\phi_\kappa(x) \approx \begin{cases} C_{-\infty}(x)\exp[+\kappa x] & \text{when } x \ll 0, \\ C_{+\infty}(x)\exp[-\kappa x] & \text{when } x \gg 0. \end{cases} \tag{4.72}$$

The ground state $\phi_0(x)$ has no nodes; i.e., it vanishes nowhere. The lowest nontrivial eigenstate $\phi_1(x)$ has precisely one node, and so on.

This concludes our illustrative example.

Okay, let us return to the eigenvalue problem (4.66). To show the dynamical stability of the soliton we must show that if $\phi_0(x)$ is a static kink or static antikink, then the associated eigenvalue problem has no negative eigenvalues. The proof is surprisingly simple. It is based upon the observation that we can in fact immediately write down the ground state!

Notice first that if $\phi_0(x)$ is a static solution with finite energy, then it satisfies the field equation (4.28)

$$\frac{d^2\phi_0}{dx^2} = U'[\phi_0(x)].$$

Differentiating this once with respect to x, we get

$$\frac{d^3\phi_0}{dx^3} = U''[\phi_0(x)]\,\frac{d\phi_0}{dx}\,.$$

This we rearrange as

$$\left\{-\frac{d^2}{dx^2} + U''[\phi_0(x)]\right\}\frac{d\phi_0}{dx} = 0.$$

A comparison with (4.66) then immediately shows that $d\phi_0/dx$ is an eigenfunction with the eigenvalue $\lambda = 0$. Recall then that the kink and the antikink are *monotonic* functions. Thus the associated zero mode has a constant sign throughout the whole x-axis. But then the zero mode $d\phi_0/dx$ has *no* nodes; i.e., it is precisely the ground state for the eigenvalue problem (4.66)! This shows that there are no negative eigenvalues.

In the above argument the presence of a zero mode $d\phi_0/dx$ associated with the finite-energy static solution $\phi_0(x)$ is very central. Its existence can be understood in the following way: The model has translational symmetry; i.e., the configurations

$$\phi_0(x) \quad \text{and} \quad \phi_0(x+a)$$

have the same static energy. For a small a we can expand the displaced configuration as

$$\phi_0(x+a) \approx \phi_0(x) + a\,\frac{\partial\phi_0(x+a)}{\partial a}\Big|_{a=0}\,.$$

Clearly,

$$\frac{\partial \phi_0(x+a)}{\partial a}\Big|_{a=0} = \frac{\partial \phi_0(x)}{\partial x} .$$

Thus we see that a perturbation along $d\phi_0/dx$ leaves the static energy invariant. According to (4.68), it must then be a zero mode, i.e., an eigenmode with the eigenvalue 0. To conclude: *The zero mode comes from translational symmetry.*

Before we leave the stability considerations, we want to mention that it is sometimes possible to prove the *instability* of a static solution by examining a particular simple deformation of the given configuration. The most important example is *Derrick's scaling argument*:

Consider a scalar field theory in D space dimensions based upon the Lagrangian density

$$L = -\frac{1}{2}\,(\partial_\mu \phi^a)(\partial^\mu \phi^a) - U[\phi^a].$$

The corresponding static energy functional is given by

$$\begin{aligned} H_{\text{static}}[\phi^a] &= \frac{1}{2}\int_{\mathbb{R}^d}\left(\frac{\partial \phi^a}{\partial x^i}\right)\left(\frac{\partial \phi^a}{\partial x^i}\right) d^D x + \int_{\mathbb{R}^d} U[\phi^a] d^D x \\ &= H_1[\phi^a] + H_2[\phi^a]. \end{aligned} \tag{4.73}$$

Suppose $\phi^a(x)$ is a static solution, and consider the scaled configuration

$$\phi^a_\lambda(x) = \phi^a(\lambda x).$$

The scaled configuration has the static energy

$$\begin{aligned} H[\phi^a_\lambda] &= \frac{1}{2}\int \lambda^2 \frac{\partial \phi^a}{\partial x^i}\,(\lambda x)\,\frac{\partial \phi^a}{\partial x^i}\,(\lambda x) d^D x + \int U[\phi^a(\lambda x)] d^D x \\ &= \frac{1}{2}\int \lambda^{2-D}\frac{\partial \phi^a}{\partial y^i}\,(y)\,\frac{\partial \phi^a}{\partial y_i}\,(y) d^D y + \lambda^{-D}\int U[\phi^a[y]] d^D y \\ &= \lambda^{2-D} H_1 + \lambda^{-D} H_2. \end{aligned}$$

If ϕ^a is stable, this must be stationary at $\lambda = 1$; i.e., we must demand that

$$0 = \frac{dH}{d\lambda}\Big|_{\lambda=1} = (2-D)H_1[\phi] - DH_2[\phi].$$

This implies the following stability condition:

$$(2-D)H_1[\phi] = DH_2[\phi]. \tag{4.74}$$

If $D > 2$, the coefficients $(2-D)$ and D have opposite signs. Since H_1 and H_2 are nonnegative, they must both vanish. Thus we conclude that *a nontrival static solution in a scalar field theory is unstable if the dimension of the space exceeds 2.*

If $D = 2$, there is a small loophole. The condition (4.74) this time reduces to

$$H_2[\phi] = 0,$$

and this can be fulfilled by a nontrivial configuration, provided that the potential energy term is absent. This is known as the exceptional case in two dimensions, and we shall return to it in Section 10.8.

If $D = 1$, the condition (4.74) finally reduces to

$$H_1[\phi] = H_2[\phi];$$

i.e., a static configuration corresponding to a single scalar field can only be stable if it satisfies the *virial theorem*:

$$\frac{1}{2}\int\left(\frac{d\phi}{dx}\right)^2 = \int U[\phi(x)]dx. \tag{4.75}$$

Interestingly enough, the kink and the antikink in fact satisfy this identity pointwise; i.e., not only are the integrals identical, but so are the integrands too! This follows immediately from (4.29).

Moral: With a single exception, Derrick's scaling argument shows that there are no stable solitons in higher-dimensional scalar field theories. To stabilize the static solutions, we must therefore extend the model somehow. This can, for example, be done by coupling the scalar field to a gauge field. We shall return to this in Sections 8.7, 8.8, and 10.9.

§ 4.8 The Particle Spectrum in Nonlinear Field Theories

Finally, we will give a brief (and naive) discussion of the particle spectrum in a $(1 + 1)$-dimensional scalar field theory with a degenerate vacuum. When we include quantum-mechanical considerations, we see that there are two basic types of elementary particles:

a. On the one hand, there are the field quanta represented by small oscillations around a classical vacuum. A field quantum has mass

$$m = \sqrt{U''[\phi_0]}, \tag{4.76}$$

where ϕ is a classical vacuum.

b. On the other hand, there are the solitons/antisolitons, which are present already on the classical level. A soliton has mass

$$M = \int_{\phi_1}^{\phi_2} \sqrt{2U[\phi]}\, d\phi, \tag{4.77}$$

where ϕ_1 and ϕ_2 are consecutive classical vacua. (Strictly speaking, there are quantum-mechanical corrections to this mass formula, but we shall neglect them.)

Apart from these elementary particles, there may also be *composite particles*, which can be interpreted as bound states of the elementary particles, i.e., the field quanta and the solitons/antisolitons. For example, we have seen in the sine-Gordon model that a soliton and an antisoliton can form a bound state—the bion.

Such a bion must then also be included in the particle spectrum. In the classical version its mass can be anywhere between zero and two soliton masses. Quantum-mechanically, the mass of the bion is quantized; i.e., it can take only a discrete set of values. We shall return to this in Section 5.6.

There is also the possibility of having bound states consisting of a field quantum bound to a soliton. Let us consider the interaction between field quanta and a single soliton in the weak field limit; i.e., we consider small "oscillations" around a kink solution

$$\phi(x,t) = \phi_+(x) + \eta(x,t).$$

As we have seen, the "oscillation" $\eta(x,t)$ can be decomposed into normal modes

$$\phi(x,t) = \sum_k \phi_k(x)\exp[-i\omega_k t],$$

where $\phi_k(x)$ is an eigenstate for the eigenvalue problem (4.66); i.e.,

$$\left\{-\frac{d^2}{dx^2} + U''[\phi_+(x)]\right\}\phi_k(x) = \omega_k^2\phi_k(x).$$

Let us introduce the "potential"

$$W(x) = U''[\phi_+(x)] - m^2 \tag{4.78}$$

and the "shifted" eigenvalue

$$\lambda = \omega_k^2 - m^2. \tag{4.79}$$

Then the eigenvalue problem reduces to

$$\left\{-\frac{d^2}{dx^2} + W(x)\right\}\phi_k(x) = \lambda\phi_k(x) \qquad \text{with} \lim_{x\to\pm\infty} W(x) = 0.$$

Thus we can take over the results from Theorem 4.1:

(a) There is a continuous spectrum consisting of nonnegative eigenvalues $0 < \lambda < \infty$. They correspond to frequencies ω_k satisfying the inequality

$$\omega_k > m.$$

Asymptotically, the fluctuation is given by

$$\eta(x,t) = \phi_k(x)\exp[-i\omega_k t] \underset{x\ll 0}{\approx} \exp\left\{i\left[x\sqrt{\omega_k^2 - m^2} - \omega_k t\right]\right\}. \tag{4.80}$$

This corresponds to a free particle with momentum p and energy E given by

$$p = \sqrt{\omega_k^2 - m^2} \quad \text{and} \quad E = \omega_k.$$

Notice that $E^2 - p^2 = m^2$; i.e., the rest mass of the particle is precisely m. Thus it is interpreted as a *field quantum that is scattered on the soliton*. When the field quantum is far way from the soliton, it behaves like a free particle, but when it is close to the soliton, they interact in a complicated way, and the approximation (4.80) correspondingly breaks down.

(b) There is also a finite discrete spectrum

$$\lambda_0 < \lambda_1 < \cdots < \lambda_n < 0.$$

As we have seen, the ground state λ_0 corresponds to a zero mode; i.e.,

$$\lambda_0 = -m^2.$$

The ground state arises from the translational symmetry of the model. It represents no increase in the energy relative to the soliton but simply reflects the fact that the kink solution, $\phi_+(x)$, is degenerate; i.e., it is just one particular member of a continuous family $\phi_+(x - x_0)$ of kink solutions. The other discrete eigenstates, if they are actually present in the model, are more interesting. In that case the fluctuation dies off exponentially when we move away from the soliton. These states are interpreted as *bound states* of a field quantum and a soliton. Notice that the total mass of the bound state lies between M and $M + m$.

The question then arises as to whether there are models with nontrivial discrete eigenstates.

Exercise 4.8.1
Problem: (a) Consider the ϕ^4-model. Show that the potential (4.78) is given by

$$W(x) = -\frac{6\alpha^2}{\cosh^2[\alpha x]} \qquad \text{with } \alpha = \frac{\mu}{\sqrt{2}}. \tag{4.81}$$

(b) Consider the sine-Gordon model. Show that the potential (4.78) is given by

$$W(x) = -\frac{2\alpha^2}{\cosh^2[\alpha x]} \qquad \text{with } \alpha = \mu. \tag{4.82}$$

From Exercise 4.8.1 we see that in our most popular models, the ϕ^4-model and the sine-Gordon model, we must determine the discrete spectrum for the *Eckhardt potential*:

$$W(x) = -\frac{\alpha^2 s(s+1)}{\cosh^2[\alpha x]}.$$

ILLUSTRATIVE EXAMPLE:
THE SPECTRUM FOR THE ECKHARDT POTENTIAL.

The eigenvalue equation is given by

$$-\frac{d^2\phi}{dx^2} - \frac{\alpha^2 s(s+1)}{\cosh^2[\alpha x]}\phi = \lambda\phi. \tag{4.83}$$

By a suitable transformation we can eliminate the transcendental function in front of ϕ. Consider first the substitution

$$\phi(x) = \cosh^{-s}[\alpha x]w(x).$$

By a straightforward computation, we get the following equivalent equation for $w(x)$:

$$-\frac{d^2w}{dx^2} + 2\alpha s \tanh[\alpha x]\frac{dw}{dx} - \alpha^2 s^2 w(x) = \lambda w(x). \tag{4.84}$$

Since the Eckhardt potential is an even function, i.e., it possesses the symmetry $x \curvearrowright -x$, we need only look for even and odd eigenfunctions $\phi(x)$. Furthermore, $\cosh[\alpha x]$ is an even function, so we need also only look for even and odd solutions to (4.84). In particular, we need only solve equation (4.84) on the positive semiaxis, i.e., for $0 \leq x < \infty$. The odd eigenfunctions then correspond to the boundary condition

$$w(0) = 0,$$

while the even eigenfunctions correspond to the boundary condition

$$\frac{dw}{dx}(0) = 0.$$

We can now make the further variable substitution

$$\xi = \sinh^2[\alpha x].$$

The eigenvalue equation (4.84) then reduces to

$$-\xi(\xi+1)\frac{d^2w}{d\xi^2} + \left[(s-1)\xi - \frac{1}{2}\right]\frac{dw}{d\xi} = \frac{1}{4}\left(\frac{\lambda}{\alpha^2} + s^2\right)w(\xi). \tag{4.85}$$

The even solutions can be expanded in a power series like

$$w_+(\xi) = \sum_{n=0}^{\infty} a_n \xi^n, \tag{4.86a}$$

while the odd solutions can be expanded in a power series like

$$w_-(\xi) = \xi^{1/2}\sum_{n=0}^{\infty} a_n \xi^n. \tag{4.86b}$$

Substituting (4.86a) into (4.85), we get the following recurrence relation:

$$\omega_+ : a_{n+1} = -\frac{n(n-s) + \frac{1}{4}\left(\frac{\lambda}{\alpha^2} + s^2\right)}{\left(n+\frac{1}{2}\right)(n+1)} a_n.$$

A discrete eigenvalue occurs when this power series terminates after a finite number of terms, i.e., when

$$\lambda_n = -4\alpha^2\left[n(n-s) + \frac{s^2}{4}\right] = -4\alpha\left(n - \frac{s}{2}\right)^2.$$

Since

$$\phi_+ = (1+\xi)^{-(1/2)s} w_+(\xi) \underset{\xi\to\infty}{\approx} (1+\xi)^{-(1/2)s}\xi^n,$$

we must furthermore demand that

$$2n < s$$

in order to get a normalizable wave function.

In the odd case we get the recurrence relation

$$\lambda_- : a_{n+1} = -\frac{\left(n+\frac{1}{2}\right)\left(n+\frac{1}{2}-s\right)+\frac{1}{4}\left(\frac{\lambda}{\alpha^2}+s^2\right)}{\left(n^2+\frac{5}{2}\,n+\frac{3}{2}\right)}\,a_n.$$

This generates a polynomial (corresponding to an eigenfunction) when

$$\lambda_n = -4\alpha^2\left(n+\frac{1}{2}-\frac{1}{2}\,s\right)^2.$$

Since

$$\phi_-(x) = (1+\xi)^{-(1/2)s}w_-(\xi) \underset{\xi\to\infty}{\approx} (1+\xi)^{-(1/2)s}\xi^{n+(1/2)},$$

we must furthermore demand that

$$2\left(n+\frac{1}{2}\right) < s.$$

The even and odd cases may be combined in the following way: The eigenvalues are given by

$$\lambda_k = -\alpha^2(s-k)^2; \qquad k \text{ an integer, } k < s. \tag{4.87}$$

Even k then correspond to even eigenfunctions, while odd k correspond to odd eigenfunctions. The associated eigenfunction is given by

$$\phi_k(x) = \begin{cases} (1+\xi)^{-(1/2)s}P_{k/2}(\xi) & \text{when } k \text{ is even,} \\ \operatorname{sgn}[x](1+\xi)^{-(1/2)s}P_{(k-1)/2}(\xi) & \text{when } k \text{ is odd,} \end{cases} \tag{4.88}$$

where P_n denotes a polynomial of degree n obtained from the above recurrence relations. Notice that $\phi_k(x)$ has precisely k nodes, in accordance with Theorem 4.1!

Okay, we can then apply the results obtained in this example to the ϕ^4-model and the sine-Gordon model.

(a) In the ϕ^4-model we get from (4.81) that $s = 2$. Consequently, there are precisely two discrete eigenstates. The corresponding eigenvalues are given by

$$\lambda_0 = -4\alpha^2 = -2\mu^2 \quad \text{and} \quad \lambda_1 = -\alpha^2 = -\frac{1}{2}\,\mu^2.$$

According to (4.9) and (4.79) they correspond to the frequencies

$$\omega_k^2 = \lambda_k + 2\mu^2; \qquad \text{i.e., } \omega_0^2 = 0 \quad \text{and} \quad \omega_1^2 = \frac{3}{2}\,\mu^2.$$

The first of these is of course just the zero mode, while the second represents a genuine bound state! We can also determine the corresponding eigenfunction

$\phi_1(x)$. It is odd and corresponds to the case where $w_-(\xi)$ reduces to $\xi^{1/2}$. Thus, $\phi_1(x)$ is proportional to

$$\mathrm{sgn}[x](1+\xi)^{-1}\xi^{1/2} = \frac{\sinh\left[\frac{\mu}{\sqrt{2}}x\right]}{\cosh^2\left[\frac{\mu}{\sqrt{2}}x\right]}.$$

(b) In the sine-Gordon model we get from (4.82) that $s = 1$. Consequently, there is precisely one discrete eigenstate (corresponding, of course, to the zero mode). In the sine-Gordon model there is consequently no bound state of a field quantum and the soliton.

Solutions to Worked Exercises

Solution to 4.5.1

(a) Using the trigonometric identity

$$\tan\left(\alpha+\beta-\frac{\pi}{2}\right) = \frac{\tan(\alpha)\tan(\beta)-1}{\tan(\alpha)+\tan(\beta)}$$

and the expression for the static kink solution

$$\tan\left(\frac{\lambda\phi_+}{4}\right) = \exp(\mu x),$$

we get

$$\begin{aligned}\tan\left(\frac{\lambda\phi}{4}\right) &= \tan\left(\frac{\lambda\phi_+}{4}\left[\gamma(x+vt+x_0)\right]\right.\\ &\qquad\left.+\frac{\lambda\phi_+}{4}\left[\gamma(x-vt-x_0)\right]-\frac{\pi}{2}\right)\\ &= \frac{\exp(2\mu\gamma x)-1}{\exp[\mu\gamma(x+vt+x_0)]+\exp[\mu\gamma(x-vt-x_0)]}\\ &= \frac{\sinh[\mu\gamma x]}{\cosh[\mu\gamma(vt+x_0)]}.\end{aligned}$$

(b) To find the asymptotic behavior as $t\to\infty$, we use that

$$\lim_{x\to\infty}\frac{\cosh(x+a)}{\cosh x} = \exp(a); \qquad \text{i.e., } \cosh(x+a) \underset{x\to\infty}{\approx} e^a\cosh(x).$$

Solution to 4.6.1

(a) If we introduce the notation

$$a = \exp(\lambda),$$

we can write down the Bäcklund equations in the following condensed form:

$$\partial_{\pm}(\phi_1 \mp \phi_0) = 2\exp[\pm\lambda_1]\sin\left[\frac{\phi_1 \pm \phi_0}{2}\right]. \tag{4.89a}$$

$$\partial_{\pm}(\phi_3 \mp \phi_1) = 2\exp[\pm\lambda_2]\sin\left[\frac{\phi_3 \pm \phi_1}{2}\right]. \tag{4.89b}$$

$$\partial_{\pm}(\phi_2 \mp \phi_0) = 2\exp[\pm\lambda_2]\sin\left[\frac{\phi_2 \pm \phi_0}{2}\right]. \tag{4.89c}$$

$$\partial_{\pm}(\phi_3 \mp \phi_2) = 2\exp[\pm\lambda_1]\sin\left[\frac{\phi_3 \pm \phi_2}{2}\right]. \tag{4.89d}$$

By adding and subtracting the first two, we get

$$\partial_{\pm}(\phi_3 \mp \phi_0) = \pm 2\exp[\pm\lambda_1]\sin\left[\frac{\phi_1 \pm \phi_0}{2}\right] + 2\exp[\pm\lambda_2]\sin\left[\frac{\phi_3 \pm \phi_1}{2}\right].$$

Similarly we obtain from the last two

$$\partial_{\pm}(\phi_3 \mp \phi_0) = \pm 2\exp[\pm\lambda_2]\sin\left[\frac{\phi_2 \pm \phi_0}{2}\right] + 2\exp[\pm\lambda_1]\sin\left[\frac{\phi_3 \pm \phi_2}{2}\right].$$

As the left-hand sides coincide, we therefore deduce the consistency relation

$$\exp[\pm\lambda_1]\left\{\sin\left[\frac{\phi_3 \pm \phi_2}{2}\right] \mp \sin\left[\frac{\phi_1 \pm \phi_0}{2}\right]\right\} = \exp[\pm\lambda_2]\left\{\sin\left[\frac{\phi_3 \pm \phi_1}{2}\right] \mp \sin\left[\frac{\phi_2 \pm \phi_0}{2}\right]\right\}.$$

Using the well-known trigonometric identity

$$\sin x \pm \sin y = 2\sin\left[\frac{x \pm y}{2}\right]\cos\left[\frac{x \mp y}{2}\right],$$

the consistency relation reduces to

$$\exp[\pm\lambda_1]\sin\left[\frac{(\phi_3 - \phi_0) \pm (\phi_2 - \phi_1)}{4}\right] = \exp[\pm\lambda_2]\sin\left[\frac{(\phi_3 - \phi_0) \mp (\phi_2 - \phi_1)}{4}\right].$$

Using the condensed notation introduced in the exercise, this can be written as

$$\exp[\pm\lambda_1]\sin[A_{\pm}] = \exp[\pm\lambda_2]\sin[A_{\mp}], \tag{4.90}$$

which is equivalent to (4.61).

(b) Differentiating the consistency relation (4.61) we get

$$\exp[\pm\lambda_1]\cos[A_{\pm}]\partial_{\pm}A_{\pm} = \exp[\pm\lambda_2]\cos[A_{\mp}]\partial_{\mp}A_{\mp}.$$

Writing out $A_\pm$, this leads to

$$\begin{aligned}\{\exp[\pm\lambda_1]\cos[A_\pm] - \exp[\pm\lambda_2]\cos[A_\mp]\}\partial_\pm\phi_3 \\ = \left\{\exp[\pm\lambda_1]\cos[A_\pm] - \exp[\pm\lambda_2]\cos[A_\mp]\right\}\partial_\pm\phi_0 \\ \pm\left\{\exp[\pm\lambda_1]\cos[A_\pm] - \exp[\pm\lambda_2]\cos[A_\mp]\right\}\partial_\pm\phi_1 \\ \mp\left\{\exp[\pm\lambda_1]\cos[A_\pm] - \exp[\pm\lambda_2]\cos[A_\mp]\right\}\partial_\pm\phi_2.\end{aligned}$$

The middle term is okay, but the first and last terms must be rearranged using (4.89a) and (4.89c) so that we can get of $\partial_\pm\phi_0$ and $\partial_\pm\phi_2$. After some algebra, we then end up with

$$\begin{aligned}&\{\exp[\pm\lambda_1]\cos[A_\pm] - \exp[\pm\lambda_2]\cos[A_\mp]\}\partial_\pm\phi_3 = \\ &\pm\left\{\exp[\pm\lambda_1]\cos[A_\pm] - \exp[\pm\lambda_2]\cos[A_\mp]\right\}\partial_\pm\phi_1 \\ &\pm 4\exp[\pm\lambda_1]\exp[\pm\lambda_2]\cos[A_\mp]\sin\left[\frac{\phi_1\pm\phi_0}{2}\right] \\ &\mp 2\left\{\exp[\pm\lambda_1]\cos[A_\pm] + \exp[\pm\lambda_2]\cos[A_\mp]\right\}\exp[\pm\lambda_2]\sin\left[\frac{\phi_2\pm\phi_0}{2}\right].\end{aligned} \tag{$*$}$$

The first term on the right-hand side is just what we want, but the remaining two terms need a little massage! Introducing the abbreviation

$$B_\pm = \frac{(\phi_3+\phi_0)\pm(\phi_2+\phi_1)}{4},$$

we have

$$\frac{\phi_1\pm\phi_0}{2} = \pm B_\pm \mp A_\pm; \qquad \frac{\phi_2\pm\phi_0}{2} = \pm B_\pm \mp A_\mp;$$
$$\frac{\phi_3\pm\phi_1}{2} = B_\pm + A_\mp.$$

If we introduce these abbreviations, the last terms in ($*$) reduce to

$$\begin{aligned}&\exp[\pm\lambda_1]\exp[\pm\lambda_2]\{2\cos[A_\mp]\sin[B_\pm]\cos[A_\pm] + 2\cos[A_\pm]\sin[A_\mp]\cos[B_\pm] \\ &\qquad - 4\cos[A_\mp]\sin[A_\pm]\cos[B_\pm]\} \\ &+ \exp[\pm\lambda_2]\exp[\pm\lambda_2]\{2\cos[A_\mp]\sin[A_\mp]\cos[B_\pm] \\ &\qquad - \cos[A_\mp]\sin[B_\pm]\cos[A_\mp]\}.\end{aligned}$$

(A broken line with arrows links the first term $\exp[\pm\lambda_1]$ to the term $-4\cos[A_\mp]\sin[A_\pm]\cos[B_\pm]$.)

Using the consistency relation (4.61) in the combination indicated by the broken line, this is further reduced to

$$\begin{aligned}&2\exp[\pm\lambda_1]\exp[\pm\lambda_2]\left\{\cos[A_\mp]\sin[B_\pm]\cos[A_\pm] + \cos[A_\pm]\sin[A_\mp]\cos[B_\pm]\right\} \\ &\quad - 2\exp[\pm\lambda_2]\exp[\pm\lambda_2]\{\cos[A_\mp]\sin[A_\mp]\cos[B_\pm] \\ &\quad + \cos[A_\mp]\sin[B_\pm]\cos[A_\mp]\}\end{aligned}$$

$$= 2\exp[\pm\lambda_2]\left\{\exp[\pm\lambda_1]\cos[A_\pm] - \exp[\pm\lambda_2]\cos[A_\mp]\right\}$$
$$\cdot\left\{\sin[B_\pm]\cos[A_\mp] + \sin[A_\mp]\cos[B_\pm]\right\}$$
$$= 2\exp[\pm\lambda_2]\left\{\exp[\pm\lambda_1]\cos[A_\pm] - \exp[\pm\lambda_2]\cos[A_\mp]\right\}\sin\left[\frac{\phi_3 \pm \phi_1}{2}\right].$$

Inserting this back into $(*)$, this finally reduces to

$$\partial_\pm\phi_3 = \pm\partial_\pm\phi_1 + 2\exp[\pm\lambda_2]\sin\left[\frac{\phi_3 \pm \phi_1}{2}\right].$$

(c) Writing out the consistency relation (4.61) explicitly as

$$a_1 \sin\left[\frac{(\phi_3 - \phi_0) + (\phi_2 - \phi_1)}{4}\right] = a_2 \sin\left[\frac{(\phi_3 - \phi_0) - (\phi_2 - \phi_1)}{4}\right],$$

it can be rearranged as

$$a_1 \sin\left[\frac{\phi_3 - \phi_0}{4}\right]\cos\left[\frac{\phi_2 - \phi_1}{4}\right] + a_1 \cos\left[\frac{\phi_3 - \phi_0}{4}\right]\sin\left[\frac{\phi_2 - \phi_1}{4}\right]$$
$$= a \sin\left[\frac{\phi_3 - \phi_0}{4}\right]\cos\left[\frac{\phi_2 - \phi_1}{4}\right]$$
$$- a \cos\left[\frac{\phi_3 - \phi_0}{4}\right]\sin\left[\frac{\phi_2 - \phi_1}{4}\right].$$

This immediately leads to

$$(a_2 + a_1)\sin\left[\frac{\phi_2 - \phi_1}{4}\right]\cos\left[\frac{\phi_3 - \phi_0}{4}\right]$$
$$= (a_2 - a_1)\sin\left[\frac{\phi_3 - \phi_0}{4}\right]\cos\left[\frac{\phi_2 - \phi_1}{4}\right],$$

form which the Bianchi identity (4.60) follows trivially.

CHAPTER 5

Path Integrals and Instantons

§ 5.1 The Feynman Propagator Revisited

In the remaining chapter of Part I we would like to include a few aspects of the quantum theory of fields and particles. To simplify the discussion, we begin our considerations with quantum mechanics of a single particle in one space dimension.

In ordinary quantum mechanics the central concept is the Feynman propagator $K(x_b; t_b \mid x_a; t_a)$. As we have seen, it denotes the probability amplitude for a particle to move, i.e., to propagate, from the space–time point (x_a, t_a) to the space–time point (x_b, t_b). The propagator governs the dynamical evolution of the Schrödinger wave function according to the rule (2.18)

$$\psi(x_b; t_b) = \int K(x_b; t_b \mid x_a; t_a)\psi(x_a; t_a)dx_a,$$

and as a consequence, all information about the quantum behavior of the particle is stored in the propagator. Furthermore, it satisfies the group property (2.25)

$$K(x_c; t_c \mid x_a; t_a) = \int K(x_c; t_c \mid x_b; t_b)K(x_b; t_b \mid x_a; t_a)dx_b.$$

It will be useful to recall what we have learned so far about the propagator. For simplicity, we consider a physical system characterized by the Lagrangian

$$L\left(x, \frac{dx}{dt}\right) = \frac{1}{2}m\left(\frac{dx}{dt}\right)^2 - V(x).$$

Then we have previously shown that

1. *The propagator can be represented as a path integral* (2.21)

$$K(x_b; t_b \mid x_a; t_a) = \int_{x(t_a)=x_a}^{x(t_b)=x_b} \exp\left\{\frac{i}{\hbar}\int_{t_a}^{t_b}\left[\frac{m}{2}\left(\frac{dx}{dt}\right)^2 - V(x)\right]dt\right\} D[x(t)],$$

where we sum over all paths connecting the space–time points $(x_a; t_a)$ *and* $(x_b; t_b)$.

2. *The propagator is the unique solution to the Schrödinger equation*

$$i\hbar \frac{\partial}{\partial t_b} K(x_b; t_b \mid x_a; t_a) = \left[-\frac{\hbar^2}{2m} \frac{\partial^2}{\partial x_b^2} + V(x_b) \right] K(x_b; t_b \mid x_a; t_a),$$

satisfying the initial condition

$$K(x_b; t_a \mid x_a; t_a) = \delta(x_b - x_a).$$

(*Cf. the discussion in* Section 2.4. *In mathematical terminology, the propagator is thus a Green's function for the Schrödinger equation.*)

3. *The propagator has the following decomposition into a complete set of eigenfunctions for the Hamiltonian operator* (2.70):

$$K(x_b; t_b \mid x_a; t_a) = \sum_{n=0}^{\infty} \overline{\phi_n(x_a)} \phi_n(x_b) \exp\left[-\frac{i}{\hbar} E_n(t_b - t_a) \right].$$

(*Cf. the worked exercise* 2.11.3.)

Of these characterizations, only the last two have an unambiguous meaning. The first one, involving the path integral, is only a formal description, since we have not yet indicated how one should actually perform such a summation over paths!

Before we discuss the properties of the propagator further, we will indicate yet another characterization of the propagator. Recall that in quantum mechanics a physical quantity T is represented by a Hermitian operator $\hat{T}$ and that this operator can have only real eigenvalues λ_n that represent the possible values of the quantity T when we try to measure it in an actual experiment. In general, a wave function can be decomposed into a complete set of eigenfunctions for the operator $\hat{T}$:

$$\psi(x) = \sum_n a_n \psi_n(x); \qquad a_n = \langle \psi_n \mid \psi \rangle = \int \overline{\psi_n(x)}\, \psi(x) dx.$$

The coefficient a_n is then interpreted as the probability amplitude for measuring the value λ_n; i.e., in an actual experiment we will measure the value λ_n with the probability $P_n = |a_n|^2$.

For example, the position x_0 is represented by the multiplication operator

$$\hat{x}_0 \psi(x) = x_0 \cdot \psi(x). \tag{5.1}$$

The corresponding eigenfunctions are given by

$$\psi_{x_0} = \delta(x - x_0). \tag{5.2}$$

Notice that these eigenfunctions are not normalizable. This is related to the fact that the position operator has a continuous spectrum, i.e., a continuum of eigenvalues. The eigenfunctions, however, still satisfy the completeness relation (2.69):

$$\int \overline{\psi_{x_0}(x_1)} \psi_{x_0} dx_0 = \int \delta(x_0 - x_1)\delta(x_0 - x_2) dx_0 = \delta(x_1 - x_2).$$

Similarly, the momentum p_0 is represented by the differential operator

$$\hat{p}_0\psi(x) = -i\hbar\,\frac{\partial\psi}{\partial x}\,. \tag{5.3}$$

The corresponding eigenfunctions are given by the plane waves

$$\psi_{k_0}(x) = \frac{1}{\sqrt{2\pi}}\,\exp[ik_0x] \qquad \text{with } p_0 = \hbar k_0. \tag{5.4}$$

(Notice that we parametrize the momentum eigenfunctions by the wave number k_0, rather than the momentum $p_0 = \hbar k_0$.) The momentum eigenfunctions are not normalizable, but like the position eigenfunctions, they satisfy the completeness relation (2.69):

$$\int \overline{\psi_{k_0}(x_1)}\psi_{k_0}(x_2)dk_0 = \frac{1}{2\pi}\int \exp[ik_0(x_2 - x_1)]dk_0 = \delta(x_2 - x_1).$$

Here we have used that the Fourier transform of Dirac's delta function is simply 1; i.e.,

$$\int \delta(x)\exp[iyx]dx = \exp[iyx]|_{x=0} = 1.$$

By Fourier's inversion formula, we thus get

$$\delta(x) = \frac{1}{2\pi}\int \exp[-iyx]dy. \tag{5.5}$$

Exercise 5.1.1
Problem: Compute the free-particle propagator by means of formula (2.70), which in this case reduces to

$$K(x_b, t_b \mid x_a; t_a) = \int_{-\infty}^{+\infty} \overline{\psi_k(x_a)}\psi_k(x_b)\exp\left[-\frac{i}{\hbar}\,\frac{\mathbf{p}^2}{2m}\,(t_b - t_a)\right]dk,$$

where $\psi_k(x)$ is given by (5.4).

Consider now the *time-evolution operator* $\exp\left[-\frac{i}{\hbar}\,\hat{H}T\right]$. It has the matrix element

$$\langle x_b|\exp\left[-\frac{i}{\hbar}\,\hat{H}T\right]|x_a\rangle,$$

where $|x_a\rangle$ is the position eigenfunction (5.2). Using a complete set of eigenfunctions for the Hamiltonian $\hat{H}$, we can now expand the position eigenstate

$$|x_a\rangle = \sum_n \langle\psi_n \mid x_a\rangle \mid \psi_n\rangle = \sum_n \overline{\psi_n}(x_a)|\psi_n\rangle.$$

If we furthermore use that

$$\exp\left[-\frac{i}{\hbar}\,\hat{H}T\right]|\psi_n\rangle = \exp\left[-\frac{i}{\hbar}\,E_nT\right]|\psi_n\rangle,$$

then the matrix element reduces to

$$\langle x_b | \exp\left[-\frac{i}{\hbar}\hat{H}T\right] | x_a \rangle = \sum_n \overline{\psi_n}(x_n)\langle x_b \mid \psi_n\rangle \exp\left[-\frac{i}{\hbar}E_n T\right]$$
$$= \sum_n \overline{\psi_n(x_a)}\psi_n(x_b)\exp\left[-\frac{i}{\hbar}E_n T\right].$$

According to the third characterization, we have thus shown

(4) *The propagator can be represented as a matrix element of the time evolution operator* $\exp\left[-\frac{i}{\hbar}\hat{H}T\right]$; *i.e.,*

$$K(x_b; T \mid x_a; 0) = \langle x_b | \exp\left[-\frac{i}{\hbar}\hat{H}T\right] | x_a\rangle. \tag{5.6}$$

As an especially important example of how one can extract information from the propagator, we will now show how one can in principle calculate the energy spectrum of a particle. For this purpose we consider the trace of the time evolution operator

$$\begin{aligned} G(T) &= \mathrm{Tr}\left\{\exp\left[-\frac{i}{\hbar}\hat{H}T\right]\right\} = \int \langle x_0 | \exp\left[-\frac{i}{\hbar}\hat{H}T\right] | x_0\rangle dx_0 \\ &= \int K(x_0; T \mid x_0; 0)dx_0. \end{aligned} \tag{5.7}$$

Using the third characterization of the propagator, this reduces to

$$G(T) = \int \sum_{n=0}^{\infty} |\phi_n(x_0)|^2 \exp\left[-\frac{i}{\hbar}E_n T\right] dx_0 = \sum_{n=0}^{\infty} \exp\left[-\frac{i}{\hbar}E_n T\right], \tag{5.8}$$

where we have used the normalization of energy eigenfunctions $\phi_n(x)$. Thus *the trace of the propagator decomposes very simply in terms of harmonic phase factors depending only upon the energy levels*!

Introducing a "Fourier transform"

$$G(E) = \frac{i}{\hbar}\int_0^{\infty} G(T)\exp\left[\frac{i}{\hbar}ET\right]dT, \tag{5.9}$$

we in fact immediately get

$$G(E) = \frac{i}{\hbar}\sum_{n=0}^{\infty}\int_0^{\infty}\exp\left[-\frac{i}{\hbar}(E_n - E)T\right]dT = \sum_{n=0}^{\infty}\frac{1}{E_n - E}. \tag{5.10}$$

Consequently, *the energy levels show up as poles in the transformed trace of the propagator*.

In the above discussion we have been focusing upon propagation in the "position space." We could as well use other physical properties, such as the momentum, as our starting point. Thus we define $K(p_b; t_b \mid p_a; t_a)$ as the probability amplitude for a particle that is released with momentum p_a at time t_a to be observed with momentum p_b at time t_b. Thus $K(p_b; t_b \mid p_a; t_a)$ denotes the probability amplitude for a particle to "propagate in momentum space" from the point (p_a, t_a) to the point $(p_b; t_b)$. According to the basic rules of quantum mechanics, the momentum propagator is now related to the position propagator as follows:

$$K(p_b; t_b \mid p_a; t_a) = \int\int \langle p_b \mid x_b\rangle K(x_b; t_b \mid x_a; t_a)\langle x_a \mid p_a\rangle dx_a dx_b,$$

where $|p_a\rangle$ is the momentum eigenfunction (5.4). Here the various terms can be interpreted as follows:

a. $\langle x_a \mid p_a\rangle$ is the probability amplitude for a particle with momentum p_a to be at the position x_a.
b. $K(x_b; t_b \mid x_a; t_a)$ is the probability amplitude for a particle to propagate from x_a to x_b.
c. $\langle p_b \mid x_b\rangle$ is the probability amplitude for a particle at position x_b to be observed with momentum p_b.

As usual, we then sum over all the alternative ways the particle can propagate from p_a to p_b; i.e., we integrate over the intermediate positions. Using that

$$\langle x_a \mid p_a\rangle = \frac{1}{\sqrt{2\pi}} \exp\left[\frac{i}{\hbar} p_a x_a\right] \text{ and } \langle p_b \mid x_b\rangle = \frac{1}{\sqrt{2\pi}} \exp\left[-\frac{i}{\hbar} p_b x_b\right],$$

the above relations can now be rearranged as

$$K(p_b; t_b \mid p_a; t_a) = \frac{1}{2\pi} \int\int \exp\left[-\frac{i}{\hbar} p_b x_b\right] K(x_b; t_b \mid x_a; t_a) \cdot \exp\left[\frac{i}{\hbar} p_a x_a\right] dx_a dx_b. \tag{5.11}$$

The momentum propagator is thus obtained from the position propagator by a "double" Fourier transformation.

Consider, e.g., the free particle propagator (2.28)

$$K(x_b; t_b \mid x_a; t_a) = \sqrt{\frac{m}{2\pi i\hbar(t_b - t_a)}} \exp\left[\frac{im}{2\hbar} \frac{(x_b - x_a)^2}{t_b - t_a}\right].$$

In momentum space this is converted to

$$K(p_b; T \mid p_a; 0) = \sqrt{\frac{m}{2i\hbar T}} \frac{1}{2\pi} \int\int \exp\left\{\frac{i}{\hbar}\left[\frac{m}{2}\frac{(x_b - x_a)^2}{T} - p_b x_b + p_a x_a\right]\right\} dx_a dx_b.$$

This is just a Gaussian integral, cf. (2.27), which can be calculated in the usual way by "completing the square." Using the identity

$$\frac{m}{2}\frac{(x_b - x_a)^2}{T} - p_b x_b + p_a x_a = \frac{m}{2T}\left[x_a + \frac{T}{m}\left(p_a - \frac{m x_b}{T}\right)\right]^2 - \frac{p_a^2}{2m} T + (p_a - p_b)x_b,$$

the x_a-integration cancels the factor $\sqrt{\frac{m}{2\pi i\hbar T}}$ in front, and we are left with the formula

$$K(p_b; T \mid p_a; 0) = \exp\left[-\frac{i}{\hbar}\frac{p_a^2}{2m}T\right]\frac{1}{2\pi}\int \exp\left[\frac{i}{\hbar}(p_a - p_b)x_b\right]dx_b .$$

But according to (5.5), the last integration just produces a δ-function! We therefore finally get

$$K(p_b; T \mid p_a; 0) = \delta\left(\frac{p_b - p_a}{\hbar}\right)\exp\left[-\frac{i}{\hbar}\frac{p_a^2}{2m}T\right]. \tag{5.12}$$

This simple result has an intuitive explanation: Since the momentum of a free particle is conserved, it follows that once we release a free particle with momentum p_a, it will necessarily still have momentum p_a once we decide to measure it. This explains the δ-factor. Furthermore, a free particle with momentum p_a will have the definite energy $E = p_a^2/2m$. Thus the propagator has the characteristic phase factor indicated by the de Broglie rule, cf. (2.32),

$$\exp\left[-\frac{i}{\hbar}ET\right] = \exp\left[-\frac{i}{\hbar}\frac{p_a^2}{2m}T\right].$$

Evidently, the free particle propagator is much simpler in momentum space. This is a general feature of quantum theory, and therefore it is the momentum propagator that is most often displayed in the literature.

§ 5.2 Illustrative Example: The Harmonic Oscillator

As an example of how to use the preceding techniques we now take a closer look at the harmonic oscillator, which is characterized by the Lagrangian

$$L = \frac{1}{2}m\left(\frac{dx}{dt}\right)^2 - \frac{1}{2}m\omega^2x^2 . \tag{5.13}$$

The corresponding equation of motion is given by

$$\frac{d^2x}{dt^2} = -\omega^2 x.$$

Notice that ω is the cyclic frequency of the oscillator.

First, we must calculate the propagator. Since the Lagrangian is quadratic, we can use the same trick as for the free particle propagator; cf. the worked Exercise 2.4.1. The propagator consequently reduces to the form

$$K(x_b; T \mid x_a; 0) = A(T)\exp\left[\frac{i}{\hbar}\int_0^t L\left(x_{\text{cl}}; \frac{dx_{\text{cl}}}{dt}\right)dt\right], \tag{5.14}$$

where $x = x_{\text{cl}}(t)$ is the classical path connecting $(x_a; 0)$ and $(x_b; T)$.

To determine the classical action, we notice that the general solution of the equation of motion is given by

$$x_{\mathrm{cl}}(t) = c \cos \omega t + b \sin \omega t.$$

We shall then adjust the parameters a and b so that the boundary conditions

$$x_{\mathrm{cl}}(0) = x_a \quad \text{and} \quad x_{\mathrm{cl}}(T) = x_b$$

are satisfied. If $\omega T \neq n\pi$, this is easily obtained, and we get the unique solution

$$a = x_a \quad \text{and} \quad b = \frac{x_b - x_a \cos \omega T}{\sin \omega T}. \tag{5.15}$$

If on the contrary $\omega T = n\pi$ (corresponding to either a half period or a full period), we are in trouble. This is because a classical particle, after a half-period, necessarily is in the opposite point. After a full period it is, similarly, necessarily in the same point. (See Figure 5.1.) For $\omega T = n\pi$ we can therefore find a classical path only if $x_b = (-1)^n x_a$, and if that is the case, *any* classical path passing through $(x_a; 0)$ will in fact also pass through $(x_b; T)$! In analogy with optics, we say that the classical path has a *caustic* when $\omega T = n\pi$. At the caustics the propagator thus becomes singular!

Okay, neglecting the caustics for a moment we can complete the calculation. The classical action is given by

$$S[x_{\mathrm{cl}}] = \frac{m\omega}{2} \left[(b^2 - a^2) \cos \omega T \sin \omega T - 2ab \sin^2 \omega T\right],$$

and substituting the values (5.15) for a and b we now get, after a lengthy but trivial calculation,

$$S[x_{\mathrm{cl}}] = \frac{m\omega}{2 \sin \omega T} \left[(x_b^2 + x_a^2) \cos \omega T - 2x_b x_a\right]. \tag{5.16}$$

As expected, this is highly singular at the caustics, i.e., when $\omega T = n\pi$.

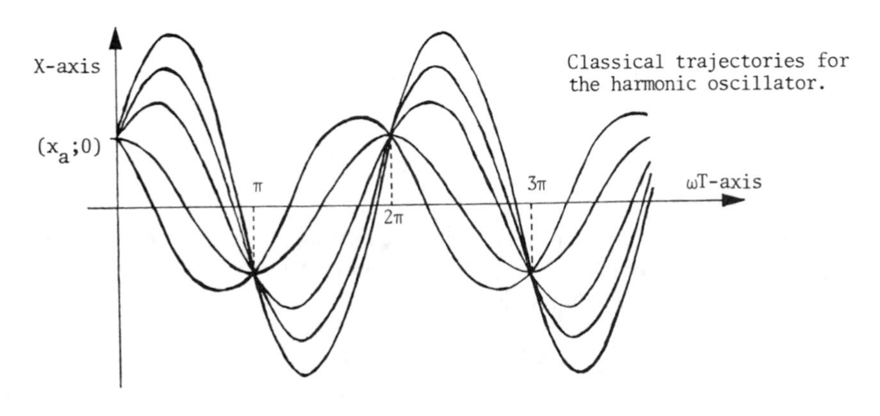

Figure 5.1.

We proceed to calculate the amplitude $A(T)$. It can be determined by the same trick as the one we used for the free particle propagator, i.e.,by using the group property (2.25). By translational invariance in time, the group property reduces to

$$K(x_3; T_2 + T_1 \mid x_1; 0) = \int K(x_3; T_2 \mid x_2; 0) K(x_2; T_1 \mid x_1; 0) dx_2.$$

Substituting the expression (5.14), we now get

$$\begin{aligned} &A(T_2 + T_1) \\ &\cdot \exp\left[\frac{im\omega}{2\hbar \sin \omega(T_2 + T_1)} \left\{(x_3^2 + x_1^2) \cos \omega(T_2 + T_1) - 2x_3 x_1\right\}\right] \\ &= A(T_2)A(T_1) \int \exp\left[\frac{im\omega}{2\hbar} \left\{\frac{(x_3^2 + x_2^2) \cos \omega T_2 - 2x_3 x_2}{\sin \omega T_2}\right.\right. \\ &\quad + \left.\left.\frac{(x_2^2 + x_1^2) \cos \omega T_1 - 2x_2 x_1}{\sin \omega T_1}\right\}\right] dx_2. \end{aligned}$$

This integral looks rather messy, but after all, it is just a harmless Gaussian integral. Using the identity

$$\begin{aligned} &\frac{im\omega}{2\hbar}\left[\frac{(x_3^2 + x_2^2) \cos \omega T_2 - 2x_3 x_2}{\sin \omega T_2} + \frac{(x_2^2 + x_1^2) \cos \omega T_1 - 2x_2 x_1}{\sin \omega T_1}\right] \\ &= \frac{im\omega \sin \omega(T_2 + T_1)}{2\hbar \sin \omega T_2 \sin \omega T_1}\left[x_2 - \frac{x_3 \sin \omega T_1 + x_1 \sin \omega T_2}{\sin \omega(T_2 + T_1)}\right]^2 \\ &\quad + \frac{im\omega}{2\hbar \sin \omega(T_2 + T_1)}\left[(x_3^2 + x^2) \cos \omega(T_2 + T_1) - 2x_3 x_1\right], \end{aligned}$$

we can immediately integrate out the x_2-dependence, whereby the right-hand side reduces to

$$\begin{aligned} &A(T_2)A(T_1)\sqrt{\frac{2\pi i\hbar \sin \omega T_1 \sin \omega T_2}{m\omega \sin \omega(T_2 + T_1)}} \\ &\cdot \exp\left[\frac{im\omega}{2\hbar \sin \omega(T_2 + T_1)} \{(x_3^2 + x_1^2) \cos \omega(T_2 + T_1) - 2x_3^2 x_1^2\}\right]. \end{aligned}$$

Most reassuring, the x_3, x_1-dependence is the same as on the left-hand side, and we can therefore factor out the phase factor completely. We then obtain the following functional equation for the amplitude $A(T)$:

$$\frac{A(T_2 + T_1)}{A(T_2)A(T_1)} = \sqrt{\frac{2\pi i\hbar \sin \omega T_2 \sin \omega T_1}{m\omega \sin \omega(T_2 + T_1)}} = \frac{\sqrt{\frac{m\omega}{2\pi i\hbar \sin \omega(T_2+T_1)}}}{\sqrt{\frac{m\omega}{2\pi i\hbar \sin \omega T_2}}\sqrt{\frac{m\omega}{2\pi i\hbar \sin \omega T_1}}}.$$

It has the general solution

$$A(T) = \exp[i\alpha T]\sqrt{\frac{m\omega}{2\pi i\hbar \sin \omega T}},$$

but as in the case of the free particle propagator we shall neglect the additional phase $\exp[i\alpha T]$, i.e., put $\alpha = 0$. (This corresponds to a normalization of the energy and has no physical consequences.) Notice that in the limit where $\omega \to 0$, the function $A(T)$ reduces to the free particle amplitude

$$A(T) = \sqrt{\frac{m}{2\pi i\hbar T}},$$

as we expect it to do!

Consequently, we have now determined completely the propagator for the harmonic oscillator. In its full glory, it is given by

$$K(x_b; T \mid x_a; 0) = \sqrt{\frac{m\omega}{2\pi i\hbar \sin\omega T}} \cdot \exp\left[\frac{im\omega}{2\hbar\sin\omega T}\left\{(x_b^2 + x_a^2)\cos\omega T - 2x_b x_a\right\}\right]. \tag{5.17}$$

Notice, too, that it also reduces to the free particle propagator in the limit of small T. Expanding the phase to the next lowest order, we in fact get

$$\exp\left[\frac{im\omega}{2\hbar\sin\omega T}\left\{(x_b^2 + x_a^2)\cos\omega T - 2x_b x_a\right\}\right] \approx \exp\left[\frac{im}{2\hbar T}(x_b - x_a)^2\right]\exp\left[-\frac{im\omega^2}{2\hbar}\left\{x_b^2 + x_a^2 + x_b x_a\right\}\frac{T}{3}\right].$$

Here the first phase factor is the free particle phase factor, while the second phase factor reproduces the interaction phase factor,

$$\exp\left[-\frac{i}{\hbar}\int_0^T V(x)dt\right],$$

integrated along the free particle path, i.e., the straight line $x = x_a + (x_b - x_a)t/T$. Thus the "infinitesimal propagator" is in accordance with our previous equation (2.53), which we used in our derivative of the Schrödinger equation. Observe also that if we had included the phase factor $\exp[i\alpha T]$, this would no longer be correct.

To be honest, the above formula (5.17) for the propagator of the harmonic oscillator is not quite correct. This is due to the fact that we have been rather careless about solving Gaussian integrals of the type

$$\int_{-\infty}^{+\infty} e^{-i\alpha x^2} dx = \sqrt{\frac{i\pi}{\alpha}}. \tag{5.18}$$

The above formula is obtained from the rigorous formula (2.27)

$$\int_{-\infty}^{+\infty} e^{-\lambda x^2} dx = \sqrt{\frac{\pi}{\lambda}}, \qquad \lambda > 0,$$

by performing an analytic continuation. But this analytic continuation is actually double-valued, since it involves the square root of a complex number. This gives rise to a phase ambiguity in the Gaussian integral (5.18). As a consequence, the propagator has a phase ambiguity (corresponding to a phase ± 1 or $\pm i$). Below the first caustic, i.e., when $T < \frac{\pi}{\omega}$, formula (5.17) is correct, simply because it is in correspondence with the free particle propagator in either of the limits $\omega \to 0$ or $T \to 0$. But once we have passed the first caustic, we no longer control the phase ambiguity. We shall return to this problem at the end of this section.

Now that we have the Feynman propagator at our disposal, we can easily calculate the energy spectrum. The trace of the propagator is given by

$$\begin{aligned} G(T) &= \int K(x_0; T \mid x_0; 0) dx_0 \\ &= \frac{m\omega}{\sqrt{2\pi i \hbar \sin \omega T}} \int \exp\left[\frac{im\omega x_0^2}{\hbar \sin \omega T} (\cos \omega T - 1)\right] dx_0 \\ &= \frac{1}{\sqrt{2(\cos \omega T - 1)}} = \frac{1}{2i \sin \frac{1}{2} \omega T}. \end{aligned}$$

But this is easily rearranged as

$$G(T) = \frac{1}{e^{i(1/2)\omega T} - e^{-i(1/2)\omega T}} = \frac{e^{-i(1/2)\omega T}}{1 - e^{-i\omega T}} = \sum_{n=0}^{\infty} e^{-i(n+(1/2))\omega T}. \tag{5.19}$$

A comparison with (5.8) then immediately permits us to read off the energy spectrum

$$E_n = \left(n + \frac{1}{2}\right) \hbar\omega. \tag{5.20}$$

In the case of the harmonic oscillator, the possible values of the energy are thus evenly spaced, and perhaps a little surprising, the ground state energy is not zero as in the classical case but is instead given by

$$E_0 = \frac{1}{2} \hbar\omega. \tag{5.21}$$

The zero-point energy is often explained by referring to *Heisenberg's uncertainty principle*, according to which the indeterminacies of the momentum and the position are bounded below by

$$\langle x \rangle \langle p \rangle \gtrsim \hbar. \tag{5.22}$$

Unlike the classical case, the particle therefore cannot be at rest in the bottom of the potential well, since this would cause both $\langle x \rangle$ and $\langle p \rangle$ to vanish. It must necessarily "vibrate a little." A rough estimate of the energy is given by

$$E = \frac{\langle p \rangle^2}{2m} + \frac{1}{2} m\omega^2 \langle x \rangle^2.$$

We want to minimize this, subject to the constraint (5.22). Replacing $\langle p \rangle$ by $\hbar / \langle x \rangle$ and minimizing the resulting expression with respect to variation in $\langle x \rangle$

produces the minimum

$$E_{\min} = \hbar\omega.$$

Except for the factor $\frac{1}{2}$, which we cannot account for by such a primitive argument, this is the same as the zero-point energy (5.21).

Once we have the propagator at our disposal, we also control the dynamical evolution of the Schrödinger wave function. As in the free particle case, we can now find an exact and especially simple solution to the Schrödinger equation. At time $t = 0$ we consider the normalized wave function

$$\psi(x, 0) = \sqrt[4]{\frac{m\omega}{\hbar\pi}} \exp\left[-\frac{m\omega}{2\hbar}(x-a)^2\right]. \tag{5.23}$$

This corresponds to a Gaussian probability distribution centered at the point $x = a$. At a later time t the wave function will, according to (2.18), evolve into

$$\psi(y, t) = \int K(y; t \mid x; 0)\psi(x, 0)dx = \sqrt[4]{\frac{m\omega}{\hbar\pi}} \sqrt[2]{\frac{m\omega}{2\pi i\hbar \sin\omega t}}$$
$$\cdot \int \exp\left[\frac{im\omega}{2\hbar\sin\omega t}\left\{(y^2+x^2)\cos\omega t - 2yx\right\} - \frac{m\omega}{2\hbar}(x-a)^2\right]dx.$$

As usual, this is just a Gaussian integral. Using the identity

$$\begin{aligned}&\frac{im\omega}{2\hbar\sin\omega t}\left[(y^2+x^2)\cos\omega t - 2yx\right] - \frac{m\omega}{2\hbar}(x-a)^2\\ &\quad= \frac{m\omega i e^{i\omega t}}{2\hbar\sin\omega t}\left[x - ie^{i\omega t}(a\sin\omega t - iy)\right]^2\\ &\qquad- \frac{m\omega}{2\hbar}\left[y^2 - 2aye^{-i\omega t} + a^2\cos\omega t e^{-i\omega t}\right],\end{aligned}$$

the integration can be immediately performed, and the wave function at time t reduces to

$$\begin{aligned}\psi(y, t) = {}&\sqrt[4]{\frac{m\omega}{\pi\hbar}} \exp\left[-\frac{m\omega}{2\hbar}(y - a\cos\omega t)^2\right]\\ &\cdot \exp\left[-\frac{i\omega t}{2}\right]\exp\left\{-\frac{im\omega}{2\hbar}\left[2ay\sin\omega t - \frac{a^2}{2}\sin 2\omega t\right]\right\}.\end{aligned} \tag{5.24}$$

Apart from a complicated phase factor, this is of the same form as (5.23). Consequently, the corresponding probability distribution

$$P(x, t) = |\psi(x, t)|^2 = \sqrt{\frac{m\omega}{\pi\hbar}} \exp\left[-\frac{m\omega}{\hbar}(x - a\cos\omega t)^2\right]$$

is still a Gaussian distribution, but this time it is centered at

$$x = a\cos\omega t.$$

This is highly interesting because this means that *the wave packet oscillates back and forth following exactly the same path as the classical particle*!

If $a = 0$, the probability distribution reduces to the stationary distribution

$$P(x, t) \equiv \sqrt{\frac{m\omega}{2\hbar}} \exp\left[-\frac{m\omega}{\hbar} x^2\right]$$

corresponding to a classical particle sitting in the bottom of the potential well. Thus it is closely related to the classical ground state. It should then not come as a great surprise that the wave function for $a = 0$, i.e.,

$$\psi(x, t) = \sqrt[4]{\frac{m\omega}{\pi\hbar}} \exp\left[-\frac{m\omega}{2\hbar} x^2\right] \exp\left[-i \frac{\omega}{2} t\right], \tag{5.25}$$

represents the quantum-mechanical ground state. Notice that it has the simple time dependence $\exp\left[-\frac{i}{\hbar} E_0 t\right]$, with E_0 precisely equal to the ground state energy $\frac{1}{2}\hbar\omega$. It must therefore, in fact, be the eigenfunction of the Hamiltonian with the eigenvalue $\frac{1}{2}\hbar\omega$.

Exercise 5.2.1

Introduction: Consider a free particle where the corresponding wave function at time $t = 0$ is given by

$$\psi(x, 0) = \frac{1}{\sqrt[4]{2\pi\sigma^2}} \exp\left[-\frac{x^2}{4\sigma^2}\right].$$

Problem: (a) Show that its wave function at a later time t is given by

$$\psi(x, t) = \frac{1}{\sqrt[4]{2\pi\sigma^2(t)}} \exp\left[-\frac{x^2}{4\sigma^2(t)}\right] \exp\left[\frac{i\hbar t}{8m\sigma^2\sigma^2(t)}\right]$$

$$\cdot \exp\left[-\frac{1}{2} \arctan \frac{\hbar t}{2m\sigma^2}\right]; \qquad \sigma^2(t) = \sigma^2 + \frac{\hbar^2 t^2}{4m^2\sigma^2}.$$

(Unlike a particle that is at "rest" in the bottom of a potential well, this wave packet thus spreads out in time!)

(b) Show that the corresponding probability amplitude in "momentum space" is given by

$$\phi(k, t) = \sqrt[4]{\frac{2\sigma^2(t)}{\pi}} \exp\left[-k^2\sigma^2(t)\right]$$

$$\cdot \exp\left[\frac{i\hbar t}{8m\sigma^2\sigma^2(t)}\right] \exp\left[-\frac{1}{2} \arctan \frac{\hbar t}{2m\sigma^2}\right],$$

and as a consequence,

$$\langle p\rangle\langle x\rangle = \frac{\hbar}{2}.$$

Also, the remaining eigenfunctions can easily be extracted from the propagator. With the third characterization of the propagator in mind, we decompose it as follows:

$$K(y; T \mid x; 0) = e^{-i\omega T/2} \sqrt{\frac{m\omega}{\pi\hbar(1 - e^{-2i\omega T})}}$$

$$\cdot \exp\left\{-\frac{m\omega}{\hbar(1 - e^{-2i\omega T})} \left[\frac{1}{2}(y^2 + x^2)(1 + e^{-2i\omega T}) - 2xye^{i\omega T}\right]\right\}.$$

If we introduce the variable $z = e^{-i\omega T}$, this can be rewritten as follows:

$$K(y;T \mid x;0) = z^{-1/2}\exp\left[-\frac{m\omega}{2\hbar}(x^2+y^2)\right]\sqrt{\frac{m\omega}{\pi\hbar(1-z^2)}}$$
$$\cdot \exp\left\{-\frac{m\omega}{\hbar}\left[\frac{(x^2+y^2)z^2-2xyz}{1-z^2}\right]\right\}.$$

This should be compared with

$$\sum_{n=0}^{\infty}\overline{\phi_n(x)}\phi_n(y)\exp\left[-\frac{i}{\hbar}E_nT\right] = z^{-1/2}\sum_{n=0}^{\infty}\phi_n(x)\phi_n(y)z^n.$$

If we put

$$\phi_n(x) = \sqrt{\frac{m\omega}{\pi h}}\exp\left[-\frac{m\omega}{2h}x^2\right](2^n n!)^{-1/2}H_n\left(\sqrt{\frac{m\omega}{h}}\,x\right)$$

and introduce the rescaled variables

$$u = \sqrt{\frac{m\omega}{\hbar}}\,x \quad\text{and}\quad v = \sqrt{\frac{m\omega}{\hbar}}\,y,$$

it follows that $H_n(u)H_n(v)$ has the generating formula

$$\sum_{n=0}^{\infty}H_n(u)H_n(v)\frac{z^n}{2^n n!} = (1-z^2)^{-1/2}\exp\left[\frac{2uvz-(u^2+v^2)z^2}{1-z^2}\right]. \tag{5.26}$$

Since the left-hand side is a Taylor series in z, we can actually find $2^{-n}H_n(u)H_n(v)$ by differentiating the right-hand side n times and thereafter putting $z = 0$. It follows trivially that $H_n(u)$ is a polynomial of degree n in u. In fact, it is a *Hermite polynomial*, and the above formula is the so-called *Mehler's formula*. We leave the details as an exercise:

Worked Exercise 5.2.2
Introduction: The Hermite polynomials $H_n(x)$ are defined by the generating formula

$$\sum_{n=0}^{n}H_n(x)\frac{z^n}{n!} = \exp[2xz - z^2]. \tag{5.27}$$

Problem: (a) Deduce Rodriques's formula

$$H_n(x) = (-1)^n e^{x^2}\frac{d^n}{dx^n}e^{-x^2}.$$

(b) Show that

$$e^{-x^2} = \frac{1}{\sqrt{\pi}}\int e^{-\xi^2+2i\xi x}d\xi,$$

and consequently,

$$H_n(x) = \frac{(-2i)^n}{\sqrt{\pi}}e^{-x^2}\int \xi^n e^{-\xi^2+2i\xi x}d\xi.$$

(c) Prove Mehler's formula

$$\sum_{n=0}^{\infty} H_n(x)H_n(y)\frac{z^n}{2^n n!} = [1-z^2]^{-1/2}\exp\left[\frac{2xyz-(x^2+y^2)z^2}{1-z^2}\right]. \tag{5.28}$$

Using Mehler's formula we can in fact recalculate the propagator. First, we must determine the eigenvalues E_n and the eigenfunctions $\phi_n(x)$ for the Hamiltonian:

$$\left(-\frac{h^2}{2m}\frac{d^2}{dx^2}+\frac{1}{2}m\omega^2x^2\right)\phi_n(x) = E_n\phi_n(x).$$

(This eigenvalue problem is treated in *any* standard text on quantum mechanics!) Once we have the eigenvalues and the eigenfunctions at our disposal, we can finally explicitly perform the summation in (2.70) by means of Mehler's formula. Notice that the phase ambiguity is still present. The Taylor series in Mehler's formula has convergence radius 1, due to the poles at $z = \pm 1$. We are especially interested in the behavior at $z = e^{-i\omega T}$. We are thus actually working directly on the boundary of the convergence domain! This boundary, i.e., the circle $|z| = 1$, is decomposed into two disjoint arcs by the poles $z = \pm 1$. Furthermore, the poles correspond precisely to the caustics $\omega T = n\pi$. At two times t_a and t_b separated by a caustic, we have thus no direct relation between the phases.

Let us finally tackle the problem of the phase ambiguity in the propagator for the harmonic oscillator. Below the first caustic we know that the exact propagator is given by (5.17)

$$K(x_b; T \mid x_a; 0) = e^{-i\pi/4}\sqrt{\frac{m\omega}{2\pi\hbar\sin\omega T}}$$
$$\cdot\exp\left\{\frac{im\omega}{2\hbar\sin\omega T}\left[(x_b^2+x_a^2)\cos\omega T - 2x_bx_a\right]\right\}; \qquad T < \frac{\pi}{\omega}.$$

When $T = \frac{\pi}{2\omega}$, this reduces to the particularly simple expression

$$K\left(x_b; \frac{\pi}{2\omega} \mid x_a; 0\right) = e^{-i\pi/4}\sqrt{\frac{m\omega}{2\pi\hbar}}\exp\left[-i\frac{m\omega}{h}x_bx_a\right].$$

From the group property (2.25) and the translational invariance in time, we know that

$$K\left(x_b; \frac{\pi}{\omega} \mid x_a; 0\right) = \int K\left(x_b; \frac{\pi}{2\omega} \mid x; 0\right) K\left(x; \frac{\pi}{2\omega} \mid x_a; 0\right) dx$$
$$= e^{-i\pi/2}\frac{1}{2\pi}\int \exp\left[-i\frac{m\omega}{\hbar}(x_b+x_a)x\right]\frac{m\omega}{h}dx,$$

which by (5.5) reduces to

$$K\left(x_b; \frac{\pi}{\omega} \mid x_a; 0\right) = e^{-i\cdot\frac{\pi}{2}}\delta(x_b+x_a).$$

This takes care of the first caustic. We now proceed in the same way with the second caustic:

$$K\left(x_b; \frac{2\pi}{\omega} \mid x_a; 0\right) = \int K\left(x_b; \frac{\pi}{\omega} \mid x; 0\right) K\left(x; \frac{\pi}{\omega} \mid x_a; 0\right) dx$$

$$= e^{-i\pi} \int \delta(x_b + x)\delta(x + x_a)dx$$

$$= e^{-i\pi}\delta(x_b - x_a).$$

Continuing in this way, we finally obtain the propagator corresponding to an arbitrary caustic:

$$K\left(x_b; \frac{n\pi}{\omega} \mid x_a; 0\right) = e^{-in\cdot\frac{\pi}{2}}\delta[x_b - (-1)^n x_a]. \tag{5.29}$$

But the propagator on a caustic serves as the initial condition for the propagator on the subsequent segment; i.e., we must demand that

$$\lim_{T\to\frac{n\pi^+}{\omega}} K(x_b; T \mid x_a; 0) = e^{-in\cdot\frac{\pi}{2}}\delta[x_b - (-1)^n x_a].$$

Consider the expression

$$e^{-i\pi/4}\sqrt{\frac{m\omega}{2\pi\hbar|\sin\omega T|}} \exp\left\{\frac{im\omega}{2\hbar\sin\omega T}\left[(x_b^2 + x_a^2)\cos\omega T - 2x_b x_a\right]\right\}.$$

(Notice that it differs from Feynman's expression (5.17) by an inclusion of the absolute value of $\sin\omega T$ under the square root.) If

$$\omega T = n\pi + \omega\epsilon \qquad (0 < \epsilon \ll 1),$$

we can approximate the above expression by

$$e^{-i\pi/4}\sqrt{\frac{m}{2\pi\hbar\epsilon}} \exp\left\{(-1)^n \frac{im}{2\hbar\epsilon}\left[(x_b^2 + x_a^2)(-1)^n - 2x_b x_a\right]\right\}$$

$$= e^{-i\pi/4}\sqrt{\frac{m}{2\pi\hbar\epsilon}} \exp\left\{\frac{im}{2\hbar\epsilon}\left[x_b - (-1)^n x_a\right]^2\right\}.$$

Comparing this with the "infinitesimal" free particle propagator, we see that it has the limit

$$\delta(x_b - (-1)^n x_a).$$

All we need to get the correct boundary condition is therefore just an inclusion of the additional phase factor

$$e^{-i\frac{\pi}{2}\,\mathrm{Int}\left[\frac{\omega T}{\pi}\right]}$$

where Int[x] represents the greatest integer below x. The correct propagator for the harmonic oscillator is consequently given by the following formula:

$$K(x_b; T \mid x_a; 0) = e^{-i\frac{\pi}{2}\left(\frac{1}{2}+\mathrm{Int}\left[\frac{\omega T}{\pi}\right]\right)} \sqrt{\frac{m\omega}{2\pi\hbar|\sin\omega T|}} \cdot \exp\left\{\frac{im\omega}{2\hbar\sin\omega T}\left[(x_b^2 + x_a^2)\cos\omega T - 2x_b x_a\right]\right\}. \tag{5.30}$$

This is the famous *Feynman–Soriau formula*, which takes into account the proper behavior of the propagator at the caustics. It was discovered by Soriau in 1974 and has an interesting geometrical–topological interpretation, which we unfortunately cannot explain within this simple framework. It has also been derived by adding a small anharmonic term, say ϵx^4, to the potential and then studying the limit of the anharmonic propagator as $\epsilon \to 0$.

§ 5.3 The Path Integral Revisited

Up to this point we have been treating the subject of path integrals in a hand-waving way. Although we are definitely not going to turn it into a rigorous concept, we shall now try to make it a bit more precise. The basic idea is to construct a limiting procedure that allows us to compute the path integral as a limit of ordinary multiple integrals (in much the same way as an ordinary integral can be obtained as the limit of Riemann sums).

There are several different ways in which this procedure can be deduced. Let us first concentrate on the same reasoning as we used when we derived the Schrödinger equation (Section 2.9). To compute

$$\int_{x(t_a)=x_a}^{x(t_b)=x_b} \exp\left[\frac{i}{\hbar}\int_{t_a}^{t_b} L\left(x; \frac{dx}{dt}\right) dt\right] D[x(t)],$$

we slice up the time interval from t_a to t_b into N equal pieces:

$$t_0 = t_a, \quad t_1 = t_a + \epsilon, \ldots,$$
$$t_{N-1} = t_a + \epsilon(N-1), \quad t_N = b, \quad \text{with } \epsilon = \frac{t_b - t_a}{N}.$$

By repeated use of the group property (2.25), we then get

$$\int_{x(t_a)=x_a}^{x(t_b)=x_b} \exp\left\{\frac{i}{\hbar} S[x(t)]\right\} D[x(t)] = \int \cdots \int K(x_N; t_N \mid x_{N-1}; t_{N-1}) \cdots K(x_1; t_1 \mid x_0; t_0) dx_1 \cdots dx_{N-1}.$$

(To simplify the notation we have put $x_a = x_0$ and $x_b = x_N$.) At this point the formula is exact, but it presupposes the knowledge of the propagator we want to calculate! In the limit where $N \to \infty$ we can, however, use that the propagators involved become infinitesimal propagators. We can therefore replace them by the approximate expression, cf. (2.53),

$$K(y; t+\epsilon \mid x; t) \approx \sqrt{\frac{m}{2\pi i\hbar\epsilon}} \exp\left[\frac{i}{\hbar}\frac{m}{2}\frac{(x-y)^2}{\epsilon}\right] \exp\left\{-\frac{i}{\hbar}V\left(\frac{x+y}{2}\right)\epsilon\right\}.$$

Here the first two terms come from the free particle propagator, while the last term comes from the interaction amplitude. We thus arrive at the following limiting procedure:

$$\begin{aligned}
&\int_{x(t_a)=x_a}^{x(t_b)=x_b} \exp\left\{\frac{i}{\hbar}\int_{t_a}^{t_b}\left[\frac{m}{2}\left(\frac{dx}{dt}\right)^2 - V(x)\right]dt\right\} D[x(t)] \\
&= \lim_{N\to\infty}\left[\sqrt{\frac{m}{2\pi i\hbar\epsilon}}\right]^N \int\cdots\int \qquad\qquad (5.31)\\
&\exp\left[\frac{i}{\hbar}\sum_{k=1}^{N}\frac{m}{2}\frac{(x_k-x_{k-1})^2}{\epsilon} - V\left(\frac{x_k+x_{k-1}}{2}\right)\epsilon\right] dx_1\cdots dx_{N-1}
\end{aligned}$$

with $\epsilon = (t_b - t_a)/N$.

In fact, this is the original formula given by Feynman. We can rederive the same formula from a slightly different point of view: Consider an arbitrary (continuous) path leading from $(x_a; t_a)$ to $(x_b; t_b)$. Using time slicing, we can approximate this path by the piecewise straight line connecting the intermediary points $(x_0; t_0), \ldots, (x_N; t_N)$. Rather than summing over all paths, we now restrict ourselves to a summation over piecewise linear paths. (See Figure 5.2.) The action of a piecewise linear path is given by

$$\begin{aligned}
\int_{t_0}^{t_N} L\left(x; \frac{dx}{dt}\right) dt = \frac{m}{2}\frac{(x_1-x_0)^2}{\epsilon} + \int_{t_0}^{t_1} V(x)dt + \cdots \\
+ \frac{m}{2}\frac{(x_N-x_{N-1})^2}{\epsilon} + \int_{t_{N-1}}^{t_N} V(x)dt.
\end{aligned}$$

The exponentiated action is therefore approximately given by

$$\exp\left\{\frac{i}{\hbar}S[x(t)]\right\} \approx \exp\left\{\frac{i}{\hbar}\sum_{k=1}^{N}\frac{m}{2}\frac{(x_i-x_{i-1})^2}{\epsilon} - V\left(\frac{x_i+x_{i-1}}{2}\right)\epsilon\right\}.$$

But since the piecewise linear path is completely characterized by its vertex points $(x_1, \ldots, x_{N-1})$, we can now sum up the contributions from the "Nth order" piecewise linear paths simply by integrating over the vertex coordinates. In the limit where $N \to \infty$, the piecewise linear paths "fill out" the whole space of paths, and

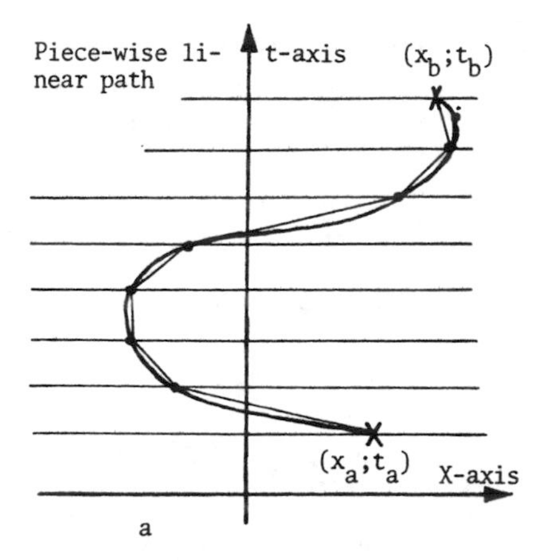

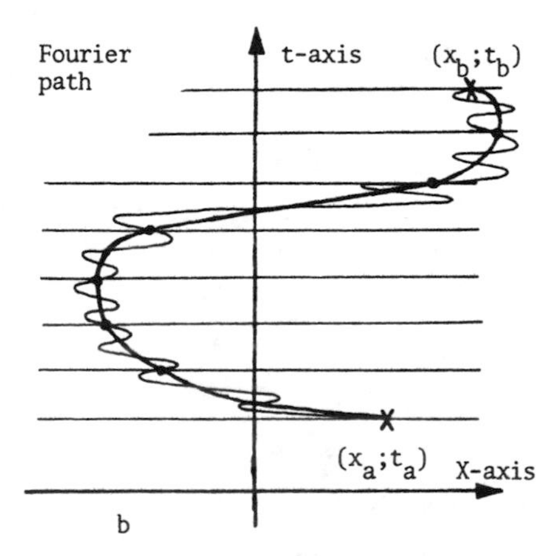

Figure 5.2.

we therefore propose the following limiting procedure:

$$\int_{x(t_a)=x_a}^{x(t_b)=x_b} \exp\left\{\frac{i}{\hbar} S[x(t)]\right\} D[x(t)] \approx \lim_{N\to\infty} \int \cdots \int$$
$$\exp\left[\frac{i}{\hbar}\sum_{k=1}^{N} \frac{m}{2}\frac{(x_k - x_{k-1})^2}{\epsilon} - V\left(\frac{x_k + x_{k-1}}{2}\right)\epsilon\right] dx_1 \cdots dx_{N-1}. \tag{5.32}$$

Comparing this with (5.30), we see that it is given by almost the same expression, except that we have "forgotten" the integration measure

$$\left(\sqrt{\frac{m}{2\pi i\hbar\epsilon}}\right)^N.$$

For various reasons, we will, however, prefer to neglect the integration measure. Rather than calculate a single path integral, we will therefore calculate the ratio of two path integrals (both involving a particle of mass m). Then the measure drops out, and we end up with the formula

$$\frac{\int_{x(t_a)=x_a}^{x(t_b)=x_b} \exp\left\{\frac{i}{\hbar} S[x(t)]\right\} D[x(t)]}{\int_{x(t_a)=x_a}^{x(t_b)=x_b} \exp\left\{\frac{i}{\hbar} S_0[x(t)]\right\} D[x(t)]} \stackrel{\text{def}}{=} \lim_{N\to\infty} \frac{\int\cdots\int \exp\left[\frac{i}{\hbar}\sum_{k=1}^{N} \frac{m}{2}\frac{(x_k-x_{k-1})^2}{\epsilon} - V\left(\frac{x_k+x_{k-1}}{2}\right)\epsilon\right] dx_1\cdots dx_{N-1}}{\int\cdots\int \exp\left[\frac{i}{\hbar}\sum_{k=1}^{N} \frac{m}{2}\frac{(x_k-x_{k-1}^2)}{\epsilon} - V_0\left(\frac{x_k+x_{k-1}}{2}\right)\epsilon\right] dx_1\cdots dx_{N-1}}. \tag{5.33}$$

Usually, S_0 is the free particle action, so that we are actually calculating a propagator relative to a free particle propagator.

When we calculate a path integral, we need not necessarily sum over piecewise straight lines. Other "complete" families of paths can be used as well. Suppose, for example, we want to calculate a propagator of the form

$$K(0; T \mid 0; 0).$$

By slicing up the time interval, we again break up an arbitrary path, $x(t)$, into segments connecting the intermediary points $x_1 = x(t_1), \ldots, x_{N-1} = x(t_{N-1})$. This time we will approximate the given path $x(t)$ by a Fourier path, i.e., a path of the form

$$\tilde{x}(t) = \sum_{k=1}^{N-1} a_k \sin \pi \frac{kt}{T}.$$

(See Figure 5.2b.) If we choose the coefficients a_k such that

$$x_j = \sum_{k=1}^{N-1} a_k \sin\left[\pi \cdot \frac{jk}{N}\right], \qquad j = 1, \ldots, N-1, \tag{5.34}$$

we evidently obtain that the Fourier path passes through the same intermediary points as the given path; cf. Figure 5.2.

Again, the approximate path is completely parametrized by the vertex coordinates $(x_1, \ldots, x_{N-1})$, and we can therefore sum up the contributions from such paths by integrating over the vertex coordinates. In practice it is, however, more convenient to integrate over the Fourier components $(a_1, \ldots, a_{N-1})$. Since the relationship (5.34) between $(x_1, \ldots, x_{N-1})$ and $(a_1, \ldots, a_{N-1})$ is one-to-one and smooth, we evidently have

$$\int\cdots\int \{\quad\} dx_1 \cdots dx_{N-1} = \int\cdots\int \{\quad\} \det\left(\frac{\partial x_i}{\partial a_j}\right) da_1 \cdots da_{N-1}.$$

But the transformation (5.34) is in fact linear, so that the Jacobian is independent of $(a_1, \ldots, a_{N-1})$! We can therefore forget about it, and we finally arrive at the formula

$$\begin{aligned}
&\frac{\int_{x(0)=0}^{x(T)=0} \exp\left\{\frac{i}{\hbar}\, S[x(t)]\right\} D[x(t)]}{\int_{x(0)=0}^{x(T)=0} \exp\left\{\frac{i}{\hbar}\, S_0[x(t)]\right\} D[x(t)]} \\
&= \lim_{N\to\infty} \frac{\int\cdots\int \exp\left\{\frac{i}{\hbar}\, S\left[\sum_{k=1}^{N-1} a_k \sin\frac{k\pi t}{T}\right]\right\} da_1\cdots da_{N-1}}{\int\cdots\int \exp\left\{\frac{i}{\hbar}\, S_0\left[\sum_{k=1}^{N-1} a_k \sin\frac{k\pi t}{T}\right]\right\} da_1\cdots da_{N-1}}\,.
\end{aligned} \tag{5.35}$$

To see how it works, we shall now return to our favorite example:

Test Case: The Harmonic Oscillator. According to Exercise 2.4.1, the path integral can be expanded around the classical path, whereby we get

$$\begin{aligned}
K(x_2; T \mid x_1; 0) &= \exp\left\{\frac{i}{\hbar}\, S[x_{\text{cl}}]\right\} \\
&\cdot \int_{x(0)=0}^{x(T)=0} \exp\left\{\frac{i}{\hbar}\int_0^T \left[\frac{m}{2}\left(\frac{dx}{dt}\right)^2 - \frac{m}{2}\,\omega^2 x^2\right] dt\right\} D[x(t)].
\end{aligned}$$

According to (5.35), this can be further rearranged as

$$\begin{aligned}
K(x_2; T \mid x_1; 0) &= \exp\left\{\frac{i}{\hbar}\, S[x_{\text{cl}}]\right\} K_0(0; T \mid 0; 0) \\
&\cdot \lim_{N\to\infty} \frac{\int\cdots\int \exp\left\{\frac{i}{\hbar}\, S\left[\sum_{k=1}^{N-1} a_k \sin\frac{k\pi t}{T}\right]\right\} da_1\cdots da_{N-1}}{\int\cdots\int \exp\left\{\frac{i}{\hbar}\, S_0\left[\sum_{k=1}^{N-1} a_k \sin\frac{k\pi t}{T}\right]\right\} da_1\cdots da_{N-1}}\,,
\end{aligned}$$

where K_0 is the free particle propagator, which is given by

$$K_0(0; T \mid 0; 0) = \sqrt{\frac{m}{2\pi i\hbar T}}\,.$$

The remaining problem is therefore the computation of the multiple integral. Using the orthogonality relations for the sine function, it is easy to calculate the action, which reduces to

$$S\left\{\sum_{k=1}^{N-1} a_k \sin\frac{k\pi t}{T}\right\} = \frac{mT}{4}\sum_{k=1}^{N-1} a_k^2 \left(\frac{k^2\pi^2}{T^2} - \omega^2\right).$$

But then it is trivial to perform the integration over Fourier components, since the exponent is not only quadratic, but even diagonal in $(a_1, \ldots, a_{N-1})$:

$$\begin{aligned}
&\int\cdots\int \exp\left\{\frac{i}{\hbar}\, S\left[\sum_{k=1}^{N-1} a_k \sin\frac{k\pi t}{T}\right]\right\} da_1\cdots da_{N-1} \\
&= \left(\sqrt{\frac{4\pi i\hbar}{mT}}\right)^{N-1} \prod_{k=1}^{N-1}\left(\frac{k^2\pi^2}{T^2} - \omega^2\right)^{-1/2}.
\end{aligned}$$

To calculate the same multiple integral for the free particle action, we simply put $\omega = 0$ in the above formula. Thus we finally get

$$K(x_2; T \mid x_1; 0) = \exp\left\{\frac{i}{\hbar} S[x_{\text{cl}}]\right\} \cdot \sqrt{\frac{m}{2\pi i \hbar T}} \left[\lim_{N\to\infty} \prod_{k=1}^{N-1} \left(1 - \frac{\omega^2 T^2}{\pi^2 k^2}\right)\right]^{-1/2}.$$

But using Euler's famous product formula for the sine function,

$$\sin(\pi x) = \pi x \left[\prod_{k=1}^{\infty} \left(1 - \frac{x^2}{k^2}\right)\right], \tag{5.36}$$

this reduces precisely to

$$K(x_2; T \mid x_1; 0) = \sqrt{\frac{m}{2\pi i \hbar \sin \omega T}} \exp\left\{\frac{i}{\hbar} S[x_{\text{cl}}]\right\}$$

in accordance with (5.17)!

What about the phase ambiguity? This can also be resolved if we are a little more careful! Consider the analytic extension of the Gaussian integral. If we carefully separate the phase from the modulus, we get

$$\int_{-\infty}^{+\infty} e^{i\lambda x^2} dx = \sqrt{\frac{i\pi}{\lambda}} = \begin{cases} \sqrt{\frac{\pi}{|\lambda|}} e^{i\cdot\frac{\pi}{4}} & \text{when } \lambda > 0, \\ \sqrt{\frac{\pi}{|\lambda|}} e^{-i\cdot\frac{\pi}{4}} & \text{when } \lambda < 0. \end{cases}$$

Now, going back to the action of the Fourier path,

$$\exp\left[\frac{i}{\hbar} S\right] = \exp\left\{\frac{imT}{4\hbar} \sum_{k=1}^{N-1} a_n^2 \left(\frac{k^2\pi^2}{T^2} - \omega^2\right)\right\}, \tag{5.37}$$

we observe that the analogue of λ is negative when $k < \omega T/\pi$, and positive when $k > \omega T/\pi$. Consequently, there are $\text{Int}[\omega T/\pi]$ "negative terms" (where $\text{Int}[x]$ denotes the integer part of x, i.e., the greatest integer smaller than or equal to x). The multiple integral is therefore actually given by

$$\int \cdots \int \exp\left[\frac{imT}{4\hbar} \sum_{k=1}^{N-1} a_k^2 \left(\frac{k^2\pi^2}{T^2} - \omega^2\right)\right] da_1 \cdots da_{N-1}$$
$$= e^{-i\frac{\pi}{2}\text{Int}\left[\frac{\omega T}{\pi}\right]} \left(e^{i\frac{\pi}{4}} \sqrt{\frac{4\pi\hbar}{mT}}\right)^{N-1} \prod_{k=1}^{N-1} \left|\frac{k^2\pi^2}{T^2} - \omega^2\right|^{-1/2}.$$

For the free particle propagator this ambiguity does not occur. The corrected formula for the propagator thus comes out as follows:

$$K(x_2; T \mid x_1; 0) = e^{-i\left(\frac{\pi}{4} + \frac{\pi}{2}\text{Int}\left[\frac{\omega T}{\pi}\right]\right)} \sqrt{\frac{m\omega}{2\pi i \hbar |\sin \omega T|}} \exp\left\{\frac{i}{\hbar} S[x_{\text{cl}}]\right\}. \tag{5.38}$$

But that is precisely the Feynman–Soriau formula!

Now that we have seen how the limiting procedure based upon Fourier paths works, let us return for a moment to the problem of the integration measure. If we want to calculate a path integral directly, i.e., we do not calculate a ratio any longer, we must necessarily incorporate an integration measure in the limiting procedure. Feynman suggested that one could use the same integration measure as for the piecewise linear path. This assumption leads to the formula

$$\int_{x(0)=0}^{x(T)=0} \exp\left\{\frac{i}{\hbar}\int_0^T \left[\frac{m}{2}\left(\frac{dx}{dt}\right)^2 - V(x)\right] dt\right\} D[x(t)]$$
$$\stackrel{?}{=} \lim_{N\to\infty}\left(\sqrt{\frac{m}{2\pi i\hbar\epsilon}}\right)^N$$
$$\int\cdots\int \exp\left\{\frac{i}{\hbar}S\left[\sum_{k=1}^{N-1} a_k \sin\frac{k\pi t}{T}\right]\right\} dx_1\cdots dx_{N-1} \tag{5.39}$$
$$= \lim_{N\to\infty}\left(\sqrt{\frac{m}{2\pi i\hbar\epsilon}}\right)^N \det\left[\frac{\partial x_i}{\partial a_j}\right]$$
$$\int\cdots\int \exp\left\{\frac{i}{\hbar}S\left[\sum_{k=1}^{N-1} a_k \sin\frac{k\pi t}{T}\right]\right\} da_1\cdots da_{N-1}.$$

But here the right-hand side diverges, as we can easily see, when we try to calculate it in the case of a free particle; cf. Exercise 5.3.1 below:

Exercise 5.3.1
Introduction: Consider the matrix $\overset{=}{A}$ involved in the linear transformation (5.34), i.e.,

$$A_{jk} = \sin\left[\pi\,\frac{jk}{N}\right]; \qquad j,k = 1,\ldots,N-1.$$

(a) Show that

$$\overset{=}{A}\overset{=}{A}^T = \frac{N}{2}\cdot\overset{=}{I} \quad\text{and}\quad \det[\overset{=}{A}] = \left[\frac{N}{2}\right]^{\frac{N-1}{2}}.$$

(b) Show that the series involved in the limiting procedure (5.39) is given by

$$\lim_{N\to\infty}\frac{1}{\Gamma(N)}\left[\frac{N}{\pi}\right]^{N-1}\sqrt{\frac{mM}{2\pi ihT}},$$

and using Stirling's formula for the Γ-function to verify that it diverges.

In fact, by considering the case of the free particle propagation, it is not difficult to construct the correct integration measure, which (when we integrate over the

Fourier components) turns out to be given by

$$\int_{x(0)=0}^{x(T)=0} \exp\left\{\frac{i}{\hbar} S[x(t)]\right\} D[x(t)] = \lim_{N\to\infty} \left(\frac{\pi}{\sqrt{2}}\right)^{N-1} \Gamma(N)$$

$$\cdot \left[\sqrt{\frac{m}{2\pi i\hbar T}}\right]^N \int \cdots \int \exp\left\{\frac{i}{\hbar} S\left[\sum_{k=1}^{N-1} a_k \sin\frac{k\pi t}{T}\right]\right\} d^{N-1}a.$$

(As a consistency check this formula can be used to rederive the propagator of the harmonic oscillator.) So now it is clear why we prefer to neglect the integration measure: Every time we introduce a new limiting procedure, i.e., a new denumerable complete set of paths, we would have to introduce a new integration measure!

§ 5.4 Illustrative Example: The Time-Dependent Oscillator

As another very important example of an exact calculation of a propagator, we now look at the quadratic Lagrangian

$$L = \frac{1}{2} m \left(\frac{dx}{dt}\right)^2 - \frac{1}{2} mW(t)x^2.$$

Since it is quadratic, we can as usual expand around a classical solution, so that we need only bother about the calculation of the following path integral:

$$\int_{x(t_a)=0}^{x(t_b)=0} \exp\left\{\frac{im}{2\hbar}\int_{t_a}^{t_b}\left[\left(\frac{dx}{dt}\right)^2 - W(t)x^2\right]dt\right\} D[x(t)].$$

Let us first rewrite the action in a more suitable form. Performing a partial integration, we get

$$S[x(t)] = -\frac{m}{2}\int_{t_a}^{t_b}\left[x\frac{d^2x}{dt^2} + W(t)x^2\right]dt.$$

Here we have neglected the boundary terms due to the boundary conditions $x(t_a) = x(t_b) = 0$. Inserting this, we see immediately that the above path integral is actually an infinite-dimensional generalization of the usual Gaussian integral, since it now takes the form

$$\int_{x(t_a)=0}^{x(t_b)=0} \exp\left\{-\frac{im}{2\hbar}\int_{t_a}^{t_b} x(t)\left[\frac{d^2}{dt^2} + W(t)\right]x(t)dt\right\} D[x(t)].$$

To compute it, we ought therefore to diagonalize the Hermitian operator

$$\frac{d^2}{dt^2} + W(t).$$

At the moment, we shall, however, proceed a little differently. By a beautiful transformation of variables we can change the action to the free particle action! Let f be a solution of second-order differential equation

$$\left\{ \frac{d^2}{dt^2} + W(t) \right\} f(t) = 0; \tag{5.40}$$

i.e., f belongs to the kernel of the above differential operator. The solution can be chosen almost completely arbitrarily within the two-dimensional solution space. The only thing we will assume is that f does not vanish at the initial endpoint:

$$f(t_a) \neq 0.$$

(Notice that $f(t)$ is not an admissible path, since it breaks the boundary conditions!) Using f, we then construct the following linear transformation, where $x(t)$ is replaced by the path $y(t)$:

$$x(t) = f(t) \int_{t_a}^{t} \frac{y'(s)}{f(s)} ds. \tag{5.41}$$

Here we assume that the transformed function $y(t)$ also satisfies the boundary condition

$$y(t_a) = 0.$$

Differentiating (5.41), we obtain

$$x'(t) = f'(t) \int_{t_a}^{t} \frac{y'(s)}{f(s)} ds + y'(t) = \frac{f'(t)}{f(t)} x(t) + y'(t),$$

so that the inverse transformation is given by

$$y(t) = x(t) - \int_{t_a}^{t} \frac{f'(s)}{f(s)} x(s) ds. \tag{5.42}$$

We can now show that the above transformation has the desired effect. Using that

$$x''(t) = f''(t) \int_{t_a}^{t} \frac{y'(s)}{f(s)} ds + \frac{f'(t)y'(t)}{f(t)} + y''(t),$$

we obtain

$$\left\{ \frac{d^2}{dt^2} + W(t) \right\} x(t) = \{f''(t) + W(t)f(t)\} \int_{t_a}^{t} \left[\frac{y'(s)}{f(s)} \right] ds$$
$$+ \frac{f'(t)y'(t)}{f(t)} + y''(t).$$

But here the first term vanishes on account of (5.40). Consequently, the action reduces to

$$S[x(t)] = -\frac{m}{2} \int_{t_a}^{t_b} [F(t)f'(t)y'(t) + F(t)f(t)y''(t)]dt$$
$$\text{with } F(t) = \int_{t_a}^{t} \frac{y'(s)}{f(s)} ds.$$

By performing a partial integration on the second term, this can be further rearranged as

$$S[x(t)] = \frac{m}{2}\int_{t_a}^{t_b}\left[\left(\frac{dy}{dt}\right)^2\right]dt - \frac{m}{2}\left[x(t)y'(t)\right]_{t_a}^{t_b}.$$

But here the boundary terms vanish due to the boundary conditions satisfied by $x(t)$. Thus we precisely end up with the free particle action in terms of the transformed path $y(t)$!

There is only one complication associated with the above transformation, and that is the boundary condition associated with the final endpoint t_b. The boundary conditions satisfied by $x(t)$ are transformed into the following conditions on the transformed path $y(t)$:

$$y(t_a) = 0; \qquad \int_{t_a}^{t_b}\frac{y'(s)}{f(s)}\,ds = 0.$$

The second boundary condition is nonlocal and therefore not easy to handle directly. We shall therefore introduce another trick! Using the identity

$$\delta(x(t_b)) = \frac{1}{2\pi}\int \exp\{-i\alpha x(t_b)\}d\alpha,$$

cf. (5.5), we can formally introduce an integration over the final endpoint:

$$\int_{x(t_a)=0}^{x(t_b)=0}\exp\left\{\frac{i}{\hbar}S[x(t)]\right\}D[x(t)]$$
$$= \frac{1}{2\pi}\int_{x(t_a)=0}^{x(t_b)\text{arbitrary}}\int_{-\infty}^{\infty}\exp\left[-i\alpha x(t_b)\right]\exp\left\{\frac{i}{\hbar}S[x(t)]\right\}d\alpha\, D[x(t)].$$

This is because the integration over α now produces a δ-function that picks up the correct boundary condition! (Notice that if we attempt to calculate the path integral by a limiting procedure, we must now also integrate over the final endpoint x_N.) Changing variables, we then get

$$\cdots = \frac{1}{2\pi}\int_{y(t_a)=0}^{y(t_b)\text{arbitrary}}\int_{-\infty}^{\infty}\exp\left\{-i\alpha f(t_b)\int_{t_a}^{t_b}\frac{y'(s)}{f(s)}ds\right\}$$
$$\cdot\exp\left\{\frac{im}{2\hbar}\int_{t_a}^{t_b}\left(\frac{dy}{dt}\right)^2dt\right\}\det\left[\frac{\delta x}{\delta y}\right]d\alpha\, D[y(t)].$$

Here the infinite-dimensional generalization of the Jacobi determinant is independent of $y(t)$ because the transformation (5.41) is linear. The remaining integral is Gaussian. "Completing the square," the whole formula therefore reduces to

$$\cdots = \frac{1}{2\pi}\det\left[\frac{\delta x}{\delta y}\right]\int_{-\infty}^{\infty}\exp\left\{-\frac{i\hbar}{2m}\alpha^2 f^2(t_b)\int_{t_a}^{t_b}\frac{dt}{f^2(t)}\right\}da$$
$$\cdot\int_{\gamma(t_a)=0}^{\gamma(t_b)\text{arbitrary}}\exp\left\{\frac{im}{2\hbar}\int_{t_a}^{t_b}\left(\frac{d\gamma}{dt}\right)^2dt\right\}D[\gamma(t)],$$

with

$$\gamma(t) = y(t) - \frac{\alpha t}{m} f(t_b) \int_{t_a}^{t} \frac{ds}{f(s)} .$$

At this point we get a pleasant surprise: We can actually carry out the α-integration! Furthermore, the remaining path integral is within our reach, since it involves only the free particle propagator. In fact, we get

$$\begin{aligned}
&\int_{\gamma(t_a)=0}^{\gamma(t_b)\text{arbitrary}} \exp\left\{ \frac{im}{2\hbar} \int_{t_a}^{t_b} \left(\frac{d\gamma}{dt} \right)^2 dt \right\} D[\gamma(t)] \\
&\quad = \int_{-\infty}^{\infty} K_0(x; t_b \mid 0; t_a) dx \\
&\quad = \sqrt{\frac{m}{2\pi i h(t_b - t_a)}} \int_{-\infty}^{\infty} \exp\left[\frac{im}{2h} \frac{x^2}{(t_b - t_a)} \right] dx = 1.
\end{aligned}$$

This should hardly come as a surprise, since by construction, the above path integral represents the probability amplitude for finding the particle *anywhere* at the time t_b. The total path integral thus collapses into the simple expression

$$\begin{aligned}
&\int_{x(t_a)=0}^{x(t_b)=0} \exp\left\{ \frac{im}{2\hbar} \int_{t_a}^{t_b} \left[\left(\frac{dx}{dt} \right)^2 - W(t)x^2 \right] dt \right\} D[x(t)] \\
&\quad = \det\left[\frac{\delta x}{\delta y} \right] \sqrt{\frac{m}{2\pi i h f^2(t_b) \int_{t_a}^{t_b} \frac{ds}{f^2(s)}}} .
\end{aligned} \tag{5.43}$$

It remains to calculate the Jacobian! We shall calculate it using a very naive approach. (The following argument is included for illustrative purposes only; it is certainly not a rigorous procedure!) As in the approximation procedure for path integrals, we discretize the linear transformation by introducing a time slicing. The paths $x(t)$ and $y(t)$ are then replaced by the multidimensional points

$$(x_0, x_1, \ldots, x_N) \text{ and } (y_0, y_1, \ldots, y_N) \text{ with } x_k = x(t_k) \text{ and } y_k = y(t_k).$$

The linear transformation (5.42) can then be approximated by

$$\begin{aligned}
y_n &= x_n - \frac{T}{N} \sum_{k=1}^{n} \frac{f'(t_k)}{f(t_k)} \frac{(x_k + x_{k-1})}{2} \\
&= \sum_{k=1}^{N} \delta_{nk} x_k - \frac{1}{2} \sum_{k=1}^{n} \frac{f'(t_k)}{f(t_k)} \frac{T}{N} x_k - \frac{1}{2} \sum_{k=0}^{n-1} \frac{f'(t_{k+1})}{f(t_{k+1})} \frac{T}{N} x_k .
\end{aligned}$$

(This is actually *the* delicate point, since the discrete approximation of the integral is by no means unique, and the Jacobi determinant turns out to be very sensitive to the choice of the approximation procedure.) Okay, so the Jacobi matrix has now been replaced by a lower diagonal matrix. The determinant thus comes exclusively from the diagonal; i.e.,

$$J_N = \det\left[\frac{\partial y_i}{\partial x_j}\right] = \prod_{k=1}^{N}\left(1 - \frac{1}{2}\frac{f'(t_k)}{f(t_k)}\frac{T}{N}\right).$$

Taking the limit $N \to \infty$, we then find that

$$\begin{aligned}
\det\left[\frac{\delta y}{\delta x}\right] &= \lim_{N\to\infty} J_N = \lim_{N\to\infty}\exp\left[\log\prod_{k=1}^{N}\left(1 - \frac{1}{2}\frac{f'(t_k)}{f(t_k)}\frac{T}{N}\right)\right] \\
&= \lim_{N\to\infty}\exp\left[\sum_{k=1}^{N}\log\left(1 - \frac{1}{2}\frac{f'(t_k)}{f(t_k)}\frac{T}{N}\right)\right] \\
&= \lim_{N\to\infty}\exp\left[-\frac{1}{2}\left(\sum_{k=1}^{N}\frac{f'(t_k)}{f(t_k)}\frac{T}{N}\right)\right] \\
&= \exp\left[-\frac{1}{2}\int_{t_a}^{t_b}\frac{f'(t)}{f(t)}\,dt\right] \\
&= \exp\left[-\frac{1}{2}\log\left\{\frac{f(t_b)}{f(t_a)}\right\}\right] = \sqrt{\frac{f(t_a)}{f(t_b)}}.
\end{aligned}$$

Consequently,

$$\det\left[\frac{\delta x}{\delta y}\right] = \det{}^{-1}\left[\frac{\delta y}{\delta x}\right] = \sqrt{\frac{f(t_b)}{f(t_a)}}.$$

Inserting this into (5.43), our formula for the path integral finally boils down to the following remarkably simple result.

The path integral corresponding to the quadratic action

$$\begin{aligned}
S[x(t)] &= \frac{m}{2}\int_{t_a}^{t_b}\left[\left(\frac{dx}{dt}\right)^2 - W(t)x^2\right]dt \\
&= -\frac{m}{2}\int_{t_a}^{t_b} x(t)\left[\frac{d^2}{dt^2} + W(t)\right]x(t)dt
\end{aligned} \tag{5.44}$$

is given by

$$\int_{x(t_a)=0}^{x(t_b)=0}\exp\left\{\frac{i}{\hbar}S[x(t)]\right\}D[x(t)] = \sqrt{\frac{m}{2\pi i\hbar f(t_a)f(t_b)\int_{t_a}^{t_b}\frac{dt}{f^2(t)}}}, \tag{5.45}$$

where $f(t)$ is an almost arbitrary solution to the differential equation

$$\left\{\frac{d^2}{dt^2} + W(t)\right\}f(t) = 0,$$

the only constraint being that $f(t_a) \neq 0$.

Notice that the above formula in fact includes the free particle propagator (with $W(t) = 0$) and the harmonic oscillator (with $W(t) = \omega^2$). One can easily check that (5.45) reproduces our previous findings in these two cases by putting $f(t) \equiv 1$, respectively $f(t) = \cos\omega(t - t_a)$.

§ 5.5 Path Integrals and Determinants

Now that we have the formula for the propagator corresponding to a quadratic Lagrangian at our disposal, we will look at it from a somewhat different point of view. Since the action is quadratic,

$$S[x(t)] = -\frac{m}{2}\int_{t_a}^{t_b} x(t)\left\{\frac{d^2}{dt^2} + W(t)\right\} x(t)dt,$$

we can "diagonalize" it. For this purpose we consider the Hermitian operator

$$-\frac{d^2}{dt^2} - W(t),$$

which acts upon the space of paths $x(t)$ all satisfying the boundary conditions $x(t_a) = x(t_b) = 0$. It possesses a complete set of normalized eigenfunctions $\phi_n(t)$:

$$\left\{-\frac{d^2}{dt^2} - W(t)\right\}\phi_n(t) = \lambda_n\phi_n(t);$$

$$\phi_n(t_a) = \phi_n(t_b) = 0; \qquad \int_{t_a}^{t_b}\phi_n^2(t)dt = 1.$$

A given path $x(t)$ can now be approximated by a linear combination

$$\tilde{x}_N(t) = \sum_{n=1}^{N} a_n\phi_n(t) \qquad \text{with } a_n = \int_{t_a}^{t_b}\phi_n(t)x(t)dt.$$

Notice that the corresponding action of the approximative path reduces to

$$S[\tilde{x}_N(t)] = \frac{m}{2}\sum_{n=1}^{N}\lambda_n a_n^2.$$

The summation over approximative paths can therefore immediately be carried out, since it is just a product of ordinary Gaussian integrals:

$$\int\cdots\int \exp\left\{\frac{i}{\hbar}\, S[x_N(t)]\right\} da_1\cdots da_N = \left(\sqrt{\frac{2\pi i\hbar}{m}}\right)^N \sqrt{\frac{1}{\lambda_1\cdots\lambda_N}}\,.$$

In the limit where $N \to \infty$, the approximative paths fill out the whole space of paths, and the path integral is therefore essentially given by

$$\left(\prod_{n=1}^{\infty} \lambda_n\right)^{-1/2}.$$

By analogy with the finite-dimensional case, we define the determinant of the Hermitian operator $-\partial_t^2 - W(t)$ to be the infinite product of eigenvalues. Of course, the determinant will in general be highly divergent, but we can "regularize" it in the usual way by calculating the ratio of two determinants. From the above calculation we then learn the following important lesson:

The path integral corresponding to a quadratic Lagrangian is essentially given by the determinant of the associated differential operator; i.e.,

$$\begin{aligned}\int_{x(t_a)=0}^{x(t_b)=0} \exp\left\{\frac{im}{2\hbar}\int_{t_a}^{t_b} x(t)\left[-\frac{d^2}{dt^2} - W(t)\right]x(t)dt\right\} D[x(t)] \\ = \Delta\left\{\det\left[-\frac{d^2}{dt^2} - W(t)\right]\right\}^{-1/2},\end{aligned} \tag{5.46}$$

where the right-hand side should actually be interpreted as a limiting procedure (*i.e., it is a shorthand version of the following expression*):

$$\begin{aligned}\lim_{N\to\infty} \Delta(N)\left[\prod_{n=1}^{N} \lambda_n\right]^{-1/2} \\ = \lim_{N\to\infty} \Delta(N) \int \cdots \int \exp\left[\frac{im}{2\hbar}\sum_{n=1}^{N} \lambda_n a_n^2\right]\left(\sqrt{\frac{m}{2\pi i\hbar}}\right)^N d^N a.\end{aligned} \tag{5.47}$$

Notice that we have included a factor $\left(\sqrt{\frac{m}{2\pi i\hbar}}\right)^N$ in the integration over the generalized Fourier components; i.e., *the proper integration measure for evaluating the determinant* is given by

$$\prod_{k=1}^{N}\left[\frac{m}{2\pi i\hbar}\, da_k\right]. \tag{5.48}$$

Apart from that, we still need a further integration measure $\Delta(N)$, since the determinant is only proportional to the path integral. The above characterization of the path integral is the one used by, for instance, Coleman (1977).

Since we already know how to compute the path integral, we can now extract a relation for the determinant. To avoid divergence problems, we calculate the ratio of two determinants. According to (5.45), it is given by

$$\frac{\det[-\partial_t^2 - W(t)]}{\det[-\partial_t^2 - V(t)]} = \frac{f_W(t_a) f_W(t_b) \int_{t_a}^{t_b} \frac{dt}{f_W^2(t)}}{f_V(t_a) f_V(t_b) \int_{t_a}^{t_b} \frac{dt}{f_V^2(t)}}. \tag{5.49}$$

In this formula it is presupposed that f_W and f_V doe not vanish at t_a. It can, however, be simplified considerably by going to the singular limit, where f_W and f_V *do* vanish at t_a! Let us denote f_W^0 the unique solution to the differential equation $\left\{-\partial_t^2 - W(t)\right\} f(t) = 0$ which satisfies the boundary conditions

$$f_W^0(t_a) = 0; \qquad \frac{d}{dt}\, f_W^0(t_a) = 1.$$

Similarly, we denote by f_W^1 the solution that satisfies the boundary conditions

$$f_W^1(t_a) = 1; \qquad \frac{d}{dt}\, f_W^1(t_a) = 0.$$

In the above identity (5.48), we can then put

$$f_W = f_W^0 + \epsilon f_W^1; \qquad f_V = f_V^0 + \epsilon f_V^1.$$

It follows that

$$\lim_{\epsilon\to 0} \frac{f_W(t_a)}{f_V(t_a)} = 1 \quad \text{and} \quad \lim_{\epsilon\to 0} \frac{f_W(t_b)}{f_V(t_b)} = \frac{f_W^0(t_b)}{f_V^0(t_b)}.$$

Finally, the limit of the integral

$$\int_{t_a}^{t_b} \frac{dt}{[f_W(t)]^2}$$

diverges, due to the vanishing of f_W^0 at t_a. But since almost all the contribution to the integral comes from an "infinitesimal neighborhood" of t_a (in the limit where $\epsilon \to 0$), it follows that it diverges like

$$\int_{t_a}^{t_b} \frac{dt}{[t - t_a]^2}.$$

Consequently,

$$\lim_{\epsilon\to 0} \left\{ \int_{t_a}^{t_b} \frac{dt}{f_W^2(t)} \right\} \Big/ \left\{ \int_{t_a}^{t_b} \frac{dt}{f_V^2(t)} \right\} = 1.$$

Thus the identity (5.49) collapses into the extremely simple determinantal relation

$$\frac{\det[-\partial_t^2 - W(t)]}{\det[-\partial_t^2 - V(t)]} = \frac{f_W^0(t_b)}{f_V^0(t_b)}. \tag{5.50}$$

If, for example, we put $V(t) = 0$ (and consequently $f_V^0(t) = t - t_a$), it follows that

$$\frac{\det[-\partial_t^2 - W(t)]}{\det[-\partial_t^2]} = \frac{f_W^0(t_b)}{t_b - t_a}.$$

This can be used to calculate a propagator (corresponding to a quadratic Lagrangian) relative to the free particles propagator K_0:

$$
\begin{aligned}
K(x_b; t_b \mid x_a; t_a) &= K(0, t_b \mid 0; t_a) \exp\left\{\frac{i}{\hbar} S[x_{\text{cl}}]\right\} \\
&= \left[\frac{\det\{-\partial_t^2 - W(t)\}}{\det\{-\partial_t^2\}}\right]^{-1/2} K_0(0; t_b \mid 0; t_a) \exp\left\{\frac{i}{\hbar} S[x_{\text{cl}}]\right\} \\
&= \sqrt{\frac{m}{2\pi i\hbar f_W^0(t_b)}} \exp\left\{\frac{i}{\hbar} S[x_{\text{cl}}]\right\}.
\end{aligned}
\tag{5.51}
$$

E.g., in the case of the harmonic oscillator we put $W(t) = \omega^2$, and consequently, $f_W(t) = \frac{1}{\omega} \sin\omega(t - t_a)$. In this way we recover the by now well-known result (5.17).

Remark. Incidentally, the investigation of determinants corresponding to linear operators has a long tradition in mathematics. For example, the basic determinantal relation (5.50) has been known at least since the twenties (Van Vleck, 1928). This is very fortuitous, since in our derivation some dirt has been swept under the rug. The passage from quotients of path integrals to quotients of determinants works only if the integration measure $\Delta(N)$ introduced in (5.47) is actually independent of the potential function $W(t)$. Since the result we have deduced, i.e., the determinantal relation (5.50), is known to be correct, we have thus now justified this assumption.

In fact, the determinantal relation (5.50) can be proved by a very general and beautiful reasoning that emphasizes its basic position. For the benefit of those who are acquainted with more advanced analysis, we include the main line of the argument:

Consider the expressions

$$
g(\lambda) = \frac{\det(-\partial_t^2 - W(t) - \lambda)}{\det(-\partial_t^2 - V(t) - \lambda)}
$$

and

$$
h(\lambda) = \frac{f^0_{(W+\lambda)}(t_b)}{f^0_{(V+\lambda)}(t_b)}.
$$

It is important for the following that λ is treated as a complex variable. It can be proven quite generally that a differential operator of the form

$$
-\partial_t^2 - W(t)
$$

has a discrete spectrum of real eigenvalues

$$
\lambda_1, \lambda_2, \ldots, \lambda_n, \ldots
$$

that are bounded below and that tend to infinity as $n \to \infty$. Each of these eigenvalues has multiplicity one; i.e., the associate eigenspace is one-dimensional.

When λ coincides with one of these eigenvalues, say $\lambda = \lambda_n$, it follows that the "shifted" operator

$$\left\{-\partial_t^2 - W(t) - \lambda_n\right\}$$

has the eigenvalue zero, so that its determinant vanishes. As a consequence, the function $g(\lambda)$ becomes a meromorphic function with simple zeros at the eigenvalues λ_k^W and simple poles at the eigenvalues λ_n^V.

Consider next the function $f^0_{W+\lambda}(t)$. By definition it is the unique solution to the differential equation

$$\left\{-\partial_t^2 - W(t) - \lambda\right\} f(t) = 0$$

that satisfies the boundary conditions

$$f(t_a) = 0; \qquad f'(t_a) = 1.$$

It follows that λ is an eigenvalue of the operator $\left\{-\partial_t^2 - W(t)\right\}$ if and only if

$$f^0_{(W+\lambda)}(t_b) = 0,$$

and when that happens $f_{(W+\lambda)}(t)$ is in fact the associated eigenfunction (except for a normalization factor). As a consequence, the function $h(\lambda)$ becomes a meromorphic function with simple zeros at λ_k^W and simple poles at λ_n^V.

The quotient function

$$g(\lambda)/h(\lambda)$$

is thus an analytic function without zeros and poles! (Thus it has a behavior similar to exp[λ], for example.) Furthermore, from general properties of determinants (respectively solutions to differential equations) it can be shown that $g(\lambda)$ (respectively $h(\lambda)$) tends to 1 as λ tends to infinity, except along the real axis. The same then holds for the quotient, which as a consequence must be a bounded analytic function. But then a famous criterion of Liouville guarantees that it is constant; i.e.,

$$\frac{g(\lambda)}{h(\lambda)} = 1.$$

Specializing to $\lambda = 0$ we precisely recover the determinantal relation (5.50).

§ 5.6 The Bohr–Sommerfield Quantization Rule

We shall now encounter the important problem of computing quantum corrections to classical quantities. In particular, we shall consider quantum corrections to the classical energy. In the case of a one-dimensional particle, we know that it can be found by studying the trace of the propagator. Inserting the path integral expression for the propagator, this trace can be reexpressed as

$$
\begin{aligned}
G(T) &= \int_{-\infty}^{\infty} \int_{x(0)=x_0}^{x(T)=x_0} \exp\left\{\frac{i}{\hbar}\, S[x(t)]\right\} D[x(t)]dx_0 \\
&= \int_{x(0)=x(T)} \exp\left\{\frac{i}{\hbar}\, S[x(t)]\right\} D[x(t)],
\end{aligned}
\tag{5.52}
$$

where we consequently sum over all paths that return to the same point. This expression is known as the *path-cum-trace integral*.

There are now in principle two different approximation procedures available for the evaluation of this path-cum-trace integral. The first method is the *weak-coupling approximation*. Let us assume that the potential has the general shape indicated in Figure 5.3. To calculate the low-lying energy states, we notice that near the bottom of the potential well we can approximate the potential by

$$V(x) \approx \frac{1}{2}\, V''(0)x^2.$$

Thus the problem is essentially reduced to the calculation of the path-cum-trace integral for the harmonic oscillator! This we have already solved; cf. (5.19–20), and the low-lying energy states are consequently given by

$$E_n \approx \left(n + \frac{1}{2}\right) h\omega \qquad \text{with } V''(0) = m\omega^2. \tag{5.53}$$

The second method is the so-called *WKB-approximation*, which can be used to find the high-lying energy states. It is thus essentially a nonperturbative method. It leads to the so-called *Bohr–Sommerfeld quantization rule*, which in its original form states that

$$\oint p\, dq = n \cdot h, \tag{5.54}$$

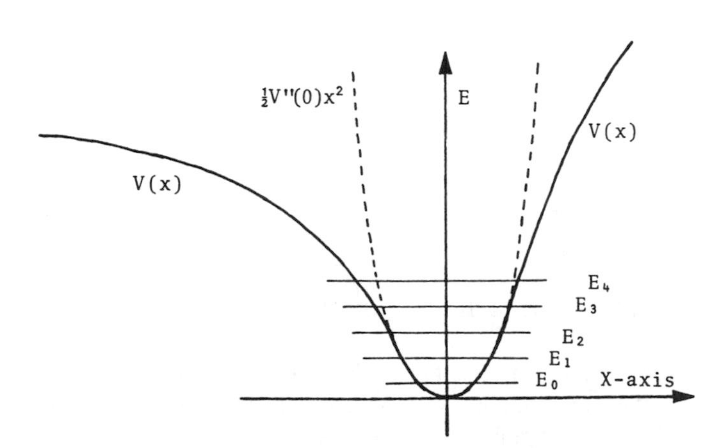

Figure 5.3.

where q is the position coordinate, p the conjugate momentum, and we integrate along a periodic orbit. Notice that the left-hand side is the area bounded by the closed orbit; cf. Figure 5.4, so that the quantization rule states that *the area enclosed in phase-space has to be an integral multiple of h*. The original WKB method, which was based upon the construction of approximative solutions to the Schrödinger equation, will not be discussed here. But before we jump out in the path integral version of the WKB method, it will be useful to take a closer look at classical mechanics and the original derivative of the quantization rule (5.54).

ILLUSTRATIVE EXAMPLE:
THE BOHR–SOMMERFELD QUANTIZATION RULE.

Let us first collect a few useful results concerning the action. Suppose $(x_a;\, t_a)$ and $(x_b;\, t_b)$ are given space–time points. Then

$$S(x_b;\, t_b \mid x_a;\, t_a)$$

will denote the action along the classical path connecting the two space–time points. The resulting function of the initial and final space–time points is known as *Hamilton's principle function*. (If there is more than one classical path connecting the space–time points $(x_a;\, t_a)$ and $(x_b;\, t_b)$, it will be a multivalued function.) All the partial derivatives of Hamilton's principle function have direct physical significance:

$$\frac{\partial S}{\partial x_a} = -p_a; \qquad \frac{\partial S}{\partial x_b} = p_b; \qquad \frac{\partial S}{\partial t_a} = E; \qquad \frac{\partial S}{\partial t_b} = -E. \tag{5.55}$$

Here p_a is the conjugate momentum at the initial point x_a, p_b the conjugate momentum at x_b, and E the energy (which is conserved along the classical path).

PROOF. A change δx_b in the spatial position of the final space–time point will cause a change $\delta x(t)$ in the classical path. (See Figure 5.5.) The corresponding

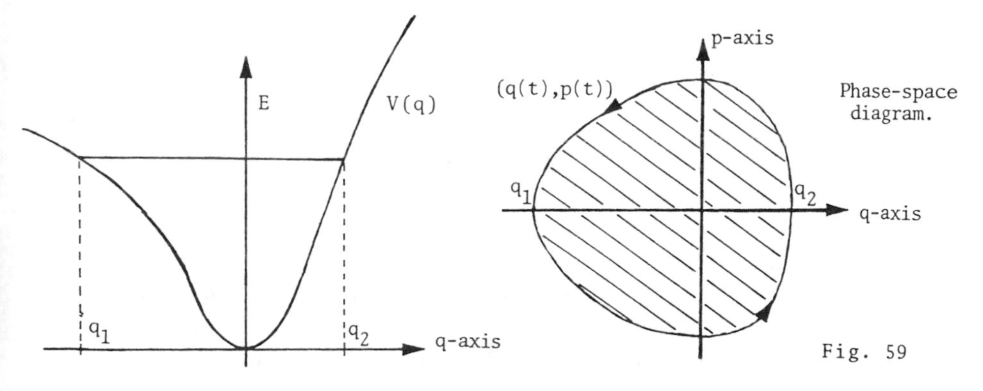

Figure 5.4.

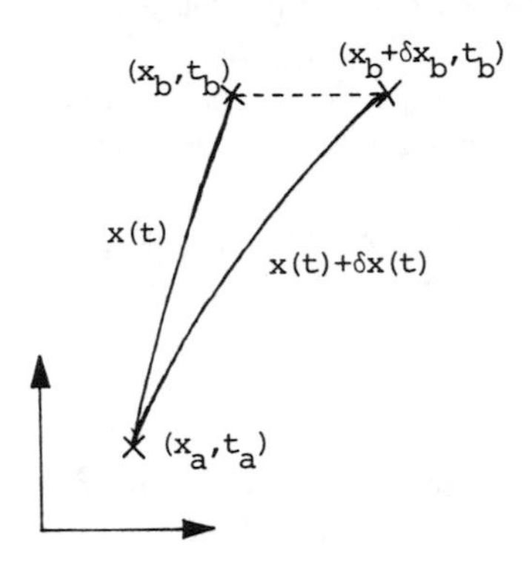

Figure 5.5.

change in the action is given by

$$\delta S = \int_{t_a}^{t_b} \left[\frac{\partial L}{\partial x} \,\delta x(t) + \frac{\partial L}{\partial \dot{x}} \,\delta \dot{x}(t) \right] dt$$
$$= \int_{t_a}^{t_b} \left[\frac{\partial L}{\partial x} - \frac{d}{dt}\frac{\partial L}{\partial \dot{x}} \right] \delta x(t) dt + \left[\frac{\partial L}{\partial \dot{x}} \delta x(t) \right]_{t_a}^{t_b} .$$

But here the integrand vanishes due to the equations of motion, and we end up with

$$\delta S = \frac{\partial L}{\partial \dot{x}} \,(t_b)\delta x(t_b) - \frac{\partial L}{\partial \dot{x}} \,(t_a)\delta x(t_a) = \frac{\partial L}{\partial \dot{x}} \,(t_b)\delta x_b = p_b \delta x_b .$$

It follows that

$$\frac{\delta S}{\delta x_b} = p_b .$$

Similarly, we may consider a change δt_b in the temporal position of the final space–time point (See Figure 5.6.) Notice first that

$$\delta \dot{x}(t_b) = -\frac{\delta x(t_b)}{\delta t_b} .$$

The corresponding change in the action is also slightly more complicated,

$$\delta S = L(t_b)\delta t_b + \int_{t_a}^{t_b} \left[\frac{\partial L}{\partial x} \,\delta x(t) + \frac{\partial L}{\partial \dot{x}} \,\delta \dot{x}(t) \right] dt ,$$

where the first term is caused by the change in the upper limit of the integration domain. As before, we can now carry out a partial integration, leaving us with the

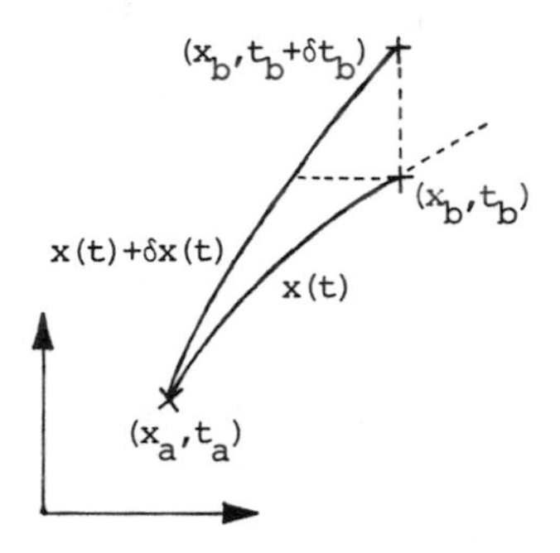

Figure 5.6.

formula

$$\delta S = L(t_b)\delta t_b + \frac{\partial L}{\partial \dot{x}}\,(t_b)\delta x(t_b) = L(t_b)\delta t_b - \frac{\partial L}{\partial \dot{x}}\,(t_b)\dot{x}(t_b)\delta t_b$$
$$= \left[L(t_b) - \frac{\partial L}{\partial \dot{x}}\,(t_b)\dot{x}(t_b)\right]\delta t_b.$$

But the energy is precisely given by

$$E = p\dot{q} - L = \frac{\partial L}{\partial \dot{x}}\,\dot{x} - L,$$

so that we end up with the identity

$$\delta S = -E\delta t_b; \qquad \text{i.e.,}\ \frac{\partial S}{\partial t_b} = -E.$$

□

Notice that the above considerations were in fact anticipated in the discussion of the Einstein–de Broglie rules; cf. the discussion in Section 2.5.

After these general considerations, we now return to the discussion of a one-dimensional particle in a potential well like the one sketched in Figure 5.4. In the general case there will exist a one-parameter family of periodic orbits, which we can label by the fundamental period T:

$$x = x_T(t).$$

Each periodic orbit will furthermore be characterized by its energy E, which thus becomes a function of the period T; i.e., $E = E(T)$. From (5.55) we learn that

$$\frac{dS}{dT} = -E, \tag{5.56}$$

where $S(T)$ is the action of the periodic orbit $x_T(t)$. We can now introduce the Legendre transformation of the action function $S(T)$. It is defined by the relation

$$W(E) = S(T) - \frac{dS}{dT}\,T = S(T) + E \cdot T,$$

where T should be considered a function of E. (Notice the similarity with the passage from the Lagrangian to the Hamiltonian.) In analogy with (5.56), we then get

$$\frac{dW}{dE} = \frac{dS}{dT}\frac{dT}{dE} + T + E \cdot \frac{dT}{dE} = -E\,\frac{dT}{dE} + T + E\,\frac{dT}{dE} = T. \tag{5.57}$$

Finally, we can get back the action function by a Legendre transformation of $W(E)$,

$$S(T) = W(E) - T \cdot E = W(E) - \frac{dW}{dE} \cdot E, \tag{5.58}$$

where E should now be considered a function of T. Notice that the Legendre transformation is in fact given by the phase integral in (5.54),

$$\begin{aligned} W(E) &= S(T) + E \cdot T \\ &= \int_0^T \left[m\left(\frac{dx}{dt}\right)^2 - V(x) \right] dt + \int_0^T \left[m\left(\frac{dx}{dt}\right)^2 + V(x) \right] dt; \end{aligned}$$

i.e.,

$$W(E) = \int_0^T m\left(\frac{dx}{dt}\right)^2 dt = \oint p\,dq, \tag{5.59}$$

since $p = m\,\frac{dx}{dt}$ for this simple type of theory. It is thus the quantity $W(E)$ that we are going to quantize!

From (5.57) we now get

$$dE = \frac{1}{T}\,dW = \frac{\omega}{2\pi}\,dW \qquad \text{with } \omega = \frac{2\pi}{T},$$

where ω is the cyclic frequency of the periodic orbit. Following Bohr, we then assume that only a discrete subset of the periodic orbits are actually allowed and that the transition from one such periodic orbit to the next results in the emission of a quantum with energy $\hbar\omega$, where ω is the frequency of the classical orbit. The last part of the assumption is Bohr's famous correspondence principle. It follows that W can take only a discrete set of values and that the difference between two neighboring values is given by $2\pi\hbar = n$. Thus the quantization rule for W can be stated as

$$W(E_n) = 2\pi(n + c)\hbar = 2\pi\hbar \cdot n + 2\pi\hbar c,$$

where c is a constant that we cannot determine from the correspondence principle. We can look upon this formula as the first two terms in a perturbation expansion for $W(E_n)$, where we expand in powers of $1/n$. Bohr and Sommerfeld simply assumed it was zero, but using the WKB method, it was found to be actually $\frac{1}{2}$. Thus the correct quantization rule for the above case is given by

$$W(E_n) = 2\pi \left(n + \frac{1}{2}\right) \hbar. \tag{5.60}$$

As an example of its application, we shall as usual consider the harmonic oscillator. This example turns out to have two remarkable features: (1) The assumption of a one-parameter family of periodic solutions labeled by the period T breaks down completely. As a consequence, the action function $S(T)$ cannot be defined for a harmonic oscillator. (2) The quantization rule (5.60) is *exact*.

Since the action function cannot be defined, we shall define $W(E)$ through the phase integral (5.59). A general periodic orbit is given by

$$x(t) = A \cos \omega t + B \sin \omega t.$$

The associated energy is consequently

$$E = m\omega^2(A^2 + B^2),$$

while the phase integral turns out to be

$$W = 2\pi m\omega(A^2 + B^2).$$

Thus

$$W(E) = \frac{2\pi}{\omega} E,$$

and the Bohr–Sommerfeld quantization rule reduces to the well-known result (5.20)

$$E_n = \left(n + \frac{1}{2}\right) \hbar\omega.$$

Now that we understand the Bohr–Sommerfeld quantization rule, we return to the path-cum-trace integral (5.52). This time we are going to expand around a classical solution

$$x(t) = x_{\text{cl}}(t) + \eta(t).$$

The potential energy is expanded to second order:

$$V[x(t)] \approx V[x_{\text{cl}}(t)] + V'[x_{\text{cl}}(t)]\eta(t) + \frac{1}{2} V''[x_{\text{cl}}(t)]\eta^2(t).$$

Inserting this, and using the classical equation of motion, the action then decomposes as follows:

$$\begin{aligned} S[x(t)] &\approx S[x_{\text{cl}}(t)] + \int_0^T \left\{ \frac{m}{2} \left(\frac{d\eta}{dt}\right)^2 - \frac{1}{2} V''[x_{\text{cl}}(t)]\eta^2(t) \right\} dt \\ &= S[x_{\text{cl}}(t)] + \frac{m}{2} \int_0^T \eta(t) \left\{ -\frac{d^2}{dt^2} - \frac{1}{m} V''[x_{\text{cl}}(t)] \right\} \eta(t)\, dt. \end{aligned}$$

Consequently, the path-cum-trace integral reduces to

$$G(T) \approx \int_{-\infty}^{\infty} \sum_{x_{\rm cl}} \exp\left\{\frac{i}{\hbar}\, S[x_{\rm cl}(t)]\right\}$$
$$\cdot \int_{\eta(0)=0}^{\eta(T)=0} \exp\left\{\frac{im}{2\hbar}\,\eta\left\{-\frac{d^2}{dt^2} - \frac{1}{m}\, V''[x_{\rm cl}(t)]\right\}\eta\, dt\right\} D[\eta(t)]dx_0, \tag{5.61}$$

where $x_{\rm cl}(0) = x_{\rm cl}(T) = x_0$. But the remaining path integral we know precisely how to handle. According to (5.45), it is given by

$$\sqrt{\frac{m}{2\pi i\hbar f(0) f(T) \int_0^T \frac{dt}{f^2(t)}}} \qquad \text{with } \left\{m\,\frac{d^2}{dt^2} + V''[x_{\rm cl}(t)]\right\} f(t) = 0.$$

In fact, we can relate f to the classical path: The classical path solves the Newtonian equations of motion:

$$m\,\frac{d^2x_{\rm cl}}{dt^2} = -V'[x_{\rm cl}(t)].$$

Differentiating once more, we thus find

$$m\,\frac{d^3x_{\rm cl}}{dt^3} = -V''[x_{\rm cl}(t)]\,\frac{dx_{\rm cl}}{dt}\;.$$

Consequently, we can put

$$f(t) = \frac{dx_{\rm cl}}{dt}\;.$$

Thereby the path-cum-trace integral reduces to

$$G(T) \approx \int_{-\infty}^{\infty} \sum_{x_{\rm cl}} \exp\left\{\frac{i}{\hbar}\, S[x_{\rm cl}(t)]\right\} \sqrt{\frac{m}{2\pi i\hbar \dot{x}_{\rm cl}(0)\dot{x}_{\rm cl}(T) \int_0^T \frac{dt}{[\dot{x}_{\rm cl}]^2}}}\; dx_0, \tag{5.62}$$

where we sum over all classical paths satisfying the constraint $x_{\rm cl}(0) = x_{\rm cl}(T) = x_0$. Despite its complicated structure, it is essentially a one-dimensional integral of the type

$$\int_{-\infty}^{+\infty} \exp\{if(x)\}g(x)dx.$$

Such an integral can be calculated approximately by using *the stationary phase approximation*. Thus we look for a point x_0 where the phase is stationary, i.e., $f'(x_0) = 0$. We can then expand around this point:

$$f(x) \approx f(x_0) + \frac{1}{2}\, f''(x_0)(x - x_0)^2,$$
$$g(x) \approx g(x_0) + g'(x_0)(x - x_0).$$

In this approximation the integral then reduces to a Gaussian integral:

$$\int_{-\infty}^{+\infty} \exp\{if(x)\}g(x)dx \approx g(x_0)\exp\{if(x_0)\}\sqrt{\frac{2\pi i}{f''(x_0)}} . \tag{5.63}$$

Using this on the path-cum-trace integral (5.62), we see that we need only include contributions from the classical paths that produce a stationary phase; i.e.,

$$\begin{aligned} 0 &= \frac{\partial}{\partial x_0} S[x_{\text{cl}}] = \frac{\partial}{\partial x_0} S[x_0; T \mid x_0; 0] \\ &= \frac{\partial S}{\partial x_b}(x_0; T \mid x_0; 0) + \frac{\partial S}{\partial x_a}(x_0; T \mid x_0; 0). \end{aligned}$$

But according to (5.55), this gives

$$0 = p_b - p_a;$$

i.e., the momenta at $t = 0$ and $t = T$ are also identical. Thus the stationary phase approximation selects for us only the purely periodic solutions! They can be parametrized by their fundamental periodic T. Corresponding to a given T we should therefore only include the contributions from the periodic orbits:

$$x_T(t);\ x_{T/2}(t);\ x_{T/3}(t);\ \ldots;\ x_{T/n}(t);\ \ldots .$$

Before we actually apply the stationary phase approximation, we can therefore restrict ourselves to a summation over these periodic orbits:

$$G(T) \approx \int_{-\infty}^{\infty} \sum_{n=1}^{\infty} \exp\left\{\frac{i}{\hbar} nS\left[\frac{T}{n}\right]\right\} \frac{1}{|\dot{x}_{T/n}(0)|} \sqrt{\frac{n}{2\pi i\hbar}} \left[\int_0^T \frac{dt}{(\dot{x}_{T/n})^2}\right] dx_0.$$

Notice that the parametrization of the periodic orbit is determined only up to a translation in t; i.e., we choose the starting point quite arbitrarily on the closed path. Each of the x-values between the turning points x_1 and x_2 occurs twice more on the path, cf. Figure 5.7. We are now ready to perform the x_0-integration. The action of the periodic orbits does not depend upon the choice of the starting point, and neither does the expression

$$\int_0^T \frac{dt}{(\dot{x})^2} .$$

Thus the only contribution to the integral comes from

$$\int_{-\infty}^{+\infty} \frac{dx_0}{|\dot{x}_{T/n}(0)|} = 2\int_{x_1}^{x_2} \frac{dx_0/dx_0}{dt} = \oint dt = \frac{T}{n} .$$

Having taken care of the x_0-integration, we must then compute the integral over the reciprocal square of the velocity. Here we get

$$\int_0^T \frac{dt}{(\dot{x}_{T/n})^2} = 2n\int_{x_1}^{x_2} \frac{dx}{\dot{x}^3} = 2n\int_{x_1}^{x_2} \left(\frac{2[E(T/n) - V(x)]^{-3/2}}{m}\right) dx.$$

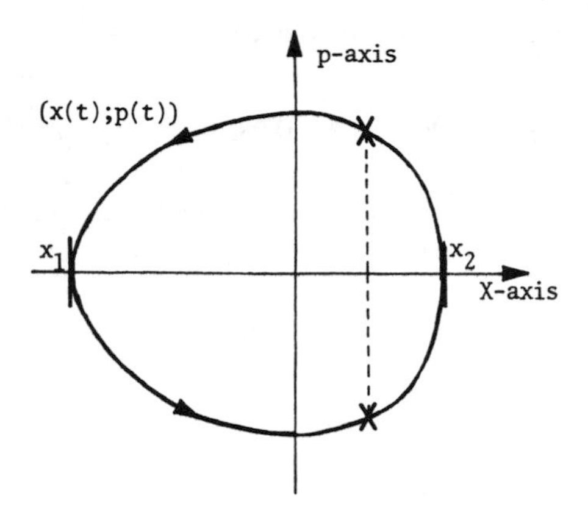

Figure 5.7.

The same integral can, however, be obtained by a different reasoning! Consider the phase integral $W(E)$, which is given by

$$W(E) = m \int_0^T (\dot{x})^2 dt = 2n \int_{x_1}^{x_2} \sqrt{2[E - V(x)]}\, dx$$

for the periodic orbit in question. Differentiating twice with respect to E, we obtain

$$\frac{dT}{dE} = \frac{d^2 W}{dE^2} = -2n \int_{x_1}^{x_2} (2[E - V(x)])^{-3/2} dx.$$

Consequently,

$$\int_0^T \frac{dt}{(\dot{x}_{T/n})^2} = -m^{3/2} \frac{dT}{dE} \, .$$

Putting all this together, the path-cum-trace integral thus finally reduces to

$$G(T) \approx \frac{1}{m} \frac{1}{\sqrt{2\pi i \hbar}} \sum_{n=1}^{\infty} \exp\left\{ \frac{i}{\hbar} nS\left[\frac{T}{n}\right]\right\} \frac{T}{n} \sqrt{\left|\frac{dE}{dT}\right|}_{T=T/n} \tag{5.64}$$

This was the hard part of the calculation! Now we can extract the energy levels. As usual, this is done by going to the transformed path-cum-trace integral and looking for the poles. The transformed path-cum-trace integral is given by, cf. (5.9),

$$G(E) = \frac{i}{\hbar} \int_0^\infty G(T) \exp\left\{ \frac{i}{\hbar} \cdot ET \right\} dE.$$

(Notice that E now represents both an integration variable and the energy of the periodic orbit. To avoid confusion, we shall write the latter as E_{cl}!) Inserting the above expression for $G(T)$, we obtain

$$G(E) \approx \frac{i}{m\hbar} \frac{1}{\sqrt{2\pi i\hbar}} \sum_{n=1}^{\infty} \int_0^{\infty} \exp\left\{\frac{i}{\hbar}\left(ET + nS\left[\frac{T}{n}\right]\right)\right\} \cdot \frac{T}{n}\sqrt{\left|\frac{dE_{\text{cl}}}{dT}\right|} dT.$$

Here it will be preferable to make a change of variables, $\tau = T/n$; i.e., τ is the fundamental period of the periodic orbit in question:

$$G(E) \approx \frac{i}{m\hbar} \frac{1}{\sqrt{2\pi i\hbar}} \sum_{n=1}^{\infty} \int_0^{\infty} \exp\left\{\frac{in}{\hbar}\left(E\tau + S[\tau]\right)\right\} \tau\sqrt{\left|\frac{dE_{\text{cl}}}{d\tau}\right|}\sqrt{n}\, d\tau.$$

The τ-integration is then performed by using the stationary phase approximation. A stationary phase requires

$$0 = \frac{\partial}{\partial \tau}\left(E\tau + S[\tau]\right) = E - \frac{\partial S}{\partial \tau} = E - E_{\text{cl}}.$$

Thus for a given value of E, we get the main contribution from the periodic orbit, which precisely has the energy E! According to (5.63), the stationary phase approximation now gives

$$G(E) \approx \frac{i}{m\hbar} \frac{1}{\sqrt{2\pi i\hbar}} \sum_{n=1}^{\infty} \sqrt{\frac{2\pi i\hbar}{n}} \frac{1}{\sqrt{\frac{d^2S}{d\tau^2}}} \exp\left\{\frac{in}{\hbar}\left(E\tau[E] + S[\tau]\right)\right\} \cdot \tau(E)\sqrt{n}\left|\frac{dE}{d\tau}\right|.$$

Using that

$$\frac{d^2S}{d\tau^2} = \frac{dE}{d\tau} \quad \text{and} \quad W(E) = S(\tau) + E\tau(E),$$

the formula finally reduces to

$$G(E) \approx \frac{i}{m\hbar}\tau(E)\sum_{n=1}^{\infty}\exp\left\{\frac{i}{\hbar}nW(E)\right\} = \frac{i}{m\hbar}\tau(E)\frac{\exp\left\{\frac{i}{\hbar}W(E)\right\}}{1-\exp\left\{\frac{i}{\hbar}W(E)\right\}}. \tag{5.65}$$

This clearly has poles when

$$W(E_n) = 2\pi n\hbar, \tag{5.66}$$

and thus we have precisely recovered the old Bohr–Sommerfeld quantization condition (without the half-integral correction term). The reason we missed the half-integral correction term is, however, very simple. It is due to the fact that we have not taken into account the phase ambiguity of the path integral. Precisely, as for the harmonic oscillator, we should pick up phase corrections. The expression

(5.62) is completely analogous to (5.38). The additional phases then come from the singularities in the integrand

$$\left[\int_0^T \frac{dt}{f^2(t)}\right]^{-1/2}.$$

These singularities correspond to the zeros of $f(t) = \dot{x}(t)$, i.e., to the turning points x_1 and x_2. In the present calculation, the turning points are thus analogous to the caustics. Each time we pass a turning point we therefore expect that we pick up an additional phase factor $e^{i\pi/2}$; cf. (5.29). Since there are $2n$ turning points in the orbit $x_{T/n}$, we consequently gain an additional phase

$$e^{-i\pi n} = (-1)^n,$$

which should be included in the formula for the propagator. The transformed path-cum-trace integral (5.65) is thereby changed into

$$\begin{aligned} G(E) &\approx \frac{i}{m\hbar}\tau(E)\sum_{n=1}^{\infty}(-1)^n \exp\left\{\frac{i}{\hbar}nW(E)\right\} \\ &= \frac{-i}{m\hbar}\tau(E)\frac{\exp\left\{\frac{i}{\hbar}W(E)\right\}}{1+\exp\left\{\frac{i}{\hbar}W(E)\right\}}. \end{aligned} \tag{5.67}$$

This causes a shift in the poles, which are now given by

$$W(E_n) = 2\pi\left(n+\frac{1}{2}\right)\hbar, \tag{5.68}$$

and that is precisely the correct quantization rule according to the original WKB calculation.

Notice too that near a pole we get the expansion

$$W(E) \approx W(E_n) + T(E)(E - E_n).$$

Using the approximation

$$\begin{aligned} 1+\exp\left\{\frac{i}{\hbar}W(E)\right\} &\approx 1+\exp\left\{\frac{i}{\hbar}W(E_n)\right\}\exp\left\{\frac{i}{\hbar}T(E)(E-E_n)\right\} \\ &= 1-\left\{1-\frac{i}{\hbar}T(E)(E-E_n)\right\} \\ &= \frac{i}{\hbar}T(E)(E_n-E), \end{aligned}$$

it follows that $G(E)$ behaves like

$$G(E) \approx \frac{1}{E_n - E}$$

when $E \approx E_n$. This should be compared with (5.10), and it clearly shows that the approximative path-cum-trace integral has the correct asymptotic behavior close to a pole.

We have been working hard to derive a result that, as we have seen, can in fact almost be deduced directly from the correspondence principle! When we are going to use the path integral technique in quantizing field theories, we will actually have to work still harder. Before we enter into these dreadful technicalities, I want to give a few examples of almost trivial applications of the preceding machinery.

The first example is concerned with the free particle. To find the energy levels, we enclose the free particle in a box of length L and use periodic boundary conditions. Thus we have effectively replaced the line by a closed curve of length L. This has the effect that the energy spectrum is discretized. We get the continuous free particle spectrum back by letting L go to infinity. The free particle can now execute periodic motions by going around and around the closed circumference: The periodic orbit x_T is given by

$$x_T(t) = \frac{Lt}{T} \, .$$

It has the following energy and action:

$$E(T) = \frac{m}{2}\left(\frac{dx}{dt}\right)^2 = \frac{mL^2}{2T^2}\, ; \qquad S(T) = ET = \frac{mL^2}{2T}\, .$$

Consequently,

$$W(E) = S(T) + ET = L\sqrt{2mE}.$$

Since there are no turning points in this problem, the quantization rule is given by (5.66); i.e.,

$$L\sqrt{2mE_n} = 2\pi n\hbar;$$

i.e.,

$$E_n = \frac{\hbar^2}{2m}\left(\frac{2n\pi}{L}\right)^2 . \tag{5.69}$$

These are the correct energy levels in the discrete version. To see this, we notice that the free particle Hamiltonian is given by

$$\tilde{H} = -\frac{\hbar^2}{2m}\frac{d^2}{dx^2}\, .$$

Since the Schrödinger wave function must be periodic, the eigenfunctions are given by

$$\psi_n(x) = \exp\left[\pm i\,\frac{2\pi n x}{L}\right],$$

and the corresponding eigenvalues precisely reproduce the above result (5.69).

The second example is concerned with a field theory. In the discussion of the sine-Gordon model we found an interesting family of periodic solutions: The bions (or breathers). In a slightly changed notation, where we emphasize the cyclic frequency ω, they are given by (cf. (4.50))

$$\phi(x,t) = \frac{4}{\lambda}\arctan\left[\frac{\eta\sin\omega t}{\cosh(\eta\omega x)}\right] \quad \text{with } \eta = \frac{\sqrt{\mu^2-\omega^2}}{\omega} \text{ and } 0<\omega<\mu. \tag{5.70}$$

The corresponding energy is given by (cf. (4.51))

$$E = 2M\sqrt{1-\frac{\omega^2}{\mu^2}}, \tag{5.71}$$

where M is the mass of a single soliton. As we have seen, it represents a classical bound state of a soliton and an antisoliton. We therefore expect that it will generate quantum bound states that should be recognizable in the energy spectrum. Since (5.70) is a one-parameter family of periodic solutions (which can easily be labeled by the period T), we can in fact use the naive Bohr–Sommerfeld quantization rule. For a classical field theory it takes the form

$$W(E) = 2\pi nh \quad \text{with } W(E) = \int_0^T\int_{-\infty}^{+\infty}\pi(x,t)\frac{\partial\phi}{\partial t}\,dt\,dx \tag{5.72}$$

(where $\pi(x,t)$ is the conjugate momentum density). Rather than calculate $W(E)$ directly from the phase integral, we shall use the basic relation (5.57)

$$\frac{dW}{dE} = T = 2\pi\omega^{-1}.$$

From (5.71) we see that ω can be replaced by E, and we get

$$\frac{dW}{dE} = 2\pi\mu^{-1}\left(1-\frac{E^2}{4M^2}\right)^{-1/2}. \tag{5.73}$$

This can trivially be integrated:

$$W(E) = 4\pi\mu^{-1}M\arcsin\left(\frac{E}{2M}\right)+C.$$

But using that $E=0$ (i.e., $\omega=\mu$) corresponds to the classical ground state, when W obviously must vanish too, we see that $C=0$. A trivial application of the naive Bohr–Sommerfeld quantization rule then gives

$$E_n = 2M\sin\left[n\cdot\frac{\mu\hbar}{2M}\right]; \qquad 0<n<\frac{\pi M}{\mu\hbar}. \tag{5.74}$$

Notice the upper limit of n; i.e., there is only a finite discrete set of energy states. This comes about in the following way: When n grows, so does the fundamental period T, which by (5.57) and (5.73) is given by

$$T(n) = \frac{dW}{dE} = \frac{2\pi}{\mu\cos\left(\frac{n\mu\hbar}{2M}\right)}.$$

But when $T \to \infty$, the corresponding soliton–antisoliton pair becomes unbound, and for n above the threshold, the classical state thus becomes unstable and breaks up into a separate soliton and antisoliton.

In the above derivation we have been using the naive Bohr–Sommerfeld quantization rule. As in the case of the single particle, we can improve this using the path integral version of WKB. In the multidimensional case, the WKB-formula, however, becomes more complicated than in the one-dimensional case, where we had just to include a "half-integral" correction term. For the special case of the sine-Gordon model it can nevertheless be shown that the energy spectrum of the bion states is still given by (5.74), with a suitable redefinition of the parameters M and μ.

§ 5.7 Instantons and Euclidean Field Theory

At this point we introduce a very important trick that will enable us to extract nontrivial information about the quantum-mechanical ground state of a system. Let us consider the Minkowski space with its indefinite metric

$$ds^2 = -dt^2 + dx^2 + dy^2 + dz^2.$$

Formally, we can turn it into a positive definite metric by using Weyl's unitary trick, i.e., by replacing the Minkowski time t with the imaginary time $\tau = it$:

$$ds^2 = d\tau^2 + dx^2 + dy^2 + dz^2.$$

Obviously, we can think of this as a kind of analytic continuation if we regard the 4-dimensional Minkowski space as being a 4-dimensional real subspace of an 8-dimensional complex space.

The above trick has been widely used in special relativity, where it allows us to avoid the distinction between upper and lower indices. But it is also widely used in quantum field theory, where it has had profound implications for our understanding of a variety of phenomena. In quantum field theory, it has further been given a special name: The analytic continuation to imaginary time is known to quantum field theoreticians as performing a *Wick rotation.*

To illustrate the applications of the Wick rotation, we consider for simplicity ordinary quantum mechanics in one space dimension. As we have seen, the most important ingredient in ordinary quantum mechanics is the Feynman propagator $K(x_b;\, t_b \mid x_a;\, t_a)$. If we assume this to be an analytic function of t_b and t_a, we can perform a Wick rotation whereby we produce the so-called *Euclidean propagator*: $K_E(x_b;\, \tau_b \mid x_a;\, \tau_a)$. For example, the free particle propagator is Wick rotated into the following Euclidean propagator:

$$K_E^0(x_b;\, \tau_b \mid x_a;\, \tau_a) = \sqrt{\frac{m}{2\pi\hbar(\tau_b - \tau_a)}} \exp\left\{-\frac{m}{2\hbar}\frac{(x_b - x_a)^2}{\tau_b - \tau_a}\right\}; \tag{5.75}$$

cf. (2.28). Similarly, the propagator for the harmonic oscillator is Wick rotated into (cf. (5.17))

$$K_E(x_b;\Upsilon \mid x_a;0) = \sqrt{\frac{m\omega}{2\pi\hbar\sin h[\omega\Upsilon]}} \cdot \exp\left[-\frac{m\omega}{2\hbar\sin h[\omega\Upsilon]}\left\{(x_b^2+x_a^2)\cosh[\omega\Upsilon]-2x_bx_a\right\}\right], \tag{5.76}$$

where the trigonometric functions have been replaced by hyperbolic functions. Notice that the caustics have disappeared and so have the phase ambiguities—the Euclidean propagator is singular only when $\tau_b = \tau_a$!

In the Minkowski space, the Feynman propagator can be reexpressed as a path integral. The same is true in the Euclidean case. To Wick rotate the path integral, we notice that the substitution $\tau \curvearrowright -i\tau$ generates the following change in the action:

$$S[x(t)] = \int_{t_a}^{t_b}\left\{\frac{m}{2}\left(\frac{dx}{dt}\right)^2 - V(x)\right\}dt \curvearrowright i\int_{\tau_a}^{\tau_b}\left\{\frac{m}{2}\left(\frac{dx}{d\tau}\right)^2 + V(x)\right\}d\tau.$$

Consequently, the Wick-rotated path integral is given by

$$\int_{x(\tau_a)=x_a}^{x(\tau_b)=x_b} \exp\left[-\frac{1}{\hbar}\int_{\tau_a}^{\tau_b}\left\{\frac{m}{2}\left(\frac{dx}{d\tau}\right)^2 + V(x)\right\}d\tau\right] D(x(\tau)),$$

where we sum over paths in the Euclidean space. Let us introduce the abbreviation

$$S_E[x(\tau)] \stackrel{\text{def}}{=} \int_{\tau_a}^{\tau_b}\left\{\frac{m}{2}\left(\frac{dx}{d\tau}\right)^2 + V(x)\right\}d\tau. \tag{5.77}$$

The quantity $S_E[x(\tau)]$ is known as the *Euclidean action*. Notice that the Euclidean action is *positive definite*. The Euclidean path integral is now given by

$$\int_{x(\tau_a)=x_a}^{x(\tau_b)=x_b} \exp\left\{-\frac{1}{\hbar}\, S_E[x(\tau)]\right\} D[x(\tau)].$$

This is one of the main achievements of the Wick rotation. Rather than an oscillatory integral, which can *never* be absolutely convergent, we have now turned the path integral into a direct generalization of a nice, decent Gaussian integral where the exponent is negative definite, and furthermore, it is at least quadratic. This puts the path integral on a much sounder footing, where we avoid completely the divergencies and phase ambiguities that plague the Minkowski version. The Euclidean path integral can be defined through limiting procedures like (5.31) or (5.33). If they are quadratic, i.e., of Gaussian type, they can also be diagonalized, so that we can use relations like (5.45) (suitably modified). In the Euclidean case it is in fact even possible to define it in terms of a rigorous integration theory on an infinite-dimensional measure space using the so-called Wiener measures.

Returning to the Euclidean propagator, we now get from (2.21) that it can be reexpressed in terms of an Euclidean path integral:

$$K_E(x_b; \tau_b \mid x_a; \tau_a) = \int_{x(\tau_a)=x_a}^{x(\tau_b)=x_b} \exp\left\{-\frac{1}{\hbar}\, S_E[x(\tau)]\right\} D[x(\tau)]. \tag{5.78}$$

The other characterizations of the Feynman propagator can also be carried over to the Euclidean version. Notice that the Wick rotation turns the Schrödinger equation into the *Heat equation*:

$$\hbar\, \frac{\partial}{\partial \tau}\, \psi(x, \tau) = \frac{\hbar^2}{2m}\, \frac{\partial^2}{\partial x^2}\, \psi(x, \tau) - V(x)\psi(x, \tau). \tag{5.79}$$

As a consequence, *the Euclidean propagator is the unique solution to the Heat equation*

$$\hbar\, \frac{\partial}{\partial \tau_b}\, K_E(x_b; \tau_b \mid x_a; \tau_a) = \left[\frac{\hbar^2}{2m}\, \frac{\partial^2}{\partial x_b^2} - V(x_b)\right] K_E(x_b; \tau_b \mid x_a; \tau_a)$$

satisfying the boundary condition

$$K_E(x_b; \tau_a \mid x_a; \tau_a) = \delta(x_b - x_a).$$

Next, we observe that the Hamiltonian operator is unaffected by a Wick rotation (since it does not contain the time). This evidently leads to the following characterization: *The Euclidean propagator has the following decomposition into a complete set of eigenfunctions for the Hamiltonian operator*:

$$K_E(x_b; \tau_b \mid x_a; \tau_a) = \sum_{n=0}^{\infty} \overline{\phi_n(x_a)}\phi_n(x_b) \exp\left[-\frac{E_n}{\hbar}\, (\tau_b - \tau_a)\right]; \tag{5.80}$$

cf. (2.70).

Finally, it can be reexpressed as a matrix element of the *Euclidean time evolution operator* $\exp\left[-\frac{1}{\hbar}\, \tilde{H}T\right]$; i.e.,

$$K_E(x_b; T \mid x_a; 0) = \langle x_b \mid \exp\left[-\frac{1}{\hbar}\, \tilde{H}T\right] \mid x_a\rangle. \tag{5.81}$$

As we have seen, the Euclidean formalism offers a chance of constructing in quantum mechanics a rigorous path integral that is not plagued by singularities and phase ambiguities. To some extent this is also true in quantum field theory. Consider the Lagrangian density

$$L[\phi; \partial_\mu\phi] = \frac{1}{2}\left(\frac{\partial\phi}{\partial t}\right)^2 - \frac{1}{2}\left(\frac{\partial\phi}{\partial \mathbf{r}}\right)^2 - U[\phi] = -\frac{1}{2}\, \eta^{\mu\nu}\partial_\mu\phi\partial_\nu\phi - U[\phi].$$

(For technical reasons, $U[\phi]$ should be at most a fourth-order polynomial in ϕ. This guarantees that the theory is renormalizable, i.e., that one can get rid of various divergent expressions in quantum field theory using a so-called renormalization procedure.) In the Euclidean formalism this Lagrangian density leads to the following Euclidean action:

$$S_E[\phi] = \int \left\{ \frac{1}{2}\left(\frac{\partial\phi}{\partial\tau}\right)^2 + \frac{1}{2}\left(\frac{\partial\phi}{\partial\mathbf{r}}\right)^2 + U[\phi] \right\} d\mathbf{r}\, d\tau$$
$$= \int \left\{ \frac{1}{2}\, \delta^{\mu\nu}\partial_\mu\phi\partial_\nu\phi + U[\phi] \right\} d\mathbf{r}\, d\tau. \tag{5.82}$$

Notice that apart from a change of sign, the Minkowski metric has been replaced by the Euclidean metric. As in quantum mechanics, the Euclidean action is positive definite, and the Euclidean path integrals

$$\int \exp\left\{-\frac{1}{\hbar}\, S_E[\phi]\right\} D[\phi]$$

are of the nice infinite-dimensional Gaussian type with a negative definite quadratic exponent. As a consequence, it has been possible to construct a rigorous integration theory for such path integrals in one and two space dimensions. In three space dimensions (corresponding to our actual world), there are, however, still unsolved technical problems.

In quantum mechanics, the Euclidean formalism, with its associated Euclidean path integrals, is a funny trick that is not really necessary (after all, it is easy to solve the original Schrödinger equation). In quantum field theory it is, on the contrary, the only way one knows to construct a rigorous theory: First one must construct the Euclidean version, and thereafter one Wick rotates back to the Minkowski space.

We now turn to a very interesting aspect of the Euclidean field theory. In a Minkowskian field theory we have seen that the path integral is dominated by classical solutions with finite energy, and this we have been using to calculate a semiclassical approximation of the path integral. In the Euclidean version we can do the same; i.e., we expect the Euclidean path integral to be dominated by classical solutions to the Euclidean equations of motion.

To investigate Euclidean configurations a little more closely we look at the space of smooth configurations with a finite Euclidean action, i.e.,

$$S_E[\phi] < \infty.$$

In analogy with the Minkowski case, a *Euclidean vacuum* is defined as a configuration with vanishing action. Since the Euclidean action is given by

$$S_E[\phi] = \int \left\{ \frac{1}{2}\left(\frac{\partial\phi}{\partial\tau}\right)^2 + \frac{1}{2}\left(\frac{\partial\phi}{\partial\mathbf{r}}\right)^2 + U[\phi] \right\} d\mathbf{r}\, d\tau, \tag{5.82}$$

this implies that

$$\frac{\partial\phi}{\partial\tau} = \frac{\partial\phi}{\partial\mathbf{r}} = U[\phi] = 0;$$

i..e, *a Euclidean vacuum* (like the Minkowski vacuum) *corresponds to a zero point of the potential energy density*:

$$\phi(\mathbf{r}, \tau) \equiv \phi_0 \qquad \text{with } U[\phi_0] = 0.$$

Next, we observe that finiteness of the Euclidean action implies that the field asymptotically approaches a Euclidean vacuum. We then look for solutions of the Euclidean equations of motion. Obviously, a Euclidean vacuum is such a solution, but there may also be nontrivial solutions with the following two properties:

a. It has a finite Euclidean action;
b. It is a local minimum of the Euclidean action (i.e., it is stable).

Since such a nontrivial solution is localized not only in space, but also in (Euclidean) time, 't Hooft has proposed to call it an *instanton*. The basic role played by the instantons is then that we expect them to dominate the Euclidean path integrals and thereby help us with the calculation of the Euclidean propagator.

At this point the reader has probably recognized a similarity with our discussion of solitons in Chapter 4. This is no accident, since there is an extremely close connection between instantons and solitons, we can formalize in the following way: Consider a Minkowskian field theory in *D space* dimensions. It is characterized by an action of the form

$$S[\phi] = \int \left\{ \frac{1}{2}\left(\frac{\partial \phi}{\partial t}\right)^2 - \frac{1}{2}\left(\frac{\partial \phi}{\partial \mathbf{r}}\right)^2 - U[\phi] \right\} d^D\mathbf{r}\, dt.$$

When studying solitons, we look for a static configuration $\phi(\mathbf{r}, t) = \phi(\mathbf{r})$ with the following two properties:

a. It has a finite static energy;
b. It is a local minimum of the static energy (i.e., it is stable).

Here the static energy is given by the expression

$$E_{\text{static}} = \int \left\{ \frac{1}{2}\left(\frac{\partial \phi}{\partial \mathbf{r}}\right)^2 + U[\phi] \right\} d^D\mathbf{r}. \tag{5.83}$$

Next, we consider a field theory in *D space–time* dimensions, based upon the same potential energy density $U[\phi]$. The Euclidean version of this theory is based upon the Euclidean action

$$S_E = \int \left\{ \frac{1}{2}\, \delta^{ab} \partial_a \phi \partial_b \phi + U[\phi] \right\} d^D x,$$

which is identical to (5.83)! As a consequence, we therefore derive the following extremely important principle:

A static soliton in D space dimensions is completely equivalent to an instanton in D space–time dimensions.

We shall apply this principle to solitons in one space dimension. They correspond to instantons in one space–time dimension, i.e., in zero space dimensions! This example is therefore somewhat degenerate. Now, a field in zero space dimensions can depend only upon time; i.e., it is of the form $\phi^a(t)$. Consequently, a field theory in zero space dimensions corresponds to a system with a finite number of degrees of freedom, i.e., to ordinary mechanics. If there is only one field component $\phi(t)$, we can identify it with the position of a particle moving in a single space dimension, i.e., put $x = \phi(t)$.

Let us concretize this: A particle moving in a one-dimensional potential $V(x)$ is characterized by the Lagrangian

$$L = \frac{1}{2} m \left(\frac{dx}{dt} \right)^2 - V(x).$$

In the Euclidean version this corresponds to the Euclidean action

$$S_E[\phi(\tau)] = \int_{-\infty}^{+\infty} \left[\frac{1}{2} m \left(\frac{d\phi}{d\tau} \right)^2 + V(\phi) \right] d\tau,$$

where we have denoted the Euclidean path by $x = \phi(\tau)$. Since τ is now a space-like variable, we can finally denote it by $\bar{x}$, whereby the Euclidean action becomes

$$S_E[\phi(\bar{x})] = \int_{-\infty}^{+\infty} \left\{ \frac{1}{2} m \left(\frac{d\phi}{d\bar{x}} \right)^2 + V[\phi(\bar{x})] \right\} d\bar{x},$$

which is precisely the same as the expression for the static energy in a $(1 + 1)$-dimensional field theory. Notice too a very pleasant surprise: Although we have started out with a *nonrelativistic* theory for a particle moving in a potential $V(x)$, we see that the corresponding Euclidean formalism corresponds to the static version of a *relativistic* theory in $(1 + 1)$-space–time dimension!

We can now take over the results obtained in Sections 4.3–4.4 concerning the kinks in $(1 + 1)$-dimensional field theories. First of all, a kink in a $(1 + 1)$-dimensional field theory corresponds to an instanton in ordinary quantum mechanics. E.g., to a particle moving in a double well

$$V(x) = \frac{\lambda^4}{4} (x^2 - a^2)^2$$

is associated an instanton given by

$$\phi(\tau) = a \tanh \left\{ a \frac{\sqrt{\lambda}}{2} (\tau - \tau_0) \right\} ; \tag{5.84}$$

cf. (4.23). Similarly, to a particle moving in the periodic potential

$$V(x) = \frac{\mu^4}{\lambda} \left[1 - \cos \left(\frac{\sqrt{\lambda}}{\mu} \phi \right) \right]$$

is associated an instanton given by

$$\phi(\tau) = \frac{4\mu}{\sqrt{\lambda}} \arctan \left\{ e^{\mu(\tau - \tau_0)} \right\} ; \tag{5.85}$$

cf. (4.24). Furthermore, we have seen that a theory with a degenerate vacuum is characterized by a topological charge Q and that the static energy is bounded below by the topological charge due to the Bogomolny decomposition. In the Euclidean version, this corresponds to a decomposition of the Euclidean action

$$S_E[\phi(\tau)] = \frac{m}{2}\int_{-\infty}^{+\infty}\left\{\frac{d\phi}{d\tau} \mp \sqrt{\frac{2V(\phi)}{m}}\right\}^2 + [Q]$$
$$\text{with } Q = \int_{\phi_-}^{\phi_+} \sqrt{2mV(\phi)}\,d\phi, \tag{5.86}$$

where $\phi(\tau)$ is a configuration interpolating between the classical vacua ϕ_- and ϕ_+. This leads to the following simple picture: The space of all smooth configurations with finite Euclidean action breaks up into disconnected sectors characterized by different values of the topological charge Q. In each of these sectors the Euclidean action is bounded below by the topological charge

$$S_E[\phi] \geq |Q|.$$

We expect the contribution from a given sector the Euclidean path integral to be dominated by a configuration that saturates the lower bound, i.e., from a configuration satisfying the differential equation

$$\frac{d\phi}{d\tau} = \pm\sqrt{\frac{2V(\phi)}{m}}\,. \tag{5.87}$$

In the lowest nontrivial sectors this leads precisely to the one-instantons or anti-instantons (like (5.84) and (5.85)). In the higher sectors we know that we cannot saturate the bound exactly, but we can still come arbitrarily close by considering multiinstanton configurations consisting of widely separated instantons and antiinstantons; cf. the discussion in Section 4.4

Due to the close relation between solitons and instantons, Polyakov has suggested that instantons be referred to as *pseudo-particles*, a label that is also currently used in the literature.

§ 5.8 Instantons and the Tunnel Effect

Having understood some of the formal properties of instantons, we now turn to the question of their physical significance. We know that solitons correspond to a particle-like excitation of the field. In a similar way, we will now show that instantons are associated with the so-called *tunnel effect* in quantum mechanics.

Let us first recall that instantons in ordinary mechanics are associated with a particle moving in a potential well that has a degenerate minimum (like the double well or a periodic potential). As a consequence, the classical vacuum is degenerate, and formally, the instantons represent *Euclidean solutions* that interpolate between two different classical vacua. We now proceed to examine the *quantum-mechanical ground state* of such a system. To be specific, let us concentrate for a moment upon

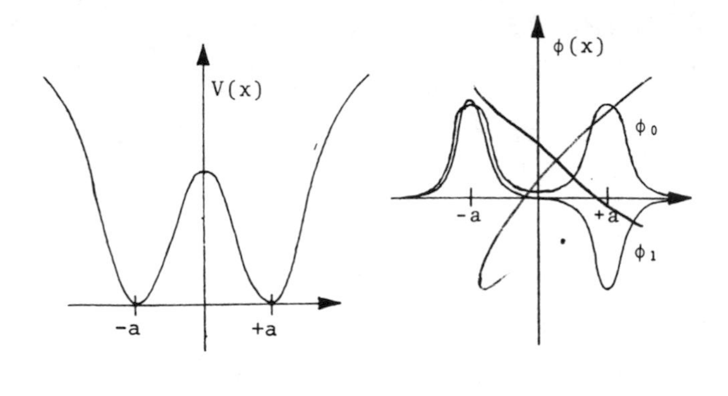

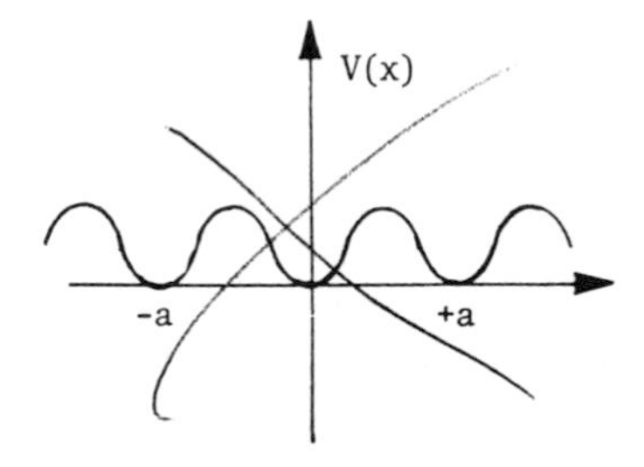

Figure 5.12.

the motion in a double well (cf. Figure 5.12). Classical, there are two ground states, represented by the configurations

$$x_+(t) \equiv a \quad \text{and} \quad x_-(t) \equiv -a,$$

corresponding to a particle sitting in the bottom of either of the wells. Quantum-mechanically, we must solve the Schrödinger equation and thereby determine the energy spectrum to see which eigenfunction has the lowest eigenvalue:

$$-\frac{\hbar^2}{2m}\frac{d^2\psi}{dx^2} + V(x)\psi(x) = E\psi(x).$$

Based upon the classical analysis, we can give the following qualitative description of the low-lying energy states: Each of the two wells corresponds to a harmonic oscillator potential. Thus the two lowest eigenstates will be related to the ground states of these two oscillators. Furthermore, the potential $V(x)$ is invariant under the inversion, $x \curvearrowright -x$, and consequently, the eigenfunctions of the Hamiltonian can be chosen to be eigenfunctions of the *parity* operator

$$\tilde{P}\psi(x) = \psi(-x);$$

i.e., they can be chosen as either even or odd functions. In the case of the double well, the two lowest eigenstates are therefore of the form

$$\psi_+(x) = \frac{1}{\sqrt{2}}\left\{\psi_0(x-a)+\psi_0(x+a)\right\} \qquad \text{and}$$

$$\psi_-(x) = \frac{1}{\sqrt{2}}\left\{\psi_0(x-a)-\psi_0(x+a)\right\},$$

where ψ_0 is the ground state wave function for a harmonic oscillator (i.e., it is given by (5.25), where we put $t = 0$).

There are now two possibilities: Either the quantum-mechanical vacuum is degenerate (like the classical vacuum), i.e., ψ_+ and ψ_- have the same energy, or the classical degeneracy has been lifted, i.e., ψ_+ and ψ_- have different energies. To find out which possibility is actually realized, we must compute the associated eigenvalues, i.e., solve the Schrödinger equation. We will not do this for a general potential, but to get a feeling for what is going on, we look at a particularly simple example:

ILLUSTRATIVE EXAMPLE: THE DOUBLE SQUARE WELL. Consider the double square well shown in Figure 5.8. Because the potential is constant "piece by piece," it is trivial to solve the Schrödinger equation. Due to the symmetry of the wave functions, we need furthermore solve it only in regions I and II. Since the potential at $x = b$ is infinitely high, the wave function must vanish at $x = b$. At $x = a$ we must furthermore demand that the wave function and its derivative be continuous. This can be simplified by demanding that the logarithmic derivatives match; i.e., that

$$\frac{1}{\psi_I}\frac{d\psi_I}{dx}\bigg|_{x=a} = \frac{1}{\psi_{II}}\frac{d\psi_{II}}{dx}\bigg|_{x=a}. \tag{5.88}$$

Okay, now for the solutions. In region I, the solution to the Schrödinger equation

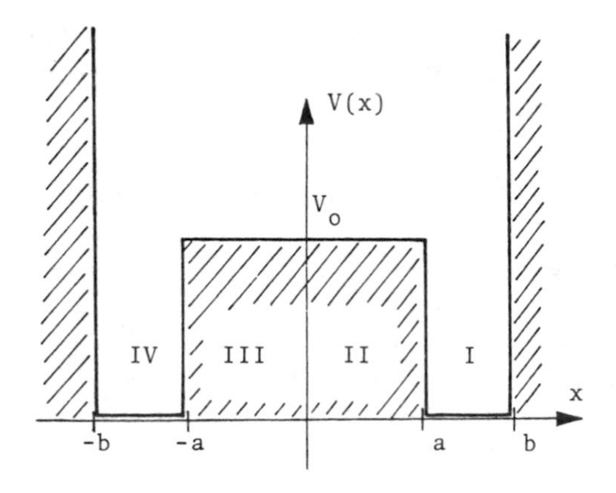

Figure 5.8.

is given by

$$\psi_I(x) = A \sin\left\{\sqrt{\frac{2mE}{\hbar^2}}\,(x-b)\right\}$$

(where A is a normalization constant that will soon drop out of the calculation). In region II, the solution depends on whether the total wave function is even or odd. For an even wave function, we get the solution

$$\psi_{II}(x) = B \cosh\left\{\sqrt{\frac{2m(V_0-E)}{\hbar^2}}\,x\right\},$$

while in the odd case we get

$$\psi_{II}(x) = B \sinh\left\{\sqrt{\frac{2m(V_0-E)}{\hbar^2}}\,x\right\}.$$

In the even case, the boundary condition (5.88) now reduces to

$$\tan\left\{\sqrt{\frac{2mE}{\hbar^2}}\,(b-a)\right\} = -\sqrt{\frac{E}{V_0-E}}\coth\left\{\sqrt{\frac{2m(V_0-E)}{\hbar^2}}\,a\right\}, \tag{5.89a}$$

while in the odd case, it reduces to

$$\tan\left\{\sqrt{\frac{2mE}{\hbar^2}}\,(b-a)\right\} = -\sqrt{\frac{E}{V_0-E}}\tanh\left\{\sqrt{\frac{2m(V_0-E)}{\hbar^2}}\,a\right\}. \tag{5.89b}$$

In the limit where $V_0 \to \infty$, i.e., where the potential barrier is infinitely high, the two conditions collapse into the single condition

$$\tan\left\{\sqrt{\frac{2mE}{\hbar^2}}\,(b-a)\right\} = 0. \tag{5.90}$$

(Notice that both $\tanh x$ and $\coth x$ tend to 1 at infinity!) In this case, the spectrum is therefore degenerate; i.e., both the even and odd wave functions have the same energy, which is given by

$$E = \frac{\hbar^2}{2m}\frac{n^2\pi^2}{(b-a)^2}; \qquad n = 1, 2, 3, \ldots \tag{5.91}$$

(cf. the free particle spectrum (5.69)).

For finite V_0 the spectra are, however, different in the two cases due to the difference between the hyperbolic tangent and cotangents functions. We can furthermore estimate the two lowest-lying energies. If V_0 is high, we expect the correct energy to be close to the "degenerate" value

$$E_0 = \frac{\hbar^2\pi^2}{2m(b-a)^2};$$

i.e., we can safely put

$$E = E_0 + \delta E \qquad (\text{where } |\delta E| \ll E_0).$$

Using the approximation

$$\sqrt{\frac{2mE}{\hbar^2}} \approx \sqrt{\frac{2mE_0}{\hbar^2}}\left(1 + \frac{\delta E}{2E_0}\right),$$

we then get

$$\tan\left\{\sqrt{\frac{2mE}{\hbar^2}}\,(b-a)\right\} = \tan\left\{\frac{\delta E}{2E_0}\sqrt{\frac{2mE_0}{\hbar^2}}\,(b-a)\right\}.$$

(Notice that $\tan\left\{\sqrt{\frac{2mE_0}{\hbar^2}}\,(b-a)\right\} = 0$, and use the addition theorem for the tangent function!) As $|\delta E| \ll E_0$, we can furthermore forget about the tangent function itself, so that the boundary conditions (5.89a) and (5.89b) finally reduce to

$$\delta E \approx \begin{cases} -\frac{2E_0\hbar}{(b-a)\sqrt{2mV_0}} \coth\left\{\sqrt{\frac{2mV_0}{\hbar^2}}\cdot a\right\} & (\text{even case}), \\ -\frac{2E_0\hbar}{(b-a)\sqrt{2mV_0}} \tanh\left\{\sqrt{\frac{2mV_0}{\hbar^2}}\cdot a\right\} & (\text{odd case}). \end{cases} \tag{5.92}$$

Thus the effect of including an additional potential well is on the one hand to produce a general shift in the ground state energy given by

$$-\frac{2E_0\hbar}{(b-a)\sqrt{2mV_0}}.$$

On the other hand, it causes a small split between the energies of the even and odd wave functions. Using the asymptotic relations

$$\coth x \underset{x\to\infty}{\approx} 1 + e^{-2x} \quad \text{and} \quad \tanh x \underset{x\to+\infty}{\approx} 1 - e^{-2x},$$

we in fact get that the even wave function has the lowest energy and that the energy split between the even and odd case is given by

$$E_- - E_+ \approx \frac{4E_0\hbar}{(b-a)\sqrt{2mV_0}} \exp\left[-\frac{1}{\hbar}\,(\sqrt{2mV_0})2a\right]. \tag{5.93}$$

Returning to the case of a general double well, we thus see that the classical degeneracy has actually been lifted. The symmetric ground state wave function suggests that there is a probability of finding it somewhere between the wells. This is related to the tunnel effect. Classically, a particle with energy E_+ cannot penetrate through the potential barrier, so it is totally unaffected by the presence of the second well. Quantum-mechanically, there is, however, a small chance for the particle to be in the region between the wells. Semiclassically, we can thus describe the ground state in the following way: Most of the time the particle is vibrating close to the bottom of either of the wells, but from time to time it leaks through the barrier and moves from one well to another; i.e., it tunnels back and forth.

Notice that the instanton solution (5.84) precisely interpolates between the two classical vacua, and we can therefore think of it as representing the tunneling between the two vacua. Similarly, we may think of the multiinstanton solutions as describing the tunneling back and forth between the two vacua.

Let us further consider the Euclidean equations of motion and the associated Euclidean energy (see Table 5.1). The Euclidean energy (which is the Wick rotated version of the Minkowski energy) is indefinite, but it is still a constant of motion. In the Minkowski case, the demand of vanishing energy leads to the classical vacua, i.e.,

$$x(t) = \pm a.$$

But in the Euclidean case it leads to the instanton equation (5.87), i.e.,

$$\frac{dx}{d\tau} = \pm\sqrt{\frac{V(x)}{2m}}\,.$$

Thus the instantons are zero-energy solutions that interpolate between the classical vacua. (The same is "almost" true for the multiinstantons.)

This is now a general aspect of the tunneling phenomena: *The existence of a Euclidean zero-energy solution that interpolates between two different classical vacua signals the presence of tunneling between these vacua, and this tunneling lifts the degeneracy of the classical vacuum.*

In light of the above interpretation of the instanton, let us return for a moment to the double square well. Here we have shown that the energy split is given by (5.93)

$$E_- - E_+ \approx \frac{4E_0\hbar}{(b-a)\sqrt{2mV_0}}\exp\left\{-\frac{1}{\hbar}\,2a\sqrt{2mV_0}\right\}.$$

This can be related to instantons in the following way: According to (5.86), the action of a single instanton is given by

$$S_0 = \int_{-a}^{+a}\sqrt{2mV(\phi)}\,d\phi = 2a\sqrt{2mV_0}\,.$$

Table 5.1.

	Minkowski version (real time)	Euclidean version (imaginary time)
Equation of motion	$m\frac{d^2x}{dt^2} = -V'(x)$	$m\frac{d^2x}{d\tau^2} = V'(x)$
Energy	$E = \frac{1}{2}m\left(\frac{dx}{dt}\right)^2 + V(x)$	$E = -\frac{1}{2}m\left(\frac{dx}{d\tau}\right)^2 + V(x)$

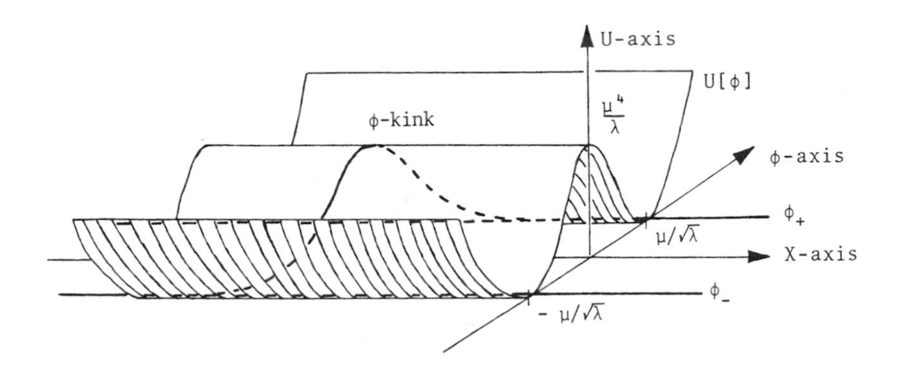

Figure 5.9.

Thus the energy split is precisely proportional to $\exp\left\{-\frac{1}{\hbar} S_0\right\}$. This strongly suggests that the energy split is related to some Euclidean path integral of the form

$$\int_{x(-\infty)=-a}^{x(\infty)=+a} \exp\left\{-\frac{1}{\hbar} S_E[x(\tau)]\right\} D[x(\tau)],$$

where we have evaluated the path integral by expanding around the instanton solution in the usual fashion. We shall make this precise in a moment.

As a final remark in this section, we consider a scalar field theory in $(1+1)$-space–time dimensions based upon a potential density $U[\phi]$ that has a degenerate minimum. This includes our favorite examples: The ϕ^4-model and the sine-Gordon model. What happens when we quantize such theories? Does the vacuum degeneracy survive the quantization, or is it lifted by instanton effects? The vacuum degeneracy persists! An instanton in this case would correspond to a soliton in two space dimensions, but the existence of such a soliton would violate Derrick's theorem (see Section 4.7). Thus a scalar field theory in $(1+1)$-dimensions cannot support instantons. It can also be argued that two different classical vacua are separated by an infinite potential barrier. To be specific, let us consider the ϕ^4-model. To be able to "visualize" the argument, let us furthermore use the string analogy of the field theory; cf. Section 3.1. The potential energy $U[\phi]$ corresponds to a double valley separated by a ridge of height $\mu^4/\sqrt{\lambda}$; see Figure 5.9. A classical vacuum corresponds to a string that lies at the bottom of one of the valleys. Now, if we want to convert one vacuum into another, we would have to lift the string across the ridge. It requires the energy $\mu^4/\lambda \cdot L$ to lift the string across a ridge of length L. Thus it requires an infinite amount of energy to lift the string across an infinitely long ridge. The same type of argument shows that the soliton is quantum-mechanically stable, i.e., that it is not allowed to decay into one of the vacua.

§ 5.9 Instanton Calculation of the Low-Lying Energy Levels

Now, that we know the general aspects of the instantons, we shall try to make an explicit calculation of the energy split in the double well. As a byproduct of this calculation we will also be able to verify the previously given description of the ground state.

Since we are not interested in the elevated energy states, we need not evaluate the total path-cum-trace integral. It suffices to calculate the asymptotic behavior of the propagator as T goes to infinity. This follows from the decomposition (2.70), which in the Euclidean formalism becomes (5.80)

$$K_E(x_b;\, \tau_b \mid x_a;\, \tau_a) = \sum_{n=0}^{\infty} \overline{\phi_n(x_a)}\phi_n(x_b) \exp\left\{-\frac{1}{\hbar}\, E_n T\right\}.$$

In the large T-limit, the right-hand side will clearly be dominated by the ground state. Also, the "amplitude" will be related to the wave function.

Consider, for example, a single potential well such as the one sketched in Figure 5.3. The classical ground state is clearly given by

$$x_0(\tau) \equiv 0,$$

which in fact is the absolute minimum of the Euclidean action, so this path will dominate the path integral. Consider a small fluctuation around the classical ground state

$$x(\tau) = \eta(\tau).$$

Expanding the potential to second order, we get

$$V[x(\tau)] \approx \frac{1}{2}\, V''(0)\eta^2(\tau) = \frac{1}{2}\, m\omega^2\eta^2(\tau) \qquad \text{with } m\omega^2 = V''(0).$$

The Euclidean propagator from 0 to 0 is therefore approximately given by

$$\begin{aligned} K_E\left(0,\, \frac{T}{2} \mid 0;\, -\frac{T}{2}\right) &\approx \exp\left\{-\frac{1}{\hbar}\, S[x_0]\right\} \\ &\cdot \int_{\eta(-T/2)=0}^{\eta(T/2)=0} \exp\left\{-\frac{m}{2\hbar}\int_{-T/2}^{T/2}\left[\left(\frac{d\eta}{d\tau}\right)^2 + \omega^2\eta^2\right] d\tau\right\} D[\eta(\tau)]. \end{aligned} \tag{5.94}$$

But the action corresponding to the classical ground state is zero, and the remaining path integral is just the Euclidean propagator for the harmonic oscillator. Consequently, we get (cf. (5.76))

$$K_E\left(0;\, \frac{T}{2} \mid 0;\, -\frac{T}{2}\right) \approx \sqrt{\frac{m\omega}{2\pi\hbar\sinh[\omega T]}}\,.$$

We then go to the large T-limit, where we can replace the hyperbolic sine by an exponential function, so that we end up with

$$K_E\left(0;\frac{T}{2}\mid 0;-\frac{T}{2}\right)\underset{T\to\infty}{\approx}\sqrt{\frac{m\omega}{\pi\hbar}}\,e^{-\frac{\omega}{2}\cdot T}. \tag{5.95}$$

On the other hand, we get from (5.80) the following asymptotic behavior:

$$K_E\left(0;\frac{T}{2}\mid 0;-\frac{T}{2}\right)\underset{T\to\infty}{\approx}|\phi_0(0)|^2\exp\left[-\frac{1}{\hbar}\,E_0T\right]. \tag{5.96}$$

A comparison therefore reveals that

$$\phi_0(0)\approx\sqrt[4]{\frac{m\omega}{\pi\hbar}}\quad\text{and}\quad E_0\approx\frac{1}{2}\,\hbar\omega. \tag{5.97}$$

This is in precise agreement with the previously obtained results concerning the harmonic oscillator (see especially (5.25)), respectively the weak coupling limit of the low-lying energy states in a single potential well (cf. (5.53)).

Having "warmed up," we now turn our attention to the double well; cf. Figure 5.12. In this case the classical ground state is degenerate,

$$x_\pm(\tau)\equiv\pm a,$$

and we shall correspondingly take a closer look at the following four propagators:

$$K\left(-a;\frac{T}{2}\mid -a;-\frac{T}{2}\right);\qquad K\left(+a;\frac{T}{2}\mid +a;-\frac{T}{2}\right);$$
$$K\left(+a;\frac{T}{2}\mid -a;-\frac{T}{2}\right);\qquad K\left(-a;\frac{T}{2}\mid +a;-\frac{T}{2}\right).$$

The first two connect a classical ground state with itself, while the last two connect two different classical ground states.

The calculation of their asymptotic behavior will be performed in two steps: In the first step we will deduce the general structure of the asymptotic behavior leaving a single parameter uncalculated. This will especially put us in a position where we can deduce the qualitative behavior of the quantum-mechanical ground state. In the second step, which will be performed in an illustrative example, we will then calculate in full detail the missing parameter. This will finally give us an explicit formula for the energy split.

Okay, to be specific, we concentrate on the propagator

$$K\left(a;\frac{T}{2}\mid -a;-\frac{T}{2}\right). \tag{5.98}$$

In this case the path integral is, in particular, dominated by the instanton solution (in the large T-limit). Besides that, we will also get contributions from strings of instantons and antiinstantons. Let us denote such a path by

$$\tilde{x}_{[n;\tau_1,\ldots,\tau_n]}(t),$$

where n is the total number of instantons and antiinstantons, while $\tau_1, \ldots, \tau_n$ are the consecutive positions of the consecutive centers. The centers consequently obey the constraint

$$-\frac{T}{2} < \tau_1 < \tau_2 < \cdots < \tau_n < \frac{T}{2} \,. \tag{5.99}$$

In order to get a string of widely separated instantons and antiinstantons, the difference between two consecutive centers should actually be at least of the order of μ (the width of the instanton). In the large T-limit the subset of configurations where some of the instantons and antiinstantons overlap will, however, be very small compared to the total set of configurations satisfying the constraint (5.99), and therefore we will neglect them. Notice too that since we propagate from $-a$ to $+a$, the integer n must necessarily be odd.

Okay, according to our general strategy for semiclassical approximations we must now expand around each such quasi-stationary path

$$x(\tau) = \tilde{x}_{[n;\tau_1,\ldots,\tau_n]}(\tau) + \eta(\tau)$$

and approximate the potential by

$$V[x(\tau)] \approx V[\tilde{x}(\tau)] + \frac{1}{2}\, V''[\tilde{x}(\tau)]\eta^2(\tau).$$

The corresponding contribution to the path integral then decomposes as

$$\exp\left\{-\frac{1}{\hbar}\, S_E[\tilde{x}_{[n;\tau_1,\ldots,\tau_n]}]\right\} \int_{\eta(-T/2)=0}^{\eta(T/2)=0} \exp\left\{-\frac{1}{\hbar}\, \tilde{S}_E[\eta(\tau)]\right\} D[\eta(\tau)],$$

where $\tilde{S}_E$ is the Euclidean action of a particle moving in the time-dependent potential

$$\tilde{U}(\eta) = \frac{1}{2}\, V''\left[\tilde{x}_{[n;\tau_1,\ldots,\tau_n]}(\tau)\right]\eta^2.$$

The remaining path integral corresponds to the Euclidean propagator of a particle moving in the above potential. Let us for notational simplicity denote this propagator by

$$\tilde{K}_n\left(0; \frac{T}{2} \mid 0; -\frac{T}{2}\right),$$

where we have suppressed the dependence upon the centers $(\tau_1, \ldots, \tau_n)$. Notice that the Euclidean action of such a string is approximately given by

$$S_E[\tilde{x}_{[n;\tau_1,\ldots,\tau_n]}] \approx nS_0,$$

where S_0 is the Euclidean action of a single instanton; cf. (4.38).

Finally, we must sum up the contributions from all the quasi-stationary paths; i.e., we must sum over n and integrate over $\tau_1, \ldots, \tau_n$. The total approximation of the propagator (5.98) is thus given by

$$K_E\left(a;\frac{T}{2}\mid -a;\frac{T}{2}\right)\approx\sum_{n\text{ odd}}\int_{-T/2}^{+T/2}\cdots\int_{-T/2}^{\tau_3}\int_{-T/2}^{\tau_2}$$
$$\cdot\exp\left[-\frac{n}{\hbar}S_0\right]\tilde{K}_n\left(0;\frac{T}{2}\mid 0;-\frac{T}{2}\right)d\tau_1 d\tau_2\cdots d\tau_n. \tag{5.100}$$

We then turn our attention to the funny propagator $\tilde{K}_n$. Consider first the case of a single instanton centered at $\tau = 0$. Notice that the potential

$$\tilde{U}[\eta]=\frac{1}{2}V''[\tilde{x}(\tau)]\eta^2$$

degenerates to a harmonic oscillator potential except when we pass right through the instanton; i.e., outside the small interval $[-\mu, \mu]$, the potential is given by

$$\tilde{U}[\eta]\approx\frac{1}{2}m\omega^2\eta^2.$$

This suggests that we should compare the distorted propagator $\tilde{K}$ with the propagator of the harmonic oscillator, which we will denote by K_ω. We therefore put

$$\Delta=\frac{\tilde{K}\left(0;\frac{T}{2}\mid 0;-\frac{T}{2}\right)}{K_\omega\left(0;\frac{T}{2}\mid 0;-\frac{T}{2}\right)} \tag{5.101}$$

where Δ is supposed to be small. Using the known expression for the Euclidean propagator of the harmonic oscillator (5.76), we then get

$$\tilde{K}\left(0;\frac{T}{2}\mid 0;-\frac{T}{2}\right)=\Delta\sqrt{\frac{m\omega}{2\pi\hbar\sinh[\omega T]}}.$$

Since we are going to investigate the asymptotic behavior in the large T-limit, we can safely approximate the hyperbolic sine by an exponential function, and we end up with

$$K\left(0;\frac{T}{2}\mid 0;-\frac{T}{2}\right)\approx\Delta e^{-\omega T/2}\sqrt{\frac{m\omega}{2\pi\hbar}}.$$

Before we proceed to the case of multiinstantons, we want to make two important remarks about the behavior of the correction term Δ in the large T-limit:

In the large T-limit

1. *Δ is independent of T, and*
2. *Δ is independent of the center of the instanton.*

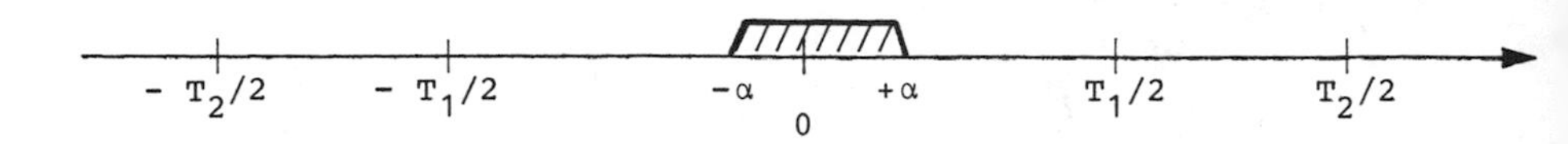

Figure 5.10.

PROOF. Suppose Δ depended upon T, and consider two different large values of T (Figure 5.10). According to the group property (2.25), we then get

$$\tilde{K}\left(0;\ \frac{T_2}{2}\ \middle|\ 0;\ -\frac{T_2}{2}\right) = \int\int \tilde{K}\left(0;\ \frac{T_2}{2}\ \middle|\ y;\ \frac{T_1}{2}\right)$$
$$\cdot\,\tilde{K}\left(y;\ \frac{T_1}{2}\ \middle|\ x;\ -\frac{T_1}{2}\right)\tilde{K}\left(x;\ -\frac{T_1}{2}\ \middle|\ 0;\ \frac{T_2}{2}\right) dx\,dy.$$

Outside $[-T_1/2;\ T_1/2]$, the curly propagator reduces to the oscillator propagator. Furthermore, $\tilde{K}(y;\ T_1/2 \mid x;\ -T_1/2)$ is exponentially small, except when x and y are close to the "classical" vacuum $\eta \equiv 0$; cf. the behavior of the ground state wave function (5.25). In the above integral we can therefore safely replace $\tilde{K}(y;\ T_1/2 \mid x;\ -T_1/2)$ by $\Delta(T_1)K_\omega(y;\ T_1/2 \mid x;\ -T_1/2)$. Using the group property once more, we therefore get

$$\tilde{K}\left(0;\ \frac{T_2}{2}\ \middle|\ 0;\ -\frac{T_2}{2}\right) \approx \Delta(T_1)K_\omega\left(0;\ \frac{T_2}{2}\ \middle|\ 0;\ -\frac{T_2}{2}\right),$$

i.e., $\Delta(T_2) = \Delta(T_1)$. The second statement is proven by a similar argument. □

Notice, too, that the correction term corresponding to an antiinstanton is the same as for an instanton, since they generate the same potential $\tilde{U}[\eta]$.

We now turn our attention to the multiinstanton case. Here the "particle" moves in the potential

$$\tilde{U}[\eta] = \frac{1}{2}\ V''[\tilde{x}_{[n;\tau_1,\ldots,\tau_n]}(\tau)]\eta^2.$$

For large T and widely separated instantons and instantons (Figure 5.11) we then get exactly as before that the potential reduces to the oscillator potential except when we pass through one of the instantons or antiinstantons. From the group

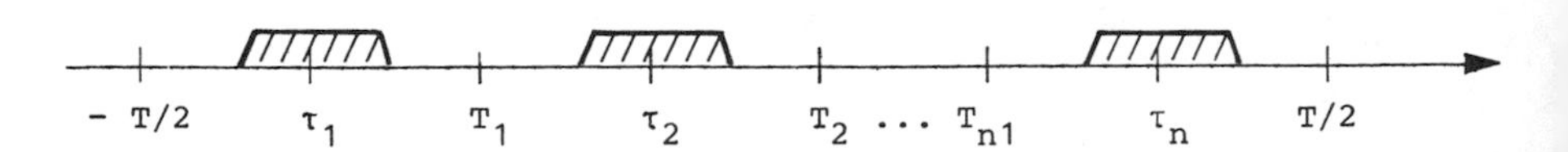

Figure 5.11.

property (2.25) we get

$$\tilde{K}_n\left(0; \frac{T}{2} \mid 0; -\frac{T}{2}\right) = \int \cdots \int \tilde{K}\left(0; \frac{T}{2} \mid x_{n-1}; T_{n-1}\right)$$
$$\cdots \tilde{K}\left(x_1; T_1 \mid 0; -\frac{T}{2}\right) dx_1 \cdots dx_{n-1}.$$

As in the above proof, we can safely replace

$$\tilde{K}\left(x_k; T_k \mid x_{k-1}; T_{k-1}\right)$$

by

$$\Delta \cdot K_\omega(x_k; T_k \mid x_{k-1}; T_{k-1}).$$

Using the group property once more, we therefore deduce that

$$\tilde{K}_n\left(0; \frac{T}{2} \mid 0; -\frac{T}{2}\right) \approx \Delta^n K_\omega\left(0; \frac{T}{2} \mid 0; -\frac{T}{2}\right) \approx \Delta^n e^{-\omega T/2}\sqrt{\frac{m\omega}{\pi\hbar}}\,.$$

Inserting this into our approximate formula for the full propagator, (5.100) now reduces to

$$K_E\left(a; \frac{T}{2} \mid -a; -\frac{T}{2}\right) \underset{T\to\infty}{\approx} \sqrt{\frac{m\omega}{\pi\hbar}}\, e^{-\omega T/2}$$
$$\cdot \sum_{n \text{ odd}} \int_{-T/2}^{+T/2} \cdots \int_{-T/2}^{\tau_3} \int_{-T/2}^{\tau_2} \exp\left[-\frac{n}{\hbar}\, S_0\right] \Delta^n d\tau_1 d\tau_2 \cdots d\tau_n.$$

But neither S_0 nor Δ depends upon the intermediary times $\tau_1, \ldots, \tau_n$. Consequently, the time integration can be trivially executed. Using a symmetry argument, we quickly get

$$\int_{-T/2}^{+T/2} \cdots \int_{-T/2}^{\tau_3} \int_{-T/2}^{\tau_2} d\tau_1 d\tau_2 \cdots d\tau_n$$
$$= \frac{1}{n!} \int_{-T/2}^{+T/2} \cdots \int_{-T/2}^{+T/2} \int_{-T/2}^{+T/2} d\tau_1 d\tau_2 \cdots t\tau_n = \frac{T^n}{n!}\,.$$

Inserting this, we can finish our calculation:

$$K_E\left(a; \frac{T}{2} \mid -a; -\frac{T}{2}\right)$$
$$\approx \sqrt{\frac{m\omega}{\pi\hbar}}\, e^{-\omega T/2} \sum_{n \text{ odd}} \frac{1}{n!}\left\{\Delta T \exp\left[-\frac{1}{\hbar}\, S_0\right]\right\}^n \tag{5.102}$$
$$= \sqrt{\frac{m\omega}{\pi\hbar}}\, e^{-\omega T/2} \sinh\left\{\Delta T \exp\left[-\frac{1}{\hbar}\, S_0\right]\right\}.$$

Had we considered the propagator from $-a$ to $-a$ itself, everything would have been the same, except that this time n should be even. We would therefore end up with the analogous formula

$$K_E\left(-a; \frac{T}{2} \mid -a; -\frac{T}{2}\right) \underset{T\to\infty}{\approx} \sqrt{\frac{m\omega}{\pi\hbar}}\, e^{-\omega T/2} \cosh\left\{\Delta T \exp\left\{-\frac{1}{\hbar} S_0\right\}\right\}. \tag{5.103}$$

Finally, the propagators are invariant under the substitution

$$a \to -a; \qquad -a \to a.$$

Putting everything together, we have thus shown that

$$\begin{aligned} K_E\left(\pm a; \frac{T}{2} \mid \pm a; -\frac{T}{2}\right) &\underset{T\to\infty}{\approx} \frac{1}{2}\sqrt{\frac{m\omega}{\pi\hbar}}\, e^{-\{\frac{1}{2}\omega - \Delta \exp\left[-\frac{1}{\hbar} S_0\right]\}T} \\ &\quad + \frac{1}{2}\sqrt{\frac{m\omega}{\pi\hbar}}\, e^{-\{\frac{1}{2}\omega + \Delta \exp\left[-\frac{1}{\hbar} S_0\right]\}T}; \end{aligned} \tag{5.104a}$$

$$\begin{aligned} K_E\left(\mp a; \frac{T}{2} \mid \pm a; -\frac{T}{2}\right) &\underset{T\to\infty}{\approx} \frac{1}{2}\sqrt{\frac{m\omega}{\pi\hbar}}\, e^{-\{\frac{1}{2}\omega - \Delta \exp\left[-\frac{1}{\hbar} S_0\right]\}T} \\ &\quad - \frac{1}{2}\sqrt{\frac{m\omega}{\pi\hbar}}\, e^{-\{\frac{1}{2}\omega + \Delta \exp\left[-\frac{1}{\hbar} S_0\right]\}T}; \end{aligned} \tag{5.104b}$$

Comparing this with the general formula (5.80), we see that the ground state has split up into two low-lying energy states ϕ_0 and ϕ_1. The energy split is given by

$$E_1 - E_0 = 2\hbar\Delta \exp\left[-\frac{1}{\hbar} S_0\right], \tag{5.105}$$

and the wave functions ϕ_0 and ϕ_1 satisfy

$$\phi_0^2(\pm a) = \phi_1^2(\pm a) = \phi_0(a)\phi_0(-a) = -\phi_1(a)\phi_1(-a) = \frac{1}{2}\sqrt{\frac{m\omega}{\pi\hbar}}.$$

Thus the ground state wave function ϕ_0 together with its close companion ϕ_1 peaks at the classical vacua $\pm a$, but the peaks are only half as great as the corresponding peak for the oscillator ground state, cf. (5.25),

$$\phi_0(+a) = \phi_0(-a) = \frac{1}{\sqrt{2}}\sqrt[4]{\frac{m\omega}{\pi\hbar}} = \phi_1(+a) = -\phi_1(-a).$$

Clearly, the true quantum-mechanical ground state ϕ_0 corresponds to an even combination of the corresponding oscillator ground states ϕ_- and ϕ_+, while ϕ_1 corresponds to an odd combination. Notice that in accordance with the node theorem in Section 4.7, the even combination has no node, while the odd combination has precisely one node. We have thus in full detail verified the qualitative description outlined in the preceding section.

Let us make some general comments about the method we have used: If we imagine that the instantons are some kind of particles (which they are *not*!), then we can think of an arbitrary distribution of instantons and antiinstantons as a kind of "instanton gas." To calculate the Euclidean propagator, we are then summing up various types of configurations in this "instanton gas," but the crucial point is that the calculation is dominated by configurations consisting of widely separated instantons and antiinstantons. For this reason the above method is often referred to as the *dilute gas approximation*.

We can now check whether the method is self-consistent, i.e., if the main contribution to the approximation (5.104) really does come from widely separated instantons and antiinstantons. Consider the intermediary result (5.102):

$$K_E\left(a; \frac{T}{2} \mid -a; -\frac{T}{2}\right) \underset{T\to\infty}{\approx} \sqrt{\frac{m\omega}{\pi\hbar}}\, e^{-\omega T/2} \sum_{n\ \text{odd}} \frac{1}{n!} \left\{\Delta T \exp\left[-\frac{1}{\hbar} S_0\right]\right\}^n .$$

Now, in a series like

$$\sum_n \frac{1}{n!} x^n$$

the main contribution comes from the terms where

$$n \approx x.$$

(As long as $n < x$, the factorial is dominated by the power and vice versa.) In our case, the main contribution thus comes from terms where

$$n \approx \Delta T \exp\left[-\frac{1}{\hbar} S_0\right],$$

i.e.,

$$\frac{n}{T} \approx \Delta T \exp\left[-\frac{1}{\hbar} S_0\right].$$

But n/T is the mean density of the instantons and antiinstantons, and since the right-hand side is supposed to be small, this mean density is very small. Thus the main contribution really comes from "dilute gas" configurations.

As another application of the dilute gas approximation we now consider a particle moving in a periodic potential, say an electron moving in a one-dimensional crystal. Thus the potential is closely analogous to the potential in the sine-Gordon model, and we should keep the sine-Gordon model in the back of our minds throughout the following discussion. In the case of a periodic potential, the classical ground state is infinitely degenerate (cf. Figure 5.12):

$$x_j(\tau) = ja; \qquad j = \ldots, -2, -1, 0, 1, 2, \ldots.$$

We shall correspondingly try to estimate the propagator that connects two such classical ground states, which we label by the integers j_- and j_+:

$$K_E\left(j_+a; \frac{T}{2} \mid j_-a; -\frac{T}{2}\right).$$

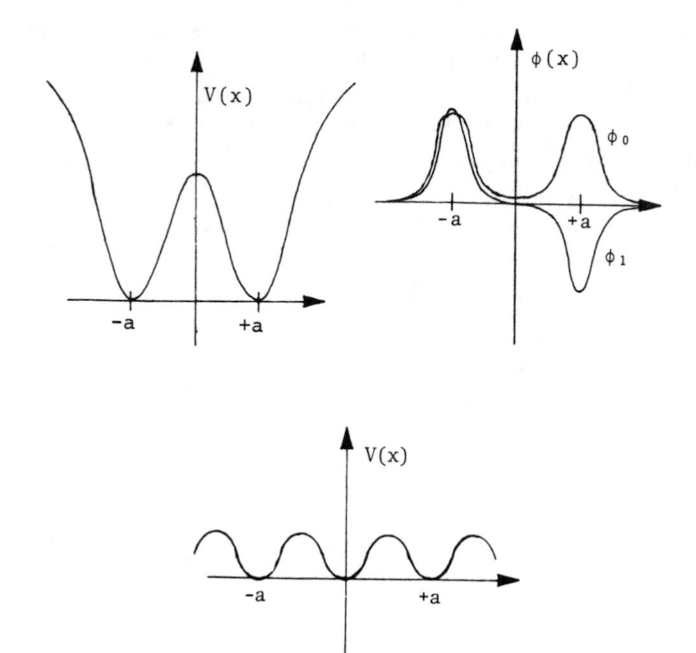

Figure 5.12.

The corresponding path integral is dominated by multiinstanton configurations, i.e., by strings of widely separated instantons and/or antiinstantons. But notice that this time there is *no* restriction on the positions of the antiinstantons relative to the instantons. If we label such a multiinstanton configuration by

$$\tilde{x}_{[n;m;\tau_1,\ldots,\tau_n;\mu_1,\ldots,\mu_m]}$$

(where n is the number of instantons, m the number of antiinstantons, etc.), then the only restrictions to be placed upon n, m, and the center positions of the instantons, respectively the antiinstantons, are accordingly given by

$$n - m = j_+ - j_-; \quad -\frac{T}{2} < \tau_1 < \cdots < \tau_n < \frac{T}{2};$$
$$-\frac{T}{2} < \mu_1 < \cdots < \mu_m < \frac{T}{2}.$$

Following the same chain of arguments as in the case of the double well, we therefore obtain the following asymptotic estimate:

$$K_E\left(j_+a;\frac{T}{2} \mid j_-a;-\frac{T}{2}\right)$$
$$\underset{T\to\infty}{\approx} \sqrt{\frac{m\omega}{\pi\hbar}}\, e^{-\omega T/2} \sum_n \sum_m \frac{1}{n!}\,\frac{1}{m!}\left\{\Delta T \exp\left[-\frac{1}{\hbar}\,S_0\right]\right\}^{n+m} \delta_{n+m-j_++j_-}. \tag{5.106}$$

Using that

$$\delta_{ab} = \frac{1}{2\pi}\int_0^{2\pi} e^{i\theta(a-b)}d\theta,$$

we can in fact get rid of the restriction on the summation over n and m, and we thereby get

$$K_E\left(j_+a;\ \frac{T}{2} \mid j_-a;\ -\frac{T}{2}\right) \underset{T\to\infty}{\approx} \sqrt{\frac{m\omega}{\pi\hbar}}\, e^{-\omega T/2}\frac{1}{2\pi}\int_0^{2\pi}\sum_n\sum_m \frac{1}{n!}\frac{1}{m!}$$
$$\cdot\left\{\Delta T \exp\left[-\frac{1}{\hbar}\,S_0\right]\exp[i\theta]\right\}^n$$
$$\cdot\left\{\Delta T \exp\left[-\frac{1}{\hbar}\,S_0\right]\exp[i\theta]\right\}^m e^{-i\theta(j_+-j_-)}d\theta.$$

But then we can explicitly perform the double summation, which in fact, decomposes into the product of two separate summations:

$$= \sqrt{\frac{m\omega}{\pi\hbar}}\, e^{-\omega T/2}\frac{1}{2\pi}\int_0^{2\pi} e^{-2\Delta T\exp\left\{-\frac{1}{\hbar}S_0\right\}\cos\theta} e^{-i\theta(j_+-j_-)}d\theta.$$

So the calculation has come to an end, and after a little rearrangement we get the final result:

$$\begin{aligned} &K_E\left(j_+a;\ \frac{T}{2} \mid j_-a;\ -\frac{T}{2}\right) \\ &\underset{T\to\infty}{\approx} \sqrt{\frac{m\omega}{\pi\hbar}}\,\frac{1}{2\pi}\int_0^{2\pi} e^{-i\theta j_+}e^{i\theta j_-}e^{-\left\{\frac{1}{2}\,\omega-2\Delta\cdot\exp\left[-\frac{1}{\hbar}\,S_0\right]\cos\theta\right\}T}d\theta. \end{aligned} \tag{5.107}$$

Now, how are we going to interpret this asymptotic behavior? If we compare it with the general expression for the asymptotic behavior of a Euclidean propagator (5.80), we see that this time the classical degenerate vacuum has in fact been split up into a *continuous* band of low-lying energy states $\phi_\theta(x)$ labeled by the angle θ. Furthermore, the energy of the eigenstate ϕ_θ (relative to the ground state $\theta = 0$) is given by

$$E(\theta) \approx 2\Delta\exp\left\{-\frac{1}{\hbar}\,S_0\right\}(\cos\theta - 1). \tag{5.108}$$

Notice also that

$$\phi_\theta(ja) = \sqrt[4]{\frac{m\omega}{\pi\hbar}}\,\frac{1}{\sqrt{2\pi}}\,e^{ij\theta}.$$

This suggests that the eigenstates ϕ_θ are a superposition of the form

$$\phi_\theta(x) \approx \frac{1}{\sqrt{2\pi}}\sum_{j=-\infty}^{+\infty}\phi(x-ja)e^{ij\theta}, \tag{5.109}$$

where $\phi(x)$ is the oscillator ground state wave function; cf. (5.25).

Notice that the periodic potential possesses a discrete symmetry, since it is invariant under the translation

$$x \to x + a.$$

If we neglect tunneling effects and correspondingly try to represent the quantum-mechanical ground state by the discrete series of wave functions

$$\phi_j(x) \approx \phi(x - ja),$$

then we break this translational symmetry. The effect of the instantons is thus, precisely as for the double well, to restore this symmetry: The correct eigenstates $\phi_\theta(x)$ are given by (5.109), which are now eigenfunctions of the translation operator

$$\phi_\theta(x + a) = e^{i\theta}\phi_\theta(x).$$

All the preceding results are in fact well known from solid state physics, where wave functions, which like ϕ_θ are eigenfunctions of the translation operator, are known as *Bloch waves*.

§ 5.10 Illustrative Example: Calculation of the Parameter Δ

So far, our discussion of the double well has been dominated by qualitative arguments. Now we will finish the calculation, so that we in principle can compute the actual figure representing the difference in energy between the two lowest-lying states.

We have already deduced the formula (5.105)

$$E_1 - E_0 \approx 2\hbar\Delta \exp\left[-\frac{1}{\hbar} S_0\right].$$

In this formula S_0 is the action of a single instanton, which according to (5.86) is given by

$$S_0 = \int_{-a}^{+a} \sqrt{2mV(x)}\, dx. \tag{5.110}$$

Thus it is a computable number once we have specified the potential. The other parameter Δ is, according to (5.101), given by

$$\Delta = \lim_{T\to\infty} \frac{\tilde{K}\left(0; \frac{T}{2} \mid 0; -\frac{T}{2}\right)}{K_\omega\left(0; \frac{T}{2} \mid 0; -\frac{T}{2}\right)}.$$

Here $\tilde{K}$ is the Euclidean propagator corresponding to the motion of a particle in the potential

$$\tilde{U}[x] = \frac{1}{2} V''[\tilde{x}_0(\tau)]x^2,$$

where $\tilde{x}_0(\tau)$ is the one-instanton solution (5.84), and K_ω is the propagator of a harmonic oscillator with frequency ω. In terms of path integrals, it is thus given by

$$\Delta = \lim_{T\to\infty} \frac{\int_{x(-T/2)=0}^{x(+T/2)=0} \exp\left\{-\frac{m}{2\hbar}\int_{-T/2}^{+T/2}\left\{\left(\frac{dx}{d\tau}\right)^2 + \frac{1}{m}V''[x(\tau)]\right\}d\tau\right\} D[x(\tau)]}{\int_{x(-T/2)=0}^{x(+T/2)=0} \exp\left\{-\frac{m}{2\hbar}\int_{-T/2}^{+T/2}\left\{\left(\frac{dx}{d\tau}\right)^2 + \omega^2 x^2\right\}d\tau\right\} D[x(\tau)]}.$$

One way of calculating this ratio is to use the relationship between path integrals and determinants. According to (5.46), we thus get

$$\Delta = \lim_{T\to\infty}\left[\frac{\det\left\{-\partial_\tau^2 - \frac{1}{m}V''[\tilde{x}(\tau)]\right\}}{\det\left\{-\partial_\tau^2 - \omega^2\right\}}\right]^{-1/2}.$$

But in the limit where $T \to \infty$ we know in fact that the differential operator

$$-\partial_\tau^2 - \frac{1}{m}V''[\tilde{x}(\tau)] \tag{5.111}$$

has a zero eigenmode given by

$$\phi_0(\tau) = \frac{d\tilde{x}_0}{d\tau}\ ; \tag{5.112}$$

i.e., by construction, ϕ_0 is an eigenfunction of the differential operator (5.111) with the eigenvalue 0. This is the old story that we encounter every time we have a symmetry in the Lagrangian that is broken by the particular solution in question; cf. the discussion of the translational mode in connection with the solitons (Section 4.7).

But the other differential operator

$$-\partial_\tau^2 - \omega^2 \tag{5.113}$$

has *no* zero eigenmode. Its lowest eigenstate is given by

$$\phi_1(\tau) = \sin\left[\frac{\pi\tau}{T}\right]; \qquad \text{i.e., } \lambda_1^2 = \omega^2 + \frac{\pi^2}{T^2}.$$

As a consequence, only the first determinant goes to zero in the large T-limit. Evidently, this implies that

$$\Delta \to \infty \qquad \text{as } T \to \infty.$$

But this is a disaster! The *whole* discussion preceding this illustrative example was based upon the assumption that Δ is a very small quantity, which is furthermore independent of T for large T! So where did we go wrong?

Fortunately, it turns out that all the conclusions in the preceding discussion are in fact correct, but that we have been a little sloppy in the introduction of Δ, where we have, in fact, committed a subtle but serious error.

To understand this, we must reconsider the whole philosophy behind our approach: To calculate a path integral, we proceed in two steps: First, we find the quasi-stationary paths, i.e., in the present case the multiinstantons

$$\tilde{x}_{[x;\tau_1,\ldots,\tau_n]}(\tau).$$

Next, we consider the fluctuations around each such configuration:

$$x(\tau) = \tilde{x}_{[n;\tau_1,\ldots,\tau_n]}(\tau) + \eta(\tau).$$

The path integral is then evaluated as a sum of contributions coming from each quasi-stationary path, i.e.,

$$\begin{aligned} &\int \exp\left\{-\frac{1}{\hbar}\, S_E[x(\tau)]\right\} D[x(\tau)] \\ &\quad\approx \sum_{n,\tau_1,\ldots,\tau_n} \int \exp\left\{-\frac{1}{\hbar}\, S_E[x(\tau)+\eta(\tau)]\right\} D[\eta(\tau)] \end{aligned}$$

(where the sum over the instanton centers $\tau_1, \ldots, \tau_n$ is actually an integration!). Each of the "shifted" path integrals is thereafter evaluated in the Gaussian approximation. The problem now arises because the set of quasi-stationary paths is labeled not only by a discrete parameter, but also by *continuous* parameters. To study this in detail, consider the one-instanton contribution. It comes from a curve in path space $\tilde{x}_{\tau_0}(\tau)$, where τ_0 labels the center of the instanton. (See Figure 5.13.) When we perform the Gaussian approximation corresponding to fluctuations around $\tilde{x}_{\tau_0}$, we in fact make a "double counting." The fluctuations *along* the instanton curve in path space include contributions from the nearby one-instanton paths

$$\tilde{x}_{\tau_0+\epsilon}(\tau),$$

but they have already been included when we "sum" over the positions τ_0! Thus

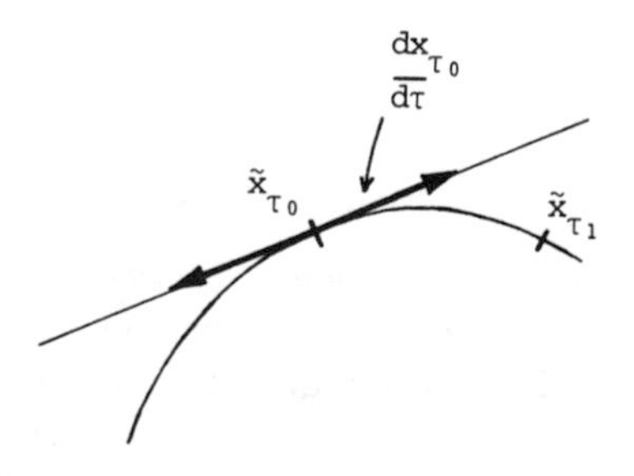

Figure 5.13.

we see that *when performing the Gaussian approximation, we ought to include only fluctuations that are perpendicular to the instanton curve in the path space*!

Notice that

$$\tilde{x}_{\tau_0+\epsilon}(\tau) = \tilde{x}_{\tau_0}(\tau - \epsilon) = \tilde{x}_{\tau_0}(\tau) - \epsilon \, \frac{d\tilde{x}_{\tau_0}}{d\tau} \, . \tag{5.114}$$

Consequently, the fluctuations along the instanton curve are precisely generated by the zero eigenmode; i.e., these fluctuations are of the form

$$\eta(\tau) = \lambda \, \frac{d\tilde{x}_{\tau_0}}{d\tau} \, .$$

This indicates a connection with our determinantal problem: On the one hand, it is the zero mode that generates fluctuations along the instanton curve (and thereby causes a "double counting" in our naive approach). On the other hand, it produces a zero eigenvalue in the determinant (as $T \to \infty$).

We can now resolve the problem in the following way: When calculating the shifted path integral in the Gaussian approximation, we must integrate only over fluctuations that are perpendicular to the instanton curve. Let us denote the resulting path integral by:

$$\int \exp\left\{-\frac{1}{\hbar} \, S_E[\tilde{x}_0(\tau) + \eta(\tau)]\right\} D[\eta(\tau)].$$

Let us furthermore consider the associated differential operator (5.111). It possesses a complete set of normalized eigenfunctions

$$\left\{-\partial_\tau^2 - \frac{1}{m} \, V''[x_0(\tau)]\right\} \phi_n(\tau) = \lambda_n \phi_n(\tau);$$

$$\int_{-T/2}^{+T/2} \phi_n(\tau)\phi_m(\tau) d\tau = \delta_{nm},$$

where we know that $\phi_0(\tau)$ is proportional to $d\tilde{x}_0/d\tau$ (in the limit where T goes to infinity). Normally, we would integrate over all the "Fourier components" in the shifted path,

$$\eta(\tau) = a_0\phi_0(\tau) + \sum_{n=1}^{\infty} a_n\phi_n(\tau),$$

but in the above path integral we have excluded integration over a_0, since this is to be replaced by an integration over the instanton center τ_0. This causes, however, a slight normalization problem. In the standard evaluation of the path integral (leading to the determinant), we use the integration measure

$$\sqrt{\frac{m}{2\pi\hbar}} \, da_0;$$

cf. (5.48). We must now reexpress this in terms of $d\tau_0$. According to (5.114), we get

$$\tilde{x}_{\tau_0 - d\tau_0}(\tau) \approx \tilde{x}_{\tau_0} + S_0^{1/2} d\tau_0 \phi_0,$$

and consequently,

$$da_0 = S_0^{1/2} d\tau_0.$$

The correct integration measure associated with the integration over the instanton center is therefore given by

$$\sqrt{\frac{mS_0}{2\pi\hbar}}\, d\tau_0. \tag{5.115}$$

When we originally performed the integration over the instanton center, we missed the factor in front of $d\tau_0$. This will now be included in the corrected definition of Δ. To sum up, the correct definition of Δ is therefore given by

$$\begin{aligned}\Delta &= \lim_{T\to\infty}\sqrt{\frac{mS_0}{2\pi\hbar}}\,\frac{\not{\tilde{K}}\left(0;\,\frac{T}{2}\mid 0;\,-\frac{T}{2}\right)}{K_\omega\left(0;\,\frac{T}{2}\mid 0;\,-\frac{T}{2}\right)}\\ &= \sqrt{\frac{mS_0}{2\pi\hbar}}\lim_{T\to\infty}\left[\frac{\not{\det}\left\{-\partial_\tau^2 - \frac{1}{m}V''[x_0(\tau)]\right\}}{\det\left\{-\partial_\tau^2 - \omega^2\right\}}\right]^{-1/2}\end{aligned} \tag{5.116}$$

(where the stroke indicates that we shall omit the lowest eigenvalue in the evaluation of the determinant!) This expression can now be computed using the following strategy. On a finite interval $[-T/2, +T/2]$ the differential operator (5.111) will not have an exact zero eigenmode (i.e., the smallest eigenvalue $\lambda_0(T)$ will vanish only in the limit as $T\to\infty$). Using the determinantal relation (5.50), we therefore get

$$\begin{aligned}\Delta &= \sqrt{\frac{mS_0}{2\pi\hbar}}\lim_{T\to\infty}\left[\frac{\det\left\{-\partial_\tau^2 - \frac{1}{m}V''[x_0(\tau)]\right\}}{\lambda_0(T)\det\left\{-\partial_\tau^2 - \omega^2\right\}}\right]^{-1/2}\\ &= \sqrt{\frac{mS_0}{2\pi\hbar}}\lim_{T\to\infty}\left[\frac{\lambda_0(T)\sinh[\omega T]}{\omega f_0\left(\frac{1}{2}T\right)}\right]^{1/2}\\ &= \sqrt{\frac{mS_0}{2\pi\hbar}}\lim_{T\to\infty}\left[\frac{\lambda_0(T)\exp[\omega T]}{2\omega f_0\left(\frac{1}{2}T\right)}\right]^{1/2}.\end{aligned} \tag{5.117}$$

It remains to calculate $f_0(T/2)$ and $\lambda_0(T)$:

(a) *Calculation of* $f_0(T/2)$:

By definition, $f_0(\tau)$ is the unique solution to the differential equation

$$\left\{-\partial_\tau^2 - \frac{1}{m}V''[\tilde{x}_0(\tau)]\right\} f_0(\tau) = 0 \tag{5.118}$$

that satisfies the boundary condition

$$f_0\left(-\frac{T}{2}\right) = 0, \qquad \partial_\tau f_0\left(-\frac{T}{2}\right) = 0;$$

cf. the discussion in Section 5.5. One particular solution to the above differential equation is

$$\phi_0(\tau) = S_0^{-1/2}\frac{d\tilde{x}_0}{d\tau}.$$

In the large T-limit this is properly normalized and must therefore have the asymptotic behavior

$$\phi_0(\tau) \underset{\tau\to+\infty}{\approx} C_0 e^{-\omega\tau},$$

since (5.111) reduces to (5.113) for large τ. Here C_0 is a constant, which by construction is extracted from the asymptotic behavior of the one-instanton solution

$$C_0 \stackrel{\text{def}}{=} \lim_{\tau\to\infty}\phi_0(\tau)e^{\omega\tau} = S_0^{-1/2}\lim_{\tau\to\infty}e^{\omega\tau}\frac{d\tilde{x}_0}{d\tau}. \tag{5.119}$$

Here the parameter S_0 is given by (5.110). Thus C_0 can be computed once we have specified the potential. Since $\tilde{x}_0$ is an odd function, it follows that ϕ_0 is an even function. Consequently, we may summarize its asymptotic behavior as follows:

$$\phi_0(\tau) \approx \begin{cases} C_0 e^{-\omega\tau} & \text{as } \tau \to +\infty, \\ C_0 e^{\omega\tau} & \text{as } \tau \to -\infty. \end{cases} \tag{5.120}$$

The second-order differential equation (5.118) has two independent solutions. The second one can be constructed explicitly from ϕ_0 as follows: Put

$$\psi_0(\tau) = \phi_0(\tau)\int_0^\tau \frac{dS}{\phi_0^2(S)}. \tag{5.121}$$

(Notice that ϕ_0 has *no* nodes and that ψ_0 is *not* proportional to ϕ_0, since the integral is τ-dependent.) Differentiating the above identity, we now get

$$\psi_0'(\tau) = \phi_0'(\tau)\int_0^\tau \frac{dS}{\phi_0^2(S)} + \frac{1}{\phi_0(\tau)}$$

and

$$\psi_0''(\tau) = \psi_0''(\tau)\int_0^\tau \frac{dS}{\phi_0^2(S)}.$$

These imply the following two identities:

$$W[\phi_0;\psi_0] = \phi_0\frac{d\psi_0}{d\tau} - \psi_0\frac{d\phi_0}{d\tau} = 1 \tag{5.122}$$

and

$$\psi_0''\phi_0 = \phi_0''\psi_0. \tag{5.123}$$

(The second one suffices to show that ψ_0 is another solution.) Since ϕ_0 is an even solution to (5.118), it follows that ψ_0 is an *odd* solution.

We are also going to need the asymptotic behavior of ψ_0. From (5.120) we get

$$\psi_0(\tau) \underset{\tau\to+\infty}{\approx} C_0 e^{-\omega\tau} \int_0^\tau \frac{dS}{C_0^2 e^{-2\omega\tau}} \approx \frac{e^{\omega\tau}}{2C_0\omega}.$$

The asymptotic behavior may thus be summarized as follows:

$$\psi_0(\tau) \approx \begin{cases} \frac{1}{2C_0\omega}\, e^{\omega\tau} & \text{as } \tau \to +\infty, \\ -\frac{1}{2C_0\omega}\, e^{-\omega\tau} & \text{as } \tau \to -\infty. \end{cases} \tag{5.124}$$

We proceed to investigate the particular solution f_0, which is the one we are really interested in. We know that it is a linear combination of ϕ_0 and ψ_0:

$$f_0(\tau) = A\phi_0(\tau) + B\psi_0(\tau).$$

Clearly, A and B can be expressed in terms of Wronskians:

$$W[\phi_0; f_0] = BW[\phi_0; \psi_0] = B$$

and

$$W[\psi_0; f_0] = AW[\psi_0; \phi_0] = -A.$$

But these Wronskians may also be computed directly using the boundary conditions satisfied by f_0:

$$W[\phi_0; f_0] = \left[\phi_0 \frac{df_0}{d\tau} - f_0 \frac{d\phi_0}{d\tau}\right]_{|\tau=-T/2} = \phi_0\left(-\frac{T}{2}\right)$$

and

$$W[\psi_0; f_0] = \left[\psi_0 \frac{df_0}{d\tau} - f_0 \frac{d\psi_0}{d\tau}\right]_{|\tau=-T/2} = \psi_0\left(-\frac{T}{2}\right).$$

Consequently,

$$f_0(\tau) = \phi_0\left(-\frac{T}{2}\right)\psi_0(\tau) - \psi_0\left(-\frac{T}{2}\right)\phi_0(\tau). \tag{5.125}$$

From this expression, which is exact, we can extract the asymptotic behavior of f_0:

$$f_0(\tau) \underset{\tau\to+\infty}{\approx} \frac{1}{2\omega}\left\{e^{-\omega\frac{T}{2}} e^{\omega\tau} + e^{\omega\frac{T}{2}} e^{-\omega\tau}\right\}.$$

In particular, we get

$$f_0\left(\frac{T}{2}\right) \approx \frac{1}{\omega} \qquad \text{in the limit for large } T. \tag{5.126}$$

(b) *Calculation of* $\lambda_0(T)$:

To compute the lowest eigenvalue, we must first determine the solution $f_\lambda(t)$ to the differential equation

$$\left\{-\partial_\tau^2 - \frac{1}{m} V''[\tilde{x}_0(\tau)]\right\} f_\lambda(\tau) = \lambda f_\lambda(\tau) \tag{5.127}$$

that satisfies the boundary conditions

$$f_\lambda\left(-\frac{T}{2}\right) = \partial_\tau f_\lambda\left(-\frac{T}{2}\right) = 0.$$

The eigenvalues λ are then determined from the subsidiary condition

$$f_\lambda\left(\frac{T}{2}\right) = 0.$$

For small values of λ we can now use the approximation

$$f_\lambda(\tau) \approx f_0(\tau) + \lambda \left.\frac{df_\lambda}{d\lambda}\right|_{\lambda=0}. \tag{5.128}$$

If we introduce the function

$$\Phi(\tau) = \left.\frac{df_\lambda}{d\lambda}\right|_{\lambda=0}(\tau),$$

we get from (5.127) that it is the unique solution to the differential equation

$$\left\{-\partial_\tau^2 - \frac{1}{m} V''[\tilde{x}_0(\tau)]\right\} \Phi(\tau) = f_0(\tau) \tag{5.129}$$

that satisfies the boundary conditions

$$\Phi\left(-\frac{T}{2}\right) = \frac{d}{d\tau}\Phi\left(-\frac{T}{2}\right).$$

The general solution to the inhomogeneous equation (5.129) is given by

$$\begin{aligned}\Phi(\tau) &= \alpha\phi_0(\tau) + \beta\psi_0(\tau) \\ &\quad + \int_{-T/2}^{\tau} [\psi_0(s)\phi_0(\tau) - \phi_0(s)\psi_0(\tau)] f_0(s) ds.\end{aligned}$$

But from the above boundary conditions it follows that both α and β vanish; i.e., Φ is given by

$$\Phi(\tau) = \int_{-T/2}^{\tau} [\psi_0(s)\phi_0(\tau) - \phi_0(s)\psi_0(\tau)] f_0(s) ds. \tag{5.130}$$

Inserting this in (5.128), we thus end up with the following approximation:

$$f_\lambda(\tau) \approx f_0(\tau) + \lambda \int_{-T/2}^{\tau} [\psi_0(s)\phi_0(\tau) - \phi_0(s)\psi_0(\tau)] f_0(s) ds.$$

But f_0 is given by (5.125). Inserting this and using the symmetry properties of ϕ_0 and ψ_0, we now get in the large T-limit,

$$f_\lambda\left(\frac{T}{2}\right) \underset{T\to\infty}{\approx} f_0\left(\frac{T}{2}\right) + \lambda \int_{-T/2}^{T/2} \left\{\phi_0^2\left(\frac{T}{2}\right)\psi_0^2(s) - \psi_0^2\left(\frac{T}{2}\right)\phi_0^2(s)\right\} ds.$$

But in the same limit,

$$\int_{-T/2}^{T/2} \phi_0^2(s)ds \approx \int_{-\infty}^{+\infty} \phi_0^2(s)ds = 1,$$

while

$$\int_{-T/2}^{T/2} \psi_0^2(s)ds \approx A \cdot e^{\omega T}.$$

Consequently, the first term is bounded (since $\phi_0^2\left(\frac{T}{2}\right)$ goes like $e^{-\omega T}$), while the second term grows exponentially. Inserting the asymptotic behavior of f_0 and ψ_0, we thus end up with

$$f_\lambda\left(\frac{T}{2}\right) \underset{T\to\infty}{\approx} \frac{1}{\omega} - \lambda \frac{1}{4C_0^2\omega^2} e^{\omega T}.$$

It follows that the lowest eigenvalue is given by

$$\lambda_0(T) \approx 4C_0^2\omega e^{-\omega T}. \tag{5.131}$$

Now that we have calculated both $f_0\left(\frac{T}{2}\right)$ and $\lambda_0(T)$, we can return to the formula for Δ. Inserting (5.126) and (5.131) into (5.117), we obtain

$$\Delta = C_0\sqrt{\frac{mS_0\omega}{\pi\hbar}}. \tag{5.132}$$

Thus the energy split (5.105) is given by

$$E_1 - E_0 \approx 2C_0\sqrt{\frac{mS_0\hbar\omega}{\pi}} \exp\left\{-\frac{1}{\hbar} S_0\right\}. \tag{5.133}$$

This ends our discussion of the double well. As an example of the application of the above machinery it is now possible to work out the detailed formula for the energy split in the double well specified by the fourth-order potential

$$V(x) = \frac{\lambda^4}{4}\left(x^2 - \frac{\mu^2}{\lambda}\right)^2.$$

Details can be found in (Gildener and Patrascioiu, 1977).

Solutions to Worked Exercises

Solution to 5.2.2

(a) From Taylor's formula we get

$$H_n(x) = \frac{\partial^n}{\partial z^n}\bigg|_{z=0} \exp[2xz - z^2] = e^{x^2} \frac{\partial^n}{\partial z^n}\bigg|_{z=0} \exp[-(x-z)^2]$$
$$= (-1)^n e^{x^2} \frac{\partial^n}{\partial x^n} \left\{\exp[-(x-z)^2]\right\}\big|_{z=0} = (-1)^n e^{x^2} \frac{\partial^n}{\partial x^n} e^{-x^2}.$$

(b) By completing the square,

$$-u^2 + 2iux = -(u-ix)^2 - x^2,$$

we can immediately compute the Gaussian integral:

$$\int_{-\infty}^{\infty} e^{-u^2+2iux} du = e^{-x^2} \int_{-\infty}^{\infty} e^{-\xi^2} d\xi = \sqrt{\pi} e^{-x^2}.$$

Using Rodriques's formula, we now get

$$H_n(x) = \frac{(-1)^n}{\sqrt{\pi}} e^{x^2} \frac{\partial^n}{\partial x^n} \int e^{-u^2+2iux} du$$
$$= \frac{(-2i)^n}{\sqrt{\pi}} e^{x^2} \int u^n e^{-u^2+2iux} du.$$

(c)

$$\sum_{n=0}^{\infty} H_n(x) H_n(y) \frac{z^n}{2^n n!} = \sum_{n=0}^{\infty} \frac{z^n}{2^n n!} \frac{(-2i)^{2n}}{\pi}$$
$$\cdot e^{x^2+y^2} \int\int u^n v^n e^{-u^2+2iux-v^2+2ivy} du\, dv$$
$$= \frac{1}{\pi} e^{x^2+y^2} \int\int \left[\sum_{n=0}^{\infty} \frac{(-2uvz)^n}{n!}\right] e^{-u^2+2iux-v^2+2ivy} du\, dv$$
$$= \frac{1}{\pi} e^{x^2+y^2} \int\int e^{-2uvz-u^2+2iux-v^2+2ivy} du\, dv.$$

We can now easily compute the Gaussian integrals, whereby we obtain

$$\sum_{n=0}^{\infty} H_n(x) H_n(y) \frac{z^n}{2^n n!} = (1-z^2)^{-1/2} \exp\left[\frac{2xyz - (x^2+y^2)z^2}{1-z^2}\right].$$

PART II

Basic Principles and Applications of Differential Geometry

General References to Part II

M. Spivak, *A Comprehensive Introduction to Differential Geometry,* Publish or Perish (1970).

S. I. Goldberg, *Curvature and Homology*, Academic Press, New York (1962).

G. deRham, *Variétés Differentiables. Formes, Courants, Formes Harmoniques*, Hermann, Paris (1960).

V. Guillemin and A. Pollack, *Differential Topology*, Prentice-Hall, Englewood Cliffs (1974).

I. M. Gelfand and G. E. Shilov, *Generalized Functions*, Academic Press, New York (1964).

L. I. Schiff, *Quantum Mechanics*, McGraw-Hill, Kogakusha Ltd. (1968).

A. Jaffe and C. H. Taubes, *Vortices and Monopoles*, Birkhäuser (1981).

H. B. Nielsen and P. Olesen, "Vortex-line models for dual strings," *Nucl. Phys.* **B61** (1973) 45.

G. S. LaRue, W. M. Fairbank, and A. F. Hebard, "Evidence for the existence of fractionally charged matter," *Phys. Rev. Lett.* **38** (1977) 1011.

P. Dirac, "Quantized singularities in the electromagnetic field," *Proc. Roy. Soc.* **A133** (1931) 60.

P. Dirac, "The theory of magnetic poles," *Phys. Rev.* **74** (1948) 817.

A. A. Belavin and A. M. Polyakov, "Metastable states of two-dimensonal isotropic ferromagnets," *JETP Lett.* **22** (1975) 245.

G. Woo, "Pseudoparticle configurations in two-dimensional ferromagnets," *J. Math. Phys.* **18** (1977) 1264.

CHAPTER 6

Differentiable Manifolds — Tensor Analysis

§ 6.1 Coordinate Systems

To simplify our discussion, we shall work entirely with subsets of Euclidean spaces! We will assume the reader to be familiar with Euclidean spaces, but to fix notation, some useful properties and definitions are collected in Figure 6.1.

Observe that once we have equipped the Euclidean space $\mathbb{R}^n$ with a metric, the rest of the definitions are standard definitions that are common for all metric spaces.

In what follows we shall especially be interested in subsets of $\mathbb{R}^n$ that might be referred to as "smooth surfaces." Consider, for instance, the unit sphere S^2 in three-dimensional Euclidean space $\mathbb{R}^3$:

$$S^2 = \{x \in \mathbb{R}^3 \mid (x^1)^2 + (x^2)^2 + (x^3)^2 = 1\}.$$

If we prefer, we may think of S^2 as a model of Earth's surface. Now we can clearly characterize a point on the surface through its Cartesian coordinates (x^1, x^2, x^3). This, on the other hand, is rather clumsy, because one of the Cartesian coordinates is superfluous: We can use the equation $(x^1)^2+(x^2)^2+(x^3)^2 = 1$ to eliminate, e.g., x^3. Because it suffices to use two coordinates when we want to characterize points on S^2, we say that S^2 is a smooth two-dimensional surface in $\mathbb{R}^3$. In geography it is customary to use latitude and longitude as coordinates. In our mathematical model these are referred to as polar coordinates (θ, ϕ), θ being the polar angle, while ϕ is the azimuthal angle. (See Figure 6.2.)

Let us try to sharpen these ideas. Let M be a subset of the Euclidean space $\mathbb{R}^N$. We want to give an exact meaning to the statement, M is an n-dimensional smooth surface.

The key concept is that of a *coordinate system*: Let U be an open subset of $\mathbb{R}^n$, and let $\phi : U \curvearrowright M$ be an injective map. As ϕ is injective, it establishes a one-to-one correspondence between points in the domain U and points in the image $\phi(U)$. Thus each point P in $\phi(U)$ is represented by exactly one set of coordinates $(x^1, \ldots, x^n)$ (see Figure 6.3),

$$P = \phi(x^1, \ldots, x^n).$$

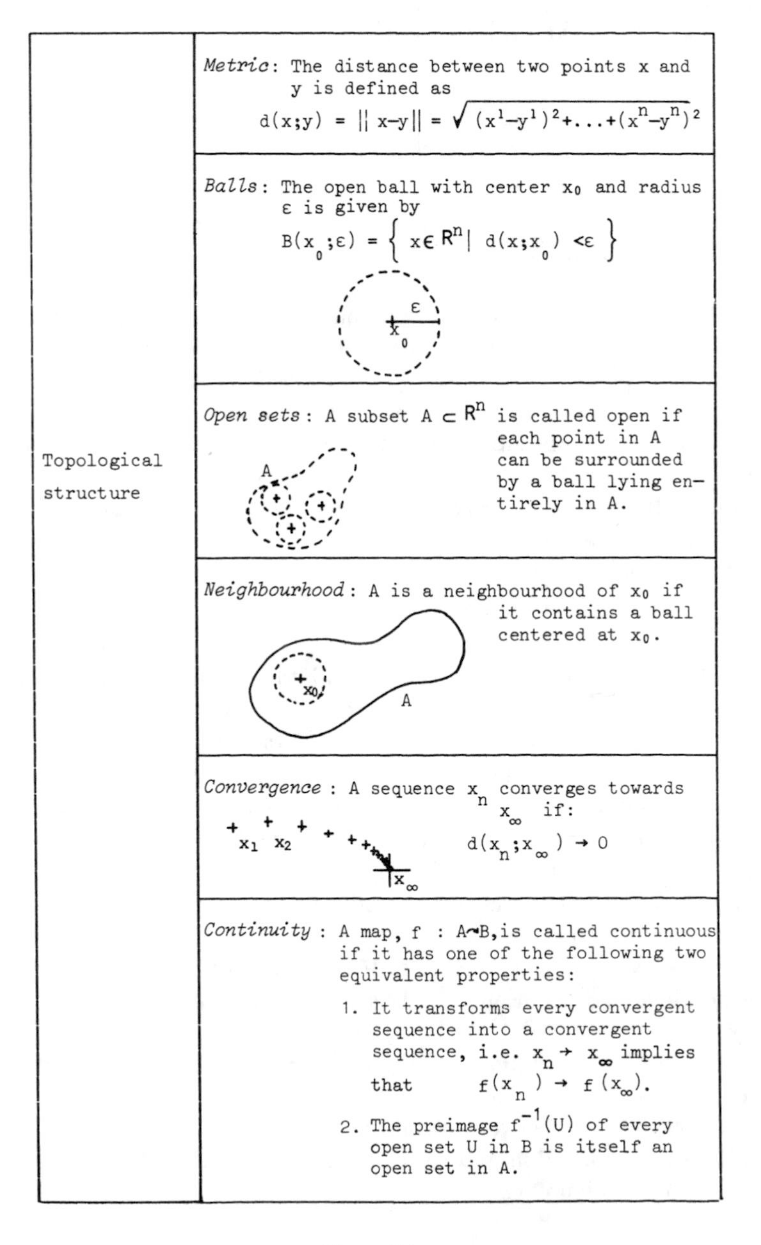

Topological structure
Metric: The distance between two points x and y is defined as
$d(x;y) = \|x-y\| = \sqrt{(x^1-y^1)^2+\ldots+(x^n-y^n)^2}$
Balls: The open ball with center x_0 and radius ε is given by
$B(x_0;\varepsilon) = \left\{ x \in R^n \mid d(x;x_0) < \varepsilon \right\}$
ε
x_0
Open sets: A subset $A \subset R^n$ is called open if each point in A can be surrounded by a ball lying entirely in A.
A
Neighbourhood: A is a neighbourhood of x_0 if it contains a ball centered at x_0.
x_0
A
Convergence: A sequence x_n converges towards x_∞ if:
$d(x_n;x_\infty) \to 0$
x_1 x_2
x_∞
Continuity: A map, $f: A \to B$, is called continuous if it has one of the following two equivalent properties:
1. It transforms every convergent sequence into a convergent sequence, i.e. $x_n \to x_\infty$ implies that $f(x_n) \to f(x_\infty)$.
2. The preimage $f^{-1}(U)$ of every open set U in B is itself an open set in A.

Figure 6.1.

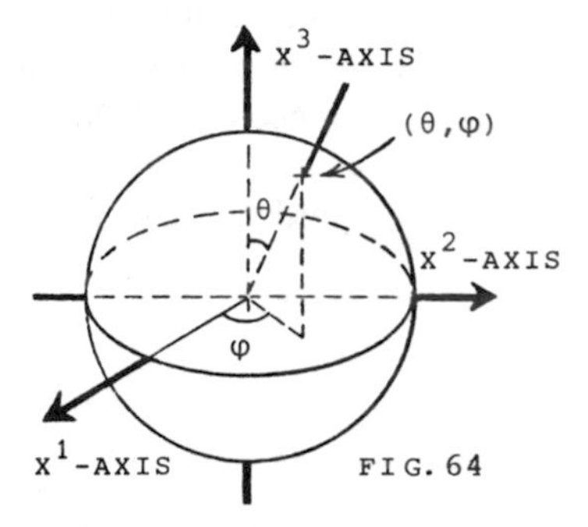

Figure 6.2.

We may therefore characterize the points in $\phi(U)$ through their coordinates. But that is not enough! To be of any interest, ϕ must respect the topology on M. Being a subset of $\mathbb{R}^N$ we know that M is equipped with a *metric*. Obviously, we must demand that points in M lie close to each other if and only if their coordinates lie close to each other! We can formalize this in the following way:

Let P be an arbitrary point in $\phi(U)$. The coordinates are expected to give information about the properties of the surface around P. But then it is necessary that the coordinates cover a neighborhood of P. We can achieve this by demanding that $\phi(U)$ should be an open subset of M. Hence, if Q is sufficiently close to P, then Q lies in $\phi(U)$, and we can represent it by a set of coordinates: $(x^1+\epsilon^1, \ldots, x^n+\epsilon^n)$.

Now, let (P_i), $i = 1, 2, \ldots$ be a sequence in $\phi(U)$ converging to a point P_∞ in $\phi(U)$, and look at the corresponding coordinates $(x_i^1, \ldots, x_i^n)$ and $(x_\infty^1, \ldots, x_\infty^n)$. We want the following to hold: P_i converges to P_∞ if and only if $(x_i^1, \ldots, x_i^n)$ converges to $(x_\infty^1, \ldots, x_\infty^n)$, but this we can achieve by demanding that ϕ be a *homeomorphism*, i.e., that ϕ and the inverse map ϕ^{-1} be continuous maps!

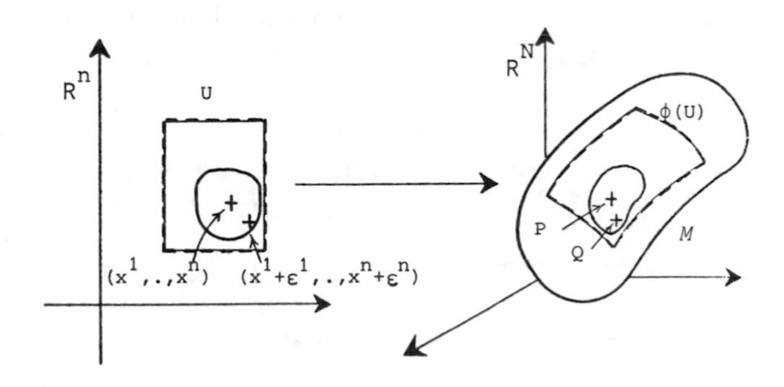

Figure 6.3.

By demanding that ϕ be a homeomorphism from an open subset $U \subset \mathbb{R}^n$ to an open subset $\phi(U) \subset M$, we have thus guaranteed that the coordinates respect the topology of M. But it would also be nice if we could express the smoothness of M in terms of the coordinate map ϕ.

First, we observe that ϕ can be regarded as a map, $U \curvearrowright \mathbb{R}^N$,

$$y^1 = \phi^1(x^1, \ldots, x^n)$$
$$\vdots$$
$$y^N = \phi^N(x^1, \ldots, x^n).$$

But then we must obviously demand that ϕ be a *smooth* map, i.e., that the components $\phi^1, \ldots, \phi^N$ have partial derivatives

$$\frac{\partial \phi^i}{\partial x^j}, \quad \frac{\partial^2 \phi^i}{\partial x^j \partial x^k}, \quad \frac{\partial^3 \phi^i}{\partial x^j \partial x^k \partial x^l}, \quad \text{etc.}$$

of arbitrarily high order! If ϕ is not smooth, we could easily be in trouble, as we can convince ourselves by considering the example $\phi(x, y) = (x, y, |x|)$. (See Figure 6.4.)

But even that is not enough! Given a smooth map $\phi : U \curvearrowright \mathbb{R}^N$, we may form the Jacobian matrix

$$\left[\frac{\partial \phi^i}{\partial x^j}\right] = \begin{bmatrix} \frac{\partial \phi^1}{\partial x^1} & \cdots & \frac{\partial \phi^1}{\partial x^n} \\ \vdots & & \\ \frac{\partial \phi^N}{\partial x^1} & \cdots & \frac{\partial \phi^N}{\partial x^n} \end{bmatrix}. \tag{6.1}$$

We say that ϕ is regular at a point $x_0 = (x_0^1, \ldots, x_0^n)$ if the Jacobian matrix $\left[\frac{\partial \phi^i}{\partial x^j}\right]\Big|_{x_0}$ has a maximal rank at that point. The maximal rank is n, and that ϕ is regular simply means that the n column vectors

$$\begin{bmatrix} \frac{\partial \phi^1}{\partial x^1} \\ \vdots \\ \frac{\partial \phi^N}{\partial x^1} \end{bmatrix}, \ldots, \begin{bmatrix} \frac{\partial \phi^1}{\partial x^n} \\ \vdots \\ \frac{\partial \phi^N}{\partial x^n} \end{bmatrix}$$

are linearly independent. We will demand that ϕ be regular everywhere in U. This is a rather technical assumption, but we will give a simple geometrical interpretation

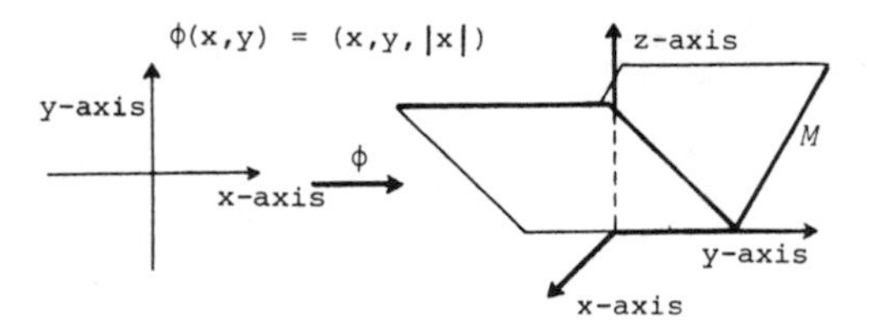

Figure 6.4.

in a moment. Let us first look at an example that shows that the assumption is necessary:

$$\phi : \mathbb{R}^2 \curvearrowright \mathbb{R}^3; \qquad \phi(x, y) = (x^2, x^3, y)$$

In this case ϕ is a smooth map, but the Jacobian matrix

$$\left[\frac{\partial \phi^i}{\partial x^j}\right] = \begin{bmatrix} 2x & 0 \\ 3x^2 & 0 \\ 0 & 1 \end{bmatrix}$$

is not of maximal rank when $x = 0$. Thus ϕ is singular along the y-axis. The corresponding points in M form the sharp edge! (See Figure 6.5)

Let us summarize the preceding discussion:

Definition 6.1. Let M be a subset of $\mathbb{R}^N$. A *coordinate system* on M is a pair (ϕ, U) where $U \subset \mathbb{R}^n$ is an open subset and $\phi : U \curvearrowright M$ is an injective map with the following two properties:

a. ϕ is a homeomorphism from the open set U onto the open set $\phi(U)$. (ϕ respects the topology of M.)
b. ϕ is a smooth regular map. (ϕ respects the smoothness of M.)

We may consider a coordinate system (ϕ, U) as a parametrization of a "smooth surface," but this parametrization need not cover the whole surface. Actually, there need not exist a single coordinate system covering a given "smooth surface." The sphere, for example, cannot be covered by a single coordinate system (this will be demonstrated in Section 6.2).

Let us look again at the Euclidean subset $M \subset \mathbb{R}^N$. Suppose that we have a family of coordinate systems (ϕ_i, U_i), $i \in I$, all of the same dimension n, and suppose furthermore that this family covers M; (see Figure 6.6),

$$M \subset \bigcup_{i \in I} \phi_i(U_i).$$

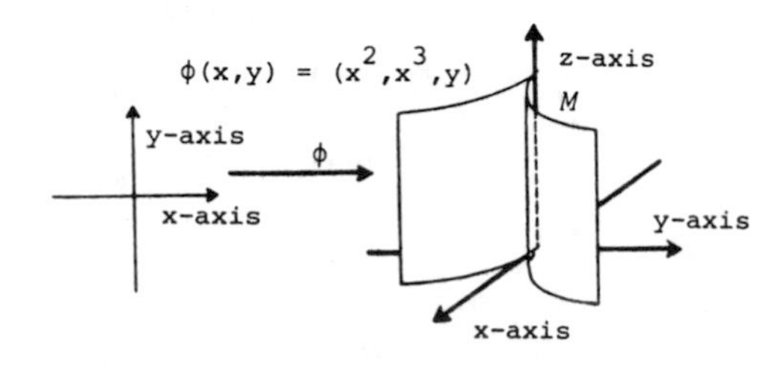

Figure 6.5.

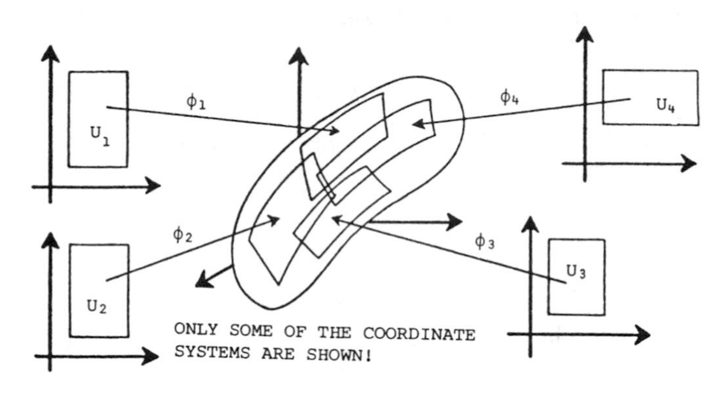

Figure 6.6.

In this case, some of the coordinate systems may, of course, overlap, which means that the same point P is represented by different coordinates, say $(x^1, \ldots, x^n)$ with respect to the coordinate system (ϕ_1, U_1) and $(y^1, \ldots, y^n)$ with respect to the coordinate system (ϕ_2, U_2); i.e.,

$$P = \phi_1(x^1, \ldots, x^n) = \phi_2(y^1, \ldots, y^n).$$

In the overlap region $\phi_1(U_1) \cap \phi_2(U_2)$ we have thus the possibility of exchanging coordinates! To investigate this more closely we now introduce the sets (see Figure 6.7)

$$\begin{aligned} U_{12} &= \phi_1^{-1}(\phi_1(U_1) \cap \phi_2(U_2)), \\ U_{21} &= \phi_2^{-1}(\phi_1(U_1) \cap \phi_2(U_2)). \end{aligned}$$

Observe that the overlap region $\phi_1(U_1) \cap \phi_2(U_2)$ is an open subset of M, and since ϕ_1 and ϕ_2 are continuous maps, we conclude that U_{12} and U_{21} are open sets. We now introduce the *transition functions* (see Figure 6.8)

$$\begin{aligned} \phi_{21} &: U_{12} \curvearrowright U_{21}, \qquad & \phi_{21} &= \phi_2^{-1} \circ \phi_1, \\ \phi_{12} &: U_{21} \curvearrowright U_{12}, \qquad & \phi_{12} &= \phi_1^{-1} \circ \phi_2. \end{aligned} \tag{6.2}$$

Suppose P is a point in the overlap region. Then P is represented by two sets of coordinates, $(x^1, \ldots, x^n)$ and $(y^1, \ldots, y^n)$, where

$$(x^1, \ldots, x^n) = \phi_1^{-1}(P) \quad \text{and} \quad (y^1, \ldots, y^n) = \phi_2^{-1}(P).$$

But then it is evident from Figure 6.8 that

$$(y^1, \ldots, y^n) = \phi_{21}(x^1, \ldots, x^n).$$

Thus the transition function ϕ_{21} has a simple meaning: *It expresses the new coordinates* $(y^1, \ldots, y^n)$ *in terms of the old ones* $(x^1, \ldots, x^n)$. In a similar way,

$$(x^1, \ldots, x^n) = \phi_{12}(y^1, \ldots, y^n);$$

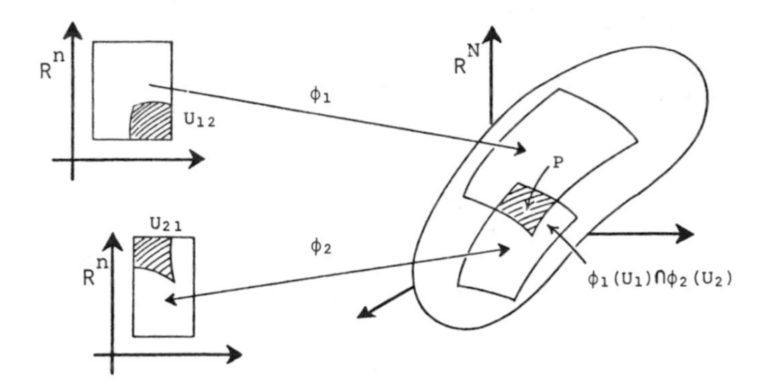

Figure 6.7.

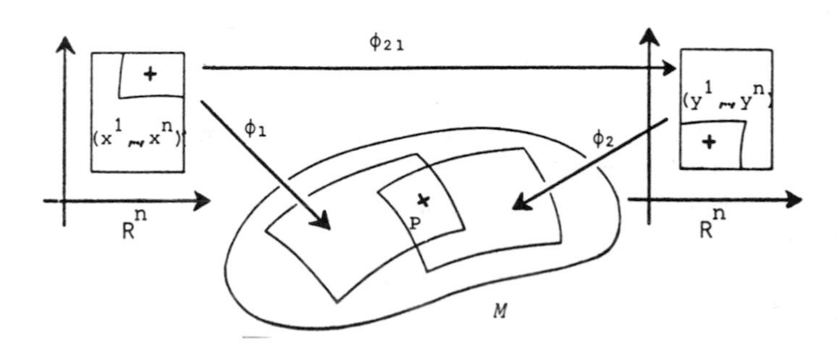

Figure 6.8.

i.e., the second transition function ϕ_{12} expresses the old coordinates in terms of the new ones! As

$$\phi_{21} = \phi_2^{-1} \circ \phi_1 \quad \text{and} \quad \phi_{12} = \phi_1^{-1} \circ \phi_2,$$

it follows that the transition functions are homeomorphisms. It will, however, be important later on that the new coordinates $(y^1, \ldots, y^n)$ depend not only continuously but even smoothly upon the old coordinates $(x^1, \ldots, x^n)$. Therefore, we demand that the transition functions be smooth functions. When the transition functions are smooth, we will also say that the two coordinate systems are *smoothly related*. The preceding discussion motivates the following definition:

Definition 6.2. Let M be a subset of $\mathbb{R}^N$. An *atlas* on M is a family of coordinate systems $(\phi_i, U_i)_{i \in I}$ with the following properties:

a. The family $(\phi_i, U_i)_{i \in I}$ covers M; i.e., $M \subset \bigcup_{i \in I} \phi_i(U_i)$.
b. Any two coordinate systems in the family are smoothly related.

§ 6.2 Differentiable Manifolds

Let us look at subset $M \subset \mathbb{R}^N$ equipped with an atlas $(\phi_i, U_i)_{i \in I}$. Suppose $f : M \to \mathbb{R}$ is a map from M to $\mathbb{R}$. We know what it means to say that f is continuous, because M is a metric space. Let P_0 be an arbitrary point in M. Now, we also want to assign a meaning to the statement, f is differentiable at the point P_0. Let us choose a coordinate system (ϕ_1, U_1) that covers $P_0 = \phi_1(x_0^1, \ldots, x_0^n)$. (See Figure 6.9.) Then we can represent f by an ordinary Euclidean function $\bar{f}$ defined on the domain of the coordinate system U_1, $y = \bar{f}(x^1, \ldots, x^n)$ with $x(^1, \ldots, x^n) \in U_1$, where $\bar{f} = f \circ \phi_1$.

But as f itself is defined on a neighborhood of P_0, we see that the Euclidean representative $\bar{f}$ is defined on a neighborhood of $(x_0^1, \ldots, x_0^n)$. Consequently, we make the following definition.

Definition 6.3. f is differentiable at the point P_0 if and only if the Euclidean representative $\bar{f}$ is differentiable in the usual sense at $(x_0^1, \ldots, x_0^n)$.

One might be afraid that this definition is meaningless, because we could choose another coordinate system (ϕ_2, U_2) covering P_0. Then f would be represented by another Euclidean function $\bar{f}'$, and what are we going to do if it turns out that $\bar{f}$ is differentiable at $(x_0^1, \ldots, x_0^n)$, while $\bar{f}'$ is *not* differentiable at $(y_0^1, \ldots, y_0^n)$?

But do not be afraid! $\bar{f}'$ is given by the expression $\bar{f}' = f \circ \phi_2$, which can be rearranged in the following way:

$$\bar{f}' = f \circ \phi_2 = f \circ (\phi_1 \circ \phi_1^{-1}) \circ \phi_2 = (f \circ \phi_1) \circ (\phi_1^{-1} \circ \phi_2) = \bar{f} \circ \phi_{12}.$$

(See Figure 6.10.) As the two coordinate systems are smoothly related, we know that the transition function ϕ_{12} is a smooth function. Thus we see that whenever $\bar{f}$ is differentiable, so is $\bar{f}'$!

Because an atlas makes it possible for us to introduce the concept of a differentiable map $f : M \to \mathbb{R}$, we say that an atlas generates *a differentiable structure on* M.

Now suppose we have been given two different atlases $(\phi_i, U_i)_{i \in I}$ and $(\psi_j, V_j)_{j \in J}$ that cover the same subset M. If all coordinate systems in the first

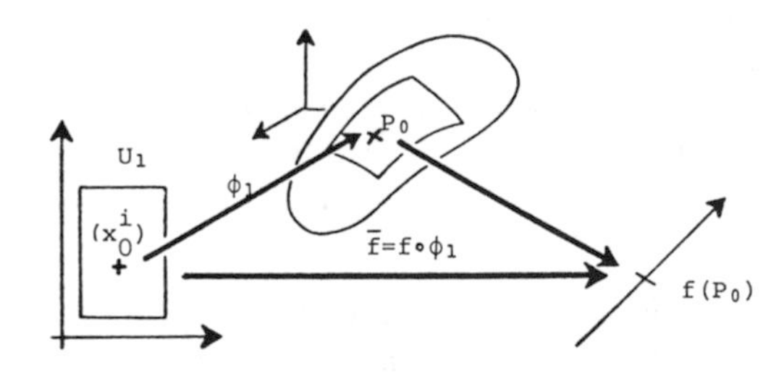

Figure 6.9.

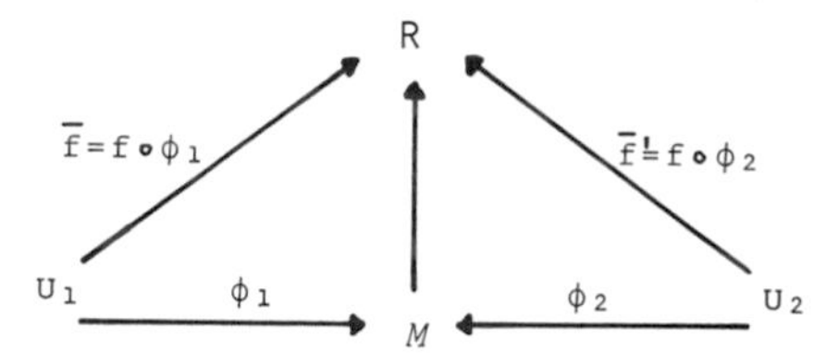

Figure 6.10.

atlas are smoothly related to all the coordinate systems in the second atlas, then they will obviously generate the same differentiable structure, i.e., a continuous map, $f : M \curvearrowright \mathbb{R}$ will be differentiable relative to the first atlas exactly when it is differentiable relative to the second atlas.

This motivates the following concept:

Definition 6.4. An atlas $(\phi_i, U_i)_{i\in I}$ on a subset M is called *maximal* if it has the following property: Whenever (ψ, V) is a coordinate system that is smoothly related to *all* the coordinates systems (ϕ_i, U_i) in our atlas, then (ψ, V) itself belongs to our atlas; i.e., $(\psi, V) = (\phi_j, U_j)$ for some $j \in J$.

Thus a maximal atlas comprises all possible smooth exchanges of the coordinates on M.

Given an atlas, say $(\phi_i, U_i)_{i\in I}$, we may easily generate a maximal atlas. Simply supply the given atlas with all possible coordinate systems that are smoothly related to the given atlas. The use of maximal atlases will be important for us later on, when we study the principle of general covariance.

We can now formalize the definition of a differentiable structure.

Definition 6.5. A differentiable structure on a Euclidean subset M is a maximal atlas on M.

Finally, we can give a precise definition of what we understand by a "smooth surface," which from now on we will refer to as a *differentiable manifold*:

Definition 6.6. A differentiable manifold M is a Euclidean subset M equipped with a differentiable structure.

Let M be a differentiable manifold. If we can cover it with a single coordinate system, we call it a *simple manifold*. In that case, a map $f : M \curvearrowright \mathbb{R}$ can be investigated through a single coordinate representative:

$$\bar{f}(x^1, \ldots, x^n).$$

In general, however, we will need several coordinate systems to cover M. A map $f : M \curvearrowright \mathbb{R}$, must then be represented by several coordinate representatives

$$\bar{f}_i(x^1, \ldots, x^n)$$

corresponding to the various coordinate systems (ϕ_i, U_i). In the overlapping regions

$$\Omega_{ij} = \phi_i(U_i) \cap \phi_j(U_j),$$

the coordinate representatives are then patched together through the relations

$$\bar{f}_i(x^1, \ldots, x^n) = \bar{f}_j(y^1, \ldots, y^n),$$

where

$$\phi_i(x^1, \ldots, x^n) = \phi_j(y^1, \ldots, y^n).$$

(See Figure 6.11.) Finally, we mention that an n-dimensional differentiable manifold M is frequently denoted by M^n, where the dimension of M is incorporated explicitly.

ILLUSTRATIVE EXAMPLE. *The sphere*

We are now armed with heavy artillery! Let us return to the sphere S^2 and see how the machinery works.

We want to construct an atlas on the sphere, and thus we must construct some coordinate systems. If U denotes the unit ball in $\mathbb{R}^2$, we may define a map $\phi : U \curvearrowright S^2$ in the following way:

$$\phi(x^1, x^2) = \left(x^1, x^2, \sqrt{1 - (x^1)^2 - (x^2)^2}\right).$$

Clearly, this map has a simple geometric interpretation: If we identify $\mathbb{R}^2$ with the $x^1 - x^2$-plane in $\mathbb{R}^3$, then ϕ is nothing but the projection along the x^3-axis! (See

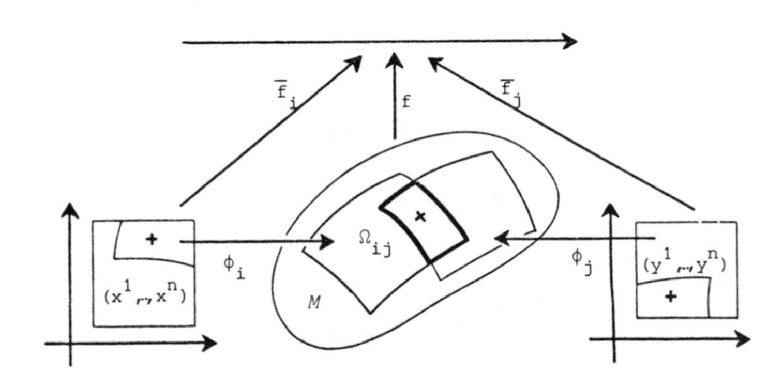

Figure 6.11.

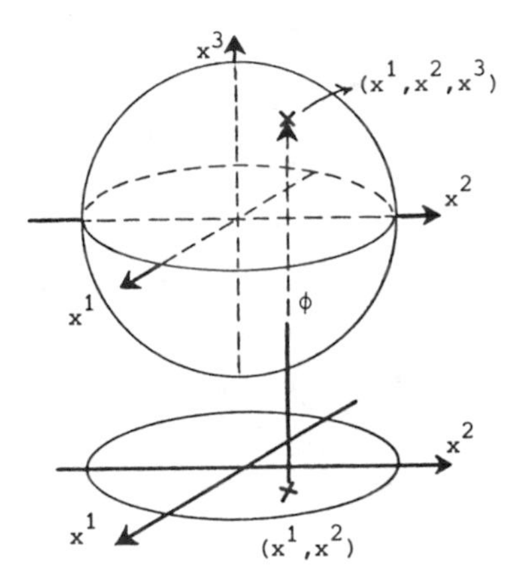

Figure 6.12.

Figure 6.12.) Let us check that (ϕ, U) is actually a coordinate system: U is obviously open, and $\phi(U)$ is the northern hemisphere

$$\phi(U) = \{x \in S^2 \mid x^3 > 0\},$$

which is open too (when regarded as a subset of the topological space S^2!) As

$$(x^1, x^2) \curvearrowright \sqrt{1 - (x^1)^2 - (x^2)^2}$$

is a continuous map, it follows that ϕ is continuous! ϕ^{-1} is nothing but the projection on the first two coordinates:

$$(x^1, x^2, x^3) \in \phi(U) : \phi^{-1}(x^1, x^2, x^3) = (x^1, x^2).$$

Hence ϕ^{-1} is continuous. Thus we have shown that ϕ *respects the topology of* S^2. Then we must check that ϕ is a smooth map:

We know that the square root function $\sqrt{t}$ is smooth when $t > 0$. (At $t = 0$ it is not smooth due to the fact that

$$\frac{d}{dt}\left(\sqrt{t}\right) = \frac{1}{2\sqrt{t}} \to \infty \qquad \text{when } t \to 0^+.)$$

But we have restricted U to values of (x^1, x^2) where $(x^1)^2 + (x^2)^2 < 1$. Thus there are no troubles, and ϕ is a smooth map.

We must check the Jacobian matrix too, but that is easily done:

$$\left[\frac{\partial\phi^i}{\partial x^j}\right] = \begin{bmatrix} \frac{\partial\phi^1}{\partial x^1} & \frac{\partial\phi^1}{\partial x^2} \\ \frac{\partial\phi^2}{\partial x^1} & \frac{\partial\phi^2}{\partial x^2} \\ \frac{\partial\phi^3}{\partial x^1} & \frac{\partial\phi^3}{\partial x^2} \end{bmatrix} = \begin{bmatrix} 1 & 0 \\ 0 & 1 \\ \frac{-x^1}{\sqrt{1-(x^1)^2-(x^2)^2}} & \frac{-x^2}{\sqrt{1-(x^1)^2-(x^2)^2}} \end{bmatrix},$$

and we see that $\left(\frac{\partial\phi^i}{\partial x^j}\right)$ always have rank two due to the upper square unit matrix. Consequently, we have also shown that ϕ *respects the smoothness of* S^2.

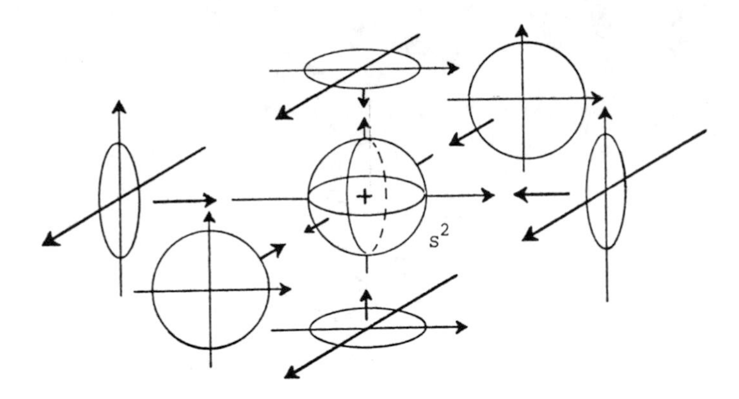

Figure 6.13.

Clearly, we may construct six coordinate systems in this way, and they obviously cover the sphere. (See Figure 6.13.)

To see that they form an atlas, we must check that they are smoothly related. (See Figure 6.14.) From the figure we read off

$$(x^1, x^2) \xrightarrow{\phi_1} \left(x^1, x^2, \sqrt{1-(x^1)^2-(x^2)^2}\right)$$
$$\xrightarrow{\phi_2^{-1}} \left(x^1, \sqrt{1-(x^1)^2-(x^2)^2}\right).$$

Hence the transition function ϕ_{21} is given by

$$y^1 = x^1,$$
$$y^2 = \sqrt{1-(x^1)^2-(x^2)^2},$$

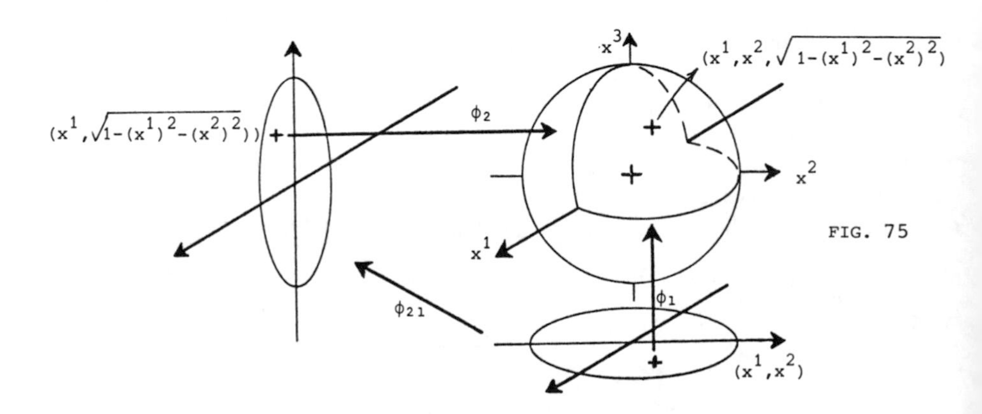

Figure 6.14.

and the corresponding domain U_{12} by

$$U_{12} = \left\{(x^1, x^2) \mid x^2 > 0, (x^1)^2 + (x^2)^2) < 1\right\}.$$

As the square root never obtains the value 0, the transition function ϕ_{12} is obviously smooth! In a similar way we can check that all the other transition functions are smooth, and thus we have shown in full detail how to construct an atlas on S^2.

Consequently, S^2 is a differentiable manifold, and we shall refer to the coordinates constructed above as *standard coordinates*.

What about the polar coordinates (θ, φ)? (See Figure 6.15.) In our new context they correspond to a coordinate system (ϕ, U), where

$$\phi(\theta, \varphi) = (\sin\theta\cos\varphi, \sin\theta\sin\varphi, \cos\theta).$$

But what can we choose as our domain U? There are two kinds of troubles:

The first trouble is not so serious. Due to the periodicity of the trigonometric functions, several values of the coordinates actually describe the same point:

$$\phi(\theta, \varphi) = \phi(\theta, \varphi + 2\pi),$$
$$\phi(\theta, \varphi) = \phi(-\theta, \varphi + \pi).$$

But the second trouble is more serious. At the north pole and the south pole the coordinate ϕ is completely indeterminate! Let us compute the Jacobian matrix:

$$\left[\frac{\partial\phi^i}{\partial x^j}\right] = \begin{bmatrix} \frac{\partial\phi^1}{\partial\theta} & \frac{\partial\phi^1}{\partial\phi} \\ \frac{\partial\phi^2}{\partial\theta} & \frac{\partial\phi^2}{\partial\phi} \\ \frac{\partial\phi^3}{\partial\theta} & \frac{\partial\phi^3}{\partial\phi} \end{bmatrix} = \begin{bmatrix} \cos\theta\cdot\cos\varphi & -\sin\theta\cdot\sin\varphi \\ \cos\theta\cdot\sin\varphi & \sin\theta\cdot\cos\varphi \\ -\sin\theta & 0 \end{bmatrix}.$$

At the north pole ($\theta = 0$) or the south pole ($\theta = \pi$) we see that $\sin\theta = 0$, and therefore we find that

$$\left[\frac{\partial\phi^i}{\partial x^j}\right]_{\text{pole}} = \begin{bmatrix} ? & 0 \\ ? & 0 \\ 0 & 0 \end{bmatrix},$$

which shows that ϕ is singular when $\theta = 0$ or $\theta = \pi$. Thus the coordinate system breaks down at the north pole and the south pole, and we will have to exclude

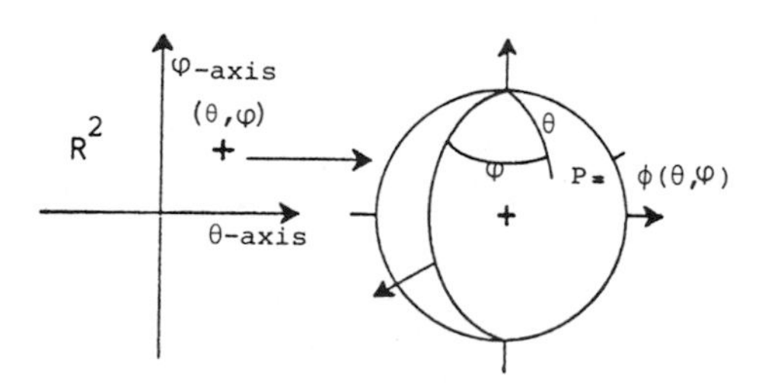

Figure 6.15.

these from the range of ϕ because they are singular points. (Let me emphasize once more: The sphere itself is perfect and smooth at the poles. The singularities arise *only* because we have chosen a "bad" coordinate system!)

The best thing we can do is therefore to put U equal to something like

$$U = \{(\theta, \varphi) \mid 0 < \theta < \pi, 0 < \varphi < 2\pi\}.$$

We see that we have missed not only the poles, but also the arc Γ joining the poles. The arc "singularity" is due to the periodicity in ϕ. Clearly, we may construct an atlas using two coordinate systems of this kind (see Figure 6.16).

Now we have constructed two different atlases on the sphere. Does it mean that we have constructed two different types of differentiable structures on the sphere? No, because if we try to express the polar coordinates in terms of the standard coordinates, we find that the polar coordinates depend smoothly upon the standard coordinates and vice versa:

$$x^1 = \sin\theta \cdot \cos\varphi, \qquad \theta = \arcsin\left[\sqrt{(x^1)^2 + (x^2)^2}\right]$$

$$x^2 = \sin\theta \cdot \sin\varphi, \qquad \varphi = \arctan\left[\frac{x^2}{x^1}\right].$$

Thus all the polar coordinate systems are smoothly related to the standard coordinate systems! So they are obviously equivalent; i.e., if a function $f : S^2 \curvearrowright \mathbb{R}$ is a differentiable function when expressed in polar coordinates, it will also be differentiable when expressed in standard coordinates. Thus it is irrelevant whether we use standard coordinates or polar coordinates on the sphere, since they will generate the same maximal atlas and consequently the same differentiable structure.

In the discussion of the sphere S^2, the reader might have wondered why it was necessary to use several coordinate systems to cover the sphere. It required six standard coordinate systems, respectively two polar coordinate systems, to cover the sphere! If we were smart enough, might we then hope to find a single coordinate system covering the whole of the sphere?

However, it is impossible to do that: *Any atlas on the sphere consists of at least two coordinate systems.*

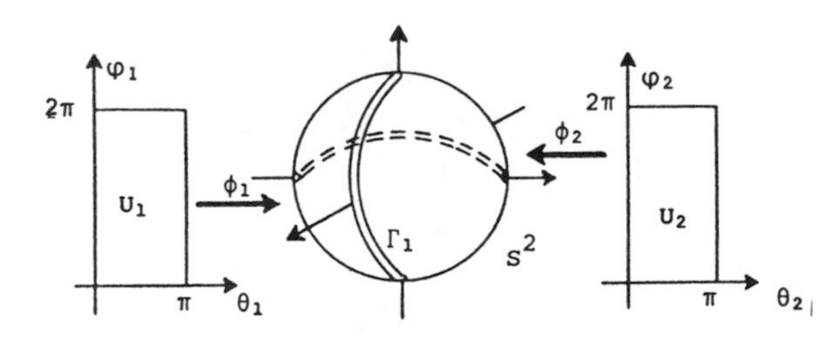

Figure 6.16.

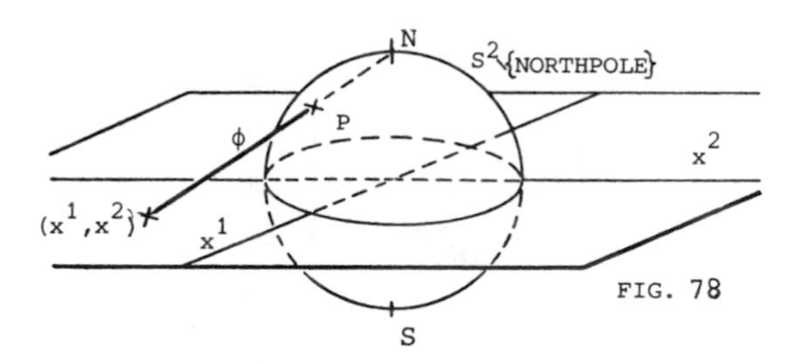

Figure 6.17.

The reason for this peculiar fact is purely *topological*. It is intimately connected with the fact that the sphere is a *compact* topological space, since it is a bounded, closed subset of $\mathbb{R}^3$! If we had a single coordinate system (ϕ, U) covering the whole of the sphere, then ϕ would be a homeomorphism of $U \subset \mathbb{R}^2$ onto the sphere S^2:

$$\phi : U \xrightarrow[\text{onto}]{\text{homeomorphic}} S^2.$$

But this forces U to be compact itself. Therefore, U is both *compact* and *open*. But the only subset of a Euclidean space that is both open and compact is the empty set. This is a contradiction! Hence the sphere S^2 is not a simple manifold.

However, if we cut out just one point from the sphere, then what remains is *not* compact, and nothing can prevent us from covering the sphere minus one point with a single coordinate system! For simplicity, we cut out the north pole. Then stereographic projection defines a nice coordinate system (see Figure 6.17)

$$\phi : \mathbb{R}^2 \underset{\text{onto}}{\longrightarrow} S^2\backslash\{\text{north pole}\}.$$

§ 6.3 Product Manifolds and Manifolds Defined by Constraints

Now consider two Euclidean manifolds $M^m \subset \mathbb{R}^p$, $N^n \subset \mathbb{R}^q$, where M^m is an m-dimensional and N^n an n-dimensional manifold. We can form the product set $M \times N$ consisting of all pairs (x, y) with $x \in M$ and $y \in N$. Clearly, $M \times N$ is a subset of the Euclidean space $\mathbb{R}^p \times \mathbb{R}^q = \mathbb{R}^{p+q}$ (See Figure 6.18). We want to construct an atlas covering $M \times N$. Let (ϕ, U) be a coordinate system on M and let (ψ, V) be a coordinate system on N. (See Figure 6.19.) Then we define a coordinate system on the product set $M \times N$ as follows:

$$\begin{aligned}&\phi \otimes \psi : U \times V \to M \times N\\&\phi \otimes \psi(x^1, \ldots, x^m; y^1, \ldots, y^n) = (\phi(x^1, \ldots, x^m); \psi(y^1, \ldots, y^n)).\end{aligned} \tag{6.3}$$

Clearly, $U \times V$ is an open set in $\mathbb{R}^r \times \mathbb{R}^m = \mathbb{R}^{n+m}$, and it can be checked

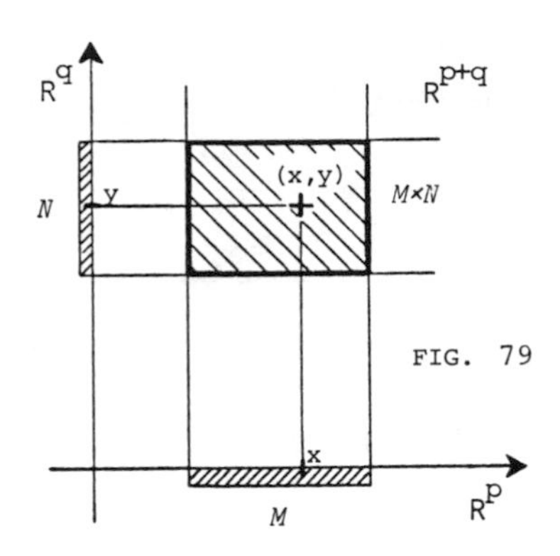

Figure 6.18.

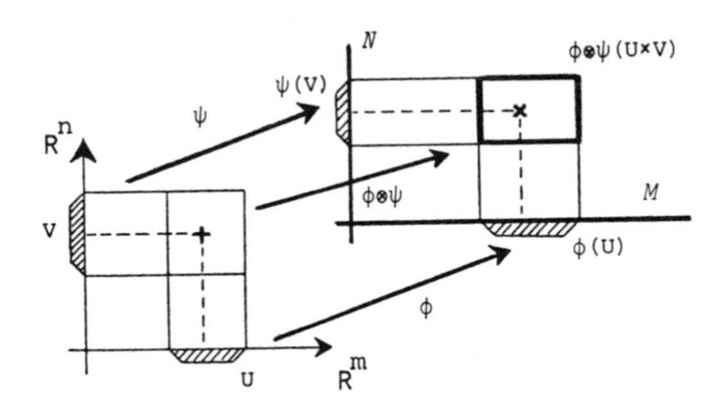

Figure 6.19.

completely that $\phi \otimes \psi$ is a homeomorphism and that it is a regular map. Thus it is a nice, respectable coordinate system. If $(\phi_i, U_i)_{i\in I}$ is an atlas for M and $(\psi_j, V_j)_{j\in J}$ is an atlas for N, then $(\phi_i \otimes \psi_j, U_i \otimes V_j)$ is an atlas for $M \times N$. We call $M \times N$ the *product manifold of M and N*. By construction, $M \times N$ becomes an $(m + n)$-dimensional manifold.

Some examples may throw some light on this rather abstract procedure:

The cylinder: $S^1 \times \mathbb{R}$, (Figure 6.20). We have used polar coordinates on S^1. The cylinder is supposed to be extended to infinity in both directions.

The torus: $S^1 \times S^1$. In Figure 6.21 the two circles have been drawn with very different radii $a > b$. (Observe that strictly speaking, $S^1 \times S^1$ should be constructed as a subset of $\mathbb{R}^2 \times \mathbb{R}^2 = \mathbb{R}^4$, but that is not so easy to visualize. Observe also that if $a \leq b$, then the torus constructed in Figure 6.21 becomes a smooth surface with self-intersections. Such a surface is not a Euclidean manifold.)

We will now sketch a very general method for constructing Euclidean manifolds. First, we recall the following problem from classical calculus:

Let $f : \mathbb{R}^{m+n} \curvearrowright \mathbb{R}^n$ be a smooth function, and consider the equation

$$f(\bar{x}, \bar{y}) = 0,$$

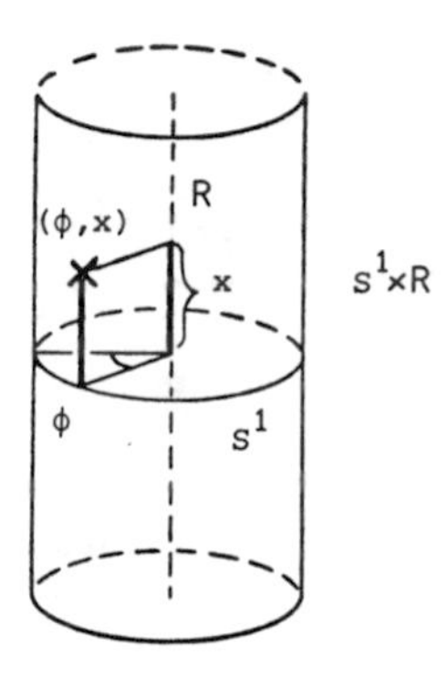

Figure 6.20.

i.e.,

$$\begin{cases} f^1(x^1, \ldots, x^m, y^1, \ldots, y^n) = 0, \\ \vdots \\ f^n(x^1, \ldots, x^m, y^1, \ldots, y^n) = 0. \end{cases}$$

Suppose $(\bar{x}_0, \bar{y}_0)$ is a solution of the equation $f(\bar{x}, \bar{y}) = 0$. Under what circumstances can we find a neighborhood U of $(\bar{x}_0, \bar{y}_0)$ in which we can solve the equation $f(\bar{x}, \bar{y}) = 0$ with respect to $\bar{y}$ and obtain $\bar{y}$ as a smooth function g of $\bar{x}$; i.e., $\bar{y} = g(\bar{x})$? (See Figure 6.22.)

This problem is well known. To motivate the solution, we study the linearized problem, where $f(\bar{x}_0 + \Delta\bar{x}; \bar{y}_0 + \Delta\bar{y})$ is replaced by

$$f(\bar{x}_0, \bar{y}_0) + \bar{\bar{D}}_x f \cdot \Delta\bar{\bar{x}}_| + \bar{\bar{D}}_y f \cdot \Delta\bar{\bar{y}}_|.$$

Thus the equation $f(\bar{x}, \bar{y}) = 0$ is replaced by the equation

$$\bar{\bar{D}}_x f \cdot \Delta\bar{\bar{x}}_| + \bar{\bar{D}}_y f \cdot \Delta\bar{\bar{y}}_| = 0,$$

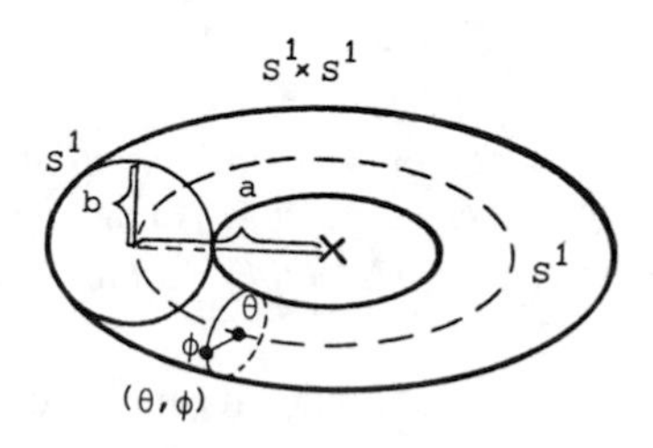

Figure 6.21.

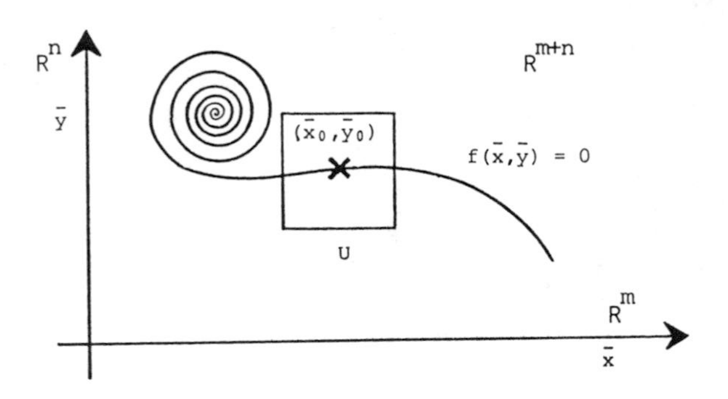

Figure 6.22.

where we have taken into account the assumption that $(\bar{x}_0, \bar{y}_0)$ is a solution; i.e., $f(\bar{x}_0, \bar{y}_0) = 0$. Now, the linearized equation can be solved with respect to $\Delta\bar{\bar{y}}_|$, provided that $\bar{\bar{D}}_y f = \left(\frac{\partial f^i}{\partial y^j}\right)$ is a regular square matrix. In that case, we furthermore get

$$\Delta\bar{\bar{y}}_| = -(\bar{\bar{D}}_y f)^{-1} \cdot (\bar{\bar{D}}_x f) \cdot \Delta\bar{\bar{x}}_|.$$

With this in mind, we now state the solution to our problem:

Theorem 6.1. *(Theorem of implicit functions)*

Let $f : \mathbb{R}^{m+n} \curvearrowright \mathbb{R}^n$ be a smooth function. If $\bar{\bar{D}}_y f$ is a regular square matrix, when it is evaluated at a solution $(\bar{x}_0, \bar{y}_0)$ to the equation $f(\bar{x}, \bar{y}) = 0$, there exists a neighborhood U of $(\bar{x}_0, \bar{y}_0)$ and a smooth function $g : \mathbb{R}^m \curvearrowright \mathbb{R}^n$ such that

1. *$\underline{f}(\bar{x}, \bar{y}) = \underline{0}$ iff $\bar{y} = \underline{g}(\bar{x})$ for all $(\bar{x}, \bar{y})$ in the neighborhood U.*
2. *$\bar{\bar{D}}_x g = -(\bar{\bar{D}}_y f)^{-1}(\bar{\bar{D}}_x f)$.*

(I.e., the partial derivatives of y with respect to x are found by implicit differentiation:

$$\frac{\partial f^i}{\partial x^j} + \frac{\partial f^i}{\partial y^k}\frac{\partial y^k}{\partial x^j} = 0.)$$

Using this cornerstone of classical analysis, we can now show:

Theorem 6.2. *Let $f : \mathbb{R}^q \to \mathbb{R}^p$ be a smooth function $(q > p)$. Let M be the set of solutions to the equation $f(\bar{z}) = 0$. If M is nonempty and if f has maximal rank throughout M, then M is a differentiable manifold of dimension $(q - p)$.*

We will not go through the proof in all details, but let us sketch how to construct coordinate systems on M. Suppose M is not empty and suppose $\bar{z}_0$ belongs to M;

i.e., $f(\bar{z}_0) = 0$. We know that

$$\bar{\bar{D}}_z f_{\big|\bar{z}=\bar{z}_0} = \begin{bmatrix} \frac{\partial f^1}{\partial z^1} & \cdots & \frac{\partial f^1}{\partial z^q} \\ \vdots & & \\ \frac{\partial f^p}{\partial z^1} & \cdots & \frac{\partial f^p}{\partial z^q} \end{bmatrix}$$

has rank p, which is the highest possible rank. Hence we can find p columns that taken together constitute a regular square matrix. To simplify , let us assume that the last p columns are independent, and let us introduce the notation

$$\bar{z} = (\bar{x}, \bar{y}); \qquad \text{i.e., } (z^1, \ldots, z^q) = (x^1, \ldots, x^{q-p}, y^1, \ldots, y^p).$$

Then $\bar{\bar{D}}_y f$ is a regular square matrix. According to the theorem of implicit functions, we can now find a neighborhood U of $\bar{z}_0 = (\bar{x}_0, y_0)$ of the form

$$U = \{(\bar{x}, \bar{y}) \mid x_0^i - \epsilon < x^i < x_0^i + \epsilon, y_0^j - \epsilon < y^j < y_0^j + \epsilon\}$$

and a smooth map $g : \mathbb{R}^{q-p} \to \mathbb{R}^p$ such that $(\bar{x}, \bar{y}) \in M \cap U$ iff $\bar{y} = g(\bar{x})$ when $\bar{x}$ belongs to $\pi(U)$, where

$$\pi(U) = \{x \mid x_0^i - \epsilon < x < x_0^i + \epsilon\};$$

i.e., $\pi(U)$ is the projection of U into $\mathbb{R}^{q-p}$; cf. Figure 6.23. But this means that restricting ourselves to U we can define a coordinate system $(\phi, \pi(U))$, $\phi : \pi(U) \curvearrowright M$ in the following way:

$$\phi(\bar{x}) = (\bar{x}, g(\bar{x})).$$

Such a coordinate system will be called a *standard coordinate system*. We must now check that ϕ really defines a coordinate system:

a. Since U is open, $M \cap U$ automatically becomes an open subset of M. As the projection map $\pi = \phi^{-1}$ is trivially continuous, we see that ϕ is a homeomorphism of the open subset $\pi(U)$ onto the open subset $M \cap U$.

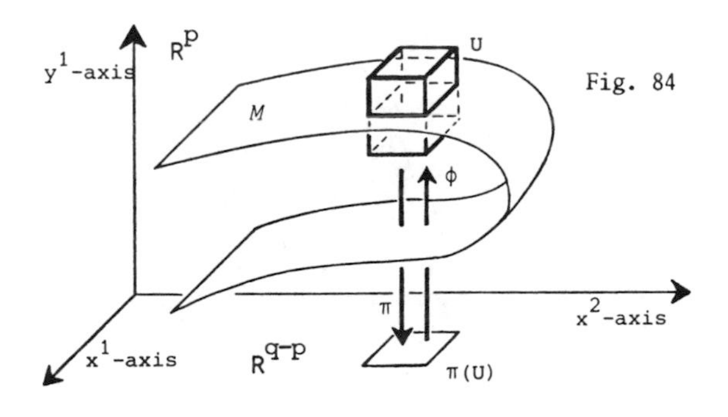

Figure 6.23.

b. ϕ is a smooth function, since g is. Furthermore, ϕ is trivially regular, since

$$\overline{\overline{D}}_x\phi = \begin{bmatrix} 1 & & 0 & \vdots & \frac{\partial \phi^1}{\partial y^1} & \cdots \\ & \ddots & & \vdots & & \\ 0 & & 1 & & & \end{bmatrix},$$

where the left part of the Jacobian matrix is the unit $(q-p) \times (q-p)$ matrix, which is trivially regular.

We can clearly cover M by standard coordinate systems, and it is not too difficult to check that they are all smoothly related. Thus we have defined an atlas, and this shows that M is a differentiable manifold of dimension $(q-p)$; see Figure 6.24.

If M is generated as the set of solutions of an equation $f(\bar{z}) = 0$, we say that *M is defined by an equation of constraint*. Clearly, the sphere S^2 is defined by the following equation of constraint:

$$(x^1)^2 + (x^2)^2 + (x^3)^2 = 1.$$

If we review the illustrative example of the sphere, we will, in fact, find that most of it was an exemplification of the above abstract discussion.

The assumption that f has maximal rank is essential. Consider for instance the following trivial example: Let $f : \mathbb{R}^2 \curvearrowright \mathbb{R}$ be the smooth map given by $f(x, y) = x^2 - y^2$. Then $\overline{\overline{D}}f = \left(\frac{\partial f}{\partial x}; \frac{\partial f}{\partial y}\right) = (2x, -2y)$ is singular at the point $(x, y) = (0, 0)$. But the point $(0, 0)$ clearly belongs to $M = \{(x, y) \mid x^2 - y^2 = 0\}$. In accordance with this, we easily find that M is not a 1-dimensional differentiable manifold. It is impossible to find a smooth map ϕ that maps an open interval bijectively onto an open neighborhood of $(0, 0)$ in M. (See Figure 6.25.)

As another illuminative example we consider the smooth map, $f : \mathbb{R}^3 \curvearrowright \mathbb{R}$, given by

$$f(x, y, z) = (x^2 + y^2 - 1)^2 + z^2.$$

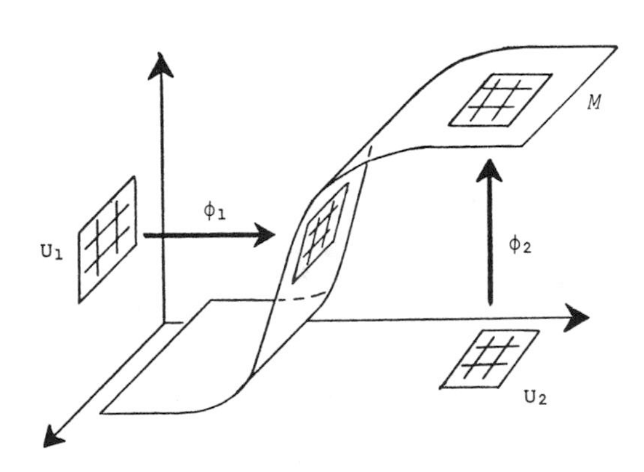

Figure 6.24.

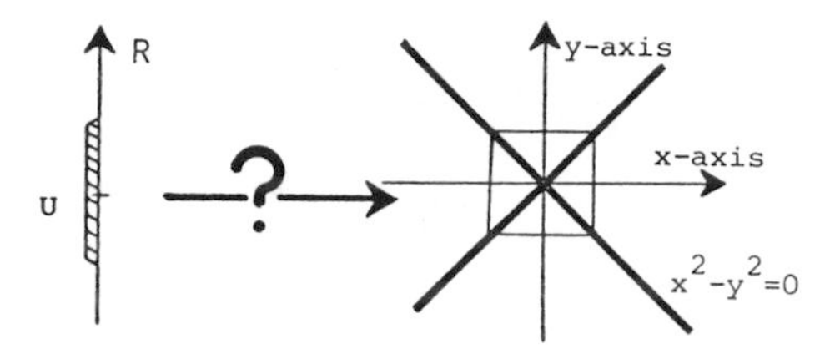

Figure 6.25.

Here

$$\bar{\bar{D}}_x f = \left(\frac{\partial f}{\partial x} ; \frac{\partial f}{\partial y} ; \frac{\partial f}{\partial z} \right) = (4x[x^2 + y^2 - 1]; 4y[x^2 + y^2 - 1]; 2z),$$

which is singular at all points on $M = \{(x, y, z) \mid f(x, y, z) = 0\} = \{(x, y, z) \mid x^2 + y^2 = 1, z = 0\}$. The subset M is, nevertheless, clearly a 1-dimensional differentiable manifold (it is isomorphic to the circle S^1). What goes wrong is the dimensionality. If f had maximal rank, then M would have been a 2-*dimensional* manifold! (See Figure 6.26.)

As a final example of differentiable manifolds defined by constraints we look at the configuration space for a system consisting of a finite number of particles in classical mechanics. Consider, for instance, a *double pendulum*: It consists of two particles, where the positions are constrained through the equations $(q^1)^2 + (q^2)^2 + (q^3)^2 = 1$ and $(q^4 - q^1)^2 + (q^5 - q^2)^2 + (q^6 - q^3)^2 = 1$. Since the smooth map $f : \mathbb{R}^6 \to \mathbb{R}^2$ given by

$$f(q^1, \ldots, q^6) = \begin{bmatrix} (q^1)^2 + (q^2)^2 + (q^3)^2 - 1 \\ (q^4 - q^1)^2 + (q^5 - q^2)^2 + (q^6 - q^3)^2 - 1 \end{bmatrix}$$

is regular on the configuration space M, we conclude that M is a 4-dimensional differentiable manifold. In fact, M is isomorphic to $S^2 \times S^2$.

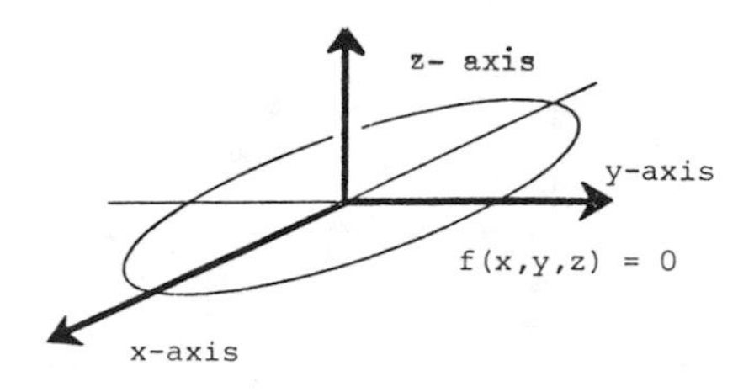

Figure 6.26.

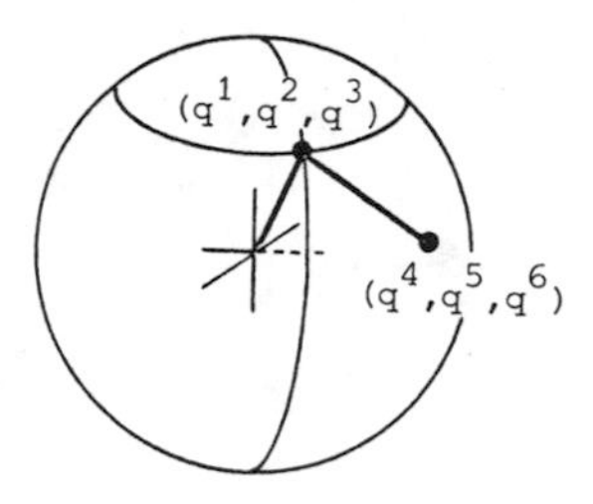

Figure 6.27.

§ 6.4 Tangent Vectors

We are now in a position to introduce tangent vectors. Consider a Euclidean manifold $M \subset \mathbb{R}^N$ (Figure 6.28).

Let P be a point in M. Then we have a lot of vectors at P in the surrounding space $\mathbb{R}^N$. Observe that a vector is defined as a directed line segment PQ. Thus a vector is characterized not only by its direction and length but also by its base point P!

Hence if two vectors have different base point P and P', then they are different vectors, even if they have the same length and direction.

Let us return to the vectors with the base point P. Not all of them are tangent vectors to the smooth surface M! Let $\lambda : \mathbb{R} \curvearrowright M$ be a smooth curve passing through P. Then λ generates a tangent vector at P, the *velocity vector*

$$\mathbf{v}_P = \frac{d\lambda}{dt}\Big|_{t=0}. \tag{6.4}$$

The set of vectors generated by *all* possible smooth curves in M passing through P forms a set of vectors called the *tangent space* and, it is denoted by

$$\mathbf{T}_P(M).$$

(See Figure 6.29.) To investigate the structure of this tangent space we introduce a coordinate system (ϕ, U), which covers a neighborhood of P. To simplify, we choose the coordinate system such that P has the coordinates $(0, \ldots, 0)$ (see

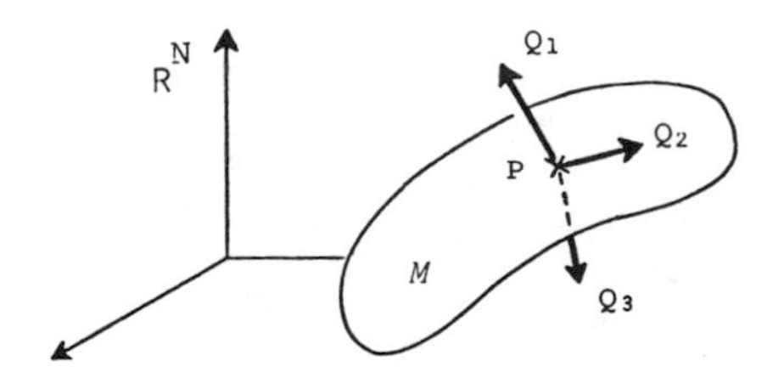

Figure 6.28.

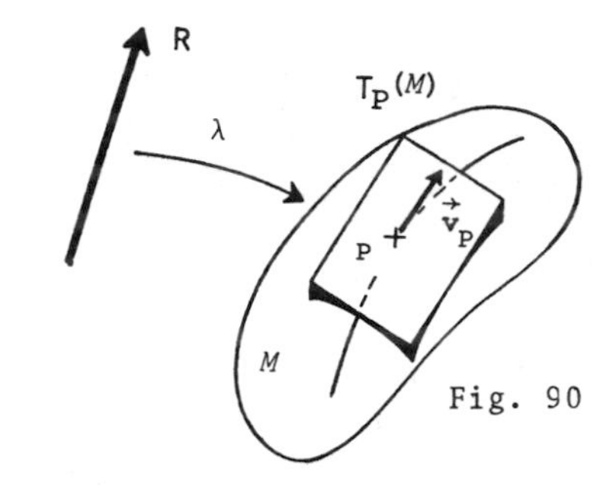

Figure 6.29.

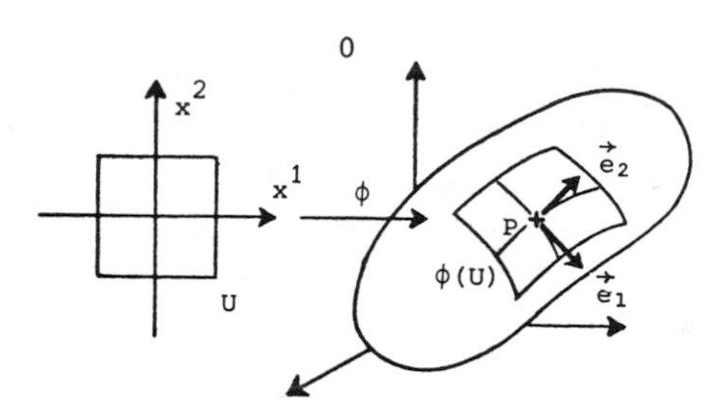

Figure 6.30.

Figure 6.30). Consider the curves:

$$\lambda_1(t) = \phi(t, 0, 0, \ldots, 0); \ldots; \lambda_n(t) = \phi(0, 0, \ldots, 0, t). \tag{6.5}$$

They are called the *coordinate lines*. By differentiation we now obtain the tangent vectors

$$\mathbf{e}_1 = \frac{d\lambda_1}{dt}\Big|_{t=0}; \ldots; \mathbf{e}_n = \frac{d\lambda_n}{dt}\Big|_{t=0}. \tag{6.6}$$

They are called the *canonical frame vectors*. To investigate them a little more closely we will write out the parametrization of the coordinate lines in their full glory:

$$\lambda_1(t) = \begin{bmatrix} \phi^1(t, 0, \ldots, 0) \\ \vdots \\ \phi^N(t, 0, \ldots, 0) \end{bmatrix}; \ldots; \lambda_n(t) = \begin{bmatrix} \phi^1(0, \ldots, 0, t) \\ \vdots \\ \phi^N(0, \ldots, 0, t) \end{bmatrix}.$$

By differentiation we get

$$\mathbf{e}_1 : \frac{d\lambda_1}{dt}\Big|_{t=0} = \begin{bmatrix} \frac{\partial\phi^1}{\partial x^1} \\ \vdots \\ \frac{\partial\phi^N}{\partial x^1} \end{bmatrix}_{\big|(0,\ldots,0)}; \ldots; \mathbf{e}_n : \frac{d\lambda_n}{dt}\Big|_{t=0} = \begin{bmatrix} \frac{\partial\phi^1}{\partial x^n} \\ \vdots \\ \frac{\partial\phi^N}{\partial x^n} \end{bmatrix}_{\big|(0,\ldots,0)}.$$

Here the coordinates are the Cartesian coordinates in $\mathbb{R}^N$, and we will refer to them as the *extrinsic coordinates* of the vector. Now take a look at the Jacobian matrix

$$\left(\frac{\partial \phi^i}{\partial x^j}\right) = \begin{bmatrix} \frac{\partial \phi^1}{\partial x^1} & \cdots & \frac{\partial \phi^1}{\partial x^n} \\ \vdots & & \vdots \\ \frac{\partial \phi^N}{\partial x^1} & \cdots & \frac{\partial \phi^N}{\partial x^n} \end{bmatrix}.$$

It is obvious that *the n column vectors in the Jacobian matrix are nothing but the extrinsic coordinates of the canonical frame vectors*. But we have previously demanded that

$$\left(\frac{\partial \phi^i}{\partial x^j}\right)$$

be of maximal rank n. Therefore, the column vectors are linearly independent! Thus, we have shown that *the canonical frame vectors are linearly independent.*

Next, we want to show that the canonical frame vectors span the whole tangent space $\mathbf{T}_p(M)$. Let $\mathbf{v}_p$ be an arbitrary tangent vector. Then $\mathbf{v}_p$ is generated by a smooth curve λ. Let us write out the parametrization of λ:

First we observe that λ has the coordinate representation

$$\bar{\lambda}(t) = (x^1(t), \ldots, x^n(t))$$

with respect to a coordinate system (ϕ, U); cf. Figure 6.31. We can then write down the parametrization of λ with respect to the Cartesian coordinates in the surrounding space $\mathbb{R}^N$:

$$\lambda(t) = \begin{bmatrix} \phi^1(x^1(t), \ldots, x^n(t)) \\ \vdots \\ \phi^N(x^1(t), \ldots, x^n(t)) \end{bmatrix}.$$

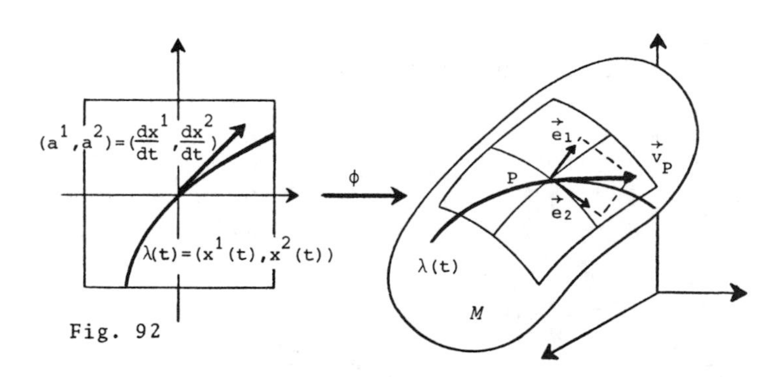

Figure 6.31.

By differentiation we get

$$\mathbf{v}_p = \frac{d\lambda}{dt}\Big|_{t=0} = \begin{bmatrix} \frac{\partial\phi^1}{\partial x^1}\frac{dx^1}{dt} + \cdots + \frac{\partial\phi^1}{\partial x^n}\frac{dx^n}{dt} \\ \vdots \\ \frac{\partial\phi^N}{\partial x^1}\frac{dx^1}{dt} + \cdots + \frac{\partial\phi^N}{\partial x^n}\frac{dx^n}{dt} \end{bmatrix} = \frac{dx^1}{dt}\mathbf{e}_1 + \cdots + \frac{dx^n}{dt}\mathbf{e}_n. \tag{6.7}$$

Consequently, we see that the vectors $\mathbf{e}_1, \ldots, \mathbf{e}_n$ really span the tangent vector $\mathbf{v}_p$. On the other hand, if $\mathbf{u}$ is a vector spanned by $\mathbf{e}_1, \ldots, \mathbf{e}_n$, i.e.,

$$\mathbf{u} = a^1\mathbf{e}_1 + \cdots + a^n\mathbf{e}_n,$$

then $\mathbf{u}$ is a tangent vector, since it is obvious that it is generated by the curve $\lambda(t)$ with the coordinate representation

$$\lambda(t) = (a^1t, \ldots, a^nt).$$

As $\mathbf{e}_1, \ldots, \mathbf{e}_n$ are linearly independent, we therefore conclude that *the tangent space $\mathbf{T}_p(M)$ is an n-dimensional vector space*. The vectors $\mathbf{e}_1, \ldots, \mathbf{e}_n$ constitute a frame for this vector space. Formula (6.7) now motivates the following definition:

Definition 6.7. Let $\mathbf{v}_p$ be a tangent vector generated by λ and let λ have the coordinate representation $\bar{\lambda}(t) = (x^1(t), \ldots, x^n(t))$. Then the numbers $a^i = \frac{dx^i}{dt}$ are called the *intrinsic* coordinates of $\mathbf{v}_p$ with respect to the coordinate system (ϕ, U).

Observe that the intrinsic coordinates a^i are simply the components of $\mathbf{v}_p$ with respect to the canonical frame $(\mathbf{e}_1, \ldots, \mathbf{e}_n)$. The numbers a^i are also referred to as the *contravariant* components of $\mathbf{v}_p$. We shall explain this mysterious name in a moment!

Now, let us see what happens if we exchange coordinates, say from $(x^1, \ldots, x^n)$-coordinates to $(y^1, \ldots, y^n)$-coordinates. (See Figure 6.32.) Then the same tangent vector $\mathbf{v}_p$ is described by two set of numbers,

$$\mathbf{v}_p : \begin{cases} (a^1_{(1)}, \ldots, a^n_{(1)}) & \text{with respect to the } x\text{-coordinates,} \\ (a^1_{(2)}, \ldots, a^n_{(2)}) & \text{with respect to the } y\text{-coordinates.} \end{cases}$$

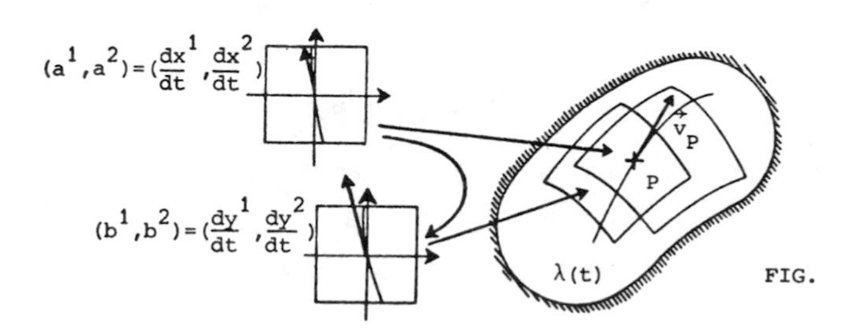

Figure 6.32.

We want to find the transformation rule that allows us to pass from x-coordinates to y-coordinates. It will be convenient to introduce the notation $\bar{\bar{D}}_{21}$ for the Jacobian matrix of the transition function; i.e.,

$$\bar{\bar{D}}_{21} = \left(\frac{\partial y^i}{\partial x^j} \right).$$

Okay, we are ready: Let $\mathbf{v}_p$ be generated by the smooth curve λ. Then λ has the coordinate representation

$$x^i = x^i(t)$$

in x-coordinates, and it has the coordinate representation

$$y^i = y^i(x^j(t))$$

in y-coordinates. Consequently, we obtain

$$a^i_{(2)} = \frac{dy^i}{dt} = \frac{\partial y^i}{\partial x^j} \frac{dx^i}{dt} = \frac{\partial y^i}{\partial x^j} a^i_{(1)}. \tag{6.8}$$

In a matrix formulation we may collect the intrinsic coordinates into a column vector

$$\bar{\bar{a}}_{(2)_|} = \begin{bmatrix} a^1_2 \\ \vdots \\ a^n_2 \end{bmatrix}; \qquad \bar{\bar{a}}_{(1)_|} = \begin{bmatrix} a^1_1 \\ \vdots \\ a^n_1 \end{bmatrix}.$$

Then we can rearrange the transformation rule in the following way:

$$\bar{\bar{a}}_{(2)_|} = \bar{\bar{D}}_{21} \bar{\bar{a}}_{(1)_|} \tag{6.9}$$

This is the most important of all the formulas involving the tangent vectors.

Let us also investigate the connection between the old canonical frame vectors $\mathbf{e}_{(1)_i}$ and the new canonical frame vectors $\mathbf{e}_{(2)_i}$. Note that $\mathbf{e}_{(1)_i}$ has the old coordinates $a^j_{(1)} = \delta^j_i = (0, \ldots, 1, \ldots, 0)$, where the number 1 is in the ith position. But now we may compute the new coordinates from the above formula:

$$a^j_{(2)} = \frac{\partial y^j}{\partial x^k} \cdot a^k_{(1)} = \frac{\partial y^j}{\partial x^k} \cdot \delta^k_i = \frac{\partial y^j}{\partial x^i}.$$

Consequently, the vector $\mathbf{e}_{(1)_i}$ has the following decomposition along the new canonical frame vectors:

$$\mathbf{e}_{(1)_i} = \mathbf{e}_{(2)_j} \cdot a^j_{(2)} = \mathbf{e}_{(2)_j} \cdot \frac{\partial y^j}{\partial x^i}. \tag{6.10}$$

Let us fix our notation: ϕ_{21} and ϕ_{12} are the transition functions with the corresponding Jacobian matrices $\bar{\bar{D}}_{21}$ and $\bar{\bar{D}}_{12}$.

$$y = \phi_{21}(x), \qquad x = \phi_{12}(y),$$
$$\bar{\bar{D}}_{21} = \left[\frac{\partial y^i}{\partial x^j} \right], \qquad \bar{\bar{D}}_{12} = \left[\frac{\partial x^i}{\partial y^j} \right]. \tag{6.11}$$

As ϕ_{21} and ϕ_{12} are inverse maps, we conclude that $\bar{\bar{D}}_{21}$ and $\bar{\bar{D}}_{12}$ are reciprocal matrices! Thus $\frac{\partial y^i}{\partial x^j}$ and $\frac{\partial x^i}{\partial y^j}$ are components of reciprocal matrices, and we may therefore rearrange formula (6.11) in the following way:

$$\mathbf{e}_{(2)_j} = \mathbf{e}_{(1)_i} \frac{\partial x^i}{\partial y^j} \,. \tag{6.12}$$

A quantity that transforms in the same way as the canonical frame vectors is said to transform *covariantly* (*covarians* = transform in the same way). Hence a covariant quantity is a quantity ω_j that transforms according to the rule

$$\omega_{(2)_j} = \omega_{(1)_i} \frac{\partial x^i}{\partial y^j} \tag{6.13}$$

when we exchange coordinates. Observe that we have put the index i in the *lower position*: ω_i. It is referred to as a covariant index. Observe that a covariant index is transformed with the matrix $\frac{\partial x^i}{\partial y^j}$ from the right.

The canonical frame vectors transform with the matrix $\frac{\partial x^i}{\partial y^j}$. A quantity that transforms with the *reciprocal* matrix $\frac{\partial y^i}{\partial x^j}$ is said to transform *contravariantly* (cotravarians = transforms in the opposite way). Hence a contravariant quantity is a quantity a^i that transforms according to the rule

$$a^i_{(2)} = \frac{\partial y^i}{\partial x^j} a^j_{(1)} \tag{6.14}$$

when we exchange coordinates. This time we have put the index i in the *upper position*: a^i. It is referred to as a contravariant index. Observe that a contravariant index is transformed with the matrix

$$\frac{\partial y^i}{\partial x^j}$$

from the left.

As we have already seen, the components of a tangent vector transforms contravariantly!

In the preceding sections we have discussed only the local aspects of functions and tangent vectors; i.e., we have focused upon the immediate neighborhood of a specific point P. We will now briefly discuss how we can introduce scalar fields and vector fields on a manifold M.

Let us look at a scalar field first. In our geometrical interpretation, a scalar field is simply a smooth function $\phi : M \curvearrowright \mathbb{R}$. If we introduce two coordinate systems, say x-coordinates and y–coordinates, we may now represent ϕ by the ordinary Euclidean functions

$$\bar{\phi}_{(1)}(x^i) \quad \text{and} \quad \bar{\phi}_{(2)}(y^i),$$

and these are related through the transformation rule

$$\bar{\phi}_{(1)}(x^i) = \bar{\phi}_{(2)}(y^i),$$

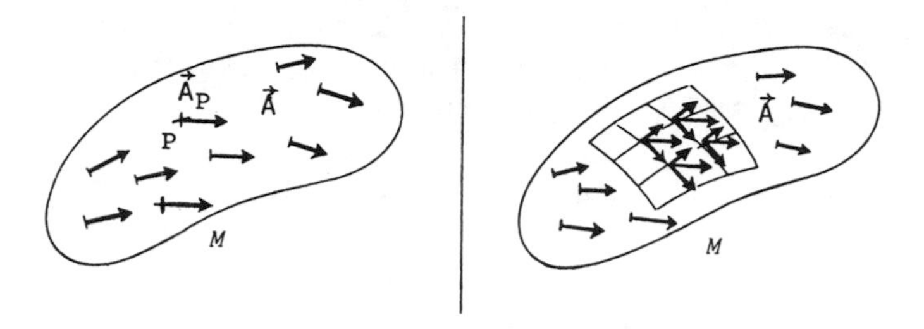

Figure 6.33.

where x^i and y^i are coordinates of the *same* space–time point P. When we investigate such a scalar field $P \to \phi(P)$, we will always introduce coordinates and use the Euclidean representative. We then "forget" the bar and write the corresponding Euclidean function as

$$x^\alpha \curvearrowright \phi(x^\alpha).$$

Next, we have vector fields. To construct a vector field $\mathbf{A}$ on an arbitrary manifold M, we attach a tangent vector, $\mathbf{A}_P$ to each point P in our manifold (see Figure 6.33). If we introduce coordinates $(x^1, \ldots, x^n)$ on M, then the tangent vector at the point $P(x^1, \ldots, x^n)$ can be decomposed along the canonical frame vectors $\mathbf{e}_1, \ldots, \mathbf{e}_n$ as follows:

$$\mathbf{A}_P = a^i(x^1, \ldots, x^n)\mathbf{e}_i.$$

We say that the vector field is *smooth* if its components $a^i(x^1, \ldots, x^n)$ are smooth functions of the coordinates. In general, we want to investigate a vector field $\mathbf{A}$ through its components $a^i(x)$, but one should be careful. If the manifold M is a nontrivial manifold, then several coordinate systems will be needed to cover it! (Compare the discussion of S^2 in Section 6.2.)

If we have two such coordinate regions Ω_1 and Ω_2 that overlap in the region Ω_{12}, then we will have to use two coordinate expressions $a^i_{(1)}(x)$ and $a^i_{(2)}(y)$ to describe the vector field $\mathbf{A}$. In the overlapping region Ω_{12} these two coordinate

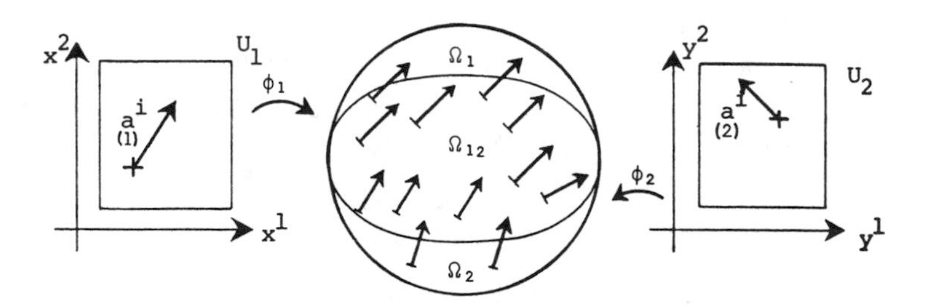

Figure 6.34.

expressions are now connected through the transformation formula (6.8):

$$a^i_{(2)}(y^1, \dots, y^n) = \frac{\partial y^i}{\partial x^j}\, a^j_{(1)}(x^1, \dots, x^n).$$

§ 6.5 Metrics

In the ordinary Euclidean space $\mathbb{R}^n$ we have an inner product $\langle \mid \rangle$,

$$\langle x \mid y \rangle = x^1 y^1 + \cdots x^n y^n,$$

which is symmetric, bilinear, and positive definite. We want to extend this concept to the tangent vectors on a manifold M. To each tangent space on the manifold M we therefore associate an inner product. This family of inner products is called a *metric*, and it is denoted by the letter $\boldsymbol{g}$. If $\mathbf{v}_P$ and $\mathbf{u}_P$ are tangent vectors belonging to the *same* tangent space $\mathbf{T}_P(M)$ we denote the value of their inner product by: $\boldsymbol{g}(\mathbf{v}_P; \mathbf{u}_P)$. (See Figure 6.35). Observe that we do *not* define the inner product of tangent vectors $\mathbf{v}_P$ and $\mathbf{u}_Q$ with different base points.

As in the Euclidean case, we shall assume that $\boldsymbol{g}$ is *symmetric*,

$$\boldsymbol{g}(\mathbf{v}_P; \mathbf{u}_P) = \boldsymbol{g}(\mathbf{u}_P; \mathbf{v}_P), \tag{6.15}$$

and *bilinear*

$$\boldsymbol{g}(\lambda \mathbf{v}_P + \mu \mathbf{u}_P; \mathbf{w}_P) = \lambda \boldsymbol{g}(\mathbf{v}_P; \mathbf{u}_P) + \mu \boldsymbol{g}(\mathbf{u}_P; \mathbf{w}_P). \tag{6.16}$$

But we shall *not* demand that the metric be positive definite. We only demand that it be *nondegenerate*; i.e., the only vector that is orthogonal to all other vectors in $\mathbf{T}_P(M)$ is the zero vector:

$$\boldsymbol{g}(\mathbf{v}_P; \mathbf{u}_P) = 0 \qquad \text{for all } \mathbf{u}_P \text{ in } T_P(M) \text{ implies } \mathbf{v}_P = \mathbf{0}. \tag{6.17}$$

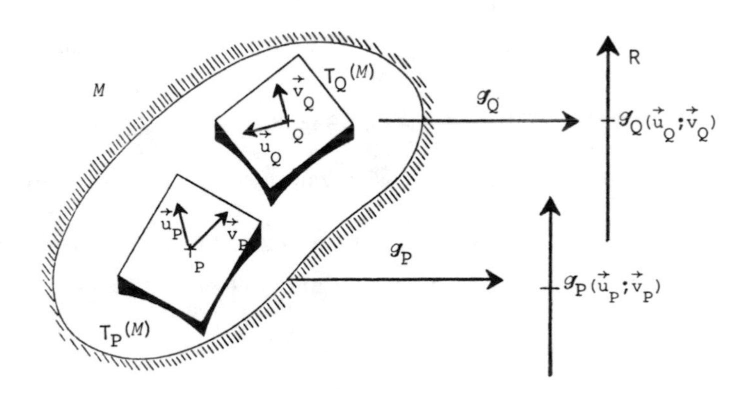

Figure 6.35.

Let us introduce coordinates, and let $\mathbf{e}_1, \ldots, \mathbf{e}_n$ be the corresponding canonical frame vectors. The numbers

$$g_{ij} = \boldsymbol{g}(\mathbf{e}_i; \mathbf{e}_j) \tag{6.18}$$

are referred to as the *metric coefficients*. Let $\mathbf{v}_P$ and $\mathbf{u}_P$ be two tangent vectors with coordinates a^i and b^i. We can then express the inner product of $\mathbf{v}_P$ and $\mathbf{u}_P$ entirely in terms of the metric coefficients and the coordinates a^i and b^i:

$$\boldsymbol{g}(\mathbf{v}_P; \mathbf{u}_P) = \boldsymbol{g}(a^i\mathbf{e}_i; b^j\mathbf{e}_j) = a^ib^j\boldsymbol{g}(\mathbf{e}_i; \mathbf{e}_j) = g_{ij}a^ib^j. \tag{6.19}$$

At each point P the metric is therefore completely characterized by its components $g_{ij}(P)$. Thus we may regard the metric coefficients as functions defined on the range of the coordinate system. If we identify P with its coordinates $(x^1, \ldots, x^n)$, we may furthermore regard g_{ij} as ordinary Euclidean functions

$$g_{ij} = g_{ij}(x^1, \ldots, x^n).$$

We say that the metric is *smooth* if g_{ij} depends smoothly upon the coordinates. In what follows we will work only with smooth metrics!

Let us summarize the discussion:

Definition 6.8. Let M be a differentiable manifold. A smooth metric $\boldsymbol{g}$ is a family of inner products, defined on each of the tangent spaces, with the following properties:

a. $\boldsymbol{g}$ is a bilinear map that is symmetric and nondegenerate.

b. The metric coefficients g_{ij} depend smoothly upon the coordinates $(x^1, \ldots, x^n)$.

The metric coefficients are often collected in a matrix $\overline{\overline{G}} = (g_{ij})$, which by construction is symmetric. Let us denote by $\overline{\overline{a}}_|$ and $\overline{\overline{b}}_|$ the coordinates of two tangent vectors $\mathbf{v}_P$ and $\mathbf{u}_P$. Then we may rearrange the formula for the inner product in the following way:

$$\boldsymbol{g}(\mathbf{v}_P; \mathbf{u}_P) = g_{ij}a^ib^j = a^ig_{ij}b^j = \overline{\overline{a}}_|^{+}\,\overline{\overline{G}}\,\overline{\overline{b}}_|.$$

Next, we want to show that the matrix $\overline{\overline{G}}$ is *regular*. Let us assume that $\overline{\overline{b}}_|$ is a column vector with the property

$$\overline{\overline{G}}\,\overline{\overline{b}}_| = \overline{\overline{0}}_|.$$

We have finished the proof as soon as we can show that $\overline{\overline{b}}_|$ is necessarily the zero vector!

Let $\overline{\overline{b}}_|$ be the components of a fixed tangent vector $\mathbf{u}_P$ and denote by $\overline{\overline{a}}_|$ the components of an arbitrary tangent vector $\mathbf{v}_P$. From $\overline{\overline{G}}\,\overline{\overline{b}}_| = \overline{\overline{0}}_|$ we conclude that

$$\overline{\overline{a}}_|^{+}(\overline{\overline{G}}\,\overline{\overline{b}}_|) = 0 \qquad \text{for all } \overline{\overline{a}}_|;$$

i.e.,

$$\boldsymbol{g}(\mathbf{v}_P; \mathbf{u}_P) = 0 \qquad \text{for all } \mathbf{v}_P.$$

Consequently, **u** vanishes because $\boldsymbol{g}$ is nondegenerate. It follows that $\overline{\overline{b}}_| = \overline{\overline{0}}_|$, as we wanted it to be.

As $\overline{\overline{G}} = (g_{ij})$ is a regular matrix, we know that it has a reciprocal matrix! We will denote the elements of the reciprocal matrix by g^{ij}. By construction, we get

$$g^{ij} g_{jk} = \delta^i_k. \tag{6.20}$$

Then we want to see how the metric coefficients transform when we pass from one coordinate system to another. But that is easily done. Let there be given two coordinate systems, x-coordinates and y-coordinates. If $g_{(1)_{\alpha\beta}}$ are the old metric coefficients corresponding to x-coordinates and $g_{(2)_{\alpha\beta}}$ are the new metric coefficients corresponding to y-coordinates, we now get

$$\begin{aligned} g_{(2)_{\alpha\beta}} &= \boldsymbol{g}\left(\mathbf{e}_{(2)_\alpha}; \mathbf{e}_{(2)_\beta}\right) = \boldsymbol{g}\left(\mathbf{e}_{(1)_\gamma} \frac{\partial x^\gamma}{\partial y^\alpha}; \mathbf{e}_{(1)_\delta} \frac{\partial x^\delta}{\partial y^\beta}\right) \\ &= \boldsymbol{g}\left(\mathbf{e}_{(1)_\gamma}; \mathbf{e}_{(1)_\delta}\right) \frac{\partial x^\gamma}{\partial y^\alpha} \frac{\partial x^\delta}{\partial y^\beta} = g_{(1)_{\gamma\delta}} \frac{\partial x^\gamma}{\partial y^\alpha} \frac{\partial x^\delta}{\partial y^\beta}, \end{aligned}$$

where we have used (6.12). So the metric coefficients transform covariantly:

$$g_{(2)_{\alpha\beta}} = g_{(1)_{\gamma\delta}} \frac{\partial x^\gamma}{\partial y^\alpha} \frac{\partial x^\delta}{\partial y^\beta}. \tag{6.21}$$

Worked Exercise 6.5.1
Problem: Let g_{ij} be the covariant components of the metric. Show that the components g^{ij} of the reciprocal matrix transform contravariantly.

Exercise 6.5.2
Problem: Consider the Euclidean space $\mathbb{R}^3$ with the usual metric. In Cartesian coordinates the metric coefficients reduce to the unit matrix

$$\overline{\overline{G}}_{\text{Cartesian}} = \begin{bmatrix} 1 & 0 & 0 \\ 0 & 1 & 0 \\ 0 & 0 & 1 \end{bmatrix}.$$

Let us introduce spherical coordinates (r, θ, φ) through the relations

$$\begin{bmatrix} x \\ y \\ z \end{bmatrix} = \begin{bmatrix} r \sin\theta \cos\varphi \\ r \sin\theta \sin\varphi \\ r \cos\theta \end{bmatrix}$$

and cylindrical coordinates (ρ, φ, z) through the relations

$$\begin{bmatrix} x \\ y \\ z \end{bmatrix} = \begin{bmatrix} \rho \cos\varphi \\ \rho \sin\varphi \\ z \end{bmatrix}.$$

Show that the metric coefficients in spherical, respectively cylindrical, coordinates are given by

$$\overline{\overline{G}}_{\text{spherical}} = \begin{bmatrix} 1 & 0 & 0 \\ 0 & r^2 & 0 \\ 0 & 0 & r^2 \sin^2\theta \end{bmatrix} \qquad \overline{\overline{G}}_{\text{cylindrical}} = \begin{bmatrix} 1 & 0 & 0 \\ 0 & \rho^2 & 0 \\ 0 & 0 & 1 \end{bmatrix}. \tag{6.22}$$

(Hint: It is preferable to work in the matrix language. Let $\overline{\overline{D}}_{12} = \frac{\partial x^i}{\partial y^j}$ be the Jacobian matrix corresponding to a coordinate transformation. Then the transformation rule

(6.21) reduces to

$$\overline{\overline{G}}_{(2)} = \overline{\overline{D}}_{12}^{+}\overline{\overline{G}}_{(1)}\overline{\overline{D}}_{12}.) \tag{6.23}$$

We will need some elementary properties of the metric. From the regularity of $\overline{\overline{G}}$ we conclude that the determinant $g = \det[\overline{\overline{G}}]$ is nonzero. From the transformation rule (6.23) we furthermore get

$$g_{(2)} = g_{(1)}[\det(\overline{\overline{D}}_{12})]^2. \tag{6.24}$$

It follows trivially that the sign of g is independent of the coordinate system we work in. Clearly, the determinant depends continuously upon the coordinates. Now let us assume that our manifold M is *connected* (i.e., any two points in M may be joined with a smooth curve within M). If g were positive at a point P and negative at a point Q, then g would have to be zero somewhere between P and Q. Since this is excluded, we obtain the following lemma:

Lemma 6.1. *g has the constant sign throughout M.*

We know that $\overline{\overline{G}}$, being symmetric, has n real eigenvalues $\lambda_1, \ldots, \lambda_n$. We also know that $\overline{\overline{G}}$ is regular; therefore, none of the eigenvalues can be zero. (Remember that $\det[G] = \lambda_1 \cdot \ldots \cdot \lambda_n$.) It can be shown that the number of positive eigenvalues is independent of the coordinate system. (The actual values of $\lambda_1, \ldots, \lambda_n$ depend strongly on the coordinate system in question!) Now let us assume again that our manifold is connected. Then we can extend the previous argument to show that the number of positive eigenvalues is constant throughout our manifold. We will especially be interested in two cases:

Definition 6.9.

a. If all the eigenvalues are strictly positive, we say that the metric is a *Euclidean metric*. In that case the metric is positive definite. A manifold with a Euclidean metric is called a *Riemannian manifold*. (If the metric is not Euclidean, we speak of a pseudo-Riemannian manifold.)
b. If one of the eigenvalues is negative and the rest of them are positive, we say that the metric is a *Minkowski metric*. In this case the metric is nondefinite. (Hence a manifold with a Minkowski metric is a special example of a pseudo-Riemannian manifold.)

Let us investigate the Minkowski metric a little more closely:

Lemma 6.2. *Let M be a manifold with a Minkowski metric. Let $\mathbf{T}_p(M)$ be the tangent space at the point P. Then we can decompose $\mathbf{T}_p(M)$ into an orthogonal sum*

$$\mathbf{T}_p(M) = V_- \oplus U_+ ;$$

where V_- is a one-dimensional subspace such that the restriction of $\mathbf{g}$ *to* V_- *is negative definite and* U_+ *is an* $(n-1)$*-dimensional subspace such that the restriction of* $\mathbf{g}$ *to* U_+ *is positive definite.*

PROOF. Choose a coordinate system covering P and let g_{ij} be the metric coefficients with respect to this coordinate system. Then $\overline{\overline{G}} = (g_{ij})$ has one negative eigenvalue λ_0 and $(n-1)$ positive eigenvalues $\lambda_1, \ldots, \lambda_{n-1}$. Let V_- be the eigenspace corresponding to λ_0 and let U_+ be the subspace spanned by the eigenspaces corresponding to $\lambda_1, \ldots, \lambda_{n-1}$.

The restriction of $\mathbf{g}$ to V_- is negative definite. To see this, let $\mathbf{v}$ be an eigenvector with eigenvalue λ_0 and coordinates a^i. Then

$$\mathbf{g}(\mathbf{v}, \mathbf{v}) = a^i g_{ij} a^j = a^i(\lambda_0 a^i) = \lambda_0(a^i a^i) < 0.$$

Hence $\mathbf{v}$ has a negative square.

Similarly, the $(n-1)$ eigenvectors $\mathbf{u}_1, \ldots, \mathbf{u}_{n-1}$ with eigenvalues $\lambda_1, \ldots, \lambda_{n-1}$ have positive squares. But they form a frame for U_+ and therefore, the restriction of $\mathbf{g}$ to U_+ is positive definite.

Finally, we must show that V is orthogonal to U. It suffices to show that $\mathbf{v}$ is orthogonal to each of the frame vectors $\mathbf{u}_1, \ldots, \mathbf{u}_{n-1}$. But

$$\mathbf{g}(\mathbf{v}; \mathbf{u}_k) = a^i g_{ij} b^j = a^i(\lambda_k b^i) = \lambda_k(a^i b^i)$$

and

$$\mathbf{g}(\mathbf{u}; \mathbf{v}) = b^i g_{ij} a^j = b^i(\lambda_0 a^i) = \lambda_0(a^i b^i).$$

As $\lambda_0 \neq \lambda_k$ (λ_k is positive and λ_0 is negative), it follows that $\mathbf{g}(\mathbf{v}; \mathbf{u}_k)$ must vanish. □

This lemma motivates the following classification:

Definition 6.10. Let M be a differentiable manifold with a Minkowski metric:
1. A vector with negative square is called *Time-like*.
2. A vector with zero square is called *null vector*.
3. A vector with positive square is called *space-like*.

The null vectors form a cone, the *light-cone*, that separates the time-like vectors from the space-like vectors. (See Figure 6.36.)

ILLUSTRATIVE EXAMPLE. *The sphere II*

Let us look at a sphere with an arbitrary radius r and let us introduce spherical coordinates (Figure 6.37):

$$\begin{bmatrix} x \\ y \\ z \end{bmatrix} = \begin{bmatrix} r\sin\theta\cos\varphi \\ r\sin\theta\sin\varphi \\ r\cos\theta \end{bmatrix}.$$

Here (θ, φ) are the *intrinsic* coordinates on the sphere and (x, y, z) are the *extrinsic* coordinates. At that point (θ_0, φ_0) we have the canonical frame vectors $\mathbf{e}_\theta$ and $\mathbf{e}_\phi$

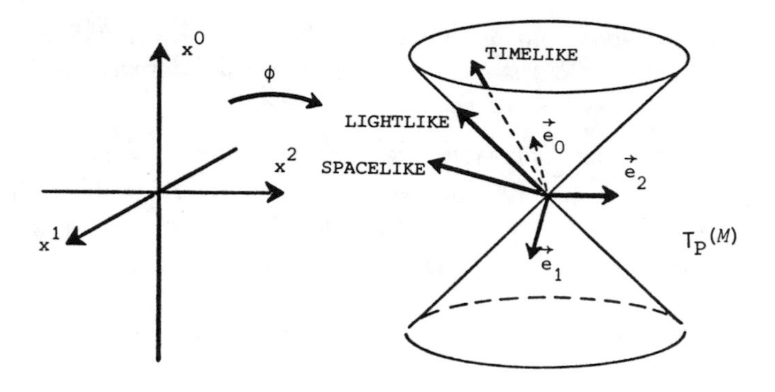

Figure 6.36.

with the extrinsic coordinates

$$\mathbf{e}_\theta = \begin{bmatrix} \frac{\partial x}{\partial \theta} \\ \frac{\partial y}{\partial \theta} \\ \frac{\partial z}{\partial \theta} \end{bmatrix} = \begin{bmatrix} r\cos\theta_0\cos\varphi_0 \\ r\cos\theta_0\sin\varphi_0 \\ -r\sin\theta_0 \end{bmatrix}; \quad \mathbf{e}_\varphi = \begin{bmatrix} \frac{\partial x}{\partial \varphi} \\ \frac{\partial y}{\partial \varphi} \\ \frac{\partial z}{\partial \varphi} \end{bmatrix} = \begin{bmatrix} -r\sin\theta_0\sin\varphi_0 \\ r\sin\theta_0\cos\varphi_0 \\ 0 \end{bmatrix}.$$

The *intrinsic* coordinates are by definition

$$\mathbf{e}_\theta = \begin{bmatrix} 1 \\ 0 \end{bmatrix} \quad \text{and} \quad \mathbf{e}_\varphi = \begin{bmatrix} 0 \\ 1 \end{bmatrix}.$$

The frame vectors $\mathbf{e}_\theta$ and $\mathbf{e}_\varphi$ are obviously orthogonal to the radial vector $\mathbf{r}$, and they are orthogonal to each other. (See Figure 6.38.) We also want to construct a metric on the sphere. But here we may use the usual inner product $\langle \mathbf{v} \mid \mathbf{u} \rangle = v^1u^1 + v^2u^2 + v^3u^3$ associated with $\mathbb{R}^3$. Thus, for any two tangent vectors $\mathbf{v}_P$ and $\mathbf{u}_P$ we put

$$\mathbf{g}(\mathbf{v}_P, \mathbf{u}_P) = \langle \mathbf{v}_P, \mathbf{u}_P \rangle. \tag{6.25}$$

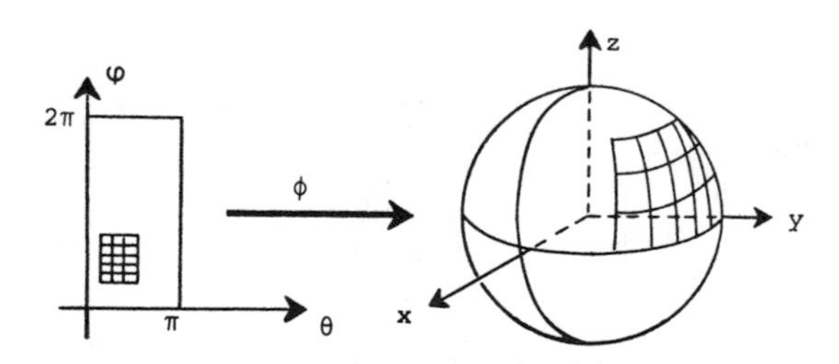

Figure 6.37.

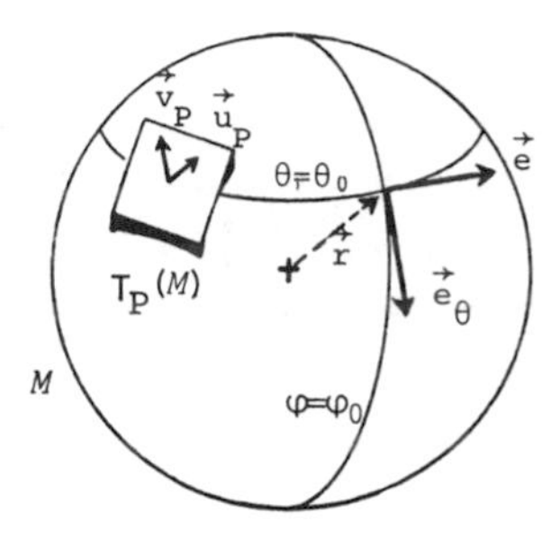

Figure 6.38.

This is called the *induced metric*. Let us compute the metric coefficients:

$$g_{ij} = \mathbf{g}(\mathbf{e}_i; \mathbf{e}_j) = \langle \mathbf{e}_i \mid \mathbf{e}_j \rangle$$

$$g_{\theta\theta} = [r\cos\theta_0\cos\varphi_0;\ r\cos\theta_0\sin\varphi_0;\ -r\sin\theta_0]\cdot\begin{bmatrix} r\cos\theta_0\cos\varphi_0 \\ r\cos\theta_0\sin\varphi_0 \\ -r\sin\theta_0\end{bmatrix} = r^2,$$

$$g_{\theta\varphi} = [r\cos\theta_0\cos\varphi_0;\ r\cos\theta_0\sin\varphi_0;\ -r\sin\theta_0]\cdot\begin{bmatrix} -r\sin\theta_0\cos\varphi_0 \\ r\sin\theta_0\cos\varphi_0 \\ -r\sin\theta_0\end{bmatrix} = 0,$$

$$g_{\varphi\varphi} = [-r\sin\theta_0\sin\varphi_0;\ r\sin\theta_0\cos\varphi_0;\ 0]\cdot\begin{bmatrix} -r\sin\theta_0\sin\varphi_0 \\ r\sin\theta_0\cos\varphi_0 \\ 0\end{bmatrix} = r^2\sin^2\theta_0.$$

Therefore, we find that

$$g_{ij}(\theta_0, \varphi_0) = r^2\begin{bmatrix} 1 & 0 \\ 0 & \sin^2\theta_0\end{bmatrix}. \tag{6.26}$$

Observe that the metric becomes singular at the north pole and the south pole,

$$g_{ij}\big|_{\text{pole}} = r^2\begin{bmatrix} 1 & 0 \\ 0 & 0\end{bmatrix},$$

which again tells us that the coordinate system breaks down at the poles!

Clearly, this example can be generalized to an arbitrary Euclidean manifold, which then becomes a Riemannian manifold, when we equip it with the induced metric. Let us emphasize, however, that the induced metric is not the only possible metric on a Euclidean manifold. We may choose the family of inner products completely arbitrarily.

Exercise 6.5.3
Problem: Consider a torus with outer radius a and inner radius b. Use the angles θ and φ as shown in Figure 6.39 to parametrize the surface. Compute the metric coefficients of the induced metric in (θ, φ)-coordinates.

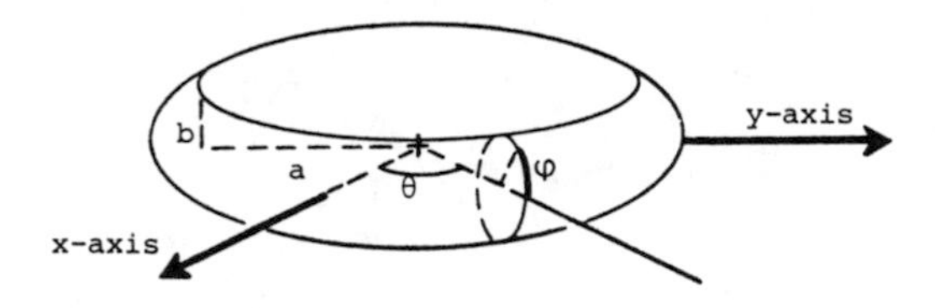

Figure 6.39.

Exercise 6.5.4

Problem: Consider the unit sphere S^{n-1} in the Euclidean space $\mathbb{R}^n$. We introduce spherical coordinates as indicated below. Show that the metric components are given as shown below.

$$
\begin{aligned}
x^1 &= r \sin\theta^1 \sin\theta^2 \cdots \sin\theta^{n-2} \cos\theta^{n-1}, \\
x^2 &= r \sin\theta^1 \sin\theta^2 \cdots \sin\theta^{n-2} \sin\theta^{n-1}, \\
x^3 &= r \sin\theta^1 \sin\theta^2 \cdots \cos\theta^{n-2}, \\
&\vdots \\
x^{n-1} &= r \sin\theta^1 \cos\theta^2, \\
x^n &= r \cos\theta^1,
\end{aligned}
$$

$$
g_{ij}(r, \theta^1, \ldots, \theta^{n-1}) =
\begin{bmatrix}
1 & & & & & 0 \\
& r^2 & & & & \\
& & r^2 \sin^2\theta^1 & & & \\
& & & r^2 \sin^2\theta^1 \sin^2\theta^2 & & \\
& 0 & & & \ddots & \\
& & & & & r^2 \sin^2\theta^1 \cdots \sin^2\theta^{n-2}
\end{bmatrix}. \tag{6.27}
$$

§ 6.6 The Minkowski Space

Consider a four-dimensional manifold M that we may identify with the four-dimensional Euclidean space $\mathbb{R}^4$, but forget about the extrinsic coordinates, as they are of no importance in our example. We shall use this manifold M as a model of our *space–time*. So, each point P in our manifold M represents an *event*. This event could be the absorption of a photon by a particle. It all happened at a specific position at a specific time!

A particle moving in ordinary space is represented by a *world line* in our model. (See Figure 6.40.) The world line is a smooth curve comprising all the space–time points occupied by the particle during its existence.

Next, we want to introduce coordinates in our manifold. To this purpose we shall use an inertial frame of reference; i.e., we imagine an observer equipped with a standard rod for the measurement of length and a standard clock for the measurement of time. Equipped with these instruments our observer can characterize

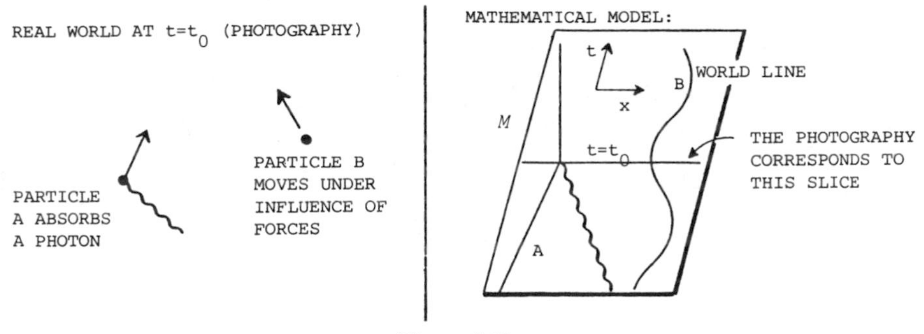

Figure 6.40.

each point in space by means of three Cartesian coordinates (x^1, x^2, x^3) and can characterize any event by its position in space (x^1, x^2, x^3) and the time x^0. Hence any event is characterized by a set of four coordinates, (x^0, x^1, x^2, x^3). In our mathematical model such an observer corresponds to a specific coordinate system, $\phi : \mathbb{R}^4 \curvearrowright M$, covering the whole space–time manifold (Figure 6.41).

What do we mean by an inertial frame of reference? Physically, an inertial frame of reference is characterized by the following property:

Definition 6.11. An inertial frame of reference is a frame of reference in which any free particle moves with constant velocity. The coordinate system corresponding to an inertial frame of reference will be referred to as an inertial frame, and we speak about inertial coordinates.

Consider the world line of a free particle. In inertial coordinates it is parametrized in the following way,

$$\begin{bmatrix} x^1 \\ x^2 \\ x^3 \end{bmatrix} = \begin{bmatrix} a^1 \\ a^2 \\ a^3 \end{bmatrix} x^0 + \begin{bmatrix} b^1 \\ b^2 \\ b^3 \end{bmatrix}; \qquad \text{i.e., } \overline{\overline{x}}_| = \overline{\overline{a}}_| x^0 + \overline{\overline{b}}_|,$$

where $\overline{\overline{a}}_|$ is the constant velocity of the particle. In the coordinate space this is simply the parametrization of a straight line; cf. Figure 6.42

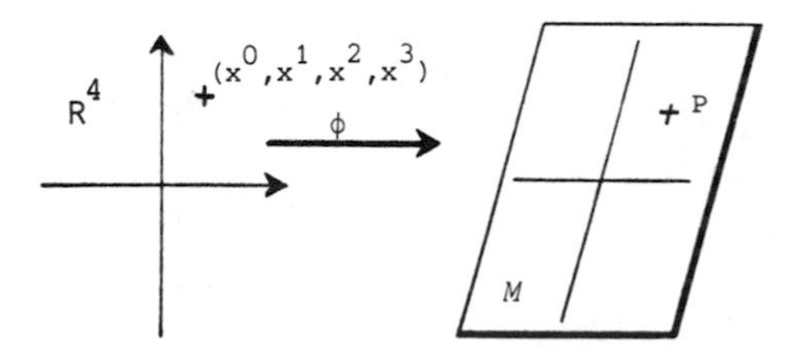

Figure 6.41.

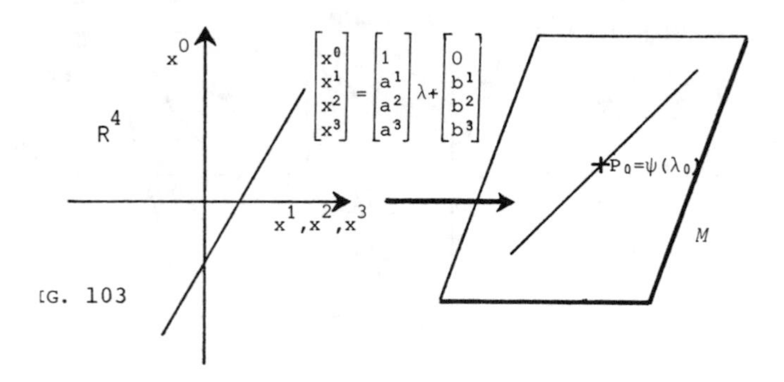

Figure 6.42.

Now, suppose we have been given two inertial frames of reference. Then we have two coordinate systems, $(\phi_1, \mathbb{R}^4)$ and $(\phi_2, \mathbb{R}^4)$, covering the space–time manifold. The transition function ϕ_{21} that exchanges the old coordinates (x^0, x^1, x^2, x^3) with the new coordinates (y^0, y^1, y^2, y^3) is referred to as a *Poincaré transformation*. What can we say about the transition function? Observe that the world line of a free particle is represented by a straight line in both coordinate systems (Figure 6.43).

Thus the function $y = \phi_{21}(x)$ maps straight lines onto straight lines, and we conclude that it is a linear function,

$$y^\alpha = A^\alpha{}_\beta x^\beta + b^\alpha .$$

To proceed, we must invoke the following assumptions:

Basic Assumptions of Special Relativity:

1. The speed of light has the same value in all directions in all inertial frames of reference; i.e., the speed of light is independent of the observer.
2. All inertial frames of reference are equivalent.

Let us introduce matrix notation. Put

$$\overline{\overline{\eta}} = \begin{bmatrix} -1 & 0 & 0 & 0 \\ 0 & 1 & 0 & 0 \\ 0 & 0 & 1 & 0 \\ 0 & 0 & 0 & 1 \end{bmatrix} \tag{6.28}$$

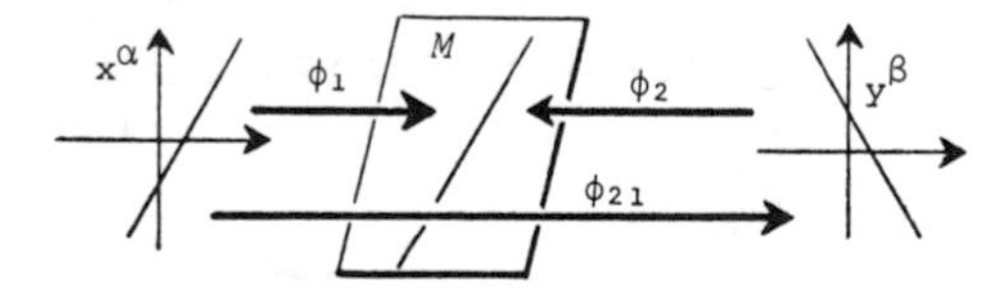

Figure 6.43.

We can then show

Theorem 6.3. *A Poincaré transformation connecting two inertial frames is given by the formula*

$$\overline{\overline{y}}_| = \overline{\overline{A}}\,\overline{\overline{x}}_| + \overline{\overline{b}}_|,$$

where $\overline{\overline{A}}$ satisfies

$$\overline{\overline{A}}^+ \overline{\overline{\eta}}\,\overline{\overline{A}} = \overline{\overline{A}}\,\overline{\overline{\eta}}\,\overline{\overline{A}}^+ = \overline{\overline{\eta}}. \tag{6.29}$$

PROOF. We have normalized our units so that the speed of light is one ($c = 1$). Suppose P and Q are two points lying on the world line of a free particle. Then the speed of this particle is given by

$$v = \frac{\sqrt{(x_Q^1 - x_P^1)^2 + (x_Q^2 - x_P^2)^2 + (x_Q^3 - x_P^3)^2}}{|x_Q^0 - x_P^0|}.$$

If we introduce the notation $\Delta\overline{\overline{x}}_| = \overline{\overline{x}}_|\big|_Q - \overline{\overline{x}}_|\big|_P$, we therefore conclude that the points P and Q lie on the world line of a photon if and only if

$$-(x_Q^0 - x_P^0)^2 + (x_Q^1 - x_P^1)^2 + (x_Q^2 - x_P^2)^2 + (x_Q^3 - x_P^3)^2 = 0;$$

i.e.,

$$(\Delta\overline{\overline{x}}_|)^+ \overline{\overline{\eta}}(\Delta\overline{\overline{x}}_|) = 0.$$

But the condition for P and Q to lie on a world line of a photon must be valid in any inertial frame. Therefore, we conclude that

$$(\Delta\overline{\overline{x}}_|)^+ \overline{\overline{\eta}}(\Delta\overline{\overline{x}}_|) = 0 \quad \text{iff} \quad (\Delta\overline{\overline{y}}_|)^+ \overline{\overline{\eta}}(\Delta\overline{\overline{y}}_|) = 0. \tag{6.30}$$

But here

$$\Delta\overline{\overline{y}}_| = \overline{\overline{y}}_|\big|_Q - \overline{\overline{y}}_|\big|_P = \overline{\overline{A}}(\overline{\overline{x}}_|\big|_Q - \overline{\overline{x}}_|\big|_P) = \overline{\overline{A}}\Delta\overline{\overline{x}}_|.$$

Thus (6.30) can be rearranged as

$$(\Delta\overline{\overline{x}}_|)^+ \overline{\overline{\eta}}(\Delta\overline{\overline{x}}_|) = 0 \quad \text{iff} \quad (\Delta\overline{\overline{x}}_|)^+ \overline{\overline{A}}^+ \overline{\overline{\eta}}\,\overline{\overline{A}}(\Delta\overline{\overline{x}}_|) = 0.$$

Hence the two symmetric matrices $\overline{\overline{\eta}}$ and $\overline{\overline{A}}^+ \overline{\overline{\eta}}\,\overline{\overline{A}}$ generate the same light cone K, where

$$K = \{(x^0, x^1, x^2, x^3) \mid -(x^0)^2 + (x^1)^2 + (x^2)^2 + (x^3)^2 = 0\}.$$

But then they must be proportional. Therefore, we conclude that

$$\overline{\overline{A}}^+ \overline{\overline{\eta}}\,\overline{\overline{A}} = \lambda\overline{\overline{\eta}}. \tag{6.31}$$

In this way we have attached a proportionality factor λ to the Poincaré transformation ϕ_{21}, which was given by

$$\overline{\overline{y}}_| = \phi_{21}(\overline{\overline{x}}_|) = \overline{\overline{A}}\,\overline{\overline{x}}_| + \overline{\overline{b}}_|.$$

The inverse Poincaré transformation ϕ_{12} is consequently given by

$$\overline{\overline{x}}_| = \phi_{12}(\overline{\overline{y}}_|) = \overline{\overline{A}}^{-1}\overline{\overline{y}}_| - \overline{\overline{A}}^{-1}\overline{\overline{b}}_|.$$

Thus ϕ_{12} is characterized by the matrix $\overline{\overline{A}}^{-1}$. A simple algebraic manipulation of equation (6.31) gives us

$$\frac{1}{\lambda}\,\overline{\overline{\eta}} = (\overline{\overline{A}}^{-1})^{+}\overline{\overline{\eta}}(\overline{\overline{A}}^{-1}),$$

so the proportionality factor corresponding to ϕ_{12} is $\frac{1}{\lambda}$. As the two inertial frames of reference are equivalent, we conclude that

$$\lambda = \frac{1}{\lambda}\,; \qquad \text{i.e., } \lambda = \pm 1.$$

But from (6.31) we also get

$$(\det[\overline{\overline{A}}])^2 = \lambda.$$

Consequently, $\lambda = +1$. We have therefore shown that

$$\overline{\overline{A}}^{+}\overline{\overline{\eta}}\,\overline{\overline{A}} = \overline{\overline{\eta}} \quad \text{and} \quad (\overline{\overline{A}}^{-1})^{+}\overline{\overline{\eta}}(\overline{\overline{A}})^{-1}) = \overline{\overline{\eta}}.$$

The last formula can easily be converted to give

$$\overline{\overline{A}}\,\overline{\overline{\eta}}^{-1}\overline{\overline{A}}^{+} = \overline{\overline{\eta}}^{-1}.$$

But $\overline{\overline{\eta}}^{-1} = \overline{\overline{\eta}}$, and therefore $\overline{\overline{A}}$ also satisfies

$$\overline{\overline{A}}\,\overline{\overline{\eta}}\,\overline{\overline{A}}^{+} = \overline{\overline{\eta}}.$$

□

This theorem motivates the following definition.

Definition 6.12. A matrix $\overline{\overline{A}}$ with the property

$$\overline{\overline{A}}^{+}\overline{\overline{\eta}}\,\overline{\overline{A}} = \overline{\overline{A}}\,\overline{\overline{\eta}}\,\overline{\overline{A}}^{+} = \overline{\overline{\eta}} \tag{6.32}$$

is called a *Lorentz-matrix*. The Lorentz matrices constitute a group called the Lorentz group, which we denote by $O(3, 1)$. A homogeneous transformation

$$\overline{\overline{y}}_| = \overline{\overline{A}}\,\overline{\overline{x}}_|,$$

where $\overline{\overline{A}}$ is a Lorentz matrix, is called a *Lorentz transformation*.

Exercise 6.6.1
Problem: (1) Show that the matrix $\overline{\overline{\eta}}$ itself is a Lorentz matrix. What is the physical interpretation of the associated Lorentz transformation?

(2) A Lorentz matrix that preserves the direction of time is called *orthochronous*. A Lorentz matrix that preserves the orientation of the spatial axes is called *proper*.

Show that a Lorentz matrix is proper and orthochronous if and only if it has the properties

$$A^0{}_0 > 0, \qquad \det \bar{\bar{A}} = 1,$$

and that the matrices with both properties constitute a subgroup of the Lorentz group (which is denoted by $SO_\uparrow(3, 1)$).

(3) Show that the spatial rotation group $SO(3)$ can be considered a subgroup of the Lorentz group.

Exercise 6.6.2

Problem: Consider the matrix:

$$\bar{\bar{A}} = \frac{1}{\sqrt{1-v^2}} \begin{bmatrix} 1 & v & 0 & 0 \\ v & 1 & 0 & 0 \\ 0 & 0 & 1 & 0 \\ 0 & 0 & 0 & 1 \end{bmatrix}.$$

Show that it is a proper orthochronous Lorentz matrix. Consider the corresponding Lorentz transformation, $\bar{\bar{y}} = \bar{\bar{A}}\bar{\bar{x}}$, and show that it corresponds to a transformation between two observers, where the second observer moves with velocity v along the x-axis of the first observer. This transformation is called the *standard Lorentz transformation*.

We will now construct a metric on our space–time manifold M. Let $\mathbf{v}_P$ and $\mathbf{u}_P$ be two tangent vectors at the same point P. Let $\bar{\bar{a}}_|$ and $\bar{\bar{b}}_|$ be their coordinates relative to an inertial frame, and consider the number

$$\bar{\bar{a}}_|^{+}\bar{\bar{\eta}}\bar{\bar{b}}_|.$$

This is actually independent of the inertial frame. To see this, we introduce another inertial frame. Let the corresponding Poincaré transformation be given by

$$\bar{\bar{y}}_| = \bar{\bar{A}}\bar{\bar{x}}_| + \bar{\bar{b}}_|.$$

Then the Jacobian matrix reduces to

$$\bar{\bar{D}}_{21} = \left(\frac{\partial y^\alpha}{\partial x^\beta}\right) = \bar{\bar{A}}.$$

According to (6.14), the new coordinates of $\mathbf{v}_P$ and $\mathbf{u}_P$ are given by $\bar{\bar{A}}\bar{\bar{a}}_|$ and $\bar{\bar{A}}\bar{\bar{b}}_|$. Hence in the new inertial frame we compute the number

$$(\bar{\bar{A}}\bar{\bar{a}}_|)^{+}\bar{\bar{\eta}}(\bar{\bar{A}}\bar{\bar{b}}_|) = \bar{\bar{a}}_|^{+}(\bar{\bar{A}}^{+}\bar{\bar{\eta}}\bar{\bar{A}})\bar{\bar{b}}_| = \bar{\bar{a}}_|^{+}\bar{\bar{\eta}}\bar{\bar{b}}_|,$$

and this shows that the number $\bar{\bar{a}}_|^{+}\bar{\bar{\eta}}\bar{\bar{b}}_|$ is really independent of the inertial frame. Thus we may make the following definition: *The inner product between two tangent vectors $\mathbf{u}_P$ and $\mathbf{v}_P$ is given by*

$$\mathbf{g}(\mathbf{u}_P; \mathbf{v}_P) = \bar{\bar{a}}_|^{+}\bar{\bar{\eta}}\bar{\bar{b}}_|,$$

where $\bar{\bar{a}}_|$ and $\bar{\bar{b}}_|$ are the components of $\mathbf{v}_P$ and $\mathbf{u}_P$ relative to an arbitrary inertial frame.

In this way we have defined a metric! We can easily read off the metric coefficients:

$$(\eta_{\alpha\beta}) = \begin{bmatrix} -1 & 0 & 0 & 0 \\ 0 & 1 & 0 & 0 \\ 0 & 0 & 1 & 0 \\ 0 & 0 & 0 & 1 \end{bmatrix}.$$

(Compare with (6.28).) They are clearly smooth functions of the coordinates. Furthermore, $\overline{\overline{\eta}}$ is symmetric and regular, and it has the eigenvalues $-1, 1, 1, 1$, so it is a nice Minkowski metric. The four-dimensional manifold $\mathbb{R}^4$ equipped with the above metric is called *Minkowski space.*

ILLUSTRATIVE EXAMPLE. *Spherical symmetry in the Minkowski space*

Very often, we study physical systems having some kind of symmetry, and it is then advantageous to choose coordinates that reflect the symmetry.

If the physical problem in question has spherical symmetry, we would naturally introduce spherical coordinates.

The transition functions from inertial coordinates (x^0, x, y, z) to spherical coordinates (t, r, θ, φ) are given by

$$\phi_{12} : x^0 = t; \quad x = r\sin\theta\cos\varphi; \quad y = r\sin\theta\sin\varphi; \quad z = r\cos\theta;$$

$$\phi_{21} : t = x^0; \quad r = \sqrt{x^2+y^2+z^2};$$

$$\theta = \arccos\left[\frac{z}{\sqrt{x^2+y^2+z^2}}\right]; \quad \varphi = \arctan\left[\frac{y}{x}\right].$$

The corresponding Jacobian matrices are given by (1) From inertial to spherical coordinates:

$$\overline{\overline{D}}_{21} = \left(\frac{\partial y^\alpha}{\partial x^\beta}\right) = \begin{bmatrix} 1 & 0 & 0 & 0 \\ 0 & \frac{x}{r} & \frac{y}{r} & \frac{z}{r} \\ 0 & -\frac{xz}{r^2\sqrt{r^2-z^2}} & -\frac{yz}{r^2\sqrt{r^2-z^2}} & -\frac{r^2-z^2}{r^2\sqrt{r^2-z^2}} \\ 0 & -\frac{y}{x^2+y^2} & \frac{x}{x^2+y^2} & 0 \end{bmatrix}. \tag{6.34}$$

(2) From spherical to inertial coordinates:

$$\overline{\overline{D}}_{12} = \frac{\partial x^\alpha}{\partial y^\beta} = \begin{bmatrix} 1 & 0 & 0 & 0 \\ 0 & \sin\theta\cos\varphi & r\cos\theta\cos\varphi & -r\sin\theta\sin\varphi \\ 0 & \sin\theta\sin\varphi & r\cos\theta\sin\varphi & r\sin\theta\cos\varphi \\ 0 & \cos\theta & -r\sin\theta & 0 \end{bmatrix}. \tag{6.35}$$

We can then find the metric coefficients in spherical coordinates. Using the transformation rule (6.23), we easily get

$$g_{\alpha\beta} = \begin{bmatrix} -1 & & & \\ & 1 & & 0 \\ 0 & & r^2 & \\ & & & r^2\sin^2\theta \end{bmatrix}. \tag{6.36}$$

Observe that the part of the metric associated with the polar angles is nothing but the Euclidean metric associated with a sphere of radius r. (See (6.26).) Observe also that the metric breaks down on the z-axis: $\theta = 0, \pi$. That is just the old coordinate singularity we found on the sphere. So strictly speaking, spherical coordinates do not cover the whole space–time manifold!

Now let us also look at a tangent vector $\mathbf{A}$ (i.e., a four-vector). Let us for a moment use inertial coordinates (A^t, A^x, A^y, A^z). Furthermore, we assume that the vector part, (A^x, A^y, A^z), is a radial vector, i.e., of the form

$$A(\sin\theta\cos\varphi, \sin\theta\sin\varphi, \cos\theta),$$

where A is the spatial length of the space vector. This is the typical form of a four-vector in problems with spherical symmetry (Figure 6.44). Let us try to find the spherical coordinates of this four-vector. Here we should use the transformation rule (6.8). Thus we get

$$\overline{\overline{A'_{|}}} = \begin{bmatrix} 1 & 0 & 0 & 0 \\ 0 & \sin\theta\cos\varphi & \sin\theta\sin\varphi & \cos\theta \\ 0 & -\frac{\cos\theta\cos\varphi}{r} & -\frac{\cos\theta\sin\varphi}{r} & \frac{\sin\theta}{r} \\ 0 & -\frac{\sin\varphi}{r\sin\theta} & \frac{\cos\varphi}{r\sin\theta} & 0 \end{bmatrix} \begin{bmatrix} A^0 \\ A\sin\theta\cos\varphi \\ A\sin\theta\sin\varphi \\ A\cos\theta \end{bmatrix} = \begin{bmatrix} A^0 \\ A \\ 0 \\ 0 \end{bmatrix}.$$

So, in spherical coordinates we get the simple expression

$$A^t = A^0; \quad A^r = A; \quad A^\theta = A^\varphi = 0.$$

Exercise 6.6.3
Problem: Introduce cylindrical coordinates (t, ρ, φ, z) (cf. Figure 6.45): Write down the transition functions. Show that the Jacobian matrices are given by

$$\overline{\overline{D}}_{21} = \begin{bmatrix} 1 & 0 & 0 & 0 \\ 0 & \cos\varphi & \sin\varphi & 0 \\ 0 & -\frac{\sin\varphi}{\rho} & \frac{\cos\varphi}{\rho} & 0 \\ 0 & 0 & 0 & 1 \end{bmatrix}, \tag{6.37}$$

$$\overline{\overline{D}}_{12} = \begin{bmatrix} 1 & 0 & 0 & 0 \\ 0 & \cos\varphi & -\rho\sin\varphi & 0 \\ 0 & \sin\varphi & \rho\cos\varphi & 0 \\ 0 & 0 & 0 & 1 \end{bmatrix}, \tag{6.38}$$

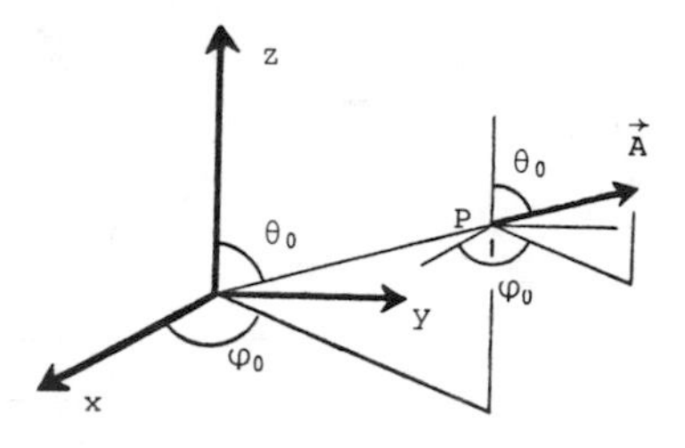

Figure 6.44.

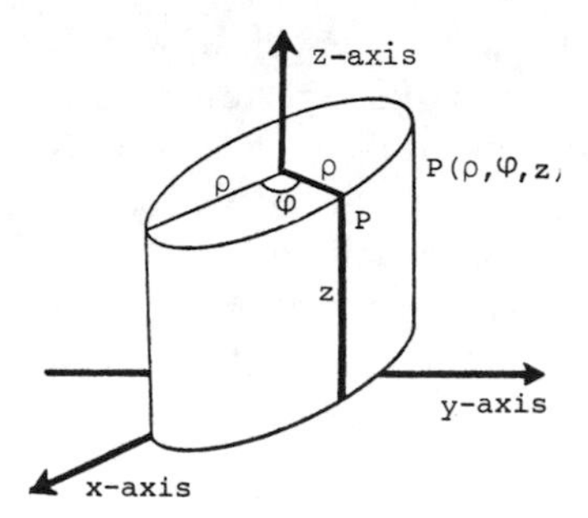

Figure 6.45.

and that the metric is given by

$$g_{\alpha\beta} = \begin{bmatrix} -1 & & & 0 \\ & 1 & & \\ & & \rho^2 & \\ 0 & & & 1 \end{bmatrix}. \tag{6.39}$$

§ 6.7 The Action Principle for a Relativistic Particle

Now, consider the world line of a particle, which need not be free. Let it have the parametrization, $x^\alpha = x^\alpha(\lambda)$ (Figure 6.46). Then the speed is given by the formula

$$v = \frac{ds}{dt} = \frac{ds/d\lambda}{dt/d\lambda} = \frac{\sqrt{(dx^1/d\lambda)^2 + (dx^2/d\lambda)^2 + (dx^3/d\lambda)^2}}{dx^0/d\lambda} < 1,$$

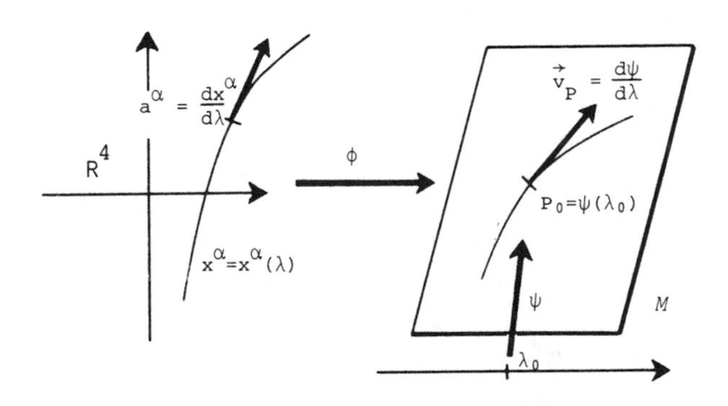

Figure 6.46.

which implies that

$$\mathbf{g}(\mathbf{v}_P;\mathbf{v}_P) = -\left(\frac{dx^0}{d\lambda}\right)^2 + \left(\frac{dx^1}{d\lambda}\right)^2 + \left(\frac{dx^2}{d\lambda}\right)^2 + \left(\frac{dx^3}{d\lambda}\right)^2 < 0.$$

Thus the tangent vector $\mathbf{v}_P$ is always a time-like. We therefore call the world line of an ordinary particle a *time-like curve.*

Now let us look a little more closely at the world line:

Let P_1 and P_2 be two points on the world line corresponding to the parameter values λ_1 and λ_2. Then we can define the arc-length of the arc $\widehat{P_1 P_2}$ by the following formula

$$\text{Arc-length} = \int_{\lambda_1}^{\lambda_2} \sqrt{-g_{\alpha\beta}(x(\lambda))\frac{dx^\alpha}{d\lambda}\frac{dx^\beta}{d\lambda}}\,d\lambda\,. \tag{6.40}$$

Observe that the arc-length is independent of the coordinate system chosen because the integrand is coordinate-independent:

$$\sqrt{-g_{\alpha\beta}(x(\lambda))\frac{dx^\alpha}{d\lambda}\frac{dx^\beta}{d\lambda}} = \sqrt{-\mathbf{g}(\mathbf{v}_P(\lambda),\mathbf{v}_P(\lambda))}.$$

Observe also that the arc-length is independent of the parametrization: If we exchange the parameter λ with a new parameter s such that $\lambda = \lambda(s)$ with $\lambda_1 = \lambda(s_1)$ and $\lambda_2 = (\lambda(s_2)$, we get

$$\begin{aligned}
\int_{\lambda_1}^{\lambda_2} \sqrt{-g_{\alpha\beta}(x(\lambda))\frac{dx^\alpha}{d\lambda}\frac{dx^\beta}{d\lambda}}\,d\lambda &= \int_{s_1}^{s_2} \sqrt{-g_{\alpha\beta}(x(s))\frac{dx^\alpha}{d\lambda}\frac{dx^\beta}{d\lambda}}\cdot\frac{d\lambda}{ds}\,ds \\
&= \int_{s_1}^{s_2} \sqrt{-g_{\alpha\beta}(x(s))\left[\frac{dx^\alpha}{d\lambda}\frac{d\lambda}{ds}\right]\left[\frac{dx^\beta}{d\lambda}\frac{d\lambda}{ds}\right]}\,ds \\
&= \int_{s=s_1}^{s=s_2} \sqrt{-g_{\alpha\beta}(x(s))\frac{dx^\alpha}{ds}\frac{dx^\beta}{ds}}\,ds.
\end{aligned}$$

Hence it is a reasonable definition. We will try to give a physical interpretation of the arc-length.

Let us look at a free particle first. As the particle is free, we may choose an inertial frame of reference that follows the particle; i.e., the particle is at *rest* in this inertial frame. Consequently, we can parametrize its world line as

$$x^0 = \lambda; \qquad x^1, x^2, x^3 = 0.$$

(See Figure 6.47.) Observe that the parameter λ represents the time measured on a standard clock that is at rest relative to the particle! In this particular simple coordinate system we get

$$\text{Arc-length} = \int_{\lambda_1}^{\lambda_2} \sqrt{-g_{00}\frac{dx^0}{d\lambda}\frac{dx^0}{d\lambda}}\,d\lambda = \int_{\lambda_1}^{\lambda_2} d\lambda = \lambda_2 - \lambda_1.$$

Thus we have shown:

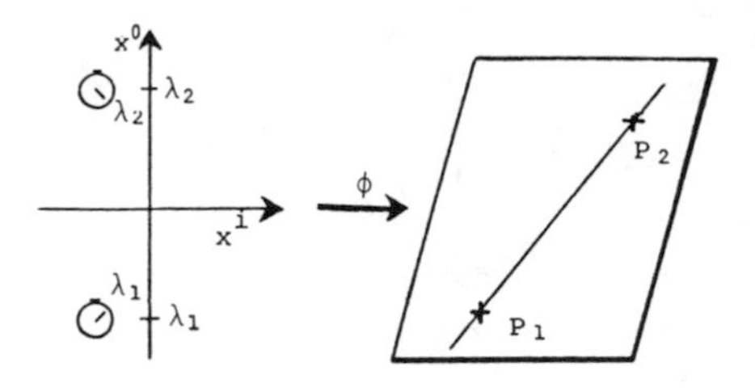

Figure 6.47.

The arc-length corresponding to a free particle is the time interval measured on a standard clock that is at rest relative to the particle!

Then we look at an arbitrary particle moving in space. When the world line is no longer straight, the above argument does not apply. Let us, however, assume that *the rate of a standard clock is unaffected by accelerations*. Then we may approximate the curved world line with a world line that is piecewise straight: The piecewise straight world line corresponds to a standard clock that is piecewise free. (See Figure 6.48.) Therefore, the above argument applies, and the arc-length of the piecewise straight world line is the time interval measured on this standard clock! Going to the limit, we conclude that

The arc-length of a world line is equal to the time measured on a standard clock following the particle. This will be referred to as the proper time τ of the particle.

Returning to the world line of a particle moving in an arbitrary way, we may now use the proper time τ of the particle as a parameter. It is connected to the old parameter λ via the formula

$$\tau = \int_{\lambda_0}^{\lambda} \sqrt{-g_{\alpha\beta}(x(\lambda))\frac{dx^\alpha}{d\lambda}\frac{dx^\beta}{d\lambda}}\, d\lambda. \tag{6.41}$$

Differentiating this formula, we get

$$\frac{d\tau}{d\lambda} = \sqrt{-g_{\alpha\beta}\cdot\frac{dx^\alpha}{d\lambda}\frac{dx^\beta}{d\lambda}};$$

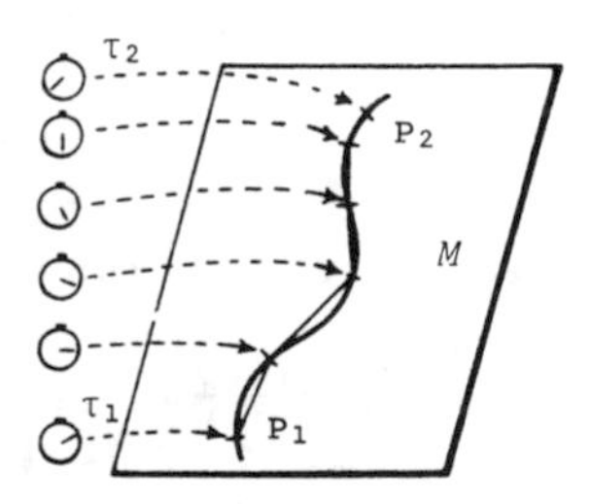

Figure 6.48.

i.e., in symbolic notation we have

$$d\tau = \sqrt{-g_{\alpha\beta}dx^\alpha dx^\beta}.$$

This formula is often "squared," whereby we get

$$d\tau^2 = -g_{\alpha\beta}dx^\alpha dx^\beta, \tag{6.42}$$

and $d\tau^2$ is then referred to as the *square of the line element* or just the *line element*.

Using the proper time τ as a parameter, we get the tangent vector

$$U^\alpha = \frac{dx^\alpha}{d\tau}, \tag{6.43}$$

which is referred to as the *four-velocity*. The four-velocity has a very simple property. If we square it, we get

$$\mathbf{g}(\mathbf{U}_P, \mathbf{U}_P) = g_{\alpha\beta}\frac{dx^\alpha}{d\tau}\frac{dx^\beta}{d\tau} = g_{\alpha\beta}\frac{dx^\alpha}{d\lambda}\frac{dx^\beta}{d\lambda}\left[\frac{d\lambda}{d\tau}\right]^2 = -1.$$

Thus the four-velocity is *a unit vector pointing in the direction of time*. (See Figure 6.49.)

Even in special relativity there is nothing sacred about inertial frames. We may introduce any coordinate system we like! We should, however, be aware of the following fact: If (ϕ, U) is not an inertial frame, then (Figure 6.50)

a. The metric coefficients no longer reduce to the trivial ones; i.e.,

$$g_{\alpha\beta} \neq \eta_{\alpha\beta}.$$

b. The world line of a *free* particle is no longer parametrized as a straight line; i.e.,

$$\frac{d^2x^\alpha}{d\tau^2} \neq 0.$$

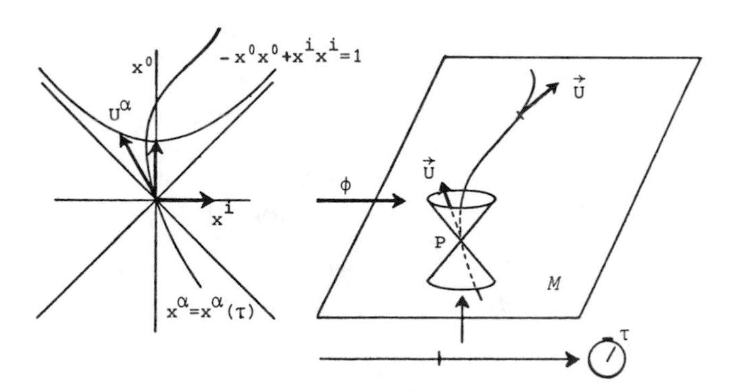

Figure 6.49.

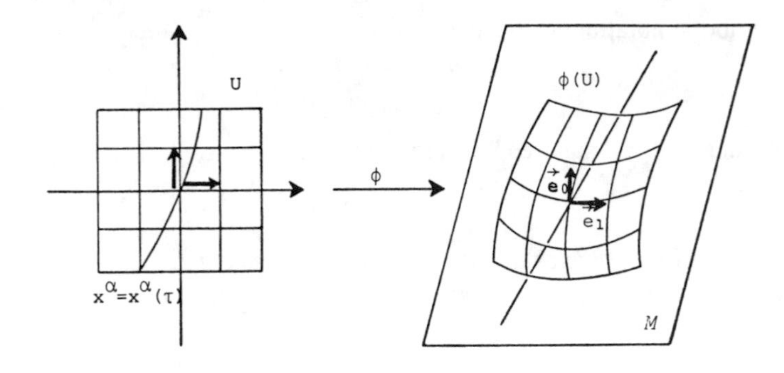

Figure 6.50.

Since a free particle has a nontrivial acceleration in curvilinear coordinates, we say that it experiences fictitious forces. We call the forces fictitious because they owe their existence to the choice of the coordinate system. If we return to the inertial frame, the fictitious forces disappear.

Let us try to find the equations of motion of a free particle expressed in arbitrary coordinates. It would be nice if we could derive these equations of motion from a Lagrangian principle; i.e., the world line should extremize an action of the form

$$S = \int_{\lambda_1}^{\lambda_2} L\left(x^\alpha; \frac{dx^\alpha}{d\lambda}\right) d\lambda.$$

Observe that the Lagrangian depends on all four space–time coordinates and their derivatives. In a relativistic formulation, time and space coordinates must be on the same footing.

In Euclidean space we know that a straight line extremizes the arc-length. This suggests the following theorem:

Theorem 6.4. *The relativistic Lagrangian for a free particle is given by*

$$L\left(x^\alpha; \frac{dx^\alpha}{d\lambda}\right) = m\sqrt{-g_{\alpha\beta}\frac{dx^\alpha}{d\lambda}\frac{dx^\beta}{d\lambda}}, \tag{6.44}$$

so the corresponding action is proportional to the proper time

$$S = -\int_{\tau_1}^{\tau_2} m d\tau = -\int_{\lambda_1}^{\lambda_2} m\sqrt{-g_{\alpha\beta}(x(\lambda))\frac{dx^\alpha}{d\lambda}\frac{dx^\beta}{d\lambda}}\, d\lambda. \tag{6.45}$$

The equations of motion can be written in the form

$$\frac{d^2x^\nu}{d\tau^2} = -\Gamma^\nu{}_{\alpha\beta}\frac{dx^\alpha}{d\tau}\frac{dx^\beta}{d\tau}, \tag{6.46}$$

where $\Gamma^\nu{}_{\alpha\beta}$, the so-called Christoffel field, is given by

$$\Gamma^\nu{}_{\alpha\beta} = \frac{1}{2}\, g^{\mu\nu}\left[\frac{\partial g_{\alpha\mu}}{\partial x^\beta} + \frac{\partial g_{\beta\mu}}{\partial x^\alpha} - \frac{\partial g_{\alpha\beta}}{\partial x^\mu}\right]. \tag{6.47}$$

PROOF. The Lagrangian principle leads to the following Euler–Lagrange equations:

$$\frac{\partial L}{dx^\mu} = \frac{d}{d\lambda}\left[\frac{\partial L}{\partial\left(\frac{dx^\mu}{d\lambda}\right)}\right]; \qquad \mu = 0, 1, 2, 3.$$

The evaluation of these equations using (6.44) is straightforward but tedious, so we leave it as an exercise:

Worked Exercise 6.7.1
Problem: Show that the Euler–Lagrange equation corresponding to the action (6.45) reduces to

$$\frac{d^2x^\nu}{d\tau^2} = -\frac{1}{2}\, g^{\mu\nu}\left[\frac{\partial g_{\mu\alpha}}{\partial x^\beta} + \frac{\partial g_{\mu\beta}}{\partial x^\alpha} - \frac{\partial g_{\alpha\beta}}{\partial x^\mu}\right]\frac{dx^\alpha}{d\tau}\frac{dx^\beta}{d\tau}.$$

To conclude the proof, we must show that the equations of motion reduce to

$$\left(\frac{d^2x^\nu}{d\tau^2}\right) = 0$$

in an inertial frame. But in an inertial frame we know that the metric coefficients are constant, $g_{\alpha\beta} = \eta_{\alpha\beta}$, so their derivatives vanish. □

If we compare this with the expression for the Lorentz force in *inertial coordinates*,

$$m\,\frac{d^2x^\nu}{d\tau^2} = qF^\nu_{\ \alpha}\,\frac{dx^\alpha}{d\tau},$$

we see that we may identify $-m\Gamma^\nu_{\ \alpha\beta}$ *with the field strengths of the fictitious forces*.

Let us look at the electromagnetic force once more. We know that in terms of inertial coordinates it may be rewritten as (1.29)

$$F^\nu_{\ \alpha} = g^{\mu\nu}\left[\frac{\partial A_\alpha}{\partial x^\mu} - \frac{\partial A_\mu}{\partial x^\alpha}\right].$$

Thus it is expressed as a combination of derivatives of the Maxwell field A^α. For this reason we say that the Maxwell field acts as *potentials* for the electromagnetic forces. Returning to curvilinear coordinates, we have seen that the Christoffel field may be expressed as a combination of derivatives of the metric coefficients:

$$\Gamma^\nu_{\ \alpha\beta} = \frac{1}{2}\, g^{\mu\nu}\left[\frac{\partial g_{\mu\alpha}}{\partial x^\beta} + \frac{\partial g_{\beta\mu}}{\partial x^\alpha} - \frac{\partial g_{\alpha\beta}}{\partial x^\mu}\right]. \tag{6.47}$$

For this reason we say that *the metric coefficients act as potentials for the fictitious forces*.

Finally, we notice that if we parametrize the world line by the frame time, i.e., put $\lambda = x^0\ (= t)$, then the relativistic action can be rearranged as

$$S = -m\int_{t_1}^{t_2}\sqrt{1-\mathbf{v}^2}\,dt. \tag{6.48}$$

For a nonrelativistic particle, i..e, when $v \ll 1$, this reduces in the lowest-order approximation to the nonrelativistic action

$$S \approx -m \int_{t_1}^{t_2} \left(1 - \frac{1}{2}\mathbf{v}^2\right) dt = \int_{t_1}^{t_2} \frac{1}{2} m\mathbf{v}^2 dt + m(t_2 - t_1)$$

(except for an irrelevant constant).

Consider an arbitrary manifold with a metric, which can be a Riemannian metric or a Minkowski metric. A curve connecting two points that extremizes the arc-length is called a *geodesic*. In the above analysis we have made no special use of the fact that we were working in Minkowski space. Consequently, geodesics on an arbitrary manifold are characterized by the *geodesic equation*

$$\frac{d^2x^\alpha}{ds^2} = -\Gamma^\alpha_{\mu\nu} \frac{dx^\mu}{ds} \frac{dx^\nu}{ds} , \tag{6.49}$$

where s is the arc-length and $\Gamma^\alpha_{\mu\nu}$ are the Christoffel fields. The Christoffel fields are defined by the formula (6.47). But unfortunately, this formula is very inconvenient for computational purposes. When we want to compute the geodesic equations, it is therefore preferable to extract them directly from a variational principle. Of course, we could use the Euler–Lagrange equations coming from extremizing the arc-length itself,

$$S = \int_{\lambda_1}^{\lambda_2} \sqrt{g_{\alpha\beta}(x(\lambda)) \frac{dx^\alpha}{d\lambda} \frac{dx^\beta}{d\lambda}}\, d\lambda,$$

but this leads to very complicated computations due to the square root. So in practice, one uses the following useful lemma:

Lemma 6.3. *Let γ be a geodesic:*
1. *Then γ extremizes the functional*

$$I = \int_{\lambda_1}^{\lambda_2} \frac{1}{2} g_{\alpha\beta}(x(\lambda)) \frac{dx^\alpha}{d\lambda} \frac{dx^\beta}{d\lambda}\, d\lambda. \tag{6.50}$$

2. *Furthermore, when $x^\alpha = x^\alpha(\lambda)$ extremizes the functional I, then*

$$g_{\alpha\beta}(x(\lambda)) \frac{dx^\alpha}{d\lambda} \frac{dx^\beta}{d\lambda}$$

is constant along the geodesic, so that we can identify the parameter λ with the arc-length s.

PROOF. Consider the Lagrangian

$$L\left(x^\alpha, \frac{dx^\alpha}{d\lambda}\right) = \frac{1}{2} g_{\mu\nu}(x(\lambda)) \frac{dx^\mu}{d\lambda} \frac{dx^\nu}{d\lambda} .$$

It leads to the Euler–Lagrange equations

$$\frac{1}{2} \partial_\alpha g_{\mu\nu} \frac{dx^\mu}{d\lambda} \frac{dx^\nu}{d\lambda} = \frac{d}{d\lambda}\left[g_{\alpha\nu} \frac{dx^\nu}{d\lambda}\right] = \partial_\mu g_{\alpha\nu} \frac{dx^\mu}{d\lambda} \frac{dx^\nu}{d\lambda} + g_{\alpha\nu} \frac{d^2x^\nu}{d\lambda^2}$$

i.e.,

$$g_{\alpha\nu}\frac{d^2x^\nu}{d\lambda^2} = -\frac{1}{2}\left[\partial_\mu g_{\alpha\nu} + \partial_\nu g_{\alpha\mu} - \partial_\alpha g_{\mu\nu}\right]\frac{dx^\mu}{d\lambda}\frac{dx^\nu}{d\lambda}$$

but they are identical to the geodesic equations if we can show property (2). This is done in the following way:

Let $x^\alpha = x^\alpha(\lambda)$ extremize the functional I. Observe that $L\left(x^\alpha; \frac{dx^\alpha}{d\lambda}\right)$ is homogeneous in the variable $dx^\alpha/d\lambda$ of degree 2. Thus it satisfies the Euler equation

$$\frac{dx^\alpha}{d\lambda}\frac{\partial L}{\partial(dx^\alpha/d\lambda)} = 2L.$$

We must show that $L\left(x^\alpha(\lambda); \frac{dx^\alpha}{d\lambda}\right)$ is constant along $x^\alpha = x^\alpha(\lambda)$. Consequently, we consider the total derivative

$$\frac{dL}{d\lambda} = \frac{\partial L}{\partial x^\alpha}\frac{dx^\alpha}{d\lambda} + \frac{\partial L}{\partial(dx^\alpha/d\lambda)}\frac{d^2x^\alpha}{d\lambda^2}.$$

Using the Euler–Lagrange equations, this is rearranged as

$$\begin{aligned}\frac{dL}{d\lambda} &= \frac{d}{d\lambda}\left\{\frac{\partial L}{\partial(dx^\alpha/d\lambda)}\right\}\frac{dx^\alpha}{d\lambda} + \frac{\partial L}{\partial(dx^\alpha/d\lambda)}\frac{d^2x^\alpha}{d\lambda^2}\\ &= \frac{d}{d\lambda}\left[\frac{\partial L}{\partial(dx^\alpha/d\lambda)}\frac{dx^\alpha}{d\lambda}\right] = 2\frac{dL}{d\lambda},\end{aligned}$$

where we have taken advantage of the Euler relation. But this implies immediately that

$$\frac{dL}{d\lambda} = 0;$$

i.e., λ must be proportional to the arc-length s. □

Observe that when we have found the geodesic equations, we can directly extract the Christoffel fields as the coefficients on the right-hand side. By an abuse of notation, the above functional is often denoted by

$$I = \frac{1}{2}\int d\tau^2.$$

Exercise 6.7.2
Problem: Consider the two-sphere S^2 with the line element (cf. (6.26))

$$ds^2 = d\theta^2 + \sin^2\theta d\varphi^2.$$

1. Determine the geodesic equations and show that the meridian $\varphi = \varphi_0$ is geodesic.
2. Compute the Christoffel field.

Worked Exercise 6.7.3
Problem: In Minkowski space the line element is given by

$$d\tau^2 = -dt^2 + dr^2 + r^2(d\theta^2 + \sin^2\theta d\varphi^2),$$

when it is expressed in terms of spherical coordinates; cf. (6.39). Compute the associated Christoffel field.

Okay, you might say, this is very nice. We can work in arbitrary coordinates if we are willing to pay the price of fictitious forces and nontrivial metric coefficients and all that, but why should we do it? After all, we have at our disposal a nice family of coordinate systems, the inertial frames, that gives a simple description of physics!

There is, however, a subtle reason for working in arbitrary coordinates that was pointed out by Einstein. Fictitious and gravitational forces have a strange property in common:

If a test particle is released in a gravitational field, its acceleration is independent of its mass. Two test particles with different masses will follow each other in the gravitational field. This is known as *Galileo's principle*, and it has been tested in modern times with an extremely high precision! (See Figure 6.51.)

But the fictitious forces have the same property: The acceleration of the test particle depends only on the field strength $-\Gamma^{\nu}{}_{\alpha\beta}$, and the field strength itself depends only on the metric coefficients. Hence if we release free particles, then their accelerations relative to an arbitrary observer, say a rotating frame of reference, will be independent of their masses. Especially if we release two free particle close to each other, then they will stay close to each other!

Okay, one might say, I accept that they have this property in common, but have they anything else in common? For instance, I can transform the fictitious forces completely away by choosing a suitable coordinate system. But I cannot transform away the gravitational field!

It is true that I cannot transform away the gravitational field completely. But I can transform it away locally! Consider a gravitational field, say that of earth. Now let us release a small box so that it is freely falling in the gravitational field. Inside the box we put an observer with a standard rod and a standard clock. Thus he can assign coordinates to all the events inside his box. (See Figure 6.52.)

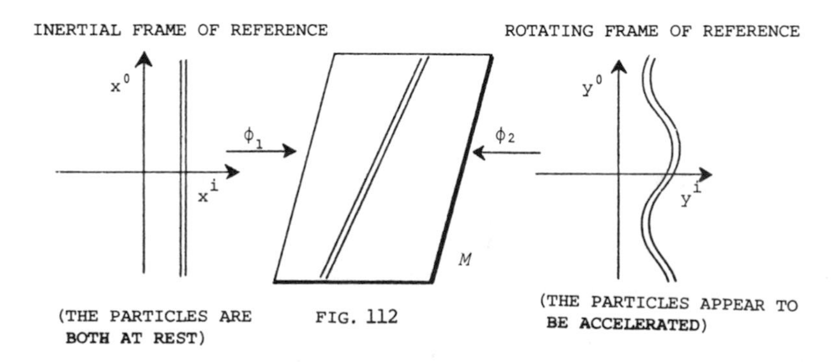

Figure 6.51.

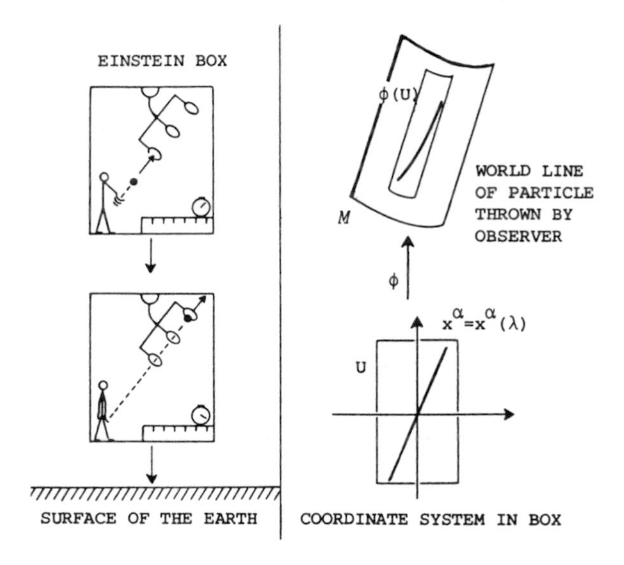

Figure 6.52.

But now the miracle happens. If he releases particles in his box, they will be freely falling too. Therefore, they will move with a constant velocity relative to his coordinate system, and they will act as free particles relative to his coordinate system! Consequently, we have succeeded in *transforming the gravitational field away inside the box by choosing a suitable coordinate system*!

When we say that we have transformed away the gravitational field inside the box, this is not the whole truth. Actually, if we were very careful, we would find that if we released two test particles as shown in Figure 6.53, they would slowly approach each other. Consequently, there is still an attractive force between the two test particles. This force is extremely weak and is known as the *Tidal force*. The Tidal force owes its existence to the inhomogeneities of the gravitational field, and it is a second-order effect, so we may loosely say that we can transform away the gravitational field to first-order inside the box!

So there is a good reason why we might be interested in studying fictitious forces: *It might teach us something about gravitational forces*!

§ 6.8 Covectors

The next concept we want to introduce on our manifold M is that of covector:

Definition 6.13. A *covector* ω_P at the point P is a linear map

$$\omega_P : \mathbf{T}_P(M) \to \mathbb{R}$$

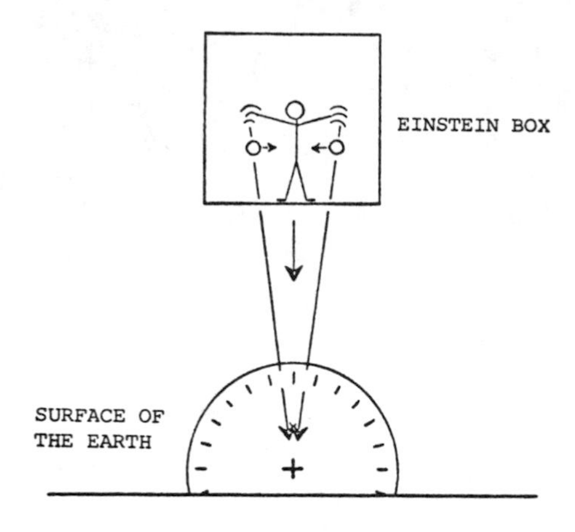

Figure 6.53.

(See Figure 6.54.) Here will use a notation introduced by Dirac. Given a tangent vector $\mathbf{v}_P$, we denote it by $|\mathbf{v}_P\rangle$ The symbol $|\rangle$ is called a *ket*. Similarly, a covector $\boldsymbol{\omega}_P$ is denoted by $\langle\boldsymbol{\omega}_P|$, where the symbol $\langle|$ is called a *bra*. Then the value of $\boldsymbol{\omega}_P$ at $\mathbf{v}_P$ is written as follows:

$$\omega(\mathbf{v}_P) = \langle\omega|\mathbf{v}_P\rangle, \tag{6.52}$$

where the symbol $\langle\,|\,\rangle$ is called a *bracket*. Using this bracket the linearity of $\boldsymbol{\omega}_P$ may be expressed in the following way:

$$\langle\boldsymbol{\omega}_P|\lambda\mathbf{v}_P + \mu\mathbf{u}_P\rangle = \lambda\langle\boldsymbol{\omega}_P|\mathbf{v}_P\rangle + \mu\langle\boldsymbol{\omega}_P|\mathbf{u}_P\rangle.$$

To justify the name *covector*, we must show that the covectors form a vector space. If ω and χ are covectors, i.e., linear maps from the tangent space, we define their sum $\omega_P + \chi_P$ as follows:

$$(\omega_P + \chi_P)(\mathbf{v}_P) = \omega_P(\mathbf{v}_P) + \chi_P(\mathbf{v}_P).$$

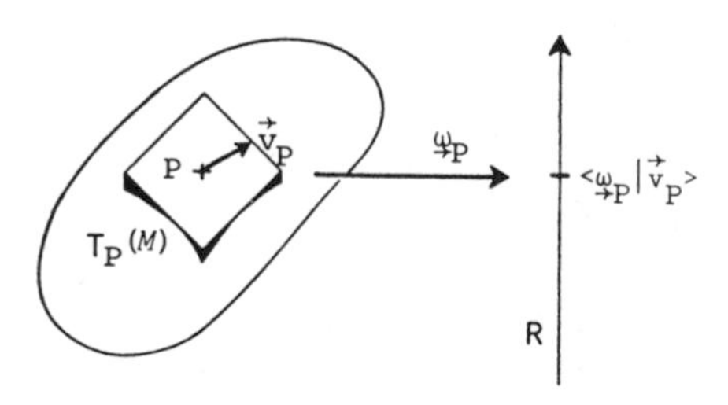

Figure 6.54.

(It can be easily verified that $\omega_P + \chi_P$ is a linear map), and similarly, we define the product of a covector ω and a scalar λ in the following way:

$$(\lambda\omega_P)(\mathbf{v}_P) = \lambda\omega(\mathbf{v}_P).$$

Using the bracket notation we may rewrite these rules as

a. Sum: $\langle\omega_P + \chi_P|\mathbf{v}_P\rangle = \langle\omega|\mathbf{v}_P\rangle + \langle\chi_P|\mathbf{v}_P\rangle$.
b. Product with a scalar: $\langle\lambda\omega_P|\mathbf{v}_P\rangle = \lambda\langle\omega_P|\mathbf{v}_P\rangle$.

Consequently, the bracket is bilinear in both of its arguments.

So the covectors at the point P form a vector space, which mathematicians call the *dual vector space* and which they denote by

$$\mathbf{T}_P^*(M).$$

Let us investigate the structure of this vector space a little more closely. Choose a covector ω_P. Introduce a coordinate system around P. (See Figure 6.55.) We can then define *coordinates* of the covector in the following way:

$$\omega_i = \langle\omega_P|\mathbf{e}_i\rangle. \tag{6.53}$$

Using the coordinates of ω_P we can express the value of ω_P at a tangent vector $\mathbf{v}_P$ in a simple way. If $\mathbf{v}_P$ has the coordinates a^i, we find that

$$\langle\omega_P|\mathbf{v}_P\rangle = \langle\omega_P|a^i\mathbf{e}_i\rangle = a^i\langle\omega_P|\mathbf{e}_i\rangle = \omega_i a^i. \tag{6.54}$$

Observe that an expression like $\omega_i a^i$ is coordinate independent because $\langle\omega_P|\mathbf{v}_P\rangle$ is defined in a purely geometrical way without any reference to the coordinate system used on M. *This is a general feature of expressions where we sum over a repeated index.*

Now let ω_P have the coordinates ω_i and χ_P have the coordinates χ_i. Then $\omega_P + \chi_P$ gets the coordinates

$$\langle\omega + \chi_P|\mathbf{e}_i\rangle = \langle\omega_P|\mathbf{e}_i\rangle + \langle\chi_P|\mathbf{e}_i\rangle = \omega_i + \chi_i.$$

In a similar way $\lambda\omega_P$ gets the coordinates

$$\langle\lambda\omega_P|e_i\rangle = \lambda\langle\omega_P|e_i\rangle = \lambda\omega_i.$$

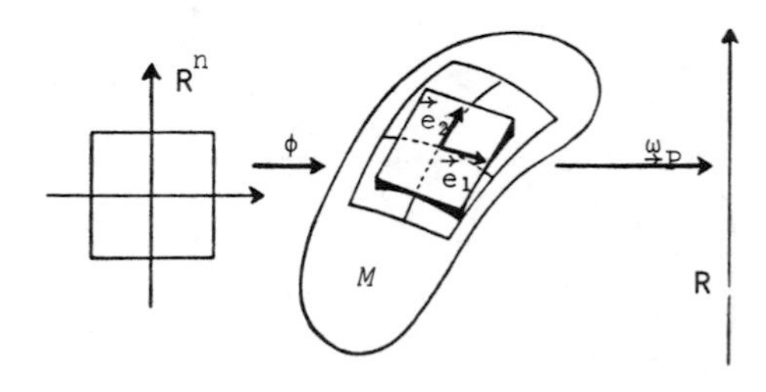

Figure 6.55.

Thus we see that once we have introduced coordinates, we may identify the dual space $\mathbf{T}^*_P(M)$ with the vector space $\mathbb{R}_n$ is the sense that the covector ω_P is represented by its coordinates $(\omega_1, \ldots, \omega_n)$. In particular, we see that $\mathbf{T}^*_P(M)$ is an n-dimensional vector space. To each point P on our manifold we have therefore now introduced two n-dimensional vector spaces,

$$\mathbf{T}_P(M) \quad \text{and} \quad \mathbf{T}^*_P(M).$$

When discussing tangent vectors, we noticed that a tangent vector $\mathbf{v}_P$ had a simple geometrical interpretation. It could be regarded as a velocity vector corresponding to a smooth curve λ passing through P. Now we want to give a similar geometrical interpretation of a covector. Consider a differentiable function, $f : M \curvearrowright \mathbb{R}$, defined in a neighborhood of P, and let λ be a smooth curve passing through P (see Figure 6.56). Then $t \curvearrowright f(\lambda(t))$ is an ordinary function from $\mathbb{R}$ into $\mathbb{R}$ merely describing the variation of f along λ. Since it is an ordinary function, we can differentiate it at $t = 0$:

$$\frac{df(\lambda(t))}{dt}\Big|_{t=0}.$$

Let us investigate this number a little more closely. It represents the rate of change of f in the direction of λ. If we introduce coordinates $(x^1, \ldots, x^n)$ around P, we can represent f by an ordinary Euclidean function $y = \bar{f}(x^1, \ldots, x^n)$, and we can furthermore parametrize the curve λ as $x^i = x^i(t)$. Then the tangent vector $\mathbf{v}_P$ has the intrinsic coordinates

$$a^i = \frac{dx^i}{dt}\Big|_{t=0}.$$

Using these parametrizations, we now obtain

$$\frac{df(\lambda(t))}{dt}\Big|_{t=0} = \frac{d}{dt}\bar{f}(x^i(t))\Big|_{t=0} = \frac{\partial \bar{f}}{\partial x^i}\frac{dx^i}{dt}\Big|_{t=0} = \frac{\partial \bar{f}}{\partial x^i}a^i. \tag{6.55}$$

This formula shows that the number $\frac{d}{dt} f(\lambda(t))$ depends only on the tangent vector $\mathbf{v}_P$ and not on the particular curve generating it. It also shows that this number

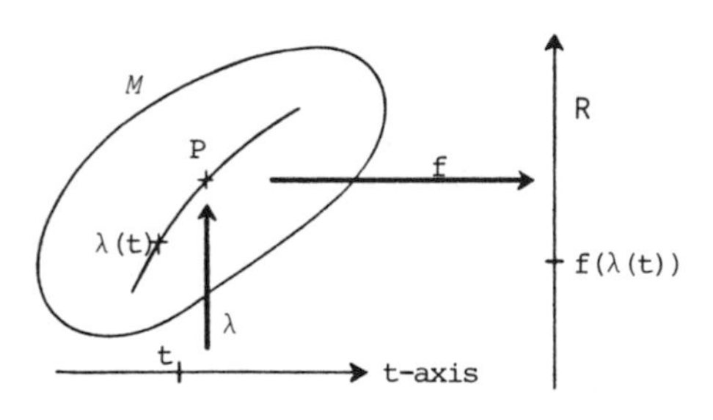

Figure 6.56.

depends linearly on the tangent vector $\mathbf{v}_P$. We therefore define a linear map on the tangent space $\mathbf{T}_P(M)$ by

$$\mathbf{v}_P \to \frac{df(\lambda(t))}{dt},$$

where λ is any smooth curve generating $\mathbf{v}_P$. But a linear map from the tangent space $\mathbf{T}_P(M)$ is nothing but a covector! This covector will be denoted by $\boldsymbol{d}f$, and we refer to this expression as the *exterior derivative of* f.

We have thus shown that

$$\langle \boldsymbol{d}f | \mathbf{v}_P \rangle = \frac{d}{dt} f(\lambda(t))\Big|_{t=0} = \frac{\partial \bar{f}}{\partial x^i} a^i. \tag{6.56}$$

If we compare this with (6.54), we can immediately read off the coordinates of $\boldsymbol{d}f$:

$$\boldsymbol{d}f : \left(\frac{\partial \bar{f}}{\partial x^i}, \frac{\partial \bar{f}}{\partial x^2}, \ldots, \frac{\partial \bar{f}}{\partial x^n} \right). \tag{6.57}$$

But then $\boldsymbol{d}f$ is nothing but a generalization of the gradient vectors from elementary vector analysis in Euclidean spaces!

Using gradient vectors we may analyze the structure of the dual space $\mathbf{T}^*_P(M)$ a little more closely. Let P_0 be a point on M, and let us introduce coordinates $(x^1, \ldots, x^n)$ around P_0. Then we have constructed a frame for the ordinary tangent space $\mathbf{T}_{P_0}(M)$ in the following way: First we introduced the coordinate lines on M (see Figure 6.57)

$$\lambda_1(t) = \phi(t, 0, \ldots, 0); \ldots; \lambda_n(t) = \phi(0, \ldots, 0, t)$$

and then the coordinate lines generated by the canonical frame vectors

$$\mathbf{e}_1 = \frac{d\lambda_1}{dt}\Big|_{t=0}; \ldots; \mathbf{e}_n = \frac{d\lambda_n}{dt}\Big|_{t=0}.$$

Now we want to do something similar. First, we observe that we can define a smooth function on M in the following way:

$$Q(x^1, \ldots, x^n) \curvearrowright x^1.$$

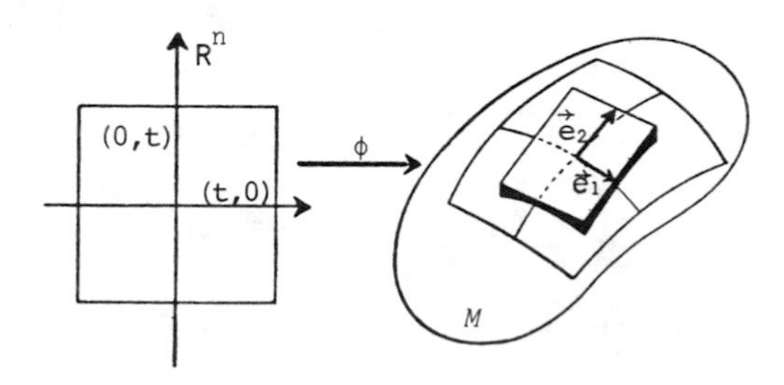

Figure 6.57.

This function is referred to as the *first* coordinate function, and for simplicity we will denote it by x^1. Thus $x^1(Q) = x_0^1$, where Q has the coordinates $(x_0^1, \ldots, x_0^n)$. In a similar way we can define the other coordinate functions x^i, $i = 2, \ldots, n$, where $x^i(Q) = x_0^i$.

Being smooth functions, these coordinate functions x^i will generate covectors $\boldsymbol{d}x^i$ at the point P_0. These covectors have particularly simple coordinates; e.g., $\boldsymbol{d}x^1$ gets the coordinates

$$\boldsymbol{d}x^1 : \left(\frac{\partial x^1}{\partial x^1}, \frac{\partial x^1}{\partial x^2}, \ldots, \frac{\partial x^1}{\partial x^n} \right) = (1, 0, \ldots, 0),$$

and similarly,

$$\boldsymbol{d}x^2 : (0, 1, \ldots, 0); \ldots ; \boldsymbol{d}x^n : (0, 0, \ldots, 1).$$

Therefore, *the covectors $\boldsymbol{d}x^1; \ldots ; \boldsymbol{d}x^n$ serve as the canonical frame vectors in the dual space $\mathbf{T}_P^*(M)$.*

If an arbitrary covector $\boldsymbol{\omega}_P$ has the coordinates ω_i, we now get

$$\boldsymbol{\omega}_P = \omega_1 \boldsymbol{d}x^1 + \cdots + \omega_n \boldsymbol{d}x^n. \tag{6.58}$$

This formula is, in particular, valid for the gradient vector $\boldsymbol{d}f$: Therefore we conclude

$$\boldsymbol{d}f = \frac{\partial \bar{f}}{\partial x^i}\, \boldsymbol{d}x^i. \tag{6.59}$$

Here we see something strange! This formula looks exactly like the well-known formula for the differential:

$$df = \frac{\partial f}{\partial x^i}\, dx^i.$$

But in modern geometry the idea of infinitesimal increments has been buried because of the logical difficulties involved in their use. The symbol $\boldsymbol{d}$ is now interpreted as an exterior derivative, and the miracle happens: Many of the formulas involving the exterior derivative look exactly like the "old" formulas involving differentials. So in modern geometry we work with the same formulas; they have just been given a *new interpretation*.

Finally, we may investigate how the coordinates of a covector $\boldsymbol{\omega}_P$ are changed if we exchange the coordinate system. Let the old coordinates be $\omega_{(1)_i}$ and the new coordinates $\omega_{(2)_i}$. Using (6.12), we then get

$$\omega_{(2)_j} = \langle \boldsymbol{\omega}_P | \mathbf{e}_{(2)_j} \rangle = \left\langle \boldsymbol{\omega}_P | \mathbf{e}_{(1)_i} \frac{\partial x^i}{\partial y^j} \right\rangle = \langle \boldsymbol{\omega}_P | \mathbf{e}_{(1)_i} \rangle \frac{\partial x^i}{\partial y^j} = \omega_{(1)_i} \frac{\partial x^i}{\partial y^j}\,.$$

So the coordinates of a covector are transformed according to the rule

$$\omega_{(2)_j} = \omega_{(1))_i} \frac{\partial x^i}{\partial y^j}\,. \tag{6.60}$$

Thus the coordinates transform covariantly, and this justifies their name: covectors!

We can also easily find out how the canonical frame vectors transform. If we introduce two coordinate systems, then the old coordinates $(x^1, \dots, x^n)$ are connected to the new coordinates $(y^1, \dots, y^n)$ via the transition functions

$$y^1 = y^1(x^1, \dots, x^n),$$
$$\vdots$$
$$y^n = y^n(x^1, \dots, x^n).$$

But here $y^i = y^i(x^1, \dots, x^n)$ is nothing but the new coordinate functions expressed in terms of the old coordinates! A simple application of (6.59) then gives

$$\boldsymbol{d}y^i = \frac{\partial y^i}{\partial x^j}\, \boldsymbol{d}x^j. \tag{6.61}$$

Up to this point we have been working with a manifold *without a metric*. At each point of the manifold we have learned how to construct two n-dimensional vector spaces $\mathbf{T}_P(M)$ and $\mathbf{T}_P^*(M)$. Let us now assume that we have also been given a metric $\boldsymbol{g}$ on the manifold M. Then we can do something strange: We can fuse the two vector spaces $\mathbf{T}_P(M)$ and $\mathbf{T}_P^*(M)$ together so that they become indistinguishable! I.e., we may construct an identification procedure where each tangent vector $\mathbf{v}_P$ is identified with a unique covector ω_P in a purely geometrical manner.

To see how this can be done, let us choose a tangent vector $\mathbf{v}_P$. Then we construct a linear map, $\mathbf{T}_P(M) \curvearrowright \mathbb{R}$, in the following way:

$$\mathbf{u}_P \curvearrowright \boldsymbol{g}(\mathbf{v}_P; \mathbf{u}_P).$$

But a linear map, $\mathbf{T}_P(M) \curvearrowright \mathbb{R}$, is nothing but a covector, and therefore our tangent vector $\mathbf{v}_P$ has generated a covector, which will be denoted by

$$\boldsymbol{g}(\mathbf{v}_P; \cdot).$$

Let us introduce a coordinate system around P. Let $\mathbf{v}_P$ have the coordinates a^i. What are the coordinates of the corresponding covector $\boldsymbol{g}(\mathbf{v}_P; \cdot)$? They are easily found. Denoting them by a_i, we get

$$a_i = \boldsymbol{g}(\mathbf{v}_P, \mathbf{e}_i) = g_{jk} a^j \delta_i^k = g_{ji} a^j. \tag{6.62}$$

Thus the tangent vector with coordinates a^i generates the covector with coordinates $a_i = g_{ij} a^j$.

The map $I : a^i \curvearrowright a_i = g_{ij} a^j$, is a vector-space isomorphism between $\mathbf{T}_P(M)$ and $\mathbf{T}_P^*(M)$ because g_{ij} is a regular matrix. Therefore, we may use this map to identify $\mathbf{T}_P(M)$ with $\mathbf{T}_P^*(M)$. This is called the *canonical identification*!

Exercise 6.8.1
Problem: Let ω_P be a covector characterized by the coordinates ω_i. Show that the corresponding tangent vector is characterized by the components ω^i, where

$$\omega^k = g^{ki}\omega_i. \tag{6.63}$$

It should be emphasized that tangent vectors and covectors were born in different vector spaces with different geometrical properties. If there is a matrix g on the manifold, we may, however, identify them. Hence we will often see that when there is a metric, one simply "forgets" the dual space, i.e., the covectors. But then we have only tangent vectors at our disposal. Given a tangent vector, we can now characterize it by two sets of numbers:

a. We can decompose $\mathbf{v}_P$ along the canonical frame vectors $\mathbf{e}_i$; i.e.,

$$\mathbf{v}_P = a^i \mathbf{e}_i,$$

where the numbers a^i are referred to as the *contravariant components*.

b. We can use the numbers

$$a_i = g(\mathbf{v}_P, \mathbf{e}_i).$$

They are referred to as the *covariant components*. If $\mathbf{e}_i$ is a unit vector, then a_i is simply the orthogonal projection of $\mathbf{v}_P$ onto $\mathbf{e}_i$! (See Figure 6.58.)

Due to the rules

$$a_i = g_{ij}a^i \quad \text{and} \quad a^i = g^{ij}a_j, \tag{6.64}$$

we often say that *the metric coefficients are used to "raise and lower indices"*!

ILLUSTRATIVE EXAMPLE. *The gauge potential outside a solenoid*

Let us consider the Minkowski space M. Here we have a metric g, so we do not have to distinguish between tangent vectors and covectors. A vector field $\mathbf{A}$ could, for instance, represent the gauge potential in electromagnetism. We have previously computed the components of the gauge potential outside a solenoid in an inertial frame (Section 1.4).

$$\mathbf{A} : \frac{B_0 a^2}{2}\left[0, -\frac{y}{x^2+y^2}, \frac{x}{x^2+y^2}, 0\right].$$

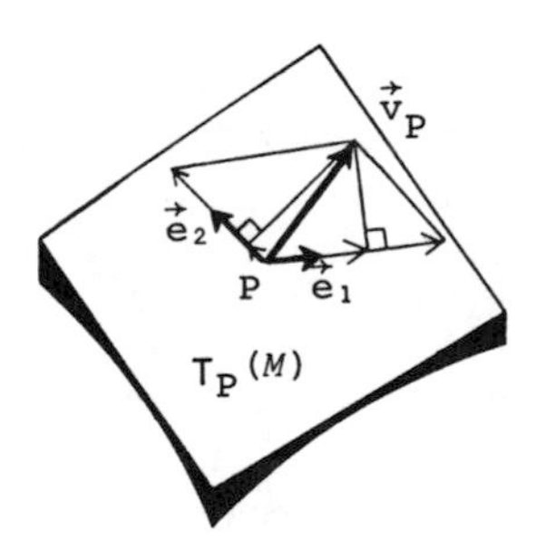

Figure 6.58.

The spatial part was circulating around the z-axis. This makes it natural to introduce cylindrical coordinates (t, ρ, φ, z). After all, there is nothing sacred about inertial coordinates in special relativity! In Exercise 6.6.3 we have computed the Jacobian matrices for the transition functions and the metric coefficients. We are therefore ready to find the contravariant components of the potential in cylindrical coordinates. Using formula (6.8), we get

$$A^{\alpha}_{(2)} = \frac{B_0 a^2}{2} \cdot \begin{bmatrix} 1 & 0 & 0 & 0 \\ 1 & \cos\varphi & \sin\varphi & 0 \\ 0 & -\frac{\sin\varphi}{\rho} & \frac{\cos\varphi}{\rho} & 0 \\ 0 & 0 & 0 & 1 \end{bmatrix} \cdot \begin{bmatrix} 0 \\ -\frac{\sin\varphi}{\rho} \\ \frac{\cos\varphi}{\rho} \\ 0 \end{bmatrix} = \frac{B_0 a^2}{2} \begin{bmatrix} 0 \\ 0 \\ \frac{1}{\rho^2} \\ 0 \end{bmatrix},$$

but as we know the metric coefficients (6.39), we can easily find the covariant components in cylindrical coordinates. Using that

$$A_\alpha = g_{\alpha\beta} A^\beta, \qquad \text{i.e., } \overline{A}_{\text{cov}} = \overline{\overline{G}} \overline{A}_{\text{contrav.}},$$

we then obtain

$$A_t = 0; \quad A_\rho = 0; \quad A_\varphi = \frac{B_0 a^2}{2}; \quad A_z = 0.$$

We can also compute the square of the gauge potential A using cylindrical coordinates:

$$\mathbf{g}(\mathbf{A}, \mathbf{A}) = A_\alpha A^\alpha = \left[\frac{B_0 a^2}{2}\right]^2 \cdot [0, 0, 1, 0] \cdot \begin{bmatrix} 0 \\ 0 \\ \frac{1}{\rho^2} \\ 0 \end{bmatrix} = \left[\frac{B_0 a^2}{2}\right]^2 \frac{1}{\rho^2}.$$

Thus $\mathbf{A}$ is a space–like vector, and its magnitude falls off like $1/\rho$.

If ϕ is a scalar field in the Minkowski space M, we may generate a covector field by taking the exterior derivative $\boldsymbol{d}\phi$. This is the gradient field corresponding to ϕ, and in terms of coordinates it is represented by $\partial_\mu \phi$. Observe that while $\partial_\mu \phi$ consists of the ordinary derivatives

$$\partial_\mu \phi = \frac{\partial \phi}{\partial x^\mu},$$

the contravariant components $\partial^\mu \phi$ have a more complicated structure involving the metric coefficients

$$\partial^\mu \phi = g^{\mu\nu} \partial_\nu \phi = g^{\mu\nu} \frac{\partial \phi}{\partial x^\nu}.$$

Thus in spherical coordinates, for instance, we get (compare Section 6.6)

$$\partial^t \phi = \frac{-\partial \phi}{\partial t}; \quad \partial^r \phi = \frac{\partial \phi}{\partial r}; \quad \partial^\theta \phi = r^2 \frac{\partial \phi}{\partial \theta}; \quad \partial^\varphi \phi = r^2 \sin^2\theta \frac{\partial \phi}{\partial \varphi}.$$

Returning to the potential $\mathbf{A}$, we can now write down a gauge transformation in a coordinate-free manner as follows:

$$\mathbf{A} \rightarrow \mathbf{A} + \boldsymbol{d}\chi, \tag{6.65}$$

where χ is an arbitrary scalar field. (If we write it out in coordinates, we get

$$A_\alpha \to A_\alpha + \partial_\alpha \chi,$$

which coincides with a gauge transformation in inertial coordinates. Therefore, it is valid in an arbitrary coordinate system.)

To see how (6.65) works, let us consider the potential outside a solenoid, and let us work in cylindrical coordinates:

$$(A_t, A_\rho, A_\varphi, A_z) = \frac{B_0 a^2}{2}\,(0, 0, 1, 0).$$

As the scalar function we choose the coordinate function φ multiplied by the constant $-\frac{B_0 a^2}{2}$; i.e.,

$$\chi(t, \rho, \varphi, z) = -\frac{B_0 a^2}{2}\,\varphi.$$

Then the "differential" $\boldsymbol{d}\chi$ gets the coordinates

$$\left(\frac{\partial \chi}{\partial t}, \frac{\partial \chi}{\partial \rho}, \frac{\partial \chi}{\partial \varphi}, \frac{\partial \chi}{\partial z}\right) = \frac{B_0 a^2}{2}\,(0, 0, 1, 0),$$

and $\mathbf{A} + \boldsymbol{d}\chi$ the coordinates

$$\frac{B_0 a^2}{2}\,(0, 0, -1, 0) + \frac{B_0 a^2}{2}\,(0, 0, 1, 0) = \mathbf{0}.$$

This does not mean that we have managed to gauge away $\mathbf{A}$ completely. The coordinate function φ is *not* defined throughout the whole of space–time. We must exclude a half-plane bounded by the x-axis, where φ makes a jump. Hence all we have shown is that if we cut away a half-plane bounded by the z-axis, then we can gauge away $\mathbf{A}$ in the remaining part of space–time. So we have arrived at the same conclusion as we did the first time (Section 1.4). At the same time, we have learned an important lesson:

Every time we use a cyclic coordinate (i.e., a periodic coordinate), there is a singularity associated with this coordinate, and the corresponding coordinate function does not *define a smooth scalar field on the whole of our manifold.*

§ 6.9 Tensors

By now we should begin to have a feeling of how to construct the geometrical concepts associated with a manifold. We consider linear maps to be the natural geometric concept associated with the tangent space. Metrics were constructed as bilinear maps on the tangent space, and covectors were constructed as linear maps on the tangent space.

We can now generalize the concepts of a metric and a covector. Consider a map $\mathbf{F}_P(.;\ldots;.)$ with k arguments. The arguments are all tangent vectors, $\mathbf{v}_P^{(1)}, \ldots, \mathbf{v}_P^{(k)}$, and $\mathbf{F}_P$ is suppose to map such a k-tuple $(\mathbf{v}_P^{(1)};\ldots;\mathbf{v}_P^{(k)})$ into a real

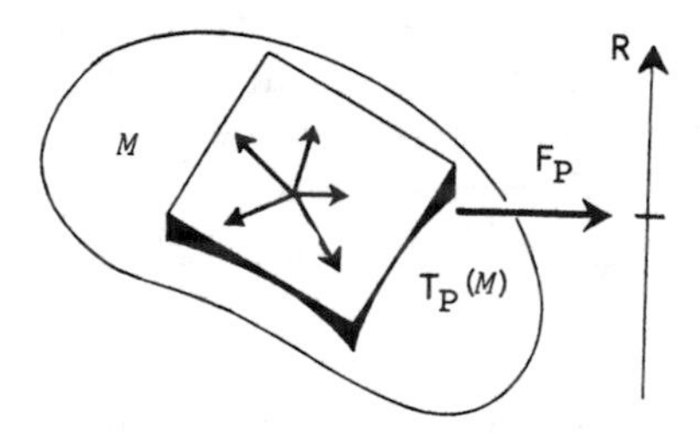

Figure 6.59.

number $\mathbf{F}_P(\mathbf{v}_P^{(1)}; \ldots, \mathbf{v}_P^{(k)})$. (See Figure 6.59.) The set of all k-tuples $(\mathbf{v}_P^{(1)}; \ldots; \mathbf{v}_P^{(k)})$ is denoted by

$$\underbrace{\mathbf{T}_P(M) \times \cdots \times \mathbf{T}_P(M)}_{k \text{ factors}},$$

and thus $\mathbf{F}_P$ is defined as a map

$$\mathbf{F}_P : \mathbf{T}_P(M) \times \cdots \times \mathbf{T}_P(M) \curvearrowright \mathbb{R}.$$

For it to be of any interest, we will, of course, assume that $\mathbf{F}_P$ is linear in each of its arguments:

$$\mathbf{F}_P(\mathbf{v}_P^{(1)}; \ldots; \lambda \mathbf{u}_P + \mu \mathbf{w}_P; \ldots; \mathbf{v}_P^{(k)}) = \lambda \mathbf{F}_P(\mathbf{v}^{(1)}; \ldots; \mathbf{u}_P; \ldots; \mathbf{v}_P^{(k)})$$
$$+ \mu \mathbf{F}_P(\mathbf{v}_P^{(1)}; \ldots; \mathbf{w}_P; \ldots; \mathbf{v}_P^{(k)}).$$

We express this property of $\mathbf{F}_P$ by saying that it is a *multilinear map*. This is the kind of map we will be interested in, and we will give it a special name:

Definition 6.14. A *cotensor* $\mathbf{F}_P$ is a multilinear map

$$\mathbf{F}_P : \mathbf{T}_P(M) \times \cdots \times \mathbf{T}_P(M) \curvearrowright \mathbb{R}.$$

If $\mathbf{F}_P$ has k arguments, we say that $\mathbf{F}_P$ is a cotensor of *rank* k, and the set of all cotensors of rank k will be denoted by $\mathbf{T}^{(0,k)}(M)$.

Clearly, a covector $\boldsymbol{\omega}_P$ is nothing but a cotensor of rank 1, and a metric $\boldsymbol{g}_P$ is a symmetric, nondegenerate cotensor of rank 2.

The next thing we must investigate is the possibility of introducing "coordinates" to facilitate computations with cotensors (and to show that these cotensors actually correspond to the tensor concept familiar from ordinary physics). For simplicity, we will illustrate the machinery with cotensors of rank 3.

Let us start by introducing coordinates $(x^1, \ldots, x^n)$ around the point P in M. If $\mathbf{S}_P$ is a cotensor associated with the tangent space $\mathbf{T}_P(M)$, we will refer to the numbers

$$S_{ijk} = \mathbf{S}_P(\mathbf{e}_i; \mathbf{e}_j; \mathbf{e}_k) \tag{6.66}$$

as *the components of* $\mathbf{S}_P$ *with respect to the coordinates* $(x^1, \ldots, x^n)$. Let $\mathbf{u}_P, \mathbf{v}_P, \mathbf{w}_P$ be three arbitrary tangent vectors with the coordinates a^i, b^j, c^k respectively. We can then easily express the value of $\mathbf{S}_P(\mathbf{u}_P; \mathbf{v}_P; \mathbf{w}_P)$ in terms of the components of $\mathbf{S}_P$ and the coordinates of $\mathbf{u}_P$, $\mathbf{v}_P$ and $\mathbf{w}_P$:

$$\begin{aligned}\mathbf{S}_P(\mathbf{u}_P; \mathbf{v}_P; \mathbf{w}_P) &= \mathbf{S}_P(a^i\mathbf{e}_i; b^j\mathbf{e}_j; c^k\mathbf{e}_k)\\ &= a^ib^jc^k\mathbf{S}_P(\mathbf{e}_i; \mathbf{e}_j; \mathbf{e}_k) = S_{ijk}a^ib^jc^k.\end{aligned}$$

Observe that although the specific components S_{ijk} and the coordinates a^i, b^j, c^k depend strongly on the coordinate system, the number $S_{ijk}a^ib^jc^k$ is *independent of the coordinate system*! This is an example of the following important rule: *Whenever we contract all upper indices and lower indices, the resulting number is an invariant; i.e., it is independent of the coordinate system employed.*

Let us check how the components of $\mathbf{S}_P$ transform under an exchange from x-coordinates to y-coordinates,

$$\begin{aligned}S_{(2)_{abc}} &= \mathbf{S}_P(\mathbf{e}_{(2)_a}; \mathbf{e}_{(2)_b}; \mathbf{e}_{(2)_c}) = \mathbf{S}_P\left(\mathbf{e}_{(1)_i}\frac{\partial x^i}{\partial y^a}; \mathbf{e}_{(1)_j}\frac{\partial x^j}{\partial y^b}; \mathbf{e}_{(1)_k}\frac{\partial x^k}{\partial y^c}\right)\\ &= \mathbf{S}_P(\mathbf{e}_{(1)_i}; \mathbf{e}_{(1)_j}; \mathbf{e}_{(1)_k})\frac{\partial x^i}{\partial y^a}\frac{\partial x^j}{\partial y^b}\frac{\partial x^k}{\partial y^c}\\ &= S_{(1)_{ijk}}\frac{\partial x^i}{\partial y^a}\frac{\partial x^j}{\partial y^b}\frac{\partial x^k}{\partial y^c},\end{aligned}$$

where we have used (6.12) and the multilinearity of $\mathbf{S}_P$. Thus we see that the components transform as covariant components (which justifies the name *co*tensor):

$$S_{(2)_{abc}} = S_{(1)_{ijk}}\frac{\partial x^i}{\partial y^a}\frac{\partial x^j}{\partial y^b}\frac{\partial x^k}{\partial y^c}. \tag{6.67}$$

We have various possibilities for manipulating tensors:

1. *the sum of two cotensors*:
   If $\mathbf{S}_P$ and $\mathbf{T}_P$ are cotensors of the same rank k, then there sum $\mathbf{S}_P + \mathbf{T}_P$ is defined as the following cotensor of the same rank k,

   $$(\mathbf{S}_P + \mathbf{T}_P)(\mathbf{u}_1; \ldots; \mathbf{u}_k) = \mathbf{S}_P(\mathbf{u}_1; \ldots; \mathbf{u}_k) + \mathbf{T}_P(\mathbf{u}_1; \ldots; \mathbf{u}_k),$$

   and the sum is characterized by the components

   $$S_{i_1\cdots i_k} + T_{i_1\cdots i_k}.$$

2. *Product of a cotensor and scalar*:
   If $\mathbf{S}_P$ is a cotensor of rank k and λ is a scalar, then we can define a new cotensor of rank k, $\lambda\mathbf{S}_P$, through the formula

   $$(\lambda\mathbf{S}_P)(\mathbf{u}_1; \ldots; \mathbf{u}_k) = \lambda\mathbf{S}_P(\mathbf{u}_1; \ldots; \mathbf{u}_k).$$

   The product of a cotensor and a scalar is characterized by the components $\lambda S_{i_1\ldots i_k}$.

3. *Tensor product*:

   If $\mathbf{S}_P$ and $\mathbf{T}_P$ are cotensors of rank s and t, then their tensor product, $\mathbf{S}_P \otimes \mathbf{T}_P$ is defined as the following cotensor of rank $s+t$:

$$(\mathbf{S}_P \otimes \mathbf{T}_P)(\mathbf{u}_1; \dots; \mathbf{u}_s; \mathbf{v}_1; \dots; \mathbf{v}_t) = \mathbf{S}_P(\mathbf{u}_1; \dots; \mathbf{u}_s)\mathbf{T}_P(\mathbf{v}_1; \dots; \mathbf{v}_t).$$

   The tensor product has the components

$$S_{i_1 \dots i_s} T_{j_1 \dots j_t}.$$

4. *Contraction*:

   If $\mathbf{T}_P$ is a cotensor of rank k and $\mathbf{v}_{(1)}, \dots, \mathbf{v}_{(\ell)}$ are tangent vectors, then their contraction is defined as the following cotensor of rank $(k-\ell)$,

$$\mathbf{T}_P(\mathbf{v}_{(1)}; \dots; \mathbf{v}_{(\ell)}; \dots),$$

   and the contraction is characterized by the components

$$T_{i_1 \dots i_\ell i_{\ell+1} \dots i_k} a^{i_1}_{(1)} \dots a^{i_\ell}_{(\ell)}.$$

It is important to observe that the cotensor space $\mathbf{T}_P^{(0,k)}(M)$ actually carries a linear structure; i.e., it is a vector space.

Once we have the tensor product at our disposal, we can discuss how to build up cotensors of arbitrary rank systematically using the covectors as building blocks.

We have previously studied the space of covectors $\mathbf{T}_P^*(M) = \mathbf{T}_P^{(0,1)}(M)$; and especially we found the canonical frame vectors $\boldsymbol{dx}^1; \dots; \boldsymbol{dx}^n$. These are the covectors that we shall use as the basic units!

Consider first an arbitrary covector $\boldsymbol{\omega}_P$. If it has the components ω_i, we can decompose it in the following way:

$$\boldsymbol{\omega}_P = \omega_i \boldsymbol{dx}^i. \tag{6.58}$$

Next, we consider a cotensor $\mathbf{F}_P$ of rank 2! Let it have the components $F_{k\ell}$. Then we can construct another cotensor of rank 2,

$$F_{ij} \boldsymbol{dx}^i \otimes \boldsymbol{dx}^j.$$

What are the components of this cotensor?

$$F_{ij} \boldsymbol{dx}^i \otimes \boldsymbol{dx}^j(\mathbf{e}_k; \mathbf{e}_\ell) = F_{ij} \boldsymbol{dx}^i(\mathbf{e}_k) \cdot \boldsymbol{dx}^j(\mathbf{e}_\ell) = F_{ij}\delta^i_k \delta^j_\ell = F_{k\ell}.$$

Hence $F_{ij}\boldsymbol{dx}^i \otimes \boldsymbol{dx}^j$ has the components $F_{k\ell}$ too, and we conclude that $\mathbf{F}_P$ and $F_{ij}\boldsymbol{dx}^i \otimes \boldsymbol{dx}^j$ are identical cotensors; i.e.,

$$\mathbf{F}_P = F_{ij} \boldsymbol{dx}^i \otimes \boldsymbol{dx}^j. \tag{6.68}$$

This is an important result: Any cotensor of rank 2 can be expressed as a linear combination of the cotensors

$$\boldsymbol{dx}^1 \otimes \boldsymbol{dx}^1; \boldsymbol{dx}^2 \otimes \boldsymbol{dx}^2; \dots; \boldsymbol{dx}^n \otimes \boldsymbol{dx}^n.$$

Therefore, the cotensors $\boldsymbol{dx}^i \otimes \boldsymbol{dx}^j$ generate the whole tensor space $\mathbf{T}_P^{(0,2)}(M)$, so we see that they act as canonical frame tensors.

Now we can guess the rest of the story! Let $\mathbf{T}_P$ be a cotensor of rank 3 with components T_{ijk}; we can then decompose it in the following way:

$$\mathbf{T}_P = T_{ijk}\boldsymbol{dx}^i \otimes \boldsymbol{dx}^j \otimes \boldsymbol{dx}^k, \qquad \text{etc!} \tag{6.69}$$

Consider the metric $\boldsymbol{g}$. It is a cotensor of rank 2. If we choose a coordinate system, then all cotensors of rank 2 can be expanded in terms of the basic cotensors $\boldsymbol{dx}^\alpha \otimes \boldsymbol{dx}^\beta$. In particular, we may expand the metric tensor

$$\boldsymbol{g} = g_{\alpha\beta}\boldsymbol{dx}^\alpha \otimes \boldsymbol{dx}^\beta. \tag{A}$$

This formula is interesting, because formally it looks very much like the "square of the line element"

$$-d\tau^2 = g_{\alpha\beta}dx^\alpha dx^\beta. \tag{B}$$

But (A) has an exact meaning: It states that two well-defined cotensors are identical, while (B) only has a symbolic meaning. It is a mnemonic rule of how to write down the arc-length of a curve:

$$\int_{\tau_1}^{\tau_2} d\tau = \int_{\lambda_1}^{\lambda_2} \sqrt{-g_{\alpha\beta}(x(\lambda))\frac{dx^\alpha}{d\lambda}\frac{dx^\beta}{d\lambda}}\,d\lambda$$

(or even worse: dt is an infinitesimal arc-length, and dt^2 is the square of an infinitesimal quantity, whatever that is).

Hence we see again that the formulas of modern differential geometry very often resemble formulas from "old days" involving "infinitesimal quantities" and other mysterious things. Using the notation $\boldsymbol{g} = g_{\alpha\beta}\boldsymbol{dx}^\alpha \otimes \boldsymbol{dx}^\beta$ we can write the metric on a 2-sphere as (6.26)

$$\boldsymbol{g} = r^2(\boldsymbol{d\theta} \otimes \boldsymbol{d\theta} + \sin^2\theta\boldsymbol{d\varphi} \otimes \boldsymbol{d\varphi})$$

and the metric on the Minkowski space expressed in spherical coordinates as (6.36)

$$\boldsymbol{g} = -\boldsymbol{dt} \otimes \boldsymbol{dt} + \boldsymbol{dr} \otimes \boldsymbol{dr} + r^2(\boldsymbol{d\theta} \otimes \boldsymbol{d\theta} + \sin^2\theta\boldsymbol{d\varphi} \otimes \boldsymbol{d\varphi}).$$

Let us look at a point P on a manifold M^n. To this point P we have attached the tangent space $\mathbf{T}_P(M)$, and the tangent space has been used to generate all the cotensor spaces. But to the point P we have also attached the cotangent space $\mathbf{T}^*_P(M)$, and that is an n-dimensional vector space with a structure very similar to that of the tangent space. We can therefore generalize the discussion of cotensors in the following way:

Definition 6.15. A mixed tensor $\mathbf{F}_P$ of type (k, ℓ) is a multilinear map

$$\mathbf{F}_P : \underbrace{\mathbf{T}^*_P(M) \times \cdots \times \mathbf{T}^*_P(M)}_{k-\text{factors}} \times \underbrace{\mathbf{T}_P(M) \times \cdots \times \mathbf{T}_P(M)}_{\ell-\text{factors}} \to \mathbb{R}$$

A mixed tensor of type (k, ℓ) has rank $k + \ell$, and the space of all mixed tensors of type (k, ℓ) is denoted $\mathbf{T}^{(k,\ell)}_P(M)$.

A mixed tensor of type $(k, 0)$ is called simply a *tensor of rank k*. Let us focus attention on the tensors of rank 1. A tensor $\mathbf{A}_P$ of rank 1 is characterized by its components $A^i = A_P(\boldsymbol{dx}^i)$, and these components transform contravariantly. Thus there seems to be an intimate connection between tensors of rank 1 and tangent vectors. In fact, we can identify them:

Let $\mathbf{v}_P$ be a tangent vector. Then the map

$$\omega_P \rightarrow \langle \omega_P | \mathbf{v}_P \rangle$$

is a linear map $\mathbf{T}^*_P(M) \curvearrowright \mathbb{R}$, since the bracket is bilinear. Therefore, $\mathbf{v}_P$ generates a unique tensor of rank 1, which we also denote by $\mathbf{v}_P$.

Exercise 6.9.1
Problem: Consider a tangent vector $\mathbf{v}_P$. Show that the components of $\mathbf{v}_P$ as a tangent vector and the components of $\mathbf{v}_P$ as a tensor of rank 1 are identical.

When we have tensors of an arbitrary type at our disposal, we can generalize the contractions. Let, for instance, $\mathbf{T}_P$ be a tensor of type $(1, 2)$. It is then characterized by its components

$$T^a_{\ bc} = \mathbf{T}_P(\boldsymbol{dx}^a; \mathbf{e}_b; \mathbf{e}_c)$$

with respect to some coordinates $(x^1, \ldots, x^n)$. Here the index a transforms contravariantly, and the indices b and c transform covariantly. But then we may contract the indices a and b, obtaining the quantity

$$T^a_{\ ac}.$$

As we sum over the index a, this quantity is characterized by only one index:

$$S_c = T^a_{\ ac}.$$

The important point is now that the quantity S_c transforms covariantly! This follows from the computation

$$\begin{aligned} S_{(2)_c} = T^a_{(2)_{ac}} &= \frac{\partial y^a}{\partial x^i} T^i_{(1)_{jk}} \frac{\partial x^j}{\partial y^a} \frac{\partial x^k}{\partial y^c} \\ &= \delta^j_i T^i_{(1)_{jk}} \frac{\partial x^k}{\partial y^c} = T^i_{(1)_{ik}} \frac{\partial x^k}{\partial y^c} = S_{(1)_k} \frac{\partial x^k}{\partial y^c}, \end{aligned}$$

where we have used that $\left(\frac{\partial y^a}{\partial x^i}\right)$ and $\left(\frac{\partial x^j}{\partial y^a}\right)$ are reciprocal matrices. But if S_c transforms covariantly, we may regard it as the components of a cotensor $\mathbf{S}_P$. We say that $\mathbf{S}_P$ is generated from $\mathbf{T}_P$ by contraction in the first two variables. This is obviously a general rule: *Whenever we contract an upper index with a lower index, the resulting quantity transforms as the components of a tensor* (where the degree, of course, is lowered by two!). The only trouble with contractions is that they are almost impossible to write in a coordinate-free manner! As contractions play an important role in applications, we will therefore write down many equations involving tensors using component notation.

In the following table we have summarized the most important properties of mixed tensors:

Components	A mixed tensor T_P of type (k, ℓ) is completely determined by its components $$T^{i_1 \dots i_k}{}_{j_1 \dots j_\ell} = T_P\left(dx^{i_1}; \dots; dx^{i_k}; e_{j_1}; \dots; e_{j_\ell}\right)$$ with respect to a coordinate system. The indices $i_1, \dots, i_k$ transform contravariantly, and the indices $j_1, \dots, j_\ell$ transform covariantly.	(6.70)
Linear structure	If S_P and T_P are mixed tensors of the same type (k, ℓ), and λ is a real scalar, then we can form the mixed tensors $S_P + T_P$ and λT_P of the same type (k, ℓ). Furthermore, these mixed tensors are characterized by the components $$S^{i_1 \dots i_k}{}_{j_1 \dots j_\ell} + T^{i_1 \dots i_k}{}_{j_1 \dots j_\ell};\ \lambda T^{i_1 \dots i_k}{}_{j_1 \dots j_\ell}.$$	(6.71)
Tensor product	If S_P and T_P are mixed tensors of type (k_1, ℓ_1) and (k_2, l_2), we can form the mixed tensor $S_P \otimes T_P$ of type $(k_1 + k_2, \ell_1 + \ell_2)$. The tensor product is characterized by the components $$S^{i_1 \dots i_k}{}_{j_1 \dots j_\ell} T^{i_1 \dots i_k}{}_{j_1 \dots j_\ell}.$$	(6.72)
Contraction	If S_P is a mixed tensor of type (k, ℓ), we can form a mixed tensor T_P of type $(k-1, \ell-1)$ by contracting a contravariant and a covariant index. The contraction is characterized by the components $$T^{i_2 \dots i_k}{}_{j_2 \dots j_\ell} = S^{i i_2 \dots i_k}{}_{i j_2 \dots j_\ell}.$$	(6.73)

In what follows we are going to deal extensively with *tensor fields*. To construct a tensor field $\mathbf{T}$ of rank 3 we attach to each point P in our manifold a tensor $\mathbf{T}_P$ of rank 3! Let us introduce coordinates $(x^1, \dots, x^n)$ on M. Then the tensor at the point $P(x^1, \dots, x^n)$ is characterized by its components

$$T^{abc}(x^1, \dots, x^n).$$

We say that the tensor field $\mathbf{T}$ is a *smooth* field if the components $T^{abc}(x^1, \dots, x^n)$ are smooth functions of the coordinates. If nothing else is stated, we will always assume the tensor fields to be smooth.

ILLUSTRATIVE EXAMPLE. *The unit tensor field*

Let M be an arbitrary manifold. Then we construct a mixed tensor $\mathbf{T}$ of type $(1, 1)$ in the following way. At each point P we consider the bilinear map

$$\mathbf{T}_P(\boldsymbol{\omega}_P; \mathbf{v}_P) = \langle \boldsymbol{\omega}_P | \mathbf{v}_P \rangle.$$

The corresponding components are

$$\mathbf{T}_P(\boldsymbol{d}x^i; \mathbf{e}_j) = \langle \boldsymbol{d}x^i | \mathbf{e}_j \rangle = \delta^i_j,$$

i.e., the Kronecker delta! It is a remarkable fact that the components of this tensor have the same values in *all* coordinate systems. As the components are constant throughout the manifold, they obviously depend smoothly upon the underlying coordinates! We therefore conclude that

The Kronecker delta $\delta^i_{\ j}$ are the components of a smooth tensor field on M, called the unit tensor field of type $(1, 1)$.

Exercise 6.9.2
Problem: Show that the Christoffel fields

$$\Gamma^{\nu}_{\ \alpha\beta} = \frac{1}{2} g^{\nu\mu} \left[\frac{\partial g_{\mu\alpha}}{\partial x^{\beta}} + \frac{\partial g_{\beta\mu}}{\partial x^{\alpha}} - \frac{\partial g_{\alpha\beta}}{\partial x^{\mu}} \right] \tag{6.74}$$

are *not* the components of a mixed tensor field. (Hint: Show that they do not transform homogeneously under a coordinate transformation.)

Exercise 6.9.3
Problem: Consider a point P on our manifold. To each coordinate system $(x^1, \ldots, x^n)$ around P we attach a quantity $T^{ij}_{\ \ k}$ with two upper indices and one lower index. A priori the upper indices need not transform contravariantly, and the lower index need not transform covariantly. Show the following:

If the quantities U^{ℓ}_k given by

$$U^{\ell}_{\ k} = T^{ij}_{\ \ k} S^{\ell}_{\ ij}$$

are the components of a mixed tensor of type $(1, 1)$ *whenever $S^{\ell}_{\ ij}$ are the components of a mixed tensor of type* $(1, 2)$, *then $T^{ij}_{\ \ k}$ are the components of a mixed tensor of type* $(2, 1)$.

The method outlined in Exercise 6.9.3 is very useful when we want to show that a given quantity transforms like a tensor. Clearly, it can be generalized to mixed tensors of arbitrary type.

Suppose now that we have attached a *metric* $\mathbf{g}$ to our manifold M. Then we have previously shown (Section 6.8) how to identify tangent vectors and covectors using the bijective linear map.

$$I : \mathbf{T}^*_P(M) \curvearrowright \mathbf{T}_P(M)$$

generated by the metric. If we write it out in components, it is given by

$$I : a_i \curvearrowright a^i = g^{ij} a_j .$$

Exercise 6.9.4
Problem: Show that $I(\boldsymbol{dx}^k)$ is characterized by the components g^{ki}.

We can now in a similar way identify all tensor spaces of the same rank. For simplicity, we sketch the idea using tensors of rank 2.

Let $\mathbf{T}$ be a cotensor of rank 2 characterized by the covariant components T_{ij}. Then we identify $\mathbf{T}$ with the following tensor of rank 2:

$$I_1(\mathbf{T})(\omega; \chi) \stackrel{\text{def}}{=} \mathbf{T}(I(\omega); \mathbf{I}(\chi)).$$

The tensor $I_1(\mathbf{T})$ is characterized by a set of contravariant components that we denote by T^{ij}. They are given by

$$T^{k\ell} = I_1(\mathbf{T})(\boldsymbol{dx}^k; \boldsymbol{dx}^\ell) = \mathbf{T}(I(\boldsymbol{dx}^k); I(\boldsymbol{dx}^\ell)) = T_{ij}g^{ki}g^{\ell j}.$$

In this way we have clearly constructed a bijective linear map

$$I_1 : \mathbf{T}_P^{(0,2)}(M) \curvearrowright \mathbf{T}_P^{(2,0)}(M).$$

We can also identify $\mathbf{T}$ with the following mixed tensor type of $(1, 1)$:

$$I_2(\mathbf{T})(\boldsymbol{\omega}; \mathbf{v}) \stackrel{\text{def}}{=} \mathbf{T}(I(\boldsymbol{\omega}); \mathbf{v}).$$

Here $I_2(\mathbf{T})$ is characterized by the components

$$T^k_\ell = I_2(\mathbf{T})(\boldsymbol{dx}^k; \mathbf{e}_\ell) = \mathbf{T}(I(\boldsymbol{dx}^k); \mathbf{e}_\ell) = T_{ij}g^{ki}\delta^j_\ell = T_{i\ell}g^{ki}.$$

In this way we have generated a bijective linear map

$$I_2 : \mathbf{T}^{(0,2)}(M) \rightarrow \mathbf{T}^{(1,1)}(M).$$

On a manifold M with a metric $\boldsymbol{g}$ we have therefore a *canonical identification* of the tensor space of rank 2:

$$\mathbf{T}_P^{(0,2)}(M) \stackrel{\boldsymbol{g}}{\simeq} \mathbf{T}_P^{(1,1)}(M) \stackrel{\boldsymbol{g}}{\simeq} \mathbf{T}_P^{(2,0)}(M).$$

Obviously, these results generalize immediately to all tensors of higher rank.

If there is a metric $\boldsymbol{g}$ on our manifold M, it is customary not to distinguish among the various types of tensors of the same rank. One simply speaks of tensors, and a given tensor $\mathbf{T}$ is then represented by various types of components. For instance, a tensor $\mathbf{T}$ of rank 2 is represented by the following four types of components:

$$T_{ij}, \quad T^i{}_j, \quad T_i{}^j, \quad T^{ij},$$

and these components are related to each other through the formulas

$$\begin{aligned} T^i{}_j &= g^{ik}T_{kj} \qquad & T_i{}^j &= g^{kj}T_{ik} \qquad & T^{ij} &= g^{ik}g^{\ell j}T_{k\ell} \\ T_{ij} &= g_{ik}T^k{}_j \qquad & T_{ij} &= g_{kj}T_i{}^k \qquad & T_{ij} &= g_{ik}g_{\ell j}T^{k\ell} \end{aligned} \tag{6.75}$$

This is known as the art of raising and lowering indices!

ILLUSTRATIVE EXAMPLE. *The metric cotensor.*

Let us apply this to the metric cotensor itself. Thus we want to find the mixed components of $\boldsymbol{g}$ and the contravariant components of $\boldsymbol{g}$. Raising one of the indices, we find that

$$g^i_j = g^{ik}g_{kj} = \delta^i_j.$$

Consequently, the mixed components are nothing but the Kronecker delta! As we have already used the notation g^{ij} to denote the reciprocal matrix, we will for a moment use $\tilde{g}^{ij}$ to denote the contravariant components of $\boldsymbol{g}$. Raising both of the indices, we find that

$$\tilde{g}^{ij} = g^{ik}g^{\ell j}g_{k\ell} = g^{ik}\delta^j_k = g^{ij}.$$

So we may relax: g^{ij} actually denotes the contravariant components of $\boldsymbol{g}$ too. Hence the metric $\boldsymbol{g}$ is characterized by each of the following components:

$$g_{ij}, \qquad \delta^i_j \quad \text{and} \quad g^{ij}. \tag{6.76}$$

where g_{ij} and g^{ij} are reciprocal matrices and δ^i_j is the unit matrix.

§ 6.10 Tensor Fields in Physics

To see how a physical quantity can be represented by a tensor, we will consider the electromagnetic field. It is characterized by the field strengths **E** and **B** measured in an inertial frame of reference. The components $E_x, \dots, B_z$ are collected into a skew-symmetric square matrix

$$F_{\alpha\beta} = \begin{bmatrix} 0 & -E_x & -E_y & -E_z \\ E_x & 0 & B_z & -B_y \\ E_y & -B_z & 0 & B_x \\ E_z & B_y & -B_x & 0 \end{bmatrix}.$$

If we exchange the inertial frame of reference $\mathbf{S}_1$, characterized by inertial coordinates x^α, with the inertial frame of reference $\mathbf{S}_2$, characterized by inertial coordinates y^β, then it is an experimental fact that the electromagnetic field strengths transform according to the rule

$$\mathbf{F}_{(2)_{\alpha\beta}} = \mathbf{F}_{(1)_{\gamma\delta}} \breve{A}^\gamma_\alpha \breve{A}^\delta_\beta, \tag{6.77}$$

where $(\breve{A})^\alpha_\beta$ is the reciprocal Lorentz matrix; i.e.,

$$(\breve{A})^\alpha_\beta = \frac{\partial x^\alpha}{\partial y^\beta}.$$

This allows us to define a cotensor field **F** of rank 2 on Minkowski space. let P be a space–time point and $\mathbf{u}_P$, $\mathbf{v}_P$ two tangent vectors. Then we define

$$\mathbf{F}_P(\mathbf{u}_P; \mathbf{v}_P) = F_{\alpha\beta} u^\alpha v^\beta, \tag{6.78}$$

where $F_{\alpha\beta}$, u^α, v^β are the components relative to an arbitrary inertial frame. That the number on the right-hand side is independent of the inertial frame follows immediately from the transformation rule (6.77). Thus in our mathematical model the electromagnetic field is represented by a skew-symmetric cotensor field **F** of rank 2.

Of course, we do not have to restrict ourselves to inertial frames. As a specific example let us consider a pure monopole field. Let us assume that the monopole is at rest at the origin of the inertial frame of reference $\mathbf{S}_1$. We want to find the components of the monopole field in terms of spherical coordinates (t, r, θ, φ). In the inertial frame the electric field strength **E** vanishes, and

$$\mathbf{B} = \frac{g}{4\pi} \cdot \frac{\mathbf{r}}{r^3},$$

where g is the strength of the monopole. Therefore,

$$F_{(1)\alpha\beta} = \frac{g}{4\pi r^2}\begin{bmatrix} 0 & 0 & 0 & 0 \\ 0 & 0 & \frac{z}{r} & -\frac{y}{r} \\ 0 & -\frac{z}{r} & 0 & \frac{x}{r} \\ 0 & \frac{y}{r} & -\frac{x}{r} & 0 \end{bmatrix}.$$

Using the transformation rule (6.67), we find

$$F_{(2)\alpha\beta} = F_{(1)\gamma\delta}\frac{\partial x^\gamma}{\partial y^\alpha}\frac{\partial x^\delta}{\partial y^\beta} = \left(\frac{\partial x^\gamma}{\partial y^\alpha}\right)F_{(1)\gamma\delta}\left(\frac{\partial x^\delta}{\partial y^\beta}\right);$$

i.e.,

$$\bar{\bar{F}}_{(2)} = (\bar{\bar{D}}_{12})^+\bar{\bar{F}}_{(1)}(\bar{\bar{D}}_{12}).$$

The Jacobian matrices were computed in Section 6.6. Inserting (6.35), we get

$$\bar{\bar{F}}_{(2)} = \frac{g}{4\pi}\begin{bmatrix} 0 & 0 & 0 & 0 \\ 0 & 0 & 0 & 0 \\ 0 & 0 & 0 & \sin\theta \\ 0 & 0 & -\sin\theta & 0 \end{bmatrix}. \tag{6.79}$$

Thus only the $F_{\theta\varphi}$-term is nonvanishing in this particularly simple coordinate system. The corresponding contravariant components become

$$F^{\alpha\beta} = g^{\alpha\gamma}g^{\delta\beta}F_{\gamma\delta} = g^{\alpha\gamma}F_{\gamma\delta}g^{\delta\beta};$$

i.e.,

$$\bar{\bar{F}}_{\text{contrav.}} = (\bar{\bar{G}}^{-1})\bar{\bar{F}}_{\text{covar.}}(\bar{\bar{G}}^{-1}).$$

We have already determined the metric coefficients in spherical coordinates; see (6.36):

$$\bar{\bar{G}} = \begin{bmatrix} -1 & 0 & 0 & 0 \\ 0 & 1 & 0 & 0 \\ 0 & 0 & r^2 & 0 \\ 0 & 0 & 0 & r^2\sin^2\theta \end{bmatrix}; \quad \text{i.e., } \bar{\bar{G}}^{-1} = \begin{bmatrix} -1 & 0 & 0 & 0 \\ 0 & 1 & 0 & 0 \\ 0 & 0 & \frac{1}{r^2} & 0 \\ 0 & 0 & 0 & \frac{1}{r^2\sin^2\theta} \end{bmatrix},$$

and from this we obtain the contravariant components of **F**:

$$F^{\alpha\beta}_{(2)} = \frac{g}{4\pi}\begin{bmatrix} 0 & 0 & 0 & 0 \\ 0 & 0 & 0 & 0 \\ 0 & 0 & 0 & \frac{1}{r^4\sin\theta} \\ 0 & 0 & \frac{1}{r^4\sin\theta} & 0 \end{bmatrix}. \tag{6.80}$$

Observe that we cannot find the "strength" of the field just by observing the covariant (or the contravariant) components of **F**. To do that we must form an invariant quantity. If we have a tensor **T** of rank 2, we can form the following two invariants:

a. The trace of $\mathbf{T} : T^\alpha_{\ \alpha}$.

b. The "square" of $\mathbf{T}$: $\frac{1}{2}\, T_{\alpha\beta}T^{\alpha\beta}$.

But the electromagnetic field tensor $\mathbf{F}$ is skew-symmetric, i.e., $\mathbf{F}(\mathbf{v}_P, \mathbf{u}_P) = -\mathbf{F}(\mathbf{u}_P, \mathbf{v}_P)$, so the trace vanishes automatically. Concerning the square, we get

$$\frac{1}{2}\, F_{\alpha\beta}F^{\alpha\beta} = \frac{1}{2}\left(\frac{g}{4\pi r^2}\right)^2 (1+1) = \left(\frac{g}{4\pi r^2}\right)^2 .$$

Therefore, the "strength" of the field depends only on the radial coordinate r, and it varies as $\frac{1}{r^2}$!

Illustrative Example. *The four-momentum and the energy–momentum tensor.*

As another example of a tensor in Minkowski space we consider the energy–momentum tensor $\mathbf{T}$. This is a symmetric tensor of rank 2. In an inertial frame the contravariant components $T^{\alpha\beta}$ have the well-known interpretation (1.37)

$$T^{\alpha\beta} = \begin{bmatrix} \text{Energy density} & \text{Energy current} \\ \text{Momentum density} & \text{Momentum currents} \end{bmatrix}$$

For a sourceless electromagnetic field $\mathbf{F}$ the energy–momentum tensor can depend only on $\mathbf{F}$; i.e., $\mathbf{T} = \mathbf{T}(\mathbf{F})$. In an inertial frame we have found the following relation (1.41):

$$T^{\alpha\beta} = -F^{\alpha\gamma}F^{\beta}_{\ \gamma} - \frac{1}{4}\,\eta^{\alpha\beta}(F^{\gamma}_{\ \delta}F^{\delta}_{\ \gamma}).$$

As $\eta^{\alpha\beta}$ are the contravariant components of the metric $\mathbf{g}$, we conclude that the general relation is

$$T^{\alpha\beta} = -F^{\alpha\gamma}F^{\beta}_{\ \gamma} - \frac{1}{4}\,g^{\alpha\beta}(F^{\gamma}_{\ \delta}F^{\delta}_{\ \gamma}).$$

(Since the left- and the right-hand sides are components of tensors that coincide in an inertial frame, they must be identical.)

By definition, the tensors are constructed as multilinear maps on the tangent space $\mathbf{T}_P(M)$ and the cotangent space $\mathbf{T}^*_P(M)$. It may seem strange to represent a physical quantity by a multilinear map, so let us look at some examples:

(A) Let us assume that we are observing a particle moving in space–time with a four-momentum $\mathbf{P}$. The four-momentum was introduced as a tangent vector along the world line

$$P^\alpha = m\,\frac{dx^\alpha}{d\tau},$$

but we can identify it with a covector $\mathbf{P}$, i.e., a cotensor of rank 1, and characterize it by its covariant components P_α. (See Figure 6.60.) As observers, we are ourselves moving through space–time along a certain world line with a certain four-velocity $\mathbf{U}$. As $\mathbf{P}$ is a linear map, $\mathbf{T}_P(M) \curvearrowright \mathbb{R}$, will map $\mathbf{U}$ into a real number $\langle \mathbf{P} \mid \mathbf{U} \rangle$. What is the physical interpretation of this number?

Let us choose an inertial frame where we are *momentarily at rest*. With respect to this coordinate system, $\mathbf{P}$ is described by the components P_α, where $-P_0$ *is the*

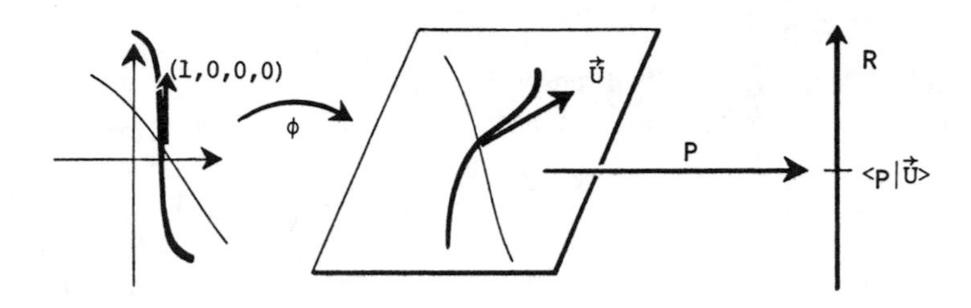

Figure 6.60.

energy that the observers will measure, because the observers are at rest relative to this inertial frame.

On the other hand, the four-velocity **U** will have the coordinate $U^\alpha = (1, 0, 0, 0)$ with respect to this inertial frame. We therefore get

$$\langle \mathbf{P}, \mathbf{U} \rangle = P_\alpha U^\alpha = P_0.$$

So we conclude that $-\langle \mathbf{P} \mid \mathbf{U} \rangle$ *is the energy of a particle with four-momentum* **P** *measured by an observer with four-velocity* **U**.

(B) Consider the energy–momentum tensor **T** of a certain field configuration. Let us assume that we are observing this field configuration and that we are moving with a certain four-velocity **U**. As **T** is a bilinear map,

$$\mathbf{T}_P(M) \times \mathbf{T}_P(M) \curvearrowright \mathbb{R},$$

it will map the pair of tangent vectors $(\mathbf{U}, \mathbf{U})$ into the real number $\mathbf{T}(\mathbf{U}, \mathbf{U})$. To find the physical interpretation of this number, we introduce again an inertial frame where we are momentarily at rest. With respect to this frame, **T** is characterized by the covariant components $T_{\alpha\beta}$, where T_{00} is the energy density of the field that *we* will measure! Our four-velocity, on the other hand, is characterized by the components $U^\alpha = (1, 0, 0, 0)$. Therefore, we find that

$$\mathbf{T}(\mathbf{U}, \mathbf{U}) = T_{\alpha\beta} U^\alpha U^\beta = T_{00},$$

and we conclude that $\mathbf{T}(\mathbf{U}, \mathbf{U})$ *is the energy density of a field with energy–momentum* **T** *measured by an observer with four-velocity* **U**.

Solutions to Worked Exercises

Solution to 6.5.1

Let $g^{ij}_{(2)}$ and $g^{ij}_{(2)}$ be the components with respect to two coordinate systems (ϕ_2, U_2). Then by definition,

$$g^{ij}_{(2)} g_{(2)jk} = \delta^i_k \qquad (\text{and } g^{ij}_{(1)} g_{(1)jk} = \delta^i_k).$$

But we know that g_{ij} transforms covariantly! Consequently,

$$g_{(2)_{jk}} = \frac{dx^\ell}{\partial y^j}\frac{\partial x^m}{\partial y^k}\, g_{(1)_{\ell m}}.$$

Substituting this, we get

$$g^{ij}_{(2)}\frac{\partial x^\ell}{\partial y^j}\frac{\partial x^m}{\partial y^k}\, g_{(1)_{\ell m}} = \delta^i_k.$$

Multiplying by the reciprocal matrix $\frac{\partial y^k}{\partial x^p}$, we get

$$g^{ij}_{(2)}\frac{\partial x^\ell}{\partial y^j}\,\delta^m_p\, g_{(1)_{\ell m}} = \delta^i_k\,\frac{\partial y^k}{\partial x^p}\,; \qquad \text{i.e., } g^{ij}_{(2)}\frac{\partial x^\ell}{\partial y^j}\, g_{(1)_{\ell p}} = \frac{\partial y^i}{\partial x^p}\,.$$

Multiplying by the reciprocal matrix $g^{pq}_{(1)}$, we further get

$$g^{ij}_{(2)}\frac{\partial x^\ell}{\partial y^j}\,\delta^q_\ell = \frac{\partial y^i}{\partial x^p}\, g^{pq}_{(1)}\,; \qquad \text{i.e., } g^{ij}_{(2)}\frac{\partial x^q}{\partial y^j} = \frac{\partial y^i}{\partial x^p}\, g^{pq}_{(1)}.$$

Multiplying by the reciprocal matrix $\partial y^k/\partial x^q$, we finally obtain

$$g^{ij}_{(2)}\delta^k_j = \frac{\partial y^i}{\partial x^p}\frac{\partial y^k}{\partial x^q}\, g^{pq}_{(1)}\,; \qquad \text{i.e., } g^{ik}_{(2)} = \frac{\partial y^i}{\partial x^p}\frac{\partial y^k}{\partial x^q}\, g^{pq}_{(1)}.$$

So finally, we get the formula we wanted ! After a lot of index gymnastique we can see with our own eyes that g^{ij} transforms contravariantly!

Solution to 6.7.1

Taking the Lagrangian

$$L\left(x^\alpha;\ \frac{dx^\alpha}{d\lambda}\right) = -m\sqrt{-g_{\alpha\beta}\frac{dx^\alpha}{d\lambda}\frac{dx^\beta}{d\lambda}}$$

as a starting point, the Euler–Lagrange equation $\frac{\partial L}{\partial x^\alpha} = \frac{d}{d\lambda}\left[\frac{\partial L}{\partial\left(\frac{dx^\alpha}{d\lambda}\right)}\right]$ reduces to

$$\begin{aligned}
\frac{\frac{\partial g_{\alpha\beta}}{\partial x^\mu}\frac{dx^\alpha}{d\lambda}\frac{dx^\beta}{d\lambda}}{2\sqrt{\dots}} &= \frac{d}{d\lambda}\left[\frac{g_{\alpha\beta}\delta^\alpha_\mu\frac{dx^\beta}{d\lambda} + g_{\alpha\beta}\frac{dx^\alpha}{d\lambda}\delta^\beta_\mu}{2\sqrt{\dots}}\right] \\
&= \frac{d}{d\lambda}\left[\frac{g_{\mu\beta}\frac{dx^\beta}{d\lambda} + g_{\alpha\mu}\frac{dx^\alpha}{d\lambda}}{2\sqrt{\dots}}\right] = \frac{d}{d\lambda}\left[\frac{g_{\alpha\mu}\frac{dx^\alpha}{d\lambda}}{2\sqrt{\dots}}\right] \\
&= \frac{\sqrt{\dots}\left[\frac{\partial g_{\alpha\mu}}{\partial x^\beta}\frac{dx^\beta}{d\lambda}\frac{dx^\alpha}{d\lambda} + g_{\alpha\mu}\frac{d^2x^\alpha}{d\lambda^2}\right] - g_{\alpha\mu}\frac{dx^\alpha}{d\lambda}\frac{d}{d\lambda}\sqrt{\dots}}{(\sqrt{\dots})^2}\,.
\end{aligned}$$

This is obviously a mess, but we should not be surprised, because even a straight line may be characterized by a complicated parametrization:

$$x^\alpha = a^\alpha \sin(e^{\tan\lambda}) + b^\alpha .$$

Hence even a simple curve like a straight line may solve a complicated differential equation, if we choose a crazy parametrization! Let us take advantage of the fact that we can choose our parameter λ as we like, so let us use the proper time τ as the parameter! Hence we put $\lambda = \tau$ in the equation of motion and use the simple relations

$$g_{\alpha\beta}\frac{dx^\alpha}{d\tau}\frac{dx^\beta}{d\tau} = 1 \Rightarrow \sqrt{\ldots} = 1, \qquad \frac{d}{dt}\sqrt{\ldots} = 0.$$

Then the equation of motion is simplified considerably!:

$$\frac{1}{2}\frac{\partial g_{\alpha\beta}}{\partial x^\mu}\frac{dx^\alpha}{d\tau}\frac{dx^\beta}{d\tau} = \frac{\partial g_{\alpha\mu}}{\partial x^\beta}\frac{dx^\alpha}{d\tau}\frac{dx^\beta}{d\tau} + g_{\alpha\mu}\frac{d^2x^\alpha}{d\tau^2}$$
$$\Rightarrow g_{\alpha\mu}\frac{d^2x^\alpha}{d\tau^2} + \frac{1}{2}\left[\frac{\partial g_{\alpha\mu}}{\partial x^\beta} + \frac{\partial g_{\beta\mu}}{\partial x^\alpha} - \frac{\partial g_{\alpha\beta}}{\partial x^\mu}\right]\frac{dx^\alpha}{d\tau}\frac{dx^\beta}{d\tau} = 0.$$

(Here we have split the term

$$\frac{\partial g_{\alpha\mu}}{\partial x^\beta}\frac{dx^\beta}{d\tau}\frac{dx^\alpha}{d\tau} = \frac{1}{2}\frac{\partial g_{\alpha\mu}}{\partial x^\beta}\frac{dx^\beta}{d\tau} + \frac{1}{2}\frac{\partial g_{\alpha\mu}}{\partial x^\beta}\frac{dx^\beta}{d\tau}\frac{dx^\alpha}{d\tau}.$$

Performing the substitutions $\alpha \to \beta, \beta \to \alpha$ in the last term, this is rearranged as

$$\frac{1}{2}\frac{\partial g_{\alpha\mu}}{\partial x^\beta}\frac{dx^\beta}{d\tau}\frac{dx^\alpha}{d\tau} + \frac{1}{2}\frac{\partial g_{\beta\mu}}{\partial x^\alpha}\frac{dx^\alpha}{d\tau}\frac{dx^\beta}{d\tau}.)$$

If we multiply this equation by $g^{\mu\nu}$, we get

$$\delta^\nu_\alpha\frac{d^2x^\alpha}{d\tau^2} + \frac{1}{2}g^{\mu\nu}\left[\frac{\partial g_{\mu\alpha}}{\partial x^\beta} + \frac{\partial g_{\beta\mu}}{\partial x^\alpha} - \frac{\partial g_{\alpha\beta}}{\partial x^\mu}\right]\frac{dx^\alpha}{d\tau}\frac{dx^\beta}{d\tau} = 0$$
$$\Rightarrow \frac{d^2x^\nu}{d\tau^2} + \frac{1}{2}g^{\mu\nu}\left[\frac{\partial g_{\mu\alpha}}{\partial x^\beta} + \frac{\partial g_{\beta\mu}}{\partial x^\alpha} - \frac{\partial g_{\alpha\beta}}{\partial x^\mu}\right]\frac{dx^\alpha}{d\tau}\frac{dx^\beta}{d\tau} = 0,$$

and we are through!

Solution to 6.7.3

If we extremize

$$s = \frac{1}{2}\int d\tau^2$$
$$= \frac{1}{2}\int\left(\frac{dt}{d\tau}\right)^2 - \left(\frac{dr}{d\tau}\right)^2 - r^2\left[\left(\frac{d\theta}{d\tau}\right)^2 + \sin^2\theta\left(\frac{d\varphi}{d\tau}\right)^2\right]d\tau,$$

we get the Euler–Lagrange equations

$$\frac{d^2t}{d\tau^2} = 0; \quad \frac{d^2r}{d\tau^2} = r\left[\left(\frac{d\theta}{d\tau}\right)^2 + \sin^2\theta\left(\frac{d\varphi}{d\tau}\right)^2\right];$$

$$\frac{d}{d\tau}\left[r^2\,\frac{d\theta}{d\tau}\right] = r^2\sin\theta\cos\theta\left(\frac{d\varphi}{d\tau}\right)^2; \quad \frac{d}{d\tau}\left[r^2\sin^2\theta\,\frac{d\varphi}{d\tau}\right] = 0,$$

which we may rearrange as

$$\begin{aligned}
\frac{d^2t}{d\tau^2} &= 0, \\
\frac{d^2r}{d\tau^2} &= r\left(\frac{d\theta}{d\tau}\right)^2 + r\sin^2\theta\left(\frac{d\varphi}{d\tau}\right)^2, \\
\frac{d^2\theta}{d\tau^2} &= -\frac{2}{r}\,\frac{dr}{d\tau}\,\frac{d\theta}{d\tau} + \sin\theta\cos\theta\left(\frac{d\varphi}{d\tau}\right)^2, \\
\frac{d^2\varphi}{d\tau^2} &= -\frac{2}{r}\,\frac{dr}{d\tau}\,\frac{d\varphi}{d\tau} - 2\cot\theta\,\frac{d\theta}{d\tau}\,\frac{d\varphi}{d\tau}.
\end{aligned} \tag{6.81}$$

These equations then comprise the equations of motion for a free particle. But we can also use them to read off the Christoffel fields. Using the formula

$$\frac{d^2x^\alpha}{d\tau^2} = -\Gamma^\alpha{}_{\mu\nu}\,\frac{dx^\mu}{d\tau}\,\frac{dx^\nu}{d\tau},$$

we immediately get the following nonvanishing components of the Christoffel fields:

$$\begin{aligned}
&\Gamma^r{}_{\theta\theta} = -r, \qquad \Gamma^r{}_{\varphi\varphi} = -r\sin^2\theta, \\
&\Gamma^\theta{}_{r\theta} = \Gamma^\theta{}_{\theta r} = \frac{1}{r}\,\Gamma^\theta{}_{\varphi\varphi} = -\sin\theta\cos\theta, \\
&\Gamma^\varphi{}_{r\varphi} = \Gamma^\varphi{}_{\varphi r} = \frac{1}{r}\,\Gamma^\varphi{}_{\theta\varphi} = \Gamma^\varphi{}_{\varphi\theta} = \cot\theta.
\end{aligned} \tag{6.82}$$

CHAPTER 7

Differential Forms and the Exterior Calculus

7.1 Introduction

We would like to extend the differential calculus to include tensors. Consider an arbitrary Euclidean manifold M and a cotensor field $\mathbf{T}$ of rank 2 on this manifold. If we introduce coordinates $(x^1, \ldots, x^n)$, we can characterize $\mathbf{T}$ by its covariant components

$$T_{\alpha\beta}(x^1, \ldots, x^n).$$

Here we might try to differentiate the components of $\mathbf{T}$ with respect to one of the coordinates x^μ; i.e., we form the components $\partial_\mu T_{\alpha\beta}(x^1, \ldots, x^n)$. The question is, Is this new quantity a cotensor field of rank 3; i.e., do the components $\partial_\mu T_{\alpha\beta}(x^1, \ldots, x^n)$ coincide with the components of some cotensor of rank 3? To investigate this we must try to show that the quantity $\partial_\mu T_{\alpha\beta}$ transforms covariantly. Therefore, we introduce new coordinates $(y^1, \ldots, y^n)$.

With respect to the old coordinates $(x^1, \ldots, x^n)$ the quantity has the components

$$\frac{\partial}{\partial x^\mu_{(1)}} T_{\alpha\beta}(x^1, \ldots, x^n),$$

and with respect to the new coordinates $(y^1, \ldots, y^n)$ the quantity has the components

$$\frac{\partial}{\partial y^\mu_{(2)}} T_{\alpha\beta}(y^1, \ldots, y^n).$$

As

$$T_{(2)_{\alpha\beta}}(y^1, \ldots, y^n) = T_{(1)_{\gamma\delta}}(x^1, \ldots, x^n) \frac{\partial x^\gamma}{\partial y^\alpha} \frac{\partial x^\delta}{\partial y^\beta},$$

we get

$$\frac{\partial}{\partial y^\mu} T_{(2)_{\alpha\beta}}(y^1, \ldots, y^n) = \frac{\partial}{\partial y^\mu} \left[T_{(1)_{\gamma\delta}}(x^1, \ldots, x^n) \frac{\partial x^\gamma}{\partial y^\alpha} \frac{\partial x^\delta}{\partial y^\beta} \right]$$

$$= \frac{\partial}{\partial x^\nu} T_{(1)_{\gamma\delta}} \frac{\partial x^\nu}{\partial y^\mu} \frac{\partial x^\gamma}{\partial y^\alpha} \frac{\partial x^\delta}{\partial y^\beta} + T_{(1)_{\gamma\delta}} \left[\frac{\partial^2 x^\gamma}{\partial y^\mu \partial y^\alpha} \frac{\partial x^\delta}{\partial y^\beta} + \frac{\partial x^\gamma}{\partial y^\alpha} \frac{\partial^2 x^\delta}{\partial y^\mu \partial y^\beta} \right].$$

The first term is exactly the term we are after, but the second term spoils everything. So the quantity with the components $\partial_\mu T_{\alpha\beta}$ is certainly *not* a cotensor!

We can even understand intuitively what went wrong! We are looking at a cotensor field $\mathbf{T}$ and trying to form something like the partial derivative of it. (See Figure 7.1.) Intuitively, we would then try to make an expression like

$$\frac{1}{\epsilon}\left[\mathbf{T}_{P(x^1;\dots;x^\mu+\epsilon;\dots;x^n)} - \mathbf{T}_{P(x^1;\dots;x^\mu;\dots,x^n)}\right].$$

But this has *no* meaning at all because we are trying to subtract two tensors with *different base points*! But they lie in different tensor spaces and have absolutely nothing to do with each other. Of course, we could still try to form the component expression

$$\frac{1}{\epsilon}\left[T_{\alpha\beta}(x^1;\dots;x^\mu+\epsilon;\dots;x^n) - T_{\alpha\beta}(x^1;\dots;x^\mu;\dots,x^n)\right].$$

This is legitimate, but observe that this expression has *no* geometrical meaning. *You cannot compare two tensors with different base points P and Q just by comparing their components*. Introducing a coordinate system that covers P and Q, we may characterize the tensors $\mathbf{T}_P$ and $\mathbf{T}_Q$ by their components

$$T_{\alpha\beta}(P) \quad \text{and} \quad T_{\alpha\beta}(Q).$$

We might hope that it would make sense to say that the two tensors $\mathbf{T}_P$ and $\mathbf{T}_Q$ are almost equal if their components $T_{\alpha\beta}(P)$ and $T_{\alpha\beta}(Q)$ are almost equal. But watch out! Let us change the coordinate system in such a way that the coordinates in the neighborhood of P are unaffected, while the coordinates in the neighborhood of Q are changed drastically (see Figure 7.2). Then $\mathbf{T}_P$ is still characterized by the components

$$T'_{\alpha\beta}(P) = T_{\alpha\beta}(P),$$

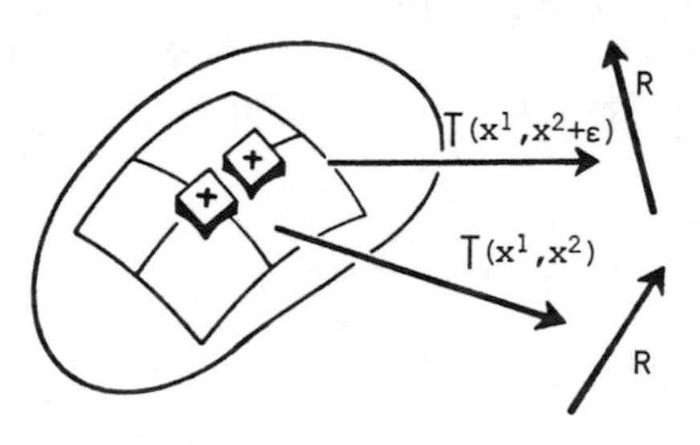

Figure 7.1.

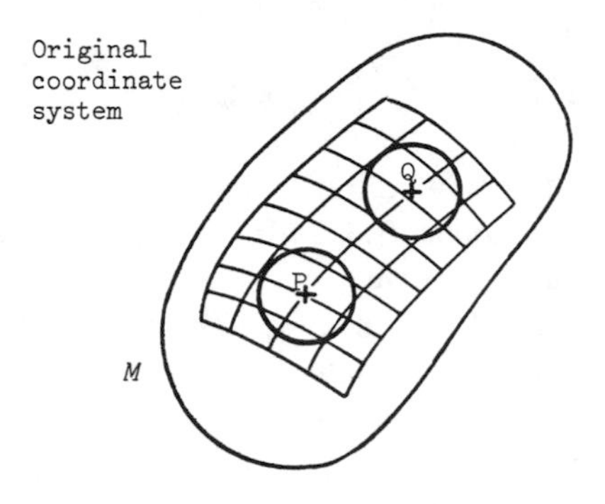

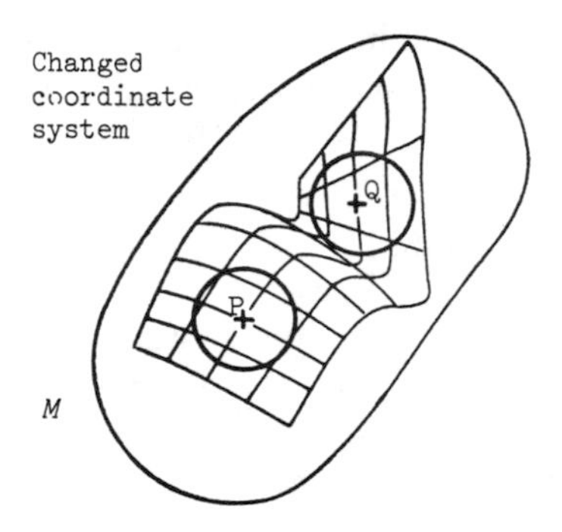

Figure 7.2.

while $\mathbf{T}_Q$ will be characterized by the new components

$$T'_{\alpha\beta}(Q) = T_{\gamma\delta}(Q)\,\frac{\partial x^\gamma}{\partial y^\alpha}\,\frac{\partial x^\delta}{\partial y^\beta}\,.$$

Thus although we have fixed the components $\mathbf{T}_P$, we can change the components of $\mathbf{T}_Q$ into almost anything!

This shows clearly that it makes no sense to compare cotensors with different base points just by comparing their components. In these notes we will not try to attack the general problem of constructing derivatives of cotensor fields but will show only that it is possible to overcome some of the difficulties if we are willing to restrict ourselves to a special kind of cotensor.

7.2 k-Forms — The Wedge Product

Now let $\mathbf{F}$ be a cotensor of arbitrary rank. We say that $\mathbf{F}$ is *skew-symmetric* if it changes sign whenever we exchange two of its arguments:

$$\mathbf{F}(\ldots;\mathbf{u}_P;\ldots;\mathbf{v}_P;\ldots) = -\mathbf{F}(\ldots;\mathbf{v}_P;\ldots;\mathbf{u}_P;\ldots).$$

If we characterize $\mathbf{F}$ by its components,

$$F_{a_1 \ldots a_k} = \mathbf{F}\left(\mathbf{e}_{a_1}; \ldots; \mathbf{e}_{a_k}\right),$$

we see that $\mathbf{F}$ is skew-symmetric whenever the components $F_{a_1 \ldots a_k}$ are skew-symmetric in the indices $a_1, \ldots, a_k$! The skew-symmetric cotensors are so important that mathematicians have given them a special name:

Definition 7.1. A *k-form* is a skew-symmetric cotensor of rank k.

Consider a specific point P on M^n. We have previously attached a whole family of cotensor spaces $\mathbf{T}_P^{(0,k)}(M)$ to this point. Now we want to extract the skew-symmetric cotensors. The set of k-forms at the point P will be denoted by $\Lambda_P^k(M)$. Let us start with some elementary remarks:

If $\mathbf{F}$ and $\mathbf{G}$ are skew-symmetric cotensors of rank k, then so are $\mathbf{F} + \mathbf{G}$ and $\lambda\mathbf{F}$. Therefore, the set of k-forms

$$\Lambda_P^k(M)$$

forms a *vector space*. If $k > n$, then $\Lambda_P^k(M)$ degenerates and will contain only the zero form:

$$\Lambda_P^k(M) = \{0\}, \qquad k > n.$$

To see this, let $\mathbf{F}$ be a skew-symmetric cotensor of rank k. Then $\mathbf{F}$ is characterized by its components

$$F_{a_1 \ldots a_k},$$

but as $k > n$, two of the indices must always coincide (since an index a_i can take only the values $1, \ldots, n$), whence $F_{a_1 \ldots a_k}$ vanishes! So in the case of k-forms, we do *not* have an infinite family of tensor spaces! We will use the convention that cotensors of rank 1 are counted as 1-forms. Of course, it has no meaning to say that a covector is skew-symmetric, but it is useful to include them among the forms. In a similar way it is useful to treat scalars as 0-forms. Consequently, the whole family of forms looks as follows:

$$\Lambda_P^0(M) = \mathbb{R}; \quad \Lambda_P^1(M) = \mathbf{T}_P^*(M); \quad \Lambda_P^2(M); \ldots; \Lambda_P^n(M);$$
$$\Lambda_P^{n+1}(M) = \{0\}; \quad \Lambda_P^{n+2}(M) = \{0\}; \ldots.$$

Working with ordinary cotensors, we have previously introduce the tensor product: If $\mathbf{F}$ and $\mathbf{G}$ are arbitrary cotensors, then the tensor product $\mathbf{F} \otimes \mathbf{G}$ defined by

$$\mathbf{F} \otimes \mathbf{G}(\mathbf{v}_1; \ldots; \mathbf{v}_k; \mathbf{u}_1; \ldots; \mathbf{u}_m) \doteq \mathbf{F}(\mathbf{v}_1; \ldots; \mathbf{v}_k)\mathbf{G}(\mathbf{u}_1; \ldots; \mathbf{u}_m)$$

is a cotensor of rank $k+m$. But if we restrict ourselves to *skew-symmetric* cotensors, this composition is no longer relevant because $\mathbf{F} \otimes \mathbf{G}$ is not necessarily skew-symmetric. (If we interchange $\mathbf{v}_i$ and $\mathbf{u}_j$, we can say nothing about what happens of $\mathbf{F}(\mathbf{v}_1; \ldots; \mathbf{v}_k)\mathbf{G}(\mathbf{u}_1; \ldots; \mathbf{u}_m)$.) We will therefore try to modify this composition:

If $\omega_{a_1\ldots a_k}$ is a quantity with indices $a_1, \ldots, a_k$, then we can construct a skew-symmetric quantity in the following way:

$$\omega_{[a_1\ldots a_k]} = \frac{1}{k!}\sum_{\pi}(-1)^{\pi}\omega_{\pi(a_1)\ldots\pi(a_k)}, \tag{7.1}$$

where we sum overall all permutations π of the indices $a_1, \ldots, a_k$ and $(-1)^{\pi}$ is the sign of the permutation; i.e.,

$$\begin{aligned} (-1)^{\pi} &= +1 \qquad \text{if } \pi \text{ is an even permutation,} \\ (-1)^{\pi} &= -1 \qquad \text{if } \pi \text{ is an odd permutation,} \end{aligned}$$

For instance, we get

$$\omega_{[ab]} = \frac{1}{2!}\left[\omega_{ab} - \omega_{ba}\right]$$

and

$$\omega_{[abc]} = \frac{1}{3!}\left[\omega_{abc} + \omega_{bca} + \omega_{cab} - \omega_{acb} - \omega_{cba} - \omega_{bac}\right].$$

The quantity $\omega_{[a_1\ldots a_k]}$ is called the skew-symmetric of $\omega_{a_1\ldots a_k}$. Observe that $\omega_{[a_1\ldots a_k]}$ is completely skew-symmetric, and that if $\omega_{a_1\ldots a_k}$ is born skew-symmetric, then

$$\omega_{[a_1\ldots a_k]} = \omega_{a_1\ldots a_k}.$$

(This is, of course, the reason why we have included the factor $\frac{1}{k!}$.)

If we introduce the abbreviation

$$\operatorname{Sgn}\begin{bmatrix} b_1 \ldots b_k \\ a_1 \ldots a_k \end{bmatrix} = \begin{cases} +1 & \text{if } (b_1 \ldots b_k) \text{ is an even permutation of} \\ & \qquad (a_1 \ldots a_k), \\ -1 & \text{if } (b_1 \ldots b_k) \text{ is an odd permutation of} \\ & \qquad (a_1 \ldots a_k), \\ 0 & \text{otherwise,} \end{cases} \tag{7.2}$$

we can write down the skew-symmetrization as an explicit summation:

$$\omega_{[a_1\ldots a_k]} = \frac{1}{k!}\operatorname{Sgn}\begin{bmatrix} b_1 \ldots b_k \\ a_1 \ldots a_k \end{bmatrix}\omega_{b_1\ldots b_k}.$$

With these preliminaries, we may return to the skew-symmetric cotensors: If **F** is a k-form characterized by components $F_{a_1\ldots a_k}$ and **G** an m-form with components $G_{b_1\ldots b_m}$, then the tensor product is characterized by the components

$$F_{a_1\ldots a_k}G_{b_1\ldots b_m}.$$

Next, we form the skew-symmetrization

$$\frac{(k+m)!}{k!m!}\,F_{[a_1\ldots a_k}G_{b_1\ldots b_m]},$$

where the factor $\frac{(k+m)!}{k!m!}$ has been included to make life easier. The statistical factor $\frac{(k+m!)}{k!m!}$ removes "double counting":

$$\frac{(k+m)!}{k!m!} F_{[a_1 \dots a_k} G_{b_1 \dots b_m]} = \\ \frac{1}{k!m!} \operatorname{Sgn} \begin{bmatrix} c_1 \dots c_k d_1 \dots d_m \\ a_1 \dots a_k b_1 \dots b_m \end{bmatrix} F_{c_1 \dots c_k} G_{d_1 \dots d_m}.$$

Observe that $\operatorname{Sgn} \begin{bmatrix} c_1 \dots c_k d_1 \dots d_m \\ a_1 \dots a_k b_1 \dots b_m \end{bmatrix}$ and $F_{c_1 \dots c_k}$ are both skew-symmetric in $(c_1 \dots c_k)$. When we perform a permutation of $(c_1 \dots c_k)$, we pick up two factors $(-1)^\pi$. All permutations of $(c_1 \dots c_k)$ thus give the same contribution to the sum. We need therefore only consider a single representative, for instance the ordered set $(c_1 \dots c_k)$ characterized by the property $c_1 < \dots < c_k$. There are $k!$ permutations of $(c_1 \dots c_k)$ giving the same contribution, and similarly there are $m!$ permutations of $(d_1 \dots d_m)$ giving the same contribution. Using this, we obtain the formula

$$\frac{(k+m)!}{k!m!} F_{[a_1 \dots a_k} G_{b_1 \dots b_m]} = \\ \sum_{c_1 < \dots < c_k} \sum_{d_1 < \dots < d_m} \operatorname{Sgn} \begin{bmatrix} c_1 \dots c_k d_1 \dots d_m \\ a_1 \dots a_k b_1 \dots b_m \end{bmatrix} F_{c_1 \dots c_k} G_{d_1 \dots d_m},$$

where we have obviously avoided "double counting." Observe that

$$\frac{(k+m)!}{k!m!} F_{[a_1 \dots a_k} G_{b_1 \dots b_m]} = \\ \frac{1}{k!m!} \sum_\pi (-1)^\pi F_{\pi(a_1) \dots \pi(a_k)} G_{\pi(b_1) \dots \pi(b_m)}.$$

Each of the terms $F_{\pi(a_1) \dots \pi(a_k)} G_{\pi(b_1) \dots \pi(b_m)}$ transforms covariantly. Therefore,

$$\frac{(k+m)!}{k!m!} F_{[a_1 \dots a_k} G_{b_1 \dots b_m]}$$

coincide with the components of a skew-symmetric cotensor of rank $k+m$. This motivates the following definition:

Definition 7.2. Let $\mathbf{F}$ be a k-form with components $F_{a_1 \dots a_k}$ and $\mathbf{G}$ an m-form with components $G_{b_1 \dots b_m}$. Then the *wedge product* $\mathbf{F} \wedge \mathbf{G}$ is the $(k+m)$-form with components

$$\frac{(k+m)!}{k!m!} F_{[a_1 \dots a_k} G_{b_1 \dots b_m]}.$$

Let us give a simple example: If $\mathbf{A}$ and $\mathbf{B}$ are covectors, i.e., 1-forms, then their wedge product $\mathbf{A} \wedge \mathbf{B}$ has the components

$$2A_{[a} B_{b]} = A_a B_b - A_b B_a = A_a B_b - B_a A_b.$$

From this we see that

$$\mathbf{A} \wedge \mathbf{B} = [\mathbf{A} \otimes \mathbf{B} - \mathbf{B} \otimes \mathbf{A}]. \tag{7.3}$$

Exercise 7.2.1
Problem: (a) Consider a 2-form $\mathbf{F}$ characterized by the components F_{ab} and a 1-form $\mathbf{B}$ characterized by the components B_c. Show that $\mathbf{F} \wedge \mathbf{B}$ is characterized by the components

$$F_{ab}B_c + F_{bc}B_a + F_{ca}B_b. \tag{7.3}$$

(b) Let $\mathbf{A}$, $\mathbf{B}$, $\mathbf{C}$ be 1-forms. Show that

$$\begin{aligned}(\mathbf{A} \wedge \mathbf{B}) \wedge \mathbf{C} = [&\mathbf{A} \otimes \mathbf{B} \otimes \mathbf{C} + \mathbf{B} \otimes \mathbf{C} \otimes \mathbf{A} + \mathbf{C} \otimes \mathbf{A} \otimes \mathbf{B} \\ &- \mathbf{B} \otimes \mathbf{A} \otimes \mathbf{C} - \mathbf{A} \otimes \mathbf{C} \otimes \mathbf{B} - \mathbf{C} \otimes \mathbf{B} \otimes \mathbf{A}].\end{aligned} \tag{7.4}$$

The wedge product $\wedge$ has some simple algebraic properties:

Theorem 7.1. *The wedge product is associative and distributive:*

$$(\mathbf{F} \wedge \mathbf{G}) \wedge \mathbf{H} = \mathbf{F} \wedge (\mathbf{G} \wedge \mathbf{H}). \tag{7.5}$$

$$\mathbf{F} \wedge (\lambda \mathbf{G} + \mu \mathbf{H}) = \lambda \mathbf{F} \wedge \mathbf{G} + \mu \mathbf{F} \wedge \mathbf{H}. \tag{7.6}$$

If $\mathbf{F}$ *is a k-form and* $\mathbf{G}$ *is an m-form, then*

$$\mathbf{F} \wedge \mathbf{G} = (-1)^{km} \mathbf{G} \wedge \mathbf{F}. \tag{7.7}$$

Observe especially that covectors anticommute!

PROOF. Associativity follows from the simple observation that $(\mathbf{F} \wedge \mathbf{G}) \wedge \mathbf{H}$ has the components

$$\begin{aligned}&\frac{(k+\ell+m)!}{(k+\ell)!m!}\left[\frac{(k+\ell)!}{k!\ell!} F_{[[a_1 \ldots a_k} G_{b_1 \ldots b_\ell]} H_{c_1 \ldots c_m]}\right] = \\ &\quad \frac{(k+\ell+m)!}{k!\ell!m!} F_{[a_1 \ldots a_k} G_{b_1 \ldots b_\ell} H_{c_1 \ldots c_m]}.\end{aligned}$$

Distributivity also follows directly when we compare the components of both sides of the equation. We omit the proof, which consists simply in keeping track of a lot of indices!

Finally, we consider the wedge products $\mathbf{F} \wedge \mathbf{G}$ and $\mathbf{G} \wedge \mathbf{F}$:

$$(\mathbf{F} \wedge \mathbf{G})_{a_1 \ldots a_k b_1 \ldots b_m} = \frac{(k+m)!}{k!m!}[F_{a_1 \ldots a_k} G_{b_1 \ldots b_m} + \cdots$$

$$(\mathbf{G} \wedge \mathbf{F})_{b_1 \ldots b_m a_1 \ldots a_k} = \frac{(k+m)!}{k!m!}[G_{b_1 \ldots b_m} F_{a_1 \ldots a_k} + \cdots$$

Consequently,

$$(\mathbf{F} \wedge \mathbf{G})_{a_1 \ldots a_k b_1 \ldots b_m} = (\mathbf{G} \wedge \mathbf{F})_{b_1 \ldots b_m a_1 \ldots a_k}.$$

Observe that it will require $k \cdot m$ transpositions to obtain $a_1 \ldots a_k b_1 \ldots b_m$ from $b_1 \ldots b_m a_1 \ldots a_k$. (It will require k transpositions to move b_m through $a_1 \ldots a_k$, it will then require another k transposition to move b_{m-1} through $a_1 \dots a_k$, etc.) Using that $\mathbf{G} \wedge \mathbf{F}$ is a skew-symmetric, we get

$$(\mathbf{G} \wedge \mathbf{F})_{b_1 \dots b_m a_1 \dots a_k} = (-1)^{km} (\mathbf{G} \wedge \mathbf{F})_{a_1 \dots a_k b_1 \dots b_m}.$$

Combining these two formulas we, finally obtain

$$(\mathbf{F} \wedge \mathbf{G})_{a_1 \dots a_k b_1 \dots b_m} = (-1)^{km} (\mathbf{G} \wedge \mathbf{F})_{a_1 \dots a_k b_1 \dots b_m},$$

showing that

$$\mathbf{F} \wedge \mathbf{G} = (-1)^{km} \mathbf{G} \wedge \mathbf{F}.$$

□

The most surprising of these rules is no doubt the special "commutation" rule (7.7). There is an easy way to remember this rule. If $\mathbf{F}$ is a form of even degree, we call it an even form, and similarly, a form $\mathbf{G}$ of odd degree is called an odd form. The rule (7.7) can now be summarized in the following form:

Even forms always commute; odd forms anticommute.

So if we consider an expression like

$$\mathbf{F}_1 \wedge \dots \wedge \mathbf{F}_n$$

and we want to interchange the order of the forms, then we only have to worry about the odd forms: Every time we interchange two odd forms, it costs a sign.

Now let us look at a 2-form $\mathbf{F}$. Introducing a coordinate system, we may decompose $\mathbf{F}$ along the basic cotensors $\boldsymbol{dx}^i \otimes \boldsymbol{dx}^j$; i.e.,

$$\mathbf{F} = F_{ij} \boldsymbol{dx}^i \otimes \boldsymbol{dx}^j,$$

where F_{ij} are the components of $\mathbf{F}$ with respect to this coordinate system (cf. (6.68)). But F_{ij} is skew-symmetric in i and j. Consequently, we can rewrite F_{ij} as $F_{ij} = \frac{1}{2}(F_{ij} - F_{ji})$. Thus we can rearrange the decomposition of $\mathbf{F}$ in the following way:

$$\mathbf{F} = \frac{1}{2} (F_{ij} \boldsymbol{dx}^i \otimes \boldsymbol{dx}^j - F_{ji} \boldsymbol{dx}^i \otimes \boldsymbol{dx}^j).$$

Interchanging the dummy indices i and j in the last term, we further get

$$\mathbf{F} = \frac{1}{2} (F_{ij} \boldsymbol{dx}^i \otimes \boldsymbol{dx}^j - F_{ij} \boldsymbol{dx}^j \otimes \boldsymbol{dx}^i) = \frac{1}{2} F_{ij} \boldsymbol{dx}^i \wedge \boldsymbol{dx}^j.$$

This may obviously be generalized to arbitrary k-forms:

Theorem 7.2. *If* $\mathbf{F}$ *has the components* $F_{i_1 \dots i_k}$, *we can expand* $\mathbf{F}$ *in the following way:*

$$\mathbf{F} = \frac{1}{k!} F_{i_1 \dots i_k} \boldsymbol{dx}^{i_1} \wedge \cdots \wedge \boldsymbol{dx}^{ik}. \tag{7.8}$$

Exercise 7.2.2
Problem: Let us introduce coordinates on a manifold M^n. The coordinate system generates canonical frames $(\mathbf{e}_1, \dots, \mathbf{e}_n)$ and $(\boldsymbol{dx}^1, \dots, \boldsymbol{dx}^n)$.

a. Show that

$$(dx^{a_1} \wedge \cdots \wedge dx^{a_k})_{b_1 \ldots b_k} = \text{Sgn}\begin{bmatrix} a_1 \ldots a_k \\ b_1 \ldots b_k \end{bmatrix};$$

i.e.,

$$dx^1 \wedge \cdots \wedge dx^k(\mathbf{e}_{i_1}, \ldots, \mathbf{e}_{i_k}) = \begin{cases} +1 & \text{if } (i_1 \ldots i_k) \text{ is an even permutation of } (1 \ldots k), \\ -1 & \text{if } (i_1 \ldots i_k) \text{ is an odd permutation of } (1 \ldots k), \\ 0 & \text{otherwise.} \end{cases}$$

b. Let $(\mathbf{u}_1 \ldots \mathbf{u}_n)$ be an n-tuple of tangent vectors, where $\mathbf{u}_j$ is characterized by the contravariant components $A^i{}_j$; i.e., $\mathbf{u}_j = \mathbf{e}_i A^i{}_j$. Show that

$$dx_1 \wedge \ldots \wedge dx^n(\mathbf{u}_1, \ldots, \mathbf{u}_n) = \det \overline{\overline{A}}. \tag{7.9}$$

Exercise 7.2.3
Problem: Let $\mathbf{T}$ be a form of maximal rank n. Observe that $\mathbf{T}$ is characterized by the single component $T_{1\ldots n}$ since all other components can be obtained from this using a permutation of the indices. Show that $\mathbf{T}$ can be decomposed as

$$\mathbf{T} = T_{1\ldots n} dx^1 \wedge \cdots \wedge dx^n. \tag{7.10}$$

In the rest of this section we will try to develop a geometrical interpretation of differential forms to help us understand more intuitively the concept of k-form.

We start by considering the ordinary 3-dimensional Euclidean space $\mathbb{R}^3$. Let us investigate the k-forms associated with the tangent space at the origin:

	Canonical frame:
0-forms	1
1-forms	dx; dy; dz
2-forms	$dx \wedge dy$; $dy \wedge dz$; $dz \wedge dx$
3-forms	$dx \wedge dy \wedge dz$

Consider the basic two-form $dx \wedge dy$. Remembering that $dx \wedge dy$ is defined as a linear map, $\mathbf{T}_P(M) \times \mathbf{T}_P(M) \curvearrowright \mathbb{R}$, we want to compute its value on a pair of tangent vectors $dx \wedge dy(\mathbf{u}, \mathbf{v})$. Observe that $dx \wedge dy$ has the components ϵ_{ij3} (cf. Exercise 7.2.2). Consequently, we get

$$dx \wedge dy(\mathbf{u}, \mathbf{v}) = \epsilon_{ij3} a^i b^j = \det \begin{bmatrix} a^1 & b^1 \\ a^2 & b^2 \end{bmatrix}.$$

But this is easy to interpret. We use that $\mathbf{u}, \mathbf{v}$ span a parallelogram, which we project onto the (x, y)-plane (see Figure 7.3). But then we conclude that $dx \wedge dy(\mathbf{u}, \mathbf{v})$ *is the area of this projection*. For this reason we say that $dx \wedge dy$ defines a *unit of area* in the (x, y)-plane. Observe also that $dx \wedge dy$ generates an orientation

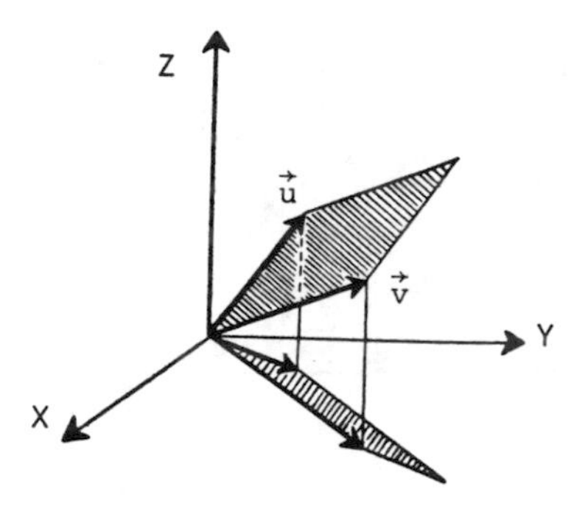

Figure 7.3.

in the (x, y)-plane. The projected area $\boldsymbol{dx} \wedge \boldsymbol{dy}(\mathbf{u}, \mathbf{v})$ is positive if and only if the projection of $(\mathbf{u}, \mathbf{v})$ defines the same orientation as $(\mathbf{e}_x, \mathbf{e}_y)$.

Exercise 7.2.4
Problem: Consider the basic one-form $\boldsymbol{dx}$. Show that $\boldsymbol{dx}$ defines a unit of length along the x-axis in the sense that $\boldsymbol{dx}(\mathbf{u})$ is the length of the projection of $\mathbf{u}$ onto the x-axis.

Exercise 7.2.5
Problem: (a) Show that $\boldsymbol{dx} \wedge \boldsymbol{dy} \wedge \boldsymbol{dz}$ defines a volume-form in $\mathbb{R}^3$ in the sense that $\boldsymbol{dx} \wedge \boldsymbol{dy} \wedge \boldsymbol{dz}(\mathbf{U}_1; \mathbf{U}_2; \mathbf{U}_3)$ is the volume of the parallelepiped spanned by $\mathbf{U}_1$, $\mathbf{U}_2$ and $\mathbf{U}_3$.

(b) Show that $\boldsymbol{dx} \wedge \boldsymbol{dy} \wedge \boldsymbol{dz}$ defines an orientation in $\mathbb{R}^3$ in the sense that $(\mathbf{U}_1, \mathbf{U}_2, \mathbf{U}_3)$ is positively oriented if and only if $\boldsymbol{dx} \wedge \boldsymbol{dy} \wedge \boldsymbol{dz}(\mathbf{U}_1; \mathbf{U}_2; \mathbf{U}_3)$ is positive.

(Hint: Decompose the triple $(\mathbf{U}_1, \mathbf{U}_2, \mathbf{U}_3)$ as

$$\mathbf{U}_j = \mathbf{e}_i A^i_{\ j}$$

and show that

$$\boldsymbol{dx} \wedge \boldsymbol{dy} \wedge \boldsymbol{dz}(\mathbf{U}_1; \mathbf{U}_2; \mathbf{U}_3) = \det(A^i_{\ j}).$$

Compare with the discussion in Section 1.1.)

Obviously, the above analysis depends on very special properties of the Euclidean space $\mathbb{R}^3$. It is, however, possible to convert it into a purely geometrical form suitable for generalizations to arbitrary Euclidean manifolds.

Consider once more the basic two-form $\boldsymbol{dx} \wedge \boldsymbol{dy}$. The coordinate functions x and y define two *stratifications* of the Euclidean space:

$$\ldots, x = -2, \quad x = -1, \quad x = 0, \quad x = 1, \quad x = 2, \ldots,$$
$$\ldots, y = -2, \quad y = -1, \quad y = 0, \quad y = 1, \quad y = 2, \ldots.$$

Together these two stratifications form what is generally known as a "honeycomb structure" (see Figure 7.4).

If we return to a parallelogram spanned by two tangent vectors $\mathbf{u}$ and $\mathbf{v}$, then it is clear that the projected area is simply equal to the number of tubes intersected

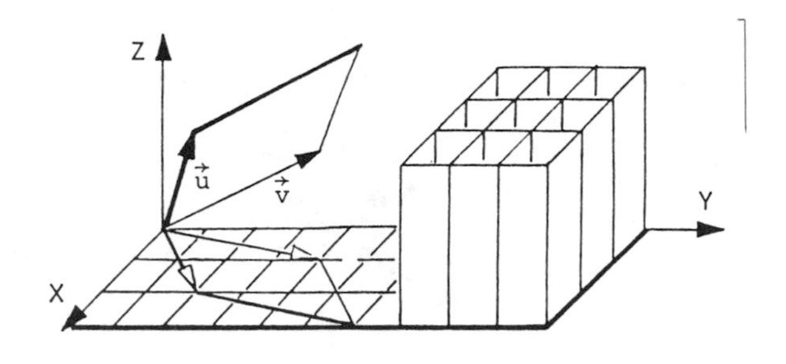

Figure 7.4.

by the parallelogram. We therefore conclude that $dx \wedge dy(\mathbf{u}, \mathbf{v})$ *is the number of x-y-tubes that are intersected by the parallelogram spanned by* $\mathbf{u}$ *and* $\mathbf{v}$.

Exercise 7.2.6
Problem: (a) Consider the basic one-form dx. The coordinate function x defines a stratification $\ldots, x = -1, x = 0, x = 1, \ldots$. Show that $dx(\mathbf{u})$ is equal to the number of hyperplanes $x = k$ that are intersected by the vector $\mathbf{u}$.

(b) Consider the basic three-form $dx \wedge dy \wedge dz$. The coordinate functions x, y, and z generate three stratifications, which together form a cell structure. Show that $dx \wedge dy \wedge dz(\mathbf{u}; \mathbf{v}; \mathbf{w})$ is equal to the number of cells contained in the parallelepiped spanned by $\mathbf{u}$, $\mathbf{v}$, and $\mathbf{w}$.

We can now transfer the above machinery to an arbitrary manifold. Consider a two-dimensional manifold M with local coordinates (x^1, x^2). Let P_0 be a point on M and consider the basic two-form $dx^1 \wedge dx^2$ at P_0. The two coordinate functions x^1 and x^2 generate two stratifications on $M : x^1 = \ldots, -1, 0, 1, \ldots$ and $x^2 = \ldots, -1, 0, 1, \ldots$. These two stratifications divide the surface into a great number of cells (see Figure 7.5).

Consider two tangent vectors $\mathbf{u}$ and $\mathbf{v}$ at P_0. This time $\mathbf{u}$ and $\mathbf{v}$ span a parallelogram that lies outside the manifold M. We must therefore use a trick. In coordinate

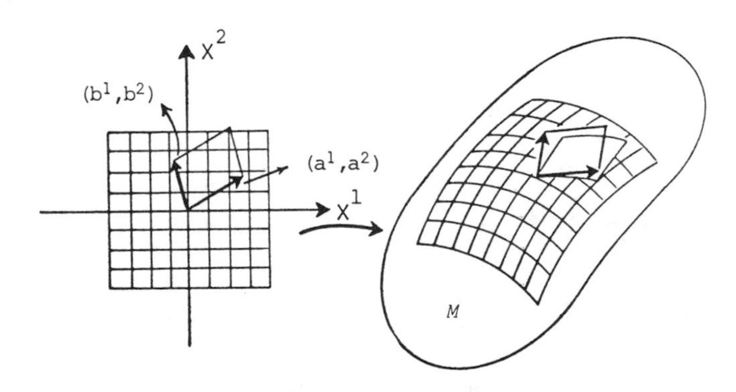

Figure 7.5.

space, (a^1, a^2) and (b^1, b^2) span an ordinary parallelogram. The number

$$\boldsymbol{dx}^1 \wedge \boldsymbol{dx}^2(\mathbf{u}; \mathbf{v})$$

is therefore equal to the number of cells contained in this parallelogram. But since this parallelogram is a subset of the coordinate domain U, we may transfer it to M. The image will be referred to as the "parallelogram" swept out by $\mathbf{u}$ and $\mathbf{v}$. Therefore, we conclude in the usual way that $\boldsymbol{dx}^1 \wedge \boldsymbol{dx}^2(\mathbf{u}; \mathbf{v})$ *denotes the number of cells contained in the parallelogram swept out by* $\mathbf{u}$ *and* $\mathbf{v}$.

7.3 The Exterior Derivative

That was a long algebraic digression. Let us return to our main problem. We want to construct a differential operator that converts a cotensor field of degree k into a cotensor field of degree $k+1$, i.e., something like

$$T_{a_1\ldots a_k}(x^1; \ldots; x^k) \to \partial_\mu T_{a_1\ldots a_k}(x^1; \ldots; x^k).$$

We now restrict ourselves to skew-symmetric cotensor fields. They play a key role in differential geometry and are called *differential forms*. If it is clear from context that we are working with a skew-symmetric cotensor field, then a differential form of rank k will be referred to simply as a k-form. The set of all smooth differential forms of rank k will be denoted by $\Lambda^k(M)$. Observe that $\Lambda^k(M)$ is an infinite-dimensional vector space for $0 \leq k \leq n$.

Now, let us consider a differential form of degree zero, i.e., a scalar field ϕ. Then we have previously introduced the differential operator $\boldsymbol{d}$ that converts the scalar field ϕ into a differential form of degree one, i.e., a covector field $\boldsymbol{d}\phi$. If we introduce coordinates, then ϕ is represented by an ordinary Euclidean function $\phi(x^1; \ldots; x^n)$, and $\boldsymbol{d}\phi$ is characterized by the components $\partial_\mu\phi(x^1; \ldots; x^n)$. It is this differential operator $\boldsymbol{d}$ we want to extend!

Consider a differential form of degree one, i.e., a covector field $\mathbf{A}$ characterized by the components

$$A_\alpha(x^1, \ldots, x^n).$$

Differentiating this we get the quantity

$$\frac{\partial}{\partial x^\mu} A_\alpha(x^1, \ldots, x^n) = \partial_\mu A_\alpha(x^1, \ldots, x^n),$$

but this is of little interest because it is not skew-symmetric in μ and α. Therefore, we antisymmetrize it and get

$$2\partial_{[\mu}A_{\alpha]} = \partial_\mu A_\alpha - \partial_\alpha A_\mu.$$

This is skew-symmetric, and if we can show that it transforms covariantly, then we have succeeded. Introducing another coordinate system, we get the new components

$$2\partial_{(2)[\mu}A_{(2)_\alpha]} = \frac{\partial}{\partial y^\mu} A_{(2)_\alpha}(y^1, \ldots, y^n) - \frac{\partial}{\partial y^\alpha} A_{(2)_\mu}(y^1, \ldots, y^n).$$

But $A_{(2)_\alpha}(y) = A_{(1)_\beta}(x)\,\frac{\partial x^\beta}{\partial y^\alpha}$. Inserting this, we get

$$\begin{aligned}
2\partial_{(2)[\mu}A_{(2)_\alpha]} &= \frac{\partial}{\partial y^\mu}\left[A_{(1)_\beta}(x)\,\frac{\partial x^\beta}{\partial y^\alpha}\right] - \frac{\partial}{\partial y^\alpha}\left[A_{(1)_\beta}(x)\,\frac{\partial x^\beta}{\partial y^\mu}\right] \\
&= \frac{\partial A_{(1)_\beta}}{\partial x^\gamma}\,\frac{\partial x^\gamma}{\partial y^\mu}\,\frac{\partial x^\beta}{\partial y^\alpha} + A_{(1)_\beta}\,\frac{\partial^2 x^\beta}{\partial y^\mu \partial y^\alpha} - \frac{\partial A_{(1)_\beta}}{\partial x^\gamma}\,\frac{\partial x^\gamma}{\partial y^\alpha}\,\frac{\partial x^\beta}{\partial y^\mu} \\
&\quad - A_{(1)_\beta}\,\frac{\partial^2 x^\beta}{\partial y^\alpha \partial y^\mu}.
\end{aligned}$$

Using the fact that partial derivatives commute, we observe that the spoiling terms cancel each other! Finally, we get

$$2\partial_{(2)[\mu}A_{(2)_\alpha]} = \partial_{(1)_\gamma}A_{(1)_\beta}\,\frac{\partial x^\gamma}{\partial x^\mu}\,\frac{\partial x^\beta}{\partial y^\alpha} - \partial_{(1)_\gamma}A_{(1)_\beta}\,\frac{\partial x^\gamma}{\partial y^\alpha}\,\frac{\partial x^\beta}{\partial y^\mu}.$$

Here we interchange the dummy indices β and γ in the last term and get

$$\begin{aligned}
2\partial_{(2)[\mu}A_{(2)_\alpha]} &= \left[\partial_{(1)_\gamma}A_{(1)_\beta} - \partial_{(1)_\beta}A_{(1)_\gamma}\right]\frac{\partial x^\gamma}{\partial x^\mu}\,\frac{\partial x^\beta}{\partial y^\alpha} \\
&= 2\partial_{(1)[\gamma}A_{(1)_\beta]}\,\frac{\partial x^\gamma}{\partial y^\mu}\,\frac{\partial x^\beta}{\partial y^\alpha}.
\end{aligned}$$

So everything works! The expression $2\partial_{[\mu}A_{\beta]}$ coincides with the components of a differential form of degree 2, which we will denote by $\boldsymbol{d}\mathbf{A}$.

This may immediately be generalized:

Definition 7.3. If $\mathbf{F}$ is a differential form of degree k characterized by the components $F_{i_1\ldots i_k}(x)$, then $\boldsymbol{d}\mathbf{F}$ is the differential form of degree $k+1$ characterized by the components

$$\frac{(k+1)!}{k!}\,\partial_{[\mu}F_{i_1,\ldots,i_k]}.$$

The differential operator $\boldsymbol{d} : N^k(M) \to N^{k+1}(M)$ is called the *exterior derivative*.

We may think of $\boldsymbol{d}$ as a kind of 1-form characterized by the components ∂_μ. In fact, ∂_μ transforms covariantly according to the chain rule:

$$\partial_{(2)_\mu} = \frac{\partial}{\partial y^\mu} = \frac{\partial}{\partial x^\nu}\,\frac{\partial x^\nu}{\partial y^\mu} = \partial_{(1)_\nu}\,\frac{\partial x^\nu}{\partial y^\mu}.$$

The construction of the exterior derivative is then formally equivalent to the "wedge product" $\boldsymbol{d} \wedge \mathbf{F}$. This also explains the statistical factor. (Compare with Definition 7.2.)

Exercise 7.3.1
Problem: Let **F** be a 2-form characterized by the components $F_{\alpha\beta}$. Show that $\boldsymbol{d}\mathbf{F}$ is characterized by the components

$$(\boldsymbol{d}\mathbf{F})_{\alpha\beta\gamma} = \partial_\alpha F_{\beta\gamma} + \partial_\beta F_{\gamma\alpha} + \partial_\gamma F_{\alpha\beta}. \tag{7.11}$$

The exterior derivative has several important properties. It is a linear operator, i.e.,

$$\boldsymbol{d}(\mathbf{F} + \mathbf{G}) = \boldsymbol{d}\mathbf{F} + \boldsymbol{d}\mathbf{G}, \qquad \boldsymbol{d}(\lambda\mathbf{F}) = \lambda\boldsymbol{d}\mathbf{F}, \tag{7.12}$$

as can be easily checked. But the most important property is the following:

Theorem 7.3. *(Poincaré's lemma)*

$$\boldsymbol{dd} = 0; \tag{7.13}$$

i.e., automatically vanishes if applied twice.

PROOF. Consider first a scalar field ϕ represented by the Euclidean function $\phi(x^1, \ldots, x^n)$. We then get $(\boldsymbol{d}\phi)_\mu = \partial_\mu\phi$, which implies that

$$(\boldsymbol{d}^2\phi)_{\mu\nu} = \partial_\nu\partial_\mu\phi - \partial_\mu\partial_\nu\phi = 0.$$

This shows the mechanism in the cancellations. We then consider an arbitrary differential form **F** of rank k. Let **F** have the components $F_{i_1\ldots i_k}$. We then get for $\boldsymbol{d}\mathbf{F}$,

$$\frac{(k+1)!}{k!}\,\partial_{[\mu}F_{i_1,\ldots,i_k]},$$

and similarly for $\boldsymbol{d}^2\mathbf{F}$,

$$\frac{(k+2)!}{(k+1)!}\,\frac{(k+1)!}{k!}\,\partial_{[\nu}\partial_{[\mu}F_{i_1\ldots i_k]]} =$$
$$\frac{(k+2)!}{k!}\,\partial_{[\nu}\partial_\mu F_{i_1\ldots i_k]} = \frac{1}{k!}\,[\partial_\nu\partial_\mu F_{i_1\ldots i_k} + \cdots].$$

But this sum vanishes, because if we write out all the terms they occur in pairs

$$\partial_\alpha\partial_\beta F_{j_1\ldots j_k} \quad \text{and} \quad \partial_\beta\partial_\alpha F_{j_1\ldots j_k}.$$

As $\beta\alpha j_1 \ldots j_k$ is generated from $\alpha\beta j_1 \ldots j_k$ by applying *one* transposition, they will have opposite signs; i.e., we may collect the two terms into

$$(\partial_\alpha\partial_\beta - \partial_\beta\partial_\alpha)F_{j_1\ldots j_k},$$

and this vanishes automatically. □

Before we derive more rules, we introduce some more notation.

Exercise 7.3.2
Problem: Let **F** be a k-form characterized by the components $F_{a_1\ldots a_n}$ and **G** an m-form characterized by the components $G_{b_1\ldots b_m}$. Show that we can decompose $\mathbf{F} \wedge \mathbf{G}$ as

$$\frac{1}{k!m!}\,F_{a_1\ldots a_k}G_{b_1\ldots b_m}\boldsymbol{d}x^{a_1} \wedge \cdots \wedge \boldsymbol{d}a^{a_k} \wedge \boldsymbol{d}x^{b_1} \wedge \cdots \wedge \boldsymbol{d}x^{b_m}. \tag{7.14}$$

(Hint: Use the distributivity of $\wedge$.)

Exercise 7.3.3
Problem: Consider a quantity $\omega_{a_1 \ldots a_k}$ that is not necessarily skew-symmetric. Show that

$$\omega_{[a_1 \ldots a_k]} \mathbf{dx}^{a_1} \wedge \cdots \wedge \mathbf{dx}^{a_k} = \omega_{a_1 \ldots a_k} \mathbf{dx}^{a_1} \wedge \cdots \wedge \mathbf{dx}^{a_k}. \tag{7.15}$$

If $\mathbf{F}$ is a differential form of degree k, we can decompose it along the basic k-forms (7.8). If we form the exterior derivative $\boldsymbol{d}\mathbf{F}$, we know that it is characterized by the components

$$\frac{(k+1)!}{k!} \partial_{[\mu} F_{i_1 \ldots i_k]}.$$

Therefore, we can decompose $\boldsymbol{d}\mathbf{F}$ in the following way

$$\begin{aligned} \boldsymbol{d}\mathbf{F} &= \frac{1}{(k+1)!} \frac{(k+1)!}{k!} \partial_{[\mu} F_{i_1 \ldots i_k]} \mathbf{dx}^{\mu} \wedge \mathbf{dx}^{i_1} \wedge \cdots \wedge \mathbf{dx}^{i_k} \\ &= \frac{1}{k!} \partial_{[\mu} F_{i_1 \ldots i_k]} \mathbf{dx}^{\mu} \wedge \mathbf{dx}^{i_1} \wedge \cdots \wedge \mathbf{dx}^{i_k}. \end{aligned}$$

But according to Exercise 7.3.3, we are allowed to forget the skew-symmetrization! In this way we obtain the formula

$$\boldsymbol{d}\mathbf{F} = \frac{1}{k!} \partial_{\mu} F_{i_1 \ldots i_k} \mathbf{dx}^{\mu} \wedge \mathbf{dx}^{i_1} \wedge \cdots \wedge \mathbf{dx}^{i_k}. \tag{7.16}$$

We are now in a position to prove a generalization of the familiar rule for differentiating a product (Leibniz's rule):

$$\frac{d}{dx}(f \cdot g) = \frac{df}{dx} \cdot g + f \cdot \frac{dg}{dx}.$$

Theorem 7.4. (*Leibniz's rule for differential forms*) *If* $\mathbf{F}$ *is a k-form and* $\mathbf{G}$ *is an* ℓ*-form, then*

$$\boldsymbol{d}(\mathbf{F} \wedge \mathbf{G}) = \boldsymbol{d}\mathbf{F} \wedge \mathbf{G} + (-1)^k \mathbf{F} \wedge \boldsymbol{d}\mathbf{G}. \tag{7.17}$$

PROOF. If we combine (7.14) an (7.16), we immediately obtain

$$\boldsymbol{d}(\mathbf{F} \wedge \mathbf{G}) = \frac{1}{k!\ell!} \partial_{\mu}(F_{i_1 \ldots i_k} G_{j_1 \ldots j_\ell}) \mathbf{dx}^{\mu} \wedge \mathbf{dx}^{i_1} \wedge \cdots \wedge \mathbf{dx}^{i_k} \wedge \mathbf{dx}^{j_1} \wedge \cdots \wedge \mathbf{dx}^{j_\ell}.$$

According to Leibniz's rule for ordinary functions, this can be rearranged as

$$\begin{aligned} &\frac{1}{k!\ell!} (\partial_{\mu} F_{i_1 \ldots i_k}) G_{j_1 \ldots j_\ell} \mathbf{dx}^{\mu} \wedge \mathbf{dx}^{i_1} \wedge \cdots \wedge \mathbf{dx}^{i_k} \wedge \mathbf{dx}^{j_1} \wedge \cdots \wedge \mathbf{dx}^{j_\ell} \\ &+ \frac{1}{k!\ell!} F_{i_1 \ldots i_k} (\partial_{\mu} G_{j_1 \ldots j_\ell}) \mathbf{dx}^{\mu} \wedge \mathbf{dx}^{i_1} \wedge \cdots \wedge \mathbf{dx}^{i_k} \wedge \mathbf{dx}^{j_1} \wedge \cdots \wedge \mathbf{dx}^{j_\ell}. \end{aligned}$$

In the last term we would like to move $\boldsymbol{dx}^\mu$ through $\boldsymbol{dx}^{i_1} \wedge \cdots \wedge \boldsymbol{dx}^{i_k}$. But covectors anticommute, and therefore we must pay with a factor $(-1)^k$:

$$\begin{aligned}\boldsymbol{d}(\mathbf{F} \wedge \mathbf{G}) = &\left(\frac{1}{k!}\partial_\mu F_{i_1\ldots i_k}\right) \boldsymbol{dx}^\mu \wedge \boldsymbol{dx}^{i_1} \wedge \cdots \\ &\wedge \boldsymbol{dx}^{i_k} \left(\frac{1}{\ell!} G_{j_1\ldots j_\ell}\right) \boldsymbol{dx}^{j_1} \wedge \cdots \wedge \boldsymbol{dx}^{j_\ell} \\ &+ (-1)^k \left(\frac{1}{k!} F_{i_1\ldots i_k}\right) \boldsymbol{dx}^{i_1} \wedge \cdots \\ &\wedge \boldsymbol{dx}^{i_k} \left(\frac{1}{\ell!}\partial_\mu G_{j_1\ldots j_\ell}\right) \boldsymbol{dx}^\mu \wedge \boldsymbol{dx}^{j_1} \wedge \cdots \wedge \boldsymbol{dx}^{j_\ell} \\ = &\ \boldsymbol{d}\mathbf{F} \wedge \mathbf{G} + (-1)^k \mathbf{F} \wedge \boldsymbol{d}\mathbf{G}.\end{aligned}$$

□

We may think of $\boldsymbol{d}$ as an *odd* form (cf. the discussion after Definition 7.3). In Leibniz's rule we have the two terms

$$\boldsymbol{d}\mathbf{F} \wedge \mathbf{G} \quad \text{and} \quad (-1)^k \mathbf{F} \wedge \boldsymbol{d}\mathbf{G}.$$

In the last term we have interchanged $\boldsymbol{d}$ and $\mathbf{F}$. This costs a sign if $\mathbf{F}$ is an odd form.

Observe that if ϕ is a differential form of degree 0, i.e., a scalar field, then $\phi \wedge \mathbf{G}$ is just a fancy way of writing $\phi(x^1, \ldots, x^n)\mathbf{G}$. In that case, Leibniz's rule degenerates to

$$\boldsymbol{d}(\phi\mathbf{G}) = \boldsymbol{d}\phi \wedge \mathbf{G} + \phi \wedge \boldsymbol{d}\mathbf{G} = \boldsymbol{d}\phi \wedge \mathbf{G} + \phi\boldsymbol{d}\mathbf{G}. \tag{7.18}$$

We can now recapture (7.16) in the following way. Consider a k-form

$$\mathbf{F} = \frac{1}{k!} F_{i_1\ldots i_k}(x)\boldsymbol{dx}^{i_1} \wedge \cdots \wedge \boldsymbol{dx}^{i_k}.$$

If we keep the coordinate system fixed, we may treat $F_{i_1\ldots i_k}(x)$ as a scalar field. Applying the exterior derivative, we therefore obtain

$$\begin{aligned}&= \frac{1}{k!}[\boldsymbol{d}F_{i_1\ldots i_k}(x)] \wedge \boldsymbol{dx}^{i_1} \wedge \cdots \wedge \boldsymbol{dx}^{i_k} \\ &+ \frac{1}{k!} F_{i_1\ldots i_k}(x) \wedge \boldsymbol{d}[\boldsymbol{dx}^{i_1} \wedge \cdots \wedge \boldsymbol{dx}^{i_k}].\end{aligned}$$

Using Leibniz's rule once more, we see that $\boldsymbol{d}[\boldsymbol{dx}^{i_1} \wedge \cdots \wedge \boldsymbol{dx}^{i_k}]$ automatically vanishes, since it involves only double derivatives (Theorem 3). We are therefore left with the first term. If we use that

$$\boldsymbol{d}F_{i_1\ldots i_k}(x) = \partial_\mu F_{i_1\ldots i_k}\boldsymbol{dx}^\mu,$$

we finally recapture (7.16).

We conclude this section with a definition of two important types of differential forms:

Definition 7.4.

a. A differential form $\mathbf{F}$ is *closed* if its exterior derivative vanishes; i.e., if $\boldsymbol{d}\mathbf{F} = 0$.

b. A differential form $\mathbf{G}$ is *exact* if it is the exterior derivative of another differential form $\mathbf{F}$; i.e., $\mathbf{G} = \boldsymbol{d}\mathbf{F}$.

Observe that an exact form $\mathbf{G}$ is automatically closed, since $\mathbf{G} = \boldsymbol{d}\mathbf{F}$ implies that

$$\boldsymbol{d}\mathbf{G} \Rightarrow \boldsymbol{d}^2\mathbf{F} = 0,$$

but the converse need not be true! (Examples will be given later on. See, e.g., Section 7.8, where the case of the monopole field is discussed.)

Exercise 7.3.4
Problem: Let $\mathbf{F}$ be a closed form and $\mathbf{G}$ an exact form. Show that $\mathbf{F} \wedge \mathbf{G}$ is exact.

A closely related type of differential form is given by

Definition 7.5. A k-form ω is called *simple* if there exist k smooth scalar fields $\phi_1 \ldots, \phi_k$ such that ω can be decomposed as

$$\omega = \boldsymbol{d}\phi_1 \wedge \cdots \wedge \boldsymbol{d}\phi_k. \tag{7.19}$$

The basic forms generated by a coordinate system and the simple forms are intimately connected. If $\phi_1, \ldots, \phi_k$ are sufficiently well behaved, then

$$\begin{aligned} y^1 &= \phi_1(x^1, \ldots, x^n), \\ &\vdots \\ y^k &= \phi_k(x^1, \ldots, x^n), \\ y^{k+1} &= x^{k+1} \\ &\vdots \\ y^n &= x^n \end{aligned}$$

is an admissible exchange of coordinates, and in the new coordinates $(y^1 \ldots, y^n)$ we can decompose ω as

$$\omega = \boldsymbol{d}y^1 \wedge \cdots \wedge \boldsymbol{d}y^k.$$

So ω is a basic form generated by the new coordinates.

Consider a simple form $\boldsymbol{d}\phi_1 \wedge \cdots \wedge \boldsymbol{d}\phi_k$. It can be interpreted geometrically in the same way as the basic forms. It generates a stratification, $\phi_1 = \ldots, -1, 0, 1, \ldots, \ldots, \phi_k = \ldots, -1, 0, 1, \ldots$, and $\boldsymbol{d}\phi_1 \wedge \cdots \wedge \boldsymbol{d}\phi_k(\mathbf{u}_1; \ldots; \mathbf{u}_k)$ is equal to the number of "tubes" intersected by the "parallelepiped" spanned by $(\mathbf{u}_1, \ldots, \mathbf{u}_n)$.

Simple forms have especially nice properties:

Lemma 7.1. *A simple form is closed and exact.*

PROOF. As $d^2 = 0$, it follows trivially that $d\phi_1 \wedge \cdots \wedge d\phi_k$ is closed. The exactness follows from the formula

$$d\phi_1 \wedge \cdots \wedge d\phi_k = d[\phi_1 d\phi_2 \wedge \cdots \wedge d\phi_k].$$

□

Worked Exercise 7.3.5
Problem: Let $f_1, \ldots, f_k : \mathbb{R} \to \mathbb{R}$ be smooth real functions. Show that

$$\omega = f_1(x^1) \ldots f_k(x^k) dx^1 \wedge \cdots \wedge dx^k \tag{7.20}$$

is a simple form.

7.4 The Volume Form

Suppose the manifold M^n is also equipped with a metric g. Then we can speak of areas, volumes, etc. in the tangent space $\mathbf{T}_P(M)$. Let us make this more precise. Consider an n-tuple $(\mathbf{u}_1, \ldots, \mathbf{u}_n)$ that generates a "parallelepiped." We want to find out how to compute the n-dimensional volume of this "parallelepiped." If we introduce coordinates $(x^1, \ldots, x^n)$, it would be tempting to put

$$\text{Vol}[\mathbf{u}_1; \ldots; \mathbf{u}_n] \stackrel{?}{=} dx^1 \wedge \cdots \wedge dx^n[\mathbf{u}_1; \ldots; \mathbf{u}_n]$$

by analogy to the lower-dimensional Euclidean cases. (See Exercise 7.2.5.) But here we should be careful, since the canonical frame generated by $(x^1, \ldots, x^n)$ need not be orthonormal.

The tangent space $\mathbf{T}_P(M)$ is simply isomorphic to the standard Euclidean space $\mathbb{R}^n$. Recall the standard definitions of a volume in n-dimensional Euclidean space. Let $(\mathbf{e}_1^0, \ldots, \mathbf{e}_n^0)$ be an orthonormal frame, and consider an arbitrary n-tuple of vectors $(\mathbf{v}_1, \ldots, \mathbf{v}_n)$. They span a "parallelepiped" with the volume

$$\text{Vol}(\mathbf{v}_1; \ldots; \mathbf{v}_n) = \det[\overline{\overline{B}}], \tag{7.21}$$

where the matrix elements $B^i_{\ j}$ are the coordinates of $\mathbf{v}_j$,

$$\mathbf{v}_j = \mathbf{e}_i^0 B^i_{\ j}.$$

Furthermore, $(\mathbf{v}_1, \ldots, \mathbf{v}_n)$ generates the same orientation as $(\mathbf{e}_1^0, \ldots, \mathbf{e}_n^0)$ if and only if $\det[\overline{\overline{B}}]$ is positive.

Now let us introduce coordinates around the point P. Then the canonical frame $(\mathbf{e}_1, \ldots, \mathbf{e}_n)$ generated from the coordinates need not be an orthonormal frame, but we can decompose it into the orthonormal frame $(\mathbf{e}_1^0, \ldots, \mathbf{e}_n^0)$:

$$\mathbf{e}_j = \mathbf{e}_i^0 E^i_{\ j}. \tag{7.22}$$

Furthermore, let $(\mathbf{u}, \ldots, \mathbf{u}_n)$ be an n-tuple of tangent vectors, where $\mathbf{u}_j$ is characterized by the contravariant components $A^i_{\ j}$. From Exercise 7.2.2 we know that (7.9)

$$dx^1 \wedge \cdots \wedge dx^n(\mathbf{u}_1, \ldots, \mathbf{u}_n) = \det(\overline{\overline{A}}).$$

We now get

$$\mathbf{u}_j = \mathbf{e}_i A^i{}_j = \mathbf{e}^0_k E^k{}_i A^i{}_j .$$

From (7.21) it then follows that

$$\text{Vol}(\mathbf{u}_1, \ldots, \mathbf{u}_n) = \det(\overline{\overline{E}}\,\overline{\overline{A}}) = \det(\overline{\overline{E}}) \cdot \boldsymbol{dx}^1 \wedge \cdots \wedge \boldsymbol{dx}^n(\mathbf{u}_1, \ldots, \mathbf{u}_n). \quad (7.23)$$

It remains to interpret $\det[\overline{\overline{E}}]$. consider the coordinate exchange

$$y^i = E^i{}_j x^j, \qquad \frac{\partial y^i}{\partial x^j} = E^i{}_j .$$

Observe that the orthonormal frame $(\mathbf{e}^0_1, \ldots, \mathbf{e}^0_n)$ is the canonical frame generated from the $(y^1, \ldots, y^n)$-coordinates. (This follows directly from (7.22) and (6.10).) The metric coefficients g_{ij} corresponding to the $(y^1, \ldots, y^n)$-coordinates therefore reduce to the Kronecker delta δ_{ij}. The transformation formula (6.21) for the metric coefficients may be rearranged as

$$g_{ij} = g^0_{k\ell} E^k{}_i E^\ell{}_j = \delta_{k\ell} E^k{}_i E^\ell{}_j = E^k{}_i E^k{}_j ;$$

i.e.,

$$\overline{\overline{G}} = \overline{\overline{E}}^\dagger \overline{\overline{E}}$$

It follows that

$$g = \det[\overline{\overline{G}}] = [\det(\overline{\overline{E}})]^2; \qquad \text{i.e., } \det[\overline{\overline{E}}] = \sqrt{g}.$$

Inserting this into (7.23), we have thus determined the volume form in the tangent space $\mathbf{T}_P(M)$:

$$\sqrt{g}\boldsymbol{dx}^1 \wedge \cdots \wedge \boldsymbol{dx}^n . \quad (7.24)$$

Later on we shall recover this formula from another point of view. The volume form also controls the orientation of the tangent space. This follows immediately from (7.9), which shows us that $(\mathbf{u}_1, \ldots, \mathbf{u}_n)$ generates the same orientation as the canonical frame $(\mathbf{e}_1, \ldots, \mathbf{e}_n)$ if and only if $\boldsymbol{dx}^1 \wedge \cdots \wedge \boldsymbol{dx}^n(\mathbf{u}_1; \ldots; \mathbf{u}_n)$ is positive.

In the preceding discussion we have been focusing upon a single tangent space. We will now try to extend the discussion of the local aspects of the volume form to its global aspects. Our final aim is to construct a global differential form ε that at every point P reduces to the local volume form given by (7.24). But as preparation for this, we must first discuss the orientability of manifolds.

Let M be a manifold and P_0 a fixed point in M. Consider the set of all coordinate systems surrounding P_0. Each coordinate system generates an orientation on the tangent space $\mathbf{T}_{P_0}(M)$.

Now consider two sets of coordinates: $(x^1, \ldots, x^n)$ and $(y^1, \ldots, y^n)$. From (6.12) we know that

$$\mathbf{e}_{(2)_j} = \mathbf{e}_{(1)_i} \frac{\partial x^i}{\partial y^j} .$$

Therefore, we conclude that $(x^1, \ldots, x^n)$ and $(y^1, \ldots, y^n)$ generate the same orientation on the tangent space $\mathbf{T}_{P_0}(M)$, provided that the Jacobian $\det\left[\frac{\partial x^i}{\partial y^j}\right]$ is positive.

Okay! Let us choose a positive orientation in our tangent space $\mathbf{T}_{P_0}(M)$. This is done by choosing a specific coordinate system (ϕ_0, U_0), which we declare to generate the positive orientation. If (ϕ, U) is an arbitrary coordinate system, then it generates a *positive* orientation if

$$\det\left[\frac{\partial x^\alpha}{\partial y^\beta}\right]$$

is positive, and it generates a *negative* orientation if its Jacobian is negative!

In what follows we will always assume that we have chosen a positive orientation in our tangent space $\mathbf{T}_{P_0}(M)$.

This was the local aspect (i.e., we have been focusing on what happens in a neighborhood of a point P_0). Now we turn to the global aspect of orientation. First, we consider a *simple* manifold, i.e., a manifold that can be covered by a single coordinate system (ϕ_0, U_0). It generates an orientation in each of the tangent spaces on M, and we declare these orientations to be the positive orientations!

Then we consider an arbitrary manifold. Suppose we have two coordinate systems (ϕ_1, U_1) and (ϕ_2, U_2) that overlap. If we are going to let them generate the same orientation in the tangent spaces, we must be careful in the overlapping region $\Omega_{12} = \phi_1(U_1) \cap \phi_2(U_2)$. We must demand that they generate the *same* orientation in the overlapping region Ω_{12}, i.e., that $\det[\partial x^\alpha / \partial y^\beta]$ be positive throughout the overlapping region. This motivates the following definition:

Definition 7.6. A manifold M is said to be *orientable* if we can choose an atlas $(\phi_i, U_i)_{i \in I}$ such that any two coordinate systems in our atlas generate the same orientation; i.e.,

$$\det\left(\frac{\partial x^\alpha}{\partial y^\beta}\right)$$

is positive throughout all overlapping regions.

When a manifold is orientable, we can choose an atlas with the above property. This atlas will then generate a specific orientation in each of the tangent spaces on M, which we arbitrarily declare to be positive!

Let us end this digression on orientability with an example of a manifold that is *not* orientable, the Möbius strip (see Figure 7.6).

Exercise 7.4.1
Problem: Let M^n be a differentiable manifold. Let us assume the existence of an n-form $\mathbf{\Omega}$ that is nowhere vanishing. If we introduce coordinates $(x^1, \ldots, x^n)$, we can decompose $\mathbf{\Omega}$ as $\mathbf{\Omega} = \Omega_{1\ldots n}(x)\boldsymbol{dx}^1 \wedge \cdots \wedge \boldsymbol{dx}^n$. That $\mathbf{\Omega}$ is nowhere vanishing implies that $\Omega_{1\ldots n}(x)$ is nowhere zero.

a. Let $(x^1, \ldots, x^n)$ denote an arbitrary set of coordinates. Show that the sign of $\mathbf{\Omega}(\mathbf{e}_1; \ldots; \mathbf{e}_n)$ is constant throughout the range of the coordinate system.

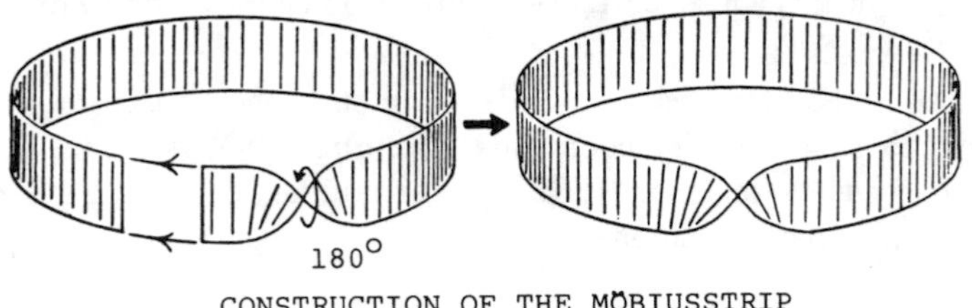

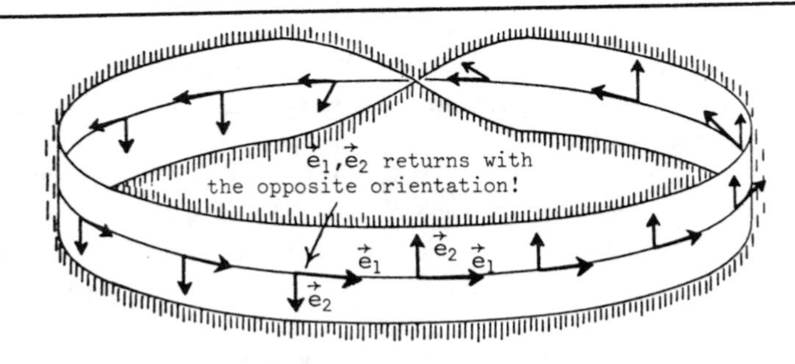

Figure 7.6.

b. Show that Ω generates an orientation of M by defining a coordinate system $(x^1, \ldots, x^n)$ to be positively oriented if and only if $\mathbf{\Omega}(\mathbf{e}_1; \ldots; \mathbf{e}_n)$ is positive.
c. Show that an n-tuple $(\mathbf{v}_1, \ldots, \mathbf{v}_n)$ is positively oriented with respect to the orientation generated by $\mathbf{\Omega}$ if and only if $\mathbf{\Omega}(\mathbf{v}_1; \ldots; \mathbf{v}_n)$ is positive.

Exercises 7.4.2
Problem: Let M^n be an n-dimensional manifold in $\mathbb{R}^{n+1}$; i.e., M is a smooth hypersurface. Let us assume the existence of a smooth normal vector field $\mathbf{n}(x)$ to M; i.e., to each point x in M we associate a normal vector $\mathbf{n}(x)$ to the tangent space. Show that M is orientable. (Hint: Let $(x^1, \ldots, x^{n+1})$ denote the extrinsic coordinates in $\mathbb{R}^{n+1}$. Let $(\mathbf{v}^1, \ldots, \mathbf{v}^n)$ be an arbitrary n-tuple tangent to M. Define a smooth n-form Ω on M in the following way:

$$\Omega(\mathbf{v}_1; \ldots; \mathbf{v}_n) = \boldsymbol{dx}^1 \wedge \cdots \wedge \boldsymbol{dx}^{n+1}(\mathbf{n}; \mathbf{v}_1; \ldots \mathbf{v}_n)$$

and apply Exercise 7.4.1.)

Consider the sphere and the Möbius strip in $\mathbb{R}^3$ (see Figures 7.7 and 7.8)! You can see with your own eyes that the sphere allows a normal field, while the Möbius strip does not!

As preparation for the construction of a global volume form, we are going to investigate two special quantities that are *not* true geometrical objects; i.e., they can be defined only in connection with a coordinate system on the manifold.

The first quantity is intimately connected with the metric $\boldsymbol{g}$ on our manifold. To be specific, let us assume that $\boldsymbol{g}$ is a Minkowski metric. Consider a point P_0 in M and a coordinate system covering P_0. We can then characterize $\boldsymbol{g}$ in terms of its

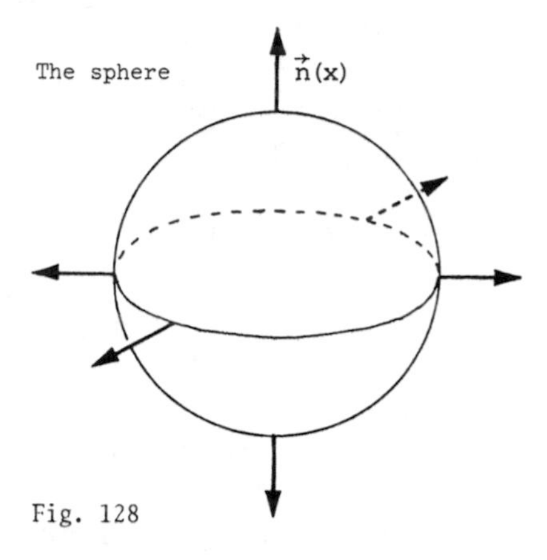

Figure 7.7.

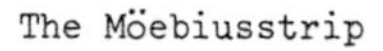

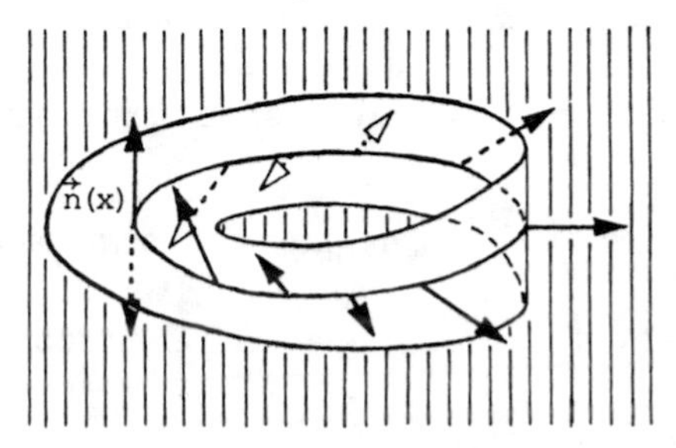

Figure 7.8.

components,

$$g_{\alpha\beta} = g_{\alpha\beta}(x_0^1, \ldots, x_0^n),$$

and we may compute the determinant

$$g = \det[g_{\alpha\beta}].$$

This determinant is necessarily negative because $\mathbf{g}$ is a Minkowski metric. It is this determinant we are going to examine. Obviously, $g(x_0)$ is *not* a scalar; i.e., it does *not* transform according to the rule $g_{(2)}(y_0) = g_{(1)}(x_0)$ when we exchange the coordinate system. To find the transformation rule, we use that $g_{\alpha\beta}$ transforms covariantly (6.21):

$$g_{(2)_{\alpha\beta}} = g_{(1)_{\gamma\delta}} \frac{\partial x^\gamma}{\partial y^\alpha} \frac{\partial x^\delta}{\partial y^\beta} \, .$$

We can rearrange this as a matrix equation (6.23):

$$\overline{\overline{G}}_{(2)} = \overline{\overline{D}}^{+}_{12}\overline{\overline{G}}_{(1)}\overline{\overline{D}}_{12}.$$

If we evaluate the determinant, we find that

$$\det[\overline{\overline{G}}_{(2)}] = \det[\overline{\overline{D}}^{+}_{12}] \cdot \det[\overline{\overline{G}}_{(1)}] \cdot \det[\overline{\overline{D}}_{12}];$$

i.e.,

$$g_{(2)}(y_0) = \left[\det\left(\frac{\partial x^\alpha}{\partial y^\beta}\right)\right]^2 g_{(1)}(x_0). \tag{7.25}$$

This transformation property makes it natural to look at $\sqrt{-g(x)}$, and we conclude:

Lemma 7.2. *On a manifold with a Minkowski metric,* $\sqrt{-g(x)}$ *transforms according to the rule*

$$\sqrt{-g_{(2)}(y)} = \left|\det\left(\frac{\partial x^\alpha}{\partial y^\beta}\right)\right| \sqrt{-g_{(1)}(x)}. \tag{7.26}$$

(*On a manifold with a Euclidean metric we just replace* $\sqrt{-g}$ *by* $\sqrt{g}$.)

Keeping this in mind, we now introduce the next quantity, the Levi-Civita symbol:

$$\epsilon_{a_1\ldots a_n} = \begin{cases} +1 & \text{if } (a_1, \ldots, a_n) \text{ is an even permutation of } (1, \ldots, n), \\ -1 & \text{if } (a_1 \ldots a_n) \text{ is an odd permutation of } (1, \ldots, n), \\ 0 & \text{otherwise.} \end{cases} \tag{7.27}$$

Again we select a point P_0 and an arbitrary coordinate system to this point. Then we attach the Levi-Civita symbol to this point P_0. This does *not* correspond to the components of a cotensor, since by definition it transforms according to the rule

$$\epsilon_{a_1\ldots a_n}(y_0) = \epsilon_{a_1\ldots a_n}(x_0). \tag{7.28}$$

However, we may use these two quantities, $\sqrt{-g}$ and $\epsilon_{a_1\ldots a_n}$ to construct a differential form. Let us choose specific coordinates $(x^1, \ldots, x^n)$ and consider the n-form $\mathbf{T}$ characterized by the components

$$T_{a_1\ldots a_n} = \sqrt{-g(x)}\epsilon_{a_1\ldots a_n},$$

with respect to this particular coordinate system.

Let us try to compute the components of $\mathbf{T}$ with respect to some other coordinates $(y^1, \ldots, y^n)$. As $T_{a_1\ldots a_n}$ transforms covariantly, we get

$$\begin{aligned} T_{(2)b_1\ldots b_n}(y_0) &= T_{(1)a_1\ldots a_n}(x_0)\frac{\partial x^{a_1}}{\partial y^{b_1}} \cdots \frac{\partial x^{a_n}}{\partial y^{b_n}} \\ &= \sqrt{-g_{(1)}(x_0)}\epsilon_{a_1\ldots a_n}\frac{\partial x^{a_1}}{\partial y^{b_1}} \cdots \frac{\partial x^{a_n}}{\partial y^{b_n}}. \end{aligned} \tag{$*$}$$

Here we should try to study

$$\epsilon_{a_1 \ldots a_n} \frac{\partial x^{a_1}}{\partial y^{b_1}} \cdots \frac{\partial x^{a_n}}{\partial y^{b_n}}$$

a little more closely. If $(b_1, \ldots, b_n) = (1, \ldots, n)$, then

$$\epsilon_{a_1 \ldots a_n} \frac{\partial x^{a_1}}{\partial y^{1}} \cdots \frac{\partial x^{a_n}}{\partial y^{n}} = \det \left[\frac{\partial x^\alpha}{\partial y^\beta} \right]$$

by the very definition of a determinant! In the general case we conclude that

$$\begin{aligned} \epsilon_{a_1 \ldots a_n} \frac{\partial x^{a_1}}{\partial y^{b_1}} \cdots \frac{\partial x^{a_n}}{\partial y^{b_n}} &= \det \begin{bmatrix} \frac{\partial x^1}{\partial y^{b_1}} & \cdots & \frac{\partial x^1}{\partial y^{b_n}} \\ \vdots & & \vdots \\ \frac{\partial x^n}{\partial y^{b_1}} & \cdots & \frac{\partial x^n}{\partial y^{b_n}} \end{bmatrix} \\ &= \begin{cases} \det \left[\frac{\partial x^\alpha}{\partial y^\beta} \right] & \text{if } (b_1, \ldots, b_n) \text{ is an even permutation of } (1, \ldots, n), \\ -\det \left[\frac{\partial x^\alpha}{\partial y^\beta} \right] & \text{if } (b_1, \ldots, b_n) \text{ is an odd permutation of } (1, \ldots, n), \\ 0 & \text{otherwise.} \end{cases} \end{aligned}$$

Consequently, we obtain the following formula:

$$\epsilon_{a_1 \ldots a_n} \frac{\partial x^{a_1}}{\partial y^{b_1}} \cdots \frac{\partial x^{a_n}}{\partial y^{b_n}} = \epsilon_{b_1 \ldots b_n} \det \left(\frac{\partial x^\alpha}{\partial y^\beta} \right). \tag{7.29}$$

Inserting this in $(*)$, we get

$$T_{(2)_{b_1 \ldots b_n}}(y_0) = \sqrt{-g_{(1)}(x_0)} \epsilon_{b_1 \ldots b_n} \det \left(\frac{\partial x^\alpha}{\partial y^\beta} \right).$$

But using (7.26), we finally obtain

$$\begin{aligned} T_{(2)_{b_1 \ldots b_n}}(y_0) &= \frac{\det \left(\frac{\partial x^\alpha}{\partial y^\beta} \right)}{\left| \det \left(\frac{\partial x^\alpha}{\partial y^\beta} \right) \right|} \sqrt{-g_{(2)}(y_0)} \epsilon_{b_1 \ldots b_n} \\ &= \operatorname{Sgn} \left(\det \left[\frac{\partial x^\alpha}{\partial y^\beta} \right] \right) \sqrt{-g_{(2)}(y_0)} \epsilon_{b_1 \ldots b_n}. \end{aligned}$$

Consequently, we see that up to a sign, $T_{(2)_{b_1 \ldots b_n}}$ has exactly the same form as $T_{(1)_{a_1 \ldots a_n}}$ This motivates the introduction of the *Levi-Civita form*:

Definition 7.7. Let M be an orientable manifold of dimension n. Then the *Levi-Civita form* associated with the chosen orientation is the n-form ε characterized by the components

$$[\varepsilon]_{a_1 \ldots a_n} = \begin{cases} \sqrt{-g(x)} \epsilon_{a_1 \ldots a_n} & \text{with respect to a positively oriented coordinate system;} \\ -\sqrt{-g(x)} \epsilon_{a_1 \ldots a_n} & \text{with respect to a negatively oriented coordinate system.} \end{cases}$$

(If the metric is Euclidean, we replace $\sqrt{-g(x)}$ by $\sqrt{g(x)}$.)

Observe that the Levi-Civita form ε is defined only after we have fixed the orientation of our manifold. If we change this orientation, then the Levi-Civita tensor is replaced by the opposite tensor; i.e.,

$$\varepsilon \rightarrow \varepsilon' = -\varepsilon.$$

For this reason, physicists refer to it as a *pseudotensor*.

Exercise 7.4.3
Problem: Consider the Levi-Civita symbol $\epsilon_{a_1\ldots a_n}$. We define

$$\epsilon^{a_1\ldots a_n} = \epsilon_{a_1\ldots a_n}.$$

(1) Show that

$$\epsilon_{a_1\ldots a_n}\epsilon^{a_1\ldots a_n} = n!. \tag{7.30}$$

$$\epsilon^{a_1\ldots a_{n-k}b_1\ldots b_k}\epsilon_{a_1\ldots a_{n-k}c_1\ldots c_k} = (n-k)!\text{Sgn}\begin{bmatrix} b_1 \ldots b_k \\ c_1 \ldots c_k \end{bmatrix}. \tag{7.31}$$

(2) Consider the case $n = 2$ and $n = 3$. Deduce the rules

$$\epsilon_{ab}\epsilon^{ab'} = \delta_b^{b'}; \quad \epsilon_{abc}\epsilon^{abc'} = 2\delta_c^{c'}; \quad \epsilon_{abc}\epsilon^{ab'c'} = \delta_b^{b'}\delta_c^{c'} - \delta_c^{b'}\delta_b^{c'}. \tag{7.32}$$

(3) Show that the contravariant components of the Levi-Civita form ε are given by

$$[\varepsilon]^{a_1\ldots a_n} = \begin{cases} -\frac{1}{\sqrt{-g(x)}}\epsilon^{a_1\ldots a_n} & \text{(positively oriented coordinate system)}, \\ +\frac{1}{\sqrt{-g(x)}}\epsilon^{a_1\ldots a_n} & \text{(negatively oriented coordinate system)}. \end{cases}$$

(4) Let $(x^1, \ldots, x^n)$ be positively oriented coordinates. Show that the Levi-Civita form can be decomposed as

$$\varepsilon = \sqrt{-g(x)}\boldsymbol{d}x^1 \wedge \cdots \wedge \boldsymbol{d}x^n. \tag{7.34}$$

Let $(x^1, \ldots, x^n)$ denote positively oriented coordinates on a Riemannian manifold M^n. From (7.34) we learn that the Levi-Civita form ε can be decomposed as

$$\varepsilon = \sqrt{g}\boldsymbol{d}x^1 \wedge \cdots \wedge \boldsymbol{d}x^n.$$

If we compare this with (7.24), we see that ε is nothing but the *volume form*. Observe that ε is globally well-defined, so that on an orientable manifold we can piece together all the volume forms on the different tangent spaces to a globally well-defined n-form. On a nonorientable manifold this is no longer possible.

Observe also that ε not only generates the volume but also the *orientation* of the tangent spaces. If $(\mathbf{v}_1 \ldots, \mathbf{v}_n)$ is an arbitrary n-tuple, then $\varepsilon(\mathbf{v}_1; \ldots; \mathbf{v}_n)$ is the volume of the parallelepiped spanned by $(\mathbf{v}_1, \ldots, \mathbf{v}_n)$, and it is positive if and only if $(\mathbf{v}_1, \ldots, \mathbf{v}_n)$ is positively oriented.

The Levi-Civita form can also be used to generalize the cross product from the ordinary Euclidean space $\mathbb{R}^3$. Let M^n be a Riemannian manifold, and consider a set of $(n-1)$ tangent vectors, $\mathbf{u}_1, \ldots, \mathbf{u}_{n-1}$. If we contract this set with the Levi-Civita form ε, we obtain a 1-form

$$\boldsymbol{n} = \varepsilon(\cdot; \mathbf{u}_1; \ldots; \mathbf{u}_{n-1}). \tag{7.35}$$

Since we work on a manifold with a metric, this 1-form is equivalent to a tangent vector $\mathbf{n}$, and it is this tangent vector $\mathbf{n}$ that generalizes the usual cross product (compare the discussion in Section 1.1):

Lemma 7.3. *The vector* $\mathbf{n}$ *is characterized by the following three properties:*

1. $\mathbf{n}$ *is orthogonal to each of the tangent vectors* $\mathbf{u}_1, \ldots, \mathbf{u}_{n-1}$.
2. *The length of* $\mathbf{n}$ *is equal to the "area" of the "parallelogram" spanned by* $\mathbf{u}_1, \ldots, \mathbf{u}_{n-1}$.
3. *The n-tuple* $(\mathbf{n}, \mathbf{u}_1, \ldots, \mathbf{u}_{n-1})$ *is positively oriented.*

PROOF. First we deduce a useful formula. Let $\mathbf{v}$ be an arbitrary tangent vector. Then

$$\begin{aligned} g(\mathbf{n}; \mathbf{v}) &= n_a v^a = \sqrt{g}\epsilon_{ab_1 \ldots b_{n-1}} u^{b^1} \ldots u^{b_{n-1}} v^a \\ &= \varepsilon(\mathbf{v}; \mathbf{u}_1; \ldots; \mathbf{u}_{n-1}). \end{aligned} \tag{7.36}$$

The rest now follows easily:

1. $g(\mathbf{n}; \mathbf{u}_i) = \varepsilon(\mathbf{u}_i; \mathbf{u}_1; \ldots; \mathbf{u}_{n-1})$, which vanishes automatically since ε is skew-symmetric
2. Consider the normalized vector $\mathbf{n}/\|\mathbf{n}\|$. It is a unit vector orthogonal to $\mathbf{u}_1, \ldots, \mathbf{u}_{n-1}$. Therefore, the volume of the parallelepiped spanned by $(\mathbf{n}/\|\mathbf{n}\|)$, $\mathbf{u}_1, \ldots, \mathbf{u}_{n-1}$ reduces to the area of the parallelogram spanned by $\mathbf{u}_1, \ldots, \mathbf{u}_{n-1}$. Consequently, we get that

$$\begin{aligned} \text{Area}[\mathbf{u}_1; \ldots; \mathbf{u}_{n-1}] &= \left(\mathbf{n}/\|\mathbf{n}\|; \mathbf{u}_1; \ldots; \mathbf{u}_{n-1}\right) \\ &= g(\mathbf{n}; \mathbf{n}/\|\mathbf{n}\|) \\ &= \frac{1}{\|\mathbf{n}\|} g(\mathbf{n}; \mathbf{n}) = \|\mathbf{n}\|. \end{aligned}$$

3. We immediately get that

$$\varepsilon(\mathbf{n}; \mathbf{u}_1; \ldots; \mathbf{u}_{n-1}) = g(\mathbf{n}, \mathbf{n}) > 0.$$

□

Lemma 7.3 motivates that we define $\mathbf{n}$ to be the cross product of $\mathbf{u}_1, \ldots, \mathbf{u}_{n-1}$, and we write it in the usual manner:

$$\mathbf{n} = \mathbf{u}_1 \times \cdots \times \mathbf{u}_{n-1}.$$

Exercise 7.4.4
Problem: Consider the Euclidean space $\mathbb{R}^{n+1}$.

a. Let $(\mathbf{v}, \mathbf{u}_1, \ldots, \mathbf{u}_n)$ be an arbitrary $(n+1)$-tuple. Show that the volume of the parallelepiped spanned by $(\mathbf{v}, \mathbf{u}_1, \ldots, \mathbf{u}_n)$ is given by the familiar formula

$$\text{Vol}[\mathbf{v}, \mathbf{u}_1, \ldots, \mathbf{u}_n] = \mathbf{v} \cdot (\mathbf{u}_1 \times \cdots \times \mathbf{u}_n). \tag{7.37}$$

b. Let M^n be an orientable n-dimensional manifold in $\mathbb{R}^{n+1}$. At each point in M we select a positively oriented orthonormal frame $(\mathbf{u}_1, \ldots, \mathbf{u}_n)$ in the corresponding tangent space. Show that the unit normal vector field

$$\mathbf{n} = \mathbf{u}_1 \times \cdots \times \mathbf{u}_n$$

is a globally smooth normal vector field on M (compare with Exercise 7.4.2).

7.5 The Dual Map

As another important application of the Levi-Civita form, we will use it to construct the dual map $*$, which allows us to identify forms of different rank. This gives a greater flexibility in the manipulations of various physical quantities.

Let $\mathbf{F}$ be a k-form with the components $F_{i_1 \ldots i_k}$. We start by considering the associated tensor with the contravariant components

$$F^{j_1 \ldots j_k} = g^{i_1 j_1} \cdots g^{i_k j_k} F_{i_1 \ldots i_k}.$$

Then we contract this tensor with the Levi-Civita form and obtain

$$\frac{1}{k!} \sqrt{g(x)} \epsilon_{a_1 \ldots a_{n-k} b_1 \ldots b_k} F^{b_1 \ldots b_k} \begin{bmatrix} \text{positively oriented} \\ \text{coordinate system!} \end{bmatrix},$$

where $\frac{1}{k!}$ is included to make life easier!

An $\epsilon_{a_1 \ldots a_n}$ is skew-symmetric, we have thus produced a form of degree $n - k$. This form is called the *dual form*, and it is denoted by $^*\mathbf{F}$, or sometimes by $\overset{*}{\mathbf{F}}$.

Definition 7.8[1]**.** Let $\mathbf{F}$ be a k-form on an orientable manifold M^n with a metric $\boldsymbol{g}$. Then the dual form $^*\mathbf{F}$ is the $(n-k)$-form characterized by the components

$$(^*\mathbf{F})_{a_1 \ldots a_{n-k}} = \begin{cases} \frac{1}{k!} \sqrt{g(x)} \epsilon_{a_1 \ldots a_{n-k} b_1 \ldots b_k} F^{b_1 \ldots b_k} & \text{(Euclidean metric),} \\ \frac{1}{k!} \sqrt{-g(x)} \epsilon_{a_1 \ldots a_{n-k} b_1 \ldots b_k} F^{b_1 \ldots b_k} & \text{(Minkowski metric)} \end{cases} \tag{7.38}$$

with respect to a positively oriented coordinate system.

Exercise 7.5.1
Problem: (a) Let ϕ be a scalar field on a Riemannian manifold and let $(x^1, \ldots, x^n)$ be positively oriented coordinates. Show that

$$^*\phi = \phi \sqrt{g} \boldsymbol{dx}^1 \wedge \cdots \wedge \boldsymbol{dx}^n. \tag{7.39}$$

(b) Show that

$$\varepsilon = {}^*1, \qquad {}^*\varepsilon = \begin{cases} +1 & \text{(Euclidean metric),} \\ -1 & \text{(Minkowski metric).} \end{cases} \tag{7.40}$$

[1] This definition differs slightly from the one generally adopted by mathematicians. See, for instance, Goldberg (1962) or de Rham (1955).

Exercise 7.5.2
Problem: Show that the *contravariant* components of $^*\mathbf{F}$ are obtained by contracting the components of $\mathbf{F}$ with the *contravariant* components of ε; i..e,

$$({}^*\mathbf{F})^{a_1\ldots a_{n-k}} = \begin{cases} \frac{1}{k!\sqrt{g}}\,\epsilon^{a_1\ldots a_{n-k}b_1\ldots b_k}F_{b_1\ldots b_k} & \text{(Euclidean metric)},\\ \frac{-1}{k!\sqrt{-g}}\,\epsilon^{a_1\ldots a_{n-k}b_1\ldots b_k}F_{b_1\ldots b_k} & \text{(Minkowski metric)}.\end{cases} \tag{7.41}$$

In this way we have constructed a map from $\Lambda^k(M)$ to $\Lambda^{n-k}(M)$. It is called the *dual map* or the *Hodge duality*. By construction, it is obviously linear:

$$^*(\mathbf{F}+\mathbf{G}) = {}^*\mathbf{F} + {}^*\mathbf{G}, \qquad {}^*(\lambda\mathbf{F}) = \lambda({}^*\mathbf{F}). \tag{7.42}$$

But a more interesting property is the following:

Theorem 7.5.

$$\begin{aligned} {}^{**}\mathbf{T} &= (-1)^{k(n-k)}\mathbf{T} \qquad \textit{(Euclidean metric)},\\ {}^{**}\mathbf{T} &= -(-1)^{k(n-k)}\mathbf{T} \quad \textit{(Minkowski metric)}.\end{aligned} \tag{7.43}$$

So up to a sign, $*$ is its own inverse. In particular, the dual map $* : \Lambda^k(M) \curvearrowright \Lambda^{n-k}(M)$ is an isomorphism.

PROOF. For simplicity we consider only the case of a Minkowski metric. Let $\mathbf{F}$ be an arbitrary k-form. Then $^*\mathbf{F}$ has the components

$$\frac{1}{k!}\sqrt{-g}\epsilon_{a_1\ldots a_{n-k}b_1\ldots b_k}F^{b_1\ldots b_k}.$$

We want to compute the components of $^{**}\mathbf{F}$.

The first thing we must do is to raise the indices of $^*\mathbf{F}$, i.e., determine the contravariant components of $^*\mathbf{F}$. According to Exercise 7.5.2, they are given by the formula (7.41)

$$({}^*\mathbf{F})^{a_1\ldots a_{n-k}} = -\frac{1}{k!\sqrt{-g}}\epsilon^{a_1\ldots a_{n-k}b_1\ldots b_k}F_{b_1\ldots b_k}.$$

Then we shall contract the Levi-Civita form with $^*\mathbf{F}$ (remembering the statistical factor!)

$$\begin{aligned}({}^{**}\mathbf{F})_{c_1\ldots c_k} &= \frac{1}{(n-k)!}\sqrt{-g}\epsilon_{c_1\ldots c_k a_1\ldots a_{n-k}}({}^*\mathbf{F})^{a_1\ldots a_{n-k}}\\ &= -\frac{1}{k!(n-k)!}\epsilon_{c_1\ldots c_k a_1\ldots a_{n-k}}\epsilon^{a_1\ldots a_{n-k}b_1\ldots b_k}F_{b_1\ldots b_k}.\end{aligned} \tag{Δ}$$

The rest of the proof is a combinatorial argument! First, we observe that

$$\epsilon_{c_1\ldots c_k a_1\ldots a_{n-k}} = (-1)^{k(n-k)}\epsilon_{a_1\ldots a_{n-k}c_1\ldots c_k}.$$

Then we use that (cf. Exercise 7.4.3)

$$\epsilon_{a_1\ldots a_{n-k}c_1\ldots c_k}\epsilon^{a_1\ldots a_{n-k}b_1\ldots b_k} = (n-k)!\mathrm{Sgn}\begin{bmatrix} b_1 & \ldots & b_k\\ c_1 & \ldots & c_k\end{bmatrix}.$$

Inserting these formulas into (Δ), the components of ($^{**}\mathbf{F}$) reduce to

$$(^{**}\mathbf{F})_{c_1\ldots c_k} = -\frac{1}{k!}\,\mathrm{Sgn}\begin{bmatrix} b_1 \ldots b_k \\ c_1 \ldots c_k \end{bmatrix} F_{b_1\ldots b_k}(-1)^{k(n-k)}. \qquad (\square)$$

But $F_{b_1\ldots b_k}$ is skew-symmetric, and therefore

$$\mathrm{Sgn}\begin{bmatrix} b_1 \ldots b_k \\ c_1 \ldots c_k \end{bmatrix} F_{b_1\ldots b_k} = k!\,F_{c_1\ldots c_k},$$

and if we insert that into ($\square$), we are through. $\square$

From Theorems 7.3 and 7.5 we learn the following important rule for dealing with the operators in an exterior algebra: "*Be wise — apply them twice.*"

We shall evaluate the sign associated with the double map $**$ so often that it pays to summarize in the Table 7.1.

Observe that the dual form $^*\mathbf{F}$ involves the Levi-Civita form. Thus it depends on the orientation of our manifold. If we exchange the orientation for the opposite one, then the dual form is replaced by the opposite form

$$^*F \curvearrowright \overset{*}{\mathbf{F}}{}' = -\overset{*}{\mathbf{F}}. \qquad (7.44)$$

For this reason physicists also call the dual form $^*\mathbf{F}$ a *pseudotensor*.

Exercise 7.5.3

Introduction: Let M^n be an orientable Riemann manifold. A differential form $\mathbf{F}$ is called *self-dual* if it satisfies

$$^*\mathbf{F} = \mathbf{F},$$

and *anti–self-dual* if

$$^*\mathbf{F} = -\mathbf{F}.$$

Problem: Show that self-dual and anti–self-dual forms can exist only in Riemannian spaces of dimension 4, 8, 12, 16, ...

Exercise 7.5.4

Problem: Let M^n be an orientable Riemannian manifold and $\mathbf{A}$ a 1-form on M. Then $\mathbf{A}$ is equivalent to a tangent vector $\mathbf{a}$. Let $(\mathbf{u}_1, \ldots, \mathbf{u}_{n-1})$ be an arbitrary $(n-1)$-tuple. Show that the volume of the parallelepiped spanned by $(\mathbf{u}_1, \ldots, \mathbf{u}_{n-1}, \mathbf{a})$ is given by

$$\mathrm{Vol}[\mathbf{u}_1; \ldots; \mathbf{u}_{n-1}; \mathbf{a}] = {}^*\mathbf{A}(\mathbf{u}_1; \ldots; \mathbf{u}_{n-1}) = \mathbf{a} \times (\mathbf{u}_1 \times \cdots \times \mathbf{u}_{n-1}). \qquad (7.45)$$

This gives a geometric characterization of the dual form $^*\mathbf{A}$.

Table 7.1.

Euclidean metric			Minkowski metric		
$^{**}F$	k even	k odd	$^{**}F$	k even	k odd
n even	F	$-F$	n even	$-F$	F
n odd	F	F	n odd	$-F$	$-F$

Exercise 7.5.5
Problem: (a) Show that

$$^*(d^{a_1} \wedge \cdots \wedge dx^{a_k}) = \frac{\sqrt{\pm g}}{(n-k)!} \epsilon_{i_1\ldots i_{n-k} j_1 \ldots j_k} g^{j_1 a_1} \ldots g^{j_k a_k} dx^{i_1} \wedge \cdots \wedge dx^{i_{n-k}}. \tag{7.46}$$

(b) Consider spherical coordinates in Euclidean space $\mathbb{R}^3$. Show that the dual map is given by

$$^*dr = r^2 \sin\theta d\theta \wedge d\varphi; \quad ^*d\theta = \sin\theta d\varphi \wedge dr; \quad ^*d\varphi = \frac{1}{\sin\theta} dr \wedge d\theta.$$

The scalar product between two 1-forms is given by $g^{ij} A_i B_j = A^j B_j$. Clearly, this is a special case of a *contraction* between a k-form **T** and an m-form **S**, where $k \geq m$:

$$T_{i_1\ldots i_{k-m} j_1 \ldots j_m} S^{j_1 \ldots j_m}.$$

The result obviously is a differential form of degree $(k - m)$. It would be nice to be able to express such contractions in a coordinate-free manner, and as we shall see, this is possible using the dual map. Thus the following pattern emerges: When we have a metric on an orientable manifold, we can use this metric to construct a dual map $*$, but once we have constructed the dual map, we can forget about the metric: It has been "coded" into the dual map.

The first thing we will construct is the scalar product between 1-forms. Let **A** and **B** be two 1-forms. Then $^*\mathbf{A}$ is an $(n-1)$-form, and therefore $^*\mathbf{A} \wedge \mathbf{B}$ is an n-form. But this can be dualized to give a scalar

$$^*(^*\mathbf{A} \wedge \mathbf{B}).$$

This scalar is a coordinate invariant formed out of A_i and B_i, and therefore it must be proportional to $A_i B^i$ (since there are no other coordinate invariants that can be made out of g_{ij}, A_i, B_i, and $\sqrt{g}\epsilon_{i_1\ldots i_n}$!) Okay, that was a rather abstract argument. Let us work out the coordinate expression for $^*(^*\mathbf{A} \wedge \mathbf{B})$ in, e.g., the Euclidean case:

$$\mathbf{C} = {}^*(^*\mathbf{A} \wedge \mathbf{B}); \qquad \text{i.e., } {}^*\mathbf{C} = {}^*\mathbf{A} \wedge \mathbf{B}.$$

Here the right-hand side is characterized by the component

$$\begin{aligned}(^*\mathbf{A} \wedge \mathbf{B})_{12\ldots n} &= \frac{1}{(n-1)!} \left[(^*\mathbf{A})_{12\ldots(n-1)} (\mathbf{B})_n + \ldots - \ldots\right] \\ &= (^*\mathbf{A})_{12\ldots n-1} (\mathbf{B})_n + (-1)^{n-1} (^*\mathbf{A})_{2\ldots n} (B)_1 + \ldots \\ &= \sqrt{g} A^n B_n + \sqrt{g} A^1 B_1 + \ldots = \sqrt{g} A^i B_i,\end{aligned}$$

whereas the left-hand side is characterized by the component

$$(^*\mathbf{C})_{1\ldots n} = \sqrt{g}\epsilon_{1\ldots n} \mathbf{C} = \sqrt{g}\mathbf{C}.$$

From this we conclude that

$$^*(^*\mathbf{A} \wedge \mathbf{B}) = \mathbf{C} = A^i B_i,$$

which justifies our claim. The expression $^*(^*\mathbf{A} \wedge \mathbf{B})$ is not manifestly symmetric in $\mathbf{A}$ and $\mathbf{B}$, but neither is $A^i B_i$! Observe that the dualization of $\mathbf{A}$ is reflected in the fact that the index of $\mathbf{A}$ has been raised, $A_i \curvearrowright A^i$. We have thus shown:

Theorem 7.6. *Let* $\mathbf{A}$ *and* $\mathbf{B}$ *be* 1*-forms. Then*

$$^*(^*\mathbf{A} \wedge \mathbf{B}) = {}^*(^*\mathbf{B} \wedge \mathbf{A}) = A^i B_i \qquad (\textit{Euclidean metric}), \tag{7.47}$$

$$^*(^*\mathbf{A} \wedge \mathbf{B}) = {}^*(^*\mathbf{B} \wedge \mathbf{A}) = -A^\mu B_\mu \quad (\textit{Minkowski metric}). \tag{7.48}$$

This theorem has an important consequence. It tells us how to reconstruct the *metric* from the dual map! Sometimes it is preferable to omit the last dualization, which in both cases produces the formula

$$^*\mathbf{A} \wedge \mathbf{B} = {}^*\mathbf{B} \wedge \mathbf{A} = A^i B_i \varepsilon. \tag{7.49}$$

The above formula suggests a generalization of scalar products to arbitrary differential forms. Before we proceed, we need a technical result:

Worked Exercise 7.5.6
Problem: Let $\mathbf{T}$, $\mathbf{U}$ be k-forms. Show that

$$^*\mathbf{T} \wedge \mathbf{U} = \frac{1}{k!} T^{i_1 \ldots i_k} U_{i_1 \ldots i_k} \varepsilon. \tag{7.50}$$

From Exercise 7.5.6 we see that $^*\mathbf{T} \wedge \mathbf{U}$ is symmetric in $\mathbf{T}$ and $\mathbf{U}$, and that it represents the coordinate invariant

$$\frac{1}{k!} T^{i_1 \ldots i_k} U_{i_1 \ldots i_k} = \sum_{i_1 < i_2 < \cdots < i_k} T^{i_1 \ldots i_k} U_{i_1 \ldots i_k}.$$

The statistical factor is very reasonable, since it removes "double counting."

Let us introduce the following abbreviation for the scalar product of two k-forms:

$$(\mathbf{T} \mid \mathbf{U}) \stackrel{\text{def}}{=} \frac{1}{k!} T^{i_1 \ldots i_k} U_{i_1 \ldots i_k}. \tag{7.51}$$

Then we have succeeded in generalizing Theorem 7.6 to arbitrary k-forms:

Theorem 7.7. *Let* $\mathbf{T}$, $\mathbf{U}$ *be* k*-forms. Then*

$$^*(^*\mathbf{T} \wedge \mathbf{U}) = (\mathbf{T} \mid \mathbf{U}) = \frac{1}{k!} T^{i_1 \ldots i_k} U_{i_1 \ldots i_k} \qquad (\textit{Euclidean metric}). \tag{7.52}$$

$$-^*(^*\mathbf{T} \wedge \mathbf{U}) = (\mathbf{T} \mid \mathbf{U}) = \frac{1}{k!} T^{\mu_1 \ldots \mu_k} U_{\mu_1 \ldots \mu_k} \quad (\textit{Minkowski metric}). \tag{7.53}$$

Exercise 7.5.7
Problem: Show that

a. $(\mathbf{d}x^a \mid \mathbf{d}x^b) = g^{ab}$.
b. $(\mathbf{d}x^a \wedge \mathbf{d}x^b \mid \mathbf{d}x^c \wedge \mathbf{d}x^d) = g^{ac} g^{bd} - g^{ad} g^{bc}$.

Remark: It can be shown in general that

$$(dx^{a_1} \wedge \cdots \wedge dx^{a_k} \mid dx^{b_1} \wedge \cdots \wedge dx^{b_k}) = \frac{1}{k!} g^{i_1 j_1} \ldots g^{i_k j_k} \operatorname{Sgn} \begin{bmatrix} a_1 \ldots a_k \\ i_1 \ldots i_k \end{bmatrix} \operatorname{Sgn} \begin{bmatrix} b_1 \ldots b_k \\ j_1 \ldots j_k \end{bmatrix}. \tag{7.54}$$

Exercise 7.5.8
Problem: Let $\mathbf{T}$ be an n-form on M^n; i.e., $\mathbf{T}$ is a differentiable form of maximal rank. Show that

$$\mathbf{T} = (\mathbf{T} \mid \varepsilon)\varepsilon.$$

Consider a single point P on our manifold. To this point we have attached the finite-dimensional vector space $\Lambda^k_P(M)$ consisting of all k-forms. By construction, the map

$$(\mathbf{T}, \mathbf{U}) \curvearrowright (\mathbf{T} \mid \mathbf{U})$$

is a symmetric bilinear map on the vector space $\Lambda^k_P(M)$. Thus there is a good chance that it actually defines a *metric* on $\Lambda^k_P(M)$. To check this, we must show that it is a nondegenerate map. Actually, the following holds:

Theorem 7.8.

a. *Let M be a Riemannian orientable manifold. Then the scalar product* $(\mathbf{T}, \mathbf{U}) \curvearrowright (\mathbf{T} \mid \mathbf{U})$ *defines a Euclidean metric.*
b. *Let M be a pseudo-Riemannian orientable manifold. Then the scalar product* $(\mathbf{T}, \mathbf{U}) \curvearrowright (\mathbf{T}|\mathbf{U})$ *defines an indefinite metric.*

PROOF. (a) We can choose a coordinate system at P_0 with metric coefficients

$$g_{ij}(P_0) = \delta_{ij}.$$

Then the scalar product $(\mathbf{T}|\mathbf{U})$ reduces to

$$(\mathbf{T}|\mathbf{U}) = \sum_{i_1 < \cdots < i_k} T_{i_1 \ldots i_k} U_{i_1 \ldots i_k},$$

which is obviously positive definite.

(b) We must show that the inner product $(\,|\,)$ is nondegenerate. Let $\mathbf{T}$ be a k-form that is perpendicular to all other k-forms; i.e.,

$$0 = (\mathbf{T}|\mathbf{U}) = \sum_{i_1 < \cdots < i_k} T^{i_1 \ldots i_k} U_{i_1 \ldots i_k} \qquad \text{for all } \mathbf{U}.$$

Choose $\mathbf{U}$ with the component $U_{12\ldots k} = 1$ and such that all other components $U_{i_1 \ldots i_k}$ where $i_1 < \ldots < i_k$ are zero. Then we deduce that $T^{12\ldots k} = 0$. In a similar way, we can show that all the other contravariant components of $\mathbf{T}$ vanish. But then $\mathbf{T} = 0$. □

Observe that the dual map $*$ maps the vector space $\Lambda^k_{P_0}(M)$ bijectively onto the vector space $\Lambda^{n-k}_{P_0}(M)$. The following theorem is therefore no great surprise:

Theorem 7.9.

a. *Let M be a Riemannian orientable manifold. The dual map $*$ is an isometry; i.e.,*

$$({}^*\mathbf{T} \mid {}^*\mathbf{U}) = (\mathbf{T} \mid \mathbf{U}) \tag{7.55}$$

for all k-forms $\mathbf{T}$ *and* $\mathbf{U}$.

b. *Let M be an orientable manifold with a Minkowski metric. Then the dual map $*$ is an antiisometry; i.e.,*

$$({}^*\mathbf{T} \mid {}^*\mathbf{U}) = -(\mathbf{T} \mid \mathbf{U}) \tag{7.56}$$

for all k-forms $\mathbf{T}$ *and* $\mathbf{U}$.

PROOF. (The Riemannian case)

$$\begin{aligned}({}^*\mathbf{T} \mid {}^*\mathbf{U}) &= {}^*({}^{**}\mathbf{T} \wedge {}^*\mathbf{U}) = (-1)^{k(n-k)*}(\mathbf{T} \wedge {}^*\mathbf{U}) \\ &= {}^*({}^*\mathbf{U} \wedge \mathbf{T}) = (\mathbf{U} \mid \mathbf{T}) = (\mathbf{T} \mid \mathbf{U}).\end{aligned}$$

Observe that we have two signs involved, one coming from the double dualization and the other coming from the exchange of $\mathbf{T}$ and ${}^*\mathbf{U}$. □

So much for the scalar product. Now we proceed to investigate the more general contractions: Let $\mathbf{T}$ be a k-form and $\mathbf{U}$ an m-form where $k \geq m$. Then we can extend the above machinery, and we immediately see that

$${}^*({}^*\mathbf{T} \wedge \mathbf{U})$$

becomes a $(k-m)$-form.

But the only general coordinate invariant $(k-m)$-form that can be made out of $T_{i_1 \ldots i_k}$ and $U_{j_1 \ldots j_m}$ is the contraction

$$\frac{1}{m!} T_{i_1 \ldots i_{k-m} j_1 \ldots j_m} U^{j_1 \ldots j_m}. \tag{7.57}$$

Thus ${}^*({}^*\mathbf{T} \wedge \mathbf{U})$ must be proportional to this contraction. In the following we will use the abbreviation $\mathbf{T} \cdot \mathbf{U}$ for the contraction; i.e., $\mathbf{T} \cdot \mathbf{U}$ is the $(k-m)$-form characterized by the components (7.57). Observe that dualization itself is a contraction between the Levi-Civita form $\boldsymbol{\varepsilon}$ and the differential form $\mathbf{U}$ we want to dualize; i.e.,

$${}^*\mathbf{U} = \boldsymbol{\varepsilon} \cdot \mathbf{U}.$$

Exercise 7.5.9
Problem: Show that

$${}^*({}^*\boldsymbol{\varepsilon} \wedge \mathbf{U}) = \begin{cases} +\boldsymbol{\varepsilon} \cdot \mathbf{U} & \text{(Euclidean metric)}, \\ -\boldsymbol{\varepsilon} \cdot \mathbf{U} & \text{(Minkowski metric)}. \end{cases}$$

Exercise 7.5.10 (For combinatorial fans!)
Problem: Show that

$${}^*({}^*\mathbf{T} \wedge \mathbf{U}) = \begin{cases} (-1)^{(k-m)(n-k)}\mathbf{T} \cdot \mathbf{U} & \text{(Euclidean metric)}, \\ -(-1)^{(k-m)(n-k)}\mathbf{T} \cdot \mathbf{U} & \text{(Minkowski metric)}. \end{cases} \tag{7.58}$$

7.6 The Codifferential and the Laplacian

Once we have the dual map at our disposal, we can construct another "natural" differential operator:

Definition 7.9. For an arbitrary k-form $\mathbf{T}$, the codifferential of $\mathbf{T}$, denoted by $\delta\mathbf{T}$, is the $(k-1)$-form defined by

$$\delta\mathbf{T} = \begin{cases} (-1)^{k(n-k+1)} {}^*\boldsymbol{d}{}^*\mathbf{T} & \text{(Euclidean metric)}, \\ -(-1)^{k(n-k+1)} {}^*\boldsymbol{d}{}^*\mathbf{T} & \text{(Minkowski metric)}. \end{cases} \tag{7.59}$$

In this way we have constructed a differential operator $\delta : \Lambda^k(M) \to \Lambda^{k-1}(M)$. It is obviously linear, since it is composed of the linear maps $*$ and $\boldsymbol{d}$. The sign of δ is a convention that makes forthcoming formulas easier to work with. Observe that δ has an important property in common with the exterior derivative $\boldsymbol{d}$:

Theorem 7.10.

$$\delta^2 = 0. \tag{7.60}$$

PROOF.

$$\delta\delta = \pm({}^*\boldsymbol{d}{}^*)({}^*\boldsymbol{d}{}^*) = \pm{}^*\boldsymbol{d}\boldsymbol{d}{}^* = 0,$$

where we have used Theorem 7.5. □

So once again we see that it pays to apply the operators twice! Note that there is no formula for δ analogous to the Leibniz rule. This is due to the fact that ${}^*(\mathbf{T} \wedge \mathbf{S})$ cannot be reexpressed in a simple way. (See, however, Exercise 7.6.1 for a possible generalization.)

Exercise 7.6.1
Problem: Prove the following generalization of Leibniz's rule: $(k \geq m+1)$:

$$\delta(\mathbf{T} \cdot \mathbf{U}) = (-1)^{n-m}\delta\mathbf{T} \cdot \mathbf{U} + (-1)^{n-k}\mathbf{T} \cdot \boldsymbol{d}\mathbf{U}. \tag{7.61}$$

(Yes, it is an ordinary exterior derivative in the last term!)

Let $\mathbf{T}$ be a differential form. We have previously seen that we can symbolically consider $\boldsymbol{d}$ as a 1-form and the exterior derivative of $\mathbf{T}$ as a kind of wedge product: "$\boldsymbol{d}\mathbf{T} = \boldsymbol{d} \wedge \mathbf{T}$." In the same spirit, we can consider the codifferential of $\mathbf{T}$ as a kind of contraction: $\delta\mathbf{T} = (-1)^k \boldsymbol{d} \cdot \mathbf{T}$. This follows from (7.58)–(7.59).

Exercise 7.6.2
Problem: Prove the following identities:

$${}^*\delta\mathbf{T} = (-1)^{n-k+1}\boldsymbol{d}{}^*\mathbf{T}. \tag{7.62}$$

$${}^*\boldsymbol{d}\mathbf{T} = (-1)^{n-k+1}\delta{}^*\mathbf{T}. \tag{7.63}$$

Observe that the above formulas hold for the cases of both a Euclidean metric and a Minkowski metric!

If we think of $\boldsymbol{d}$ as generalizing the gradient, then we may think of $\boldsymbol{\delta}$ as generalizing the *divergence*. To see this, we will work out what $\boldsymbol{\delta}$ is in a few cases. For simplicity, we work with a Euclidean metric:

a. Let ϕ be a 0-form. Then $\delta\phi = 0$. This is obvious because $\boldsymbol{\delta}$ lowers the degree of a form.

b. Let $\mathbf{A}$ be a 1-form. If we put $B = \boldsymbol{\delta}\mathbf{A}$, then B is the scalar given by

$$B = (-1)^{n\,*}\boldsymbol{d}^*\mathbf{A}; \qquad \text{i.e., } {}^*B = (-1)^n \boldsymbol{d}^*\mathbf{A}.$$

If $\mathbf{A}$ has the components A_i; then ${}^*\mathbf{A}$ is an $(n-1)$-form characterized by the components

$$({}^*\mathbf{A})_{12\ldots(n-1)} = \sqrt{g}\epsilon_{12\ldots(n-1)i}A^i = \sqrt{g}A^n$$

(where the other components are obtained by cyclic permutation). Taking the exterior derivative, we get an n-form characterized by the component

$$\begin{aligned}(\boldsymbol{d}^*\mathbf{A})_{12\ldots n} &= \frac{1}{(n-1)!}\left[\partial_1({}^*\mathbf{A})_{2\ldots n} - \cdots + \cdots\right] \\ &= \partial_1({}^*\mathbf{A})_{2\ldots n} + (-1)^{n-1}\partial_2({}^*\mathbf{A})_{3\ldots n1} + \partial_3({}^*\mathbf{A})_{4\ldots n12} + \ldots \\ &= (-1)^{n-1}\partial_1\left(\sqrt{g}A^1\right) + (-1)^{n-1}\partial_2\left(\sqrt{g}A^2\right) + \cdots \\ &= (-1)^{n-1}\partial_i\left(\sqrt{g}A^i\right).\end{aligned}$$

Consequently, $(-1)^n\boldsymbol{d}^*\mathbf{A}$ is characterized by the component

$$[(-1)^n\boldsymbol{d}^*\mathbf{A}]_{12\ldots n} = -\partial_i\left(\sqrt{g}A^i\right).$$

On the other hand, ${}^*B$ is characterized by the component

$$[{}^*B]_{12\ldots n} = \sqrt{g}B.$$

We thus get

$$\boldsymbol{\delta}\mathbf{A} = -\frac{1}{\sqrt{g}}\,\partial_i\left(\sqrt{g}A^i\right).$$

Observe the sign, which is conventional: $\boldsymbol{\delta}$ really is *minus* the divergence!

c. Let $\mathbf{F}$ be a two-form. If we put $\mathbf{G} = \boldsymbol{\delta}\mathbf{F}$, then $\mathbf{G}$ is a one-form:

$$\mathbf{G} = {}^*\boldsymbol{d}^*\mathbf{F}; \qquad \text{i.e., } {}^*\mathbf{G} = (-1)^{n-1}\boldsymbol{d}^*\mathbf{F}.$$

Here ${}^*\mathbf{F}$ is an $(n-2)$-form characterized by the components

$$\begin{aligned}({}^*\mathbf{F})_{12\ldots n-2} &= \frac{1}{2!}\sqrt{g}\epsilon_{12\ldots(n-2)_{ij}}F^{ij} = \frac{1}{2!}\sqrt{g}\left[F^{(n-1)n} - F^{n(n-1)}\right] \\ &= \sqrt{g}F^{(n-1)n}.\end{aligned}$$

In the rest of the discussion we will put $n = 4$. This makes it easy to see what is going on, and the calculations can easily be generalized. $\boldsymbol{d}^*\mathbf{F}$ is then a 3-form characterized by the components

$$\begin{aligned}(\boldsymbol{d}^*\mathbf{F})_{123} &= \frac{1}{2!}\left[\partial_1(^*\mathbf{F})_{23} - \partial_1(^*\mathbf{F}_{32}) + \cdots - \cdots\right] \\ &= \partial_1(^*\mathbf{F})_{23} + \partial_2(^*\mathbf{F})_{31} + \partial_3(^*\mathbf{F})_{12} \\ &= \partial_1\left(\sqrt{g}F^{14}\right) + \partial_2\left(\sqrt{g}F^{24}\right) + \partial_3\left(\sqrt{g}F^{34}\right).\end{aligned}$$

But here we can artificially introduce the term $\partial_4\left(\sqrt{g}F^{44}\right)$, since it vanishes automatically due to the skew-symmetry of F^{ij}. Thus we have obtained

$$(\boldsymbol{d}^*\mathbf{F})_{123} = \partial_i\left(\sqrt{g}F^{i4}\right)$$

and this formula is immediately generalized to

$$(\boldsymbol{d}^*\mathbf{F})_{12\ldots(n-1)} = (-1)^n\partial_i\left(\sqrt{g}F^{in}\right).$$

If $\mathbf{G}$ is a one-form characterized by components $G_{1\ldots n}$ we get

$$(^*\mathbf{G})_{12\ldots(n-1)} = \sqrt{g}\epsilon_{12\ldots(n-1)i}G^i = \sqrt{g}G^n.$$

Consequently we have

$$(\boldsymbol{\delta}\mathbf{F})^n = -\frac{1}{\sqrt{g}}\,\partial_i\left[\sqrt{g}F^{in}\right]$$

which is generalized to

$$(\boldsymbol{\delta}\mathbf{F})^j = -\frac{1}{\sqrt{g}}\,\partial_i\left(\sqrt{g}F^{ij}\right).$$

We collect the results obtained in Table 7.2.

A general coordinate-invariant expression for the components of $\boldsymbol{\delta}\mathbf{T}$ is hard to derive. The details are left as an exercise for those who are especially interested in combinatorics:

Worked Exercise 7.6.3
Problem: Let $\mathbf{T}$ be a k-form. Show that the contravariant components of $\boldsymbol{\delta}\mathbf{T}$ are given by

$$(\boldsymbol{\delta}\mathbf{T})^{a_1\ldots a_{k-1}} = -\frac{1}{\sqrt{g}}\,\partial_b\left(\sqrt{g}G^{ba_1\ldots a_{k-1}}\right). \tag{7.64}$$

Table 7.2.

	Euclidean metric	Minkowski metric
0-form: ϕ	$\delta_\phi = 0$	$\delta_\phi = 0$
1-form: A	$\delta_A = -\frac{1}{\sqrt{g}}\,\partial_i(\sqrt{g}A^i)$	$\delta_A = -\frac{1}{\sqrt{-g}}\,\partial_\mu(\sqrt{-g}A^\mu)$
2-form: F	$(\delta_F)^j = -\frac{1}{\sqrt{g}}\,\partial_i(\sqrt{g}F^{ij})$	$(\delta_F)^\nu = -\frac{1}{\sqrt{-g}}\,\partial_\mu(\sqrt{-g}F^{\mu\nu})$

Once we control the codifferential $\boldsymbol{\delta}$, we can now generalize the Laplacian to arbitrary Riemannian manifolds. For a scalar field ϕ this is particularly easy. Consider a Euclidean space with the usual Cartesian coordinates. Then the Laplacian is defined as

$$\Delta\phi = \nabla \cdot (\nabla\phi) = \sum_{i=1}^{n} \frac{d^2\phi}{\partial(x^i)^2} \, . \tag{7.65}$$

In the language of differential forms this is immediately generalized to

$$\Delta\phi = -\boldsymbol{\delta d}\phi = \frac{1}{\sqrt{g}}\,\partial_i\left(\sqrt{g}\partial^i\phi\right). \tag{7.66}$$

Consequently, (7.66) is the covariant generalization of the expression $\partial_i\partial^i\phi$, which is valid only for Cartesian coordinates.

This suggests that we could generalize the Laplacian for arbitrary differential forms in the following way:

$$-\boldsymbol{\Delta} \stackrel{?}{=} \boldsymbol{\delta d},$$

but unfortunately this is not correct.

To see this, we consider a one-form $\mathbf{A}$, Then $\boldsymbol{\delta d}\mathbf{A}$ is a 1-form characterized by the components

$$\begin{aligned}(\boldsymbol{\delta d}\mathbf{A})^j &= -\frac{1}{\sqrt{g}}\,\partial_i\left[\sqrt{g}(\partial^i A^j - \partial^j A^i)\right] \\ &= -\frac{1}{\sqrt{g}}\,\partial_i\left[\sqrt{g}\partial^i A^j\right] + \frac{1}{\sqrt{g}}\,\partial_i\left[\sqrt{g}\partial^j A^i\right].\end{aligned}$$

It is the last term that spoils the game! On the other hand, we may consider the differential operator $\boldsymbol{d\delta}$. When it operates on $\mathbf{A}$, we know that $\boldsymbol{\delta}\mathbf{A}$ is the 0-form

$$-\frac{1}{\sqrt{g}}\,\partial_i\left(\sqrt{g}A^i\right),$$

and therefore

$$(\boldsymbol{d\delta}\mathbf{A})^j = -\partial^j\left(\frac{1}{\sqrt{g}}\,\partial_i\left(\sqrt{g}A^i\right)\right).$$

In a Cartesian coordinate system, $\sqrt{g} = 1$, and the two expressions reduce to

$$\begin{aligned}(\boldsymbol{\delta d}\mathbf{A})^j &= -\partial_i\partial^i A^j + \partial^j\partial_i A^i, \\ (\boldsymbol{d\delta}\mathbf{A})^j &= \qquad\qquad\quad - \partial^j\partial_i A^i.\end{aligned}$$

Adding the two expressions, we therefore get that $(\boldsymbol{\delta d} + \boldsymbol{d\delta})\mathbf{A}$ is characterized by the *contravariant* components

$$-\partial_i\partial^i A^j.$$

Observe that $\boldsymbol{d\delta}$ automatically vanishes for scalar fields! The correct generalization is therefore apparently given by the following symmetric expression

Definition 7.10[1]**.**

$$-\boldsymbol{\Delta} = \boldsymbol{\delta d} + \boldsymbol{d\delta} \tag{7.67}$$

Worked Exercise 7.6.4
(For combinatorial fans) Problem: Show that in a Euclidean space with Cartesian coordinates, the coordinate expression for the Laplace operator reduces to

$$(\boldsymbol{\Delta}\mathbf{T})_{i_1 \dots i_k} = (\partial_j \partial^j) T_{i_1 \dots i_k}.$$

Exercise 7.6.5
Problem: Show that the Laplacian commutes with the dual map, the exterior derivative, and the codifferential; i.e.,

$$\boldsymbol{\Delta}^* = {}^*\boldsymbol{\Delta}, \qquad \boldsymbol{d\Delta} = \boldsymbol{\Delta d}, \qquad \boldsymbol{\delta\Delta} = \boldsymbol{\Delta\delta}.$$

If we proceed to consider a pseudo-Riemannian manifold with a Minkowski metric, then the above considerations are still valid, except that $-(\boldsymbol{\delta d} + \boldsymbol{d\delta})$ here represents the d'Alembertian $\Box$. In an inertial frame it is given by the expression

$$\Box = \partial_\mu \partial^\mu = \frac{-\partial^2}{\partial t^2} + \Delta,$$

and its covariant generalization to arbitrary differential forms is

$$-\Box = \boldsymbol{d\delta} + \boldsymbol{\delta d}. \tag{7.68}$$

Okay, let us return to manifolds with arbitrary metrics. In analogy with the exterior derivative, we can now introduce two important types of differential forms.

Definition 7.11.

a. A differential form $\mathbf{F}$ is *coclosed* if its codifferential vanishes; i.e., $\boldsymbol{\delta}\mathbf{F} = 0$.
b. A differential form $\mathbf{G}$ is *coexact* if it is the codifferential of another differential form $\mathbf{F}$; i.e., $\mathbf{G} = \boldsymbol{\delta}\mathbf{F}$.

Observe that a coexact form is automatically coclosed, since $\mathbf{G} = \boldsymbol{\delta}\mathbf{F}$ implies that

$$\boldsymbol{\delta}\mathbf{G} = \boldsymbol{\delta}^2\mathbf{F} = 0.$$

Exercise 7.6.6
Problem:

a. Show that $\mathbf{F}$ is coclosed if and only if ${}^*\mathbf{F}$ is closed.
b. Show that $\mathbf{G}$ is coexact if and only if ${}^*\mathbf{G}$ is exact.

---

[1] This definition differs in sign from the one generally adopted by mathematicians. See, for instance, Goldberg (1962) or de Rham (1955).

In the rest of this section we consider only Riemannian manifolds. The generalization of the Laplacian to a Riemannian manifold suggests the following definition:

Definition 7.12. A differential form $\mathbf{T}$ is called *harmonic* if $\mathbf{\Delta T} = 0$.

Exercise 7.6.7
Problem:

a. Show that the constant function 1 and the Levi-Civita form ε are harmonic forms.
b. Let $\mathbf{T}$ be a harmonic form. Show that $^*\mathbf{T}$, $\boldsymbol{d}\mathbf{T}$ and $\delta\mathbf{T}$ are harmonic forms too.

A very important class of harmonic forms are those where both the exterior derivative and the codifferential vanish:

Definition 7.13. A differential form $\mathbf{T}$ is called primitively harmonic if it is both closed and coclosed; i.e., $\boldsymbol{d}\mathbf{T} = \delta\mathbf{T} = 0$.

A primitively harmonic form is obviously harmonic, but the converse need not be true. This is well known already for scalar functions. Consider $M = \mathbb{R}^3 \setminus \{\mathbf{0}\}$ and put $\phi(\mathbf{x}) = \frac{1}{r}$. Then it is not constant, and therefore $\boldsymbol{d}\phi \neq 0$. Consequently, ϕ is not primitively harmonic. On the other hand, it is harmonic, as can be trivially verified. In the above example, it is important to notice that M is not compact. For a compact Riemannian manifold we shall show later on that all harmonic forms are automatically primitively harmonic (see Chapter 8, Theorem 8.8).

Exercise 7.6.8
Problem: Let $\mathbf{A}$ be an exact 1-form, $\mathbf{A} = \boldsymbol{d}\phi$. Show that $\mathbf{A}$ is primitively harmonic if and only if ϕ is harmonic.

Worked Exercise 7.6.9
("Complex calculus on manifolds") Notation: Let $M \subseteq \mathbb{R}^2$ be a two-dimensional manifold, i.e., an open subset of $\mathbb{R}^2$. We identify $\mathbb{R}^2$ with the complex plain $\mathbb{C}$ through the usual identification

$$(x, y) \leftrightarrow z = x + iy.$$

We introduce complex-valued differential forms by allowing the components to be complex-valued functions; i.e., a complex-valued differential form is a skew-symmetric multilinear map

$$\mathbf{F} : \mathbf{T}_p(M) \times \cdots \times \mathbf{T}_p(M) \to \mathbb{C}.$$

Consider the complex-valued 1-form

$$\omega = [f_1(x, y) + ig_1(x, y)]\boldsymbol{d}x + [f_2(x, y) + ig_2(x, y)]\boldsymbol{d}y.$$

If we introduce the basic complex differentials

$$\boldsymbol{d}z = \boldsymbol{d}x + i\boldsymbol{d}y, \qquad \boldsymbol{d}\bar{z} = \boldsymbol{d}x - i\boldsymbol{d}y,$$

then we can expand ω as

$$\omega = h_1(z)\boldsymbol{d}z + h_2(z)\boldsymbol{d}\bar{z},$$

where

$$h_1(z) = \frac{1}{2}[f_1 + g_2] + \frac{i}{2}[g_1 - f_2] \quad \text{and} \quad h_2(z) = \frac{1}{2}[f_1 - g_2] + \frac{i}{2}[g_1 + f_2].$$

If $h(z)$ is a complex-valued function, we can introduce the partial derivatives $\frac{\partial h}{\partial z}$ and $\frac{\partial h}{\partial \bar{z}}$ through the expansion

$$\boldsymbol{d}h = \frac{\partial h}{\partial z}\boldsymbol{d}z + \frac{\partial h}{\partial \bar{z}}\boldsymbol{d}\bar{z};$$

$$\text{i.e.,} \quad \frac{\partial}{\partial z} = \frac{1}{2}\left(\frac{\partial}{\partial x} - i\frac{\partial}{\partial y}\right) \quad \text{and} \quad \frac{\partial}{\partial \bar{z}} = \left(\frac{\partial}{\partial x} + i\frac{\partial}{\partial y}\right).$$

We take it as well known that an analytic (holomorphic) function in M is characterized by the following two equivalent characterizations:

1. h can be expanded as a power series around each point z_0 in M; i.e.,

$$h(z) = \sum_n a_n (z - z_0)^n$$

   for z sufficiently close to z_0.
2. $\frac{\partial h}{\partial \bar{z}} = 0$ (Cauchy–Riemann equations).

Problem: Consider the 1-form $\omega = h(z)\boldsymbol{d}z$:

a. Show that the real part is the dual of the imaginary part.
b. Show that $h(z)$ is analytic (holomorphic) if and only if ω is closed.
c. Let us assume that $h(z)$ is analytic (holomorphic). Show that the real part and the imaginary part of $h(z)$ are harmonic functions.
d. Let $h(z)$ be holomorphic. Show that $\frac{\partial h}{\partial z}$ is holomorphic too.
e. Show that $h(z)$ is holomorphic if and only if $\boldsymbol{d}h$ is anti–self-dual in the sense that

$$^*\boldsymbol{d}h = -i\boldsymbol{d}h.$$

(Compare with Exercise 7.5.3.)

7.7 Exterior Calculus in 3 and 4 Dimensions

We start by looking at the ordinary Euclidean space $\mathbb{R}^3$. Here we have the *standard vector analysis*: The only admissible coordinate systems are the *Cartesian* coordinate systems. A vector **a** is characterized by its Cartesian components

$$(a_x, a_y, a_z).$$

(See Figure 7.9.) We do not distinguish between covariant and contravariant components, as the metric coefficients are given by the unit matrix

$$[g_{ij}] = \begin{bmatrix} 1 & 0 & 0 \\ 0 & 1 & 0 \\ 0 & 0 & 1 \end{bmatrix}.$$

(This is because we restrict ourselves to Cartesian coordinates.)

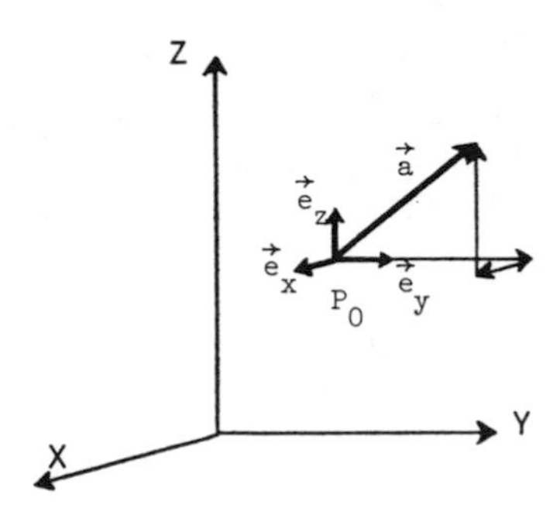

Figure 7.9.

There are two kinds of products: the *scalar product*

$$\mathbf{a} \cdot \mathbf{b} = a_x b_x + a_y b_y + a_z b_z$$

and the *cross product*

$$\mathbf{a} \times \mathbf{b} \text{ with coordinates } [a_y b_z - a_z b_y, a_z b_x - a_x b_z, a_x b_y - a_y b_x]$$

with a lot of well known algebraic properties. Then there is a differential operator

$$\nabla = \left(\frac{\partial}{\partial x} ; \frac{\partial}{\partial y} ; \frac{\partial}{\partial z} \right) .$$

Using this differential operator we can attack a scalar field ϕ, producing a vector field

$$\text{the } \textit{gradient} \quad \nabla\phi \quad \text{with coordinates } \left(\frac{\partial\phi}{\partial x} ; \frac{\partial\phi}{\partial y} ; \frac{\partial\phi}{\partial z} \right) ;$$

and we can attack a vector field $\mathbf{a}$, producing a scalar field

$$\text{the } \textit{divergence} \quad \nabla \cdot \mathbf{a} = \frac{\partial a_x}{\partial x} + \frac{\partial a_y}{\partial y} + \frac{\partial a_z}{\partial z}$$

or a vector field

$$\text{the } \textit{curl} \quad \nabla \times \mathbf{a} \text{ with coordinates } \left[\frac{\partial a_z}{\partial y} - \frac{\partial a_y}{\partial z} ; \frac{\partial a_x}{\partial z} - \frac{\partial a_z}{\partial x} ; \frac{\partial a_y}{\partial x} - \frac{\partial a_x}{\partial y} \right] .$$

This is the scheme we are going to generalize. In the *exterior* algebra we can use any coordinate system: Cartesian coordinates, spherical coordinates, cylindrical coordinates, etc. For simplicity, we shall, however, restrict ourselves to positively orientated coordinate systems. This fixes the components of the Levi-Civita form ε to be $\sqrt{g}\epsilon_{ijk}$. We distinguish between covariant and contravariant components, and we use the metric coefficients to lower and raise the indices.

In exterior algebra we play with forms. As $n = 3$, we have 0-forms, 1-forms, 2-forms, and 3-forms. The scalars are represented by 0-forms, but via the dual map we can also represent them by 3-forms! The vector fields are represented by 1-forms, but via the dual map we can also represent these by 2-forms. Thus an arbitrary form in $\mathbb{R}^3$ is either associated with a scalar or a vector.

To investigate the dual map we use formula (7.38). The results are shown in the following scheme:

The dual map in R^3		
Covariant components of F	Covariant components of ${}^*F$	
0-form: ϕ	3-form: ${}^*\phi : ({}^*\phi)_{123} = \sqrt{g}\phi$	
1-form: $E : (E_1, E_2, E_3)$	2-form: ${}^*E : \sqrt{g}\begin{bmatrix} 0 & E^3 & -E^2 \\ -E^3 & 0 & E^1 \\ E^2 & -E^1 & 0 \end{bmatrix}$	(7.69)
2-form: $B : \begin{bmatrix} 0 & B_{12} & -B_{31} \\ -B_{12} & 0 & B_{23} \\ B_{31} & -B_{23} & 0 \end{bmatrix}$	1-form: ${}^*B : \sqrt{g}\left[B^{23}, B^{31}, B^{12}\right]$	
3-form: $G : (G)_{123} = G$	0-form: ${}^*G : \frac{1}{\sqrt{g}} G$	

Most of the scheme is self-evident, but we should be careful when we dualize a three-form. A three-form is completely characterized by *just* one component, which we choose to be the 123-component. Applying formula (7.38), we get

$$ {}^*\mathbf{G} = \frac{1}{3!} \sqrt{g}\epsilon_{ijk} G^{ijk}. $$

Now we use that the contravariant components of ε are given by $\frac{1}{\sqrt{g}}\epsilon^{ijk}$ (see Exercise 7.6.1). This is where the mysterious factor $\frac{1}{\sqrt{g}}$ comes from. Raising the indices of ε and lowering those of $\mathbf{G}$ produce the equivalent formula

$$ {}^*\mathbf{G} = \frac{1}{3!} \frac{1}{\sqrt{g}} \epsilon^{ijk} G_{ijk}. $$

But here all permutations of (ijk) give the same contribution, so that the formula reduces to

$$ {}^*\mathbf{G} = \frac{1}{\sqrt{g}} \epsilon^{123} G_{123} = \frac{1}{\sqrt{g}} G. $$

We should also observe that in Euclidean spaces of odd dimension we always have $^{**}\mathbf{F} = \mathbf{F}$ for a form of arbitrary rank. (Compare this with the scheme (7.44.).)

Armed with the dual map, we can now investigate the vector analysis. In the following, **A**, **B**, and **C** will denote 1-forms.

Let us first consider the wedge product: $\mathbf{A} \wedge \mathbf{B}$ is a 2-form with the components $(A_i B_j - A_j B_i)$. This obviously generalizes the usual cross product. Although it is a 2-form, $\mathbf{A} \wedge \mathbf{B}$ actually represents a vector, which we can find by dualizing:

$$^*(\mathbf{A} \wedge \mathbf{B}) : \sqrt{g}[A^2 B^3 - A^3 B^2;\ A^3 B^1 - A^1 B^3;\ A^1 B^2 - A^2 B^1].$$

If we use Cartesian coordinates, then $\sqrt{g} = 1$, and the expression above reduces to the components of the usual cross product $\mathbf{a} \times \mathbf{b}$. Consequently, $^*(\mathbf{A} \wedge \mathbf{B})$ is the strict generalization of the usual cross product $\mathbf{a} \times \mathbf{b}$. Due to the fact that 1-forms are odd forms, i.e., they anticommute, we immediately recover the characteristic properties of the cross product:

$$\mathbf{a} \times \mathbf{b} = -\mathbf{b} \times \mathbf{a}, \qquad [\mathbf{A} \wedge \mathbf{B} = -\mathbf{B} \wedge \mathbf{A}],$$
$$\mathbf{a} \times \mathbf{a} = 0, \qquad [\mathbf{A} \wedge \mathbf{A} = 0].$$

Next we consider the scalar product between two 1-forms:

$$(\mathbf{A} \mid \mathbf{B}) = A_i B^i.$$

This obviously generalizes the usual scalar product $\mathbf{a} \cdot \mathbf{b}$. More generally, we can apply contractions between forms. Let **F** be a 2-form. Then $\mathbf{F} \cdot \mathbf{A}$ is the 1-form characterized by the components $F_{ij} A^j$. Here **F** represents a vector: $\mathbf{F} = {}^*\mathbf{B}$; i.e.,

$$F_{ij} = \sqrt{g}\epsilon_{ijk} B^k,$$

and therefore $\mathbf{F} \cdot \mathbf{A}$ is actually characterized by the components

$$\sqrt{g}\epsilon_{ijk} B^k A^j = \sqrt{g}\epsilon_{ijk} A^j B^k,$$

which are the covariant components of $\mathbf{a} \times \mathbf{b}$.

Exercise 7.7.1

Problem: (a) Let **F** and **G** be 2-forms representing the vectors **a** and **b**. Show that the scalar product

$$(\mathbf{F} \mid \mathbf{G}) = \frac{1}{2} F_{ij} G^{ij}$$

represents the ordinary scalar product $\mathbf{a} \cdot \mathbf{b}$.

(b) Let **F** be a 2-form and consider the wedge product $\mathbf{F} \wedge \mathbf{B}$. Let **F** represent the vector **a** and show that the 3-form $\mathbf{F} \wedge \mathbf{B}$ represents the ordinary scalar product $\mathbf{a} \cdot \mathbf{b}$.

Then we consider the triple product $(\mathbf{A} \wedge \mathbf{B}) \wedge \mathbf{C}$, which is a 3-form; i.e., it represents a scalar. We already know that $\mathbf{A} \wedge \mathbf{B}$ represents the cross product $\mathbf{a} \times \mathbf{b}$. To find the meaning of $(\mathbf{A} \wedge \mathbf{B}) \wedge \mathbf{C}$, we use Exercise 7.2.1, from which we get the 123-component

$$[(\mathbf{A} \wedge \mathbf{B}) \wedge \mathbf{C}]_{123} = (\mathbf{A} \wedge \mathbf{B})_{12} C_3 + (\mathbf{A} \wedge \mathbf{B})_{23} C_1 + (\mathbf{A} \wedge \mathbf{B})_{31} C_2.$$

But $\mathbf{A} \wedge \mathbf{B}$ is the dual of the cross product; i.e., $(\mathbf{A} \wedge \mathbf{B})_{12} = \sqrt{g}(\mathbf{a} \times \mathbf{b})^3$, etc. We thus find that

$$[(\mathbf{A} \wedge \mathbf{B}) \wedge \mathbf{C}]_{123} = \sqrt{g}[(\mathbf{a} \times \mathbf{b})^3 C_3 + (\mathbf{a} \times \mathbf{b})^1 C_1 + (\mathbf{a} \times \mathbf{b})^2 C_2].$$

Dualizing this, we finally get the scalar

$$^*[(\mathbf{A} \wedge \mathbf{B}) \wedge \mathbf{C}] = (\mathbf{a} \times \mathbf{b})^i C_i,$$

and we see that $[(\mathbf{A} \wedge \mathbf{B}) \wedge C]$ generalizes the triple product $(\mathbf{a} \times \mathbf{b}) \cdot \mathbf{c}$.

But $\mathbf{A} \wedge \mathbf{B}$ is a 2-form, and therefore it *commutes* with $\mathbf{C}$. We thus get

$$\mathbf{A} \wedge \mathbf{B} \wedge \mathbf{C} = \mathbf{C} \wedge \mathbf{A} \wedge \mathbf{B} = \mathbf{B} \wedge \mathbf{C} \wedge \mathbf{A}.$$

If we then dualize, this corresponds to the rule

$$(\mathbf{a} \times \mathbf{b}) \cdot \mathbf{c} = (\mathbf{c} \times \mathbf{a}) \cdot \mathbf{b} = (\mathbf{b} \times \mathbf{c}) \cdot \mathbf{a}.$$

These rules are often stated in the following way: *In a triple product it is allowed to interchange dot and cross.*

Suppose now we have three 1-forms: $\mathbf{A}$, $\mathbf{B}$, and $\mathbf{C}$. Then we can form another triple product:

$$(\mathbf{A} \wedge \mathbf{B}) \cdot \mathbf{C}.$$

This is another one-form! $\mathbf{A} \wedge \mathbf{B}$ is the dual of the cross product $\mathbf{a} \times \mathbf{b}$, but then $[^*(\mathbf{a} \times \mathbf{b})] \cdot \mathbf{C}$ is still another cross product, according to the preceding discussion! Thus $(\mathbf{A} \wedge \mathbf{B}) \cdot \mathbf{C}$ generalizes

$$-(\mathbf{a} \times \mathbf{b}) \times \mathbf{c} = \mathbf{c} \times (\mathbf{a} \times \mathbf{b}).$$

It is well known that the triple product $(\mathbf{a} \times \mathbf{b}) \times \mathbf{c}$ satisfies the rule

$$(\mathbf{a} \times \mathbf{b}) \times \mathbf{c} = (\mathbf{a} \cdot \mathbf{c})\mathbf{b} - \mathbf{a}(\mathbf{b} \cdot \mathbf{c}).$$

In the exterior algebra, this follows straightforwardly when we work out the components of $(\mathbf{A} \wedge \mathbf{B}) \cdot \mathbf{C}$:

$$(A_i B_j - A_j B_i) C^j = A_i (B_j C^j) - B_i (A_j C^j).$$

From this we immediately read off the formula

$$(\mathbf{A} \wedge \mathbf{B}) \cdot \mathbf{C} = \mathbf{A}(\mathbf{B} \mid \mathbf{C}) - (\mathbf{A} \mid \mathbf{C})\mathbf{B}.$$

If we compare this with the proofs in conventional vector analysis, we learn to appreciate the exterior algebra!

Let us collect the preceding results in the following scheme:

	Scalar product and wedge product		(7.70)
		Forms	Components
Exterior Algebra	Scalar product	$(A\|B)$	$A_i B^i$
	Contraction	$F \cdot A$	$F_{ij} A^j$
	Wedge product	$A \wedge B$	$A_i B_j - A_j B_i$
	Wedge product	$F \wedge A$	$F_{ij} A_k + F_{jk} A_i + F_{ki} A_j$

	Conventional vector analysis	Exterior algebra
Dictionary	$\mathbf{a} \cdot \mathbf{b}$	$(A\|B)$
	$\mathbf{a} \times \mathbf{b}$	$A \wedge B$
	$(\mathbf{a} \times \mathbf{b}) \cdot \mathbf{c}$	$(A \wedge B) \wedge C$
	$(\mathbf{a} \times \mathbf{b}) \times \mathbf{c}$	$-(A \wedge B) \cdot C$
	$(\mathbf{a} \times \mathbf{b}) \cdot \mathbf{c} = (\mathbf{c} \times \mathbf{a}) \cdot \mathbf{b} = (\mathbf{b} \times \mathbf{c}) \cdot \mathbf{a}$	$A \wedge B \wedge C = C \wedge A \wedge B = B \wedge C \wedge A$
	$(\mathbf{a} \times \mathbf{b}) \times \mathbf{c} = (\mathbf{a} \cdot \mathbf{c})\mathbf{b} - \mathbf{a}(\mathbf{b} \cdot \mathbf{c})$	$(A \wedge B) \cdot C = A(B\|C) - (A\|C)B$

We now proceed by investigating the differential calculus. In the exterior algebra we have the exterior derivative $\boldsymbol{d}$. It converts a scalar field ϕ into a one-form $\boldsymbol{d}\phi$ with the components

$$\left(\frac{\partial \phi}{\partial x^1}, \frac{\partial \phi}{\partial x^2}, \frac{\partial \phi}{\partial x^3} \right).$$

Thus $\boldsymbol{d}\phi$ generalizes the gradient $\nabla \phi$. It converts a 1-form $\mathbf{A}$ into the 2-form $\boldsymbol{d}\mathbf{A}$ with components

$$\begin{bmatrix} 0 & \partial_1 A_2 - \partial_2 A_1 & \cdots \\ \cdots & 0 & \partial_2 A_3 - \partial_3 A_2 \\ \partial_3 A_1 - \partial_1 A_3 & \cdots & 0 \end{bmatrix}.$$

But this 2-form obviously represents the curl $\nabla \times \mathbf{a}$.

Finally, $\boldsymbol{d}$ converts a 2-form into a 3-form. If the 2-form represents a vector field, i.e., we consider the 2-form $^*\mathbf{A}$, then the 3-form $\boldsymbol{d}^*\mathbf{A}$ represents a scalar. It is this scalar we want to examine. As $\boldsymbol{d}^*\mathbf{A}$ is characterized by the components

$$\partial_i ({}^*\mathbf{A})_{jk} + \partial_j ({}^*\mathbf{A})_{ki} + \partial_k ({}^*\mathbf{A})_{ij},$$

the 123-component is given by

$$(\boldsymbol{d}^*\mathbf{A})_{123} = \partial_1(\sqrt{g}A^1) + \partial_2(\sqrt{g}A^2) + \partial_3(\sqrt{g}A^3).$$

Dualizing this, we find that $^*\mathbf{A}$ represents the scalar

$$^*(\boldsymbol{d}^*\mathbf{A}) = \frac{1}{\sqrt{g}} \partial_i(\sqrt{g}A^i).$$

If we recall that $\sqrt{g} = 1$ in Cartesian coordinates, we immediately see that $\boldsymbol{d}^*\mathbf{A}$ generalizes the divergence $\nabla \cdot \mathbf{a}$. Notice, however, that we may express the divergence more directly by means of the codifferential $\boldsymbol{\delta}$. If $\mathbf{A}$ is a 1-form, we know that

$$-\delta\mathbf{A} = \frac{1}{\sqrt{g}}\,\partial_i\left(\sqrt{g}A^i\right).$$

(Compare Table 7.2).) This is in agreement with the result above, according to (7.59), we have $-\delta = {}^*d^*$.

Exercise 7.7.2
Problem: (a) Let $\mathbf{F}$ be a 2-form representing the vector $\mathbf{a}$. Show that $\delta\mathbf{F}$ represent the curl $\nabla \times \mathbf{a}$. (b) Let $\mathbf{G}$ be a 3-form representing the scalar field ϕ. Show that $\delta\mathbf{G}$ represents the gradient $\nabla\phi$.

We know that the exterior derivative d has some simple properties:

$$d^2 = 0, \tag{7.13}$$

$$d(\mathbf{F}\wedge\mathbf{G}) = d\mathbf{F}\wedge\mathbf{G} + (-1)^{\deg \mathbf{F}}\mathbf{F}\wedge d\mathbf{G}. \tag{7.17}$$

Almost all of the identities involving grad, div, and curl are special cases of these two rules; see the scheme (7.72).

The exterior derivative in R^3 (7.71)

	0-form: ϕ	1-form: $d\phi : (\partial_1\phi, \partial_2\phi, \partial_3\phi)$
Exterior Algebra	1-form: $A : (A_1, A_2, A_3)$	2-form: $dA : \begin{bmatrix} 0 & \partial_1A_2 - \partial_2A_1 & xxx \\ xxx & 0 & \partial_2A_3 - \partial_3A_2 \\ \partial_3A_1 - \partial_1A_3 & xxx & 0 \end{bmatrix}$
	2-form: ${}^*A : \sqrt{g}\begin{bmatrix} 0 & A^3 & -A^2 \\ -A^3 & 0 & A^1 \\ A^2 & -A^1 & 0 \end{bmatrix}$	3-form: $(d^*A)_{123} = \partial_i(\sqrt{g}A^i)$
	3-form: $G : (G)_{123} = G$	4-form (Zero) $dG = 0$

	Conventional Vector analysis	Exterior algebra	Covariant formulas
Dictionary	Gradient: $\nabla\phi$	$d\phi$	$\partial_i\phi$
	Divergence: $\nabla\cdot\mathbf{a}$	$-\delta A$	$\frac{1}{\sqrt{g}}\partial_i(\sqrt{g}A^i)$
	Curl: $\nabla\times\mathbf{a}$	dA	$\sqrt{g}\varepsilon_{ijk}\partial^jA^k$
	Laplacian: $\Delta\phi$	$-\delta d\phi$	$\frac{1}{\sqrt{g}}\partial_i(\sqrt{g}\partial^i\phi)$

(A) $d^2 = 0$

1. If we apply (A) to a scalar field ϕ, we get

$$(*) \quad d(d\phi) = 0.$$

As $d\phi$ represents the gradient of ϕ, the second d represents the curl. Thus (*) generalizes the rule

$$\nabla \times (\nabla\phi) = 0.$$

Applications

2. If we apply (A) to a vector field A, we get

$$(\square) \quad d(dA) = 0.$$

As $d\mathbf{A}$ represents the curl of $\mathbf{A}$, the second d represents the divergence. Thus $(\square)$ generalizes the rule

$$\nabla \cdot (\nabla \times \mathbf{a}) = 0.$$

(B) $d(\mathbf{F} \wedge \mathbf{G}) = d\mathbf{F} \wedge \mathbf{G} + (-1)^{\deg \mathbf{F}}\mathbf{F} \wedge d\mathbf{G}$

1. If we apply (B) to scalar fields, we get

$$d(\phi\psi) = (d\phi)\psi + \psi(d\phi),$$

which generalizes

$$\nabla(\phi\psi) = (\nabla\phi)\psi + \phi(\nabla\psi).$$

Applications

2. If we apply (B) to a scalar field ϕ and a 1-form $\mathbf{A}$, we get

$$d(\phi\mathbf{A}) = (d\phi) \wedge \mathbf{A} + \phi d\mathbf{A},$$

which generalizes

$$\nabla \times (\phi\mathbf{a}) = (\nabla\phi) \times \mathbf{a} + \phi\nabla \times \mathbf{a}.$$

3. If we apply (B) to a scalar field ϕ and a 2-form ${}^*\mathbf{A}$, we get

$$d(\phi^*\mathbf{A}) = d\phi \wedge ({}^*\mathbf{A}) + \phi d^*\mathbf{A},$$

which generalizes

$$\nabla \cdot (\phi\mathbf{a}) = (\nabla\phi) \cdot \mathbf{a} + \phi\nabla \times \mathbf{a}.$$

4. If we apply (B) to 1-forms, we get

$$d(\mathbf{A} \wedge \mathbf{B}) = d\mathbf{A} \wedge \mathbf{B} - \mathbf{A} \wedge d\mathbf{B},$$

which generalizes

$$\nabla \cdot (\mathbf{a} \times \mathbf{b}) = (\nabla \times \mathbf{a}) \cdot \mathbf{b} - \mathbf{a} \cdot (\nabla \times \mathbf{b}).$$

The preceding discussion of the dual map, the scalar product, the wedge product, and the exterior derivative should convince us that exterior algebra is capable of doing almost anything that can be done in conventional vector analysis. But exterior algebra is not just another way of saying the same thing. It is a much more powerful machinery for at least two reasons:

1. It works in any coordinate system; i.e., it is a covariant formalism.
2. It works in any number of dimensions.

The only drawback in comparison with the conventional vector analysis arises when we discuss the equations of motion for particles. In conventional vector analysis, Newton's equation of motion for a particle moving in a potential $U(\mathbf{x})$ is simply given by

$$m\,\frac{d^2\mathbf{r}}{dt^2} = -\nabla U. \tag{7.72}$$

This equation of motion can be solved very elegantly in various cases using vector calculus. It corresponds, however, to the coordinate expression

$$m\,\frac{d^2x^i}{dt^2} = -\frac{\partial U}{\partial x^i}\,, \tag{7.73}$$

which is valid *only* in Cartesian coordinates. To geometrize the formula (7.72), we must first extend the above coordinate expression to a *covariant* formula!

Exercise 7.7.3
Problem: Show that the covariant generalization of (7.73) is given by

$$m\,\frac{d^2x^k}{dt^2} = -m\Gamma^k_{\ ij}\,\frac{dx^i}{dt}\,\frac{dx^j}{dt} - g^{ki}\,\frac{\partial U}{\partial x^i}\,,$$

where $\Gamma^k_{\ ij}$ are the Christoffel fields in the Euclidean space $\mathbb{R}^3$. They should not be confused with the Christoffel fields in Minkowski space! (Hint: Use that the equation of motion (7.73) extremizes the action

$$S = \int_{t_1}^{t_2}\left[\frac{1}{2}\,mg_{ij}(x)\,\frac{dx^i}{dt}\,\frac{dx^j}{dt} - U(x)\right]dt$$

and copy the discussion in Section 6.7.)

But a covariant formula explicitly containing the Christoffel field falls beyond the scope of the exterior algebra.

Worked Exercise 7.7.4
Problem: Compute the Laplacian in spherical coordinates.

In the case of the Minkowski space, the dimension is $n = 4$. Thus we have 0-forms, 1-forms, 2-forms, 3-forms, and 4-forms. Scalars are represented by 0-forms and 4-forms. Vectors are represented by 1-forms and 3-forms. Finally, we have 2-forms at our disposal. They represent a new kind of concept that did not exist in conventional vector analysis but that plays an extremely important role in 4-dimensional geometry.

The effect of the dual map is shown in the following scheme:

The dual map in Minkowski space:	
0-form: ϕ	4-form: $^{**}\phi = -\phi$ $^{*}\phi : (^{*}\phi)_{0123} = \sqrt{-g}\phi$
1-form: $\mathbf{A} : (A_0, A_1, A_2, A_3)$	3-form: $^{**}\mathbf{A} = \mathbf{A}$ $^{*}\mathbf{A} : (^{*}\mathbf{A})_{012} = \sqrt{-g}A^3$, etc. (7.74)
2-form: $\mathbf{F} : \begin{bmatrix} 0 & F_{01} & F_{02} & F_{03} \\ -F_{01} & 0 & F_{12} & -F_{31} \\ -F_{02} & -F_{12} & 0 & F_{23} \\ -F_{03} & F_{31} & -F_{23} & 0 \end{bmatrix}$	2-form: $^{**}\mathbf{F} = -\mathbf{F}$ $^{*}\mathbf{F} : \sqrt{-g} \begin{bmatrix} 0 & F^{23} & F^{31} & F^{12} \\ -F^{23} & 0 & F^{03} & -F^{02} \\ -F^{31} & -F^{03} & 0 & F^{01} \\ -F^{12} & F^{02} & -F^{01} & 0 \end{bmatrix}$
3-form: $\mathbf{G} : (\mathbf{G})_{\alpha\beta\gamma}$	1-form: $^{**}\mathbf{G} = \mathbf{G}$ $^{*}\mathbf{G} : \sqrt{-g}[G^{123}, -G^{230}, G^{301}, -G^{012}]$
4-form: $\mathbf{H} : (\mathbf{H})_{0123} = H$	0-form: $^{**}\mathbf{H} = -\mathbf{H}$ $^{*}\mathbf{H} : -\frac{1}{\sqrt{-g}} H$

The only thing we should be carefully about is dualizing a four-form. The discussion of this is left to the reader (compare the discussion following the scheme (7.69)). The rest of the scheme should be self-evident. The minus signs associated with the double dualization are characteristic of Minkowski spaces; and we should be careful about them!

Next, we will investigate the differential calculus. First, we look at a scalar field ϕ. It is converted into a vector field

$$\boldsymbol{d}\phi : (\partial_0\phi, \partial_1\phi, \partial_2\phi, \partial_3\phi),$$

which we may naturally call the gradient.

Then we look at a vector field. We can represent it by a 1-form $\mathbf{A}$. Then it is converted into a 2-form

$$\boldsymbol{d}\mathbf{A} : (\partial_\mu A_\nu - \partial_\nu A_\mu),$$

which we may naturally call the curl. But we may also represent it by a 3-form, $^{*}\mathbf{A}$. Then it is converted into a 4-form, i.e., essentially a scalar field. Dualizing this 4-form, we get

$$\boldsymbol{d}^{*}\mathbf{A} = -\delta\mathbf{A} = \frac{1}{\sqrt{-g}}\,\partial_\alpha\left(\sqrt{-g}A^\alpha\right).$$

Thus $\boldsymbol{d}^{*}\mathbf{A}$ essentially represents the divergence.

Finally, we can look at a 2-form $\mathbf{F}$, which we can also represent by its dual form $^{*}\mathbf{F}$. The exterior derivative of $\mathbf{F}$ is a 3-form, which we may call the curl,

$$\boldsymbol{d}\mathbf{F} : \partial_\mu F_{\alpha\beta} + \partial_\alpha F_{\beta\mu} + \partial_\beta F_{\mu\alpha}.$$

The exterior derivative of ${}^*\mathbf{F}$ is similarly related to the divergence of $\mathbf{F}$:

$$ {}^*\boldsymbol{d}{}^*\mathbf{F} = -\delta\mathbf{F}. $$

The exterior derivative in Minkowski space (7.75)

0-form: ϕ	1-form: $\boldsymbol{d}\phi : (\partial_0\phi, \partial_1\phi, \partial_2\phi, \partial_3\phi)$
1-form: $\mathbf{A} : (A_0, A_1, A_2, A_3)$	2-form: $\boldsymbol{d}\mathbf{A} : (\partial_\alpha A_\beta - \partial_\beta A_\alpha)$
2-form: $\mathbf{F} : (F_{\alpha\beta})$; ${}^*\mathbf{F} : \frac{1}{2}\sqrt{-g}\epsilon_{\alpha\beta\gamma\delta}F^{\gamma\delta}$	3-form: $\boldsymbol{d}\mathbf{F} : (\partial_\alpha F_{\beta\gamma} + \partial_\beta F_{\gamma\alpha} + \partial_\gamma F_{\alpha\beta})$; $(\boldsymbol{d}{}^*\mathbf{F})_{012} = \partial_\alpha\left(\sqrt{-g}F^{\alpha 3}\right)$, etc.
3-form: $({}^*\mathbf{A})_{012} = \sqrt{-g}A^3$	4-form: $(\boldsymbol{d}{}^*\mathbf{A})_{0123} = -\partial_\alpha\left(\sqrt{-g}A^\alpha\right)$

Concept	Exterior algebra	Covariant formula
Gradient	$\boldsymbol{d}\phi$	$\partial_\alpha\phi$
Curl	$\boldsymbol{d}\mathbf{A}$	$\partial_\alpha A_\beta - \partial_\beta A_\alpha$
	$\boldsymbol{d}\mathbf{F}$	$\partial_\alpha F_{\beta\gamma} + \partial_\beta F_{\gamma\alpha} + \partial_\gamma F_{\alpha\beta}$
Divergence	$-\delta\mathbf{A}$	$\frac{1}{\sqrt{-g}}\partial_\alpha\left(\sqrt{-g}A^\alpha\right)$
	$-\delta\mathbf{F}$	$\frac{1}{\sqrt{-g}}\partial_\alpha\left(\sqrt{-g}F^{\alpha\beta}\right)$
D'Alembertian	$-\delta\boldsymbol{d}\phi$	$\frac{1}{\sqrt{-g}}\partial_\alpha\left(\sqrt{-g}\partial^\alpha\phi\right)$

7.8 Electromagnetism and the Exterior Calculus

In this section we will first exemplify the manipulations of the exterior calculus by studying electromagnetism in the Euclidean space $\mathbb{R}^3$. The electric field strength will be represented by the 1-form

$$ \mathbf{E} : (E_1, E_2, E_3), $$

whereas the magnetic field strength will be represented by the 2-form

$$ \mathbf{B} : \begin{bmatrix} 0 & B_3 & -B_2 \\ -B_3 & 0 & B_1 \\ B_2 & -B_1 & 0 \end{bmatrix}. $$

There are several reasons for this, but let us just mention one of them. In Minkowski space the field strengths are represented by the 2-form

$$\mathbf{F} : \begin{bmatrix} 0 & -E_1 & -E_2 & -E_3 \\ E_1 & 0 & B_3 & -B_2 \\ E_2 & -B_3 & 0 & B_1 \\ E_3 & B_2 & -B_1 & 0 \end{bmatrix},$$

and here we clearly see that the electric field strength becomes a 1-form, while the magnetic field strength becomes a 2-form when we decompose $\mathbf{F}$ into space and time components:

$$\mathbf{F} : \begin{bmatrix} \text{space scalar} & \text{space 1-form} \\ \text{space 1-form} & \text{space 2-form} \end{bmatrix} = \begin{bmatrix} 0 & -\mathbf{E} \\ \mathbf{E} & \mathbf{B} \end{bmatrix}.$$

If we recall that $\mathbf{B}$ is a 2-form, it is not difficult to translate Maxwell's equations:

Dictionary

Conventional vector analysis	Exterior calculus	
$\nabla \cdot \mathbf{B} = 0$	$d\mathbf{B} = 0$	(7.76)
$\dfrac{\partial \mathbf{B}}{\partial t} + \nabla \times \mathbf{E} = \mathbf{0}$	$\dfrac{\partial \mathbf{B}}{\partial t} + d\mathbf{E} = 0$	(7.77)
$\nabla \cdot \mathbf{E} = \dfrac{1}{\epsilon_0}\rho$	$-\delta\mathbf{E} = \dfrac{1}{\epsilon_0}\rho$	(7.78)
$\dfrac{\partial \mathbf{E}}{\partial t} - c^2\nabla \times \mathbf{B} = \dfrac{-1}{\epsilon_0}\mathbf{j}$	$\dfrac{\partial \mathbf{E}}{\partial t} - c^2\delta\mathbf{B} = \dfrac{-1}{\epsilon_0}\mathbf{J}$	(7.79)

We can now reshuffle the Maxwell equations in the usual way, as shown in the following exercise:

Worked Exercise 7.8.1
Problem: Use the exterior algebra to reexamine the introduction of gauge potentials, the equation of continuity, and the equations of motion for the gauge potentials.

ILLUSTRATIVE EXAMPLE. *The magnetic field outside a wire.*

Suppose we have a uniform current $\mathbf{j}$ in a wire, which we identify with the z-axis. It is well known that the current produces a static magnetic field $\mathbf{B}$ circulating around the wire. (See Figure 7.10.) According to Biot and Savart's law, we get

$$\mathbf{B} = \frac{j}{2\pi\epsilon_0 c^2}\left(\frac{-y}{x^2+y^2}\,;\,\frac{x}{x^2+y^2}\,;\,0\right).$$

At this level we do not have to distinguish between covariant and contravariant

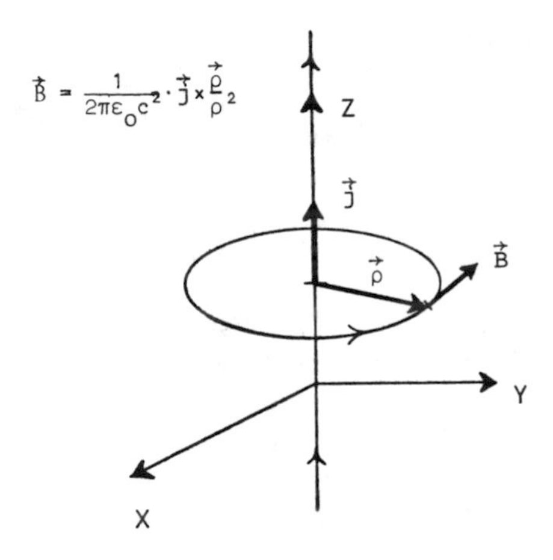

Figure 7.10.

components. In the exterior algebra, the magnetic field is described by the 2-form

$$\mathbf{B} : \frac{j}{2\pi\epsilon_0 c^2}\begin{bmatrix} 0 & 0 & \frac{-x}{x^2+y^2} \\ 0 & 0 & \frac{y}{x^2+y^2} \\ \frac{x}{x^2+y^2} & \frac{-y}{x^2+y^2} & 0 \end{bmatrix},$$

or, if it is preferable, by its dual form

$$^*\mathbf{B} : \frac{j}{2\pi\epsilon_0 c^2}\left[-\frac{y}{x^2+y^2}\ ;\ \frac{x}{x^2+y^2}\ ;0\right],$$

which reproduces the original vector field directly. It is convenient to decompose $\mathbf{B}$ and $^*\mathbf{B}$ into basic forms:

$$\mathbf{B} = \frac{j}{2\pi\epsilon_0 c^2}\left[\frac{-y}{x^2+y^2}\,\boldsymbol{dy}\wedge\boldsymbol{dz} + \frac{x}{x^2+y^2}\,\boldsymbol{dz}\wedge\boldsymbol{dx}\right],$$

$$^*\mathbf{B} = \frac{j}{2\pi\epsilon_0 c^2}\left[\frac{-y}{x^2+y^2}\,\boldsymbol{dx} + \frac{x}{x^2+y^2}\,\boldsymbol{dy}\right].$$

Then we introduce a new coordinate system, cylindrical coordinates, which reflects the symmetry of the problem. First, we notice that Cartesian coordinates and cylindrical coordinates are related by

$$x = \rho\cos\varphi,$$
$$y = \rho\sin\varphi,$$
$$z = z;$$

i.e.,

$$\boldsymbol{dx} = \cos\varphi\,\boldsymbol{d\rho} - \rho\sin\varphi\,\boldsymbol{d\varphi},$$
$$\boldsymbol{dy} = \sin\varphi\,\boldsymbol{d\rho} + \rho\cos\varphi\,\boldsymbol{d\varphi},$$
$$\boldsymbol{dz} = \boldsymbol{dz}.$$

Inserting these formulas, we get

$$\mathbf{B} = \frac{j}{2\pi\epsilon_0 c^2}\,\frac{1}{\rho}\,dz \wedge \boldsymbol{d}\rho,$$

and similarly,

$$^*\mathbf{B} = \frac{j}{2\pi\epsilon_0 c^2}\,\boldsymbol{d}\varphi.$$

Using cylindrical coordinates, we have consequently succeeded in writing **B** and ***B** as *simple* forms. To find a vector potential that generates the magnetic field, i.e., $\mathbf{B} = \boldsymbol{d}\mathbf{A}$, we simply rearrange the expression for **B** (compare Exercise 7.3.5):

$$\mathbf{B} = \frac{j}{2\pi\epsilon_0 c^2}\,dz \wedge \boldsymbol{d}(\ln\rho) = \frac{j}{2\pi\epsilon_0 c^2}\,\boldsymbol{d}(z\boldsymbol{d}\ln\rho).$$

From this we read off

$$\mathbf{A} = \frac{j}{2\pi\epsilon_0 c^2}\,z\boldsymbol{d}\ln\rho = \frac{j}{2\pi\epsilon_0 c^2}\,\frac{z}{\rho}\,\boldsymbol{d}\rho.$$

Outside the wire we have a static magnetic field. Thus we conclude that outside the wire, the Maxwell equations reduce to

$$\boldsymbol{d}\mathbf{B} = 0 \quad \text{and} \quad \delta\mathbf{B} = 0.$$

Consequently, **B** is a primitively harmonic form.

Worked Exercise 7.8.2
Problem: Show that **B** is almost coexact; i.e., determine a three-form **S** such that

$$\mathbf{B} = \delta\mathbf{S}.$$

Show that **S** is necessarily singular along a Dirac sheet, i.e., a half-plane bounded by the z-axis. Show that **S** can be chosen as a harmonic form.

Finally, we can investigate the geometrical structure of **B** and ***B**. Here **B** is a simple form generated by the functions z and $\ln\rho$ (for simplicity, we forget $\frac{j}{2\pi\epsilon_0 c^2}$ for the moment!) These functions produce the honeycomb structure shown in Figure 7.11. Similarly, ***B** is also a simple form generated by the function φ, which, however, makes a jump somewhere between 0 and 2π (cf. Figure 7.11). This example concludes our discussion of electromagnetism and three-dimensional geometry.

In the rest of this section we return to our main interest: electromagnetic fields in Minkowski space. The field strengths are represented by a skew-symmetric cotensor field of rank 2,

$$\mathbf{F} = \frac{1}{2}\,F_{\alpha\beta}\boldsymbol{d}x^\alpha \wedge \boldsymbol{d}x^\beta.$$

Therefore, **F** is nothing but a differential form of degree 2. We may decompose it into the basic 2-forms associated with an inertial frame:

$$\begin{aligned}\mathbf{F} = E_x\boldsymbol{d}x \wedge \boldsymbol{d}t + E_y\boldsymbol{d}y \wedge \boldsymbol{d}t + E_z\boldsymbol{d}z \wedge \boldsymbol{d}t \\ + B_z\boldsymbol{d}x \wedge \boldsymbol{d}y + B_x\boldsymbol{d}y \wedge \boldsymbol{d}z + B_y\boldsymbol{d}z \wedge \boldsymbol{d}x.\end{aligned} \tag{7.80}$$

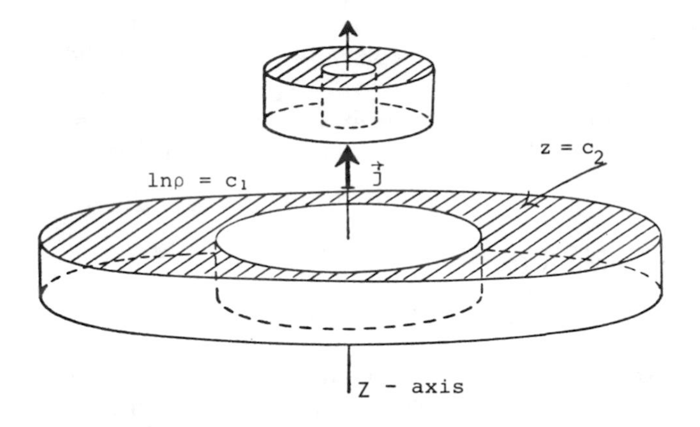

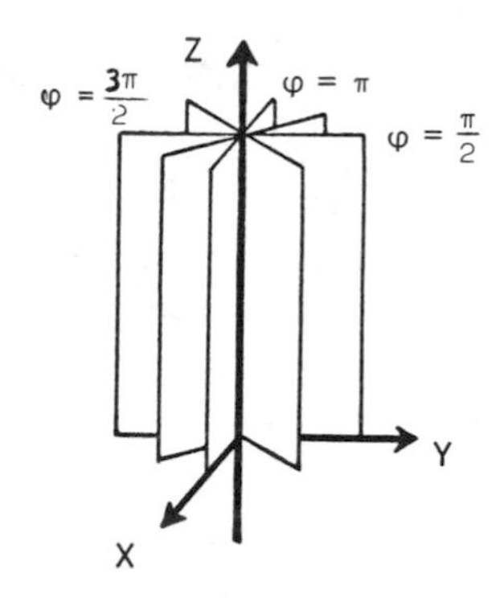

Figure 7.11.

The *gauge potential* is represented by a covector field **A**. Therefore, **A** is nothing but a differential form of degree 1. We can decompose it into the basic 1-forms associated with an inertial frame:

$$\mathbf{A} = -\phi \boldsymbol{d}t + A_x \boldsymbol{d}x + A_y \boldsymbol{d}y + A_z \boldsymbol{d}z. \tag{7.81}$$

Now we want to compute the exterior derivatives $\boldsymbol{d}\mathbf{A}$ and $\boldsymbol{d}\mathbf{F}$. We will use an *arbitrary* coordinate system. Then $\boldsymbol{d}\mathbf{A}$ is characterized by the components $\partial_\alpha A_\beta - \partial_\beta A_\alpha$. Previously we introduced the gauge potential A_α through the relation (1.30)

$$F_{\alpha\beta} = \partial_\alpha A_\beta - \partial_\beta A_\alpha,$$

valid in an inertial frame. But this can be geometrized as

$$\mathbf{F} = \boldsymbol{d}\mathbf{A}. \tag{7.82}$$

Now, the components of **F** and ***d*A** must coincide in any coordinate system. Thus we learn that (1.30) holds not only for inertial coordinates but is valid in an arbitrary coordinate system!

Furthermore, ***d*F** is characterized by the components $\partial_\alpha F_{\beta\gamma} + \partial_\beta F_{\gamma\alpha} + \partial_\gamma F_{\alpha\beta}$. In an inertial frame we know that $F_{\alpha\beta}$ solves the Maxwell equation (1.27)

$$\partial_\alpha F_{\beta\gamma} + \partial_\beta F_{\gamma\alpha} + \partial_\gamma F_{\alpha\beta} = 0.$$

This can be geometrized as

$$\boldsymbol{d}\mathbf{F} = 0. \tag{7.83}$$

So **F** is a closed form. Again we see that (1.27) not only holds for inertial coordinates but is valid for arbitrary coordinates. The fact that the equation (7.82) solves the Maxwell equation (7.83) now become a trivial consequence of the rule $\boldsymbol{d}^2 = 0$.

To finish the discussion of the electromagnetic field, we should also try to give a geometrical interpretation of the second Maxwell equation (1.28)

$$\partial_\beta F^{\alpha\beta} = J^\alpha.$$

As it involves the divergence of **F**, we should look at $\boldsymbol{\delta}\mathbf{F}$. In an *arbitrary* coordinate system this is a 1-form, characterized by the *contravariant* components

$$-(\boldsymbol{\delta}\mathbf{F})^\beta = \frac{1}{\sqrt{-g}}\,\partial_\alpha\left(\sqrt{-g}\,F^{\alpha\beta}\right).$$

In inertial coordinates, $\sqrt{-g} = 1$. Therefore, we see that the contravariant components of $\boldsymbol{\delta}\mathbf{F}$ and **J** coincide in an inertial frame. But then they are identical. Consequently, the second Maxwell equation simply states that

$$\boldsymbol{\delta}\mathbf{F} = \mathbf{J}. \tag{7.84}$$

Observe that as we know the components of $\boldsymbol{\delta}\mathbf{F}$ and **J** in an arbitrary coordinate system, we can write (7.84) in a *covariant* form:

$$\frac{1}{\sqrt{-g}}\,\partial_\beta\left(\sqrt{-g}\,F^{\alpha\beta}\right) = J^\alpha. \tag{7.85}$$

If we take the codifferential of both sides of (7.84), we immediately get

$$0 = \boldsymbol{\delta}\mathbf{J}. \tag{7.86}$$

But $-\boldsymbol{\delta}\mathbf{J}$ is the divergence of **J**, so this equation is equivalent to

$$\frac{1}{\sqrt{-g}}\,\partial_\alpha\left(\sqrt{-g}\,J^\alpha\right) = 0, \tag{7.87}$$

which reduces to the usual equation of continuity in an inertial frame. We have therefore succeeded in condensing the structure of the electromagnetic field into the following elegant form:

Theory of electromagnetism		(7.88)
	Geometrical form	Covariant form
Maxwell's first equation	$\boldsymbol{d}\mathbf{F} = 0$	$\partial_\alpha F_{\beta\gamma} + \partial_\beta F_{\gamma\alpha} + \partial_\gamma F_{\alpha\beta} = 0$
Maxwell's second equation	$\boldsymbol{\delta}\mathbf{F} = \mathbf{J}$	$\frac{1}{\sqrt{-g}}\,\partial_\beta\left(\sqrt{-g}F^{\alpha\beta}\right) = J^\alpha$
Continuity equation	$\boldsymbol{\delta}\mathbf{J} = 0$	$\frac{1}{\sqrt{-g}}\,\partial_\alpha\left(\sqrt{-g}J^\alpha\right) = 0$
Equation of gauge potential	$\mathbf{F} = \boldsymbol{d}\mathbf{A}$	$F_{\alpha\beta} = \partial_\alpha A_\beta - \partial_\beta A_\alpha$
Gauge transformation	$\mathbf{A} \to \mathbf{A} + \boldsymbol{d}\chi$	$A_\alpha \to A_\alpha + \partial_\alpha\chi$

Illustrative Example. *The monopole field.*

Observe first that we can now work in arbitrary coordinates. A relation like $F_{\alpha\beta} = \partial_\alpha A_\beta - \partial_\beta A_\alpha$ is also valid in spherical coordinates, or even in a rotating coordinate system if we prefer!

We have previously computed the field strengths of a monopole expressed in spherical coordinates (Section 6.10); cf. equation (6.79). Therefore, we can decompose the monopole field as

$$\mathbf{F} = \frac{g}{4\pi}\sin\theta\,\boldsymbol{d}\theta \wedge \boldsymbol{d}\varphi. \tag{7.89}$$

This is a simple form, which may be rewritten as

$$\mathbf{F} = \frac{-g}{4\pi}\,\boldsymbol{d}(\cos\theta) \wedge \boldsymbol{d}\varphi.$$

As it is a static field, we can suppress the time coordinate for a moment; i.e., we look at space at a specific time t_0. Apart from the normalization factor, $\mathbf{F}$ is generated by the functions $\cos\theta$ and φ. These two functions generate the honeycomb structure shown in Figure 7.12, (i.e., $\cos\theta = 0, \epsilon, 2\epsilon, \ldots$ and $\varphi = 0, \epsilon, 2\epsilon, \ldots$). The fundamental tubes are defined by $\cos\theta = +1, 0, -1$ and $\varphi = 0, 1, 2, 3,$ and

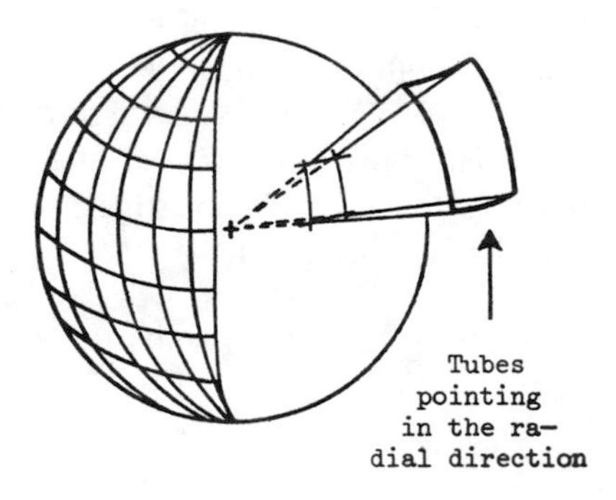

Figure 7.12.

the number of fundamental tubes emanating from the origin is $2 \times 2\pi$. Thus the number of fundamental tubes that cross a sphere around the origin is 4π. Observe also that the ϵ-tubes all intersect the unit sphere in the *same* area. To see this, we use the well-known fact that a region $\Omega = \{(\theta, \varphi) \mid \theta_1 \leq \theta \leq \theta_2; \varphi_1 \leq \varphi \leq \varphi_2\}$ covers the area

$$\int_{\theta_1}^{\theta_2} \int_{\varphi_1}^{\varphi_2} \sin\theta \, d\theta \, d\varphi = \int_{-\cos\theta_1}^{-\cos\theta_2} \int_{\varphi_1}^{\varphi_2} d(-\cos\theta) d\varphi.$$

Therefore, an arbitrary ϵ-tube covers the area

$$\int_{u=k\epsilon-\epsilon}^{u=-k\epsilon} \int_{\varphi=m}^{\varphi=m\epsilon+\epsilon} du \, d\varphi = [-k\epsilon + k\epsilon + \epsilon][m\epsilon + \epsilon - m\epsilon] = \epsilon^2.$$

This, of course, again reflects the fact that the monopole field is a spherically symmetric field!

Now observe that because we have written the monopole field as the simple form

$$\mathbf{F} = \frac{g}{4\pi} \sin\theta \, \mathbf{d}\theta \wedge \mathbf{d}\varphi,$$

we can immediately find a gauge potential $\mathbf{A}$ generating the monopole field:

$$\mathbf{F} = \frac{-g}{4\pi} \mathbf{d}(\cos\theta) \wedge \mathbf{d}\varphi = -\mathbf{d}\left(\frac{g}{4\pi} \cos\theta \, \mathbf{d}\varphi\right).$$

Thus we can use the gauge potential

$$\mathbf{A} = \frac{-g}{4\pi} \cos\theta \, \mathbf{d}\varphi. \tag{7.90}$$

Now, we should be very careful, because we have previously shown that the monopole field cannot be generated by a gauge potential throughout the whole space–time manifold (Section 1.3). Therefore, we conclude that

$$\mathbf{A} = \frac{-g}{4\pi} \cos\theta \, \mathbf{d}\varphi$$

cannot define a smooth vector field throughout our manifold. The component $\frac{g}{4\pi} \cos\theta$ is certainly a nice smooth function, so this must be related to the fact that the coordinate system itself is singular! We have several times seen that this coordinate system in fact breaks down at the z-axis, and therefore the gauge potential (7.90) is not defined on the z-axis. To investigate this a little more closely we return to Cartesian coordinates. Using that

$$\cos\theta = \frac{z}{r}; \qquad \varphi = \arctan\frac{y}{x},$$

we easily find that

$$\mathbf{A} = \frac{-g}{4\pi} \left[-\frac{zy}{r(x^2 + y^2)} \mathbf{d}x + \frac{zx}{r(x^2 + y^2)} \mathbf{d}y \right],$$

so that the gauge potential has the *Cartesian* components

$$-\frac{g}{4\pi}\left[0, -\frac{zy}{r(x^2+y^2)}, +\frac{zx}{r(x^2+y^2)}, 0\right].$$

But these components become highly singular when we approach the z-axis! If we approach the point $(0, 0, z_0)$ where $z_0 > 0$, we can safely put: $\frac{z}{r} \approx 1$. Thus A_α varies like

$$A_\alpha \approx \frac{-g}{4\pi}\left[0, -\frac{y}{x^2+y^2}, \frac{x}{x^2+y^2}, 0\right],$$

and we see that

$$|A_x|, |A_y| \rightarrow \infty$$

when we approach the z-axis! Consequently, $\mathbf{A}$ is highly singular on the z-axis in accordance with our general result: *There can exist no global gauge potential generating the monopole field.* However, if we are willing to accept a string, where the gauge potential becomes singular, then we have just found such a gauge potential. This string is, of course, the famous "Dirac string"! (Observe that in the preceding discussion we have "excluded" the origin from the space–time manifold, because at the origin the monopole field itself is singular.) We should also observe the following: The spherical coordinates not only break down at the z-axis, but the cyclic coordinate φ also has a singularity at a half-plane extending from the z-axis. This is connected with the fact that φ has to make a jump somewhere between 0 and 2π. Thus $\boldsymbol{d}\varphi$ is *not* defined on the half-plane where φ makes a jump. But if we return to Cartesian coordinates, we find that

$$\boldsymbol{d}\varphi = -\frac{y}{x^2+y^2}\,\boldsymbol{d}x + \frac{x}{x^2+y^2}\,\boldsymbol{d}y.$$

Therefore, $\boldsymbol{d}\varphi$ is perfectly well behaved (except on the z-axis), and we conclude

1. φ itself is a coordinate function that makes a jump at a half-plane, and therefore it *cannot* be extended to a smooth function on the whole space surrounding the z-axis (compare Section 1.4).
2. $\boldsymbol{d}\varphi$ does *not* make a jump at a half-plane, and consequently it can be extended to a smooth 1-form on the whole space surrounding the z-axis.

In what follows we will let $\boldsymbol{d}\varphi$ denote the smooth 1-form defined on the whole space surrounding the z-axis.

The singular gauge potential $\mathbf{A} = \frac{-g}{4\pi}\cos\theta\,\boldsymbol{d}\varphi$ is obviously not the only gauge potential generating $\mathbf{F}$. To get another example, we observe that $\boldsymbol{d}(\boldsymbol{d}\varphi) = 0$. Therefore, we consider

$$\mathbf{A}' = \mathbf{A} + \frac{g}{4\pi}\,\boldsymbol{d}\varphi = \frac{-g}{4\pi}\cos\theta\,\boldsymbol{d}\varphi + \frac{g}{4\pi}\,\boldsymbol{d}\varphi.$$

Although it formally looks like a gauge transformation, it is a *singular* gauge transformation because φ is a *singular* coordinate function. Nevertheless, $\mathbf{A}'$ generates the same field strengths. Observe that

$$\mathbf{A}' = \frac{-g}{4\pi}(\cos\theta - 1)\boldsymbol{d\varphi} \tag{7.91}$$

and that the coefficient $\cos\theta - 1$ vanishes on the positive z-axis. This suggests that $\mathbf{A}'$ is singular *only* on the negative z-axis. To confirm this we return to Cartesian coordinates:

$$\mathbf{A}' = \frac{-g}{4\pi}\left(\frac{z}{r} - 1\right)\left(-\frac{y}{x^2+y^2}\,\boldsymbol{dx} + \frac{x}{x^2+y^2}\,\boldsymbol{dy}\right).$$

If we approach the point $(0, 0, z_0)$ where $z_0 > 0$, we can use the approximation

$$r = \sqrt{x^2 + y^2 + z^2} \approx z + \frac{x^2 + y^2}{2z}$$

to get the asymptotic behavior of $\frac{z}{r} - 1$:

$$\frac{z}{r} - 1 \approx -\frac{x^2+y^2}{2z_0}\,.$$

It follows that $\mathbf{A}'$ varies like

$$\mathbf{A}' \approx \frac{-g}{4\pi}\left(\frac{y}{2z_0^2}\,\boldsymbol{dx} - \frac{x}{2z_0^2}\,\boldsymbol{dy}\right),$$

and the singularity on the positive z-axis has disappeared.

Thus we are left with an semi-infinite Dirac string: *The negative z-axis*. This is the best result we can obtain! (If we take any closed surface surrounding the monopole, then it *must* contain at least one singularity of the gauge field. Otherwise it would contradict Theorem 1.1 of Section 1.3.)

Solutions to Worked Exercises

Solution to 7.5.6

Since ${}^*\mathbf{T} \wedge \mathbf{U}$ is an n-form, we only need to find the $(12\ldots n)$-component. This is given by the expression

$$\begin{aligned}
&\frac{n!}{(n-k)!k!}\,({}^*\mathbf{T})_{[12\ldots(n-k)}(\mathbf{U})_{(n-k+1)\ldots n]} \\
&= \frac{1}{(n-k)!k!}\,\epsilon^{a_1\ldots a_{n-k}b_1\ldots b_k}({}^*\mathbf{T})_{a_1\ldots a_{n-k}}U_{b_1\ldots b_k} \\
&= \frac{\sqrt{g}}{(n-k)!k!k!}\,\epsilon^{a_1\ldots a_{n-k}b_1\ldots b_k}\epsilon_{a_1\ldots a_{n-k}c_1\ldots c_k}T^{c_1\ldots c_k}U_{b_1\ldots b_k}.
\end{aligned} \tag{7.92}$$

Now observe that

$$\epsilon^{a_1\ldots a_{n-k}b_1\ldots b_k}\epsilon_{a_1\ldots a_{n-k}c_1\ldots c_k}$$

vanishes unless $(b_1 \ldots b_k)$ is a permutation of $(c_1 \ldots c_k)$. Clearly, the following relation holds (cf. Exercise 7.4.3):

$$\frac{1}{(n-k)!} \epsilon^{a_1 \ldots a_{n-k} b_1 \ldots b_k} \epsilon_{a_1 \ldots a_{n-k} c_1 \ldots c_k} = \operatorname{Sgn} \begin{bmatrix} b_1 \ldots b_k \\ c_1 \ldots c_k \end{bmatrix},$$

where we have used that there are $(n-k)!$ permutations of $(a_1 \ldots a_{n-k})$, all contributing with the same value. Furthermore, we can use that $T^{c_1 \ldots c_k}$ is skew-symmetric, so that we get

$$\frac{1}{k!} \operatorname{Sgn} \begin{bmatrix} b_1 \ldots b_k \\ c_1 \ldots c_k \end{bmatrix} T^{c_1 \ldots c_k} = T^{b_1 \ldots b_k}.$$

Inserting these formulas into (7.92), the expression for $(^*\mathbf{T} \wedge \mathbf{U})_{1 \ldots n}$ reduces to

$$(^*\mathbf{T} \wedge \mathbf{U})_{1 \ldots n} = \frac{\sqrt{g}}{k!} T^{b_1 \ldots b_k} U_{b_1 \ldots b_k}.$$

We have thus shown

$$^*\mathbf{T} \wedge \mathbf{U} = \frac{1}{k!} T^{b_1 \ldots b_k} U_{b_1 \ldots b_k} \sqrt{g} \mathbf{dx}^1 \wedge \ldots \wedge \mathbf{dx}^n = \frac{1}{k!} T^{b_1 \ldots b_k} U_{b_1 \ldots b_k} \boldsymbol{\varepsilon}.$$

Solution to 7.5.10

For simplicity, we discuss only the Riemannian case. We want to compare $^*(\mathbf{T} \cdot \mathbf{U})$ and $^*\mathbf{T} \wedge \mathbf{U}$. For definiteness, we compare only their $1 \ldots (n-k+m)$ components, but the argument can easily be generalized to arbitrary components. The contraction $\mathbf{T} \cdot \mathbf{U}$ is characterized by the components

$$\frac{1}{m!} T_{e_1 \ldots e_{k-m} f_1 \ldots f_m} U^{f_1 \ldots f_m}.$$

The dual form is then characterized by the components

$$\frac{\sqrt{g}}{(k-m)!} \frac{1}{m!} \epsilon_{c_1 \ldots c_{n-k} g_1 \ldots g_m e_1 \ldots e_{k-m}} T^{e_1 \ldots e_{k-m} f_1 \ldots f_m} U_{f_1 \ldots f_m}.$$

If we specialize to the $(c_1 \ldots c_{n-k} g_1 \ldots g_m) = (1 \ldots n-k+m)$ component, we finally get

$$^*(\mathbf{T} \cdot \mathbf{U})_{1 \ldots (n+k+m)} = \frac{\sqrt{g}}{m!} T^{(n-k+m+1) \ldots n f_1 \ldots f_m} U_{f_1 \ldots f_m}. \tag{7.93}$$

$^*\mathbf{T} \wedge \mathbf{U}$ is characterized by the components

$$\begin{aligned}
&\frac{(n-k+m)!}{(n-k)!m!} (^*\mathbf{T})_{[c_1 \ldots c_{n-k}} U_{g_1 \ldots g_m]} \\
&= \frac{1}{(n-k)!m!} \operatorname{Sgn} \begin{bmatrix} a_1 \ldots a_{n-k} b_1 \ldots b_m \\ c_1 \ldots c_{n-k} g_1 \ldots g_m \end{bmatrix} (^*\mathbf{T})_{a_1 \ldots a_{n-k}} U_{b_1 \ldots b_m} \\
&= \frac{\sqrt{g}}{(n-k)!m!k!} \operatorname{Sgn} \begin{bmatrix} a_1 \ldots a_{n-k} b_1 \ldots b_m \\ c_1 \ldots c_{n-k} g_1 \ldots g_m \end{bmatrix} \epsilon_{a_1 \ldots a_{n-k} e_1 \ldots e_{k-m} f_1 \ldots f_m} \\
&\quad \times T^{e_1 \ldots e_{k-m} f_1 \ldots f_m} U_{b_1 \ldots b_m}.
\end{aligned}$$

If we specialize to the $(c_1, \ldots c_{n-k} g_1 \ldots g_m) = (1 \ldots n-k+m)$ component, we get

$$(^*\mathbf{T} \wedge \mathbf{U})_{1\ldots(n-k+m)} =$$

$$\frac{\sqrt{g}}{(n-k)!m!k!} \operatorname{Sgn}\begin{bmatrix} a_1 \ldots a_{n-k} \; b_1 \ldots b_m \\ 1 \ldots n-k \; n-k+1 \ldots n-k+m \end{bmatrix} \tag{7.94}$$

$$\times\, \epsilon_{a_1 \ldots a_{n-k} e_1 \ldots e_{k-m} f_1 \ldots f_m} \cdot T^{e_1 \ldots e_{k-m} f_1 \ldots f_m} U_{b_1 \ldots b_m}.$$

We must then show that (7.94) reduces to (7.93) except for a sign. This is done by brute force! We attack (7.94):

Observe first that $(a_1 \ldots a_{n-k})$ is an ordered subset of $(1 \ldots n-k+m)$ and that $(e_1 \ldots e_{k-m} f_1 \ldots f_m)$ is an ordered subset that is complementary to $(a_1 \ldots a_n)$; i.e., they are mutually disjoint, and together they exhaust $(1 \ldots n)$. Especially, $(e_1 \ldots e_{k-m} f_1 \ldots f_m)$ must contain the numbers $(n-k+m+1 \ldots n)$.

Now fix $(b_1 \ldots b_m)$ for a moment. Then $(a_1 \ldots a_{n-k})$ is determined up to a permutation. All these permutations have equal contributions when we perform the summation, and we can therefore pick out a single representative, which we denote by $(a_1(b), \ldots, a_{n-k}(b))$, since it depends on $\{b_1, \ldots, b_m\}$.

Now that $(a_1(b), \ldots, a_{n-k}(b))$ has been fixed, we observe that $(e_1 \ldots e_{k-m} f_1 \ldots f_m)$ is determined up to a permutation. Again, all these permutations give the same contribution, and we need only pick up a single representative. Since $(e_1 \ldots e_{k-m} f_1 \ldots f_m)$ contains $(n-k+m+1, \ldots, n)$; we can simply put

$$e_1 = n-k+m+1, \ldots, e_{k-m} = n.$$

Furthermore, we choose a special representative for $f_1, \ldots, f_m$, which we denote by $(f_1(b), \ldots, f_m(b))$, since it too depends on $\{b_1 \ldots, b_m\}$. With these preparations, (7.94) now reduces to

$$\frac{\sqrt{g}}{m!} \operatorname{Sgn}\begin{bmatrix} a_1(b) \cdots a_{n-k}(b) b_1 \ldots b_m \\ 1 \ldots n-k \; n-k+1 \ldots n-k+m \end{bmatrix}$$

$$\times\, \epsilon_{a_1(b) \ldots a_{n-k}(b)(n-k+m+1) \ldots n f_1(b) \ldots f_m(b)}$$

$$\times\, T^{(n-k+m+1) \ldots n f_1(b) \ldots f_m(b)} U_{b_1 \ldots b_m},$$

where we sum only over b. Observe that $(a_1(b) \ldots a_{n-k}(b) b_1 \ldots b_m)$ is an ordered subset of $(1, \ldots, n-k+m)$, and similarly, $(a_1(b) \ldots a_{n-k}(b) f_1(b) \ldots f_m(b))$ is an ordered subset of $(1, \ldots, n-k+m)$. But then $(f_1(b), \ldots, f_m(b))$ is a permutation of $(b_1 \ldots b_m)$, and we can safely put $f_1(b) = b_1, \ldots, f_m(b) = b_m$. This means that (7.94) can be further reduced to

$$\frac{\sqrt{g}}{m!} \operatorname{Sgn}\begin{bmatrix} a_1(b) \ldots a_{n-k}(b) b_1 \ldots b_m \\ 1 \ldots n-k \; n-k+1 \ldots n-k+m \end{bmatrix}$$

$$\times\, \epsilon_{a_1(b) \ldots a_{n-k}(b)(n-k+m+1) \ldots n b_1 \ldots b_m} T^{(n-k+m+1) \ldots n b_1 \ldots b_m} U_{b_1 \ldots b_m}.$$

Let us focus on the statistical factor (no summation!):

$$\operatorname{Sgn}\begin{bmatrix} a_1(b) \ldots a_{n-k}(b) b_1 \ldots b_m \\ 1 \ldots n-kn-k+1 \ldots n-k+m \end{bmatrix}$$

$$\times\, \epsilon_{a_1(b) \ldots a_{n-k}(b)(n-k+m+1) \ldots n b_1 \ldots b_m}.$$

This statistical factor is easily evaluated if we rearrange the indices of the Levi-Civita symbol

$$a_1(b)\dots a_{n-k}(b)(n-k+m+1)\dots nb_1\dots b_m \rightarrow$$
$$a_1(b)\dots a_{n-k}(b)b_1\dots b_m(n-k+m+1)\dots n.$$

This costs a factor $(-1)^{m(k-m)}$, but apart from this sign, the combinatorial factor is now recognized as a complicated way of writing the number 1! We have thus finally reduced (7.94) to the form

$$\frac{(-1)^{m(k-m)}\sqrt{g}}{m!}\,T^{(n-k+m+1)\dots nb_1\dots b_m}U_{b_1\dots b_m}.$$

If we compare this with (7.93), we deduce the following formula:

$$^*\mathbf{T}\wedge\mathbf{U} = (-1)^{m(k-m)*}(\mathbf{T}\cdot\mathbf{U}), \tag{7.95}$$

which is in fact valid for both Euclidean metrices and Minkowski metrices. The desired formula now follows by performing a dualization on both sides.

Solution to 7.6.3

Let us work it out for a Euclidean metric. Putting

$$\mathbf{S} = \delta\mathbf{T} = (-1)^{k(n-k+1)*}\boldsymbol{d}^*\mathbf{T},$$

we obtain

$$^*\mathbf{S} = (-1)^{n-k+1}\boldsymbol{d}^*\mathbf{T}. \tag{7.96}$$

(Observe that this actually holds for Minkowski metrics too.) Here $\boldsymbol{d}^*\mathbf{T}$ is an $(n-k+1)$-form characterized by the components

$$\begin{aligned}
(\boldsymbol{d}^*\mathbf{T})_{ji_1\dots i_{n-k}} &= \frac{(n-k+1)!}{(n-k)!}\,\partial_{[j}(^*\mathbf{T})_{i_1\dots i_{n-k}]}\\
&= \frac{1}{(n-k)!}\,\mathrm{Sgn}\begin{bmatrix} ab_1\dots b_{n-k}\\ ji_1\dots i_{n-k}\end{bmatrix}\partial_a(^*\mathbf{T})_{b_1\dots b_{n-k}}\\
&= \frac{1}{(n-k)!k!}\,\mathrm{Sgn}\begin{bmatrix} ab_1\dots b_{n-k}\\ ji_1\dots i_{n-k}\end{bmatrix}\\
&\quad\times\epsilon_{b_1\dots b_{n-k}c_1\dots c_k}\partial_a\left(\sqrt{g}T^{c_1\dots c_k}\right).
\end{aligned}$$

For simplicity, we investigate only the $(1,\dots,n-k+1)$-component, but the results are easily generalized:

$$\begin{aligned}
&(\boldsymbol{d}^*\mathbf{T})_{1\dots(n-k+1)}\\
&= \frac{1}{(n-k)!k!}\,\mathrm{Sgn}\begin{bmatrix} ab_1\dots b_{n-k}\\ 12\dots(n-k+1)\end{bmatrix}\epsilon_{b_1\dots b_{n-k}c_1\cdots c_k}\partial_a\left(\sqrt{g}T^{c_1\dots c_k}\right).
\end{aligned} \tag{7.97}$$

For a fixed a, the ordered set $(b_1\dots b_{n-k})$ is determined up to a permutation, since a and $(b_1\dots b_{n-k})$ are complementary subsets of $(1,\dots,n-k+1)$. All these permutations give the same contribution, so we need only specify a single representative

$$(b_1 \dots b_{n-k}) = (b_1(a), \dots, b_{n-k}(a)).$$

Once $(b_1 \dots b_{n-k})$ is fixed, the ordered set $(c_1 \dots c_k)$ is also determined up to a permutation. All these permutations give the same contribution too, and we need only specify a single representative. Since $b_1 \dots b_{n-k}$ are all different from a, we conclude that a coincides with one of the numbers $c_1 \dots c_k$. We can therefore always achieve that

$$c_1 = a.$$

The rest then can be chosen as

$$c_2 = n - k + 2, \dots, c_k = n.$$

With these preparations, (7.97) now reduces to

$$(\boldsymbol{d}^*\mathbf{T})_{1\dots(n-k+1)} = \mathrm{Sgn}\begin{bmatrix} a & b_1(a) \dots b_{n-k}(a) \\ 1 & 2 \dots (n-k+1) \end{bmatrix} \times \epsilon_{b_1(a)\dots b_{n-k}(a)a(n-k+2)\dots n}\partial_a\left(\sqrt{g}F^{a(n-k+2)\dots n}\right).$$

Using that

$$\epsilon_{a(n-k+2)\dots n b_1(a)\dots b_{n-k}(a)} = (-1)^{n-k}\epsilon_{ab_1(a)\dots b_{n-k}(a)(n-k+2)\dots n},$$

we finally obtain

$$(\boldsymbol{d}^*\mathbf{T})_{1\dots(n-k+1)} = (-1)^{n-k}\partial_a\left(\sqrt{g}T^{a(n-k+2)\dots n}\right). \tag{7.98}$$

Similarly, $^*\mathbf{S}$ is an $(n-k+1)$-form characterized by $(1, \dots, n-k+1)$-component

$$\begin{aligned}(^*\mathbf{S})_{1\dots(n-k+1)} &= \frac{\sqrt{g}}{(k-1)!}\epsilon_{1\dots(n-k+1)j_1\dots j_{k-1}}S^{j_1\dots j_{k-1}} \\ &= \sqrt{g}S^{(n-k+2)\dots n}.\end{aligned} \tag{7.99}$$

If we compare (7.98) and (7.99) and use the relation (7.96), we finally obtain the following result:

$$S^{(n-k+2)\dots n} = -\frac{1}{\sqrt{g}}\,\partial_a\left(\sqrt{g}T^{a(n-k+2)\dots n}\right),$$

which we can immediately generalize to the result (7.64).

Solution to 7.6.4

When we use Cartesian coordinates in a Euclidean space, we do not have to distinguish between covariant and contravariant components. Furthermore, we can put

$\sqrt{g} = 1$. If $\mathbf{T}$ is a k-form, the derivative $\boldsymbol{d}\mathbf{T}$ is then the $(k+1)$-form characterized by the components

$$\frac{1}{k!}\,\mathrm{Sgn}\begin{bmatrix} c & d_1 \dots d_k \\ j & i_1 \dots i_k \end{bmatrix} \partial_c T_{d_1 \dots d_k}.$$

Thus $-\boldsymbol{\delta d}\,\mathbf{T}$ is the k-form characterized by the components

$$(-\boldsymbol{\delta d}\,\mathbf{T})_{a_1 \dots a_k} = \frac{1}{k!}\,\partial_b \mathrm{Sgn}\begin{bmatrix} c & d_1 \dots d_k \\ b & a_1 \dots a_k \end{bmatrix} \partial_c T_{d_1 \dots d_k}.$$

For a fixed c, the ordered set $(d_1 \dots d_k)$ is determined up to a permutation. As all permutations give the same contribution, we need only choose a single representative

$$d_1 = d_1(c), \dots, d_k = d_k(c),$$

and the above expression reduces to

$$(-\boldsymbol{\delta d}\,\mathbf{T})_{a_1 \dots a_k} = \mathrm{Sgn}\begin{bmatrix} c & d_1(c) \dots d_k(c) \\ b & a_1 \dots a_k \end{bmatrix} \partial_b \partial_c T_{d_1(c) \dots d_k(c)},$$

where we sum only over b and c. Clearly, this sum decomposes in the following way:

$$\partial_b \partial_b T_{a_1 \dots a_k} + \sum_{b \neq c} \mathrm{Sgn}\begin{bmatrix} c & d_1(c) \dots d_k(c) \\ b & a_1 \cdots a_k \end{bmatrix} \partial_b \partial_c T_{d_1(c) \dots d_k(c)}.$$

It is the last term we must now get rid of. Since $c \neq b$, we can fix $d_1(c) = b$. Making one transposition, we get

$$\begin{aligned} \mathrm{Sgn}\begin{bmatrix} c & b & d_2(c) \dots d_k(c) \\ b & a_1 & a_2 \dots a_k \end{bmatrix} &= -\mathrm{Sgn}\begin{bmatrix} b & c & d_2(c) \dots d_k(c) \\ b & a_1 & a_2 \dots a_k \end{bmatrix} \\ &= -\mathrm{Sgn}\begin{bmatrix} c & d_2(c) \dots d_k(c) \\ a_1 & a_2 \dots a_k \end{bmatrix}. \end{aligned}$$

Inserting this, we have shown that

$$(-\boldsymbol{\delta d}\,\mathbf{T})_{a_1 \dots a_k} = \partial_b \partial_b T_{a_1 \dots a_k} - \mathrm{Sgn}\begin{bmatrix} c & d_2(c) \dots d_k(c) \\ a_1 & a_2 \dots a_k \end{bmatrix} \partial_b \partial_c T_{b d_2(c) \dots d_k(c)}.$$

On the other hand, $-\boldsymbol{d\delta}\mathbf{T}$ is the $(k-1)$-form characterized by the components

$$(-\boldsymbol{d\delta}\mathbf{T})_{a_1 \dots a_k} = \frac{1}{(k-1)!}\,\mathrm{Sgn}\begin{bmatrix} c & d_2 \dots d_k \\ a_1 & a_2 \dots a_k \end{bmatrix} \partial_c \partial_b T_{b d_2 \dots d_k}.$$

For fixed c, the ordered set $(d_2 \dots d_k)$ is determined up to a permutation. All permutations give the same contribution, and we need only choose a single representative

$$d_2 = d_2(c), \dots, d_k = d_k(c).$$

The expression above then reduces to

$$(-\boldsymbol{d}\boldsymbol{\delta}\mathbf{T})_{a_1\ldots a_k} = \operatorname{Sgn}\begin{bmatrix} c\ d_2(c)\ldots d_k(c) \\ a_1 a_2 \ldots a_k \end{bmatrix} \partial_c \partial_b T_{b d_2(c)\ldots d_k(c)}.$$

Adding the two results obtained, we finally get

$$(\boldsymbol{\Delta}\mathbf{T})_{a_1\ldots a_k} = (-\boldsymbol{d}\boldsymbol{\delta}\mathbf{T})_{a_1\ldots a_k} + (-\boldsymbol{\delta}\boldsymbol{d}\,\mathbf{T})_{a_1\ldots a_k} = \partial_b \partial_b T_{a_1\ldots a_k}.$$

Solution to 7.6.9

a.

$$\omega = [f + ig][\boldsymbol{d}x + i\boldsymbol{d}y] = [f\boldsymbol{d}x - g\boldsymbol{d}y] + i[g\boldsymbol{d}x + f\boldsymbol{d}y].$$

If we use that ${}^*\boldsymbol{d}y = \boldsymbol{d}x$, ${}^*\boldsymbol{d}x = -\boldsymbol{d}y$, we immediately get

$${}^*[g\boldsymbol{d}x + f\boldsymbol{d}y] = [-g\boldsymbol{d}y + f\boldsymbol{d}x].$$

b.

$$\boldsymbol{d}\omega = \boldsymbol{d}h \wedge \boldsymbol{d}z = \frac{\partial h}{\partial z}\,\boldsymbol{d}z \wedge \boldsymbol{d}z + \frac{\partial h}{\partial z}\,\boldsymbol{d}\bar{z} \wedge \boldsymbol{d}z.$$

Here the first term vanishes automatically because $\boldsymbol{d}z$ is a 1-form, so that

$$\boldsymbol{d}\omega = \frac{\partial h}{\partial \bar{z}}\,\boldsymbol{d}\bar{z} \wedge \boldsymbol{d}z.$$

Consequently, ω is closed and if and only if $\frac{\partial h}{\partial \bar{z}} = 0$.

c. If $h(z)$ is holomorphic, we get

$$\boldsymbol{d}h = \frac{\partial h}{\partial z}\,\boldsymbol{d}z,$$

which implies that

$$0 = \boldsymbol{d}^2 h = \boldsymbol{d}\left(\frac{\partial h}{\partial z}\right) \wedge \boldsymbol{d}z = \frac{\partial^2 h}{\partial z^2}\,\boldsymbol{d}z \wedge \boldsymbol{d}z + \frac{\partial^2 h}{\partial \bar{z}\partial z}\,\boldsymbol{d}\bar{z} \wedge \boldsymbol{d}z$$

$$= \frac{\partial^2 h}{\partial \bar{z}\partial z}\,\boldsymbol{d}\bar{z} \wedge \boldsymbol{d}z.$$

But

$$\frac{\partial^2}{\partial \bar{z}\partial z} = \frac{1}{4}\left(\frac{\partial}{\partial x} + i\,\frac{\partial}{\partial y}\right)\left(\frac{\partial}{\partial x} - i\,\frac{\partial}{\partial y}\right) = \frac{1}{4}\left(\frac{\partial^2}{\partial x^2} + \frac{\partial^2}{\partial y^2}\right).$$

Thus h is a harmonic function and so are its real and imaginary parts.

d. Let $h(z)$ be a holomorphic function. Then $\boldsymbol{d}h = \frac{\partial h}{\partial z}\,\boldsymbol{d}z$, and that is a closed 1-form. According to (b), this means that $\frac{\partial h}{\partial z}$ is a holomorphic function. We can also use that h is a harmonic function:

$$0 = \frac{\partial^2 h}{\partial \bar{z}\partial z} = \frac{\partial}{\partial z}\left(\frac{\partial h}{\partial z}\right),$$

but this is exactly the Cauchy–Riemann equation for the function $\frac{\partial h}{\partial z}$.

e. If we once more use that $^*\boldsymbol{d}x = -\boldsymbol{d}y$ and $^*\boldsymbol{d}y = \boldsymbol{d}x$, we immediately get

$$^*\boldsymbol{d}z = -i\boldsymbol{d}z, \qquad ^*\boldsymbol{d}\bar{z} = i\boldsymbol{d}\bar{z}.$$

Consequently, the decomposition

$$\boldsymbol{d}h = \frac{\partial h}{\partial z}\,\boldsymbol{d}z + \frac{\partial h}{\partial \bar{z}}\,\boldsymbol{d}\bar{z}$$

is a decomposition of $\boldsymbol{d}h$ in an anti–self dual and a self-dual part. It follows that $\boldsymbol{d}h$ is anti–self dual if and only if $\frac{\partial h}{\partial \bar{z}} = 0$.

Solution to 7.7.4

We want to compute the Laplacian in spherical coordinates. We have previously computed the metric coefficients (cf. (6.36))

$$g_{ij} = \begin{bmatrix} 1 & 0 & 0 \\ 0 & r^2 & 0 \\ 0 & 0 & r^2\sin\theta \end{bmatrix}; \qquad \sqrt{g} = r^2\sin\theta.$$

We also need the contravariant components of $\boldsymbol{d}f$, $\partial^i f = g^{ij}\partial_j f$; i.e.,

$$[\partial^i f] = \begin{bmatrix} 1 & 0 & 0 \\ 0 & \frac{1}{r^2} & 0 \\ 0 & 0 & \frac{1}{r^2\sin^2\theta} \end{bmatrix}\begin{bmatrix} \partial_r f \\ \partial_\theta f \\ \partial_\varphi f \end{bmatrix} = \begin{bmatrix} & \partial_r f \\ \frac{1}{r^2} & \partial_\theta f \\ \frac{1}{r^2\sin^2\theta} & \partial_\varphi f \end{bmatrix}.$$

We thus get

$$\begin{aligned} \Delta f &= \frac{1}{\sqrt{g}}\,\partial_i\left(\sqrt{g}\,\partial^i f\right) \\ &= \frac{1}{r^2\sin\theta}\left[\frac{\partial}{\partial r}\left(r^2\sin\theta\,\frac{\partial f}{\partial r}\right) + \frac{\partial}{\partial\theta}\left(\sin\theta\,\frac{\partial f}{\partial\theta}\right)\right. \\ &\qquad \left. + \frac{\partial}{\partial\varphi}\left(\frac{1}{\sin\theta}\,\frac{\partial f}{\partial\varphi}\right)\right]. \end{aligned} \tag{7.100}$$

Solution to 7.8.1

The first two Maxwell equations can be solved using gauge potentials. The equation $\boldsymbol{d}\mathbf{B} = 0$ is solved by the assumption that $\mathbf{B}$ can be written in the form

$$\mathbf{B} = \boldsymbol{d}\mathbf{A},$$

where $\mathbf{A}$ is a vector field! Substituting this into the second Maxwell equation (7.77), we get

$$\boldsymbol{d}\left(\frac{\partial\mathbf{A}}{\partial t} + \mathbf{E}\right) = 0,$$

and this equation can be solved by the assumption that $\frac{\partial \mathbf{A}}{\partial t} + \mathbf{E}$ can be written in the form

$$\frac{\partial \mathbf{A}}{\partial t} + \mathbf{E} = -\boldsymbol{d}\phi,$$

where ϕ is a scalar field. Thus we have rediscovered the gauge potentials ϕ, $\mathbf{A}$:

$$\mathbf{E} = -\frac{\partial \mathbf{A}}{\partial t} - \boldsymbol{d}\phi; \qquad \mathbf{B} = \boldsymbol{d}\mathbf{A}. \tag{7.101}$$

The last two Maxwell equations (7.78) and (7.79) can be used to find the equation of continuity. Rearranging them, we get

$$-\boldsymbol{\delta}\mathbf{E} = \frac{1}{\epsilon_0}\,\rho; \qquad \frac{\partial}{\partial t}\,\boldsymbol{\delta}\mathbf{E} = -\frac{1}{\epsilon_0}\frac{\partial \rho}{\partial t}$$

and

$$\frac{\partial \mathbf{E}}{\partial t} - c^2\boldsymbol{\delta}\mathbf{B} = -\frac{1}{\epsilon_0}\,\mathbf{J}; \qquad \frac{\partial}{\partial t}\,\boldsymbol{\delta}\mathbf{E} = -\frac{1}{\epsilon_0}\,\boldsymbol{\delta}\mathbf{J},$$

from which we deduce that

$$\frac{\partial \rho}{\partial t} = \boldsymbol{\delta}\mathbf{J}; \qquad \text{i.e., } \frac{\partial \rho}{\partial t} - \boldsymbol{\delta}\mathbf{J} = 0,$$

which is the proper generalization of the equation of continuity.

Finally, we deduce the equations of motion for the gauge potentials. Substituting equation (7.101) into the two remaining Maxwell equations, we get

$$\begin{cases} \boldsymbol{\delta}\left(\dfrac{\partial \mathbf{A}}{\partial t} + \boldsymbol{d}\phi\right) = +\dfrac{1}{\epsilon_0}\,\rho \\[2ex] \dfrac{\partial}{\partial t}\left(\dfrac{\partial \mathbf{A}}{\partial t} + \boldsymbol{d}\phi\right) + c^2\boldsymbol{\delta d}\mathbf{A} = +\dfrac{1}{\epsilon_0}\,\mathbf{J} \end{cases}$$

i.e.,

$$\begin{cases} \left(\dfrac{1}{c^2}\dfrac{\partial^2}{\partial t^2} + \boldsymbol{\delta d}\right)\rho - \dfrac{\partial}{\partial t}\left(\dfrac{1}{c^2}\dfrac{\partial \phi}{\partial t} - \boldsymbol{\delta}\mathbf{A}\right) = \dfrac{1}{\epsilon_0}\,\rho \\[2ex] \left(\dfrac{1}{c^2}\dfrac{\partial^2}{\partial t^2} + \boldsymbol{\delta d} + \boldsymbol{d\delta}\right)\mathbf{A} + \boldsymbol{d}\left(\dfrac{1}{c^2}\dfrac{\partial \phi}{\partial t} - \boldsymbol{\delta}\mathbf{A}\right) = \dfrac{1}{\epsilon_0 c^2}\,\mathbf{J}, \end{cases}$$

which can be further rearranged as

$$\begin{cases} \left(\dfrac{1}{c^2}\dfrac{\partial^2}{\partial t^2} - \boldsymbol{\Delta}\right)\phi - \dfrac{\partial}{\partial t}\left(\dfrac{1}{c^2}\dfrac{\partial \phi}{\partial t} - \boldsymbol{\delta}\mathbf{A}\right) = \dfrac{1}{\epsilon_0}\,\rho \\[2ex] \left(\dfrac{1}{c^2}\dfrac{\partial^2}{\partial t^2} - \boldsymbol{\Delta}\right)\mathbf{A} + \boldsymbol{d}\left(\dfrac{1}{c^2}\dfrac{\partial \phi}{\partial t} - \boldsymbol{\delta}\mathbf{A}\right) = \dfrac{1}{\epsilon_0 c^2}\,\mathbf{J}, \end{cases}$$

If $(\phi, \mathbf{A})$ satisfies the Lorenz condition

$$\frac{\partial \phi}{\partial t} - \delta \mathbf{A} = 0,$$

we thus recover the wave equations

$$\left(\frac{1}{c^2}\frac{\partial^2}{\partial t^2} - \mathbf{\Delta}\right)\phi = \frac{1}{\epsilon_0}\rho; \qquad \left(\frac{1}{c^2}\frac{\partial^2}{\partial t^2} - \mathbf{\Delta}\right)\mathbf{A} = \frac{1}{\epsilon_0 c^2}\mathbf{J}.$$

Solution to 7.8.2

First, we observe that

$$\mathbf{B} = \delta \mathbf{S} \quad \text{iff} \quad {}^*\mathbf{B} = -\boldsymbol{d}{}^*\mathbf{S}. \tag{$*$}$$

Therefore, $\mathbf{B}$ is coexact in a space region Ω if and only if ${}^*\mathbf{B}$ is exact in the same region. But

$${}^*\mathbf{B} = \frac{j}{2\pi\epsilon_0 c^2}\,\boldsymbol{d}\varphi. \tag{$**$}$$

Apart from an irrelevant constant, this is our old friend $\boldsymbol{d}\varphi$ (see Section 2.4), which is only almost exact. Consequently, $\mathbf{B}$ is only almost coexact. From $(*)$, $(**)$ we also see that

$$\mathbf{S} = -\frac{j}{2\pi\epsilon_0 c^2}\,{}^*\varphi = -\frac{j}{2\pi\epsilon_0 c^2}\,\varphi\varepsilon.$$

It remains to determine the Levi-Civita form ε. In cylindrical coordinates we know that $\sqrt{g} = \rho$ (compare Exercise 6.6.3). Thus

$$\varepsilon = \rho\boldsymbol{d}\rho \wedge \boldsymbol{d}\varphi \wedge \boldsymbol{d}z,$$

so we have finally found $\mathbf{S}$;

$$\mathbf{S} = -\frac{j}{2\pi\epsilon_0 c^2}\,\rho\varphi\boldsymbol{d}\rho \wedge \boldsymbol{d}\varphi \wedge \boldsymbol{d}z,$$

and at the same time we have managed to write $\mathbf{S}$ as a simple form. In particular, $\mathbf{S}$ is closed, and it follows that $\mathbf{S}$ is a harmonic 3-form,

$$\mathbf{\Delta} = -\boldsymbol{d}\delta\mathbf{S} - \delta\boldsymbol{d}\mathbf{S} = -\boldsymbol{d}\mathbf{B} = 0.$$

CHAPTER 8

Integral Calculus on Manifolds

8.1 Introduction

The next problem we must attack is how we can extend the integral calculus to differential forms. Before we proceed to give precise definitions, we would like to give an *intuitive* feeling of the integral concept we are going to construct. In the familiar theory of Riemann integrals we consider a function of, say, two variables $f(x, y)$ defined on a "nice" subset U of $\mathbb{R}^2$. Then we divide this region U into cells Δ_i with area ϵ^2, and in each cell, Δ_i, we choose a point (x_i, y_i). (See Figure 8.1.) We can now form the Riemann sum:

$$\sum_i f(x_i, y_i)\epsilon^2.$$

If f is "well behaved" and U is a sufficiently "nice" region, then this sum will converge to a limit as $\epsilon \rightarrow 0$, and this limit will be independent of the subdivision

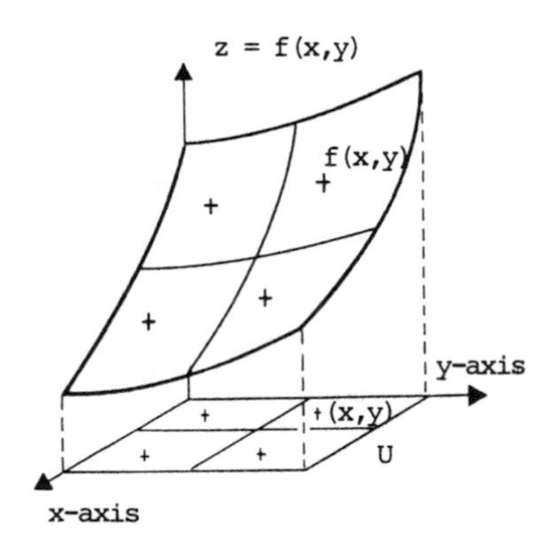

Figure 8.1.

in cells. It is this limit we define to be the Riemann integral:

$$\int_U f(x, y)dx\, dy = \lim_{\epsilon \to 0} \sum_i f(x_i, y_i)\epsilon^2.$$

Next, we consider a two-dimensional smooth surface Ω in the ordinary Euclidean space $\mathbb{R}^3$, i.e., a differential manifold of dimension 2 embedded in $\mathbb{R}^3$. Similarly, we consider a two-form $\mathbf{F}$ defined in $\mathbb{R}^3$. As Ω is a two-dimensional surface, we may introduce coordinates (λ^1, λ^2), which we assume cover Ω; i.e., we parametrize the surface Ω. (See Figure 8.2).

Then we make a subdivision of the coordinate domain U into cells Δ_i of area ϵ^2. In each of these cells we choose a point $(\lambda^1_{(i)}, \lambda^2_{(i)})$. At the corresponding point $P_i = \phi(\lambda^1_{(i)}; \lambda^2_{(i)})$ in Ω we have two basic tangent vectors, $\mathbf{e}_{(i)_1}$ and $\mathbf{e}_{(i)_2}$. As $\mathbf{F}$ is a two-form, it maps $(\mathbf{e}_{(i)_1}, \mathbf{e}_{(i)_2})$ into a real number $\mathbf{F}(\mathbf{e}_{(i)_1}, \mathbf{e}_{(i)_2})$. So we can form the sum

$$\sum_i \mathbf{F}(\mathbf{e}_{(i)_1}, \mathbf{e}_{(i)_2})\epsilon^2.$$

If $\mathbf{F}$ is "well-behaved" and Ω is a "nice" surface, then this will converge to a limit as $\epsilon \to 0$, and this limit will be independent of the subdivision into cells. It is this limit we define to be the integral of $\mathbf{F}$:

$$\int_\Omega \mathbf{F} \stackrel{\text{def}}{=} \lim_{\epsilon \to 0} \sum_i \mathbf{F}(\mathbf{e}_{(i)_1}, \mathbf{e}_{(i)_2})\epsilon^2. \tag{8.1}$$

At this point one might object that $\int_\Omega \mathbf{F}$ cannot be a geometric object, because we have been extensively using the special coordinates (λ^1, λ^2) in our definitions. However, it will be shown that $\int_\Omega \mathbf{F}$ does not depend on which coordinate system we choose, provided that it defines the same orientation on Ω. Of course, the above formula is of little value in practical calculations. Therefore, we should rearrange it!

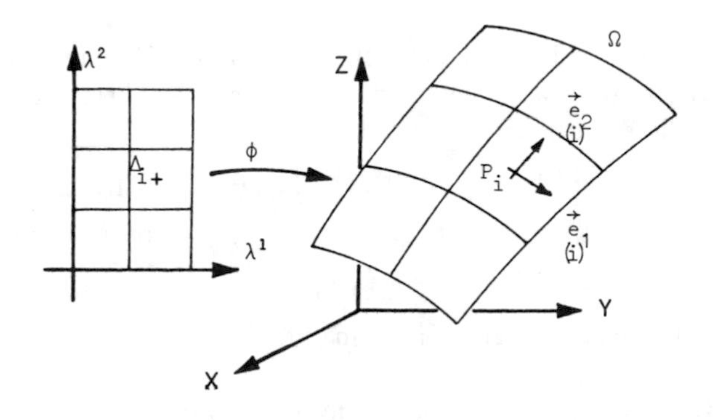

Figure 8.2.

To each point (λ^1, λ^2) there corresponds a set of canonical frame vectors $(\mathbf{e}_1, \mathbf{e}_2)$, and therefore $\mathbf{F}$ generates the ordinary function $U \curvearrowright \mathbb{R}$ given by

$$(\lambda^1, \lambda^2) \curvearrowright \mathbf{F}(\mathbf{e}_1, \mathbf{e}_2).$$

Using this function, we can convert the integral of $\mathbf{F}$ into an ordinary Riemann integral:

$$\int_\Omega \mathbf{F} = \lim_{\epsilon \to 0} \sum_i \mathbf{F}(\mathbf{e}_{(i)_1}, \mathbf{e}_{(i)_2})\epsilon^2 = \int_U \mathbf{F}(\mathbf{e}_1, \mathbf{e}_2) d\lambda^1 d\lambda^2. \tag{8.2}$$

We can also express the integral of $\mathbf{F}$ in terms of the components of $\mathbf{F}$. Here we should observe that $\mathbf{F}$ is defined as a two-form on the whole of the Euclidean space $\mathbb{R}^3$. Let $(\mathbf{i}_1, \mathbf{i}_2, \mathbf{i}_3)$ denote the canonical frame vectors corresponding to a coordinate system on $\mathbb{R}^3$. Then $\mathbf{F}$ is characterized by the components

$$F_{ab} = \mathbf{F}(\mathbf{i}_a, \mathbf{i}_b).$$

The canonical frame vectors $(\mathbf{e}_1, \mathbf{e}_2)$ on Ω are related to $(\mathbf{i}_1, \mathbf{i}_2, \mathbf{i}_3)$ in the following way:

$$\mathbf{e}_1 = \mathbf{i}_a \frac{\partial x^a}{\partial \lambda^1} \quad \text{and} \quad \mathbf{e}_2 = \mathbf{i}_b \frac{\partial x^b}{\partial \lambda^2}.$$

Using this, we can rearrange (8.2) in the following way:

$$\mathbf{F}(\mathbf{e}_1, \mathbf{e}_2) = \mathbf{F}(\mathbf{i}_a, \mathbf{i}_b) \frac{\partial x^a}{\partial \lambda^1} \frac{\partial x^b}{\partial \lambda^2} = F_{ab}(\lambda^1, \lambda^2) \frac{\partial x^a}{\partial \lambda^1} \frac{\partial x^b}{\partial \lambda^2};$$

i.e.,

$$\int_\Omega \mathbf{F} = \int_U F_{ab}(\lambda^1, \lambda^2) \frac{\partial x^a}{\partial \lambda^1} \frac{\partial x^b}{\partial \lambda^2} d\lambda^1 d\lambda^2. \tag{8.3}$$

We shall reconstruct these formulas from another point of view in a moment.

8.2 Submanifolds—Regular Domains

In the general theory we have an n-dimensional differentiable manifold M. We want to characterize what we understand by a k-dimensional surface Ω in M. To do this we must fix some notations:

$$\mathbb{R}^{(n,k)} = \{x \in \mathbb{R}^n \mid x = (x^1, x^2, \ldots, x^k, 0, \ldots, 0)\};$$
$$\mathbb{R}_-^{(n,k)} = \{x \in \mathbb{R}^n \mid x = (x^1, x^2, \ldots, x^k, 0, \ldots, 0), ; x^1 \leq 0\}.$$

(See Figure 8.3.)

We can then define the concept of a submanifold:

Definition 8.1. A subset $\Omega \subseteq M$ is said to be a *k-dimensional submanifold* if each point P in Ω has the following property: There exists an open neighborhood

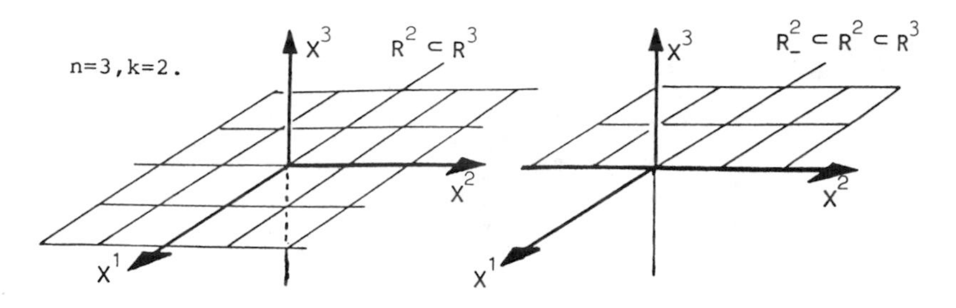

Figure 8.3.

V of P and a coordinate system (ϕ, U) such that ϕ maps $U \cap \mathbb{R}^{(n,k)}$ bijectively onto $V \cap \Omega$. (See Figure 8.4.)

A coordinate system (ϕ, U) with the above property is said to be *adapted* to Ω. According to the definition, the adapted coordinate systems actually over Ω, and hence they constitute an atlas. This justifies calling Ω a submanifold. As Ω itself is a manifold, we can introduce arbitrary coordinate systems on Ω, as long as they are compatible with the adapted coordinate systems. Clearly, the k-dimensional submanifolds are the exact counterparts of the loose concept: a k-dimensional smooth surface in M. As a specific example we may consider $M = S^2$. Then the equator S^1 is a one-dimensional submanifold. As adapted coordinate systems, we may use the standard coordinates on the sphere. (Compare the discussion in Section 6.2. See Figure 8.5 too.)

Next we should try to characterize the domains of integration. From the standard theory of integration we know that these domains should be chosen carefully, otherwise we will get into trouble with the integral. If $f : \mathbb{R} \curvearrowright \mathbb{R}$ is a continuous function, then $\int_a^b f(x)dx$ is well-defined, while $\int_{-\infty}^{+\infty} f(x)dx$ is somewhat dubious. In standard integration theory we would therefore start by restricting ourselves to closed, bounded intervals $[a, b]$. Then we will have no trouble, because these

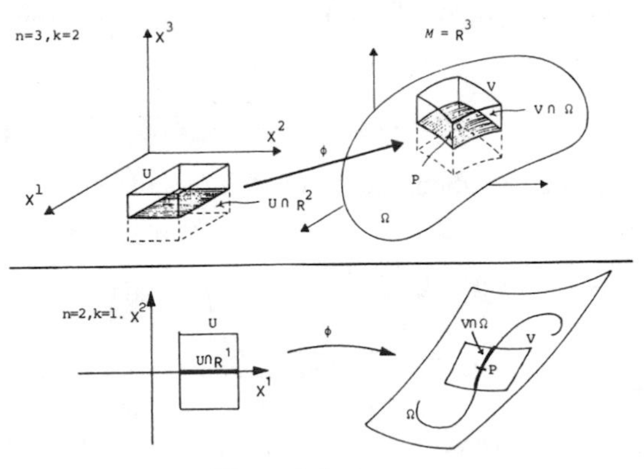

Figure 8.4.

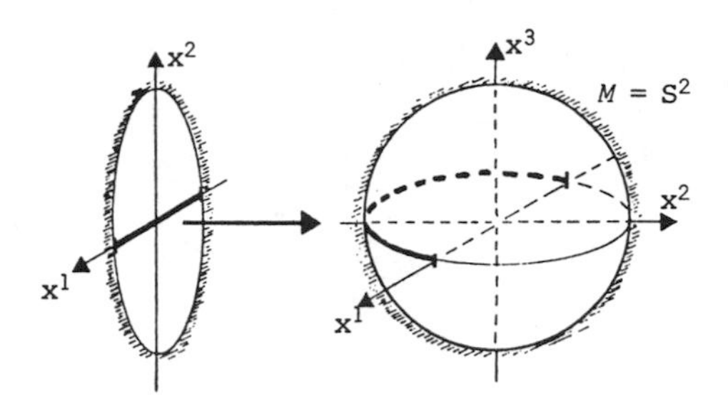

Figure 8.5.

are compact subsets of $\mathbb{R}$, and continuous functions on compact intervals have many nice properties; in particular, they are bounded and uniformly continuous. Furthermore we know that there is an intimate connection between integration and differentiation (the fundamental theorem of calculus):

$$\int_a^b \frac{df}{dx}\,dx = f(b) - f(a).$$

This relates the integral of $\frac{df}{dx}$ to the values of f at the points a and b, which constitute the boundary of the interval $[a, b]$.

Back to the general theory! To avoid trouble with convergence, we will consider only *compact* domains. (If $M \subseteq \mathbb{R}^N$, then a subset of M is compact if and only if it is closed and bounded in $\mathbb{R}^N$.) Furthermore, we should consider domains where the boundary itself is a nice "smooth surface." Otherwise, we will not be able to generalize the fundamental theory of calculus. This motivates the following definition: (See Figure 8.6 too.)

Definition 8.2. A compact subset Ω of an n-dimensional manifold M is called a k-dimensional *regular domain* if each point P in Ω has one of the following two properties:

1. There exist an open neighborhood V of P in M and a coordinate system (ϕ, U) on M such that

$$\phi : U \cap \mathbb{R}^{(n,k)} \xrightarrow{\text{bijectively}} V \cap \Omega.$$

2. There exist an open neighborhood V of P in M and a coordinate system (ϕ, U) on M such that

$$\phi : U \cap \mathbb{R}_-^{(n,k)} \xrightarrow{\text{bijectively}} V \cap \Omega \quad \text{and} \quad x^1(P) = 0.$$

The points with property 1 are called *interior* points of Ω. The set of interior points of Ω is denoted by int Ω. Clearly, it is a k-dimensional submanifold.

The points with property 2 are called boundary points of Ω. The set of boundary points is denoted by $\partial\Omega$, and it clearly is a $(k-1)$-dimensional submanifold, called

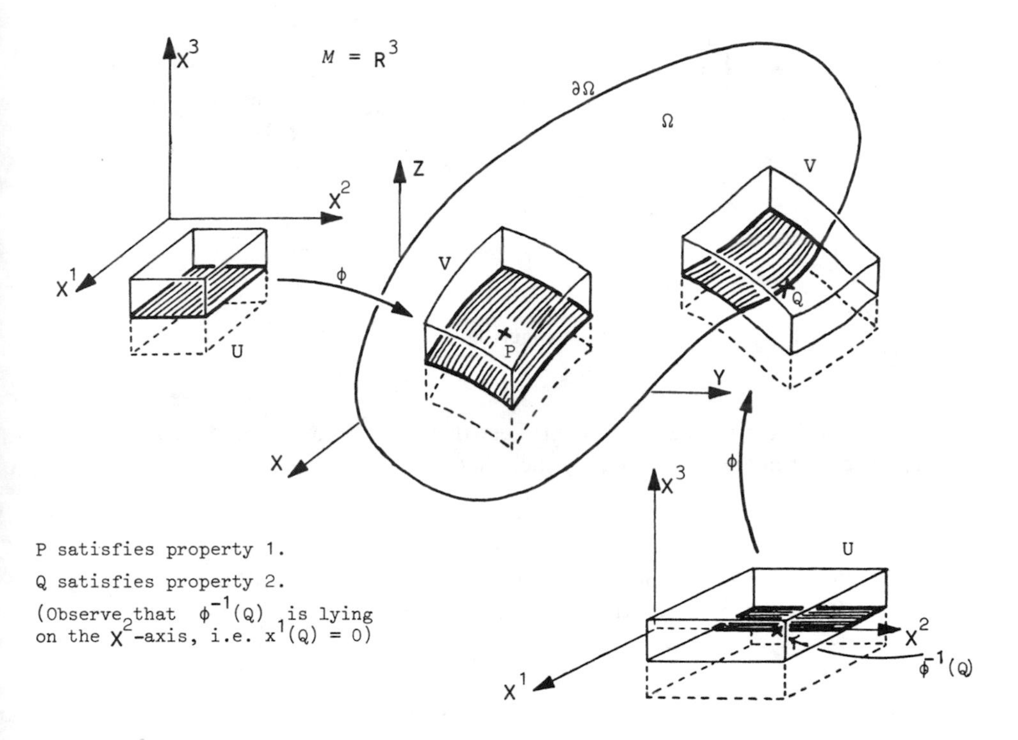

Figure 8.6.

the *boundary* of Ω. If P is a boundary point and (ϕ, U) is a coordinate system with property 2, then we may use

$$(x^2, \ldots, x^k)$$

as local coordinates on $\partial\Omega$.

Thus we have decomposed a k-dimensional regular domain into two disjoint components: int Ω and $\partial\Omega$. The boundary $\partial\Omega$ has an extremely important property! It is itself a *compact* subset:

Theorem 8.1. *$\partial\Omega$ is a compact $(k-1)$-dimensional submanifold.*

PROOF. The proof of this is somewhat technical and may be skipped.

As Ω is compact, it suffices to show that $\partial\Omega$ is closed in M. Let $(P_n)_{n\in\mathbb{N}}$ be a sequence of points in $\partial\Omega$, and assume that this sequence converges to P, i.e., $P_n \to P$. We have finished if we can show that $P \in \partial\Omega$.

As Ω is compact, it is closed in M, and therefore we know that $P \in \Omega$. But then P is either an interior point or a boundary point! Thus we must rule out the possibility of P being an interior point. (See Figure 8.7.)

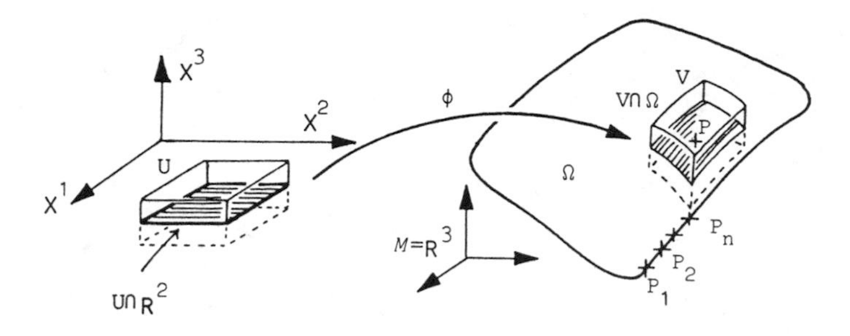

Figure 8.7.

Assume that P is an interior point. Then there exist an open neighborhood V of P and a coordinate system (ϕ, U) such that

$$\phi : U \cap \mathbb{R}^{(n,k)} \xrightarrow{\text{bijectively}} V \cap \Omega.$$

It follows that all the points in $V \cap \Omega$ are interior points. But $P_n \in \Omega$ and $P_n \to P$; therefore, there exists an integer N such that $n > N$ implies that

$$P_n \in V \cap \Omega.$$

Thus we conclude that P_n is an interior point if $n \geq N$, and that is a contradiction! □

This property of $\partial\Omega$ has important consequences. As $\partial\Omega$ is a compact submanifold, all the points in $\partial\Omega$ have property 1. Therefore, we see that $\partial\Omega$ itself is a regular domain! But as all the points in the regular domain $\partial\Omega$ are interior points, it follows that $\partial\Omega$ has *no* boundary. Consequently, $\partial(\partial\Omega)$ is empty:

Theorem 8.2. (*Cartan's lemma*) *For a regular domain* Ω, *the boundary* $\partial\Omega$ *is also a regular domain. But* $\partial\Omega$ *has vanishing boundary*

$$\partial(\partial\Omega) = \emptyset \tag{8.4}$$

We shall include a point as a zero-dimensional regular domain, although it really falls outside the scope of our definitions. We do *not* assign any boundary to a point, so that the rule $\partial\partial\Omega = \emptyset$ still holds in this very special case!

At this point we had better look at some specific examples, all using $M = \mathbb{R}^3$.

EXAMPLE 8.1 (See Figure 8.8). If Ω is the closed unit ball

$$\Omega = \{x \mid \langle x \mid x\rangle \leq 1\},$$

then int Ω is the open unit ball

$$\text{int } \Omega = \{x \mid \langle x \mid x\rangle < 1\},$$

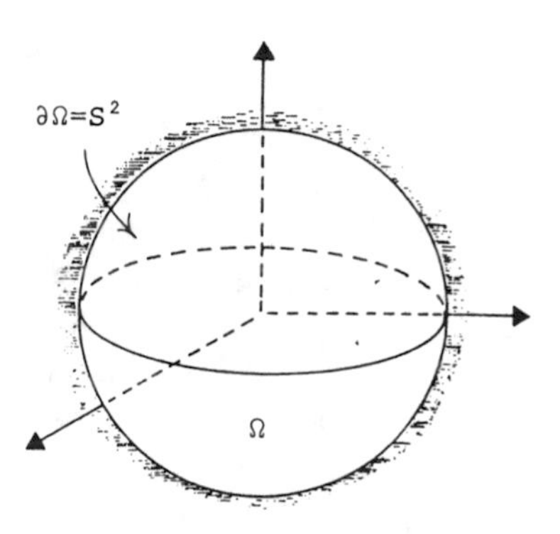

Figure 8.8.

and the boundary of Ω is the two-sphere

$$\partial\Omega = \{x \mid \langle x \mid x \rangle = 1\};$$

i.e., $\partial\Omega = S^2$. Observe that S^2 is itself a two-dimensional regular domain with boundary: $\partial\partial\Omega = \emptyset$.

EXAMPLE 8.2 (See Figure 8.9). If Ω is the northern hemisphere of S^2,

$$\Omega = \{x \in \mathbb{R}^3 \mid \langle x \mid x \rangle = 1, \ x^3 \geq 0\},$$

then int Ω is given by

$$\text{int } \Omega = \{x \in \mathbb{R}^3 \mid \langle x \mid x \rangle = 1, x^3 > 0\},$$

and the boundary of Ω is the unit circle

$$\partial\Omega = \{x \in \mathbb{R}^3 \mid \langle x \mid x \rangle = 1, x^3 = 0\};$$

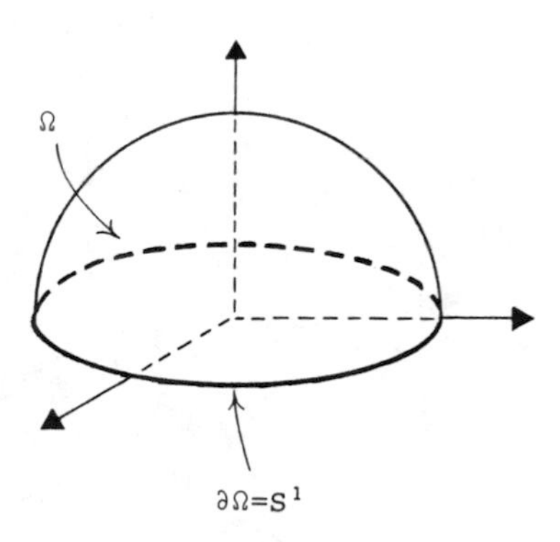

Figure 8.9.

i.e., $\partial\Omega = S^1$, where S^1 is itself a one-dimensional regular domain without boundary: $\partial\partial\Omega = \emptyset$.

Finally, we observe that the rule $\partial(\partial\Omega) = \emptyset$ reminds us of the rule $\boldsymbol{d}(\boldsymbol{d}\mathbf{F}) = 0$. They are, in fact, very intimately connected. If a differential form $\mathbf{F}$ had the property $\boldsymbol{d}\mathbf{F} = 0$, we called it a *closed* form. In a similar way we will call a regular domain Ω with the property $\partial\Omega = \emptyset$ a *closed domain*. This is in agreement with the familiar expressions "a closed curve" and "a closed surface."

At this point we should begin to have a feeling for the regular domains, which are going to be the domains of integration! There is, however, still a problem that we will have to face, and that is the orientability of Ω. If we return to our intuitive discussion of how to integrate a two-form on a surface, it is clear that if we interchange the coordinates λ^1 and λ^2, then we also change the sign of the integral, because $\mathbf{F}(\mathbf{e}_1^i, \mathbf{e}_2^i) = -\mathbf{F}(\mathbf{e}_2^i, \mathbf{e}_1^i)$; i.e., $\sum_i \mathbf{F}(\mathbf{e}_1^i, \mathbf{e}_2^i)\epsilon^2 = -\sum_i \mathbf{F}(\mathbf{e}_2^i, \mathbf{e}_1^i)\epsilon^2$.

But if we interchange λ^1 and λ^2, this simply means that we construct a new coordinate system with the opposite orientation! This clearly shows that the integral depends on the orientation.

Consider now a regular domain Ω in a manifold M. Then int Ω is a k-dimensional submanifold that can be covered by adapted coordinate systems. We say that *the regular domain Ω is orientable, provided that* int Ω *is an orientable manifold.* (Compare with the discussion in Section 7.4.) The interesting point is that whenever Ω is an orientable regular domain, so is $\partial\Omega$! In fact, the orientation chosen on int Ω induces in a canonical manner an orientation on $\partial\Omega$. To explain this, we choose an orientation on int Ω. Now, let (ϕ, U) be an *adapted* coordinate system on $\partial\Omega$. Then

$(x^1, \ldots, x^k, \ldots, x^n)$	serves as local coordinates on M;
$(x^1, \ldots, x^k, 0, \ldots, 0) \quad x^1 < 0$	serves as local coordinates on int Ω;
$(0, x^2, \ldots, x^k, 0, \ldots, 0)$	serves as local coordinates on $\partial\Omega$.

This makes the following definition reasonable: *The local coordinates* $(0, x^2, \ldots, x^k, 0, \ldots, 0)$ *generate a positive orientation on* $\partial\Omega$ *if and only if the local coordinates* $(x^1, x^2, \ldots, x^k, 0, \ldots, 0)$ *generate a positive orientation*

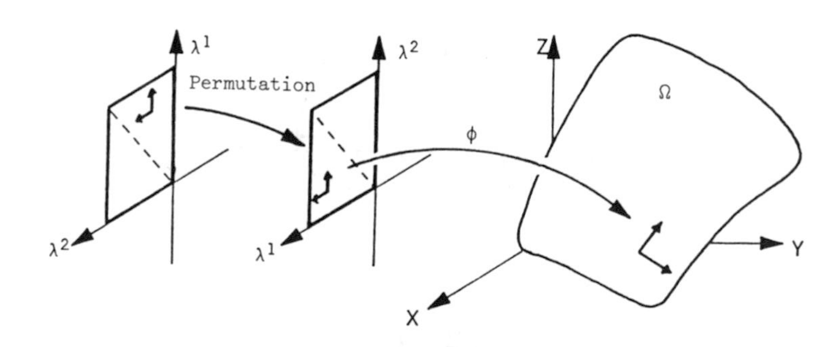

Figure 8.10.

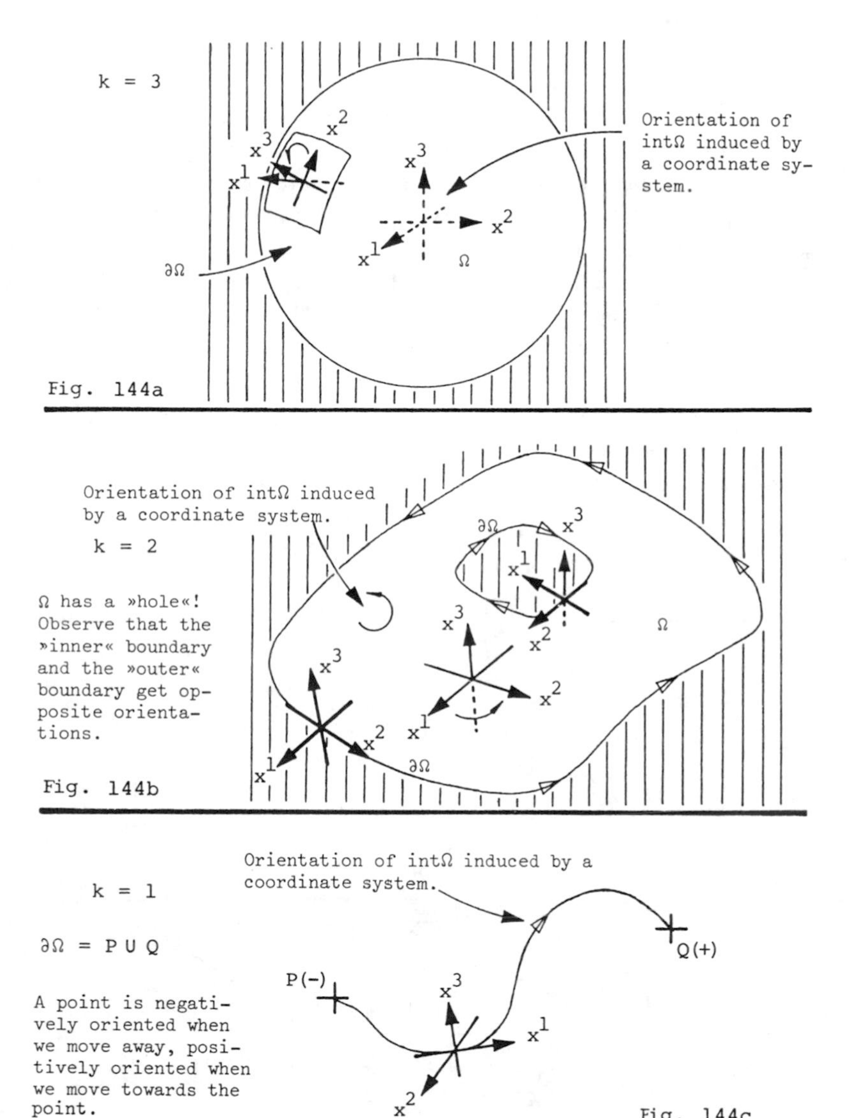

k = 3
Orientation of intΩ induced by a coordinate system.
∂Ω
Ω
Fig. 144a
Orientation of intΩ induced by a coordinate system.
k = 2
Ω has a »hole«! Observe that the »inner« boundary and the »outer« boundary get opposite orientations.
∂Ω
Ω
Fig. 144b
Orientation of intΩ induced by a coordinate system.
k = 1
∂Ω = P ∪ Q
A point is negatively oriented when we move away, positively oriented when we move towards the point.
P(−)
Q(+)
Fig. 144c

Figure 8.11.

on int Ω. We have tried to exemplify this in Figure 8.11.

Observe that in the case where $\partial\Omega$ reduces to a finite set of points, we will have to construct a special convention, as we cannot introduce coordinate systems on a single point! (See Figure 8.11c.)

8.3 The Integral of Differential Forms

With these preparations we are ready to define integrals of forms. So let Ω be an orientable k-dimensional regular domain in M^n and let $\mathbf{F}$ be a differential form of degree k defined on M^n. We assume for simplicity that int Ω is a *simple* manifold that can be covered by a single coordinate system (not necessarily adapted!). We denote the coordinates of int Ω by $(\lambda^1, \ldots, \lambda^k)$ and those of M by $(x^1, \ldots, x^k, \ldots, x^n)$. (See Figure 8.12.)

The coordinates $(\lambda^1, \ldots, \lambda^k)$ are assumed to be positively oriented. To each point $(\lambda^1, \ldots, \lambda^k)$ in U there corresponds a set of canonical frame vectors

$$\mathbf{e}_1, \ldots, \mathbf{e}_k.$$

As $\mathbf{F}$ has degree k, it maps $(\mathbf{e}_1, \ldots, \mathbf{e}_k)$ into a single real number: $\mathbf{F}(\mathbf{e}^1, \ldots, \mathbf{e}^k)$. Thus we may consider $\mathbf{F}(\mathbf{e}_1, \ldots, \mathbf{e}_k)$ as an ordinary function on U, and therefore we can define the integral as

$$\int_\Omega \mathbf{F} \stackrel{\text{def}}{=} \int_U \mathbf{F}(\mathbf{e}_1, \ldots, \mathbf{e}_k) d\lambda^1 \ldots d\lambda^k. \tag{8.5}$$

We may rearrange this formula by introducing the canonical frame vectors

$$\mathbf{i}_1, \ldots, \mathbf{i}_n$$

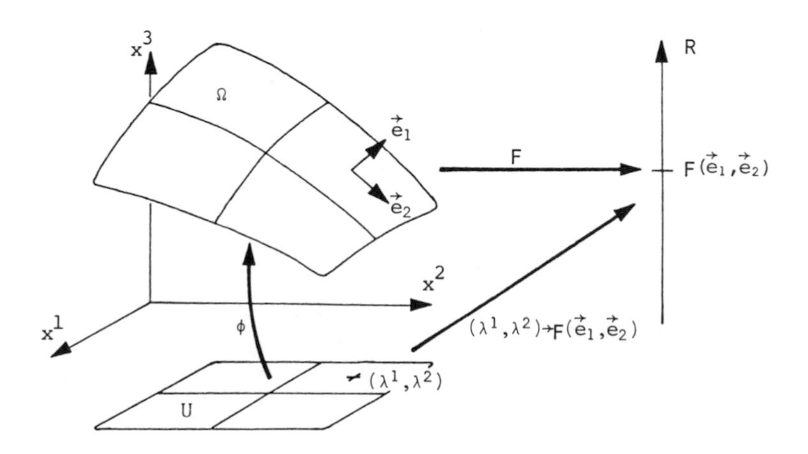

Figure 8.12.

on M.

$$\mathbf{e}_1 = \mathbf{i}_{a_1} \frac{\partial x^{a_1}}{\partial \lambda^1} \; ; \ldots ; \mathbf{e}_k = \mathbf{i}_{a_k} \frac{\partial x^{a_k}}{\partial \lambda^k} \; . \tag{8.6}$$

Inserting this, we get

$$\begin{aligned} \mathbf{F}(\mathbf{e}_1, \ldots, \mathbf{e}_k) &= \mathbf{F}(\mathbf{i}_{a_1}; \ldots ; \mathbf{i}_{a_k}) \frac{\partial x^{a_1}}{\partial \lambda^1} \cdots \frac{\partial x^{a_k}}{\partial \lambda^1} \\ &= F_{a_1 \ldots a_k} \frac{\partial x^{a_1}}{\partial \lambda^1} \cdots \frac{\partial x^{a_k}}{\partial \lambda^k} \; . \end{aligned}$$

Therefore, we finally obtain

Definition 8.3. Let $\Omega \subseteq M^n$ be an orientable k-dimensional regular domain parametrized by coordinates $(\lambda^1, \ldots, \lambda^k)$. Let $\mathbf{F}$ be a differential form of degree k characterized by the components $F_{a_1 \ldots a_k}$ with respect to coordinates $(x^1, \ldots, x^n)$ on M. Then we define the integral of $\mathbf{F}$ through the formula

$$\begin{aligned} \int_\Omega \mathbf{F} &= \int_U \mathbf{F}(\mathbf{e}_1, \ldots, \mathbf{e}_k) d\lambda^1 \ldots d\lambda^k \\ &= \int_U F_{a_1 \ldots a_k} \frac{\partial x^{a_1}}{\partial \lambda^1} \cdots \frac{\partial x^{a_k}}{\partial \lambda^k} \, d\lambda^1 \ldots d\lambda^k. \end{aligned} \tag{8.7}$$

Formula (8.5) clearly shows that the integral is independent of the coordinate system $(x^1, \ldots, x^n)$ on M, but we still have to show that it is independent of the coordinate system $(\lambda^1, \ldots, \lambda^k)$ on int Ω. Therefore we consider two sets of coordinates $(\lambda^1, \ldots, \lambda^k)$ and $(\mu^1, \ldots, \mu^k)$ that both cover int Ω and that specify the same orientation on int Ω; i.e., the Jacobian det $\left(\frac{\partial \mu^i}{\partial \lambda^j} \right)$ is always positive. First we observe that

$$\begin{aligned} \mathbf{F}\left(\mathbf{e}_{(2)_1}; \ldots ; \mathbf{e}_{(2)_k}\right) &= \mathbf{F}\left(\mathbf{e}_{(1)_{i_1}} \frac{\partial \lambda^{i_1}}{\partial \mu^1} \; ; \ldots ; \; \mathbf{e}_{(1)_{i_k}} \frac{\partial \lambda^{i_k}}{\partial \mu^k} \right) \\ &= \mathbf{F}\left(\mathbf{e}_{(1)_{i_1}}; \ldots ; \mathbf{e}_{(1)_{i_k}}\right) \frac{\partial \lambda^{i_k}}{\partial \mu^1} \cdots \frac{\partial \lambda^{i_k}}{\partial \mu^k} \, . \end{aligned}$$

But $\mathbf{F}$ is a k-form and $i_1, \ldots, i_k$ can only take the values $1, \ldots, k$. Therefore, we conclude that

$$\mathbf{F}\left(\mathbf{e}_{(1)_{i_1}}; \ldots ; \mathbf{e}_{(1)_{i_k}}\right) = \epsilon_{i_1 \ldots i_k} \mathbf{F}\left(\mathbf{e}_{(1)_1}; \ldots ; \mathbf{e}_{(1)_k}\right).$$

Inserting this, we get

$$\begin{aligned} \mathbf{F}\left(\mathbf{e}_{(2)_1}; \ldots ; \mathbf{e}_{(2)_k}\right) &= \mathbf{F}\left(\mathbf{e}_{(1)_1}; \ldots ; e_{(1)_k}\right) \epsilon_{i_1 \ldots i_k} \frac{\partial \lambda^{i_1}}{\partial \mu^1} \cdots \frac{\partial \lambda^{i_k}}{\partial \mu^k} \\ &= \mathbf{F}\left(\mathbf{e}_{(1)_1}; \ldots ; \mathbf{e}_{(1)_k}\right) \det \left[\frac{\partial \lambda^i}{\partial \mu^j} \right]. \end{aligned}$$

Using the transformation rule for Riemann integrals, we then obtain

$$\int_{U_2} \mathbf{F}(\mathbf{e}_{(2)_1}; \ldots; \mathbf{e}_{(2)_k}) d\mu^1 \ldots d\mu^k$$

$$= \int_{U_1} \mathbf{F}(\mathbf{e}_1; \ldots; \mathbf{e}_k) \left| \det \left(\frac{\partial \mu^i}{\partial \lambda^j} \right) \right| d\lambda^1 \ldots d\lambda^k$$

$$= \int_{U_1} \mathbf{F}\left(\mathbf{e}_{(1)_1}; \ldots; \mathbf{e}_{(1)_k}\right) \det \left(\frac{\partial \lambda^i}{\partial \mu^j} \right) \left| \det \left(\frac{\partial \mu^i}{\partial \lambda^j} \right) \right| d\lambda^1 \ldots d\lambda^k.$$

But $\det\left(\frac{\partial \mu^i}{\partial \lambda^j}\right)$ is assumed to be positive, and therefore we can forget about the absolute value. We then finally obtain

$$\int_{U_2} \mathbf{F}\left(\mathbf{e}_{(2)_1}; \ldots; \mathbf{e}_{(2)_k}\right) d\mu^1 \ldots d\mu^k = \int_{U_1} \mathbf{F}\left(\mathbf{e}_{(1)_1}; \ldots; \mathbf{e}_{(1)_k}\right) d\lambda^1 \ldots d\lambda^k, \tag{8.8}$$

and this shows that $\int_\Omega \mathbf{F}$ is independent of the coordinate system on int Ω, as long as it generates a positive orientation!

Up to this point we have considered only the situation where int Ω is a simple manifold; i.e., it can be covered by a single coordinate system. We now turn to the general situation, where we have to use an atlas $(\phi_i, U_i)_{i \in I}$ consisting of overlapping coordinate systems that together cover int Ω. We want to construct the integral $\int_\Omega \mathbf{F}$ of a differential form over Ω. Now, there is one case where the answer is obvious: Suppose there exists a special coordinate system (ϕ_0, U_0) such that the restriction of $\mathbf{F}$ to Ω vanishes outside $\phi_0(U_0)$. (See Figure 8.13.) Then, of course, we define the integral as

$$\int_\Omega \mathbf{F} = \int_{U_0} \mathbf{F}(\mathbf{e}_1; \ldots; \mathbf{e}_k) d\lambda^1 \ldots d\lambda^k.$$

Next we want to consider the general case of an arbitrary differential form $\mathbf{F}$. This requires, however, the introduction of some technical but important machinery:

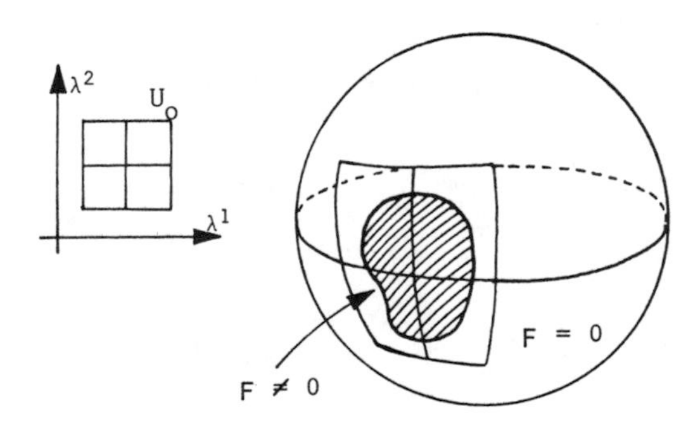

Figure 8.13.

Recall that an open covering of a topological space M is a family of open subsets $(O_i)_{i\in I}$ that covers M; i.e.,

$$M = \bigcup_{i\in I} O_i .$$

(Open coverings play an important role in the characterization of *compact* spaces. A topological space M is compact if given an arbitrary open covering of M, M can be covered by a finite subcovering.)

Then we consider a family of functions $(\psi_i)_{i\in I}$ on M,

$$\psi_i : M \curvearrowright \mathbb{R}.$$

We say that such a family is *locally finite* when at an arbitrary point $P \in M$ only a *finite* number of the functions ψ_i have a nonzero value. Observe that this property guarantees that the sum

$$\sum_{i\in I} \psi_i$$

is well-defined in a completely trivial sense, even if the index set I is uncountable. (At each point P the sum reduces effectively to a finite sum.)

Consider once more a family of functions $(\psi_i)_{i\in I}$ on our topological space M. We say that it is a *partition of unity* if it is a locally finite family of nonnegative functions such that

$$\sum_{i\in I} \psi_i = 1$$

(i.e., for any point P we have $\sum_i \psi_i(P) = 1$).

Now we can finally introduce *paracompact* spaces. As might be expected, they are characterized by a suitable property of their open coverings: *A topological space M is paracompact if to an arbitrary open covering $(O_i)_{i\in I}$ we can find a partition of unity $(\psi_i)_{i\in I}$ such that ψ_i vanishes outside O_i*. (We say that such a partition of unity is *subordinate* to the open covering. Observe that the partition of unity and the open covering have the same index set I.)

To get used to the concepts involved, the reader should try to solve the following useful exercise:

Exercise 8.3.1
Introduction: Let $(\phi_i)_{i\in I}$ be a partition of unity subordinate to the open covering $(U_i)_{i\in I}$ and let $(\psi_j)_{j\in J}$ be a partition of unity subordinate to the open covering $(O_j)_{j\in J}$. Problem:

(a) Show that $(U_i \cap O_j)_{(i,j)\in I\times J}$ is an open covering.
(b) Show that $(\phi_i \cdot \psi_j)_{(i,j)\in I\times J}$ is a partition of unity subordinate to the open covering $(U_i \cap O_j)_{(i,j)\in I\times J}$.

Now, what has all this to do with Euclidean manifolds? Well, it turns out that they satisfy the following important theorem:

Theorem 8.3. *A Euclidean manifold M is paracompact; i.e., to every open covering we can find a corresponding partition of unity.*

(For a discussion of Theorem 8.3 including a proof, see Spivak (1970).)

Okay, let us use this heavy artillery! Consider an atlas $(\phi_i, U_i)_{i\in I}$ covering int Ω. Then int Ω is a submanifold and $[\phi_i(U_i)]_{i\in I}$ an open covering. Thus we can find a partition of unity $(\psi_i)_{i\in I}$ subordinate to this open covering. We use it to cut $\mathbf{F}$ into the small pieces:

$$\mathbf{F}_i = \psi_i \mathbf{F}.$$

Here $\psi_i \mathbf{F}$ vanishes completely outside the coordinate system $\phi_i(U_i)$. Consequently, the integral

$$\int_\Omega \psi_i \mathbf{F},$$

is trivially well-defined. And furthermore, we have

$$\mathbf{F} = \left(\sum_i \psi_i\right) \mathbf{F} = \sum_i (\psi_i \mathbf{F}) \qquad \text{on int } \Omega.$$

Thus we can finally define:

Definition 8.4.

$$\int_\Omega \mathbf{F} \stackrel{\text{def}}{=} \sum_i \int_\Omega \psi_i \mathbf{F}. \tag{8.9}$$

This is a legal definition provided that we can show that the number $\sum_i \int_\Omega \psi_i \mathbf{F}$ is independent of the atlas $(\phi_i, U_i)_{i\in I}$ we choose to cover int Ω and also independent of the partition of unity. This requires lengthy but trivial calculations, where we make extensive use of Exercise 8.3.1. For details see Spivak (1970). So now we can in principle calculate the integral of a differential form over an arbitrary regular domain.

Suppose that we want to integrate a differential form over a surface that is *unbounded*. Provided that the surface is sufficiently nice, this may still be done in the usual way. Sufficiently nice means that Ω is *closed* (in the *topological* sense) and that *all points in* Ω *have either property* 1 *or property* 2 *from* Definition 8.2, Section 8.2. Then we can still divide Ω in two disjoint submanifolds: int Ω and $\partial\Omega$.

If int Ω is a simple manifold, we can choose coordinates $(\lambda^1, \ldots, \lambda^k)$ that cover int Ω. We can then formally write down the integral

$$\int_\Omega \mathbf{F} \stackrel{\text{def}}{=} \int_U \mathbf{F}(\mathbf{e}_1, \ldots, \mathbf{e}_k) d\lambda^1 \ldots d\lambda^k. \tag{8.10}$$

If the integrand $\mathbf{F}(\mathbf{e}_1, \ldots, \mathbf{e}_k)$ falls off sufficiently rapidly at "infinity," the Riemann integral will converge, and everything is nice and beautiful. This is especially the case, if $\mathbf{F}$ has *compact support*; i.e., $\mathbf{F}$ vanishes outside a compact subset of M. Otherwise the integral is divergent.

If int Ω is not simple, we must again use a partition of unity $(\psi_i)_{i\in I}$. This can be chosen such that $\psi_i \mathbf{F}$ has compact support on int Ω. Therefore, the integral $\int_\Omega \psi_i \mathbf{F}$

is trivially well-defined, and the complete integral $\int_\Omega \mathbf{F}$ is well-defined, provided that

$$\sum_i \int_\Omega \psi_i \mathbf{F}$$

is absolutely convergent; i.e., $\sum_i |\int_\Omega \psi_i \mathbf{F}| < \infty$. In that case, the value of the sum is independent of which partition of unity we use, and we can therefore define

$$\int_\Omega \mathbf{F} = \sum_i \int_\Omega \psi_i \mathbf{F}.$$

We can also use the integral concept on nonregular compact domains. Consider, e.g., the unit cube in $\mathbb{R}^3 : \Omega = \{(x, y, z) \mid 0 \leq x \leq 1, 0 \leq y \leq 1, 0 \leq z \leq 1\}$. The interior of the cube consists of all points with $0 < x < 1, 0 < y < 1$, $0 < z < 1$. The boundary consists of the remaining points, i.e., of the six faces $\Delta_1 = \{(x, y, z) \mid x = 0,\ 0 < y < 1,\ 0 < z < 1\}$ etc. Notice that the points with property (2) in Definition 8.2 do not exhaust $\partial\Omega$. We still have the points on the *edges*, which form the skeleton of the cube. (See Figure 8.14.) The interior is a nice submanifold, but the boundary is not a submanifold. The integral over $\partial\Omega$ is therefore a priori not well-defined. However, $\partial\Omega$ is piecewise smooth, and the edges are of a lower dimensionality than the faces. Therefore, we can neglect their contribution to the integral. We can then define

$$\int_{\partial\Omega} \mathbf{T} = \sum_{i=1}^{6} \int_{\Delta_i} \mathbf{T}.$$

If Ω is a regular domain, then it is composed of two parts: int Ω and $\partial\Omega$. Let us concentrate on the interior for a moment. It is a k-dimensional submanifold. To each point P on int Ω we have attached a k-dimensional tangent space $\mathbf{T}_p(\text{int }\Omega)$. If we introduce coordinates $(\lambda^1, \ldots, \lambda^k)$ on int Ω, they generate a

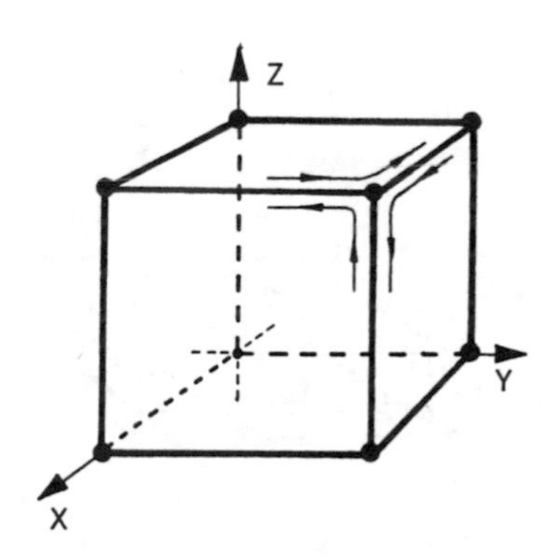

Figure 8.14.

canonical frame $(\mathbf{e}_1, \ldots, \mathbf{e}_k)$ for the tangent space $\mathbf{T}_p(\text{int }\Omega)$. Clearly, $\mathbf{T}_p(\text{int }\Omega)$ is a k-dimensional subspace of the n-dimensional tangent space $\mathbf{T}_p(M)$. (See Figure 8.15.) If $(x^1, \ldots, x^n)$ denote coordinates on M, then they generate a canonical frame $(\mathbf{i}_1, \ldots, \mathbf{i}_n)$ for the full tangent space $\mathbf{T}_p(M)$. The frame $(\mathbf{e}_1, \ldots, \mathbf{e}_k)$ is related to the frame $(\mathbf{i}_1, \ldots, \mathbf{i}_n)$ in the following way:

$$\mathbf{e}_i = \mathbf{i}_a \frac{\partial x^a}{\partial \lambda^i} . \tag{8.6}$$

Now, if $\mathbf{F}$ is a differential form of degree k defined on M, then $\mathbf{F}$ is a multilinear map on each of the tangent spaces $\mathbf{T}_p(M)$. But we may also consider $\mathbf{F}$ as a differential form defined on int Ω simply because $\mathbf{F}$ defines a multilinear map on each of the tangent spaces $\mathbf{T}_p(\text{int }\Omega) \subseteq \mathbf{T}_p(M)$. This *new* differential form is called the *restriction of* $\mathbf{F}$ *to* int Ω, and it is convenient to denote it by the same symbol $\mathbf{F}$.

When $\mathbf{F}$ is considered to be a differential form on M, it is characterized by its components

$$F_{a_1 \ldots a_k} = \mathbf{F}(\mathbf{i}_{a_1}, \ldots, \mathbf{i}_{a_k}); \qquad 1 \leq a_j \leq n,$$

and we may decompose it in the following way (7.8):

$$\mathbf{F} = \frac{1}{k!} F_{a_1 \ldots a_k} \boldsymbol{dx}^{a_1} \wedge \ldots \wedge \boldsymbol{dx}^{a_k}.$$

In a similar way the *restriction of* $\mathbf{F}$ *to* int Ω can be characterized by the *new* components

$$\mathbf{F}(\mathbf{e}_{i_1}; \ldots; \mathbf{e}_{i_k}); \qquad 1 \leq i_j \leq k,$$

and we can decompose the restriction of $\mathbf{F}$ as

$$\mathbf{F} = \mathbf{F}(\mathbf{e}_1; \ldots; \mathbf{e}_k) \boldsymbol{d\lambda}^1 \wedge \ldots \wedge \boldsymbol{d\lambda}^k, \tag{8.11}$$

since it is a k-form on a k-dimensional manifold. (Compare Exercise 7.2.3.) This is very useful when we want to integrate $\mathbf{F}$! By definition

$$\begin{aligned} \int_\Omega \mathbf{F} &= \int_\Omega \mathbf{F}(\mathbf{e}_1; \ldots; \mathbf{e}_k) \boldsymbol{d\lambda}^1 \wedge \ldots \wedge \boldsymbol{d\lambda}^k \\ &\overset{\text{def.}}{=} \int_U \mathbf{F}(\mathbf{e}_1; \ldots; \mathbf{e}_k) d\lambda^1 \ldots d\lambda^k. \end{aligned} \tag{8.12}$$

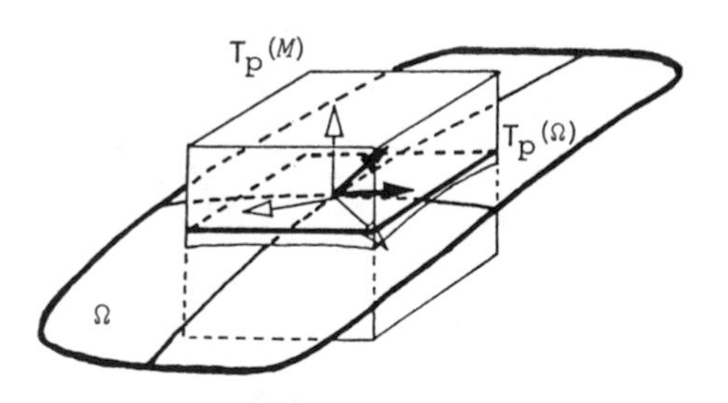

Figure 8.15.

Thus the integration is performed simply by replacing $\boldsymbol{d}\lambda^1 \wedge \ldots \wedge \boldsymbol{d}\lambda^k$ by the ordinary Riemannian volume element $d\lambda^1 \ldots d\lambda^k$! Observe that $\boldsymbol{d}\lambda^1 \wedge \ldots \wedge \boldsymbol{d}\lambda^k$ itself can be considered as a volume element in each of the tangent spaces $\mathbf{T}_p(\text{int}\Omega)$. (Compare the discussion in Section 7.4.) Consequently, the geometrical integral,

$$\int_\Omega \mathbf{F}(\mathbf{e}_1; \ldots; \mathbf{e}_k)\boldsymbol{d}\lambda^1 \wedge \ldots \wedge \boldsymbol{d}\lambda^k$$

may be considered as a generalization of the Riemann integral defined for ordinary continuous functions on ordinary Euclidean spaces!

To use this point of view in practical calculations, we must have a method to find the restriction of $\mathbf{F}$ to int Ω. One way of doing this is of course to compute $\mathbf{F}(\mathbf{e}_1, \ldots, \mathbf{e}_k)$ directly using the formula

$$\mathbf{F}(\mathbf{e}_1; \ldots; \mathbf{e}_k) = F_{a_1 \ldots a_k} \frac{\partial x^{a_1}}{\partial \lambda^1} \cdots \frac{\partial x^{a_k}}{\partial \lambda^k} .$$

But in many applications it is preferable to use the decomposition

$$\mathbf{F} = \frac{1}{k!} F_{a_1 \ldots a_k} \boldsymbol{d}x^{a_1} \wedge \ldots \wedge \boldsymbol{d}x^{a_k}$$

as a starting point. The question is then, What is the restriction of $\boldsymbol{d}x^i$ to int Ω?

Exercise 8.3.2
Problem: (a) Show that the restriction of $\boldsymbol{d}x^i$ to int Ω is given by

$$\boldsymbol{d}x^i = \frac{\partial x^i}{\partial \lambda^a} \boldsymbol{d}\lambda^a. \tag{8.13}$$

(b) Insert the result obtained in (a) in the decomposition (7.8) of $\mathbf{F}$, and show once again, by rearranging the formula obtained that the restriction of $\mathbf{F}$ to int Ω is given by

$$\mathbf{F} = \mathbf{F}(\mathbf{e}_1, \ldots, \mathbf{e}_k)\boldsymbol{d}\lambda^1 \wedge \ldots \wedge \boldsymbol{d}\lambda^k. \tag{8.11}$$

The calculations are especially easy if we use an adapted coordinate system; i.e., $(x^1, \ldots, x^k, \ldots, x^n)$ parametrized M and $(x^1, \ldots, x^k)$ parametrizes int Ω. Then the restrictions of $\boldsymbol{d}x^1, \ldots, \boldsymbol{d}x^k, \ldots, \boldsymbol{d}x^n$ to int Ω are simply given by $\boldsymbol{d}x^1 = \boldsymbol{d}x^1, \ldots, \boldsymbol{d}x^k = \boldsymbol{d}x^k$ and $\boldsymbol{d}x^{k+1} = 0, \ldots, \boldsymbol{d}x^n = 0$. So if we want to find the restriction of

$$\mathbf{F} = \frac{1}{k!} F_{a_1 \ldots a_k} \boldsymbol{d}x^{a_1} \wedge \ldots \wedge \boldsymbol{d}x^{a_k}$$

to int Ω, we simply remove all terms containing $\boldsymbol{d}x^{k+1}, \ldots, \boldsymbol{d}x^n$!

We conclude with one more remark on how to calculate the integral in practice. Suppose Ω is a nice orientable regular domain where int Ω is not a simple manifold. Then int Ω cannot be covered by a single coordinate system. In that case we ought to use overlapping coordinate systems, find a partition of unity, etc. This is very technical and very complicated. In practice we will compute the integral using our common sense! Consider, for instance, the case, $\Omega = S^2$, which is its own interior; i.e.,

$$\text{int}\,\Omega = S^2.$$

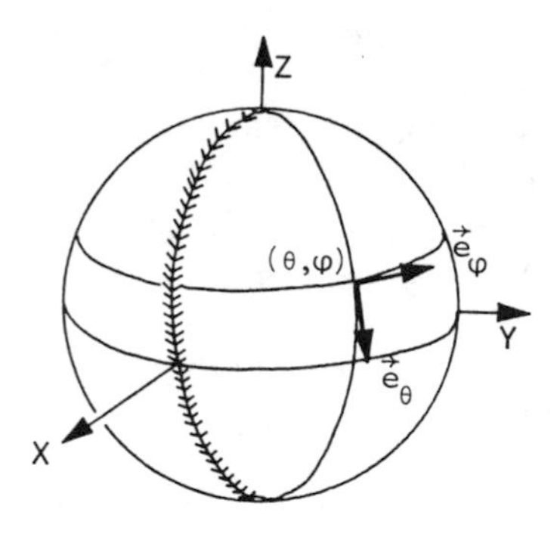

Figure 8.16.

Thus int Ω is not simple. But let us introduce spherical coordinates anyway! Then we exclude an arc joining the North Pole and the South Pole. (See Figure 8.16.) But it constitutes a subset of lower dimensionality, and therefore it does not contribute to the integral anyway. Thus we compute the integral in the following simple way:

$$\int_\Omega \mathbf{F} = \int_{\theta=0}^{\pi} \int_{\varphi=0}^{2\pi} \mathbf{F}(\mathbf{e}_\theta; \mathbf{e}_\varphi) d\theta \, d\varphi.$$

Worked Exercise 8.3.3
Problem: (a) Let $\Omega = S_+^2$ be the northern hemisphere of the unit sphere in $\mathbb{R}^3$. Show that

$$\int_{S_+^2} \boldsymbol{d}x \wedge \boldsymbol{d}y = \pi.$$

(b) Let $\Omega = B^3$ be the closed unit ball in $\mathbb{R}^3$. Show that

$$\int_{B^3} \boldsymbol{d}x \wedge \boldsymbol{d}y \wedge \boldsymbol{d}z = \frac{4}{3}\pi.$$

8.4 Elementary Properties of the Integral

The integral has, of course, several simple algebraic properties. It is obviously linear:

$$\int_\Omega (\mathbf{F} + \mathbf{G}) = \int_\Omega \mathbf{F} + \int_\Omega \mathbf{G}; \qquad \int_\Omega \lambda\mathbf{F} = \lambda \int_\Omega \mathbf{F}. \tag{8.14}$$

Now, we have been very careful in our discussion of regular domains. To each orientable k-dimensional regular domain Ω we have attached an orientable $(k-1)$-dimensional regular domain $\partial\Omega$. Suppose $\mathbf{F}$ is a differential form of degree $k-1$; then the exterior derivative $\boldsymbol{d}\mathbf{F}$ is a differential form of degree k. Thus we can form two integrals:

$$\int_{\partial\Omega} \mathbf{F} \quad \text{and} \quad \int_{\Omega} d\mathbf{F}.$$

The question is then, What is the connection between these two integrals?

Exercise 8.4.1
Problem: Let Ω be the unit cube in $\mathbb{R}^3$, and let $\mathbf{B}$ be an arbitrary smooth two-form. Show that

$$\int_{\Omega} d\mathbf{B} = \int_{\partial\Omega} \mathbf{B}.$$

It follows from Exercise 8.4.1 that the two integrals are always identical when Ω is a cube. This is a special case of the following famous theorem:

Theorem 8.4. (*Stokes's theorem*) *Let Ω be an orientable k-dimensional regular domain and let* $\mathbf{F}$ *be a differential form of degree $k-1$, then*

$$\int_{\Omega} d\mathbf{F} = \int_{\partial\Omega} \mathbf{F}. \tag{8.15}$$

For a complete proof see Spivak (1970), but the formula can also be derived from the following naive argument. The regular domain Ω can be approximated by a family of cubes: $(\Omega_i)_{i\in I}$; cf. Figure 8.17. The boundary of the cubes consists of two kinds of faces:

a. The face lies at the boundary of Ω.
b. The face lies in the interior of Ω.

In case (b) the face always coincides with a face from a neighboring cube, but they will have opposite orientations, and their contributions to the integral therefore automatically cancel. When we perform the integration, we can therefore replace

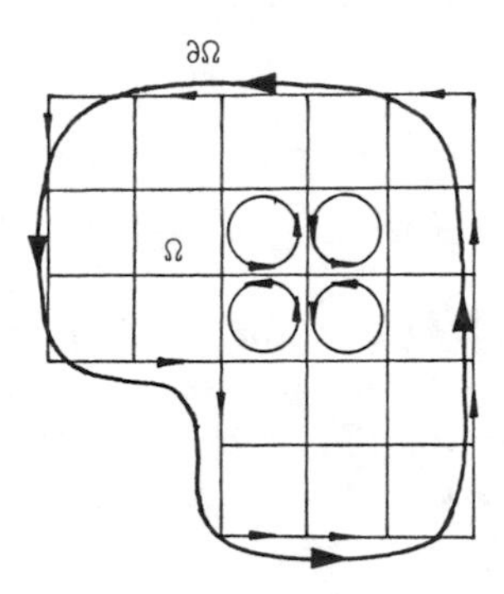

Figure 8.17.

the boundary of Ω with the boundaries of the cubes:

$$\int_{\Omega} d\mathbf{F} \approx \sum_i \int_{\Omega_i} d\mathbf{F} = \sum_i \int_{\partial\Omega_i} \mathbf{F} \approx \int_{\partial\Omega} \mathbf{F}.$$

If the manifold M is equipped with a metric, we obtain the following important corollary:

Theorem 8.5. (*Corollary to Stokes's theorem*) *Let Ω be an orientable $(n-k+1)$-dimensional regular domain and let* $\mathbf{F}$ *be a smooth k-form. Then*

$$\int_{\Omega} {}^*\delta\mathbf{F} = (-1)^{n-k+1} \int_{\partial\Omega} {}^*\mathbf{F}. \tag{8.16}$$

PROOF. From (7.62) we learn that

$${}^*\delta\mathbf{F} = (-1)^{n-k+1} d^*\mathbf{F},$$

and Theorem 8.5 is now a trivial consequence of Stokes's theorem. □

Finally, we have the rule of integration by parts. We have generalized the Leibniz rule in the following way:

$$d(\mathbf{F} \wedge \mathbf{G}) = d\mathbf{F} \wedge \mathbf{G} + (-1)^k \mathbf{F} \wedge d\mathbf{G}, \tag{7.17}$$

where $\mathbf{F}$ and $\mathbf{G}$ are differential forms of degree k and ℓ. Let Ω be a $(k+\ell+1)$-dimensional orientable regular domain. Then

$$\int_{\Omega} d(\mathbf{F} \wedge \mathbf{G}) = \int_{\Omega} d\mathbf{F} \wedge \mathbf{G} + (-1)^k \int_{\Omega} \mathbf{F} \wedge d\mathbf{G}.$$

But we may transform the left-hand side using Stokes's theorem, whereby we get

$$\int_{\partial\Omega} \mathbf{F} \wedge \mathbf{G} = \int_{\Omega} d\mathbf{F} \wedge \mathbf{G} + (-1)^k \int_{\Omega} \mathbf{F} \wedge d\mathbf{G}.$$

This we can rearrange in the following way:

Theorem 8.6. (*Theorem of integration by parts*)

$$\int_{\Omega} d\mathbf{F} \wedge \mathbf{G} = \int_{\partial\Omega} \mathbf{F} \wedge \mathbf{G} - (-1)^k \int_{\Omega} \mathbf{F} \wedge d\mathbf{G}. \tag{8.17}$$

This, of course, is a direct generalization of the familiar rule

$$\int_a^b \frac{df}{dx} g(x)dx = [f(x)g(x)]_a^b - \int_a^b f(x)\frac{dg}{dx}\,dx.$$

In many application we will be able to throw away the boundary term, either because Ω has no boundary (i.e., $\partial\Omega = \emptyset$) or because one of the differential forms vanishes on $\partial\Omega$!

Let us discuss this new integral concept in a few simple cases: Consider first *scalar fields*. A scalar field $\phi : M \curvearrowright \mathbb{R}$ can be represented as a zero-form. It can be integrated over a 0-dimensional regular domain. But a 0-dimensional regular domain consists simply of a point (or a finite collection of points). So in this case the integral reduces to

$$\int_P \phi = \phi(P);$$

i.e., it just reproduces the value of the function ϕ. This is the most trivial and most uninteresting case.

A scalar field ϕ can also be represented as the n-form (7.39)

$$^*\phi = \phi\varepsilon = \phi(x)\sqrt{g}\,\boldsymbol{dx}^1 \wedge \ldots \wedge \boldsymbol{dx}^n.$$

But then it can be integrated over n-dimensional domains including M itself. Usually such a domain Ω can be parametrized by the coordinates $(x^1, \ldots, x^n)$ on M itself, and the integral reduces to

$$\begin{aligned}\int_\Omega {}^*\phi &= \int_\Omega \phi\sqrt{g}\,\boldsymbol{dx}^1 \wedge \ldots \wedge \boldsymbol{dx}^n \\ &= \int_U \phi(x)\sqrt{g}\,dx^1 \ldots dx^n,\end{aligned} \tag{8.18}$$

where U is the coordinate domain corresponding to Ω. This shows that we can generalize the ordinary Riemann integral in Euclidean space

$$\int_\Omega \phi(x)dx^1 \ldots dx^n,$$

which is valid only for Cartesian coordinates, into the covariant expression

$$\int_\Omega \phi(x)\sqrt{g}\,dx^1 \ldots dx^n,$$

which is valid in arbitrary coordinates. All we have to do is to replace the Cartesian volume element with the covariant volume element $\sqrt{g}\,dx^1 \ldots dx^n$. When we integrate a scalar field ϕ on a manifold M with metric $\boldsymbol{g}$, we always use the covariant integral (8.18).

As an application we can choose $\phi = 1$. The integral

$$\int_\Omega \varepsilon = \int_\Omega \sqrt{g}\,dx^1 \ldots dx^n$$

then represents the volume of Ω since $\varepsilon = \sqrt{g}\,\boldsymbol{dx}^1 \wedge \ldots \wedge \boldsymbol{dx}^n$ is the volume element in the tangent space (compare the discussion in Section 7.4):

$$\text{Vol}[\Omega] = \int_\Omega \sqrt{g}\,dx^1 \ldots dx^n. \tag{8.19}$$

Here Ω is restricted to be an n-dimensional domain, but the formula (8.19) can easily be generalized to cover the "volume" of arbitrary k-dimensional domains, i.e., the length of a curve, the area of a two-dimensional surface, etc. To see how this can be done we consider an orientable k-dimensional regular domain Ω in our n-dimensional manifold M. The metric g on M induces a metric on Ω, and in this way Ω becomes k-dimensional manifold with metric. Let $(\lambda^1, \ldots, \lambda^k)$ be parametrization of Ω. The intrinsic coordinates $(\lambda^1, \ldots, \lambda^k)$ generate a canonical frame $(\mathbf{e}_1, \ldots, \mathbf{e}_k)$ in each of the tangent spaces (compare Figure 8.15). The induced metric is characterized by the intrinsic components

$$\boldsymbol{g}(\mathbf{e}_i; \mathbf{e}_j).$$

The k-dimensional volume of Ω is therefore given by

$$\mathrm{Vol}_k[\Omega] = \int_\Omega \sqrt{\det[\boldsymbol{g}(\mathbf{e}_i; \mathbf{e}_j)]}d\lambda^1 \ldots d\lambda^k. \tag{8.20}$$

This can also be reexpressed in terms of the extrinsic components of $\boldsymbol{g}$:

$$\boldsymbol{g}(\mathbf{e}_i; \mathbf{e}_j) = \boldsymbol{g}(\mathbf{i}_a; \mathbf{i}_b)\frac{\partial x^a}{\partial \lambda^i}\frac{\partial x^b}{\partial \lambda^j} = g_{ab}\frac{\partial x^a}{\partial \lambda^i}\frac{\partial x^b}{\partial \lambda^j}\,.$$

The volume formula is then rearranged as

$$\mathrm{Vol}_k[\Omega] = \int_\Omega \sqrt{\det\left[g_{ab}\frac{\partial x^a}{\partial \lambda^i}\frac{\partial x^b}{\partial \lambda^j}\right]}d\lambda^1 \ldots d\lambda^k. \tag{8.21}$$

Exercise 8.4.2
Problem: Consider a curve Γ connecting two points P and Q. Show that $\mathrm{Vol}_1[\Gamma]$ reproduces the usual formula (6.40) for the length of a curve.

Next we consider vector fields. Let $\mathbf{A}$ be a vector field. We can represent $\mathbf{A}$ by a one-form $\mathbf{A}$. It can then be integrated over a one-dimensional domain, i.e., a curve Γ. If we parameterize this curve Γ with the parameter λ, (cf. Figure 8.18), the integral reduces to

$$\int_\Gamma \mathbf{A} = \int_{\lambda_1}^{\lambda_2} \langle \mathbf{A} \mid \mathbf{e}\rangle d\lambda = \int_{\lambda_1}^{\lambda_2} A_i \frac{dx^i}{d\lambda}\, d\lambda. \tag{8.22}$$

Observe that in a Euclidean space,

$$A_i \frac{dx^i}{d\lambda} = \mathbf{A} \cdot \frac{d\mathbf{r}}{d\lambda}\,,$$

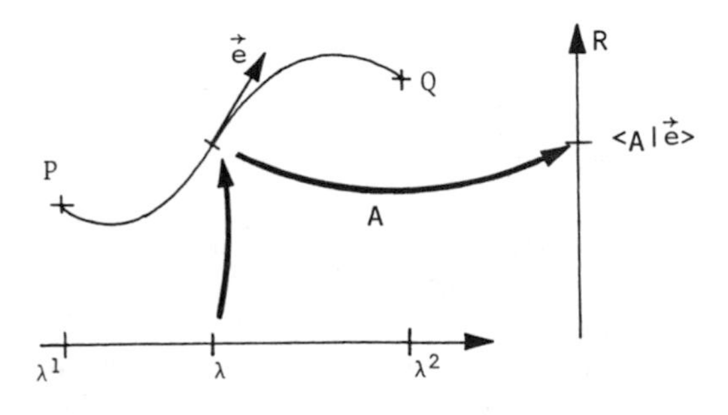

Figure 8.18.

and the integral (8.22) therefore generalizes the usual line integral

$$\int_{\Gamma} \mathbf{A} \cdot d\mathbf{r} = \int_{\lambda_1}^{\lambda_2} \mathbf{A} \cdot \frac{d\mathbf{r}}{d\lambda}\, d\lambda.$$

We shall also refer to $\int_{\Gamma} \mathbf{A}$ as a *line integral.*

Suppose $\mathbf{A}$ is generated by a scalar field ϕ; i.e., $\mathbf{A} = \boldsymbol{d}\phi$. Then Stokes's theorem gives

$$\int_{\Gamma} \boldsymbol{d}\phi = \int_{\partial\Gamma} \phi.$$

The boundary $\partial\Gamma$ consists of the points P and Q, where P is negatively oriented and Q is positively oriented. By convention we therefore put

$$\int_{\partial\Gamma} \phi = \phi(Q) - \phi(P).$$

Stokes's theorem consequently produces the formula

$$\int_{\Gamma} \boldsymbol{d}\phi = \phi(Q) - \phi(P), \tag{8.23}$$

which generalizes the usual theorem of line integrals (Section 1.1). It does not, however, use the full strength of Stokes's theorem. If we use that

$$\int_{\Gamma} \boldsymbol{d}\phi = \int_{\lambda_1}^{\lambda_2} \frac{\partial \phi}{\partial x^i} \frac{dx^i}{d\lambda}\, d\lambda = \int_{\lambda_1}^{\lambda_2} \frac{d\phi}{d\lambda}\, d\lambda,$$

we immediately see that it is a simple reformulation of the fundamental theorem of calculus.

Worked Exercise 8.4.3

Problem: We use the notation of Exercise 7.6.9. In particular, M is a two-dimensional manifold in $\mathbb{R}^2$, which we identify with $\mathbb{C}$.

a. Let Γ be a curve in M. Show that

$$\int_{\Gamma} h(z)\boldsymbol{d}z = \int_{\Gamma} h(z) dz,$$

where the integral on the right-hand side is the usual complex line integral.

b. Consider an analytic (holomorphic) function $h(z)$. Let Γ be a smooth curve connecting A and B. Show that

$$\int_{\Gamma} h'(z)\boldsymbol{d}z = h(B) - h(A).$$

c. Consider a two-dimensional regular domain Ω bounded by the closed curve Γ. Let $h(z)$ be an analytic (holomorphic) function in M. Prove the following version of Cauchy's integral theorem:

$$\int_{\Gamma} h(z)\boldsymbol{d}z = 0.$$

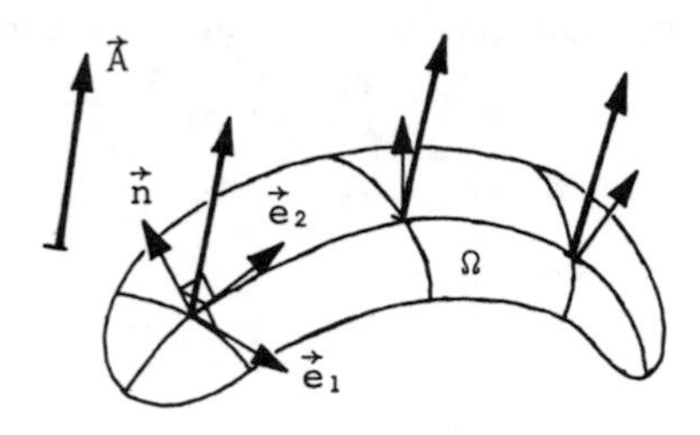

Figure 8.19.

A vector $\mathbf{A}$ can also be represented by an $(n-1)$-form $^*\mathbf{A}$. Then it can be integrated over an $(n-1)$-dimensional domain. An $(n-1)$-dimensional surface Ω is usually called a *hypersurface*. (See Figure 8.19.) If Ω is an orientable hypersurface, then it allows a normal vector field $\mathbf{n}$; i.e., at each point P on Ω the normal vector $\mathbf{n}(P)$ is orthogonal to the tangent space $\mathbf{T}_p(\Omega)$. Let us parameterize the surface Ω with the parameters $(\lambda^1, \ldots, \lambda^{n-1})$. They generate the canonical frame vectors $(\mathbf{e}_1, \ldots, \mathbf{e}_{n-1})$. Furthermore, we let $\mathbf{n}$ denote the normal vector $\mathbf{n} = \mathbf{e}_1 \times \ldots \times \mathbf{e}_{n-1}$ (compare the discussion in Section 7.4). According to Exercise 7.5.4, we have

$$^*\mathbf{A}(\mathbf{e}_1, \ldots, \mathbf{e}_{n-1}) = \mathbf{A} \cdot (\mathbf{e}_1 \times \ldots \times \mathbf{e}_{n-1}) = \mathbf{A} \cdot \mathbf{n}.$$

The integral of $^*\mathbf{A}$ therefore reduces to

$$\begin{aligned} \int_\Omega {}^*\mathbf{A} &= \int_U {}^*\mathbf{A}(\mathbf{e}_1, \ldots, \mathbf{e}_{n-1}) d\lambda^1 \ldots d\lambda^{n-1} \\ &= \int_U \mathbf{A} \cdot \mathbf{n} d\lambda^1 \ldots d\lambda^{n-1}. \end{aligned} \tag{8.24}$$

Thus $\int_\Omega {}^*\mathbf{A}$ *simply represents the flux of the vector field* $\mathbf{A}$ *through the hypersurface* Ω. For this reason we refer to the integral of $^*\mathbf{A}$ as a *flux integral*.

Having introduced flux integrals, we can now explain why Theorem 8.4 is called Stokes's theorem. If S is a surface in $\mathbb{R}^3$ bounded by the curve Γ, Theorem 8.4 gives us

$$\int_S d\mathbf{A} = \int_\Gamma \mathbf{A}.$$

The right-hand side is the line integral of $\mathbf{A}$ along Γ. We also have

$$d\mathbf{A} = {}^*(\nabla \times \mathbf{A}).$$

(Compare the discussion in Section 7.7.) The left-hand side therefore represents the flux of the curl $\nabla \times \mathbf{A}$. Consequently, Theorem 8.4 generalizes the classical Stokes's theorem from the conventional vector calculus; cf. (1.16).

Let Ω be an n-dimensional regular domain bounded by the hypersurface $\partial\Omega$. If $\mathbf{A}$ is a vector field, then $-\boldsymbol{\delta}\mathbf{A}$ represents the divergence of $\mathbf{A}$. We can integrate this divergence over Ω and get the volume integral

$$-\int_{\Omega} {}^*\delta\mathbf{A}.$$

According to the corollary to Stokes's theorem, this can be rearranged as

$$-\int_{\Omega} {}^*\delta\mathbf{A} = (-1)^{n-1}\int_{\partial\Omega} {}^*\mathbf{A}. \tag{8.25}$$

Up to a sign, we have therefore shown that *the integral of the divergence is equal to the flux through the boundary*. This is the generalization of Gauss's theorem to an arbitrary manifold. We can also work out the corresponding covariant coordinate expression. Using (8.18), (8.24), and (7.64), we get

$$\int_{\Omega} \partial_i\left(\sqrt{g}A^i\right) d^n x = \int_{\partial\Omega} A^i n_i d\lambda^1 \ldots d\lambda^{n-1}, \tag{8.26}$$

where the normal vector $\mathbf{n}$ is characterized by the covariant components

$$n_i = \sqrt{g}\,\epsilon_{ij_1\ldots j_{n-1}} \frac{\partial x^{j_1}}{\partial\lambda^1} \cdots \frac{\partial x^{j_{n-1}}}{\partial\lambda^{n-1}}. \tag{8.27}$$

We can also reproduce the integral theorems from the two-dimensional vector calculus. Consider, for instance, a region Ω bounded by a smooth curve Γ (see Figure 8.8). Then it is well known that the area of Ω is given by the line integral

$$\text{Area}[\Omega] = \frac{1}{2}\int_{\Gamma} {}^*\mathbf{r}\cdot d\mathbf{r}. \tag{8.28}$$

(Here ${}^*\mathbf{r}$ is the vector orthogonal to $\mathbf{r}$ and with the same length. Remember also that $\frac{1}{2}\,{}^*\mathbf{r}\cdot\Delta\mathbf{r}$ represents the area of the triangle spanned by $\mathbf{r}$ and $\mathbf{r}+\Delta\mathbf{r}$.) (See Figure 8.20.) It is convenient to write out this formula in Cartesian coordinates:

$$\text{Area}[\Omega] = \frac{1}{2}\int x\,dy - y\,dx. \tag{8.29}$$

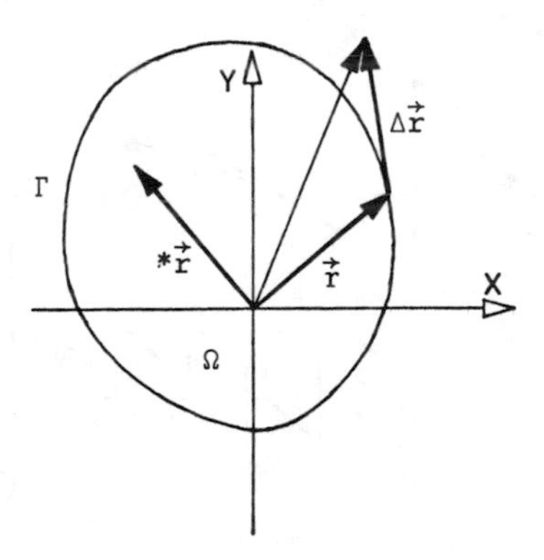

Figure 8.20.

To deduce this formula, we use that $\boldsymbol{dx} \wedge \boldsymbol{dy}$ is the "volume form" in $\mathbb{R}^2$. Consequently,

$$\text{Area}[\Omega] = \int_\Omega \boldsymbol{dx} \wedge \boldsymbol{dy} = -\int_\Omega \boldsymbol{dy} \wedge \boldsymbol{dx};$$

i.e.,

$$\text{Area}[\Omega] = \frac{1}{2}\int_\Omega \boldsymbol{dx} \wedge \boldsymbol{dy} - \boldsymbol{dy} \wedge \boldsymbol{dx} = \frac{1}{2}\int_\Omega \boldsymbol{d}(x\boldsymbol{dy} - y\boldsymbol{dx}).$$

An application of Stokes's theorem now gives

$$\text{Area}[\Omega] = \frac{1}{2}\int_\Gamma x\boldsymbol{dy} - y\boldsymbol{dx},$$

which we immediately recognize as being equivalent to (8.29).

We conclude this section with a discussion of simple forms, where we can give a naive interpretation of the integral. For simplicity, we consider the three-dimensional manifold $M = \mathbb{R}^3$. If we choose a coordinate system on M, then the coordinates (x^1, x^2, x^3) generate several simple forms, for instance, $\boldsymbol{dx}^1, \boldsymbol{dx}^1 \wedge \boldsymbol{dx}^2, \boldsymbol{dx}^1 \wedge \boldsymbol{dx}^2 \wedge \boldsymbol{dx}^3$.

Consider first the basic one-form $\boldsymbol{dx}^1$. Let Γ be a smooth curve from P to Q. (See Figure 8.21.) We want to interpret the integral $\int_\Gamma \boldsymbol{dx}^1$. We divide $[a, b]$ into unit intervals $[a, a+1], [a+1, a+2], \ldots$. Then the integral is roughly given by

$$\int_\Gamma \boldsymbol{dx}^1 \approx \sum_i \langle \boldsymbol{dx}^1 \mid \mathbf{e}_1^i \rangle.$$

The coordinate x^1 generates a stratification characterized by the surfaces

$$\ldots, x^1 = -1, \quad x^1 = 0, \quad x^1 = +1, \ldots.$$

According to our analysis in Section 7.2, the number $\langle \boldsymbol{dx}^1 \mid \mathbf{e}_1^i \rangle$ can be interpreted as the number of surfaces intersected by the curve segment corresponding to the ith

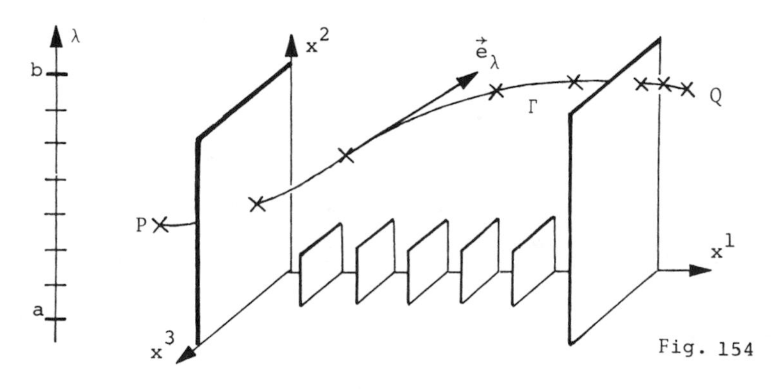

Figure 8.21.

unit interval. Therefore, the total integral can be naively interpreted as *the number of surfaces intersected by* Γ. This can be demonstrated rigorously using Stokes's theorem. It immediately gives us

$$\int_{\Omega} \boldsymbol{dx}^1 = \int_{\partial\Gamma} x^1 = x^1(Q) - x^1(P),$$

and the number $x^1(Q) - x^1(P)$ clearly represents the number of surfaces intersected by Γ. Observe that if Γ is a closed curve, then each surface is intersected twice from opposite directions. Therefore,

$$\oint_{\Gamma} \boldsymbol{dx}^1 = 0.$$

This is also in accordance with Stokes's theorem:

$$\oint_{\Gamma} \boldsymbol{dx}^1 = \int_{\partial\Gamma=\emptyset} x^1 = 0.$$

Next, we consider the basic form $\boldsymbol{dx}^1 \wedge \boldsymbol{dx}^2$. Let Ω be a smooth surface in M. (See Figure 8.22.)

We want to interpret the integral

$$\int_{\Omega} \boldsymbol{dx}^1 \wedge \boldsymbol{dx}^2.$$

We divide U into unit squares. Then the integral is roughly given by

$$\int_{\Omega} \boldsymbol{dx}^1 \wedge \boldsymbol{dx}^2 \approx \sum_i \boldsymbol{dx}^1 \wedge \boldsymbol{dx}^2 \left(\mathbf{e}_1^i; \mathbf{e}_2^i\right).$$

The coordinates (x^1, x^2) generate a honeycomb structure in M. According to our analysis in Section 7.2, the number $\boldsymbol{dx}^1 \wedge \boldsymbol{dx}^2\ (\mathbf{e}_1^i; \mathbf{e}_2^i)$ can be interpreted as the total number of tubes intersected by the surface segment corresponding to the ith square. Therefore, the integral can be naively interpreted as *the number of tubes intersected by* Ω.

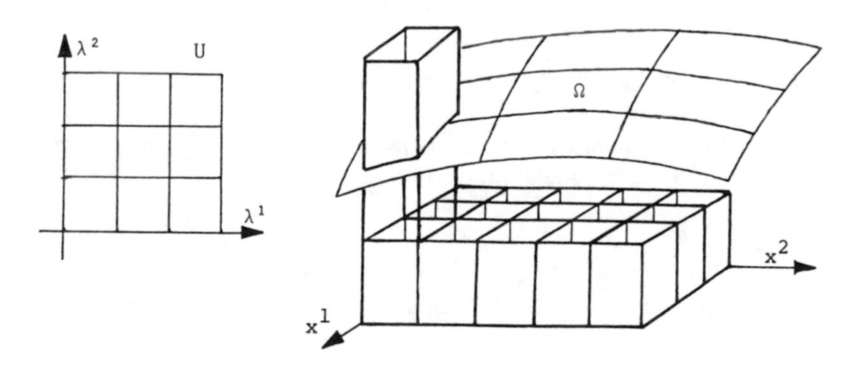

Figure 8.22.

This can also be justified more rigorously by the following argument: For simplicity, we assume that we can use (x^1, x^2) as adapted coordinates on Ω; i.e., Ω is parametrized on the form $x^3 = x^3(x^1, x^2)$. Then we get

$$\int_\Omega \boldsymbol{dx}^1 \wedge \boldsymbol{dx}^2 = \int_U dx^1 dx^2.$$

Here the coordinate domain U is an open subset of the (x^1, x^2)-plane. Tubes in M correspond to unit cells in the (x^1, x^2)-plane. The number of tubes intersected by Ω is equal to the number of unit cells contained in U. On the other hand, $\int_U dx^1 dx^2$ is equal to the area of U, and this number also represents the number of unit cells contained in U.

If Ω is a *closed* surface, then

$$\int_\Omega \boldsymbol{dx}^1 \wedge \boldsymbol{dx}^2 = 0,$$

because each tube is intersected twice with opposite orientations. This illustrates Stokes's theorem for the basic form $\boldsymbol{dx}^1 \wedge \boldsymbol{dx}^2$:

$$\int_\Omega \boldsymbol{dx}^1 \wedge \boldsymbol{dx}^2 = \int_\Omega \boldsymbol{d}(x^1 \boldsymbol{dx}^2) = \int_{\partial\Omega=\emptyset} x^1 \boldsymbol{dx}^2 = 0.$$

Exercise 8.4.4
Problem: Let $\Omega = S^2_+$ be the northern hemisphere of the unit sphere in $\mathbb{R}^3$. Show that the "number" of tubes intersected by S^2_+ is π. Compare this with Exercise 8.3.3.

Exercise 8.4.5
Problem: Consider the basic three-form $\boldsymbol{dx}^1 \wedge \boldsymbol{dx}^2 \wedge \boldsymbol{dx}^3$. The coordinate functions x^1, x^2 and x^3 generate a cell structure in M. Let Ω be a 3-dimensional regular domain in M. Show that

$$\int_\Omega \boldsymbol{dx}^1 \wedge \boldsymbol{dx}^2 \wedge \boldsymbol{dx}^3$$

can be interpreted as the number of cells contained in Ω.

8.5 The Hilbert Product of Two Differential Forms

Consider two differential forms $\mathbf{T}$ and $\mathbf{U}$ of the same degree k. We have previously (Section 7.5) introduced the scalar product between $\mathbf{T}$ and $\mathbf{U}$; cf. (7.52–7.53):

$$(\mathbf{T} \mid \mathbf{U}) = \frac{1}{k!} T^{i_1 \ldots i_k} U_{i_1 \ldots i_k} = \pm {}^*({}^*\mathbf{T} \wedge \mathbf{U}).$$

This is the relevant scalar when we look at a specific point P, i.e., when we want to introduce a metric in the finite-dimensional vector space $\lambda^k_P(M)$.

Globally, the differential forms of degree k span an infinite-dimensional vector space $\Lambda^k(M)$, and we would like to associate an *inner product* with this space.

To motivate it, we consider real-valued smooth functions on a closed interval $[a, b]$. Here we have the natural inner product

$$\langle f \mid g \rangle = \int_a^b f(x)g(x)dx = \int_{[a,b]} {}^*f \wedge g.$$

In analogy with this, we consider the n-form (7.50)

$${}^*\mathbf{T} \wedge \mathbf{U} = \frac{1}{k!}\, T^{i_1 \ldots i_k} U_{i_1 \ldots i_k} \varepsilon.$$

If it is integrable, we define the inner product in the following way:

$$\langle \mathbf{T} \mid \mathbf{U} \rangle \overset{\text{def.}}{=} \int_M {}^*\mathbf{T} \wedge \mathbf{U} = \int_M \frac{1}{k!}\, T^{i_1 \ldots i_k} U_{i_1 \ldots i_k} \sqrt{g}\, dx^1 \ldots dx^n. \tag{8.30}$$

We shall refer to this inner product as the Hilbert product. If M is compact, then the Hilbert product is always well-defined. If M is not compact, the integral is not necessarily convergent. It is, however, always well-defined if one of the differential forms has compact support.

Consider a Riemannian manifold. Let $\Lambda_0^k(M)$ denote the vector space of k-forms with compact support. The Hilbert product defines a positive definite metric in this infinite-dimensional vector space. Thus $\Lambda_0^k(M)$ is a pre-Hilbert space. We can complete it and obtain a conventional Hilbert space

$$L_k^2(M)$$

called the Hilbert space of square-integrable k-forms. An element of $L_k^2(M)$ is of the form

$$\mathbf{T} = \frac{1}{k!}\, T_{i_1 \ldots i_k}(x) \mathbf{d}x^{i_1} \wedge \ldots \wedge \mathbf{d}x^{i_k}$$

with measurable components $T_{i_1 \ldots i_k}(x)$ such that

$$\int \frac{1}{k!}\, T^{i_1 \ldots i_k}(x) T_{i_1 \ldots i_k}(x) \sqrt{g}\, d^n x$$

is convergent.

For a pseudo-Riemannian manifold things are slightly different. Here the scalar product $(\mathbf{T} \mid \mathbf{U})$ is indefinite, and therefore the Hilbert product is indefinite too. If we complete $\Lambda_0^k(M)$, we therefore get *a Hilbert space with indefinite metric*.

In the following, we shall always assume that all the differential forms we are considering have a well-defined Hilbert product. Observe first that up to a sign the dual map is a unitary operator; i.e., it preserves the inner product between two differential forms:

Theorem 8.7.

a. *The dual map $*$ is a unitary operator on a Riemannian manifold:* $\langle {}^*\mathbf{T} \mid {}^*\mathbf{U} \rangle = \langle \mathbf{T} \mid \mathbf{U} \rangle$.

b. *The dual map $*$ is an antiunitary operator on a manifold with Minkowski metric:* $\langle {}^*\mathbf{T} \mid {}^*\mathbf{U} \rangle = -\langle \mathbf{T} \mid \mathbf{U} \rangle$.

PROOF. This follows immediately from the corresponding local property (Theorem 7.9, Section 7.5):

$$\langle {}^*\mathbf{T} \mid {}^*\mathbf{U}\rangle = \int_M ({}^*\mathbf{T} \mid {}^*\mathbf{U})\varepsilon = \pm \int_M (\mathbf{T} \mid \mathbf{U})\varepsilon = \pm\langle \mathbf{T} \mid \mathbf{U}\rangle.$$

□

Then we finally arrive at a most important relationship between the exterior derivative $\boldsymbol{d}$ and the codifferential $\boldsymbol{\delta}$. Let $\mathbf{T}$ be a $(k-1)$-form, $\mathbf{U}$ a k-form, and consider the inner product

$$\langle \mathbf{U} \mid \boldsymbol{d}\mathbf{T}\rangle = \int_M {}^*\mathbf{U} \wedge \boldsymbol{d}\mathbf{T}.$$

This can be rearranged using a "partial integration":

$$\boldsymbol{d}({}^*\mathbf{U} \wedge \mathbf{T}) = \boldsymbol{d}{}^*\mathbf{U} \wedge \mathbf{T} + (-1)^{n-k*}\mathbf{U} \wedge \boldsymbol{d}\mathbf{T}.$$

From Exercise 7.6.1 we know that

$$\boldsymbol{d}{}^*\mathbf{U} = (-1)^{n-k+1*}\boldsymbol{\delta}\mathbf{U}.$$

Furthermore,

$$\int_M \boldsymbol{d}({}^*\mathbf{U} \wedge \mathbf{T}) = \int_{\partial M} {}^*\mathbf{U} \wedge \mathbf{T} = 0$$

either because M is compact without boundary or because ${}^*\mathbf{U} \wedge \mathbf{T}$ "vanishes sufficiently fast at infinity." Therefore, we obtain

$$\int {}^*\mathbf{U} \wedge \boldsymbol{d}\mathbf{T} = \int {}^*\boldsymbol{\delta}\mathbf{U} \wedge \mathbf{T}; \quad \text{i.e., } \langle \mathbf{U} \mid \boldsymbol{d}\mathbf{T}\rangle = \langle \boldsymbol{\delta}\mathbf{U} \mid \mathbf{T}\rangle. \tag{8.31}$$

Thus we see that due to our sign convention, $\boldsymbol{\delta}$ is simply the adjoint operator of $\boldsymbol{d}$. It should be emphasized that this holds both for Riemannian manifolds and manifolds with a Minkowski metric.

This has important consequences for the Laplacian operator. Let M be a Riemannian manifold. Using (8.31) we can rearrange the Laplacian as

$$-\boldsymbol{\Delta} = \boldsymbol{d\delta} + \boldsymbol{\delta d} = \boldsymbol{\delta}^\dagger\boldsymbol{\delta} + \boldsymbol{d}^\dagger\boldsymbol{d},$$

where $\boldsymbol{\delta}^\dagger$ denotes the adjoint operator. Consequently, $-\boldsymbol{\Delta}$ is a *positive Hermitian operator*. The above argument applies to arbitrary k-forms. In the special case of scalar fields, it is well known from elementary quantum mechanics, where the Hamiltonian $H = -\frac{\hbar^2}{2m}\,\boldsymbol{\Delta}$ is a positive Hermitian operator reflecting the positivity of energy!

We can also use (8.31) to deduce an important property of harmonic forms on a compact manifold.

Theorem 8.8. *Let M be a compact Riemannian manifold. A k-form $\mathbf{T}$ is harmonic if and only if it is primitively harmonic; i.e.,*

$$\boldsymbol{\Delta}\mathbf{T} = 0 \quad \textit{iff} \quad \boldsymbol{d}\mathbf{T} = \boldsymbol{\delta}\mathbf{T} = 0.$$

PROOF. Let us first observe that

$$\langle -\mathbf{\Delta T} \mid \mathbf{T}\rangle = \langle \boldsymbol{d}\delta\mathbf{T} \mid \mathbf{T}\rangle + \langle \delta\boldsymbol{d}\mathbf{T} \mid \mathbf{T}\rangle = \langle \delta\mathbf{T} \mid \delta\mathbf{T}\rangle + \langle \boldsymbol{d}\mathbf{T} \mid \boldsymbol{d}\mathbf{T}\rangle.$$

If $\mathbf{T}$ is harmonic, i.e., $\mathbf{\Delta T} = 0$, then the left-hand side vanishes automatically. But the right-hand side consists of two nonnegative terms! Hence they must both vanish:

$$\langle \delta\mathbf{T} \mid \delta\mathbf{T}\rangle = \langle \boldsymbol{d}\mathbf{T} \mid \boldsymbol{d}\mathbf{T}\rangle = 0.$$

But as the inner product is positive definite, this implies that

$$\delta\mathbf{T} = \boldsymbol{d}\mathbf{T} = 0.$$

□

Consider, for instance, a zero-form, i.e., a smooth function $\phi : M \curvearrowright \mathbb{R}$. Then $\delta\phi$ vanishes automatically, and $\boldsymbol{d}\phi$ vanishes if and only if it is constant. *Consequently, ϕ is harmonic if and only if it is constant.* This is a generalization of Liouville's theorem in complex analysis.

If M is not compact, the above consideration breaks down (unless $\mathbf{T}$ itself has compact support). Then we should be more careful! E.g., in potential theory we are interested in the electrostatic potential ϕ generating a static electric field $\mathbf{E}$. The equations of motion for ϕ reduce to the Laplace equation (cf. (1.19))

$$\mathbf{\Delta}\phi = 0$$

in space regions where there are no electrical sources. A typical problem starts with a space region, which we represent as a three-dimensional regular domain $\Omega \subseteq \mathbb{R}^3$. We have been given an electrostatic potential ϕ on the boundary of Ω, and we want to reconstruct ϕ in the interior of Ω. But observe that int Ω is not compact! Furthermore, since ϕ does not necessarily vanish on the boundary, we cannot neglect the boundary term. We will not go into any details, but the starting point is Green's identities, which the reader can have fun working out in the following exercise:

Exercise 8.5.1
Problem: Let Ω be an n-dimensional orientable domain in an n-dimensional Riemannian manifold Ω. Deduce Green's identities:

1.

$$\langle \boldsymbol{d}\phi \mid \boldsymbol{d}\psi\rangle + \langle \phi \mid \mathbf{\Delta}\psi\rangle = (-1)^{n-1}\int_{\partial\Omega} {}^*[\phi \boldsymbol{d}\psi]. \tag{8.32}$$

2.

$$\langle \phi \mid \mathbf{\Delta}\psi\rangle - \langle \mathbf{\Delta}\phi \mid \psi\rangle = (-1)^{n-1}\int_{\partial\Omega} {}^*[\phi \boldsymbol{d}\psi - \psi \boldsymbol{d}\phi]. \tag{8.33}$$

Use Green's identity to show the following property of harmonic functions: If $\phi = 0$ on $\partial\Omega$, then $\phi = 0$ in Ω.

We can also investigate singular solutions to the Laplace equation in $\mathbb{R}^3$. For instance, we know that $\phi(\mathbf{x}) = \frac{1}{r}$ is a harmonic function (generating the Coulomb field!). But it is not primitively harmonic, as it is not constant. Thus, it "violates" Theorem 8.7. This is no disaster, since it is only well-defined on $M = \mathbb{R}^3 \setminus \{\mathbf{0}\}$, which is *not* compact. Observe also that ϕ and $\boldsymbol{d}\phi$ are not square-integrable on M. The Hilbert product

$$\langle \mathbf{E} \mid \mathbf{E} \rangle = \langle \boldsymbol{d}\phi \mid \boldsymbol{d}\phi \rangle = \int_{\mathbb{R}^3 \setminus \{\mathbf{0}\}} \frac{1}{r^4}\, dx\, dy\, dz = 4\pi \int_0^\infty \frac{dr}{r^2} = \infty$$

diverges at the origin, reflecting the infinite self-energy of a point charge! If we try to repair the proof of Theorem 8.8 through a regularization procedure, i.e., put

$$M_\epsilon = \{\mathbf{x} \mid \|\mathbf{x}\| > \epsilon\},$$

then we should observe that the "partial integration" leading from $\langle \boldsymbol{\delta d}\phi \mid \phi \rangle$ to $\langle \boldsymbol{d}\phi \mid \boldsymbol{d}\phi \rangle$ is no longer valid, because the differential forms no longer vanish "sufficiently fast."

Worked Exercise 8.5.2
Problem: Let ϕ be a smooth scalar field on M_ϵ.

a. Show that

$$\langle \boldsymbol{\Delta}\phi \mid \phi \rangle = -\langle \boldsymbol{d}\phi \mid \boldsymbol{d}\phi \rangle + \int_{\partial M_\epsilon} \phi {}^*\boldsymbol{d}\phi.$$

b. Put $\phi(\mathbf{x}) = \frac{1}{r}$. Show by explicit computation that

$$\langle \boldsymbol{d}\phi \mid \boldsymbol{d}\phi \rangle = \int_{\partial M_\epsilon} \phi {}^*\boldsymbol{d}\phi = \frac{4\pi}{\epsilon}\ .$$

Moral: Whenever you work on a noncompact manifold, you should look for "boundary" contributions.

8.6 The Lagrangian Formalism and the Exterior Calculus

We will now investigate the *Lagrangian formalism* as an important example of how to apply the integral formalism.

In a field theory, the equations of motion are determined by the principle of least action. The action itself is constructed from a Lagrangian density L, which is a scalar field:

$$S = \int_\Omega L\varepsilon = \int_U L(x)\sqrt{-g}\, dx^0 dx^1 dx^2 dx^3.$$

This scalar field L is constructed from the fields and their derivatives. In the simplest case of a scalar field ϕ, it is thus a function of ϕ and $\partial_\mu\phi$. Using the covariant action, we can now construct the *covariant equations of motion*: They are obtained using the now familiar variational technique: We replace the scalar field ϕ by $\phi + \epsilon\psi$, where ψ vanishes on the boundary of Ω. This generates the new action

$$S(\epsilon) = \int_U L(\phi + \epsilon\psi, \partial_\mu\phi + \epsilon\partial_\mu\psi)\sqrt{-g}\, d^4x,$$

which is extremal when $\epsilon = 0$; i.e.,

$$0 = \frac{dS}{d\epsilon}\Big|_{\epsilon=0} = \int_U \left(\frac{\partial L}{\partial \phi}\, \psi + \frac{\partial L}{\partial(\partial_\mu \phi)}\, \partial_\mu \psi \right) \sqrt{-g}\, d^4x$$
$$= \int_U \left[\left(\sqrt{-g}\, \frac{\partial L}{\partial \phi} \right) \psi + \left(\sqrt{-g}\, \frac{\partial L}{\partial(\partial_\mu \phi)} \right) \partial_\mu \psi \right] d^4x.$$

Using partial integration on the last term, we get

$$0 = \int_U \left[\sqrt{-g}\, \frac{\partial L}{\partial \phi} - \partial_\mu \left(\sqrt{-g}\, \frac{\partial L}{\partial(\partial_\mu \phi)} \right) \right] \psi d^4x.$$

As ψ is arbitrary, this is only consistent if

$$\frac{\partial L}{\partial \phi} = \frac{1}{\sqrt{-g}}\, \partial_\mu \left(\sqrt{-g}\, \frac{\partial L}{\partial(\partial_\mu \phi)} \right). \tag{8.34}$$

If we have several fields ϕ_a, then each of the components has to satisfy the appropriate Euler equation! Thus we have shown:

Theorem 8.9. *The covariant equations of motion for a collection of scalar fields ϕ_a are given by*

$$\frac{\partial L}{\partial \phi_a} = \frac{1}{\sqrt{-g}}\, \partial_\mu \left(\sqrt{-g}\, \frac{\partial L}{\partial(\partial_\mu \phi_a)} \right). \tag{8.35}$$

Although it is much more dubious, the covariant equations of motions (8.35) are actually valid for vector fields too, when we interpret the index a as a space–time index!

Exercise 8.6.1
Problem: Consider the Lagrangian density for the electromagnetic field

$$L = -\frac{1}{4}\, F_{\mu\nu} F^{\mu\nu}.$$

Rederive the covariant Maxwell equation (7.85).

The geometrical point of view can in fact be used to throw light on one of the more subtle points in the derivative of the Euler equations.

To pass from

$$\int_U \left(\sqrt{-g}\, \frac{\partial L}{\partial(\partial_\mu \phi)} \right) \partial_\mu \psi d^4x$$

to

$$-\int_U \partial_\mu \left(\sqrt{-g}\, \frac{\partial L}{\partial(\partial_\mu \phi)} \right) \psi d^4x$$

we use a "partial integration," but we neglected the boundary terms, because "ψ vanishes at the boundary." This is justified in the following exercise:

Worked Exercise 8.6.2
Problem: (a) Show that $\frac{\partial L}{\partial(\partial_\mu \phi)}$ are the contravariant components of a vector field $\mathbf{A}$.
(b) Let ψ be a scalar field vanishing on the boundary of Ω. Show that

$$\int_\Omega \left(\sqrt{-g}A^\alpha\right)\partial_\alpha \psi d^4x = -\int_\Omega \partial_\alpha\left(\sqrt{-g}A^\alpha\right)\psi d^4x. \tag{8.36}$$

As we have seen in Section 6.7, we can also write down the equations of motion for a free particle in a covariant form. Consider now the case of an electrically charged particle moving in an electromagnetic field. Combining (1.26) with (6.46), we are led to the following covariant equations of motion:

$$m\frac{d^2x^\alpha}{d\tau^2} = qF^\alpha{}_\beta\frac{dx^\beta}{d\tau} - m\Gamma^\alpha{}_{\mu\nu}\frac{dx^\mu}{d\tau}\frac{dx^\nu}{d\tau}. \tag{8.37}$$

We would like to derive it from a Lagrangian. Now, we have previously determined the nonrelativistic action for an electrically charged particle interacting with the electromagnetic field (2.13):

$$S = \int_{t_1}^{t_2}\frac{1}{2}m\mathbf{v}^2dt - g\int_{t_1}^{t_2}[\phi - \mathbf{A}\cdot\mathbf{v}]dt.$$

This suggests that for a system consisting of an electrically charged particle and the electromagnetic field, we should use the following *relativistic* action $S = S_P + S_I + S_F$, with

$$\begin{cases} S_P = -m\displaystyle\int_{\lambda_1}^{\lambda_2}\sqrt{-g_{\alpha\beta}\frac{dx^\alpha}{d\lambda}\frac{dx^\beta}{d\lambda}}\,d\lambda, \\ S_I = q\displaystyle\int_{\lambda_1}^{\lambda_2}A_\mu\frac{dx^\mu}{d\lambda}\,d\lambda, \\ S_F = -\dfrac{1}{4}\displaystyle\int_\Omega F_{\mu\nu}F^{\mu\nu}\sqrt{-g}\,d^4x. \end{cases} \tag{8.38}$$

Here the dynamical variables to be varied consist of

a. The position of the particle: $x^\mu(\lambda)$;
b. The gauge potential: $A_\mu(x)$.

We leave it as an exercise to the reader to verify that the invariance of the action (8.38) under these variations actually leads to the desired covariant equations of motion!

Worked Exercise 8.6.3
Problem: (a) Show that the interaction term can be rearranged as

$$S_I = \int_\Omega A_\mu J^\mu\sqrt{-g}\,d^4x. \tag{8.39}$$

(Compare this with Exercise 3.10.2.)

(b) Show that the relativistic action (8.38) leads to the following equations of motion:

$$m\,\frac{d^2x^\alpha}{d\tau^2} = qF^\alpha_{\ \beta}\,\frac{dx^\beta}{d\tau} - m\Gamma^\alpha_{\ \mu\nu}\,\frac{dx^\mu}{d\tau}\,\frac{dx^\nu}{d\tau}\,. \tag{8.37}$$

$$\frac{1}{\sqrt{-g}}\,\partial_\nu\left(\sqrt{-g}F^{\mu\nu}\right) = J^\mu. \tag{7.85}$$

It is also interesting to see that the Lagrangian formalism can be thrown into a purely geometrical form. This means that we can discuss the equations of motion completely without introducing a coordinate system! Of course, the first step consists in reexpressing the action in a purely geometrical form. Let us look at some specific examples:

The Klein–Gordon field: This is a scalar field ϕ, where the action is based upon the Lagrangian density (3.49). The covariant action is then given by

$$S = \int_\Omega L\epsilon = \int_\Omega \left[-\frac{1}{2}\,(\partial_\mu\phi)(\partial^\mu\phi) - \frac{1}{2}\,m^2\phi^2\right]\sqrt{-g}\,d^4x,$$

which we rearrange as

$$S = -\frac{1}{2}\,\langle \boldsymbol{d}\phi \mid \boldsymbol{d}\phi\rangle - \frac{1}{2}\,m^2\langle\phi \mid \phi\rangle = \int_\Omega -\frac{1}{2}\,{}^*\boldsymbol{d}\phi\wedge\boldsymbol{d}\phi - \frac{1}{2}\,m^{2*}\phi\wedge\phi. \tag{8.39}$$

(Compare the discussion in Section 8.5.) This was the first step. Next we perform a variation,

$$\phi \curvearrowright \phi + \epsilon\psi,$$

where ψ is a scalar field that vanishes on the boundary of Ω. We then get

$$\begin{aligned} S(\epsilon) &= -\frac{1}{2}\,\langle \boldsymbol{d}\phi + \epsilon\boldsymbol{d}\psi \mid \boldsymbol{d}\phi + \epsilon\boldsymbol{d}\psi\rangle - \frac{1}{2}\,m^2\langle\phi+\epsilon\psi \mid \phi+\epsilon\psi\rangle \\ &= -\frac{1}{2}\,\langle \boldsymbol{d}\phi \mid \boldsymbol{d}\phi\rangle - \frac{1}{2}\,m^2\langle\phi\mid\phi\rangle + \epsilon\left[-\langle\boldsymbol{d}\phi \mid \boldsymbol{d}\psi\rangle - m^2\langle\phi\mid\psi\rangle\right] \\ &\quad + \frac{1}{2}\,\epsilon^2\left[-\langle \boldsymbol{d}\psi \mid \boldsymbol{d}\psi\rangle - m^2\langle\psi\mid\psi\rangle\right]. \end{aligned}$$

From this we immediately deduce

$$\begin{aligned} 0 = \frac{dS}{d\epsilon}\Big|_{\epsilon=0} &= -\langle\boldsymbol{d}\phi \mid \boldsymbol{d}\psi\rangle - m^2\langle\phi\mid\psi\rangle \\ &= -\langle\boldsymbol{\delta d}\phi \mid \psi\rangle - m^2\langle\phi\mid\psi\rangle = \langle -\boldsymbol{\delta d}\phi - m^2\phi \mid \psi\rangle. \end{aligned}$$

(Here we have used that ψ vanishes on the boundary to throw away the boundary term coming from the partial integration.) But this is consistent only if ϕ satisfies the equation

$$-\boldsymbol{\delta d}\phi = m^2\phi, \tag{8.40}$$

which is the geometrical form of the Klein–Gordon equation!

The Maxwell field: This is a one-form $\mathbf{A}$, and the action is based upon the Lagrangian density (3.50), which leads to a covariant action given by

$$S = -\frac{1}{4}\int_\Omega F_{\mu\nu}F^{\mu\nu}\sqrt{-g}\,d^4x; \qquad F_{\mu\nu} = \partial_\mu A_\nu - \partial_\nu A_\mu.$$

This is rearranged as

$$S = -\frac{1}{2}\langle \boldsymbol{dA} \mid \boldsymbol{dA}\rangle = -\frac{1}{2}\int_\Omega {}^*\boldsymbol{dA}\wedge\boldsymbol{dA}. \tag{8.41}$$

Then we perform a variation

$$\mathbf{A} \curvearrowright \mathbf{A} + \epsilon\mathbf{U},$$

where $\mathbf{U}$ is a one-form that vanishes on the boundary of Ω:

$$\begin{aligned} S(\epsilon) &= -\frac{1}{2}\langle \boldsymbol{dA} + \epsilon\boldsymbol{dU} \mid \boldsymbol{dA} + \epsilon\boldsymbol{dU}\rangle \\ &= -\frac{1}{2}\langle \boldsymbol{dA} \mid \boldsymbol{dA}\rangle - \epsilon\langle \boldsymbol{dA} \mid \boldsymbol{dU}\rangle - \frac{\epsilon^2}{2}\langle \boldsymbol{dU} \mid \boldsymbol{dU}\rangle. \end{aligned}$$

From this we immediately get

$$0 = \frac{dS}{d\epsilon}\Big|_{\epsilon=0} = -\langle \boldsymbol{dA} \mid \boldsymbol{dU}\rangle = -\langle \boldsymbol{\delta dA} \mid \mathbf{U}\rangle.$$

(Here we have used that $\mathbf{U}$ vanishes on the boundary of Ω to throw away the boundary term coming from the partial integration.)

But this is consistent only if

$$-\boldsymbol{\delta dA} = 0; \qquad \text{i.e.,} \ -\boldsymbol{\delta}\mathbf{F} = 0, \tag{8.42}$$

which is nothing but the Maxwell equations!

Exercise 8.6.4

Problem: (a) Consider the massive vector field $\mathbf{A}$. Show that the action (3.51) is given by

$$S = -\frac{1}{2}\langle \mathbf{F} \mid \mathbf{F}\rangle - \frac{1}{2}m^2\langle \mathbf{A} \mid \mathbf{A}\rangle; \qquad \mathbf{F} = \boldsymbol{dA}. \tag{8.43}$$

(b) Perform the variation $\mathbf{A} \curvearrowright \mathbf{A} + \epsilon\mathbf{U}$ and deduce the following equations of motion:

$$-\boldsymbol{\delta dA} = m^2\mathbf{A}. \tag{8.44}$$

(c) Show that they are equivalent to

$$\Box\mathbf{A} = m^2\mathbf{A}; \qquad -\boldsymbol{\delta}\mathbf{A} = 0. \tag{8.45}$$

We may summarize the preceding discussion Table 8.1.

8.7 Integral Calculus and Electromagnetism

As another example of how to apply the integral calculus, we will use it to reexpress in a geometrical form various electromagnetic quantities like the electric and mag-

Table 8.1.

Field	Action	Equation of motion	
Klein–Gordon	$S = \int_\Omega -\frac{1}{2}(^*\boldsymbol{d}\phi \wedge \boldsymbol{d}\phi) - \frac{1}{2}m^2(^*\phi \wedge \phi)$	$-\delta \boldsymbol{d}\phi = m^2\phi$	(8.39)
ϕ			(8.40)
Maxwell	$S = -\int_\Omega \frac{1}{2}(^*\boldsymbol{d}\mathbf{A}) \wedge \boldsymbol{d}\mathbf{A}$	$-\delta \boldsymbol{d}\mathbf{A} = 0$	(8.41)
$\mathbf{A}$			(8.42)
Massive vector	$S = \int_\Omega -\frac{1}{2}(^*\boldsymbol{d}\mathbf{A} \wedge \boldsymbol{d}\mathbf{A}) - \frac{1}{2}m^2(^*\mathbf{A} \wedge \mathbf{A})$	$-\delta \boldsymbol{d}\mathbf{A} = m^2\mathbf{A}$	(8.43)
$\mathbf{A}$			(8.44)

netic flux through a surface and the electric charge contained in a 3-dimensional regular domain.

We start out peacefully in 3-space to get some feeling for the new formalism. Remember that $\mathbf{E}$ is a 1-form, but $\mathbf{B}$ is a 2-form. Essentially, $\mathbf{B}$ is the dual of the conventional magnetic field. Now let Ω be a 3-dimensional regular domain. From the discussion of flux integrals (8.24) we get

$$\int_{\partial\Omega} \mathbf{B} = \text{The magnetic flux through the closed surface } \partial\Omega. \tag{8.46}$$

$$\int_{\partial\Omega} {}^*\mathbf{E} = \text{The electric flux through the closed surface } \partial\Omega. \tag{8.47}$$

$$\int_{\Omega} {}^*\rho = \text{The electric charge contained in } \Omega. \tag{8.48}$$

We can now use the integral calculus to deduce some well-known elementary properties:

EXAMPLE 8.1. *If Ω contains no singularities, then the magnetic flux Φ through the closed surface $\partial\Omega$ vanishes.*

This follows from an application of Stokes's theorem:

$$\Phi = \int_{\partial\Omega} \mathbf{B} = \int_{\Omega} \boldsymbol{d}\mathbf{B} = 0$$

(due to (7.76)).

EXAMPLE 8.2 (Gauss's theorem). *The electric flux through the closed surface $\partial\Omega$ is $\frac{1}{\epsilon_0}$ times the electric charge contained in Ω.*

This follows from an application of Theorem 8.5 (Corollary to Stokes's theorem). From the Maxwell equation (7.78) we get

$$\frac{1}{\epsilon_0} \times [\text{the charge}] = \frac{1}{\epsilon_0}\int_{\Omega} {}^*\rho = -\int_{\Omega} {}^*\delta\mathbf{E} = \int_{\partial\Omega} {}^*\mathbf{E}.$$

EXAMPLE 8.3. This time we consider a static situation. Let S be a surface with boundary Γ (Figure 8.23). Then *the flux of current through S is equal to $\epsilon_0 c^2$ times the circulation of the magnetic field along* Γ.

Observe first that the Maxwell equation (7.79) reduces to

$$\boldsymbol{\delta}\mathbf{B} = \frac{1}{\epsilon_0 c^2}\mathbf{J}$$

for a static configuration. The proposition then follows from an application of Theorem 8.5 (Corollary to Stokes's theorem):

$$\int_S {}^*\mathbf{J} = \epsilon_0 c^2 \int_S {}^*\boldsymbol{\delta}\mathbf{B} = \epsilon_0 c^2 \int_{\partial S=\Gamma} {}^*\mathbf{B}.$$

We conclude the discussion of electromagnetism in 3-space with the explicit computation of two important integrals:

(a) Consider the spherically symmetric monopole field.

Let S^2 be the closed surface of a sphere with radius r (Figure 8.24). Then polar coordinates (r, θ, φ) are adapted to the sphere, and we can choose (θ, φ) to parametrize it! We can now compute the magnetic flux through the closed surface S. First, we observe that the monopole field $\mathbf{B}$ is given by (cf. (7.89))

$$\mathbf{B} = \frac{g}{4\pi}\sin\theta\, \boldsymbol{d}\theta \wedge \boldsymbol{d}\varphi.$$

Then we immediately get

$$\int_{S^2} \mathbf{B} = \frac{g}{4\pi}\int_{\varphi=0}^{2\pi}\int_{\theta=0}^{\pi} \sin\theta\, d\theta\, d\varphi = g. \tag{8.49}$$

(b) This time we consider the magnetic field around a wire with current $\mathbf{J}$. Let Γ be a circle around the wire with radius ρ (Figure 8.25). Then the cylinder coordinates (ρ, φ, z) are adapted to the circle, and we can choose φ to parametrize the circle. We can now evaluate the line integral of the magnetic field along the closed

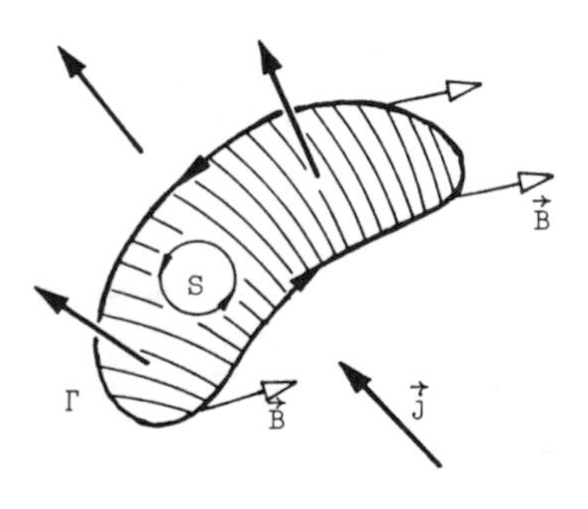

Figure 8.23.

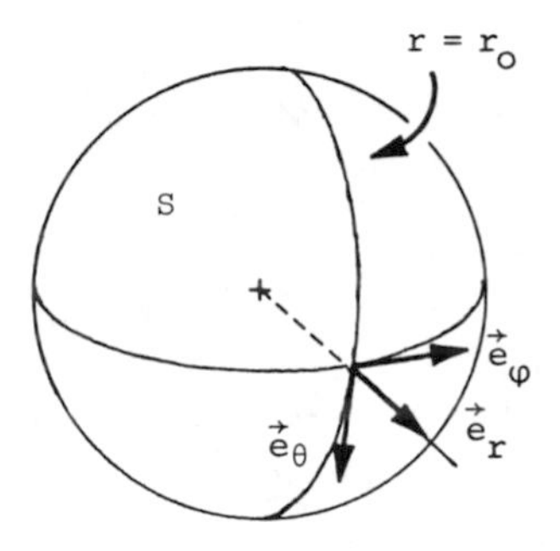

Figure 8.24.

curve Γ. For this purpose we must find the dual form ${}^*\mathbf{B}$, which is the one-form representing the magnetic field. This has been done previously (cf. Section 7.8):

$$ {}^*\mathbf{B} = \frac{j}{2\pi\epsilon_0 c^2}\, d\varphi. $$

Thus we get

$$ \int_\Gamma {}^*\mathbf{B} = \frac{j}{2\pi\epsilon_0 c^2}\int_0^{2\pi} d\varphi = \frac{j}{\epsilon_0 c^2}\,. $$

Then we proceed to consider electromagnetism in Minkowski space, i.e., 4-dimensional space–time. Here it is more complicated to express suitable quantities, so we shall adopt the following terminology:

Suppose we have chosen an inertial frame S. Let (x^0, x^1, x^2, x^3) denote the corresponding inertial coordinates. We say that the three-dimensional submanifold

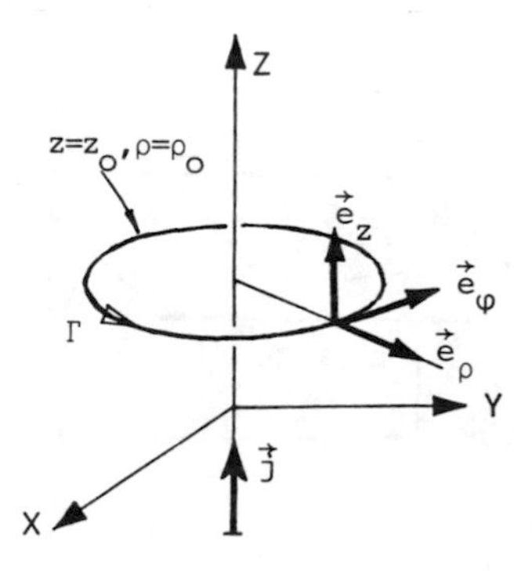

Figure 8.25.

ξ is a *space slice* if it is a subset of the form

$$\xi = \{x \in M \mid x^0 = t^0\};$$

i.e., ξ consists of all the *spatial points* at a specific time t^0. Observe that the spatial coordinates (x^1, x^2, x^3) are adapted to ξ.

The fundamental quantities describing the properties of the electromagnetic field are the field strengths **F** and ***F**, the Maxwell field **A**, and the current **J**. Now let Ω be a 3-dimensional regular domain contained in a space slice relative to the inertial frame S. Then we can form the following integrals that we want to interpret:

$$\int_{\partial\Omega} \mathbf{F}, \quad \int_{\partial\Omega} {}^*\mathbf{F}, \quad \text{and} \quad \int_{\Omega} {}^*\mathbf{J}.$$

Observe first that the restriction of $\boldsymbol{dx}^0$ to Ω vanishes. Consequently, the integrals involve only the space components of the integrands. But **F** and ***F** are decomposed as

$$\mathbf{F} = \begin{bmatrix} 0 & -\mathbf{E} \\ \mathbf{E} & \mathbf{B} \end{bmatrix} \quad \text{and} \quad {}^*\mathbf{F} = \begin{bmatrix} 0 & {}^*\mathbf{B} \\ -{}^*\mathbf{B} & {}^*\mathbf{E} \end{bmatrix}. \tag{8.50}$$

So the restriction of **F** (respectively ***F**) to $\partial\Omega$ is given by **B** (respectively ***E**). Similarly, the dual current has the following restriction to int Ω:

$${}^*\mathbf{J}_{|\Omega} = ({}^*\mathbf{J})_{123}\boldsymbol{dx}^1 \wedge \boldsymbol{dx}^2 \wedge \boldsymbol{dx}^3 = -J^0\boldsymbol{dx}^1 \wedge \boldsymbol{dx}^2 \wedge \boldsymbol{dx}^3.$$

We can now generalize the results obtained in (8.46)–(8.48) to the following:

Lemma 8.1. *Let Ω be a 3-dimensional regular domain contained in a space slice. Then*

$$\int_{\partial\Omega} \mathbf{F} = \textit{The magnetic flux through } \partial\Omega. \tag{8.51}$$

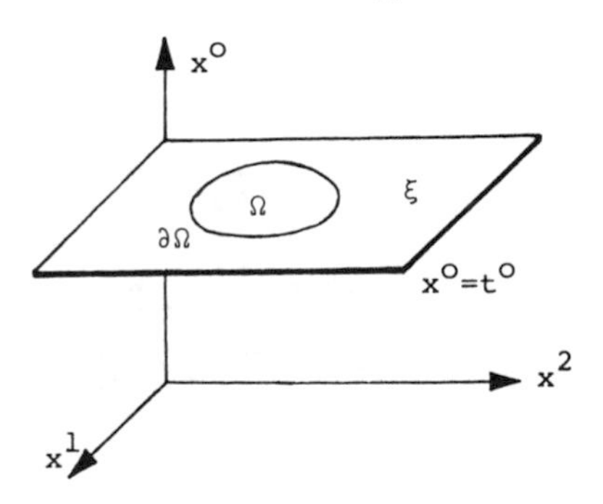

Figure 8.26.

$$\int_{\partial\Omega} {}^*\mathbf{F} = \textit{The electric flux through } \partial\Omega. \tag{8.52}$$

$$-\int_{\Omega} {}^*\mathbf{J} = \textit{The electric charge contained in } \Omega. \tag{8.53}$$

Exercise 8.7.1
Introduction: Let Ω be a 3-dimensional regular domain obtained in a space slice and let $\mathbf{F}$ be smooth throughout Ω. Problem: Use the 4-dimensional integral formalism to reexamine the following well-known results:

a. The magnetic flux through $\partial\Omega$ is zero.
b. The electric flux through $\partial\Omega$ is equal to the electric charge contained in Ω. (As usual, we have put $\epsilon_0 = c = 1$.)

ILLUSTRATIVE EXAMPLE: MAGNETIC STRINGS IN A SUPERCONDUCTOR.
We have earlier been discussing some of the features of superconductivity, especially flux quantization (see Section 2.12). Recall that in the superconducting state of a metal, the electrons will generate Cooper pairs. These Cooper pairs act as bosons, and we can therefore characterize the superconducting state by a macroscopic wave function ψ called the *order parameter* of the superconducting metal. The square of the order parameter, $|\psi|^2$, represents the density of the Cooper pairs.

We want now to study equilibrium states in a superconductor. Consider a static configuration $\psi(\mathbf{x})$, where $\psi(\mathbf{x})$ is a slowly varying spatial function. In the *Ginzburg–Landau theory* one assumes that the static energy density is given in the form

$$\begin{aligned} H &= \frac{1}{2}\,(\boldsymbol{d}\psi \mid \boldsymbol{d}\psi) + \gamma + \frac{1}{2}\,\alpha|\psi|^2 + \frac{\beta}{4}\,|\psi|^4 \\ &\left(= \frac{1}{2}\,|\nabla\psi|^2 + \gamma + \frac{1}{2}\,\alpha|\psi|^2 + \frac{\beta}{4}\,|\psi|^4\right). \end{aligned} \tag{8.54}$$

Here α is a temperature-dependent constant

$$\alpha = \alpha(T) = a\,\frac{T - T_c}{T_c} \qquad \text{(with } a \text{ positive)}, \tag{8.55}$$

where T_c is the so-called critical temperature. The constant γ is just inserted to normalize H to be zero at its global minima. The equilibrium states are found by minimizing the static energy. This leads to the Ginzburg–Landau equation

$$\Delta\psi = (\beta|\psi|^2 + \alpha)\psi. \tag{8.56}$$

Now consider the potential

$$U(\psi) = \gamma + \frac{1}{2}\,\alpha|\psi|^2 + \frac{\beta}{4}\,|\psi|^4. \tag{8.57}$$

It has the well-known shape shown in Figure 8.27. Above T_c the vacuum configuration is given by

$$\psi = 0;$$

i.e., the metal is expected to be found in its normal state where the Cooper pairs are absent. Below T_c the configuration $\psi = 0$ becomes unstable, and we get a degenerate vacuum, the *superconducting vacuum*, with a temperature-dependent density of Cooper pairs given by

$$|\psi|^2 = -\frac{\alpha}{\beta} = \frac{a}{\beta} \cdot \frac{T_c - T}{T_c} .$$

All this is in good accordance with experiment!

The coherence length: Consider a specimen, a semi-infinite slab, bounded by the (y, z)-plane. For negative x we are in the normal region, where $\psi = 0$, and for positive x we are in the superconducting region (Figure 8.28). So for sufficiently large x we expect the order parameter to be in its vacuum state

$$|\psi| = \sqrt{\frac{-\alpha}{\beta}} .$$

The coherence length $\xi(T)$ is the characteristic length it takes the order parameter to rise from its normal vacuum $\psi = 0$ at the boundary to its superconducting vacuum $|\psi| = \sqrt{\frac{-\alpha}{\beta}}$ inside the slab. The problem is essentially one-dimensional,

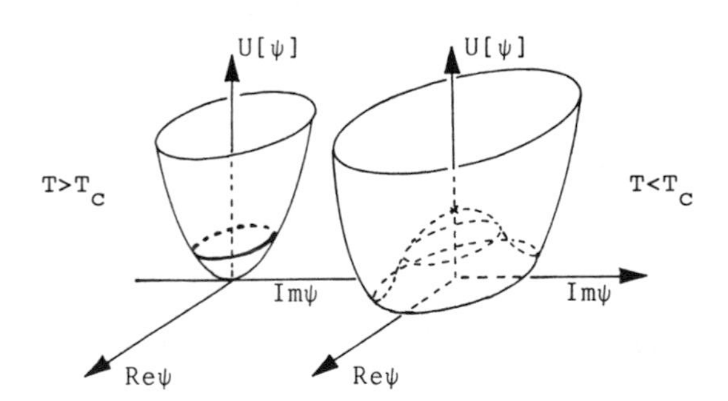

Figure 8.27.

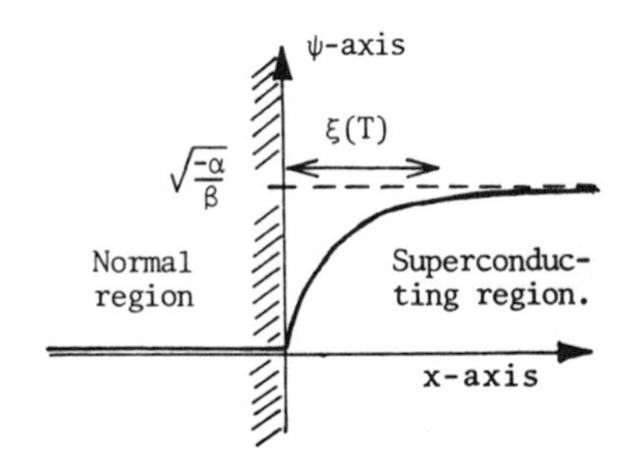

Figure 8.28.

so the Ginzburg–Landau equation (8.56) reduces to

$$\frac{d^2\psi}{dx^2} = \alpha\psi + \beta\psi^3,$$

with the boundary conditions

$$\psi(0) = 0; \qquad \psi(\infty) = \sqrt{\frac{-\alpha}{\beta}}\,.$$

The solution to this problem is simply a "half-kink"; cf. (4.23):

$$\psi(x) = \begin{cases} \sqrt{\frac{-\alpha}{\beta}}\tanh\left\{x\sqrt{\frac{-\alpha}{2}}\right\} & \text{if } x > 0, \\ 0 & \text{if } x < 0. \end{cases}$$

Consequently, the coherence length is given by

$$\xi(T) = \frac{1}{\sqrt{-\alpha}}\,. \tag{8.58}$$

Then we consider what happens when we add a magnetic field $\mathbf{B}$. The order parameter will now couple minimally to the magnetic vector potential; i.e., we must exchange the exterior derivative $\boldsymbol{d}$ with the gauge covariant exterior derivative

$$\boldsymbol{D} = \boldsymbol{d} - i\,\frac{q}{\hbar}\,\mathbf{A}$$

(where $\mathbf{A}$ is the magnetic vector potential).

Exercise 8.7.2
Problem: Show that

$$\boldsymbol{D}^2\psi = -i\,\frac{q}{\hbar}\,\mathbf{B}\psi. \tag{8.59}$$

Notice that it follows from Exercise 8.7.2 that the square of the gauge covariant exterior derivative does *not* vanish in general. When including a magnetic field, the static energy density (8.54) is modified to

$$H = \frac{1}{2}\,(\mathbf{B} \mid \mathbf{B}) + \frac{1}{2}\,(\boldsymbol{D}\psi \mid \boldsymbol{D}\psi) + \gamma + \frac{1}{2}\,\alpha|\psi|^2 + \frac{\beta}{4}\,|\psi|^4. \tag{8.60}$$

Exercise 8.7.3
Problem: Show that the equations of equilibrium configurations are given by

$$\boldsymbol{D}^*\boldsymbol{D}\psi = (\beta|\psi|^2 + \alpha)\psi\varepsilon; \qquad \text{i.e., } \frac{1}{\sqrt{g}}\,D_i\left(\sqrt{g}\,D^i\psi\right) = (\beta|\psi|^2 + \alpha)\psi; \tag{8.61a}$$

$$\begin{aligned} -\delta\mathbf{B} &= \frac{iq}{2\hbar}\,\left[\bar{\psi}D\psi - \psi\overline{D\psi}\right]; \\ \text{i.e., } \frac{1}{\sqrt{g}}\,\partial_i\left(\sqrt{g}B^{ij}\right) &= \frac{iq}{2\hbar}\,\left[\bar{\psi}\boldsymbol{D}^i\psi - \psi\overline{\boldsymbol{D}^i\psi}\right]. \end{aligned} \tag{8.61b}$$

Let us now assume that the order parameter is in its ground state (characterized by a vanishing energy density); i.e.,

$$\boldsymbol{D}\psi = 0; \qquad U(\psi) = 0. \tag{8.62}$$

Using Exercise 8.7.2, we then immediately get

$$0 = \boldsymbol{D}^2\psi = -i\ \frac{q}{\hbar}\ \mathbf{B}\psi; \tag{8.63}$$

i.e., either **B** vanishes or ψ vanishes! In the normal state $\psi = 0$, so here the magnetic field can propagate freely, but in the superconducting vacuum $|\psi| = \sqrt{\frac{-\alpha}{\beta}}$, and **B** has to vanish. Consequently, the magnetic field is expelled from any region where the order parameter is in the superconducting vacuum! This, of course, is the famous *Meissner effect.*

Observe, too, that if the order parameter is in its vacuum state, then the supercurrent, given by

$$\mathbf{J} = \frac{iq}{2\hbar}\ \left[\bar{\psi}\boldsymbol{D}\psi - \psi\overline{\boldsymbol{D}\psi}\right], \tag{8.64}$$

automatically vanishes.

The penetration length:

We can now introduce the second characteristic length in the superconductivity. Again we consider a semi-infinite slab bounded by the (y, z)-plane. We now apply an external magnetic field **B**, and we want to find out how far into the superconducting region the magnetic field penetrates. (See Figure 8.29.) This time we use the Ginzburg-Landau equation for the magnetic field:

$$-\delta\mathbf{B} = \frac{iq}{2\hbar}\ \left[\bar{\psi}\boldsymbol{D}\psi - \psi\overline{\boldsymbol{D}\psi}\right] = \frac{iq}{2\hbar}\ \left[\bar{\psi}\boldsymbol{d}\psi - \psi\overline{\boldsymbol{d}\psi}\right] + \frac{q^2}{\hbar^2}\ |\psi|^2\mathbf{A}.$$

In this equation we can approximate $|\psi|$ with its equilibrium value $\sqrt{\frac{-\alpha}{\beta}}$. This does not mean that the order parameter is in its superconducting vacuum, since the gauge covariant derivative $\boldsymbol{D}\psi$ need not vanish. In particular, we can still have superconducting currents floating around. Taking the exterior derivative on both sides, we now get

$$-\boldsymbol{d}\delta\mathbf{B} = -\frac{\alpha q^2}{\beta\hbar^2}\ \mathbf{B}.$$

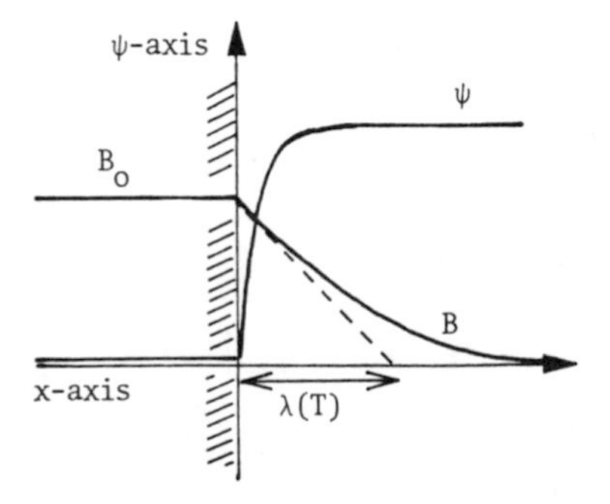

Figure 8.29.

Here we have used that ψ is proportional to $\exp\{i\varphi\}$, so that $\bar{\psi}\boldsymbol{d}\psi$ is proportional to $\boldsymbol{d}\varphi$; i.e., it is a closed form.

Since furthermore, $\boldsymbol{d}\mathbf{B}$ vanishes on account of Maxwell's equations, we arrive at the Laplace equation (cf. (2.77))

$$\boldsymbol{\Delta}\mathbf{B} = -\frac{\alpha q^2}{\beta \hbar^2}\,\mathbf{B}. \tag{8.65}$$

Again, the problem is essentially one-dimensional, and the Laplace equation therefore reduces to

$$\frac{d^2 B^{ij}}{dx^2}(x) = -\frac{\alpha q^2}{\beta \hbar^2}\,B^{ij}(x).$$

Since $\mathbf{B}$ is a closed form, we furthermore have

$$\frac{dB^{23}}{dx} = 0.$$

This leads to the following solution:

$$B_{23}(x) = 0; \qquad B_{31}(x) = B_{31}(0)\exp\left[-\sqrt{\frac{-\alpha}{\beta}}\,\frac{q}{\hbar}\,x\right];$$

$$B_{12}(x) = B_{12}(0)\exp\left[-\sqrt{\frac{-\alpha}{\beta}}\,\frac{q}{\hbar}\,x\right].$$

Consequently, the penetration length $\lambda(T)$ is given by (cf. (2.78))

$$\lambda(T) = \sqrt{\frac{-\beta}{\alpha}}\,\frac{\hbar}{q}\,. \tag{8.66}$$

(Notice that the above considerations are, strictly speaking, valid only when $\xi(T) \ll \lambda(T)$.)

We have now introduced two characteristic lengths, the coherence length $\xi(T)$ and the penetration length $\lambda(T)$, both of which depend on α and hence on the temperature. We can now form a third parameter, their ratio

$$\kappa = \frac{\lambda(T)}{\xi(T)} = \frac{\hbar\sqrt{\beta}}{q}\,, \tag{8.67}$$

which is temperature independent and is called the *Ginzburg–Landau parameter*.

Recall that there are two different types of superconductors, type I and type II. If we apply an external magnetic field, we know that the superconductivity is destroyed for sufficiently strong magnetic fields. In a type I superconductor there will be a critical field strength, B_c, above which superconductivity breaks down and the magnetic field is uniformly distributed throughout the metal. In a type II superconductor there will be two critical field strengths, B_{c_1} and B_{c_2}. When we pass B_{c_1}, superconductivity will not be completely destroyed, but rather the magnetic field will penetrate into the metal in the form of thin magnetic strings, *vortices*. Only when we pass B_{c_2} will the superconducting regions break down completely.

We want now to investigate the structure of a magnetic string in a type II superconductor. We will assume that $\xi(T) \ll \lambda(T)$. Thus a single string consists of a hard core of radius $\xi(T)$ where the density of Cooper pairs vanishes. (See Figure 8.30.) Inside this hard core we have therefore re-established the normal vacuum. Outside the hard core, the magnetic field falls off, and it vanishes essentially in a typical distance $\lambda(T)$. Thus we have also a soft core: $\xi < \rho < \lambda$. Observe that we have circulating supercurrents in the soft core that prevent the magnetic field from being spread out (the Meissner effect). Finally, we reach the superconducting vacuum outside the soft core, where neither the magnetic field nor the order parameter contributes to the energy.

Remark. Consider a magnetic string with flux Φ_0 and let us for simplicity assume that we have a constant magnetic field inside the string. Then

$$\Phi_0 = B_0 \pi \rho^2,$$

where ρ is the radius of the string. Thus the magnetic energy stored per unit length is given by

$$J = \frac{1}{2} B_0^2 \pi \rho^2 = \frac{\Phi_0^2}{2\pi \rho^2} . \tag{8.68}$$

Thus we can diminish the energy of a magnetic string with a constant flux by spreading out the string. If nothing prevents it, a magnetic string will thus grow fat. In the case of a solenoid, it is the mechanical wire, in which an electric current flows, that prevents the magnetic field from spreading out. In a superconductor it is the Meissner effect that prevents the magnetic field from spreading out.

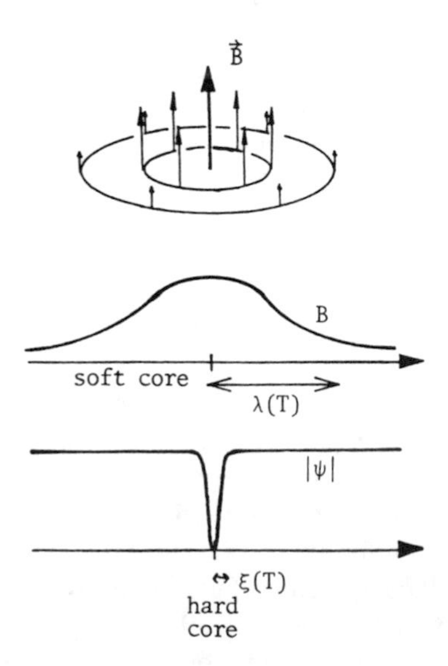

Figure 8.30.

Now let us try to compute the energy per unit length in a single vortex line. We shall neglect the hard core, which by assumption is very thin. Thus we concentrate exclusively on the soft core. Let us denote the region outside the hard core by Ω. We start by rearranging the expression for the static energy:

$$H_\Omega = \frac{1}{2}\langle \mathbf{B} \mid \mathbf{B}\rangle_\Omega + \frac{1}{2}\langle \boldsymbol{D}\psi \mid \boldsymbol{D}\psi\rangle_\Omega + \frac{\beta}{4}\int_\Omega \left(|\psi|^2 + \frac{\alpha}{\beta}\right)^2 \epsilon.$$

But $\psi = \sqrt{\frac{-\alpha}{\beta}}\exp[i\varphi]$ outside the hard core, so we can neglect the potential energy. Furthermore, the supercurrent reduces to

$$\mathbf{J} = -\frac{\alpha q}{\beta\hbar}\left[d\varphi - \frac{q}{\hbar}\mathbf{A}\right] \qquad (= \delta\mathbf{B}),$$

and the gauge covariant exterior derivative of the order parameter similarly reduces to

$$\boldsymbol{D}\psi = \sqrt{\frac{-\alpha}{\beta}}\, i\exp\{i\varphi\}\left[d\varphi - \frac{q}{\hbar}\mathbf{A}\right].$$

We therefore get, by direct inspection, that

$$\frac{1}{2}\langle \boldsymbol{D}\psi \mid \boldsymbol{D}\psi\rangle_\Omega = \frac{1}{2}\lambda^2(T)\langle \mathbf{J} \mid \mathbf{J}\rangle_\Omega = \frac{1}{2}\lambda^2(T)\langle \delta\mathbf{B} \mid \delta\mathbf{B}\rangle_\Omega.$$

As in the discussion of the penetration length, we also have

$$-\boldsymbol{d}\delta\mathbf{B} = \frac{1}{\lambda^2(T)}\mathbf{B}.$$

Finally, the expression for the static energy therefore reduces to

$$\begin{aligned} H &= \frac{1}{2}\lambda^2(T)\left[\langle \mathbf{B} \mid -\boldsymbol{d}\delta\mathbf{B}\rangle_\Omega + \langle \delta\mathbf{B} \mid \delta\mathbf{B}\rangle_\Omega\right] \\ &= \frac{1}{2}\lambda^2(T)\left[\int_\Omega -{}^*\mathbf{B}\wedge \boldsymbol{d}\delta\mathbf{B} + {}^*\delta\mathbf{B}\wedge\delta\mathbf{B}\right] \\ &= \frac{1}{2}\lambda^2(T)\int_\Omega \boldsymbol{d}[{}^*\mathbf{B}\wedge\delta\mathbf{B}]. \end{aligned}$$

Using Stokes's theorem, this is rearranged as

$$H = \frac{1}{2}\lambda^2(T)\int_{\partial\Omega} {}^*\mathbf{B}\wedge\delta\mathbf{B}. \tag{8.69}$$

Thus all we have got to know is the magnetic field and its derivative on the boundary of the hard core!

To proceed, we must determine the field configuration more accurately; i.e., we must in principle solve the Ginzburg–Landau equations (8.61a–b). It turns out to be impossible to write down an explicit solution representing a magnetic string, so we shall be content with an approximative solution. We shall concentrate on a cylindrical symmetrical string.

Notice that the flux is necessarily quantized (2.80):

$$\Phi = n\Phi_0,$$

where n is related to the jump in the phase of the order parameter when we go once around the string. For a string with n flux quanta we therefore assume that ψ has the following form

$$\psi(\mathbf{r}) = \sqrt{\frac{-\alpha}{\beta}}\,\phi(\rho)\exp[in\varphi], \qquad \text{with} \begin{cases} \lim_{\rho\to 0}\phi(\rho) = 0, \\ \lim_{\rho\to\infty}\phi(\rho) = 1. \end{cases} \tag{8.70a}$$

(Notice that the boundary condition at $\rho = 0$, which is in accordance with our general description of a magnetic string as shown in Figure 8.30, actually removes the singularity of $\exp[in\varphi]$ at the origin.) The exterior derivative of the order parameter is given by

$$\boldsymbol{d}\psi = \psi\left\{\frac{\phi'(\rho)}{\phi(\rho)}\,\boldsymbol{d}\rho + in\boldsymbol{d}\varphi\right\}.$$

Outside the soft core, it is therefore approximately given by

$$\boldsymbol{d}\psi \approx \psi\{in\boldsymbol{d}\varphi\}.$$

Since the gauge covariant exterior derivative of the order parameter must vanish outside the soft core, the magnetic vector potential is asymptotically given by

$$\mathbf{A} \approx n\,\frac{\hbar}{q}\,\boldsymbol{d}\varphi.$$

This suggests the following assumption for the structure of the magnetic vector potential

$$\mathbf{A} = n\,\frac{\hbar}{q}\,A(\rho)\boldsymbol{d}\varphi, \qquad \text{with} \begin{cases} \lim_{\rho\to 0}A(\rho) = 0, \\ \lim_{\rho\to\infty}A(\rho) = 1. \end{cases} \tag{8.70b}$$

(The boundary condition at $\rho = 0$ removes the singularity of $\boldsymbol{d}\varphi$ at the origin.)

We proceed to determine the static energy density for this type of configuration. Notice that

$$(\boldsymbol{d}\rho \mid \boldsymbol{d}\rho) = 1; \qquad (\boldsymbol{d}\rho \mid \boldsymbol{d}\varphi) = 0;$$

$$(\boldsymbol{d}\varphi \mid \boldsymbol{d}\varphi) = \frac{1}{\rho^2}; \qquad (\boldsymbol{d}\rho\wedge\boldsymbol{d}\varphi \mid \boldsymbol{d}\rho\wedge\boldsymbol{d}\varphi) = \frac{1}{\rho^2}$$

(cf. Exercises 6.6.3 and 7.5.7). Using that

$$\boldsymbol{D}\psi = \psi\left\{\frac{\phi'(\rho)}{\phi(\rho)}\,\boldsymbol{d}\rho + in[1 - A(\rho)]\boldsymbol{d}\varphi\right\}, \tag{8.71a}$$

$$\mathbf{B} = \frac{n\hbar}{q}\,A'(\rho)\boldsymbol{d}\rho\wedge\boldsymbol{d}\varphi, \tag{8.71b}$$

the expression (8.60) for the static energy reduces to

$$H = \frac{1}{2}\, n^2 \frac{\hbar^2}{q^2} \cdot \left(\frac{A'}{\rho}\right)^2 - \frac{1}{2}\frac{\alpha}{\beta} \cdot (\phi')^2 - \frac{1}{2}\frac{\alpha}{\beta}\, n^2\, \frac{(1-A)^2}{\rho^2}\, \phi^2 + \frac{\alpha^2}{4\beta}\, [\phi^2 - 1]^2.$$

(Notice that α is negative!) As expected, this depends only upon ρ, so that it is manifestly cylindrically symmetric. It follows that the static energy per unit length is given by

$$\begin{aligned} J[\phi, A] &= 2\pi \int_0^\infty J[\phi, A] d\rho \\ &= 2\pi \int_0^\infty \left\{ \frac{1}{2}\, n^2 \frac{\hbar^2}{q^2} \cdot \frac{(A')^2}{\rho} - \frac{1}{2}\frac{\alpha}{\beta}\, \rho \cdot (\phi')^2 \right. \\ &\quad \left. - \frac{1}{2}\, n^2 \frac{\alpha}{\beta}\, \frac{(1-A)^2}{\rho}\, \phi^2 + \frac{\alpha^2}{4\beta}\, \rho \cdot [\phi^2 - 1]^2 \right\} d\rho. \end{aligned} \tag{8.72}$$

Next we want to find the equations of motion for the unspecified functions $\phi(\rho)$ and $A(\rho)$. The safest but most complicated method is to plug the equation (8.70a–b) into the Ginzburg–Landau equations (8.61a–b). Notice first that

$$^*\boldsymbol{d}\rho = \rho \boldsymbol{d}\varphi \wedge \boldsymbol{d}z; \quad ^*\boldsymbol{d}\varphi = \frac{1}{\rho}\, \boldsymbol{d}z \wedge \boldsymbol{d}\rho; \quad ^*(\boldsymbol{d}\rho \wedge \boldsymbol{d}\varphi) = \frac{1}{\rho}\, \boldsymbol{d}z; \quad ^*(\boldsymbol{d}z \wedge \boldsymbol{d}\rho) = \rho \boldsymbol{d}\varphi$$

(cf. Exercises 6.6.3 and 7.5.5). It follows from (8.71a–b) that

$$\begin{aligned} -\delta \mathbf{B} &= -^*\boldsymbol{d}^*\mathbf{B} = \frac{n\hbar}{q}\, \rho \cdot \left[\frac{A'}{\rho}\right]' \boldsymbol{d}\varphi \\ \mathbf{D}^*\mathbf{D}\psi &= \psi \left\{ \rho \cdot \left(\frac{\phi'}{\phi}\right)^2 + \left[\frac{\phi'}{\phi}\, \rho\right]' - n^2\, \frac{(1-A)^2}{\rho} \right\} \varepsilon. \end{aligned}$$

Thus we obtain the following equations of motion:

$$\rho \left[\frac{A'}{\rho}\right]' = -\frac{1}{\lambda^2(T)}\, \phi^2[1 - A]; \tag{8.73a}$$

$$(\rho\phi')' = -\alpha[\phi^2 - 1]\phi + n^2\, \frac{(1-A)^2}{\rho^2}\, \phi. \tag{8.73b}$$

There is, however, a quicker way to obtain these equations. We know that the cylindrical symmetric equilibrium configuration must extremize the energy functional $J[\phi, A]$ given by (8.72). The corresponding Euler–Lagrange equations are given by

$$\frac{\partial J}{\partial \phi} = \frac{d}{d\rho}\left\{\frac{\partial J}{\partial \phi'}\right\}; \qquad \frac{\partial J}{\partial A} = \frac{d}{d\rho}\left\{\frac{\partial J}{\partial A'}\right\}.$$

Thereby we recover the equations of motion (8.73a–b). One should, however, be very careful when using this strategy. It works only because in this particular case the equations (8.70a–b) represent in fact the most general cylindrical symmetric equations. In general, a variation among a *restricted* subset of configurations need not extremize the action against *arbitrary* variations.

We must then show that the second-order differential equation (8.73) possesses a solution with the appropriate boundary condition. This is a somewhat difficult task. Using advanced analysis (Sobolev space techniques), it can be shown that the energy functional $J[\phi, A]$, which is positive definite, has a smooth minimum configuration satisfying the appropriate boundary conditions. (For details see Jaffe and Jaubes (1981).) In practice one is often contented with computer simulations.

Notice that

$$\phi(\rho) \equiv 1, \qquad A(\rho) \equiv 1$$

is a trivial solution to equation (8.73) although it breaks the boundary condition at the origin. As a consequence, the corresponding string is singular at the origin. Nevertheless, the energy density vanishes outside the origin. It is called a *vacuum texture* and obviously represents an infinitely thin string carrying n flux quanta.

Although we cannot solve the equation of motion explicitly, we can easily determine the asymptotic behavior. In the limit of large ρ, the equations of motion simplify considerably. If we introduce the functions

$$f(\rho) = \frac{1 - A(\rho)}{\sqrt{\rho}} \quad \text{and} \quad g(\rho) = \sqrt{\frac{-\alpha\rho}{\beta}}\,\{1 - \phi(\rho)\},$$

it is easy to show that asymptotically they solve the differential equations

$$f''(\rho) \underset{\rho\to\infty}{\approx} \frac{1}{\lambda^2(T)}\, f(\rho) \quad \text{and} \quad g''(\rho) \underset{\rho\to\infty}{\approx} \frac{1}{2\xi^2(T)}\, g(\rho).$$

This implies the following asymptotic behavior:

$$f(\rho) \underset{\rho\to\infty}{\approx} C_1 \exp\left[-\frac{\rho}{\lambda(T)}\right] \quad \text{and} \quad g(\rho) \underset{\rho\to\infty}{\approx} C_2 \exp\left[-\frac{\rho}{\xi(T)\sqrt{2}}\right],$$

in accordance with our previous discussion of the coherence and penetration length.

Okay, at this point we return to our estimate (8.69) for the static energy. Using the equation of motion (8.73a), it follows that

$$^*\mathbf{B} \wedge \delta\mathbf{B} = \frac{n^2h^2}{q^2\lambda^2(T)}\,\frac{1}{\rho}\,A'(\rho)[1 - A(\rho)]\phi^2(\rho)\mathbf{d}z \wedge \mathbf{d}\varphi.$$

Consequently, the energy per unit length is given by

$$J \approx \pi n^2 \frac{\hbar^2}{q^2} \left\{\frac{1}{\rho}\,A'(\rho)[1 - A(\rho)]\phi^2(\rho)\right\}\Big|_{\rho=\xi(T)}. \tag{8.74}$$

To proceed further, we notice that the equation of motion for $A(\rho)$ does not contain n explicitly! (It does contain n implicitly through the function $\phi(\rho)$.) If we impose the condition $\phi(\rho) \equiv 1$, we get a vacuum texture for the order parameter, but we can still retain the boundary conditions for $A(\rho)$. The solution to the reduced equation of motion

$$\rho\left[\frac{A'}{\rho}\right]' = -\frac{1}{\lambda^2(T)}\,[1-A] \quad \text{with } \lim_{\rho\to 0} A(\rho) = 0 \quad \text{and} \quad \lim_{\rho\to\infty} A(\rho) = 1 \tag{8.75}$$

is then likely to reproduce the vortex outside the hard core. In this approximation, $A(\rho)$ is completely independent of n. It follows that J is proportional to the square of n:

$$J_n \approx n^2 J_1. \tag{8.76}$$

As a consequence, a *vortex string with a multiple flux is unstable*. E.g., a single vortex string with a double flux quantum has twice the energy of two widely separated vortex strings with unit fluxes. This shows that when the magnetic field penetrates into the superconducting region, it will generate a uniformly distributed array of vortices all carrying a single flux quantum.

Remark. Actually, equation (8.75) can be solved explicitly using Bessel functions. The expression (8.74) can then be estimated using the known asymptotic behavior of the Bessel function in the limit of small ρ. In this way one obtains

$$J \approx \frac{n^2 \ln[\kappa]}{q^2\lambda^2(T)}, \tag{8.77}$$

where κ is the Ginzburg–Landau parameter (8.67). For details see de Gennes (1966).

8.8 The Nambu String and the Nielsen–Olesen Vortex

The soliton was introduced as a smooth, extended version of a relativistic point particle. Now we will show that a similar interpretation is possible for a vortex string. First, we must introduce a suitable generalization of the relativistic point particle known as a *relativistic string* or a *Nambu string*.

A point particle is characterized by its position $\mathbf{x}$ at time t. In four-dimensional space–time it sweeps out a time-like curve, the *world line*; cf. Figure 8.31a. The action of free particle is particularly simple, being proportional to the arc-length of the world line:

$$S = -m\int\sqrt{-g_{\mu\nu}\frac{dx^\mu}{d\lambda}\frac{dx^\nu}{d\lambda}}\,d\lambda.$$

Varying the world line, we then obtain the equation of motion, which in an inertial frame reduces to

$$m\,\frac{d^2x^\alpha}{d\tau^2} = 0.$$

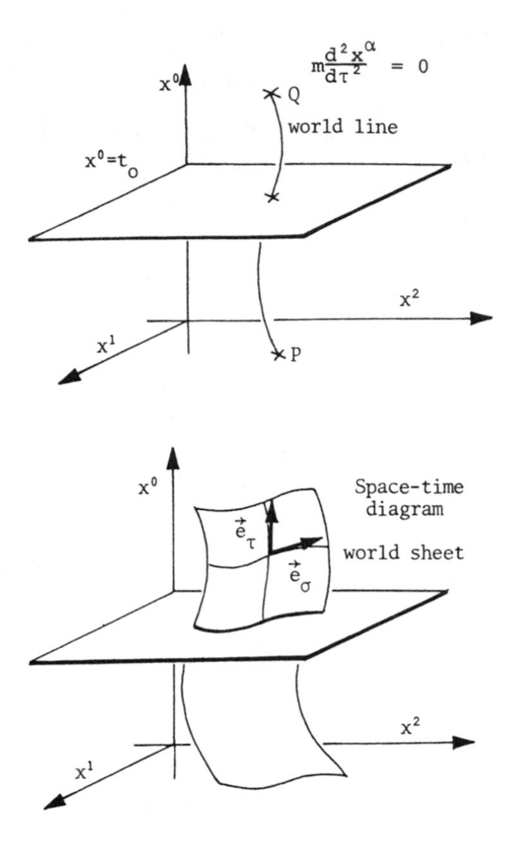

Figure 8.31.

In a similar way, a relativistic string will be characterized by its position $\mathbf{x}(\sigma)$ at time t, where $\mathbf{x} = \mathbf{x}(\sigma)$ parametrizes the spatial curve representing the string. It will be convenient to assume that the string is finite and that its endpoints correspond to $\sigma = 0$ and $\sigma = \pi$. In four-dimensional space–time, the string sweeps out a time-like sheet, the *world sheet*; cf. Figure 8.31b. The action will be chosen to be proportional to the area of the sheet. If we parametrize the sheet by

$$x^\mu = x^\mu(\sigma; \tau),$$

where σ is a space-like and τ a time-like parameter, it follows that the induced metric on the sheet is characterized by the components

$$h_{\sigma\sigma} = \mathbf{g}(\mathbf{e}_\sigma; \mathbf{e}_\sigma) = g_{\mu\nu}\frac{dx^\mu}{d\sigma}\frac{dx^\nu}{d\sigma};$$
$$h_{\sigma\tau} = \mathbf{g}(\mathbf{e}_\sigma; \mathbf{e}_\tau) = g_{\mu\nu}\frac{dx^\mu}{d\sigma}\frac{dx^\nu}{d\tau}, \quad \text{etc.}$$

From the induced metric we then get the area element

$$\sqrt{-h} = \sqrt{-\det\begin{bmatrix} h_{\sigma\sigma} & h_{\sigma\tau} \\ h_{\tau\sigma} & h_{\tau\tau} \end{bmatrix}} = \sqrt{(h_{\sigma\tau})^2 - h_{\sigma\sigma}h_{\tau\tau}}\,. \tag{8.77}$$

We then define the action to be

$$S = -\frac{1}{2\pi\alpha}\int_{\tau_1}^{\tau_2}\int_{\sigma=0}^{\sigma=\pi}\sqrt{(h_{\sigma\tau})^2 - h_{\sigma\sigma}h_{\tau\tau}}\,d\sigma\,d\tau\,. \tag{8.78}$$

From the action we get the equation of motion by varying the world sheet:

$$x^\mu(\sigma,\tau) \curvearrowright x^\mu(\sigma,\tau) + \epsilon y^\mu(\sigma,\tau), \qquad \text{where } y^\mu(\sigma,\tau_1) = y^\mu(\sigma,\tau_2) = 0.$$

Notice that there are no restrictions at the endpoints $\sigma = 0$ and $\sigma = \pi$. Thus we get not only equations of motion for the string but also boundary conditions for the endpoints. The general covariant equations of motion are extremely complicated, so we shall restrict ourselves to an inertial frame where the metric coefficients reduce to

$$g_{\mu\nu}(x) = \eta_{\mu\nu}.$$

Then the Lagrangian density depends only upon

$$\frac{dx^\mu}{d\sigma} = x'^\mu \quad \text{and} \quad \frac{dx^\mu}{d\tau} = \dot{x}^\mu.$$

We now obtain the displaced action

$$S(\epsilon) = \int_{\tau_1}^{\tau_2}\int_{\sigma=0}^{\sigma=\pi} L\left(\frac{\partial x^\mu}{\partial\sigma} + \epsilon\frac{\partial y^\mu}{\partial\sigma}\,;\,\frac{\partial x^\mu}{\partial\tau} + \epsilon\frac{\partial y^\mu}{\partial\tau}\right)d\sigma\,d\tau.$$

We must then demand that

$$\begin{aligned} 0 = \frac{dS}{d\epsilon}\Big|_{\epsilon=0} &= \int_{\tau_1}^{\tau_2}\int_{\sigma=0}^{\sigma=\pi}\left[\frac{\partial L}{\partial x'^\mu}\frac{\partial y^\mu}{\partial\sigma} + \frac{\partial L}{\partial x^\mu}\frac{\partial y^\mu}{\partial\tau}\right]d\sigma\,d\tau \\ &= -\int_{\tau_1}^{\tau_2}\int_{\sigma=0}^{\sigma=\pi}\left\{\frac{\partial}{\partial\sigma}\frac{\partial L}{\partial x'^\mu} + \frac{\partial}{\partial\tau}\frac{\partial L}{\partial x^\mu}\right\}y^\mu d\sigma\,d\tau \\ &\quad + \int_{\tau_1}^{\tau_2}\left[\frac{\partial L}{\partial x'^\mu}\,y^\mu\right]_{\sigma=0}^{\sigma=\pi}d\tau. \end{aligned}$$

As y is arbitrary, this leads on the one hand to the equations of motion

$$\frac{\partial}{\partial\sigma}\frac{\partial L}{\partial x'^\mu} + \frac{\partial}{\partial\tau}\frac{\partial L}{\partial x^\mu} = 0. \tag{8.79}$$

On the other hand, it leads to the edge conditions

$$\frac{\partial L}{\partial x'^\mu}\Big|_{\sigma=0} = \frac{\partial L}{\partial x'^\mu}\Big|_{\sigma=\pi} = 0. \tag{8.80}$$

Let us introduce the following abbreviations for the conjugate momenta:

$$P_\tau^\mu = \frac{\partial L}{\partial \dot{x}^\mu} = \frac{1}{2\pi\alpha}\frac{x'^2\dot{x}^\mu - (\dot{x} \mid x')x'^\mu}{\sqrt{(\dot{x} \mid x')^2 - \dot{x}^2 x'^2}}\,;$$
$$P_\sigma^\mu = \frac{\partial L}{\partial x'^\mu} = \frac{1}{2\pi\alpha}\frac{\dot{x}^2 x'^\mu - (\dot{x} \mid x')\dot{x}^\mu}{\sqrt{(\dot{x} \mid x')^2 - \dot{x}^2 x'^2}}\,. \tag{8.81}$$

Then the equations of motion supplied with the boundary conditions simplify to

$$\frac{\partial}{\partial\tau}P_\tau^\mu + \frac{\partial}{\partial\sigma}P_\sigma^\mu = 0; \qquad P_\sigma^\mu(\sigma = 0) = P_\sigma^\mu(\sigma = \pi) = 0. \tag{8.82}$$

Observe that we also have the following identities at our disposal:

$$P_\tau^\mu x'_\mu = P_\sigma^\mu \dot{x}_\mu = P_\tau^2 + \frac{x'^2}{4\pi^2\alpha^2} = P_\sigma^2 + \frac{\dot{x}^2}{4\pi^2\alpha^2} = 0, \tag{8.83}$$

which are trivial consequences of the explicit expressions for P_τ^μ and P_σ^μ. Notice especially that the edge conditions imply that $P_0^\mu = 0$ at the edge. As a consequence, $\dot{x}^2$ vanishes at the edge, so *the endpoints move with the speed of light.*

We may also introduce the four-momentum of the string:

$$P^\mu = \int_{\sigma=0}^{\sigma=\pi} P_\tau^\mu d\sigma \qquad (\tau \text{ fixed}). \tag{8.84}$$

The four-momentum is conserved (as it ought to be):

$$\frac{dP^\mu}{d\tau} = \int_{\sigma=0}^{\sigma=\pi}\frac{\partial P_\tau^\mu}{\partial\tau}d\sigma = -\int_{\sigma=0}^{\sigma=\pi}\frac{\partial P_\sigma^\mu}{\partial\sigma}d\sigma$$
$$= \{P_\sigma^\mu(\sigma = 0) - P_\sigma^\mu(\sigma = \pi)\} = 0.$$

Similarly, we may introduce the angular momentum of the string:

$$J^{\mu\nu} = \int_{\sigma=0}^{\sigma=\pi}\left[x^\mu P_\tau^\nu - x^\nu P_\tau^\mu\right]d\sigma. \tag{8.85}$$

Exercise 8.8.1
Problem: Show that the angular momentum is conserved; i.e.,

$$\frac{d}{d\tau}J^{\mu\nu} = 0.$$

Exercise 8.8.2
Introduction: Consider a rigid rotating string parametrized by

$$x^0 = \tau; \qquad x^1 = A\left(\sigma - \frac{1}{2}\pi\right)\cos\omega\tau;$$

$$x^2 = A\left(\sigma - \frac{1}{2}\pi\right)\sin\omega\tau; \qquad x^3 = 0.$$

Problem: (a) Show that it solves the equation of motion, provided that

$$1 = \frac{1}{2}A\omega\tau.$$

(Hint: Show that this represents the edge condition.) (b) Show that the spinning string has momentum

$$P^0 = \frac{A\pi}{4\alpha}; \qquad P^1 = P^2 = P^3 = 0.$$

(c) Show that the spinning string has angular momentum

$$J^{12} = \frac{A^2\pi^2}{16\alpha}; \qquad J^{23} = J^{31} = 0.$$

From the above exercise it follows that a spinning string behaves like a particle with rest mass

$$M = \frac{A\pi}{4\alpha}$$

and intrinsic spin

$$J = \frac{A^2\pi^2}{16\alpha} = \alpha M^2.$$

Notice especially that *the spin grows linearly with the square of the mass.*

Now, what is the relevance of such relativistic strings in high-energy physics? If we plot baryons (i.e., strongly interacting particles) in a diagram with the mass square on one axis and the spin on the other axis, then the baryons with the same isospin, strangeness, etc. fall on straight lines! (See Figure 8.32.) These are the so-called *Regge trajectories*. This remarkable property suggests that we consider the baryons to be somehow composed of relativistic strings.

Okay, with this simpleminded remark I hope to have convinced you that relativistic strings, *Nambu strings*, are very interesting objects. As in the case of point particles one would now be interested in smooth extended solutions to field theories that behave like thin strings.

ILLUSTRATIVE EXAMPLE: THE NIELSEN–OLESEN VORTEX.

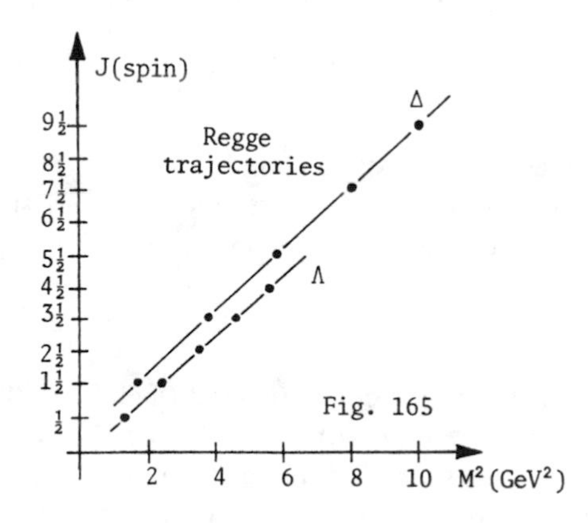

Figure 8.32.

The first example of a string-like solution in a classical field theory was found by my respected teachers, Holger Bech Nielsen and Poul Olesen (1973).

They considered a relativistic field theory based upon a complex charged scalar field coupled minimally to electromagnetism. In analogy with the ϕ^4-model, they furthermore included a nontrivial potential energy density given by

$$U(\phi) = \frac{\lambda}{4}\left(|\phi|^2 - \frac{\mu^2}{\lambda}\right)^2.$$

Models with such a potential term are generally referred to as *Higgs models*, and the corresponding scalar fields are known as *Higgs fields*. The model based upon the Lagrangian density

$$L = -\frac{1}{2}(\mathbf{F} \mid \mathbf{F}) - \frac{1}{2}(\boldsymbol{D}\phi \mid \boldsymbol{D}\phi) - U(\phi), \qquad \text{with } \boldsymbol{D}\phi = \boldsymbol{d}\phi - ie\mathbf{A}\phi, \quad (8.86)$$

where the Higgs field ϕ is coupled to an abelian gauge field, is specifically referred to as the *abelian Higgs model*. The associated field equations are highly nonlinear:

$$\begin{aligned} \boldsymbol{\delta}\mathbf{F} &= \frac{ie}{2}\left[\phi\overline{\boldsymbol{D}\phi} - \bar{\phi}\boldsymbol{D}\phi\right]; \\ &\text{i.e.,}\ \frac{1}{\sqrt{-g}}\,\partial_\nu\left(\sqrt{-g}F^{\mu\nu}\right) = \frac{ie}{2}\left[\phi\overline{D^\mu\phi} - \bar{\phi}D^\mu\phi\right] \end{aligned} \qquad (8.87a)$$

$$\boldsymbol{D}^*\boldsymbol{D}\phi = (\lambda|\phi|^2 - \mu^2)\phi\varepsilon; \quad \text{i.e.,}\ \frac{1}{\sqrt{-g}}\,D_\nu\left(\sqrt{-g}\,D^\nu\phi\right) = (\lambda|\phi|^2 - \mu^2)\phi. \qquad (8.87b)$$

In this model we now proceed to investigate the purely static configurations, where furthermore, the electric field is absent. Such a configuration is represented by a spatial function $\phi(\mathbf{r})$ and a spatial one-form $\mathbf{A} = \mathbf{A}(\mathbf{r}) \cdot \boldsymbol{d}\mathbf{r}$. Furthermore, it corresponds to a solution of the equations of motion (8.87a–b) precisely when it extremizes the static energy, which in the present model is given by

$$H = \frac{1}{2}\langle\mathbf{B} \mid \mathbf{B}\rangle + \frac{1}{2}\langle\boldsymbol{D}\phi \mid \boldsymbol{D}\phi\rangle + \int U(\phi)\varepsilon. \qquad (8.88)$$

As a consequence, the theory of static equilibrium configurations in the abelian Higgs model is completely equivalent to the Ginzburg–Landau theory for superconductivity, provided that we make the identification

$$e \curvearrowright \frac{q}{\hbar}; \qquad \lambda \curvearrowright \beta; \qquad \mu^2 \curvearrowright -\alpha.$$

(This follows immediately from a comparison of (8.60) with (8.88).)

The first important observation is that the abelian Higgs model possesses *no* soliton solutions in the strict sense; i.e., there are no nontrivial stable, static finite-energy configurations. This follows from an argument that is typical for gauge theories:

A static finite-energy configuration must satisfy the boundary conditions

$$\lim_{r\to\infty} \mathbf{B} = 0; \qquad \lim_{r\to\infty} \boldsymbol{D}\phi = 0; \qquad \lim_{r\to\infty} |\phi| = \frac{\mu}{\sqrt{\lambda}} . \tag{8.89}$$

Asymptotically, the Higgs field is therefore completely characterized by its phase factor

$$\phi(\mathbf{r}) \underset{r\to\infty}{\approx} \frac{\mu}{\sqrt{\lambda}} \exp[i\varphi(\mathbf{r})].$$

In general, the phase $\varphi(\mathbf{r})$ need not be single-valued but can make a quantized jump of $2\pi n$. But in the present case such a jump is *not* allowed!

To see this, suppose the phase makes a jump when we go once around a distant closed curve Γ. This distinct closed curve can be shrunk to a distant point, while the shrinking curve remains distant! (Topologically, it is a closed curve outside a ball, and such a curve can clearly be contracted without intersecting the ball.) Notice that by continuity, the phase must also make the jump $2\pi n$ when we go once around the shrinking curve. As the curve shrinks to a point, we thus produce a discontinuity in the phase factor $\exp[i\varphi(\mathbf{r})]$. But such a discontinuity in the phase factor contradicts the smoothness of the Higgs field!

Since the asymptotic phase $\varphi(\mathbf{r})$ is single-valued, the phase factor

$$\exp[i\varphi(\mathbf{r})] = \frac{\phi(\mathbf{r})}{|\phi(\mathbf{r})|}$$

is in fact trivial; i.e., it is single-valued throughout the whole space. Thus we can remove this phase factor completely by a gauge transformation. To conclude, we have therefore shown the existence of a gauge where ϕ is *real*!

In this particular gauge the boundary conditions (8.89) reduce to

$$\lim_{r\to\infty} A(\mathbf{r}) = 0; \qquad \lim_{r\to\infty} \phi(\mathbf{r}) = 0. \tag{8.90}$$

Consider now the deformed configuration

$$A_\epsilon(\mathbf{r}) = \epsilon A(\mathbf{r}); \qquad \phi_\epsilon(\mathbf{r}) = \epsilon \frac{\mu}{\sqrt{\lambda}} + (1-\epsilon)\phi(\mathbf{r}).$$

It satisfies the boundary condition (8.90) for any choice of ϵ. At $\epsilon = 0$ it reduces to the vacuum configuration, while for $\epsilon = 1$ it reduces to the given configuration. *Consequently, we can find a one-parameter family of finite-energy configurations that interpolate between the vacuum configuration and the given configuration.* This shows that a static finite-energy configuration cannot be stable, since we can "press" it down to the vacuum.

To obtain interesting configurations, we must therefore relax the boundary conditions (8.89). Rather than looking for point-like configurations, we will now look for string-like configurations. Consequently, we will concentrate on configurations that are independent of z. This time we therefore impose the boundary condition that the energy per unit length along the z-axis be finite. This restriction implies the following boundary conditions:

$$\lim_{\rho\to\infty} \mathbf{B} = 0; \qquad \lim_{\rho\to\infty} \boldsymbol{D}\phi = 0; \qquad \lim_{\rho\to\infty} |\phi| = \frac{\mu}{\sqrt{\lambda}}, \tag{8.91}$$

where ρ is the distance from the z-axis. As before, we conclude that asymptotically, the Higgs field is characterized by its phase factor:

$$\phi(\mathbf{r}) \underset{\rho\to\infty}{\approx} \frac{\mu}{\lambda} \exp[i\varphi(\mathbf{r})].$$

This time, however, there is nothing to prevent the phase of the Higgs field from making a jump of $2\pi n$ when we go once around a distance closed curve. (A distant closed curve cannot be shrunk to a point without intersecting the z-axis.) Notice, however, that if the phase makes a nontrivial jump, then the Higgs field must necessarily vanish somewhere inside the string corresponding to a discontinuity in the phase factor

$$\exp[i\varphi(\mathbf{r})] = \frac{\phi(\mathbf{r})}{|\phi(\mathbf{r})|}.$$

Since the gauge covariant exterior derivative of the Higgs field vanishes outside the string, we get, as usual,

$$\boldsymbol{d}\varphi - e\mathbf{A} \underset{\rho\to\infty}{\approx} 0.$$

Consequently, the jump of φ is related to the magnetic flux in the string:

$$\Phi = \lim_{\rho_0\to\infty} \int_{\rho<\rho_0} \mathbf{B} = \lim_{\rho_0\to\infty} \oint_{\rho=\rho_0} \mathbf{A} = \lim_{\rho_0\to\infty} e \oint_{\rho=\rho_0} \boldsymbol{d}\varphi = n\,\frac{2\pi}{e}.$$

This is also in accordance with the equivalence between the abelian Higgs model and superconductivity. The number n can consequently be identified with the number of flux quanta in the string.

Since the flux is quantized, it is impossible this time to interpolate between the vacuum configuration and a configuration with a nontrivial flux. A string with a nontrivial flux is thus *topologically stable*!

We proceed to look for configurations that minimize the static energy per unit length. As in the Ginzburg–Landau theory, we shall assume that the penetration length is considerably larger than the coherence length; i.e., in the present case we assume

$$\lambda \gg e^2. \tag{8.92}$$

We can then carry over the conclusions obtained in the previous section concerning magnetic strings in a type II superconductor. In the sector consisting of configurations carrying a single flux quantum, the ground state is the cylindrical symmetrical configuration. In the abelian Higgs model this is known as a *Nielsen–Olesen string*. In a sector consisting of configurations carrying multiple flux quanta there can be no exact ground state. This is because a cylindrical symmetrical configuration in unstable, since its energy grows with the *square* of the number of flux quanta; cf.

(8.76). As in the sine-Gordon model, one can, however, construct approximative ground states consisting of n widely separated Nielsen–Olesen vortices.

Notice that a string-like excitation such as the Nielsen–Olesen vortex cannot itself represent a physical particle, since it has infinite energy due to its infinite length. So if the Nielsen–Olesen vortex is going to be physically relevant, we must find a way to terminate it. This can be done by including additional point particles in the model, which then sit at the endpoints of the string. As the Nielsen–Olesen vortex carries a magnetic flux, these additional particles must necessarily be magnetic monopoles (or anti-monopoles). Furthermore, the magnetic charge must be quantized, since the flux carried away by the string is necessarily quantized. As we shall see in the next chapter, that is fine: Magnetic charges *are* in fact quantized according to the rule

$$g = n\,\frac{2\pi}{e}\,.$$

Suppose then we introduce point-monopoles in the abelian Higgs model. Notice first that it is impossible to introduce just a single monopole. This is because a single monopole is characterized by a long-range magnetic field, and that cannot exist in the abelian Higgs model due to the Meissner effect.

Let us clarify this point: One might object that the Meissner effect could just squeeze the magnetic field into a thin string extending from the monopole to infinity. But being infinitely long, the string would carry out infinite energy, and that is *not* physically acceptable. (The Coulomb field created by an ordinary charged particle is acceptable because the energy stored in the field outside a ball containing the particle is always finite. The infinite self-energy of the electron comes form the immediate neighborhood of the electron, *not* from infinity.)

This means that we can only introduce monopoles in terms of monopole–antimonopole pairs. A magnetic field line extending from the monopole can then be absorbed by the antimonopole.

Suppose then that we try to separate such a monopole–antimonopole pair. This will necessarily create a thin magnetic string between the monopole and the anti-monopole. The string has a typical thickness given by the penetration length; i.e., the radius of the string, r_0, is

$$r_0 \approx \frac{\sqrt{\lambda}}{e}\,.$$

Since the string carries the magnetic flux g, where g is the magnetic charge of the monopole, the magnetic field strength will be of the order

$$B \approx \frac{g}{\pi r_0^2}\,.$$

Thus the magnetic energy stored in the string is of the order

$$E \approx \frac{1}{2}\,B^2(\pi r_0^2 \ell) = \frac{1}{2}\,\frac{g^2}{\pi r_0^2}\,\ell,$$

where ℓ is the distance between the monopole and the antimonopole. It will therefore require infinite energy to separate the monopole–antimonopole; i.e., it is physically impossible to separate them and thereby produce free monopoles (or antimonopoles). One says that monopoles are *confined* in the abelian Higgs model. This is to be contrasted with ordinary electrodynamics. There one may also consider bound states consisting of two oppositely charged particles, say hydrogen or positronium. But in that case we can easily knock off the electron and thereby produce free electrons.

To summarize:

a. In standard electrodynamics the binding energy for a bound state consisting of oppositely charged particles is given by the *Coulomb potential*

$$V(r) = -\frac{q^2}{4\pi\epsilon_0}\frac{1}{r},$$

and it requires only a finite amount of energy to separate them.

b. In the abelian Higgs model, the binding energy for a bound state consisting of a monopole–antimonopole pair is given by the *linear potential*

$$V(r) = \frac{1}{2}\frac{g^2}{\pi r_0^2}\,r,$$

and it requires an infinite amount of energy to separate them; i.e., the monopoles will be permanently confined.

Returning to the monopole–antimonopole pair in the abelian Higgs model, we see that the potential generates an attractive force (which is independent of the distance for large distances). To prevent the pair from collapsing, we must therefore put it into rapid rotation. In this way we produce a composite particle consisting of a spinning magnetic string with a monopole at one end and an antimonopole at the other. For a sufficiently long rapidly spinning string we have thus come back to the Nambu string.

Now, what is the relevance of these considerations in high-energy physics? For various theoretical reasons, hadrons (i.e., strongly interacting particles) are generally thought to be composite particles, the constituents of which are called *quarks*. This so-called quark model has had considerable success in explaining the observed spectrum of hadrons (including some predictions of hitherto unknown hadrons). But there is one great puzzle concerning the quark model: *A free quark has never been observed as the outcome of a scattering experiment in high-energy physics*. (Some famous solid-state physicists claim to have observed free quarks in a very beautiful but delicate experiment (LaRue, Fairbanks, and Hebard, 1977), but their observations have not been confirmed by other groups, and they are not generally trusted.) Thus quarks seem to be confined.

How can we explain this quark confinement? One possible idea is as follows: The quarks interact through the strong interactions, and it is customary to introduce a field, the so-called *color field*, that is responsible for this interaction (in the same way as the electromagnetic field is responsible for the electromagnetic interaction).

In analogy with the electromagnetic field, the color field comes in two species, known as the color electric field and the color magnetic field. The quarks carry color electric charges and thus act as sources for the color electric fields. One then adds as a basic hypothesis that in the quantized theory the color vacuum will act like a superconducting vacuum, but in contrast to the abelian Higgs model, we are supposed to reverse the role of electric and magnetic fields; i.e., this time it is the color electric field that is expelled due to the Meissner effect. (Of course, this must eventually be proven directly from basic principles if we are really going to trust the argument. This is the really hard part of the game, and only little progress has been made due to great technical difficulties in the quantum theory of colored fields.)

Okay, suppose the above hypothesis concerning the colored vacuum is correct. Then we can verbally take over the arguments concerning magnetic monopoles in the abelian Higgs model. Free quarks cannot exist because the color electric field is squeezed into a finite region. It is, however, possible to collect them in a quark–antiquark pair, and due to a greater complexity in the structure of colored fields, it is also possible to put three quarks together in such a way that the net color charge is zero. (Incidentally, this is where the name "color" comes from: There are three basic colors in nature—red, green, and blue—and if we "add" them, we get white; i.e., the net color vanishes!)

Thus the following picture arises: Space is divided into two regions. The dominating region filling up space almost everywhere consists of the superconducting color vacuum. But here and there we find small regions filled with quarks and color electric fields. These "normal" regions, where the superconductivity breaks down, are called *bags*, and they act as prisons for the quarks. Thus a hadron is an extended object consisting of a bag containing either a quark–antiquark pair or a couple of three quarks. (See Figure 8.33.)

Consider a quark–antiquark pair. If we try to separate the quark from the antiquark, we necessarily produce a thin color electric string between them, and just as in the abelian Higgs model, it can be shown in a completely elementary way that the energy stored in the string is proportional to the length of the string. Thus we arrive at the typical linear *quark potential*:

$$V(r) = kr.$$

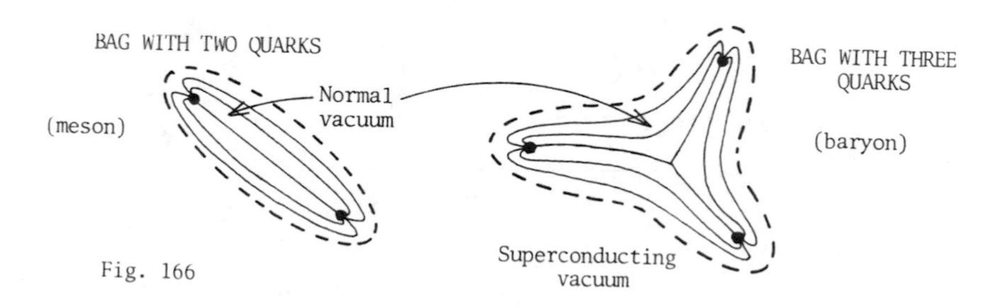

Figure 8.33.

Finally, we can excite hadrons by rotating the bags. Such rapidly spinning bags will be elongated, and we therefore expect them to behave like spinning Nambu strings!

So I hope that the reader now sees the importance of the abelian Higgs model. It serves as a theoretical laboratory where in an elementary way we can test the consistence of various properties of the model before we try to examine these properties in the more complex models that are thought to describe systems, that are actually found in nature.

We conclude this section with yet another argument that supports the interpretation of a Nielsen–Olesen vortex as a smooth extended Nambu string.

First, we rearrange the action of the Nambu string slightly. Consider an inertial frame, and let us identify the parameter τ with the corresponding time in the inertial frame; i.e., we put $\tau = t$. The string is therefore parametrized as

$$\mathbf{x} = \mathbf{x}(\sigma, t).$$

The arc-length of the string is given by

$$ds = \sqrt{\frac{\partial \mathbf{x}}{\partial \sigma} \frac{\partial \mathbf{x}}{\partial \sigma}}\, d\sigma.$$

The tangent vector $\partial \mathbf{x}/\partial s$ thus becomes a unit vector. The velocity of a point on the string is given by

$$\mathbf{v} = \frac{\partial \mathbf{x}}{\partial t}\,.$$

In our case, only the component perpendicular to the string is physically relevant. Evidently, it is given by

$$\mathbf{v}_\perp = \frac{\partial \mathbf{x}}{\partial t} - \frac{\partial \mathbf{x}}{\partial s}\left(\frac{\partial \mathbf{x}}{\partial s}\frac{\partial \mathbf{x}}{\partial t}\right); \qquad \mathbf{v}_\perp^2 = \left(\frac{\partial \mathbf{x}}{\partial t}\right)^2 - \left(\frac{\partial \mathbf{x}}{\partial s}\frac{\partial \mathbf{x}}{\partial t}\right)^2.$$

Next, we observe that metric coefficients are given by

$$h_{\sigma\sigma} = \frac{\partial \mathbf{x}}{\partial \sigma}\frac{\partial \mathbf{x}}{\partial \sigma} = \left(\frac{\partial s}{\partial \sigma}\right)^2; \quad h_{\sigma\tau} = \frac{\partial \mathbf{x}}{\partial \sigma}\frac{\partial \mathbf{x}}{\partial t} = \frac{\partial s}{\partial \sigma}\left(\frac{\partial \mathbf{x}}{\partial s}\frac{\partial \mathbf{x}}{\partial t}\right);$$

$$h_{\tau\tau} = -1 + \frac{\partial \mathbf{x}}{\partial t}\frac{\partial \mathbf{x}}{\partial t}\,.$$

Using this we can rearrange the area element as follows:

$$\sqrt{-h} = \sqrt{(h_{\sigma\tau})^2 - h_{\sigma\sigma}h_{\tau\tau}} = \frac{\partial s}{\partial \sigma}\sqrt{1 - \mathbf{v}_\perp^2}\,.$$

Consequently, we finally obtain (cf. (6.49))

$$S = -\frac{1}{2\pi\alpha}\int_{t_1}^{t_2} \sqrt{1 - \mathbf{v}_\perp^2}\, ds\, dt. \tag{8.93}$$

Next we consider a thin Nielsen-Olesen vortex string moving around in space. It could be a static vortex that we have "boosted" into a uniform motion, or it could be a more complicated motion specified by some appropriate initial data. The vortex will be a solution to the equations of motion derived from the Lagrangian density (8.86), but we are not going to specify the potential in this argument. All we use is that the model allows string-like solutions, where the fields are almost in the vacuum state outside the thin string. The deviations from the vacuum state will be exponentially small, and we shall simply neglect them. Thus the Lagrangian density vanishes outside the string, and the Lagrangian density therefore acts as a smeared-out δ-function.

We can now find an approximative expression for the action of the string. First, we introduce a rest frame for a small portion of the string. The rest frame moves with velocity $\mathbf{v}$, i.e., perpendicular to the string. The coordinates in the rest frame will be denoted by t^0, x^0, y^0, z^0, and notice that z^0 is simply the arc-length for the string! In the rest frame the string reduces to a static string. The Lagrangian density thus becomes equal to minus the energy density:

$$L = -H.$$

If J denotes the energy per unit length in the rest frame, we have therefore shown that

$$L(x^0, y^0, z^0, t^0) = -J\delta(x^0)\delta(y^0).$$

But then we get the following expression for the action per unit length:

$$\begin{aligned} S &= \int L dx^0 dy^0 dz^0 dt^0 = \int L\sqrt{1 - \mathbf{v}_\perp^2}\, dx^0 dy^0 ds\, dt \\ &= -J \int \sqrt{1 - \mathbf{v}_\perp^2}\, ds\, dt, \end{aligned}$$

where we have used the transformation formula

$$dt^0 = dt\sqrt{1 - \mathbf{v}_\perp^2}.$$

Consequently, we have deduced the following approximative expression for the total action:

$$S_{\text{vortex}} \approx -J \int \sqrt{1 - \mathbf{v}_\perp^2}\, ds\, dt. \tag{8.94}$$

But that is precisely the action of the Nambu string! (See Figure 8.34.) Furthermore, it allows us to reexpress the slope of the Regge trajectories α in terms of the energy density per unit length J:

$$\alpha = \frac{1}{2\pi J}.$$

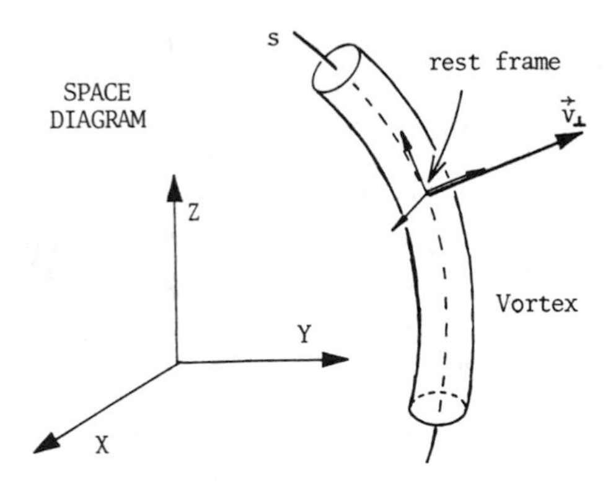

Figure 8.34.

8.9 Singular Forms

We can also extend the exterior algebra to include *singular differential forms.* A complete discussion falls beyond the scope of these notes. First, we will discuss the naive point of view generally adopted by physicists, and then we will give a brief introduction to the framework of distributions, which is a concise formalism adopted by mathematicians.

In the naive point of view, a differential form of degree k,

$$\mathbf{T} = \frac{1}{k!} T_{i_1 \dots i_k}(x) \boldsymbol{dx}^{i_1} \wedge \dots \wedge \boldsymbol{dx}^{i_k},$$

is called a *singular form* if the coordinate functions $T_{i_1, \dots, i_k}(x)$ are not smooth, i.e., they possess singularities like the discontinuity in the Heaviside function or the singularity of the δ-function.

We will assume that the usual theory of exterior calculus can be extended in a reasonable way to include singular forms, so that we can use rules like $\boldsymbol{d}^2 = 0$ or Stokes's theorem even if singular forms are involved. The first thing we will generalize is the δ-function:

Definition 8.5. If M is a manifold with metric $\mathbf{g}$, then the δ-function, δ_{p_0}, peaked at the point p_0, is defined to be the singular scalar field

$$\delta_{p_0} : M \curvearrowright \mathbb{R}$$

characterized by the property that

$$\langle \delta_{p_0} \mid \phi \rangle = \int_M \phi \delta_{p_0} \sqrt{g}\, \boldsymbol{dx}^1 \wedge \dots \wedge \boldsymbol{dx}^n = \phi(p_0) \tag{8.95}$$

for any smooth scalar field ϕ.

We can easily work out the coordinate expression for δ_{p_0}. In coordinates, the scalar field δ_{p_0} is represented by a singular Euclidean function, which we denote by $\underline{\delta}(x)$. From the integral property of δ_{p_0} we then get

$$\int_U \phi(x)\underline{\delta}(x)\sqrt{g(x)}dx^1 \ldots dx^n = \phi(x_0).$$

But then

$$\underline{\delta}(x)\sqrt{g(x)} = \delta^n(x - x_0); \qquad \text{i.e., } \underline{\delta}(x) = \frac{1}{\sqrt{g(x)}}\,\delta^n(x - x_0),$$

where $\delta^n(x - x_0)$ is the usual Euclidean δ-function.

Using singular forms, we can now associate an electric current $\mathbf{J}$ to a point particle. This is a highly singular one-form, which vanishes outside the world line of the particle. Using an inertial frame, we saw in Section 1.6 that the electric current was characterized by the contravariant components (1.34)

$$J^\mu(x) = q\int \delta^4(x - x(\tau))\,\frac{dx^\mu}{d\tau}\,d\tau.$$

The covariant coordinate expression for the δ-function suggests that we define the components of the current to be

$$J^\mu(x) = q\int \frac{1}{\sqrt{-g(x)}}\,\delta^4(x - x(\tau))\,\frac{dx^\mu}{d\tau}\,d\tau, \tag{8.96}$$

where this expression is valid in an arbitrary coordinate system. To justify that this is the covariant expression representing a singular vector field, we must show that $J^\mu(x)$ transforms contravariantly.

Observe that a general expression like

$$\int \phi(x - x(\tau))\,\frac{dx^\mu}{d\tau}\,d\tau,$$

where ϕ is a scalar field, does *not* transform contravariantly! That everything works out all right is due to very special properties of the δ-function. So let us introduce new coordinates $(y^1, \ldots, y^n)$ and see what happens. In the new coordinates the current is characterized by the components

$$J^\mu_{(2)}(y_0) = q\int \frac{1}{\sqrt{-g(y_0)}}\,\delta^4(y_0 - y(\tau))\,\frac{dy^\mu}{d\tau}\,d\tau.$$

But

$$\frac{1}{\sqrt{-g(y_0)}}\,\delta^4(y_0 - y(\tau)) = \frac{1}{\sqrt{-g(x_0)}}\,\delta^4(x_0 - x(\tau)),$$

since it represents a scalar field, and therefore we get

$$J^\mu_{(2)}(y_0) = q\int \frac{1}{\sqrt{-g(x_0)}}\,\delta^4(x_0 - x(\tau))\,\frac{\partial y^\mu}{\partial x^\nu}\,(x(\tau))\,\frac{dx^\nu}{d\tau}\,d\tau.$$

Now we use that $\delta^4(x_0 - x(\tau))$ is peaked at x_0. Consequently, we can replace $\frac{\partial y^\mu}{\partial x^\nu}(x(\tau))$ by $\frac{\partial y^\mu}{\partial x^\nu}(x_0)$ and then move it outside the integral, whereby we obtain

$$\begin{aligned} J^\mu_{(2)}(y_0) &= \frac{\partial y^\mu}{\partial x^\nu}(x_0) q \int \frac{1}{\sqrt{-g(x_0)}} \delta^4(x_0 - x(\tau)) \frac{dx^\nu}{d\tau} \, d\tau \\ &= \frac{\partial y^\nu}{\partial x^\nu}(x_0) J^\nu_{(1)}(x_0). \end{aligned}$$

So everything is okay!

Worked Exercise 8.9.1
Problem: Show that the singular current $\mathbf{J}$ with components (8.96) is conserved; i.e., $\delta\mathbf{J} = 0$.

In a similar way we can introduce a singular 2-form $\mathbf{F}$ representing the electromagnetic field generated by an electrically charged point particle. The singular forms $\mathbf{F}$ and $\mathbf{J}$ then obey the Maxwell equations

$$d\mathbf{F} = 0; \qquad \delta\mathbf{F} = \mathbf{J}.$$

Illustrative Example: How to Handle Singularities. We might wonder how we can control the singularity of $\mathbf{F}$. It is infinite at the position of the particle. How can we differentiate a function that is infinite at a single point like the function $\frac{1}{x}$? Usually, this is done by a *regularization procedure.*

Consider the static Coulomb field as an example. For simplicity, we work in the ordinary Euclidean space $\mathbb{R}^3$ and use Cartesian coordinates, etc. The Coulomb field is characterized by the components

$$E^i = \frac{q}{4\pi\epsilon_0} \cdot \frac{x^i}{r^3} \; .$$

We *regularize* the field making the substitution

$$r \curvearrowright \sqrt{r^2 + \epsilon^2}.$$

In this way we get the regularized field strength

$$E^i(\epsilon) = \frac{q}{4\pi\epsilon_0} \frac{x^i}{(r^2 + \epsilon^2)^{3/2}} \, ,$$

which is now smooth throughout the complete Euclidean space. If we differentiate this, we get

$$\nabla \cdot \mathbf{E}(\epsilon) = \frac{3q}{4\pi\epsilon_0} \cdot \frac{\epsilon^2}{(r^2 + \epsilon^2)^{5/2}} \; .$$

Consequently, the regularized electric field corresponds to a smooth charge distribution

$$\rho(\epsilon) = \frac{3q}{4\pi} \cdot \frac{\epsilon^2}{(r^2 + \epsilon^2)^{5/2}} \, ,$$

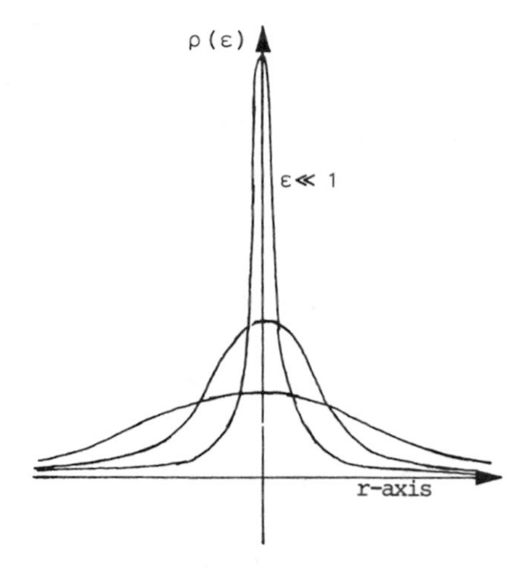

Figure 8.35.

which asymptotically vanishes like $\epsilon^2 r^{-5}$. For small ϵ it is peaked around $r = 0$, where it varies like ϵ^{-3}. (See Figure 8.35.) Observe that the total charge

$$Q(\epsilon) = 4\pi \int_0^\infty \rho(\epsilon) r^2 dr = q \int_0^\infty \frac{3\epsilon^2 r^2 dr}{(r^2 + \epsilon^2)^{5/2}} = q$$

is independent of ϵ. (Use $r = \epsilon x$.)

Thus we finally obtain the desired result

$$\lim_{\epsilon \to 0} \rho(\epsilon) = q\delta^3(\mathbf{x}),$$

from which we conclude that

$$\nabla \cdot \mathbf{E} = \frac{q}{\epsilon_0} \delta^3(\mathbf{x}).$$

Observe that the electric field $\mathbf{E}(\epsilon)$ in this example is not accompanied by a magnetic field $\mathbf{B}(\epsilon)$. The magnetic field can be obtained from the equation

$$-\frac{\partial \mathbf{B}}{\partial t} = \nabla \times \mathbf{E}(\epsilon).$$

But the curl vanishes automatically as a consequence of the spherical symmetry.

Finally, we will sketch how to introduce singular forms within the framework of distributions. Consider first the case of ordinary functions on the Euclidean space $\mathbb{R}^n$. Here a function f is characterized as a map $\mathbb{R}^n \curvearrowright \mathbb{R}$; i.e., each point x in $\mathbb{R}^n$ is mapped into a real number $f(x)$. We want now to construct singular (or generalized) functions. This is done in the following way:

First we choose a *test space*. In the theory of generalized functions, this test space consists of all smooth functions with compact support, and it is denoted by

$D(\mathbb{R}^n)$. The functions in the test space are called the *test functions*. We can now formulate the following definition.

Definition 8.6. A generalized function (i.e., a *distribution*) T is a linear functional on the test space; i.e., T is a linear map:

$$T : D(\mathbb{R}^n) \curvearrowright \mathbb{R}.$$

Now, what is the connection between ordinary functions and generalized functions? Apparently, they are defined in two completely different ways. To understand this connection we observe that any ordinary continuous function $f : \mathbb{R}^n \curvearrowright \mathbb{R}$ generates in a canonical fashion a linear functional $D(\mathbb{R}^n) \curvearrowright \mathbb{R}$, which we denote by T_f to avoid confusion. This linear functional is defined through the Hilbert product

$$T_f[\phi] = \int_{\mathbb{R}^n} f(\mathbf{x})\phi(\mathbf{x})d^n x, \tag{8.97}$$

where ϕ is any test function. The integral is well-defined because ϕ has compact support. Thus any continuous function f generates a generalized function T_f, which by abuse of notation is usually denoted by f too. But the converse is not true. Consider, e.g., the δ-function. It is defined as the linear functional

$$\delta[\phi] = \phi(\mathbf{0}). \tag{8.98}$$

In analogy with (8.97), this is often rewritten in the more informal way

$$\int \delta(x)\phi(\mathbf{x})d^n x = \phi(\mathbf{0}).$$

But there exists no ordinary function "$\delta(x)$" satisfying this integral identity, so strictly speaking, it makes no sense!

Okay, so much for the generalized functions. Consider a manifold M with metric g. We want to introduce singular k-forms. As the test space $D^k(M)$ we use the set of all smooth k-forms with compact support. They are called *test forms*. We can then define:

Definition 8.7. A weak k-form $\mathbf{T}$ is a linear functional on the test space $D^k(M)$; i.e., $\mathbf{T}$ is a linear map

$$\mathbf{T} : D^k(M) \curvearrowright \mathbb{R}.$$

Let us first check that an arbitrary smooth k-form $\mathbf{T}$ can be represented as a linear functional, so that the weak k-forms include the smooth k-forms. Let $\mathbf{T}$ be a smooth k-form. Then the linear functional is defined through the Hilbert product

$$\mathbf{T}[\mathbf{U}] \stackrel{\text{def}}{=} \langle \mathbf{T} \mid \mathbf{U} \rangle = \int_M \frac{1}{k} T^{i_1 \ldots i_k}(x) U_{i_1 \ldots i_n}(x) \sqrt{g}\, d^n x, \tag{8.99}$$

where the Hilbert product is well-defined because the test form $\mathbf{U}$ has compact support.

Let us discuss some specific examples of weak forms. First, we consider the electric current **J** associated with a point particle in Minkowski space. In the naive approach it was characterized by the contravariant components

$$J^{\mu}(x) = q \int \frac{1}{\sqrt{-g(x)}} \delta^4(x - x(\tau)) \frac{dx^{\mu}}{d\tau} \, d\tau.$$

Let **U** be a test form. Then formally we have:

$$\begin{aligned} \langle \mathbf{J} \mid \mathbf{U} \rangle &= \int_M q \int \frac{1}{\sqrt{-g(x)}} \delta^4(x - x(\tau)) \frac{dx^{\mu}}{d\tau} \, d\tau U_{\mu}(x) \sqrt{-g(x)} \, d^4x \\ &= q \int \frac{dx^{\mu}}{d\tau} \left[\int_M U_{\mu}(x) \delta^4(x - x(\tau)) d^4x \right] d\tau \\ &= q \int \frac{dx^{\mu}}{d\tau} \, U_{\mu}(x(\tau)) d\tau = q \int_{\Gamma} \mathbf{U}. \end{aligned}$$

Within the framework of distributions we therefore define the singular current as the linear functional

$$\mathbf{J}[\mathbf{U}] \stackrel{\text{def}}{=} q \int_{\Gamma} \mathbf{U}, \tag{8.100}$$

where Γ is the world line of the particle!

Next, we consider the singular Coulomb field in the ordinary Euclidean space $\mathbb{R}^3$. Its components are infinite at the origin, and we must therefore define the Hilbert product through a limit procedure. Let us put $M_{\epsilon} = \{\mathbf{x} \mid \|\mathbf{x}\| > \epsilon\}$. Then we represent **E** by the linear functional

$$\mathbf{E}[\mathbf{U}] \stackrel{\text{def}}{=} \lim_{\epsilon \to 0} \frac{q}{4\pi\epsilon_0} \int_{M_{\epsilon}} \frac{x^i}{r^3} \, U_i(\mathbf{x}) d^3x. \tag{8.101}$$

(That the limit is well-defined follows easily if we work out the integral in spherical coordinates!)

Now, if the weak forms are going to be of any use, we must be able to extend at least part of the exterior calculus. The differential calculus is the easiest to extend:

(a) *The exterior derivative*: Suppose first that **T** is a smooth k-form. Then $\boldsymbol{d}\mathbf{T}$ is represented by the linear functional

$$\boldsymbol{d}\mathbf{T}[\mathbf{U}] = \langle \boldsymbol{d}\mathbf{T} \mid \mathbf{U} \rangle = \langle \mathbf{T} \mid \delta \mathbf{U} \rangle,$$

where we can throw away the boundary term because the test form **U** has compact support. But test forms are always smooth, so if **T** is a weak form, we can immediately generalize the above result and define $\boldsymbol{d}\mathbf{T}$ to be the linear functional

$$\boldsymbol{d}\mathbf{T}[\mathbf{U}] \stackrel{\text{def}}{=} \mathbf{T}[\delta \mathbf{U}]. \tag{8.102}$$

(b) *The codifferential*: Suppose first that **T** is a smooth k-form. Then $\delta\mathbf{T}$ is represented by the linear functional

$$\delta\mathbf{T}[\mathbf{U}] = \langle \delta\mathbf{T} \mid \mathbf{U} \rangle = \langle \mathbf{T} \mid \boldsymbol{d}\mathbf{U} \rangle.$$

When $\mathbf{T}$ is a weak form, we can therefore define $\boldsymbol{\delta}\mathbf{T}$ to be the linear functional

$$\boldsymbol{\delta}\mathbf{T}[\mathbf{U}] \stackrel{\text{def}}{=} \mathbf{T}[\boldsymbol{d}\,\mathbf{U}]. \tag{8.103}$$

Exercise 8.9.2
Problem: Let M be a Riemannian manifold. Show that the Laplacian of a weak form $\mathbf{T}$ is represented by the linear functional

$$\boldsymbol{\Delta}\mathbf{T}[\mathbf{U}] \stackrel{\text{def}}{=} \mathbf{T}[\boldsymbol{\Delta}\mathbf{U}]. \tag{8.104}$$

Exercise 8.9.3
Problem: Show that we can extend the dual map to weak forms in the following way: If $\mathbf{T}$ is a weak k-form, then $^*\mathbf{T}$ is the weak $(n-k)$-form represented by the linear functional

$$^*\mathbf{T}[\mathbf{U}] \stackrel{\text{def}}{=} \begin{cases} (-1)^{k(n-k)}\mathbf{T}[^*\mathbf{U}] & \text{on a Riemannian manifold;} \\ -(-1)^{k(n-k)}\mathbf{T}[^*\mathbf{U}] & \text{on a manifold with Minkowski metric.} \end{cases} \tag{8.105}$$

Having generalized the differential operators, we can easily check some of their fundamental properties. For instance, we get

$$\boldsymbol{d}^2\mathbf{T} = 0 \quad \text{and} \quad \boldsymbol{\delta}^2\mathbf{T} = 0,$$

even if $\mathbf{T}$ is weak. This follows immediately from definitions (8.102) and (8.103) and the corresponding properties of the test forms. (E.g., we obtain

$$\boldsymbol{d}^2\mathbf{T}[\mathbf{U}] = \mathbf{T}[\boldsymbol{\delta}^2\mathbf{U}] = \mathbf{T}[0] = 0,$$

so that $\boldsymbol{d}^2\mathbf{T}$ is the zero functional, and similarly for $\boldsymbol{\delta}^2\mathbf{T}$.)

Worked Exercise 8.9.4
Problem: (a) Let $\mathbf{J}$ be the singular current (8.100) associated with a point particle in Minkowski space. Show within the framework of distributions that it is conserved; i.e., $\boldsymbol{\delta}\mathbf{J} = 0$.
(b) Let $\mathbf{E}$ be the singular Coulomb field (8.101) in ordinary Euclidean space $\mathbb{R}^3$. Show within the framework of distributions that

$$-\boldsymbol{\delta}\mathbf{E} = \frac{q}{\epsilon_0}\,\delta(x)\delta(y)\delta(z).$$

Observe that the space of weak k-forms incorporates not only all smooth k-forms, but also all orientable regular k-dimensional domains! If Ω is a k-dimensional orientable regular domain, we represent it as the linear functional

$$\Omega[\mathbf{U}] = \int_\Omega \mathbf{U}. \tag{8.106}$$

Since test forms have compact support, we can even allow Ω to be an unbounded noncompact domain. This is a very powerful generalization of the usual δ-function: A 0-dimensional regular domain Ω consisting of the single point P_0 is represented by the linear functional

$$\Omega[\phi] = \int_{P_0} \phi = \phi(P_0);$$

i.e., it generates the usual δ-function. Using this notation, we observe that the singular four-current (8.100) associated with a point particle is essentially identical to the world line of the point particle, since

$$\Gamma[\mathbf{U}] = \int_{\Gamma} \mathbf{U}; \qquad \text{i.e., } \mathbf{J} = q\Gamma.$$

Observe that for regular domains we have introduced the *boundary* operator $\partial : \Omega \curvearrowright \partial\Omega$. This can now be interpreted as a differential operator! The boundary of Ω is represented by the linear functional

$$\partial\Omega[\mathbf{U}] = \int_{\partial\Omega} \mathbf{U}.$$

Using Stokes's theorem and (8.102), this can be rearranged as

$$\partial\Omega[\mathbf{U}] = \int_{\Omega} \boldsymbol{d}\,\mathbf{U} = \Omega[\boldsymbol{d}\,\mathbf{U}] = \boldsymbol{\delta}\Omega[\mathbf{U}],$$

so that the boundary operation coincides with the codifferential! The important property $\partial\partial\Omega = \phi$ is now seen as a special case of the rule $\boldsymbol{\delta}^2\mathbf{T} = 0$.

Having come this far, we might think that everything can be done using singular forms. But that is not true! When we try to extend the wedge product or the integral calculus, we run into trouble:

Consider the wedge product of a smooth k-form $\mathbf{T}$ and a smooth m-form $\mathbf{S}$. It is represented by the linear functional

$$\mathbf{T} \wedge \mathbf{S}[\mathbf{U}] \overset{\text{def.}}{=} \langle \mathbf{T} \wedge \mathbf{S} \mid \mathbf{U} \rangle = \int_M {}^*\mathbf{U} \wedge \mathbf{T} \wedge \mathbf{S} = (-1)^{km} \int {}^*\mathbf{U} \wedge \mathbf{S} \wedge \mathbf{T}.$$

Using (7.58) this can be rewritten as

$$\mathbf{T} \wedge \mathbf{S}[\mathbf{U}] = \int {}^*(\mathbf{U} \cdot \mathbf{S}) \wedge \mathbf{T} = \langle \mathbf{T} \mid \mathbf{U} \cdot \mathbf{S} \rangle = \mathbf{T}[\mathbf{U} \cdot \mathbf{S}].$$

This shows immediately that we can generalize the wedge product to the case where one of the factors is weak and the other smooth. If $\mathbf{T}$ is a weak k-form and $\mathbf{S}$ a smooth m-form, then we define $\mathbf{T} \wedge \mathbf{S}$ by the linear functional

$$\mathbf{T} \wedge \mathbf{S}[\mathbf{U}] \overset{\text{def.}}{=} \mathbf{T}[\mathbf{U} \cdot \mathbf{S}]. \tag{8.107}$$

But if $\mathbf{S}$ is singular too, then this will not work, because $\mathbf{U} \cdot \mathbf{S}$ is then no longer a test form.

If we introduce a suitable topology on the test space $D^k(M)$ and use limit procedures, then one can sometimes extend the wedge product to a pair of weak forms or similarly extend the integral to an integral of a weak form over a regular domain; but it is not always possible! These deficiencies of the extended exterior calculus should not be underestimated. "Distribution" is not a magic word that can be used to justify any calculation we want to perform with singular quantities. Consider, e.g., the derivation of the electromagnetic energy momentum tensor in Section 1.6. Here we discussed the electromagnetic field generated by a collection of point particles. The equation (1.40)

$$\partial_\beta T^{\alpha\beta}_{\text{field}} = -F^{\alpha}_{\ \gamma} J^{\gamma}$$

had a central position in the argument. Consider the right-hand side. Here the electromagnetic field strength $F^{\alpha}_{\ \gamma}$ is singular at the position of a particle, and so is the current J^{γ}. But then their product is not well-defined, not even in the sense of distributions! Consequently, the argument in Section 1.6 is only a heuristic argument of didactic importance, not a proof in the strict sense.

Weak forms were introduced by de Rham. Among other things, he was motivated by the electric current associated with a point particle. He therefore used the name "currents" for weak forms in general. That is, however, misleading in a physical context, and I have therefore adopted the name "weak form." The reader who wants a more rigorous treatment should consult de Rham (1955) or Gelfand and Shilov (1964).

Solutions to Worked Exercises

Solution to 8.3.3

(a) We use spherical coordinates adapted to S^2_+:

$$S^2_+ = \left\{(r, \theta, \varphi) \mid r = 1; 0 \leq \theta \leq \frac{\pi}{2} ; 0 \leq \varphi \leq 2\pi\right\},$$

$$x = r \sin\theta \cos\varphi, \qquad y = r \sin\theta \sin\varphi,$$

$$\boldsymbol{dx} = \sin\theta \cos\varphi\, \boldsymbol{dr} + r\cos\theta\cos\varphi\, \boldsymbol{d\theta} - r\sin\theta\sin\varphi\, \boldsymbol{d\varphi},$$

$$\boldsymbol{dy} = \sin\theta \sin\varphi\, \boldsymbol{dr} + r\cos\theta\sin\varphi\, \boldsymbol{d\theta} + r\sin\theta\cos\varphi\, \boldsymbol{d\varphi}.$$

When we restrict $\boldsymbol{dx}$ and $\boldsymbol{dy}$ to S^2_+, we put r equal to 1 and $\boldsymbol{dr}$ equal to zero, whereby we get

$$\boldsymbol{dx} = \cos\theta\cos\varphi\, \boldsymbol{d\theta} - \sin\theta\sin\varphi\, \boldsymbol{d\varphi},$$

$$\boldsymbol{dy} = \cos\theta\sin\varphi\, \boldsymbol{d\theta} + \sin\theta\cos\varphi\, \boldsymbol{d\varphi};$$

i.e.,

$$\boldsymbol{dx} \wedge \boldsymbol{dy} = \cos\theta\sin\theta\, \boldsymbol{d\theta} \wedge \boldsymbol{d\varphi}.$$

Then we replace the geometrical volume element $\boldsymbol{d\theta} \wedge \boldsymbol{d\varphi}$ with the "Riemannian" volume element $d\theta\, d\varphi$ and finally get

$$\int_{S^2_+} \boldsymbol{dx} \wedge \boldsymbol{dy} = \int_{\theta=0}^{\pi/2} \int_{\varphi=0}^{2\pi} \cos\theta\sin\theta\, d\theta\, d\varphi = \pi.$$

(b) Spherical coordinates are also adapted to B^3:

$$B^3 = \{(r, \theta, \varphi) \mid r \leq 1, 0 \leq \theta \leq \pi, 0 \leq \varphi \leq 2\pi\}.$$

Observe that $\boldsymbol{dx} \wedge \boldsymbol{dy} \wedge \boldsymbol{dz}$ is the Levi-Civita form in $\mathbb{R}^3$. In terms of spherical coordinates, we can therefore rearrange it as

$$\boldsymbol{dx} \wedge \boldsymbol{dy} \wedge \boldsymbol{dz} = \sqrt{g}\, \boldsymbol{dr} \wedge \boldsymbol{d\theta} \wedge \boldsymbol{d\varphi}.$$

Using (6.36) we now get

$$g = r^4 \sin^2\theta; \qquad \text{i.e., } \sqrt{g} = r^2 \sin\theta.$$

When we replace the geometrical volume element $\boldsymbol{dr} \wedge \boldsymbol{d\theta} \wedge \boldsymbol{d\varphi}$ with the "Riemannian" volume element $dr\, d\theta\, d\varphi$, we finally arrive at

$$\int_{B^3} \boldsymbol{dx} \wedge \boldsymbol{dy} \wedge \boldsymbol{dz} = \int_{r=0}^{1} \int_{\theta=0}^{\pi} \int_{\varphi=0}^{2\pi} r^2 \sin\theta\, dr\, d\theta\, d\varphi = \frac{4}{3}\pi,$$

which we recognize as the volume of the unit ball.

Solution to 8.4.1

A smooth 2-form $\mathbf{B}$ can be decomposed as

$$\mathbf{B} = B_3(x, y, z)\boldsymbol{dx} \wedge \boldsymbol{dy} + B_1(x, y, z)\boldsymbol{dy} \wedge \boldsymbol{dz} + B_2(x, y, z)\boldsymbol{dz} \wedge \boldsymbol{dx}.$$

Due to linearity, it suffices to verify Stokes's theorem term by term. We have thus reduced the problem to the consideration of a smooth 2-form of the form

$$\mathbf{B} = B(x, y, z)\boldsymbol{dx} \wedge \boldsymbol{dy}.$$

We then get

$$\boldsymbol{d}\mathbf{B} = \frac{\partial B}{\partial z}\, \boldsymbol{dz} \wedge \boldsymbol{dx} \wedge \boldsymbol{dy}.$$

Cartesian coordinates are adapted to the unit cube, and the two integrals can now easily be computed

$$\text{(1)} \qquad \int_\Omega \boldsymbol{d}\mathbf{B} = \int_\Omega \frac{\partial B}{\partial z}\, \boldsymbol{dz} \wedge \boldsymbol{dx} \wedge \boldsymbol{dy} = \int_{x=0}^{1} \int_{y=0}^{1} \int_{z=0}^{1} \frac{\partial B}{\partial z}\, dz\, dx\, dy$$

$$= \int_{x=0}^{1} \int_{y=0}^{1} [B(x, y, 1) - B(x, y, 0)]dx\, dy.$$

The boundary $\partial\Omega$ consists of six surfaces, but only two of them contribute, since $\boldsymbol{dx} = 0$ (or $\boldsymbol{dy} = 0$) along surfaces where x (or y) is constant. The bottom and the top of the cube are oppositely oriented: The Cartesian coordinates (x, y) are positively oriented on the top but negatively oriented on the bottom (see Figure 8.36). We now get

$$\text{(2)} \qquad \int_{\partial\Omega} \mathbf{B} = \int_{\partial\Omega} B(x, y, z)\boldsymbol{dx} \wedge \boldsymbol{dy}$$

$$= \int_{x=0}^{1} \int_{y=0}^{1} B(x, y, 1)dx\, dy - \int_{x=0}^{1} \int_{y=0}^{1} B(x, y, 0)dx\, dy,$$

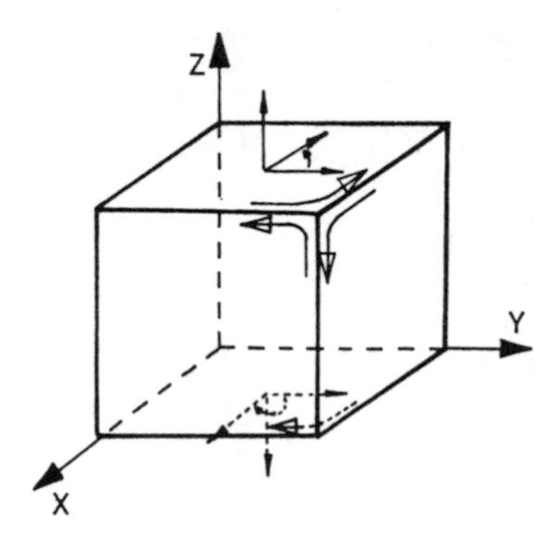

Figure 8.36.

form which we see immediately that (1) and (2) are identical.

Solution to 8.4.3

a. Let Γ have the parametrization $z(t) = x(t) + iy(t)$, $a \leq t \leq b$. When we evaluate the complex line integral, we must make the substitution

$$dz \curvearrowright \left[\frac{\partial x}{\partial t} + i \frac{\partial y}{\partial t} \right] dt.$$

When we want to integrate the one-form $h(z)\boldsymbol{dz}$, we must find the restriction of $\boldsymbol{dz}$ to Γ,

$$\boldsymbol{dz} \curvearrowright \frac{\partial z}{\partial t}\, \boldsymbol{dt} = \left[\frac{\partial x}{\partial t} + i \frac{\partial y}{\partial t} \right] \boldsymbol{dt},$$

and then exchange the geometrical volume element $\boldsymbol{dt}$ with the "Riemannian" volume element dt. Both integrals therefore reduce to

$$\int_b^a [f(x, y) + ig(x, y)] \left[\frac{\partial x}{\partial t} + i \frac{\partial y}{\partial t} \right] dt.$$

b. Notice that

$$\boldsymbol{dh}(z) = \frac{\partial h}{\partial z}\, \boldsymbol{dz} + \frac{\partial h}{\partial \bar{z}}\, \boldsymbol{d\bar{z}},$$

but $h(z)$ is holomorphic, and therefore $\frac{\partial h}{\partial \bar{z}} = 0$ (Cauchy–Riemann equation). Consequently, we get

$$\boldsymbol{dh}(z) = \frac{\partial h}{\partial z}\, \boldsymbol{dz}.$$

The formula is now a simple consequence of Stokes's theorem:

$$\int_\gamma h'(z)\boldsymbol{d}z = \int_\Gamma \boldsymbol{d}h(z) = \int_{\partial\Gamma} h(z) = h(B) - h(A).$$

c. According to Exercise 7.6.9, $h(z)\boldsymbol{d}z$ is a closed form. Therefore, we get

$$\int_\Gamma h(z)\boldsymbol{d}z = \int_{\partial\Omega} h(z)\boldsymbol{d}z = \int_\Omega \boldsymbol{d}[h(z)\boldsymbol{d}z] = 0.$$

Solution to 8.5.1

$$\begin{aligned}
\int_{\partial\Omega} {}^*[\phi\boldsymbol{d}\psi] &= \int_{\partial\Omega} \phi \wedge {}^*\boldsymbol{d}\psi = \int_\Omega \boldsymbol{d}[\phi \wedge {}^*\boldsymbol{d}\psi] \\
&= \int_\Omega \boldsymbol{d}\phi \wedge {}^*\boldsymbol{d}\psi + \int_\Omega \phi \wedge \boldsymbol{d}^*\boldsymbol{d}\psi \\
&= \int_\Omega \boldsymbol{d}\phi \wedge {}^*\boldsymbol{d}\psi - \int_\Omega \phi \wedge {}^*\boldsymbol{\Delta}\psi \\
&= (-1)^{n-1}\left[\int_\Omega {}^*\boldsymbol{d}\psi \wedge \boldsymbol{d}\phi + \int_\Omega {}^*\boldsymbol{\Delta}\psi \wedge \phi\right] \\
&= (-1)^{n-1}[\langle \boldsymbol{d}\psi \mid \boldsymbol{d}\phi\rangle + \langle \boldsymbol{\Delta}\psi \mid \phi\rangle],
\end{aligned}$$

which shows (8.32). If we interchange ϕ and ψ, we get

$$\int_{\partial\Omega} {}^*[\psi\boldsymbol{d}\phi] = (-1)^{n-1}[\langle \boldsymbol{d}\phi \mid \boldsymbol{d}\psi\rangle + \langle \boldsymbol{\Delta}\phi \mid \psi\rangle].$$

Then we subtract these two identities and get (8.33):

$$\int_{\partial\Omega} {}^*[\phi\boldsymbol{d}\psi] - \int_{\partial\Omega} {}^*[\psi\boldsymbol{d}\phi] = (-1)^{n-1}\left[\langle \boldsymbol{\Delta}\psi \mid \phi\rangle - \langle \psi \mid \boldsymbol{\Delta}\phi\rangle\right].$$

If we put $\phi = \psi$ in (8.32), we finally get

$$\langle \boldsymbol{d}\phi \mid \boldsymbol{d}\phi\rangle + \langle \phi \mid \boldsymbol{\Delta}\phi\rangle = (-1)^{n-1}\int_{\partial\Omega} {}^*[\phi\boldsymbol{d}\phi].$$

If ϕ is a harmonic function that vanishes on the boundary, we conclude that

$$\langle \boldsymbol{d}\phi \mid \boldsymbol{d}\phi\rangle = 0.$$

Therefore, $\boldsymbol{d}\phi = 0$, so that ϕ is constant; but since it vanishes on the boundary, it must vanish throughout Ω.

Solution to 8.5.2

a. This is imply the result of Exercise 8.5.1 formula (8.32) when we put $\psi = \phi$!

b.

$$d\phi = -\frac{1}{r^2}\left(\frac{x}{r}\,dx + \frac{y}{r}\,dy + \frac{z}{r}\,dz\right);$$
$$^*d\phi = -\frac{1}{r^2}\left(\frac{x}{r}\,dy\wedge dz + \frac{y}{r}\,dz\wedge dx + \frac{z}{r}\,dx\wedge dy\right).$$

Using this we get

$$\phi^* d\phi = -\frac{1}{r^4}\,[x dy\wedge dz + y\,dz\wedge dx + z\,dx\wedge dy] = -\frac{1}{r}\,\sin\theta\,d\theta\wedge d\varphi.$$

The integrals are now easily computed:

$$\langle d\phi \mid d\phi\rangle = \int_{M_\epsilon}\frac{1}{r^4}\,dx\wedge dy\wedge dz = 4\pi\int_{r=\epsilon}^{\infty}\frac{dr}{r^2} = \frac{4\pi}{\epsilon};$$

$$\int_{\partial M_\epsilon}\phi^* d\phi = -\frac{1}{\epsilon}\int_{\partial M_\epsilon}\sin\theta\,d\theta\wedge d\varphi = \frac{1}{\epsilon}\int_{\theta=0}^{\pi}\int_{\varphi=0}^{2\pi}\sin\theta\,d\theta\,d\varphi = \frac{4\pi}{\epsilon}.$$

Solution to 8.6.2

a. The Lagrangian density is a scalar field. Consequently,

$$\frac{\partial L}{\partial\epsilon}\Big|_{\epsilon=0} = \frac{\partial L}{\partial\phi}\,\phi + \frac{\partial L}{\partial(\partial_\mu\phi)}\,\partial_\mu\phi$$

is a scalar field too. From Exercise 6.9.3 it now follows that $\frac{\partial L}{\partial\phi}$ is a scalar field, while $\frac{\partial L}{\partial(\partial_\mu\phi)}$ are the contravariant components of a vector field.

b. The left-hand side is the Hilbert product of $\mathbf{A}$ and $d\psi$:

$$\int_\Omega A^\alpha(\partial_\alpha\psi)\sqrt{-g}\,d^4x = \langle \mathbf{A}\mid d\psi\rangle = \int_\Omega {}^*\mathbf{A}\wedge d\psi.$$

Using the generalized theorem of partial integration (Section 8.5), this is rearranged as

$$= -\int_{\partial\Omega}{}^*\mathbf{A}\wedge\psi + \int_\Omega d^*\mathbf{A}\wedge\psi.$$

But here the boundary term vanishes, since $\psi = 0$ on the boundary $\partial\Omega$. Using that

$$d^*\mathbf{A} = {}^*\delta\mathbf{A}$$

(see (7.62)), we finally get

$$\int_\Omega A^\alpha(\partial_\alpha\psi)\sqrt{-g}\,d^4x = \langle\mathbf{A}\mid d\psi\rangle = \langle\delta\mathbf{A}\mid\psi\rangle$$
$$= -\int_\Omega \partial_\alpha(\sqrt{-g}\mathbf{A}^\alpha)\psi d^4x.$$

Moral: When one of the factors vanishes on the boundary, then "$\boldsymbol{\delta}$ and $\boldsymbol{d}$ are adjoint operators."

Solution to 8.6.3

a.

$$S_I = q\int_{\lambda_1}^{\lambda_2} A_\mu \frac{dx^\mu}{d\lambda}\,d\lambda = q\int_{\lambda_1}^{\lambda_2}\left[\int_\Omega A_\mu(x)\delta^4(x - x(\lambda))d^4x\right]\frac{dx^\mu}{d\lambda}\,d\lambda$$
$$= \int_\Omega A_\mu(x)\left[q\int_{\lambda_1}^{\lambda_2}\frac{dx^\mu}{d\lambda}\frac{1}{\sqrt{-g}}\,\delta^4(x - x(\lambda))d\lambda\right]\sqrt{-g}\,d^4x$$
$$= \int_\Omega A_\mu(x)J^\mu(x)\sqrt{-g}\,d^4x.$$

b. The dynamical quantities to be varied are the position of the particle $x^\mu(\lambda)$ and the gauge potential $A_\mu(x)$:

$$x^\mu(\lambda) \to x^\mu(\lambda) + \epsilon y^\mu(\lambda); \qquad y^\mu(\lambda_1) = y^\mu(\lambda_2) = 0.$$
$$A_\mu(x) \to A_\mu(x) + \epsilon B_\mu(x); \qquad B_\mu = 0 \text{ on } \partial\Omega.$$

We then proceed in 4 steps:

1. *Variation of S_P*: This has been discussed in Section 6.7:

$$\frac{dS_P}{d\epsilon}\Big|_{\epsilon=0} = \int_{\tau_1}^{\tau_2}\left(m\,\frac{d^2x^\alpha}{d\tau^2} + m\Gamma^\alpha_{\ \mu\nu}\frac{dx^\mu}{d\tau}\frac{dx^\nu}{d\tau}\right)y_\alpha(\tau)d\tau.$$

2. *Variation of S_F*: This has been discussed in Exercise 8.6.1:

$$\frac{dS_F}{d\epsilon}\Big|_{\epsilon=0} = \int_\Omega \partial_\mu\left[\sqrt{-g}F^{\mu\nu}\right]B_\nu(x)d^4x.$$

3. *Variation of S_I with respect to the particle trajectory*:

$$S_I(\epsilon) = q\int_{\lambda_1}^{\lambda_2} A_\mu[x(\lambda) + \epsilon y(\lambda)]\left(\frac{dx^\mu}{d\lambda} + \epsilon\,\frac{dy^\mu}{d\lambda}\right)d\lambda;$$

i.e.,

$$\frac{dS_I}{d\epsilon}\Big|_{\epsilon=0} = q\int_{\lambda_1}^{\lambda_2}\left(\left[\frac{\partial A_\mu}{\partial x^\nu}\,y^\nu\right]\frac{dx^\mu}{d\lambda} + A_\mu\,\frac{dy^\mu}{d\lambda}\right)d\lambda$$
$$= q\int_{\lambda_1}^{\lambda_2}\left(\left[\frac{\partial A_\mu}{\partial x^\nu}\,y^\nu\right]\frac{dx^\mu}{d\lambda} - \frac{\partial A_\mu}{\partial x^\nu}\frac{dx^\nu}{d\lambda}\,y^\mu\right)d\lambda$$

(where we have performed a partial integration on the last term)

$$= q\int_{\lambda_1}^{\lambda_2}\left[\frac{\partial A_\mu}{\partial x^\nu} - \frac{\partial A_\nu}{\partial x^\mu}\right]\frac{dx^\mu}{d\lambda}\,y^\nu d\lambda$$
$$= q\int_{\lambda_1}^{\lambda_2} F_{\nu\mu}\frac{\partial x^\mu}{\partial\lambda}\,y^\nu d\lambda = q\int_{\tau_1}^{\tau_2} F^\alpha_{\ \beta}\frac{dx^\beta}{d\tau}\,y_\alpha d\tau,$$

where we have reintroduced the proper time τ.

4. *Variation of S_I with respect to the field*: Using the result obtained in (a), we easily get

$$\frac{dS_I}{d\epsilon}\Big|_{\epsilon=0} = \int_\Omega \sqrt{-g}\, J^\nu(x) B_\nu(x) d^4x.$$

Collecting all the results, we finally obtain

$$\begin{aligned} 0 &= \frac{dS}{d\epsilon}\Big|_{\epsilon=0} \\ &= \int_{\tau_1}^{\tau_2} -\left[m\, \frac{d^2x^\alpha}{d\tau^2} + m\Gamma^\alpha_{\mu\nu}\, \frac{dx^\mu}{d\tau}\, \frac{dx^\nu}{d\tau} - qF^\alpha_{\ \beta}\, \frac{dx^\beta}{d\tau}\right] y_\alpha d\tau \\ &\quad + \int_\Omega \left[\partial_\mu\left(\sqrt{-g} F^{\mu\nu}\right) + \sqrt{-g}\, J^\nu(x)\right] B_\nu d^4x. \end{aligned}$$

But y_α and B_ν are arbitrary, so this is consistent only if (8.37) and (7.85) are satisfied.

Solution to 8.7.3

a. To get the equations of motion for **B** we perform the variation

$$\mathbf{A} \curvearrowright \mathbf{A} + \epsilon \mathbf{U}$$

i.e.,

$$\mathbf{B} \curvearrowright \mathbf{B} + \epsilon\, \boldsymbol{d}\,\mathbf{U} \qquad \boldsymbol{D}\psi \curvearrowright \boldsymbol{D}\psi - i\, \frac{q}{\hbar}\, \mathbf{U}\psi.$$

We then obtain

$$\begin{aligned} H(\epsilon) = &\frac{1}{2}\, \langle \mathbf{B} \mid \mathbf{B}\rangle + \epsilon \langle \boldsymbol{d}\,\mathbf{U} \mid \mathbf{B}\rangle + \frac{1}{2}\, \epsilon^2 \langle \boldsymbol{d}\,\mathbf{U} \mid \boldsymbol{d}\,\mathbf{U}\rangle + \frac{1}{2}\, \langle \boldsymbol{D}\psi \mid \boldsymbol{D}\psi\rangle \\ &+ \frac{iq}{2\hbar}\, \epsilon \langle \mathbf{U}\psi \mid \boldsymbol{D}\psi\rangle - \frac{iq}{2\hbar}\, \epsilon \langle \boldsymbol{D}\psi \mid \mathbf{U}\psi\rangle \\ &+ \frac{1}{2}\, \frac{q^2}{\hbar^2}\, \epsilon^2 \langle \mathbf{U}\psi \mid \mathbf{U}\psi\rangle. \end{aligned}$$

Thus we must demand that

$$0 = \frac{dH}{d\epsilon}\Big|_{\epsilon=0} = \langle \boldsymbol{d}\,\mathbf{U} \mid \mathbf{B}\rangle + \frac{iq}{2\hbar}\, \left[\langle \mathbf{U}\psi \mid \boldsymbol{D}\psi\rangle - \langle \boldsymbol{D}\psi \mid \mathbf{U}\psi\rangle\right].$$

If we remember that $\langle \mathbf{U}\psi \mid \boldsymbol{D}\psi\rangle = \int U_i \bar{\psi} \boldsymbol{D}^i \psi \sqrt{g}\, d^3x$, etc., we can easily rearrange this as follows:

$$0 = \left\langle \mathbf{U} \mid \delta \mathbf{B} + \frac{iq}{2\hbar}\, (\bar{\psi} \boldsymbol{D}\psi - \psi \overline{\boldsymbol{D}\psi})\right\rangle.$$

This leads to the equations of motion

$$-\delta \mathbf{B} = \frac{iq}{2\hbar} (\bar{\psi} \boldsymbol{D}\psi - \psi \overline{\boldsymbol{D}\psi}).$$

b. To get the equations of motion for ψ, we perform the variation

$$\psi \curvearrowright \psi + \epsilon\phi;$$

i.e.,

$$\boldsymbol{D}\psi \curvearrowright \boldsymbol{D}\psi + \epsilon \boldsymbol{D}\phi; \qquad |\psi|^2 \curvearrowright |\psi|^2 + \epsilon\bar{\psi}\phi + \epsilon\psi\bar{\phi} + \epsilon^2|\phi|^2.$$

We then obtain

$$H(\epsilon) + \frac{1}{2} \langle \mathbf{B} \mid \mathbf{B}\rangle + \frac{1}{2} \langle \boldsymbol{D}\psi \mid \boldsymbol{D}\psi\rangle + \frac{1}{2} \epsilon\langle \boldsymbol{D}\phi \mid \boldsymbol{D}\psi\rangle$$
$$+ \frac{1}{2} \epsilon\langle \boldsymbol{D}\psi \mid \boldsymbol{D}\phi\rangle + \frac{1}{2} \epsilon^2\langle \boldsymbol{D}\phi \mid \boldsymbol{D}\phi\rangle + \int U[\psi + \epsilon\phi]\varepsilon.$$

Thus we must demand that

$$\begin{aligned}
0 &= \frac{dH}{d\epsilon}\Big|_{\epsilon=0} \\
&= \frac{1}{2} \langle \boldsymbol{D}\psi \mid \boldsymbol{D}\phi\rangle + \frac{1}{2} \langle \boldsymbol{D}\phi \mid \boldsymbol{D}\psi\rangle + \int \frac{\partial U}{\partial |\psi|^2} (\bar{\psi}\phi + \psi\bar{\phi})\epsilon \\
&= \frac{1}{2} \langle \boldsymbol{D}\psi \mid \boldsymbol{D}\phi\rangle + \frac{1}{2} \langle \boldsymbol{D}\phi \mid \boldsymbol{D}\psi\rangle + \left\langle \frac{\partial U}{\partial |\psi|^2} \psi \mid \phi \right\rangle \\
&\quad + \left\langle \phi \mid \frac{\partial U}{\partial |\psi|^2} \psi \right\rangle.
\end{aligned}$$

Here half of the terms involve ϕ and the other half involve $\bar{\phi}$. As usual, we may formally treat them as independent variations. We therefore get

$$0 = \langle \boldsymbol{D}\phi \mid \boldsymbol{D}\psi\rangle + \left\langle \phi \mid 2 \frac{\partial U}{\partial |\psi|^2} \psi \right\rangle.$$

Now we are almost through. But we should be a little careful about the first term:

$$\langle \boldsymbol{D}\phi \mid \boldsymbol{D}\psi\rangle = \left\langle \boldsymbol{d}\phi - i \frac{q}{\hbar} \mathbf{A}\phi \mid \boldsymbol{D}\psi \right\rangle = \langle \boldsymbol{d}\phi \mid \boldsymbol{D}\psi\rangle + i \frac{q}{\hbar} \langle \mathbf{A}\phi \mid \boldsymbol{D}\psi\rangle.$$

Here we use that

$$\langle \mathbf{A}\phi \mid \boldsymbol{D}\psi\rangle = \int A_i\bar{\phi}\boldsymbol{D}^i\psi\sqrt{g}\, d^3x,$$

so that we can rearrange it as

$$\langle \boldsymbol{D}\phi \mid \boldsymbol{D}\psi\rangle = \langle \phi \mid \delta\boldsymbol{D}\psi\rangle + i \frac{q}{\hbar} \langle \phi \mid {}^*(\mathbf{A} \wedge {}^*\boldsymbol{D}\psi)\rangle.$$

Thus we end up with the condition

$$0 = \left\langle \phi \mid \boldsymbol{\delta D}\psi + i\,\frac{q}{\hbar}\,{}^*(\mathbf{A} \wedge {}^*\boldsymbol{D}\psi) + 2\frac{\partial U}{\partial|\psi|^2}\,\psi \right\rangle.$$

This leads to the equation of motion

$$-\boldsymbol{\delta D}\psi - i\,\frac{q}{\hbar}\,{}^*(\mathbf{A} \wedge {}^*\boldsymbol{D}\psi) = 2\,\frac{\partial U}{\partial|\psi|^2}\,\psi.$$

Dualizing it, we finally get

$$2\,\frac{\partial U}{\partial|\psi|^2}\,\psi\varepsilon = \boldsymbol{d}^*\boldsymbol{D}\psi - i\,\frac{q}{\hbar}\,\mathbf{A} \wedge {}^*\boldsymbol{D}\psi = \left(\boldsymbol{d} - i\,\frac{q}{\hbar}\,\mathbf{A}\right){}^*\boldsymbol{D}\psi = \boldsymbol{D}^*\boldsymbol{D}\psi.$$

Solution 8.9.1

The verification of the current-conservation is done by a brute force calculation:

$$\begin{aligned}\frac{1}{\sqrt{-g}}\,\partial_\mu\left(\sqrt{-g}J^\mu\right) &= \frac{q}{\sqrt{-g}}\,\partial_\mu \int \frac{dx^\mu}{d\tau}\,\delta^4(x - x(\tau))d\tau \\ &= -\frac{q}{\sqrt{-g}} \int \frac{dx^\mu}{d\tau}\,\frac{\partial}{\partial x^\mu(\tau)}\,\delta^4(x - x(\tau))d\tau,\end{aligned}$$

but using the chain rule, this can be rearranged as

$$\begin{aligned}&= -\frac{q}{\sqrt{-g}} \int \frac{d}{d\tau}\,\delta^4(x - x(\tau))d\tau \\ &= -\frac{q}{\sqrt{-g}}\left[\delta^4(x - x(\tau))\right]_{\tau=-\infty}^{\tau=+\infty} = 0,\end{aligned}$$

since x is fixed and $|x^0(\tau)| \to \infty$ as $|\tau| \to \infty$.

Solution 8.9.4

a. Let ϕ be a test form of degree 0 (i.e., a smooth scalar field with compact support). Using (9.63) and (9.66), we get

$$\boldsymbol{\delta}\mathbf{J}[\phi] = \mathbf{J}[\boldsymbol{d}\phi] = q\int_\Gamma \boldsymbol{d}\phi = q[\phi(\tau = +\infty) - \phi(\tau = -\infty)] = 0,$$

since ϕ has compact support. Consequently, $\boldsymbol{\delta}\mathbf{J}$ is represented by the 0-functional.

b. Let ϕ be a test form of degree 0. Using (8.101) and (8.103), we get

$$-\boldsymbol{\delta}\mathbf{E}[\phi] = -\mathbf{E}[\boldsymbol{d}\phi] = -\lim_{\epsilon\to 0}\langle \mathbf{E} \mid \boldsymbol{d}\phi\rangle_{M_\epsilon}. \qquad (*)$$

But observe that

$$\mathbf{E} = \frac{q}{4\pi\epsilon_0}\,\frac{x^i}{r^3} = -\frac{q}{4\pi\epsilon_0}\,\boldsymbol{d}\psi,$$

where $\psi(\mathbf{x}) = \frac{1}{r}$. Inserting this into (*), we get

$$-\delta \mathbf{E}[\phi] = \frac{q}{4\pi\epsilon_0} \lim_{\epsilon \to 0} \langle \boldsymbol{d}\psi \mid \boldsymbol{d}\phi \rangle_{M_\epsilon}.$$

Now we use Green's identity (8.32) and rearrange it as

$$= \frac{q}{4\pi\epsilon_0} \lim_{\epsilon \to 0} \left[\int_{\partial M_\epsilon} {}^*\phi \boldsymbol{d}\psi - \langle \boldsymbol{\Delta}\psi \mid \phi \rangle \right].$$

But ψ is harmonic in M_ϵ, so the last term vanishes. In the first term we integrate over a sphere shrinking to a point. In the limit we can therefore safely replace $\phi(\mathbf{x})$ with $\phi(\mathbf{0})$:

$$= \frac{q}{4\pi\epsilon_0} \phi(\mathbf{0}) \lim_{\epsilon \to 0} \int_{\partial M_\epsilon} {}^*\boldsymbol{d}\psi.$$

Here we obtain

$${}^*\boldsymbol{d}\psi = -\frac{1}{r^3}\,(x\boldsymbol{d}y \wedge \boldsymbol{d}z + y\boldsymbol{d}z \wedge \boldsymbol{d}x + z\boldsymbol{d}x \wedge \boldsymbol{d}y) = -\sin\theta\, \boldsymbol{d}\theta \wedge \boldsymbol{d}\phi,$$

so that we get (M_ϵ is *outside* the sphere, so it gets the opposite orientation)

$$\int_{\partial M_\epsilon} {}^*\boldsymbol{d}\psi = 4\pi.$$

But then we have shown that

$$-\delta \mathbf{E}[\phi] = \frac{q}{\epsilon_0}\, \phi(\mathbf{0}),$$

which means that $-\delta\mathbf{E}$ is represented by $\frac{q}{\epsilon_0}$ times the δ-functional.

CHAPTER 9

Dirac Monopoles

9.1 Magnetic Charges and Currents

Previously, we have been considering only conventional electrodynamics, according to which magnetic charges are excluded. This is in accordance with experiments, where electric fields are always generated by electrically charged particles, whereas magnetic fields are generated from electric currents. Nevertheless, there is a priori no reason to exclude magnetic charges, and as shown by Dirac (1931) their existence would have interesting theoretical consequences. Especially, the magnetic charge will necessarily be quantized due to quantum-mechanical effects. Current ideas about the fundamental forces and the origin of the universe also strongly suggest that magnetic monopoles were in fact created in the early history of the universe. Most of these monopoles would have annihilated each other again (in the same way as matter and antimatter annihilate each other), but a small fraction may have survived, although they will be extremely difficult to detect, mainly due to their large mass. Even if magnetic monopoles do not exist, the underlying mathematical model is very interesting, and it has had a great impact on our understanding of gauge theories.

Let us consider a point particle that serves as a source for the electromagnetic field. We expect a singularity at the position of the particle (like the singularity of the Coulomb field). Therefore, we cut out the trajectory from Minkowski space; i.e., if Γ is the world line of the point particle, then the basic manifold is $L = M \setminus \Gamma$. We know then that the electromagnetic field is represented by a smooth 2-form $\mathbf{F}$ on L. Furthermore, $\mathbf{F}$ satisfies Maxwell's equations (7.88)

$$\boldsymbol{d}\mathbf{F} = 0; \qquad \boldsymbol{\delta}\mathbf{F} = \mathbf{J}.$$

Consider now the magnetic flux through a closed surface surrounding the point particle. To fix notation, let $\mathbf{S}$ be an inertial frame and consider a three-dimensional regular domain Ω contained in a space slice relative to $\mathbf{S}$; cf. Figure 9.1. We have then previously shown that the magnetic flux Φ through the boundary $\partial\Omega$ is given by $\int_{\partial\Omega} \mathbf{F}$ (cf. (8.51)).

We want to show that as a consequence of the Maxwell equations, the magnetic flux has the following three basic properties:

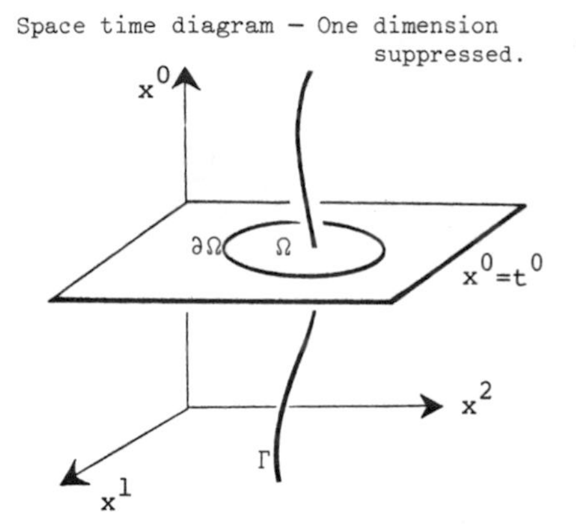

Figure 9.1.

1. *It is the same for all closed surfaces surrounding the point particle. As a consequence, it can be interpreted as the magnetic charge of the point particle* (*in the same way that the electric flux through a closed surface represents the electric charge within the surface*).
2. *It is independent of time; i.e., the magnetic charge is conserved.*
3. *It is independent of the observer; i.e., the magnetic charge is a Lorentz scalar* (*or to be precise, a pseudoscalar, since the flux depends upon the orientation*).

To show property (1), we consider two closed surfaces Ω_1 and Ω_2 contained in the same space slice; cf. Figure 9.2. The region between the two surfaces will be denoted by W. Notice that its boundary, ∂W, consists of the two surfaces Ω_1 and Ω_2, but that W induces opposite orientations on Ω_1 and Ω_2. From Stokes's theorem we now get

$$0 = \int_W \boldsymbol{d}\mathbf{F} = \int_{\partial W} \mathbf{F} = \int_{\Omega_2} \mathbf{F} - \int_{\Omega_1} \mathbf{F}; \qquad \text{i.e., } \Phi_1 = \Phi_2.$$

By repeating the same argument, we can similarly show properties (2) and (3). This requires the consideration of two closed surfaces contained in different space slices, as shown in Figure 9.2.

A point particle can, of course, in principle carry both an electric charge q and a magnetic charge g. If the particle carries both electric and magnetic charge, it is called a *dyon*.

We want to enlarge our considerations and consider a system where we have included *smooth* distribution of magnetic charges and currents. We know that the electric charges and currents may be combined in a smooth 1-form that acts as a source for the electromagnetic field through the Maxwell equation $\boldsymbol{d}^*\mathbf{F} = -^*\mathbf{J}$. This equation has two important consequences:

a. Taking the exterior derivative of both sides, we get, $\boldsymbol{d}^*\mathbf{J} = 0$, which is the continuity equation, i.e., the conservation of electric charge.

Space time diagram – One dimension suppressed

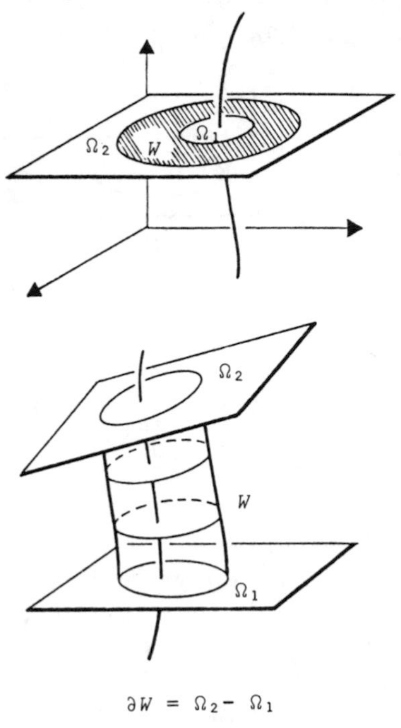

Figure 9.2.

b. If Ω is a three-dimensional volume contained in a space slice, then *the electric flux through $\partial\Omega$ is equal to the electric charge contained in Ω*:

$$\mathbf{Q} = -\int_{\Omega} {}^*\mathbf{J} = \int_{\Omega} \boldsymbol{d}^*\mathbf{F} = \int_{\partial\Omega} {}^*\mathbf{F}.$$

With this in mind, we now introduce a *magnetic* four-current $\mathbf{K}$, where k^0 is the magnetic charge density and $\mathbf{k} = (k^1, k^2, k^3)$ the magnetic current. The magnetic four-current is going to act as a source for the electromagnetic field. Consequently, we must modify the ordinary Maxwell equation, $\boldsymbol{d}\mathbf{F} = 0$. Now, if we insist that the modified equation should guarantee the conservation of magnetic charge and the identity of the magnetic charge contained in a volume with the magnetic flux through its surface, then we are forced to give it the following form:

$$\boldsymbol{d}\mathbf{F} = -{}^*\mathbf{K}. \tag{9.1}$$

In this form the symmetric role played by the electric and magnetic charges is displayed very clearly.

Exercise 9.1.1
Problem: Prove the following equivalences

$$\frac{1}{\sqrt{-g}}\,\partial_\alpha\left(\sqrt{-g}\,\overset{*}{F}{}^{\alpha\beta}\right) = -k^\beta$$
$$\text{iff}\quad \partial_\alpha F_{\beta\gamma} + \partial_\beta F_{\gamma\alpha} + \partial_\gamma F_{\alpha\beta} = \sqrt{-g}\,\epsilon_{\alpha\beta\gamma\delta}k^\delta. \tag{9.2}$$

$$\frac{1}{\sqrt{-g}}\,\partial_\alpha\left(\sqrt{-g}\,F^{\alpha\beta}\right) = j^\beta$$
$$\text{iff}\quad \partial_\alpha \overset{*}{F}_{\beta\gamma} + \partial_\beta \overset{*}{F}_{\gamma\alpha} + \partial_\gamma \overset{*}{F}_{\alpha\beta} = \sqrt{-g}\,\epsilon_{\alpha\beta\gamma\delta}j^\delta. \tag{9.3}$$

According to Exercise 9.1.1, we can translate (9.1) into the equivalent covariant formula:

$$\frac{1}{\sqrt{-g}}\,\partial_\mu\left(\sqrt{-g}\,\overset{*}{F}{}^{\mu\nu}\right) = K^\nu. \tag{9.4}$$

To summarize, we have established the following scheme for the extended Maxwell equations involving both electric and magnetic charges:

Conventional vector analysis	Covariant expressions	Geometric formula	
$\nabla \cdot \mathbf{B} = \rho_m$ $\frac{\partial \mathbf{B}}{\partial t} + \nabla \times \mathbf{E} = -\mathbf{k}$	$\partial_\alpha F_{\beta\gamma} + \partial_\beta F_{\gamma\alpha} + \partial_\gamma F_{\alpha\beta} = -\sqrt{-g}\varepsilon_{\alpha\beta\gamma\delta}K^\delta$ $\left[\frac{1}{\sqrt{-g}}\partial_\alpha(\sqrt{-g}\overset{*}{F}{}^{\alpha\beta}) = K^\beta\right]$	$dF = -{}^*K$ $\left[\delta\overset{*}{F} = -K\right]$	(9.5)
$\nabla \cdot \mathbf{E} = \frac{1}{\varepsilon_o}\rho$ $\frac{\partial \mathbf{E}}{\partial t} - c^2\nabla \times \mathbf{B} = -\frac{1}{\varepsilon_o}\mathbf{j}$	$\partial_\alpha \overset{*}{F}_{\beta\gamma} + \partial_\beta \overset{*}{F}_{\gamma\alpha} + \partial_\gamma \overset{*}{F}_{\alpha\beta} = -\sqrt{-g}\varepsilon_{\alpha\beta\gamma\delta}J^\delta$ $\left[\frac{1}{\sqrt{-g}}\partial_\beta(\sqrt{-g}F^{\alpha\beta}) = J^\alpha\right]$	$d^*F = -{}^*J$ $\left[\delta F = J\right]$	(9.6)

Using singular differential forms, we can also associate electric and magnetic currents with point particles. These currents are then represented by singular 1-forms **J** and **K** defined throughout the whole of Minkowski space. Similarly, the electromagnetic field is represented by a singular differential form **F** satisfying the modified Maxwell equations

$$-\boldsymbol{d}\mathbf{F} = {}^*\mathbf{K}, \qquad -\boldsymbol{d}^*\mathbf{F} = {}^*\mathbf{J}$$

on the whole of Minkowski space. The components of the singular currents are combined in the following scheme:

Conventional vector analysis	Covariant expression	
$\rho = q\delta^3(\mathbf{x} - \mathbf{x}(t))$ $\mathbf{j} = q\mathbf{v}\delta^3(\mathbf{x} - \mathbf{x}(t))$	$J^\mu = q \int \frac{dx^\mu}{d\tau} \frac{1}{\sqrt{-g}} \delta^4(x - x(\tau))d\tau$	(9.7)
$\rho_M = g_M\delta^3(\mathbf{x} - \mathbf{x}(t))$ $\mathbf{k} = g_M\mathbf{v}\delta^3(\mathbf{x} - \mathbf{x}(t))$	$K^\mu = g_M \int \frac{dx^\mu}{d\tau} \frac{1}{\sqrt{-g}} \delta^4(x - x(\tau))d\tau$	(9.8)

Within the framework of distributions, the currents are simply given by

$$\mathbf{J} = q\Gamma \quad \text{and} \quad \mathbf{K} = g_M\Gamma; \tag{9.9}$$

where Γ is the world line of the particle.

Next we consider the dynamics of charged point particles. They interact with the electromagnetic field and thus experience forces. To simplify calculations, we will restrict ourselves to inertial coordinates in what follows. We then already know that a charged particle experiences the Lorentz force, so that its equation of motion is given by (1.26)

$$m\frac{d^2x^\alpha}{d\tau^2} = qF^\alpha_{\ \beta}\frac{dx^\beta}{d\tau}.$$

But if the particle is magnetically charged too, it should experience a force analogous to the Lorentz force. It is this force we are going to determine. Although we might guess it correctly using a symmetry argument (see the illustrative example at the end of this section), we will derive it from a more general argument. *The interaction between the magnetically charged particles and the field should be constructed in such a way that the energy and momentum of the total system, i.e., fields and particles, are conserved.* (Compare the discussion in Section 1.6.)

Therefore, we consider a system consisting of a field and N particles. The nth particle is supposed to carry the electric charge q_n and the magnetic charge g_n (so we admit the possibility of dyons). We have previously determined the energy–momentum tensor of this system (Section 1.6). The contribution from the particles is given by (1.36)

$$T_P^{\alpha\beta} = \sum_{n=1}^N \int P_n^\alpha(\tau)\frac{dx_n^\beta}{d\tau}\,\delta^4(x - x_n(\tau))d\tau,$$

and the contribution from the field is given by (1.41)

$$T_F^{\alpha\beta} = -F^{\alpha\gamma}F_\gamma^{\ \beta} - \frac{1}{4}\,\eta^{\alpha\beta}\left[F^{\gamma\delta}F_{\gamma\delta}\right].$$

We must choose the interaction in such a way that

$$\partial_\beta T_P^{\alpha\beta} = -\partial_\beta T_F^{\alpha\beta}.$$

Okay! Let us get started. As in Section 1.6, we get (1.39)

$$\partial_\beta T_P^{\alpha\beta} = \sum_{n=1}^{N} \int \delta^4(x - x_n(\tau)) \frac{dP_n^\alpha}{d\tau} \, d\tau.$$

Then we look at the contribution from the field:

$$\begin{aligned} -\partial_\beta T_F^{\alpha\beta} &= \left(\partial_\beta F^\alpha_{\ \gamma}\right) F^{\gamma\beta} + F^\alpha_{\ \gamma}(\partial_\beta F^{\gamma\beta}) + \frac{1}{4}\,\eta^{\alpha\beta}\partial_\beta(F^{\gamma\delta}F_{\gamma\delta}) \\ &= F^\alpha_{\ \gamma} J^\gamma + (\partial_\beta F^\alpha_{\ \gamma})F^{\gamma\beta} + \frac{1}{4}\,\eta^{\alpha\beta}\partial_\beta(F^{\gamma\delta}F_{\gamma\delta}), \end{aligned} \qquad (*)$$

where we exchanged $\partial_\beta F^{\gamma\beta}$ with J^γ according to the Maxwell equations.

Worked Exercise 9.1.2
Problem: Show that the generalized Maxwell equations imply that

$$(\partial_\beta F^\alpha_{\ \gamma})F^{\gamma\beta} + \frac{1}{4}\,\eta^{\alpha\beta}\partial_\beta[F^{\gamma\delta}F_{\gamma\delta}] = -\overset{*}{F}{}^\alpha_{\ \beta} K^\beta.$$

According to Exercise 9.1.2, we can rearrange the expression $(*)$ as

$$-\partial_\beta T_F^{\alpha\beta} = F^\alpha_{\ \beta} J^\beta - \overset{*}{F}{}^\alpha_{\ \beta} K^\beta.$$

We then insert the expressions (9.7), (9.8) for the electric and magnetic currents and finally arrive at

$$-\partial_\beta T_F^{\alpha\beta} = \sum_{n=1}^{N} \int \left[q_n F^\alpha_{\ \beta} \frac{dx_n^\beta}{d\tau} - g_n \overset{*}{F}{}^\alpha_{\ \beta} \frac{dx_n^\beta}{d\tau} \right] \delta^4(x - x_n(\tau)) d\tau.$$

Comparing this with $\partial_\beta T_P^{\alpha\beta}$, we see that energy–momentum is conserved, provided that the following identity holds:

$$\frac{dP_n^\alpha}{d\tau} = q_n F^\alpha_{\ \beta} \frac{dx_n^\beta}{d\tau} - g_n \overset{*}{F}{}^\alpha_{\ \beta} \frac{dx_n^\beta}{d\tau}. \qquad (9.10)$$

Thus magnetic charge gives rise to a force very similar to the Lorentz force, except that we have interchanged the roles of electric and magnetic fields.

We may summarize this in the following scheme:

Forces arising from the interaction between charged particles and the fields Conventional vector analysis Lorentz invariant equations of motion

Forces arising from the interaction between charged particles and the fields		
Conventional vector analysis	Lorentz invariant equations of motion	
$\mathbf{F} = q(\mathbf{E} + \mathbf{v} \times \mathbf{B})$	$m\frac{d^2x^\alpha}{d\tau^2} = qF^\alpha_{\ \beta}\frac{dx^\beta}{d\tau}$	(9.11)
$\mathbf{F} = g(\mathbf{B} - \mathbf{v} \times \mathbf{E})$	$m\frac{d^2x^\alpha}{d\tau^2} = g\overset{*}{F}{}^\alpha_{\ \beta}\frac{dx^\beta}{d\tau}$	(9.12)

We can also write down the equations of motion (9.11), (9.12) in covariant form. As we have seen in Section 6.7, Theorem 6.4, this requires the introduction of Christoffel fields corresponding to the fictitious forces present in a noninertial frame. Thus we are led to the following covariant equation of motion:

$$m \frac{d^2 x^\alpha}{d\tau^2} = \left(q F^\alpha_{\ \beta} - g \overset{*}{F}{}^\alpha_{\ \beta}\right) \frac{dx^\beta}{d\tau} - m\Gamma^\alpha_{\ \mu\nu} \frac{dx^\mu}{d\tau} \frac{dx^\nu}{d\tau} . \tag{9.13}$$

(Compare this with (6.46).)

Illustrative Example: Charge Rotations. Having completed the discussion of magnetic charges, we should observe a curious fact. The equations of motion for the fields and particles exhibit a special kind of symmetry: They are invariant under the combined transformations

$$\begin{bmatrix} \mathbf{F} \\ {}^*\mathbf{F} \end{bmatrix} \mapsto \begin{bmatrix} \mathbf{F}' \\ {}^*\mathbf{F}' \end{bmatrix} = \begin{bmatrix} \cos\alpha & -\sin\alpha \\ \sin\alpha & \cos\alpha \end{bmatrix} \begin{bmatrix} \mathbf{F} \\ {}^*\mathbf{F} \end{bmatrix} \tag{9.14}$$

$$\begin{bmatrix} g_m \\ q \end{bmatrix} \to \begin{bmatrix} g'_m \\ q' \end{bmatrix} = \begin{bmatrix} \cos\alpha & -\sin\alpha \\ \sin\alpha & \cos\alpha \end{bmatrix} \begin{bmatrix} g_m \\ q \end{bmatrix} \tag{9.15}$$

(Observe that ${}^*\mathbf{F}'$ is really the dual of $\mathbf{F}'$! Observe also that the transformation rule for the charges is a consequence of the transformation rule for the field, when we identify charge with flux!)

To check this symmetry we first rearrange the equations of motion for the particles:

$$m \frac{d^2 x^\alpha}{d\tau^2} = [g_m; q] \begin{bmatrix} 0 & -1 \\ 1 & 0 \end{bmatrix} \begin{bmatrix} F^\alpha_{\ \beta} \\ {}^*F^\alpha_{\ \beta} \end{bmatrix} \frac{dx^\beta}{d\tau} .$$

We can now easily check the invariance by brute force. Then we observe that because the world lines of the particles are unaffected by the transformation, the currents will transform in the same way as the charges:

$$\begin{bmatrix} \mathbf{K} \\ \mathbf{J} \end{bmatrix} \to \begin{bmatrix} \mathbf{K}' \\ \mathbf{J}' \end{bmatrix} = \begin{bmatrix} \cos\alpha & -\sin\alpha \\ \sin\alpha & \cos\alpha \end{bmatrix} \begin{bmatrix} \mathbf{K} \\ \mathbf{J} \end{bmatrix} .$$

When we now rearrange the Maxwell equations in the form

$$\boldsymbol{d} \begin{bmatrix} \mathbf{F} \\ {}^*\mathbf{F} \end{bmatrix} = \begin{bmatrix} -{}^*\mathbf{K} \\ -{}^*\mathbf{J} \end{bmatrix} = -{}^* \begin{bmatrix} \mathbf{K} \\ \mathbf{J} \end{bmatrix} ,$$

we immediately see that this is invariant too because $\begin{bmatrix} \mathbf{F} \\ {}^*\mathbf{F} \end{bmatrix}$ and $\begin{bmatrix} \mathbf{K} \\ \mathbf{J} \end{bmatrix}$ transform with the same matrix.

This symmetry is usually referred to as *charge symmetry*. Decomposing the tensor **F**, we obtain the following charge transformation of the field strengths **E** and **B**:

$$\begin{bmatrix} \mathbf{B} \\ \mathbf{E} \end{bmatrix} \to \begin{bmatrix} \mathbf{B}' \\ \mathbf{E}' \end{bmatrix} = \begin{bmatrix} \cos\alpha & -\sin\alpha \\ \sin\alpha & \cos\alpha \end{bmatrix} \begin{bmatrix} \mathbf{B} \\ \mathbf{E} \end{bmatrix} . \tag{9.16}$$

Observe that the energy and momentum of the system are unaffected too by charge transformations. It is enough to check this for the electromagnetic field, but here it is obvious, as the energy and momentum densities are given by (1.42)

$$\epsilon = \frac{\epsilon_0}{2}(\mathbf{E}^2 + \mathbf{B}^2) \quad \text{and} \quad \mathbf{g} = \epsilon_0 \mathbf{E} \times \mathbf{B}.$$

The meaning of charge symmetry is presently not well understood, but it has one amusing consequence, which is in fact responsible for its name. On the classical level it is completely impossible to distinguish between electric and magnetic charge. What we call electric and what we call magnetic is only a matter of convention. We can redefine all charges by performing a charge transformation. If, e.g., we choose $\alpha = \pi/2$ we find that

$$q_n \to q_n' = g_n, \qquad g_n \to g_n' = -q_n;$$

i.e., what was before called electric charge is not called magnetic charge and vice versa.

Exercise 9.1.3
Problem: Introduce the complex-valued differential forms

$$\begin{aligned} F &= \mathbf{F} + i\overset{*}{\mathbf{F}} && \text{(Complex field strength)}, \\ J &= \mathbf{K} + i\mathbf{J} && \text{(Complex current)}, \\ Q &= g + iq && \text{(Complex charge)}. \end{aligned} \tag{9.17}$$

(a) Show that the complex field strength is anti–selfdual; i.e.,

$$^*F = -iF. \tag{9.18}$$

(b) Show that the equations of motion can be rearranged as

$$\boldsymbol{d}F = -{}^*J; \qquad m\frac{d^2x^\alpha}{d\tau^2} = \text{Re}(\bar{Q}iF^\alpha{}_\beta)\frac{dx^\beta}{d\tau}. \tag{9.19}$$

(c) Show that the charge symmetry transformations reduce to

$$F \to F' = e^{i\alpha}F; \qquad J \to J' = e^{i\alpha}J; \qquad Q \to Q' = e^{i\alpha}Q. \tag{9.20}$$

Show also that the Lagrangian of the free electromagnetic field is *not* invariant under charge rotations. (This is another example of a symmetry of the equations of motion that is not present at the Lagrangian level. Cf. the discussion in Section 2.2.)

9.2 The Dirac String

When we want to include magnetic monopoles in the Lagrangian formalism, we immediately run into difficulties. The electromagnetic field strength **F** is no longer derived from a global gauge potential **A**; i.e., it is not longer exact. If we insist on introducing gauge potentials, it can be done only at the expense of singularities in the gauge potentials (i.e., Dirac strings; cf. the discussion in Section 7.8), which were not present in the electromagnetic field strength. Somehow, we must learn how to handle these singularities.

ILLUSTRATIVE EXAMPLE: THE DIRAC STRING AS A PHYSICAL STRING. Consider once more the pure monopole field

$$\mathbf{B} = \frac{g_M}{4\pi r^3}\begin{bmatrix} x \\ y \\ z \end{bmatrix}. \tag{9.21}$$

It can be derived from the gauge potential (cf. (7.91)):

$$q\mathbf{A} = -\kappa(\cos\theta + 1)\nabla\varphi = \kappa \begin{bmatrix} \frac{y}{r(r-z)} \\ -\frac{x}{r(r-z)} \\ 0 \end{bmatrix} \qquad \text{with } \kappa = \frac{qg_M}{4\pi}. \tag{9.22}$$

However, $\mathbf{A}$ is singular at the origin ($r = 0$) and on the positive z-axis ($r = z$). The singularity at the origin reflects the singularity in the pure monopole field. The singularity on the positive z-axis constitutes the *string*, which was *not* present in the pure monopole field. We will now investigate this singularity and propose a physical interpretation.

To be able to control the singularities, we regularize the gauge potential

$$q\mathbf{A} \to q\mathbf{A}_\epsilon = \kappa \begin{bmatrix} \frac{y}{R(R-z)} \\ -\frac{x}{R(R-z)} \\ 0 \end{bmatrix} \qquad \text{with } R = \sqrt{r^2 + \epsilon^2}\,.$$

Observe that $\mathbf{A}$ is smooth throughout the whole of space. However, it has a tendency to "peak" at the origin and on the string. The corresponding electromagnetic field is given by

$$\mathbf{E}_\epsilon = -\frac{\partial \mathbf{A}_\epsilon}{\partial t} = \mathbf{0}, \qquad \mathbf{B}_\epsilon = \nabla \times \mathbf{A}_\epsilon.$$

They satisfy Maxwell's equations, provided that we introduce a current $\mathbf{j}_\epsilon$,

$$\nabla \times \mathbf{B}_\epsilon = \frac{1}{\epsilon_0 c^2}\,\mathbf{j}_\epsilon.$$

We may consider this to be an (extremely idealized!) model of a semi-infinite solenoid generating the magnetic field! (See Figure 9.3.)

In the limit $\epsilon \to 0$, the semi-infinite solenoid becomes *infinitely thin*, and we have a string of concentrated magnetic flux, which is spread out at the origin as the monopole field. Thus we have modified the originally pure monopole field by superimposing the string with the magnetic flux. The singularity of the $\mathbf{A}$-field at the position of the string now reflects the singularity of the magnetic field in *the modified monopole field*! Observe that if we compute the magnetic flux through a closed surface, we now get zero, because the flux of the monopole field is exactly compensated by the singular flux through the string. (See Figure 9.4.)

Worked Exercise 9.2.1
Problem: (a) Show that the regularized magnetic field is given by

$$\mathbf{B}_\epsilon = \frac{g_M}{4\pi}\left[\frac{\mathbf{r}}{R^3} - \frac{\epsilon^2(2R - z)}{R^3(R - z)^2}\,\hat{\mathbf{k}}\right].$$

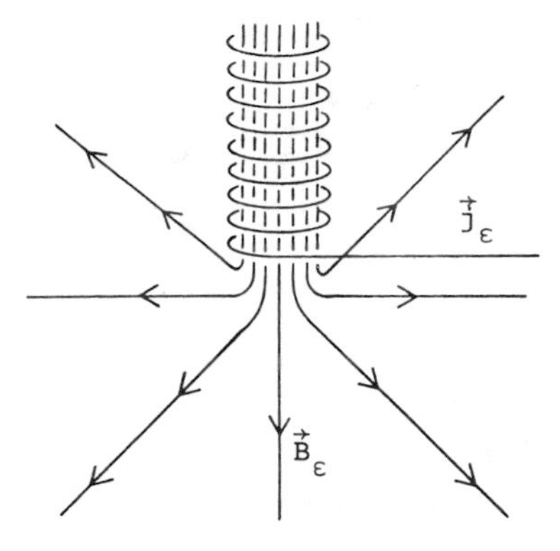

Figure 9.3.

(b) Show that the modified monopole field is given by

$$\mathbf{B} = \lim_{\epsilon \to 0} \mathbf{B}_\epsilon = \frac{g_M}{4\pi} \frac{\mathbf{r}}{r^3} - g_M(\delta(x)\delta(y)\Theta(z)\hat{\mathbf{k}}, \tag{9.23}$$

where $\Theta(z)$ is the Heaviside step function and $\hat{\mathbf{k}}$ is the unit vector along the z-axis.

The associated current $\mathbf{j}$ that is responsible for the concentrated magnetic flux along the string is extremely singular:

$$\frac{1}{\epsilon_0 c^2} \mathbf{j} = \nabla \times \mathbf{B} = -g_M \begin{bmatrix} \delta(x)\delta'(y) \\ -\delta'(x)\delta(y) \\ 0 \end{bmatrix} \cdot \Theta(z),$$

where $\delta'(x)$ is the derivative of the δ-function.

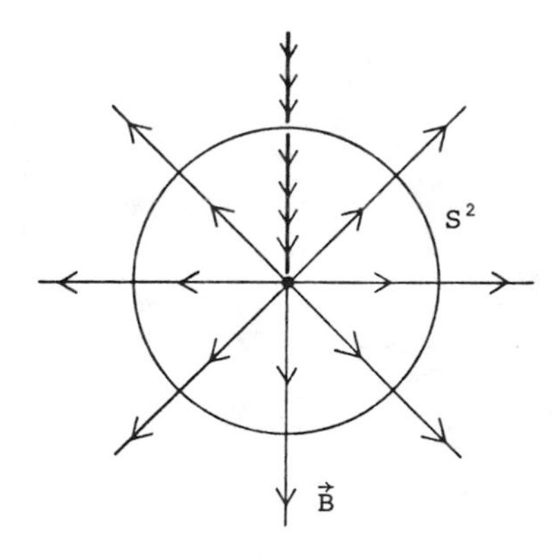

Figure 9.4.

The magnetic field associated with the string has some nice properties. It corresponds to a *negative* magnetic charge $-g_M$ at the origin, since

$$\mathbf{B}_{\text{STRING}} = -g_M \delta(x)\delta(y)\Theta(z)\hat{\mathbf{k}}$$

implies

$$\nabla \cdot \mathbf{B}_{\text{STRING}} = -g_M \delta(x)\delta(y)\frac{\partial \Theta(z)}{\partial z} = -g_M \delta(x)\delta(y)\delta(z).$$

This cancels exactly the positive magnetic charge g_M associated with the pure monopole field! Thus when we consider the modified monopole field (9.23) we see that the magnetic monopole has disappeared.

Consider the electromagnetic field $\mathbf{F}$ produced by a magnetic monopole and an electrically charged particle. This field has the following properties:

$$-\boldsymbol{d}\mathbf{F} = {}^*\mathbf{K} \qquad \text{(where } \mathbf{K} \text{ is the magnetic current)}, \tag{9.24a}$$

$$\int_\Omega \mathbf{F} = g_M \qquad \text{(where } g_M \text{ is the magnetic charge)}, \tag{9.24b}$$

where Ω is any sphere surrounding the monopole. Both of these properties make it impossible to generate $\mathbf{F}$ from a global gauge potential $\mathbf{A}$. Dirac cured these diseases in much the same manner as in the preceding example: First he chose a string extending from the monopole to infinity. This string sweeps out a two-dimensional sheet Σ in space–time. (See Figure 9.5.) The string is chosen completely arbitrarily, except that *it is never allowed to cross the trajectory of the charged particle.* This is known as *Dirac's veto*, and it will play a crucial role later on! To cure the diseases of $\mathbf{F}$ mentioned above, Dirac now introduces a singular electromagnetic field, which lives *only* on the string. From a physical point of view, Dirac's veto is then clear. If the string crosses the trajectory of a charged particle, the particle would be strongly influenced by the singular electromagnetic field associated to the string. But the string is a fictitious object, which must not disturb the electric charged particles.

Now, although the sheet Σ is *not* compact and consequently not a regular domain in the strict sense, it can still be decomposed into an interior region and a boundary. The boundary is simply the trajectory of the magnetic monopole. The interior region is a two-dimensional submanifold, int Σ, and the boundary is a closed one-dimensional submanifold, $\partial\Sigma = \Gamma_M$.

Fix a coordinate system (x^0, x^1, x^2, x^3) on Minkowski space. Then we may consider the coordinates as scalar functions $\tilde{x}^\alpha$ on int Σ,

$$P \curvearrowright \tilde{x}^\alpha(P).$$

Furthermore, we may consider the δ-function

$$\frac{1}{\sqrt{-g}}\,\delta^4(x - \tilde{x}(P))$$

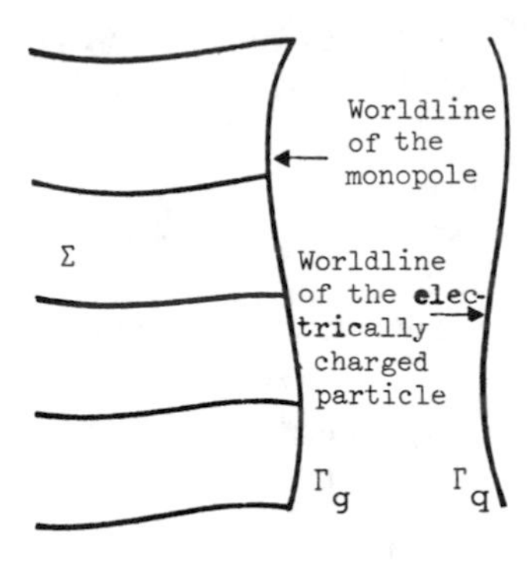

Figure 9.5.

as a scalar function on int Σ. (See Figure 9.6.) Using this we can construct the following two-form on int Σ:

$$-g_M \frac{1}{\sqrt{-g}} \delta^4(x - \tilde{x}(P)) d\tilde{x}^\alpha \wedge d\tilde{x}^\beta.$$

It being a two-form, we can then integrate it over the two-dimensional domain int Σ and obtain the quantity

$$\overset{*}{S}^{\alpha\beta}(x) = -g_M \int_\Sigma \frac{1}{\sqrt{-g}} \delta^4(x - \tilde{x}(P)) d\tilde{x}^\alpha \wedge d\tilde{x}^\beta. \tag{9.25}$$

The integral obviously depends on the point x in the δ-function and on the indices α and β. We have written the quantity in the form $\overset{*}{S}^{\alpha\beta}(x)$ because it forms the

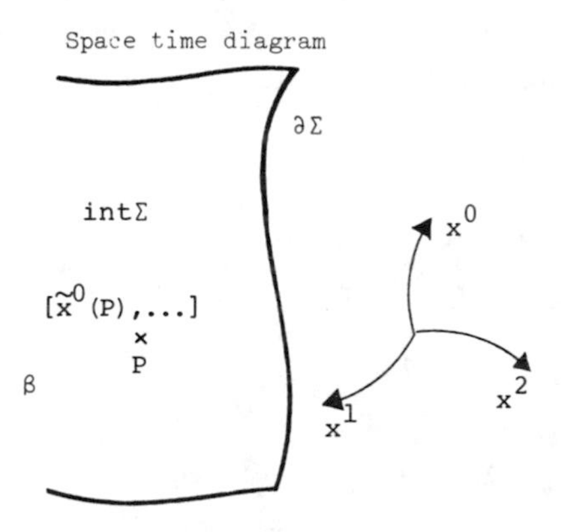

Figure 9.6.

contravariant components of a tensor. This is shown in exactly the same way as for the singular current (8.59) of Section 8.8.

The integral

$$-g_M \int_\Sigma \frac{1}{\sqrt{-g}}\, \delta^4(x - \tilde{x}(P))\boldsymbol{d}\tilde{x}^\alpha \wedge \boldsymbol{d}\tilde{x}^\beta$$

is a geometrical quantity, independent of the coordinate system we choose on int Σ. Usually we parametrize the sheet by a space-like parameter λ^1, where $-\infty < \lambda^1 < 0$, and a timelike parameter λ^2, where $-\infty < \lambda^2 < \infty$. Using such a parametrization, we can rearrange the integral as

$$\begin{aligned}\overset{*}{S}{}^{\alpha\beta}(x) = -g_M &\int_U \frac{1}{\sqrt{-g}}\, \delta^4(x - \tilde{x}(\lambda^1, \lambda^2)) \\ &\times \left[\frac{\partial\tilde{x}^\alpha}{\partial\lambda^1}\frac{\partial\tilde{x}^\beta}{\partial\lambda^2} - \frac{\partial\tilde{x}^\alpha}{\partial\lambda^2}\frac{\partial\tilde{x}^\beta}{\partial\lambda^1}\right] d\lambda^1 d\lambda^2\end{aligned} \tag{9.26}$$

with $U = \{(\lambda^1, \lambda^2) \mid \lambda^1 < 0\}$. We will use such an explicit representation of the integral whenever it is preferable.

Clearly, the tensor field with components $\overset{*}{S}{}^{\alpha\beta}(x)$ is a skew-symmetric tensor field. Consequently, $\overset{*}{S}{}^{\alpha\beta}(x)$ are the contravariant components of a 2-form. We have written this 2-form as $^*\mathbf{S}$, i.e., as the dual of another 2-form $\mathbf{S}$. As $-\mathbf{S} = {}^*({}^*\mathbf{S})$, this 2-form $\mathbf{S}$ is characterized by the covariant components:

$$\begin{aligned}S_{\alpha\beta}(x) &= -\frac{1}{2}\sqrt{-g}\,\epsilon_{\alpha\beta\gamma\delta}\,\overset{*}{S}{}^{\alpha\beta}(x) \\ &= g_M \int_U \delta^4(x - \tilde{x}(\lambda^1, \lambda^2))\epsilon_{\alpha\beta\gamma\delta}\,\frac{\partial\tilde{x}^\gamma}{\partial\lambda^1}\frac{\partial\tilde{x}^\delta}{\partial\lambda^2}\, d\lambda^1 d\lambda^2.\end{aligned} \tag{9.27}$$

Remark. Within the framework of distributions the sheet Σ itself can be considered a singular 2-form. This 2-form is closely associated with $\mathbf{S}$. If $\mathbf{U}$ denotes an arbitrary test form, we get through a formal computation

$$\begin{aligned}\langle {}^*\mathbf{S} \mid \mathbf{U}\rangle &= \frac{1}{2}\int_{\mathbf{R}^4} \overset{*}{S}{}^{\alpha\beta}(x)U_{\alpha\beta}(x)\sqrt{-g}\, d^4x \\ &= -\frac{1}{2}\int_{\mathbf{R}^4} g_M \int_U \delta^4(x - \tilde{x}(\lambda^1, \lambda^2))\left[\frac{\partial\tilde{x}^\alpha}{\partial\lambda^1}\frac{\partial\tilde{x}^\beta}{\partial\lambda^2} - \frac{\partial\tilde{x}^\alpha}{\partial\lambda^2}\frac{\partial\tilde{x}^\beta}{\partial\lambda^1}\right] \\ &\quad \times U_{\alpha\beta}(x)d\lambda^1 d\lambda^2 d^4x \\ &= -g_M \int_U \frac{\partial\tilde{x}^\alpha}{\partial\lambda^1}\frac{\partial\tilde{x}^\beta}{\partial\lambda^2}\left[\int_{\mathbf{R}^4} \delta^4(x - \tilde{x}(\lambda^1, \lambda^2))U_{\alpha\beta}(x)d^4x\right] d\lambda^1 d\lambda^2 \\ &= -g_M \int_U U_{\alpha\beta}(\tilde{x}(\lambda^1, \lambda^2))\frac{\partial\tilde{x}^\alpha}{\partial\lambda^1}\frac{\partial\tilde{x}^\beta}{\partial\lambda^2}\, d\lambda^1 d\lambda^2 = -g_M \int_\Sigma U.\end{aligned}$$

Within the framework of distributions we therefore have

$$ {}^*S = -g_M \Sigma, \qquad (\mathbf{S} = g_M {}^*\Sigma). \tag{9.28}$$

We summarize the main properties of the singular form $\mathbf{S}$ in the following lemma.

Lemma 9.1. (*Dirac's lemma*)
The singular form $\mathbf{S}$ *has the following properties:*

1. *It vanishes outside the string.*
2. $\boldsymbol{d}\mathbf{S} = {}^*\mathbf{K}$. (9.29)
3. $\int_\Omega \mathbf{S} = -g_M$ *for any closed surface* Ω *surrounding the monopole.* (9.30)

PROOF.
1. If x lies outside the sheet Σ, then the δ-function $\delta^4(x - \tilde{x}(\lambda^1, \lambda^2))$ automatically vanishes.
2. To check the relation (9.29), we use that it is equivalent to

$$\frac{1}{\sqrt{-g}}\,\partial_\alpha\left(\sqrt{-g}\overset{*}{\mathbf{S}}{}^{\alpha\beta}\right) = K^\beta.$$

Next we work our way through the left-hand side:

$$\begin{aligned}
\frac{1}{\sqrt{-g}}\,\partial_\alpha\left(\sqrt{-g}\overset{*}{\mathbf{S}}{}^{\alpha\beta}(x)\right) &= -\frac{g_M}{\sqrt{-g}}\int_U \left[\partial_\alpha \delta^4(x - \tilde{x}(\lambda^1, \lambda^2))\right] \\
&\quad \times \left[\frac{\partial \tilde{x}^\alpha}{\partial \lambda^1}\frac{\partial \tilde{x}^\beta}{\partial \lambda^2} - \frac{\partial \tilde{x}^\alpha}{\partial \lambda^2}\frac{\partial \tilde{x}^\beta}{\partial \lambda^1}\right] d\lambda^1 d\lambda^2 \\
&= \frac{g_M}{\sqrt{-g}}\int_U \left[\frac{\partial}{\partial \tilde{x}^\alpha}\,\delta^4(x - \tilde{x}(\lambda^1, \lambda^2))\right] \\
&\quad \times \left[\frac{\partial \tilde{x}^\alpha}{\partial \lambda^1}\frac{\partial \tilde{x}^\beta}{\partial \lambda^2} - \frac{\partial \tilde{x}^\alpha}{\partial \lambda^2}\frac{\partial \tilde{x}^\beta}{\partial \lambda^1}\right] d\lambda^1 d\lambda^2 \\
\text{(chain-rule)} &= \frac{g_M}{\sqrt{-g}}\int_U \left[\frac{\partial}{\partial \lambda^1}\,\delta^4(x - \tilde{x}(\lambda^1, \lambda^2))\,\frac{\partial \tilde{x}^\beta}{\partial \lambda^2}\right. \\
&\quad \left. - \frac{\partial}{\partial \lambda^2}\,\delta^4(x - \tilde{x}(\lambda^1, \eta^2))\,\frac{\partial \tilde{x}^\beta}{\partial \lambda^1}\right] d\lambda^1 d\lambda^2 \\
&= \frac{g_M}{\sqrt{-g}}\int_\Sigma \boldsymbol{d}\delta^4(x - \tilde{x}) \wedge \boldsymbol{d}\tilde{x}^\beta \\
&= \frac{g_M}{\sqrt{-g}}\int_\Sigma \boldsymbol{d}\left[\delta^4(x - \tilde{x}(\lambda^1, \lambda^2)\boldsymbol{d}x^\beta\right]
\end{aligned}$$

$$\text{(Stokes's theorem)} = \frac{g_M}{\sqrt{-g}} \int_{\partial\Sigma} \delta^4(x - \tilde{x}(\lambda^1, \lambda^2)) d\tilde{x}^\beta$$

$$= \frac{g_M}{\sqrt{-g}} \int_{-\infty}^{+\infty} \delta^4(x - \tilde{x}(\lambda^2)) \frac{\partial \tilde{x}^\beta}{\partial \lambda^2} d\lambda^2 = K^\beta.$$

3. Finally, we must compute the flux through a closed surface surrounding the monopole. But a closed surface Ω surrounding the monopole is the boundary of a regular domain W containing the monopole. Consequently, we get using (9.29)

$$\int_\Omega \mathbf{S} = \int_{\partial W} \mathbf{S} = \int_W d\mathbf{S} = -\int_W {}^*\mathbf{K} = -g_M.$$

□

Worked Exercise 9.2.2
Problem: Prove (9.30) by an explicit computation of the integral.

Remark. We can also reformulate (9.29) as

$$\delta^*\mathbf{S} = \mathbf{K}. \tag{9.31}$$

Within the framework of distributions, the boundary operation coincides with the codifferential, and we therefore get

$$\delta^*\mathbf{S} = -g_M \delta\Sigma = -g_M \partial\Sigma = -g_M \Gamma_M = -\mathbf{K}.$$

Using Dirac's lemma, we can now "cure the diseases" of $\mathbf{F}$. It was generated by a magnetic monopole and an electrically charged particle and therefore had the properties (9.24)

$$d\mathbf{F} = -{}^*\mathbf{K}, \qquad \int_\Omega \mathbf{F} = g_M.$$

We now choose an arbitrary string extending from the monopole to infinity. Associated with this string we have a weak form $\mathbf{S}$ with the properties (9.29), (9.30)

$$d\mathbf{S} = {}^*\mathbf{K}, \qquad \int_\Omega \mathbf{S} = -g_M.$$

Consequently, we see that $\mathbf{F} + \mathbf{S}$ is exact, although singular, and we therefore can find a global gauge potential $\mathbf{A}$ that generates $\mathbf{F} + \mathbf{S}$. The gauge potential $\mathbf{A}$ will, of course, be singular too. It will be singular at the position of the monopole, reflecting the singularity of $\mathbf{F}$, and it will be singular at the string, reflecting the singularity of $\mathbf{S}$.

Formally, $\mathbf{S}$ represents a concentration magnetic flux flowing towards the monopole. Hence we may formally interpret $\mathbf{F} + \mathbf{S}$ as a concentrated magnetic flux flowing towards the monopole position along the string and then spreading

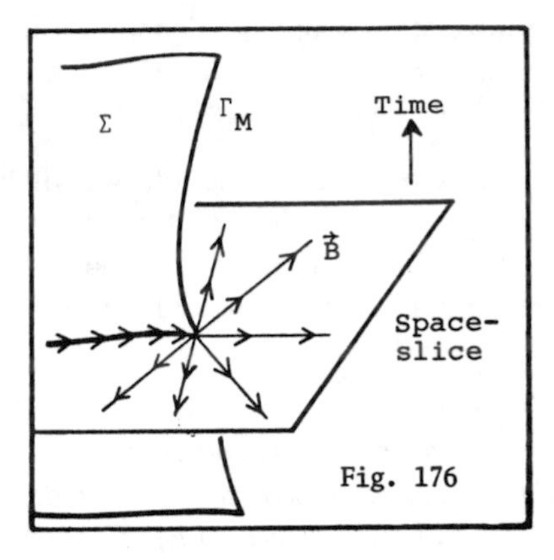

Figure 9.7.

out to produce the monopole field. (See Figure 9.7.) However, it should be emphasized that **S** has no physical meaning. The position of the string can be chosen completely arbitrarily, and the introduction of **S** is a purely formal trick that cures the diseases of **F**!

9.3 Dirac's Lagrangian Principle for Magnetic Monopoles

We are now in a position where we can state the Lagrangian principle of Dirac (Dirac, 1948). Dirac's *action* consists of three pieces:

$$S = S_{\text{PARTICLES}} + S_{\text{INTERACTION}} + S_{\text{FIELD}}, \tag{9.32}$$

where

$$S_{\text{PARTICLES}} = -m_e \int_{\Gamma_q} \sqrt{-g_{\alpha\beta} \frac{d\tilde{x}_e^\alpha}{d\lambda} \frac{d\tilde{x}_e^\beta}{d\lambda}}\, d\lambda$$

$$- m_m \int_{\Gamma_m} \sqrt{-g_{\alpha\beta} \frac{d\tilde{x}_m^\alpha}{d\lambda} \frac{d\tilde{x}_m^\beta}{d\lambda}}\, d\lambda.$$

$$S_{\text{INTERACTION}} = q \int_{\Gamma_q} A_\alpha \frac{d\tilde{x}_e^\alpha}{d\lambda}\, d\lambda = \int_{\Gamma_q} A_\alpha(x) J^\alpha(x) \sqrt{-g}\, d^4x.$$

$$S_{\text{FIELD}} = -\frac{1}{4} \int F_{\mu\nu} F^{\mu\nu} \sqrt{-g}\, d^4x.$$

Notice that the interaction term only contains a coupling between the electrically charged particle and the field! However, if we look a little more closely at S_{FIELD}, we see that it contains information about the monopole trajectory because

$$F_{\mu\nu} = \partial_\mu A_\nu - \partial_\nu A_\mu - S_{\mu\nu}. \tag{9.33}$$

Hence when we vary the trajectory of the monopole, we will have to vary the sheet, which terminates on the trajectory. But that will force $S_{\mu\nu}$ to vary, and thus S_{FIELD} contains the coupling between the monopole and the field. That the above action in fact gives the expected equations is the content of the following famous theorem:

Theorem 9.1. (*Dirac's theorem*)
All equations of motion for a system consisting of monopoles, electrically charged particles, and the electromagnetic field can be derived from Dirac's action, provided that you respect Dirac's veto; i.e., the magnetic strings are never allowed to cross the world line of an electrically charged particle.

PROOF. The proof is long and technically complicated and may be skipped on first reading.

First, we list the equations of motion that we are going to derive:

(1)
$$d\mathbf{F} = -{}^*\mathbf{K}$$

(2)
$$d^*\mathbf{F} = -{}^*\mathbf{J}$$

(3)
$$m_e\left[\frac{d^2\tilde{x}_e^\alpha}{d\tau^2} + \Gamma^\alpha{}_{\mu\nu}\frac{d\tilde{x}_e^\mu}{d\tau}\frac{d\tilde{x}_e^\nu}{d\tau}\right] = qF^\alpha{}_\beta\frac{d\tilde{x}_e^\beta}{d\tau}.$$

(4)
$$m_m\left[\frac{d^2\tilde{x}_m^\alpha}{d\tau^2} + \Gamma^\alpha{}_{\mu\nu}\frac{d\tilde{x}_m^\mu}{d\tau}\frac{d\tilde{x}_m^\nu}{d\tau}\right] = -g_m\overset{*}{F}{}^\alpha{}_\beta\frac{d\tilde{x}_m^\beta}{d\tau}.$$

Then we list the dynamical variables to be varied:

a. $\tilde{x}_e^\alpha(\lambda)$: Trajectory of electrically charged particle.
b. $\tilde{x}_m^\alpha(\lambda)$: Trajectory of monopole.
c. $A_\mu(x)$: Gauge potential.

Observe that the string coordinates $\tilde{x}(\lambda^1, \lambda^2)$ are *not* considered as dynamical variables. They are fictitious coordinates and should be completely eliminated by the end of the calculations.

Okay! Let us go to work: Equation (1) is *not* a dynamical equation. It is a purely kinematical equation that is built into the model from the beginning:

$$\mathbf{F} + \mathbf{S} = d\mathbf{A} \Rightarrow d\mathbf{F} + d\mathbf{S} = 0 \Rightarrow d\mathbf{F} = -d\mathbf{S}$$

but by construction we have: $d\mathbf{S} = {}^*\mathbf{K}$.

Equation (2). Performing the variation $A_\mu \to A_\mu + \epsilon B_\mu$, we get

$$\frac{dS_I}{d\epsilon}\Big|_{\epsilon=0} = \int_\Omega \sqrt{-g}\, J^\alpha(x) B_\alpha(x) d^4x$$

$$\frac{dS_F}{d\epsilon}\Big|_{\epsilon=0} = -\frac{1}{2}\int_\Omega F^{\mu\nu}(x)\left[\frac{\partial B_\nu}{\partial x^\mu} - \frac{\partial B_\mu}{\partial x^\nu}\right]\sqrt{-g}\, d^4x$$

$$= \int_\Omega \partial_\mu\left(\sqrt{-g}F^{\mu\nu}\right) B_\nu(x) d^4x.$$

Thus we have

$$0 = \frac{dS_I}{d\epsilon}\Big|_{\epsilon=0} + \frac{dS_F}{d\epsilon}\Big|_{\epsilon=0} = \int_\Omega \left[\partial_\mu\left(\sqrt{-g}F^{\mu\nu}\right) + \sqrt{-g}J^\nu(x)\right] B_\nu(x) d^4x,$$

which leads to the desired equation of motion:

$$\frac{1}{\sqrt{-g}}\,\partial_\mu\left(\sqrt{-g}F^{\mu\nu}\right) = -J^\mu(x).$$

Equation (3) follows from the variation of the trajectory $\tilde{x}_e^\mu(x)$. Performing the variation $\tilde{x}^\mu(\lambda) \to \tilde{x}^\mu(\lambda) + \epsilon\tilde{y}^\mu(\lambda)$, we get from a previous calculation (Exercise 8.6.3)

$$\frac{dS_P}{d\epsilon}\Big|_{\epsilon=0} = -\int_{\lambda_1}^{\lambda_2} m_e g_{\alpha\beta}\left(\frac{d^2\tilde{x}^\beta}{d\tau^2} + \Gamma^\beta_{\mu\nu}\frac{d\tilde{x}^\mu}{d\tau}\frac{d\tilde{x}^\nu}{d\tau}\right)\tilde{y}^\alpha d\tau.$$

On the other hand,

$$\frac{dS_I}{d\epsilon}\Big|_{\epsilon=0} = q\int_{\lambda_1}^{\lambda_2}\left(\left[\frac{\partial A_\alpha}{\partial x^\beta}\tilde{y}^\beta\right]\frac{d\tilde{x}^\alpha}{d\lambda} + A_\alpha\frac{d\tilde{y}^\alpha}{d\lambda}\right)d\lambda$$

$$= q\int\left[\frac{\partial A_\alpha}{\partial x^\beta}\frac{d\tilde{x}^\alpha}{d\lambda}\tilde{y}^\beta - \frac{\partial A_\alpha}{\partial x^\beta}\frac{d\tilde{x}^\beta}{d\lambda}\tilde{y}^\alpha\right]d\lambda$$

$$= q\int\left(\frac{\partial A_\alpha}{\partial x^\beta} - \frac{\partial A_\beta}{\partial x^\alpha}\right)\frac{d\tilde{x}^\alpha}{d\lambda}\tilde{y}^\beta d\lambda.$$

Consequently, we get

$$0 = \frac{dS_P}{d\epsilon}\Big|_{\epsilon=0} + \frac{dS_I}{d\epsilon}\Big|_{\epsilon=0}$$

$$= -\int\Bigg[m_e g_{\alpha\beta}\left(\frac{d^2\tilde{x}^\beta}{d\tau^2} + \Gamma^\beta_{\mu\nu}\frac{d\tilde{x}^\mu}{d\tau}\frac{d\tilde{x}^\nu}{d\tau}\right)$$

$$+ q\left(\frac{\partial A_\alpha}{\partial x^\beta} - \frac{\partial A_\beta}{\partial x^\alpha}\right)\frac{d\tilde{x}^\beta}{d\tau}\Bigg]\tilde{y}^\alpha d\tau,$$

from which we conclude:

$$m_e g_{\alpha\beta}\left(\frac{d^2\tilde{x}^\beta}{d\tau^2}+\Gamma^\beta_{\ \mu\nu}\frac{d\tilde{x}^\mu}{d\tau}\frac{d\tilde{x}^\nu}{d\tau}\right)=q\left(F_{\alpha\beta}+S_{\alpha\beta}\right)\frac{d\tilde{x}^\beta}{d\tau},$$

so we got any extra term. This is the first time we will use *Dirac's veto*: According to Dirac's veto, the sheet will never cross the trajectory of the electrically charged particle. Thus $S_{\alpha\beta}$ vanishes on the trajectory of the electric charged particle and we simply throw it away!

Equation (4) follows from the variation of the monopole trajectory and it is by far the most complicated step.

Performing the variation $\tilde{x}^\mu_m(\lambda) \to \tilde{x}^\mu_m(\lambda)+\epsilon\tilde{y}^\mu(\lambda)$ we get

$$\frac{dS_P}{d\epsilon}\Big|_{\epsilon=0}=-\int m_m g_{\alpha\beta}\left(\frac{d^2\tilde{x}^\beta_m}{d\tau^2}+\Gamma^\beta_{\ \mu\nu}\frac{d\tilde{x}^\mu_m}{d\tau}\frac{d\tilde{x}^\nu_m}{d\tau}\right)\tilde{y}^\alpha d\tau.$$

So this caused no problems! But then we must investigate $\frac{dS_F}{d\epsilon}\big|_{\epsilon=0}$: To do this we must proceed carefully. First we observe that when we vary the monopole trajectory, we must vary the sheet too:

$$\tilde{x}^\mu(\lambda^1,\lambda^2)\to\tilde{x}^\mu(\lambda^1,\lambda^2)+\epsilon\tilde{y}^\mu(\lambda^1,\lambda^2).$$

The variation $\tilde{y}^\mu(\lambda^1,\lambda^2)$ should satisfy the boundary conditions

a) $0=\tilde{y}^\mu(\lambda^1,\lambda^2)$, $\tilde{y}^\mu(0,\lambda^2)=\tilde{x}^\mu(\lambda^2)$ on the initial and final space slice;
b) $\tilde{y}^\mu(0,\lambda^2)=\tilde{x}^\mu(\lambda^2)$ i.e., the deformed sheet terminates on the deformed monopole trajectory!

(See Figure 9.8.) But apart from these boundary conditions, we can choose $\tilde{y}^\mu(\lambda^1,\lambda^2)$ arbitrarily.

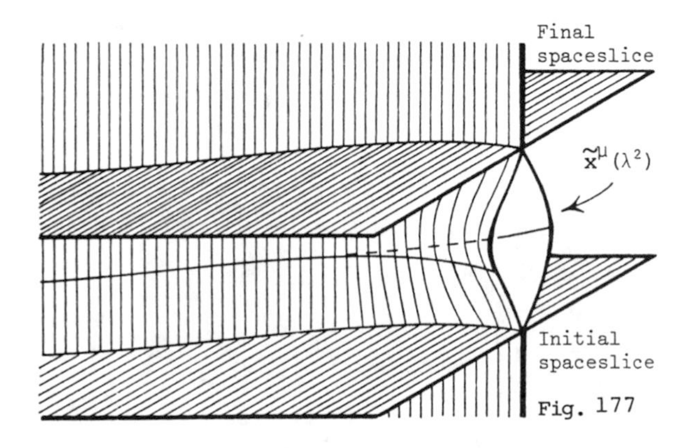

Figure 9.8.

It will also be convenient to rearrange the expression for S_{FIELD}! We have previously found (cf. (7.56)) that

$$F^{\mu\nu}F_{\mu\nu} = -\overset{*}{F}{}^{\mu\nu}\overset{*}{F}_{\mu\nu}.$$

So we my rearrange the field action as follows:

$$S_F = \frac{1}{4}\int \overset{*}{F}_{\mu\nu}\overset{*}{F}{}^{\mu\nu}\sqrt{-g}\,d^4x,$$

which implies that

$$\frac{dS_F}{d\epsilon}\Big|_{\epsilon=0} = \frac{1}{2}\int \overset{*}{F}_{\mu\nu}\frac{d\overset{*}{F}{}^{\mu\nu}}{d\epsilon}\Big|_{\epsilon=0}\sqrt{-g}\,d^4x$$
$$= -\frac{1}{2}\int \overset{*}{F}_{\mu\nu}\frac{d\overset{*}{S}{}^{\mu\nu}}{d\epsilon}\Big|_{\epsilon=0}\sqrt{-g}\,d^4x.$$

Using that

$$\frac{d\overset{*}{S}{}^{\mu\nu}}{d\epsilon}\Big|_{\epsilon=0}$$

has the form $T^{\mu\nu} - T^{\nu\mu}$, the calculation is simplified to the computation of

$$\frac{dS_F}{d\epsilon}\Big|_{\epsilon=0}$$
$$= \int \overset{*}{F}_{\mu\nu}\frac{d}{d\epsilon}\Big|_{\epsilon=0}$$
$$\times\left[g_m\int \delta^4(x-\tilde{x}-\epsilon\tilde{y})\left(\frac{\partial(\tilde{x}^\mu+\epsilon\tilde{y}^\mu)}{\partial\lambda^1}\,\frac{\partial(\tilde{x}^\nu+\epsilon\tilde{y}^\nu)}{\partial\lambda^2}\right)\right]$$
$$\times\, d\lambda^1 d\lambda^2 d^4x = \Delta_1+\Delta_2+\Delta_3,$$

where

$$\Delta_1 = -g_m\int \overset{*}{F}_{\mu\nu}\int \frac{\partial\delta^4(x-\tilde{x})}{\partial x^\alpha}\,\tilde{y}^\alpha\,\frac{\partial\tilde{x}^\mu}{\partial\lambda^1}\,\frac{\partial\tilde{x}^\nu}{\partial\lambda^2}\,d\lambda^1 d\lambda^2 d^4x,$$
$$\Delta_2 = g_m\int \overset{*}{F}_{\mu\nu}\int \delta^4(x-\tilde{x})\frac{\partial\tilde{y}^\mu}{\partial\lambda^1}\,\frac{\partial\tilde{x}^\nu}{\partial\lambda^2}\,d\lambda^1 d\lambda^2 d^4x,$$
$$\Delta_3 = g_m\int F_{\mu\nu}\int \delta^4(x-\tilde{x})\frac{\partial\tilde{x}^\mu}{\partial\lambda^1}\,\frac{\partial\tilde{y}^\nu}{\partial\lambda^2}\,d\lambda^1 d\lambda^2 d^4x.$$

In the first term Δ_1, we rearrange the integral

$$\Delta_1 = -g_m\int\left[\int \overset{*}{F}_{\mu\nu}\frac{\partial\delta^4(x-\tilde{x})}{\partial x^\alpha}\,d^4x\right]\frac{\partial\tilde{x}^\mu}{\partial\lambda^1}\,\frac{\partial\tilde{x}^\nu}{\partial\lambda^2}\,\tilde{y}^\alpha d\lambda^1 d\lambda^2.$$

By "partial integration," we then get

$$\int_\Omega \overset{*}{F}_{\mu\nu} \frac{\partial \delta^4(x-\tilde{x})}{\partial x^\alpha} d^4x = -\int_\Omega \frac{\partial \overset{*}{F}_{\mu\nu}}{\partial x^\alpha} \delta^4(x-\tilde{x}) d^4x = -\frac{\partial \overset{*}{F}_{\mu\nu}}{\partial x^\alpha}(\tilde{x}).$$

Consequently, we end up with

$$\Delta_1 = g_m \int \frac{\partial \overset{*}{F}_{\mu\nu}}{\partial x^\alpha}(\tilde{x}) \frac{\partial \tilde{x}^\mu}{\partial \lambda^1}\frac{\partial \tilde{x}^\nu}{\partial \lambda^2} \tilde{y}^\alpha d\lambda^1 d\lambda^2. \tag{*}$$

We then attack Δ_2:

$$\begin{aligned}\Delta^2 &= g_m \int \left[\int \overset{*}{F}_{\mu\nu}(x)\delta^4(x-\tilde{x})d^4x\right] \frac{\partial \tilde{y}^\mu}{\partial \lambda^1}\frac{\partial \tilde{x}^\nu}{\partial \lambda^2} d\lambda^1 d\lambda^2 \\ &= g_m \int \overset{*}{F}_{\mu\nu}(\tilde{x}) \frac{\partial \tilde{y}^\mu}{\partial \lambda^1}\frac{\partial \tilde{x}^\nu}{\partial \lambda^2} d\lambda^1 d\lambda^2.\end{aligned}$$

Using that

$$\frac{\partial}{\partial \lambda^1}\left[\overset{*}{F}_{\mu\nu}\tilde{y}^\mu\right] = \frac{\partial \overset{*}{F}_{\mu\nu}}{\partial x^\alpha}\frac{\partial \tilde{x}^\alpha}{\partial \lambda^1}\tilde{y}^\mu + \overset{*}{F}_{\mu\nu}\frac{\partial \tilde{y}^\mu}{\partial \lambda^1},$$

we may rearrange this as follows (observe that we cannot use partial integration due to the term $\frac{d\tilde{x}^\nu}{\partial \lambda^2}$!):

$$\begin{aligned}\Delta_2 = g_m &\int \frac{\partial}{\partial \lambda^1}\left(\overset{*}{F}_{\mu\nu}\tilde{y}^\mu\right)\frac{d\tilde{x}^\nu}{\partial \lambda^2} d\lambda^1 d\lambda^2 \\ &- g_m \int \frac{\partial \overset{*}{F}_{\mu\nu}}{\partial x^\alpha}\frac{\partial \tilde{x}^\alpha}{\partial \lambda^1}\tilde{y}^\mu \frac{\partial \tilde{x}^\nu}{\partial \lambda^2} d\lambda^1 d\lambda^2.\end{aligned} \tag{**}$$

In exactly the same way, we can rearrange the last term

$$\begin{aligned}\Delta_3 = g_m &\int \frac{\partial}{\partial \lambda^2}\left(\overset{*}{F}_{\mu\nu}\tilde{y}^\mu\right)\frac{\partial \tilde{x}^\mu}{\partial \lambda^1} d\lambda^1 d\lambda^2 \\ &- g_m \int \frac{\partial \overset{*}{F}_{\mu\nu}}{\partial x^\alpha}\frac{\partial \tilde{x}^\alpha}{\partial \lambda^2}\tilde{y}^\nu \frac{\partial \tilde{x}^\mu}{\partial \lambda^1} d\lambda^1 d\lambda^2.\end{aligned} \tag{***}$$

Consequently, Δ_1, Δ_2, and Δ_3 have been split into two groups of similar terms. Three of the terms obviously match together. Performing the following index substitutions,

Second term (**)	$\alpha \to \mu$;	$\nu \to \nu$;	$\mu \to \alpha$,
Third term (***)	$\mu \to \mu$;	$\alpha \to \nu$;	$\nu \to \alpha$,

these may be rearranged as

$$g_m \int \left(\partial_\alpha \overset{*}{F}_{\mu\nu} + \partial_\mu \overset{*}{F}_{\nu\alpha} + \partial_\nu \overset{*}{F}_{\alpha\mu}\right)\frac{\partial \tilde{x}^\mu}{\partial \lambda^1}\frac{\partial \tilde{x}^\nu}{\partial \lambda^2}\tilde{y}^\alpha d\lambda^1 d\lambda^2.$$

But according to the Maxwell equation already established in (2), we know that

$$\partial_\alpha \overset{*}{F}_{\mu\nu} + \partial_\mu \overset{*}{F}_{\nu\alpha} + \partial_\nu \overset{*}{F}_{\alpha\nu} = -\sqrt{-g}\,\epsilon_{\alpha\mu\nu\beta}J^\beta.$$

But J^β vanishes on the string due to *Dirac's veto*! Thus the three terms cancel automatically, and we are left with two remaining terms:

$$\begin{aligned}
\frac{dS_F}{d\epsilon}\Big|_{\epsilon=0} &= g_m \int \frac{\partial}{\partial\lambda^1}\left(\overset{*}{F}_{\mu\nu}\tilde{y}^\mu\right)\frac{\partial\tilde{x}^\nu}{\partial\lambda^2}\,d\lambda^1 d\lambda^2 \\
&\quad + g_m \int \frac{\partial}{\partial\lambda^2}\left(\overset{*}{F}_{\mu\nu}\tilde{y}^\nu\right)\frac{\partial\tilde{x}^\mu}{\partial\lambda^1}\,d\lambda^1 d\lambda^2 \\
&= g_m \int \left\{\frac{\partial}{\partial\lambda^1}\left[\overset{*}{F}_{\mu\nu}\tilde{y}^\mu\right]\frac{\partial\tilde{x}^\nu}{\partial\lambda^2} - \frac{\partial}{\partial\lambda^2}\left[\overset{*}{F}_{\mu\nu}\tilde{y}^\mu\right]\frac{\partial\tilde{x}^\nu}{\partial\lambda^1}\right\} d\lambda^1 d\lambda^2 \\
&= g_m \int_\Sigma \boldsymbol{d}\left(\overset{*}{F}_{\mu\nu}\tilde{y}^\mu\right)\wedge \boldsymbol{d}\tilde{x}^\nu = g_m \int_\Sigma \boldsymbol{d}\left[\overset{*}{F}_{\mu\nu}\tilde{y}^\mu \boldsymbol{d}\tilde{x}^\nu\right] \\
&= g_m \int_{\partial\Sigma} \overset{*}{F}_{\mu\nu}\tilde{y}^\mu \boldsymbol{d}\tilde{x}^\nu = g_m \int_{\Gamma_m} \overset{*}{F}_{\mu\nu}\frac{\partial\tilde{x}^\nu}{\partial\lambda}\,\tilde{y}^\mu d\lambda
\end{aligned}$$

because the boundary of the sheet is nothing but the monopole trajectory. Believe it or not, we are through!

Combining the above expression for

$$\frac{dS_F}{d\epsilon}\Big|_{\epsilon=0}$$

with the previous one for

$$\frac{dS_P}{d\epsilon}\Big|_{\epsilon=0},$$

we get

$$\begin{aligned}
0 &= \frac{dS_P}{d\epsilon}\Big|_{\epsilon=0} + \frac{dS_F}{d\epsilon}\Big|_{\epsilon=0} \\
&= -\int \left[m_m g_{\alpha\beta}\left(\frac{d^2\tilde{x}^\beta}{d\tau^2} + \Gamma^\beta_{\ \mu\nu}\frac{d\tilde{x}^\mu}{d\tau}\frac{d\tilde{x}^\nu}{d\tau}\right) + g_m \overset{*}{F}_{\alpha\beta}\frac{d\tilde{x}^\beta}{d\tau}\right]\tilde{y}^\alpha d\tau,
\end{aligned}$$

from which we deduce

$$m_m g_{\alpha\beta}\left[\frac{d^2\tilde{x}^\beta}{d\tau^2} + \Gamma^\beta_{\ \mu\nu}\frac{d\tilde{x}^\mu}{d\tau}\frac{d\tilde{x}^\nu}{d\tau}\right] = -g_m \overset{*}{F}_{\alpha\beta}\frac{d\tilde{x}^\beta}{d\tau}.$$

This concludes the proof of Dirac's famous theorem. □

9.4 The Angular Momentum Due to a Monopole Field

Consider a system consisting of a pure monopole and a pure electric charge and suppose that they are at rest at some time $t = 0$ (Figure 9.9). Then they will stay at rest forever! The electrically charged particle will experience the Lorentz force (9.11)

$$\mathbf{F} = q(\mathbf{E} + \mathbf{v} \times \mathbf{B})$$

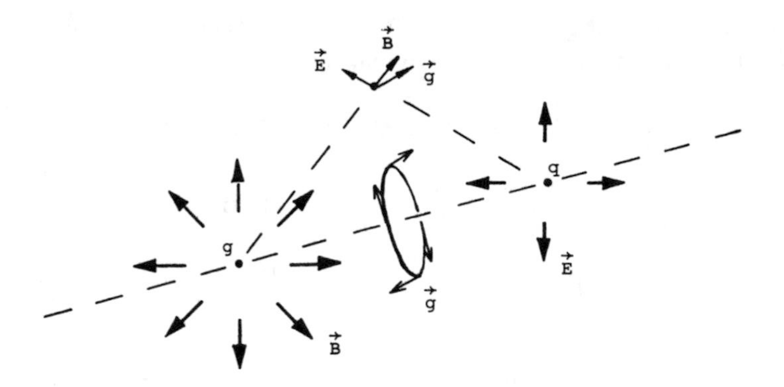

Figure 9.9.

where **E**, **B** are the field strengths produced by the monopole. But $\mathbf{E} = 0$ and $\mathbf{v} \times \mathbf{B} = \mathbf{0}$ because the particle is at rest. Similarly, the monopole will experience the force (9.12)

$$\mathbf{F} = g_M(\mathbf{B} - \mathbf{v} \times \mathbf{E}),$$

which vanishes for the same reason. Next we observe that the charge–monopole pair create an electromagnetic field with a nonvanishing momentum density (1.42)

$$\mathbf{g} = \epsilon_0 \mathbf{E} \times \mathbf{B}.$$

In fact, $\mathbf{E} \times \mathbf{B}$ is circulating around the axis connecting the monopole and the electrically charged particle. For a monopole at rest at the origin, the momentum density is responsible for an angular momentum density

$$\mathbf{j} = \mathbf{r} \times \mathbf{g},$$

and consequently, the electromagnetic field created by the charge–monopole pair carries a total angular momentum given by

$$\mathbf{J} = \int \mathbf{j} d^3\mathbf{r} = \int \mathbf{r} \times \mathbf{g} d^3\mathbf{r}.$$

(In passing, we observe that the angular momentum is independent of the choice of the origin as long as it is placed on the axis connecting the monopole–electric charge pair. This is due to the fact that **g** is perpendicular to the axis.)

If we introduce a Cartesian coordinate system with the monopole at the origin and the z-axis pointing towards the electrically charged particle, then a computation, which we leave as an exercise, shows that the components of this angular momentum are given by

$$J^1 = 0; \qquad J^2 = 0; \qquad J^3 = \frac{-g_M q}{4\pi}. \tag{9.33}$$

Worked Exercise 9.4.1
Problem: Compute the angular momentum of the electromagnetic field created by a charge-monopole pair.

So the electromagnetic field does carry a finite angular momentum along the z-axis, and the size of the angular momentum is independent of the separation between the two particles. This is the first hint that quantum-mechanically, the magnetic charge g_M and the electric charge q are *not* independent of each other. In fact, we expect something like the following naive quantum-mechanical argument to be valid:

The size of the angular momentum along the z-axis should be quantized as $\frac{n}{2}\hbar$, where n is an integer. (At this point we cannot exclude half-integer spin, because the angular momentum carried by the field is not of the conventional orbital angular momentum type.) Thus we expect

$$\frac{g_M q}{4\pi} = \frac{n}{2} \cdot \hbar ; \qquad n \text{ an integer (Dirac's quantization rule).} \tag{9.34}$$

But this is very beautiful, because this could provide us with an explanation of why electric charges are quantized in multiples of electron charges. If there exists just one monopole, then, continuing our very speculative line of argument, every electrically charged particle would interact with this monopole. Due to the quantization of the angular momentum created by the charge–monopole pair, the electric charge must therefore be an integer multiple of $\frac{2\pi\hbar}{g_M}$.

The above considerations can be extended to dyons too. Since the angular momentum is invariant under charge rotations, we can rotate one of the dyon charges into a pure monopole charge (cf. the Illustrative Example in Section 9.1):

$$(g_1, q_1) \to (g_1', q_1') = (g_M, 0) \qquad (g_1, q_2) \to (g_2', q_2').$$

It follows that the angular momentum density comes from the magnetic field $\mathbf{B}_1$ created by the first dyon and the electric field $\mathbf{E}_2$ created by the second dyon. As above, we therefore get

$$\mathbf{J} = \frac{g_1' q_2'}{4\pi} \hat{\mathbf{r}}_0$$

where $\hat{\mathbf{r}}_0$ is the unit vector pointing from the first to the second dyon. In terms of the original variables, the angular momentum is given by

$$\mathbf{J} = \frac{g_1 q_2 - g_2 q_1}{4\pi} \hat{\mathbf{r}}_0,$$

and we therefore expect the combination $(g_1 q_2 - g_2 q_1)$ to be quantized.

Now we want to investigate a little more closely the angular momentum carried by the electromagnetic field. For this purpose we consider a static spherically symmetric monopole field. We may think of it as being created by an "infinitely" heavy monopole placed at the origin. We then want to investigate the motion of an electrically charged particle in the monopole field. We shall treat the situation in the nonrelativistic approximation because then we can later on quantize the motion of the electrically charged particle using the Schrödinger equation (cf. the discussion in Chapter 2).

Newton's equation of motion is given by

$$m \frac{d^2\mathbf{r}}{dt^2} = q\mathbf{v} \times \mathbf{B} = \frac{qg_M}{4\pi} \mathbf{v} \times \frac{\mathbf{r}}{r^3} .$$

Introducing the parameter, $\kappa = \frac{qg_M}{4\pi}$, which eventually will become the angular momentum of the electromagnetic field, we rearrange this as

$$m \frac{d^2\mathbf{r}}{dt^2} = \kappa\mathbf{v} \times \frac{\mathbf{r}}{r^3} . \tag{9.35}$$

We proceed to analyze the consequences of equation (9.35) in five steps.

1. Step 1: In the first step we look for constants of motion. We multiply both sides of equation (9.35) by $\mathbf{v}$, thus obtaining

$$m\mathbf{v} \cdot \frac{d\mathbf{v}}{dt} = 0; \qquad \text{i.e.,} \ \frac{d}{dt}\left(\frac{1}{2} m\mathbf{v}^2\right) = 0;$$

i.e., the kinetic energy is conserved. This is because the monopole field performs no work on the particle, since the force $\mathbf{B}$ is always perpendicular to the velocity. Thus we have found the following *constants of motion*:

$$v \ (= \text{speed}) \quad \text{and} \quad T = \frac{1}{2} mv^2 \ (= \text{the kinetic energy}). \tag{9.36}$$

2. Step 2: Then we multiply both sides of equation (9.35) by $\mathbf{r}$, whereby we obtain

$$m\mathbf{r} \cdot \frac{d\mathbf{v}}{dt} = 0.$$

But now we can use that

$$\frac{d^2(r^2)}{dt^2} = \frac{d^2\mathbf{r}^2}{dt^2} = \frac{d}{dt}(2\mathbf{r} \cdot \mathbf{v}) = 2\mathbf{v}^2 + 2\mathbf{r}\frac{d\mathbf{v}}{dt} = 2v^2.$$

Integrating this, we get

$$r^2 = v^2t^2 + \alpha t + \beta. \tag{9.37}$$

Let us fix the time scale so that at $t = 0$ the particle is as close as possible at the origin; i.e.,

$$r(0) = d(= \text{distance of closest approach}).$$

Then equation (9.37) reduces to

$$r^2 = v^2t^2 + d^2. \tag{9.38}$$

This has an important consequence. If the speed $v \neq 0$, then the particle comes from infinity and returns to infinity. Thus there are *no bound states*. [The case $v = 0$, where the electric charge stays at rest, is a very special situation. In fact, it is unstable: The slightest perturbation and the electrically charged particle will move to infinity! This case is therefore irrelevant when we investigate the scattering of charged particles in monopole fields.]

3. Step 3: Then we consider the orbital angular momentum

$$\ell = \mathbf{r} \times m\mathbf{v}.$$

Since the monopole field is spherically symmetric, we expect that there should be a "total angular momentum" that is conserved. First, we observe that the orbital angular momentum is not conserved:

$$\frac{d}{dt}(\mathbf{r}\times m\mathbf{v}) = \mathbf{v}\times m\mathbf{v} + \mathbf{r}\times m\,\frac{d\mathbf{v}}{dt}$$
$$= \kappa\mathbf{r}\times\left(\mathbf{v}\times\frac{\mathbf{r}}{r^3}\right) = \kappa\,\frac{r\mathbf{v}-(\hat{\mathbf{r}}\cdot\mathbf{v})\mathbf{r}}{r^2}\,,$$

where $\hat{\mathbf{r}} = \frac{\mathbf{r}}{r}$ is the unit vector pointing towards the charged particle. To rearrange the right-hand side, we observe that

$$\frac{d\hat{\mathbf{r}}}{dt} = \frac{d}{dt}\left(\frac{\mathbf{r}}{r}\right) = \frac{r\mathbf{v}-\mathbf{r}\left(\frac{dr}{dt}\right)}{r^2} = \frac{r\mathbf{v}-\mathbf{r}(\hat{\mathbf{r}}\cdot\mathbf{v})}{r^2}$$

so that we finally obtain

$$\frac{d}{dt}(\mathbf{r}\times m\mathbf{v}) = \kappa\,\frac{d\hat{\mathbf{r}}}{dt}\,.$$

But then we have found the following constant of motion:

$$\mathbf{J} = \boldsymbol{\ell} - \kappa\hat{\mathbf{r}} = \mathbf{r}\times m\mathbf{v} - \kappa\hat{\mathbf{r}}. \tag{9.39}$$

This means that the total angular momentum of the system must be identified with $\mathbf{J}$. The second term, $-\kappa\hat{\mathbf{r}}$ is, of course, nothing but the angular momentum of the electromagnetic field.

4. Step 4: Since we have decomposed $\mathbf{J}$ into two orthogonal components $\boldsymbol{\ell}$ and $\kappa\hat{\mathbf{r}}$, we immediately get

$$\mathbf{J}^2 = \ell^2 + \kappa^2. \tag{9.40}$$

But κ is a constant, and J is conserved. Consequently, ℓ is conserved too. So, although the orbital angular momentum $\boldsymbol{\ell}$ is not conserved, we find that its size is conserved! This has a curious consequence: Consider the system at $t = -\infty$ and $t = 0$. Then we get

$$\ell(-\infty) = mv(-\infty)b \quad\text{and}\quad \ell(0) = mv(0)d,$$

where b is the *impact parameter*. (See Figure 9.10.) But ℓ and v are conserved. Thus we finally obtain

$$b = d. \tag{9.41}$$

Observe that if we point directly towards the monopole, i.e., $b = 0$, we will hit it! But this is a very exceptional situation: If we point directly towards the monopole, the electrically charged particle will move freely, because

$$\mathbf{v}\times\mathbf{B} = 0.$$

But the slightest perturbation and the particle will react to the force from the monopole field and no longer hit the monopole. Consequently, this is an unstable situation, and we shall neglect it.

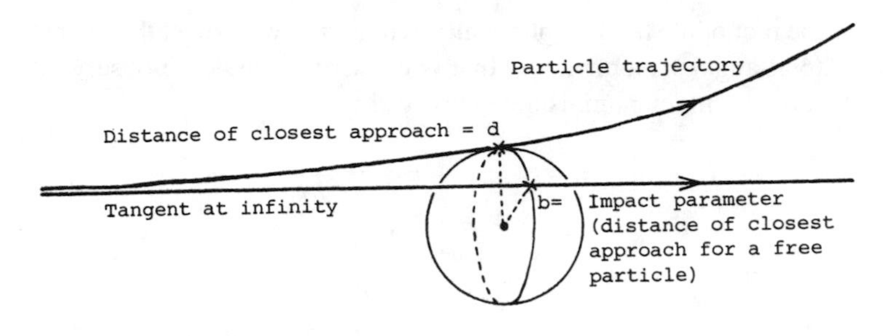

Figure 9.10.

5. Step 5: Finally, we observe that multiplying both sides of equation (9.39) by $\mathbf{r}$, we get

$$\mathbf{J} \cdot \hat{\mathbf{r}} = -\kappa; \qquad \text{i.e., } \cos(\mathbf{J}, \hat{\mathbf{r}}) = \frac{-\kappa}{J}\,. \tag{9.42}$$

But J is conserved and κ is a constant. Thus the angle between $\mathbf{J}$ and $\mathbf{r}$ is conserved, and therefore *the particle is constrained to move on a cone with* $\mathbf{J}$ *as axis* (Figure 9.11):

Worked Exercise 9.4.2
Problem: Compute the differential scattering cross section $\frac{d\sigma}{d\Omega}$ for the scattering of electrically charged particles in a monopole field.

9.5 Quantization of the Angular Momentum

As we have verified, the magnetic strength g_M of the monopole enters into the expression (9.39) for the angular momentum of a charged particle in a monopole field. When we quantize the motion of the charged particle, we know that the angular momentum becomes quantized, and we expect this to limit the possible

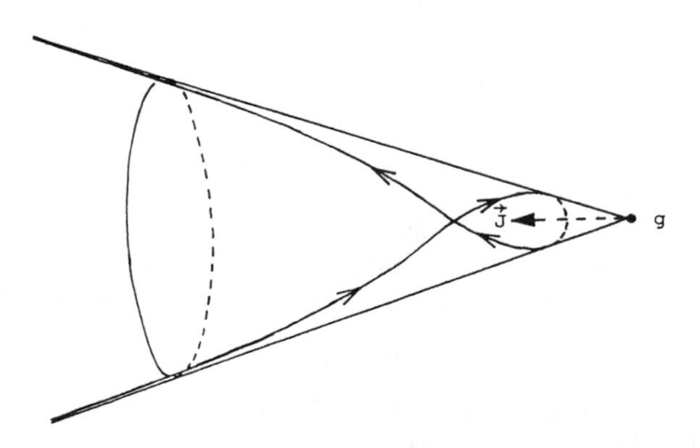

Figure 9.11.

values of κ. To determine the possible values of κ we must therefore determine the operators of angular momentum: $\hat{J}_1$, $\hat{J}_2$, $\hat{J}_3$.

For a charged particle moving in a magnetic field we have previously determined the Hamiltonian (cf. equation (2.65)):

$$\hat{H} = \frac{1}{2m}(-i\hbar\nabla - q\mathbf{A})^2, \tag{9.43}$$

where $\mathbf{A}$ is the gauge potential generating the magnetic field. Consequently, we must first find a gauge potential $\mathbf{A}$ producing the monopole field. But here we run into the well-known trouble that no globally defined gauge potential can produce the monopole field; i.e., $\mathbf{A}$ will necessarily contain singularities. However, as we have seen in Section 9.2, we may concentrate the singularities on the z-axis. We can still make a choice: The singularities form a string that can be either infinite or semi-infinite. As we shall see, the actual choice of the string has consequences on the quantum-mechanical level, and we shall therefore work out both cases. The semi-infinite string was originally introduced by Dirac, an we shall speak about the *Dirac formalism*. The infinite string has been especially advocated by Schwinger, and we shall speak about the *Schwinger formalism*. Let us collect the appropriate formulas in Figure 9.12:

Observe that the Hamiltonian (9.43) becomes a singular operator, but there is nothing to do about this for the moment. We will have to live for some time with the singularities along the z-axis, and we will have to accept similar singularities in the wave function $\psi(\mathbf{r}, t)$.

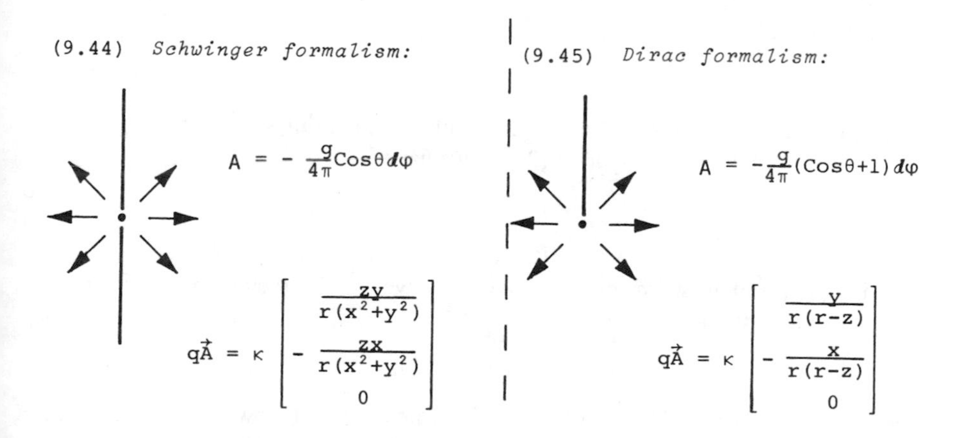

Figure 9.12.

Next we construct the operator corresponding to angular momentum. On the classical level we know that

$$\mathbf{J} = \mathbf{r} \times m\mathbf{v} - \kappa\hat{\mathbf{r}} = \mathbf{r} \times (\mathbf{p} - q\mathbf{A}) - \kappa\hat{\mathbf{r}}. \tag{9.46}$$

This corresponds to the operator

$$\hat{\mathbf{J}} = \mathbf{r} \times (i\hbar\nabla - q\mathbf{A}) - \kappa\hat{\mathbf{r}} = -i\hbar\mathbf{r} \times \nabla - (\mathbf{r} \times q\mathbf{A} + \kappa\hat{\mathbf{r}}). \tag{9.47}$$

Consequently, we have found the following candidates for the angular momentum operators:

Schwinger formalism: (9.48)	Dirac formalism: (9.49)
$\hat{J}_1 = -i\hbar\left(y\frac{\partial}{\partial z} - z\frac{\partial}{\partial y}\right) - \kappa\frac{xr}{x^2+y^2}$	$\hat{J}_1 = -i\hbar\left(y\frac{\partial}{\partial z} - z\frac{\partial}{\partial y}\right) - \kappa\frac{x}{r-z}$
$\hat{J}_2 = -i\hbar\left(z\frac{\partial}{\partial x} - x\frac{\partial}{\partial z}\right) - \kappa\frac{yr}{x^2+y^2}$	$\hat{J}_2 = -i\hbar\left(z\frac{\partial}{\partial x} - x\frac{\partial}{\partial z}\right) - \kappa\frac{y}{r-z}$
$\hat{J}_3 = -i\hbar\left(x\frac{\partial}{\partial y} - y\frac{\partial}{\partial x}\right)$	$\hat{J}_3 = -i\hbar\left(x\frac{\partial}{\partial y} - y\frac{\partial}{\partial x}\right) + \kappa$

Before we continue the discussion, we recall some general aspects of the angular momentum. According to Dirac, the components of the angular momentum are always represented by three Hermitian operators $\hat{J}_1$, $\hat{J}_2$, $\hat{J}_3$ satisfying Dirac's commutation relation:

$$[\hat{J}_i;\, \hat{J}_j] = i\hbar\epsilon_{ijk}\hat{J}_k. \tag{9.50}$$

If the system is spherically symmetric, they must furthermore commute with the Hamiltonian; i.e.,

$$[\hat{H};\, \hat{J}_i] = 0. \tag{9.51}$$

This guarantees the conservation of the angular momentum, since according to the quantum-mechanical analogue of (2.61), we have

$$i\hbar\frac{d\hat{J}_i}{dt} = [\hat{H};\, \hat{J}_i] = 0.$$

In many problems with spherical symmetry we can use the operators of *orbital* angular momentum

$$\ell = i\hbar(\mathbf{r} \times \nabla), \tag{9.52}$$

which trivially satisfy Dirac's commutation rules (9.50). But we cannot use them as angular momentum operators in this particular problem because they do not commute with the Hamiltonian (9.43). On the contrary, the above operators (9.48)–(9.49) not only satisfy Dirac's commutation rules (9.50), but they commute with the Hamiltonian (9.43) as well. The verification of this is left as an exercise.

Exercise 9.5.1
Problem:

(a) Let $f(x) \cdot \frac{\partial}{\partial x}$ be a differential operator. Show that

$$\left[f(x)\,\frac{\partial}{\partial x}\ ;\, g(x) \right] = f(x)\,\frac{\partial g}{\partial x}\ ;$$

i.e., the commutator is a multiplication operator. (Hint: Let both sides operate on a test function $\phi(x)$.)

(b) Show that the operators of angular momentum ((9.48)–(9.49)) satisfy the commutation rules

$$[\hat{J}_i, \hat{P}_j] = i\hbar\epsilon_{ijk}\hat{P}_k \tag{9.53}$$

with $\hat{P}_i = -i\hbar\,\frac{\partial}{\partial x^i} - qA_i$.

(c) Show that the operators of angular momentum ((9.48)–(9.49)) satisfy the commutation rules (9.50) and (9.51) as required by Dirac.

We proceed to investigate the spectrum of the angular momentum. Using the commutation relations (9.50), one can determine the possible eigenvalues for the angular momentum operators. (For details see, e.g., Schiff (1968)). First we introduce the square of the total angular momentum:

$$\hat{J}^2 = \hat{J}_1^2 + \hat{J}_2^2 + \hat{J}_3^2. \tag{9.54}$$

It follows that $\hat{J}^2$ commutes with $\hat{J}_1$, $\hat{J}_2$, and $\hat{J}_3$, and consequently we can diagonalize $\hat{J}^2$ and $\hat{J}_3$ simultaneously. If we label the eigenstates ψ_{jm}, we get

$$\hat{J}_3\psi_{jm} = m\hbar\psi_{jm} \tag{9.55a}$$

$$\hat{J}^2\psi_{jm} = j(j+1)\hbar^2\psi_{jm}. \tag{9.55b}$$

Furthermore, it follows that the possible eigenvalues are given by

$$m = -j, -j+1, \ldots, j-1, j,$$
$$j = 0, \frac{1}{2}, 1, \frac{3}{2}, 2, \frac{5}{2} \ \ldots. \tag{9.56}$$

Finally, it is preferable to introduce the raising and lowering operators

$$\hat{J}_+ = \hat{J}_1 + i\hat{J}_2 \qquad \hat{J}_- = \hat{J}_- - i\hat{J}_2. \tag{9.57}$$

They satisfy the commutation rules

$$[J_3; J_+] = \hbar J_+; \qquad [J_3; J_-] = \hbar J_-; \qquad [J_+; J_-] = 2\hbar J_3, \tag{9.58}$$

and as a consequence, the operators J_+ and J_- connect the different eigenfunctions with a fixed j:

$$\hat{J}_+\psi_{jm} = \hbar\sqrt{j(j+1) - m(m+1)}\,\psi_{jm},$$
$$\hat{J}_-\psi_{jm} = \hbar\sqrt{j(j+1) - m(m-1)}\,\psi_{jm}. \tag{9.59}$$

In the particular example of a charged particle in a monopole field, we can now use the explicit form (9.50)–(9.51) for the angular momentum operator to determine which of the possible eigenvalues (9.56) are actually realized. Recall that classically,

$$\mathbf{J}^2 = \ell^2 + \kappa^2. \tag{9.40}$$

This equation has a quantum-mechanical analogue. We have decomposed the quantum operators in the following way:

$$\hat{\mathbf{J}} = \mathbf{r} \times (-i\hbar\nabla - q\mathbf{A}) - \kappa\hat{\mathbf{r}} = \hat{\mathbf{L}} - \kappa\mathbf{r}.$$

But then we can square this operator relation:

$$\hat{\mathbf{J}}^2 = (\hat{\mathbf{L}} - \kappa\hat{\mathbf{r}}) \cdot (\hat{\mathbf{L}} - \kappa\hat{\mathbf{r}}) = \hat{\mathbf{L}}^2 - \kappa\hat{\mathbf{r}} \cdot \hat{\mathbf{L}} - \kappa\hat{\mathbf{L}} \cdot \hat{\mathbf{r}} + \kappa^2.$$

Worked Exercise 9.5.2
Problem: Let $\hat{\mathbf{L}}$ be the operator

$$\hat{\mathbf{L}} = \hat{\mathbf{r}} \times (-i\hbar\nabla - q\mathbf{A}).$$

Show that the operator products

$$\hat{\mathbf{L}} \cdot \hat{\mathbf{r}} \quad \text{and} \quad \hat{\mathbf{r}} \cdot \hat{\mathbf{L}}$$

both vanish.

According to Exercise 9.5.2, both operator products $\hat{\mathbf{r}} \cdot \hat{\mathbf{L}}$ and $\hat{\mathbf{L}} \cdot \hat{\mathbf{r}}$ automatically vanish. The following operator relation has therefore been established:

$$\hat{\mathbf{J}}^2 = \hat{\mathbf{L}}^2 + \kappa^2. \tag{9.60}$$

Here κ is a C-number, and $\hat{\mathbf{J}}^2$ and $\hat{\mathbf{L}}^2$ therefore have the same eigenfunctions:

$$\hat{\mathbf{L}}^2\psi_{jm} = \left(\hat{\mathbf{J}}^2_{\text{op.}} - \kappa^2\right)\psi_{jm} = [j(j+1)\hbar^2 - \kappa^2]\psi_{jm}.$$

But $\hat{\mathbf{L}}^2 = \hat{L}_1^2 + \hat{L}_2^2 + \hat{L}_3^2$ is a *positive* operator, because $\hat{L}_1$, $\hat{L}_2$ and $\hat{L}_3$ are Hermitian operators. Thus the eigenvalues of $\hat{\mathbf{L}}^2$ are positive, and we conclude that

$$j(j+1)\hbar^2 - \kappa^2 \geq 0; \qquad \text{i.e., } j(j+1) \geq \left(\frac{\kappa}{\hbar}\right)^2. \tag{9.61}$$

This is our main result because this shows that in a monopole field $j = 0$ is not allowed. Consequently, the minimal value of j is nontrivial, and the system therefore possesses an *intrinsic spin*!

To determine the exact spectrum, i.e., the allowed values of j and m together with the eigenfunctions Y_{jm}, we must now use the explicit expressions for $\hat{\mathbf{J}}^2$ and $\hat{J}_3$. It is convenient to use spherical coordinates. After a long but trivial computation, we obtain the following explicit expressions:

Schwinger formalism:

$$\hat{J}^2 = \hbar^2\left[\frac{1}{\sin\theta}\frac{\partial}{\partial\theta}\left(\sin\theta\frac{\partial}{\partial\theta}\right) + \frac{1}{\sin^2\theta}\frac{\partial^2}{\partial\varphi^2}\right] - 2i\hbar\kappa\frac{\cos\theta}{\sin^2\theta}\frac{\partial}{\partial\varphi} + \frac{\kappa^2}{\sin^2\theta}, \tag{9.62}$$

$$\hat{J}_3 = -i\hbar\frac{\partial}{\partial\varphi}.$$

Dirac formalism:

$$\hat{J}^2 = \hbar^2\left[\frac{1}{\sin\theta}\frac{\partial}{\partial\theta}\left(\sin\theta\frac{\partial}{\partial\theta}\right) + \frac{1}{\sin^2\theta}\frac{\partial^2}{\partial\varphi^2}\right] - 2i\hbar\kappa\frac{1}{1-\cos\theta}\frac{\partial}{\partial\varphi} + 2\kappa^2\frac{1}{1-\cos\theta}, \tag{9.63}$$

$$\hat{J}_3 = -i\hbar\frac{\partial}{\partial\varphi} + \kappa.$$

The Schwinger formalism is the most complicated one. But let us start with it to get a feeling for the general machinery.

We look for eigenfunctions of the form $Y_{jm}(\theta, \varphi)$. First we investigate the eigenvalue equation (9.55a),

$$\hat{J}_3\psi_{jm} = \hbar m\psi_{jm}$$

i.e.,

$$-i\hbar\frac{\partial}{\partial\varphi}Y_{jm}(\theta,\varphi) = m\hbar Y_{jm}(\theta,\varphi).$$

But this shows that we can factorize Y_{jm} in the following way:

$$Y_{jm}(\theta,\varphi) = P_{jm}(\cos\theta)e^{im\varphi}. \tag{9.64}$$

Since $Y_{jm}(\theta, \varphi)$ must be a smooth function, we demand that m be an integer, so that $e^{im\varphi}$ is periodic with the period 2π. It should be observed that $e^{im\varphi}$ is still singular on the z-axis, where it is discontinuous. (If we approach the z-axis along the line $\varphi = 0$, then $e^{im\varphi} = 1$, but if we approach the z-axis along the line $\varphi = \frac{\pi}{m}$, then $e^{im\varphi} = -1$.) Usually, we get rid of this singularity by demanding that $0 = P_{jm}(\cos\theta)$ on the z-axis, because then Y_{jm} itself will be smooth. In our models, however, we have singularities on the z-axis form the beginning, since the gauge potential **A** itself is singular on the z-axis. We shall therefore neglect the problem. Because m can take only integer values, we see that *only bosonic states are possible*.

Then we must use the second eigenvalue equation (9.55b). Inserting the expression (9.64) and introducing $x = \cos\theta$, we obtain

$$\left[-\frac{\partial}{\partial x}\left[(1-x^2)\frac{\partial}{\partial x}\right] + \frac{m^2 + 2m\frac{\kappa}{\hbar}x + \left(\frac{\kappa}{\hbar}\right)^2}{1-x^2} - j(j+1)\right]P_{jm}(x) = 0. \tag{9.65}$$

We should then look for regular normalizable solutions of this equation on the interval $[-1, +1]$. It can be shown that equation (9.65) has regular normalizable solutions if and only if m and $\frac{\kappa}{\hbar}$ are either both integers or both half-integers. Furthermore j is constrained through the relation

$$j \geq \frac{|\kappa|}{\hbar}, \tag{9.66}$$

in agreement with the previously obtained result (9.61).

Worked Exercise 9.5.3
Problem: Consider the differential equation (9.65) on $[-1, 1]$. Determine under what circumstances it has regular normalizable solutions and show that such solutions are given by

$$P_{jm}(x) = N_{jm} \cdot (1-x)^{-\frac{1}{2}\left(m+\frac{\kappa}{\hbar}\right)}(1+x)^{-\frac{1}{2}\left(m-\frac{\kappa}{\hbar}\right)} \times \frac{d^{j-m}}{dx^{j-m}}\left[(1-x)^{\left(j+\frac{\kappa}{\hbar}\right)}(1+x)^{\left(j-\frac{\kappa}{\hbar}\right)}\right]. \tag{9.67}$$

In the Schwinger formalism we know from the beginning that m is integer valued. Consequently, $\frac{\kappa}{\hbar}$ must be integer values too. To summarize, we have obtained the following results:

Theorem 9.2. *The* Schwinger *formalism leads to a* bosonic *spectrum:*

$$m,\ j,\ \frac{\kappa}{\hbar} \qquad \textit{are all integers.}$$

Furthermore, $|\kappa|$ is the intrinsic spin of the state. The eigenfunctions are given by the formula

$$Y_{jm}(\theta, \varphi) = P_{jm}(\cos\theta)e^{im\varphi}, \tag{9.64}$$

where

$$P_{jm}(x) = N_{jm}(1-x)^{-\frac{1}{2}\left(m+\frac{\kappa}{\hbar}\right)}(1+x)^{-\frac{1}{2}\left(m-\frac{\kappa}{\hbar}\right)} \times \frac{d^{j-m}}{dx^{j-m}}\left[(1-x)^{\left(j+\frac{\kappa}{\hbar}\right)}(1+x)^{\left(j-\frac{\kappa}{\hbar}\right)}\right]. \tag{9.67}$$

Now we briefly discuss the Dirac formalism. Here the spectrum is more easily obtained. From the eigenvalue equation (9.55a) we get

$$-i\frac{\partial}{\partial\varphi}Y_{jm}(\theta, \varphi) = \left(m - \frac{\kappa}{\hbar}\right)Y_{jm}(\theta, \varphi).$$

But this shows that we can factorize Y_{jm} as follows:

$$Y_{jm}(\theta, \varphi) = P_{jm}(\cos\theta)e^{i\left(m-\frac{\kappa}{\hbar}\right)\varphi}. \tag{9.68}$$

Since $e^{i\left(m-\frac{\kappa}{\hbar}\right)\varphi}$ has to be periodic, we conclude that $\left(m - \frac{\kappa}{\hbar}\right)$ is integer valued. But from the general theory, we know that m is either half-integer valued or integer valued! We have now two possibilities: Either $\left(j, m, \frac{\kappa}{\hbar}\right)$ are half-integer valued, or $\left(j, m, \frac{\kappa}{\hbar}\right)$ are all integer valued. Thus we have the possibility of a fermionic spectrum!

To check that all the listed combinations are admissible, we must investigate the second eigenvalue equation (9.55b). Inserting the expression (9.68) and introducing $x = \cos\theta$, we obtain the differential equation

$$\left[-\frac{\partial}{\partial x}\left[(1-x^2)\frac{\partial}{\partial x}\right] + \frac{m^2 + 2m\frac{\kappa}{\hbar}x + \left(\frac{\kappa}{\hbar}\right)^2}{1-x^2} - j(j+1)\right] P_{jm}(x) = 0. \tag{9.69}$$

But that is exactly the same equation as the one we analyzed in the Schwinger formalism! (See equation (9.65)). But there we found that there existed regular normalizable solutions, provided that m and $\frac{\kappa}{\hbar}$ were both half-integers or both integers! Thus they are all admissible, and we obtain the following theorem.

Theorem 9.3. *The* Dirac *formalism has both a* fermionic *and a* bosonic *spectrum:*

$$j, m, \frac{\kappa}{\hbar} \qquad \text{are all integers or all half-integers.}$$

Furthermore, $|\kappa|$ is the intrinsic spin of the state. The eigenfunctions are given by the formula

$$Y_{jm}(\theta, \varphi) = P_{jm}(\cos\theta)e^{i\left(m-\frac{\kappa}{\hbar}\right)\varphi}; \tag{9.68}$$

where

$$P_{jm}(x) = N_{jm}(1-x)^{-\frac{1}{2}\left(m+\frac{\kappa}{\hbar}\right)}(1+x)^{-\frac{1}{2}\left(m-\frac{\kappa}{\hbar}\right)} \times \frac{d^{j-m}}{dx^{j-m}}\left[(1-x)^{\left(j+\frac{\kappa}{\hbar}\right)}(1+x)^{\left(j-\frac{\kappa}{\hbar}\right)}\right]. \tag{9.67}$$

9.6 The Gauge Transformation as a Unitary Transformation

As we have seen, the choice of the gauge potential $\mathbf{A}$ has consequences on the quantum-mechanical level. This may be somewhat surprising: Usually the choice of a gauge potential is unique up to a gauge transformation, and gauge transformations leave physics unchanged. When discussing monopole fields, however, we should be careful! The crucial point is that here different choices of the gauge potential need not be related through a global gauge transformation. E.g., the spherically symmetric monopole field has been represented by the two gauge potentials

$$\mathbf{A}_1 = \frac{-g}{4\pi}\cos\theta\,\boldsymbol{d}\varphi \quad \text{and} \quad \mathbf{A}_2 = \frac{-g}{4\pi}(\cos\theta + 1)\boldsymbol{d}\varphi. \qquad (9.44)\text{–}(9.45)$$

Formally, we have

$$\mathbf{A}_2 = \mathbf{A}_1 - \frac{g}{4\pi}\,\boldsymbol{d}\varphi,$$

but φ is *not* smooth throughout space–time. It makes a jump somewhere between 0 and 2π, and therefore $\mathbf{A}_1$ and $\mathbf{A}_2$ are *not* related through a global gauge transformation.

On the quantum-mechanical level things behave a little differently. Here gauge transformations are represented by *unitary* transformations, as we will now explain.

Suppose the state of a quantum-mechanical system is represented by the Schrödinger wave functions $\psi(\mathbf{r}, t)$, and the various physical quantities are represented by Hermitian operators $\tilde{P}_1, \tilde{P}_2, \tilde{P}_3$, etc. If $\tilde{U}$ is a unitary operator, then the same state can be represented by the transformed wave function

$$\psi'(\mathbf{r}, t) = \tilde{U}\psi(\mathbf{r}, t),$$

and the various physical quantities by the transformed operators

$$\tilde{P}_1' = \tilde{U}\tilde{P}_1\tilde{U}^{-1}, \qquad \tilde{P}_2' = \tilde{U}\tilde{P}_2\tilde{U}^{-1}, \ldots.$$

The transformation

$$\psi \curvearrowright \tilde{U}\psi; \qquad \tilde{P} \curvearrowright \tilde{U}\tilde{P}\tilde{U}^{-1} = \tilde{U}\tilde{P}\tilde{U}^{\dagger} \tag{9.70}$$

is called a *unitary transformation*, and it leaves all matrix elements invariant:

$$\langle\psi_2|\tilde{P}|\psi_1\rangle \curvearrowright \langle\psi_2'|\tilde{P}'|\psi_1'\rangle = \langle\psi_2|\tilde{U}^{\dagger}\tilde{U}\tilde{P}\tilde{U}^{\dagger}\tilde{U}|\psi_1\rangle = \langle\psi_2|\tilde{P}|\psi_1\rangle.$$

Consequently, a unitary transformation leaves physics unchanged, which justifies that $\psi' = \tilde{U}\psi$ represents the same state.

Consider the Hamiltonian

$$\tilde{H} = \frac{1}{2m}\left(-i\hbar\nabla - q\mathbf{A}\right)^2 + q\phi$$

representing a charged particle moving in an electromagnetic field.

We now introduce the following unitary operator:

$$\tilde{U} = e^{i\frac{q}{\hbar}\chi(\mathbf{r},t)}. \tag{9.71}$$

It generates a unitary transformation which transforms the wave function in the following way

$$\psi(\mathbf{r}, t) \curvearrowright \psi'(\mathbf{r}, t) = e^{i\frac{q}{\hbar}\chi(\mathbf{r},t)}\psi(\mathbf{r}, t). \tag{9.72}$$

Furthermore, the operator $\mathbf{P} = -i\hbar\nabla$, representing the conjugate momentum transforms according to the rule

$$\begin{aligned}\mathbf{P} \curvearrowright \tilde{U}\mathbf{P}\tilde{U}^{-1} &= e^{i\frac{q}{\hbar}\chi(\mathbf{r},t)}(-i\hbar\nabla)e^{-i\frac{q}{\hbar}\chi(\mathbf{r},t)} \\ &= -q\nabla\chi - i\hbar\nabla = \mathbf{P} - q\nabla\chi.\end{aligned} \tag{9.73}$$

Finally, the operator $i\hbar\frac{\partial}{\partial t}$ representing the energy transforms according to the rule

$$\begin{aligned}i\hbar\frac{\partial}{\partial t} \curvearrowright \tilde{U}\left(i\hbar\frac{\partial}{\partial t}\right)\tilde{U}^{-1} &= e^{i\frac{q}{\hbar}\chi(\mathbf{r},t)}\left(i\hbar\frac{\partial}{\partial t}\right)e^{-i\frac{q}{\hbar}\chi(\mathbf{r},t)} \\ &= q\frac{\partial\chi}{\partial t} + i\hbar\frac{\partial}{\partial t}.\end{aligned} \tag{9.74}$$

Using this, we see that the *transformed* wave function $\psi'(\mathbf{r}, t)$ solves the transformed Schrödinger wave equation:

$$i\hbar \frac{\partial}{\partial t} \psi' = \left\{ \frac{1}{2m} \left[-i\hbar\nabla - q(\mathbf{A} + \nabla\chi) \right]^2 + q\left(\phi + \frac{\partial \chi}{\partial t} \right) \right\} \psi'.$$

But this is obtained from the old Schrödinger equation, provided that we perform the substitutions

$$\mathbf{A} \curvearrowright \mathbf{A} + \nabla\chi \quad \text{and} \quad \phi \curvearrowright \phi - \frac{\partial \chi}{\partial t} . \tag{9.75}$$

This justifies our claim that quantum-mechanically, gauge transformations correspond to unitary transformations! There is, however, one important difference between the classical situation and the quantum-mechanical situation: Classically, we have always demanded that $\chi(\mathbf{r}, t)$ should be a smooth function. Quantum-mechanically, this is no longer relevant. All we should demand is that $\tilde{U}(\mathbf{r}, t) = e^{i \frac{q}{\hbar} \chi(\mathbf{r}, t)}$ is a smooth function. Therefore, χ need not be single-valued, but can make jumps, provided that the jump of $\frac{q}{\hbar} \chi$ is proportional to 2π! Thus the unitary transformation is slightly more general than its classical counterpart.

With this in mind, we return to our discussion of the spherically symmetric monopole field. The choice of the two gauges

$$\text{Schwinger formalism:} \quad \mathbf{A}_1 = \frac{-g}{4\pi} \cos\theta \boldsymbol{d\varphi}, \tag{9.44}$$

$$\text{Dirac formalism:} \quad \mathbf{A}_2 = \frac{-g}{4\pi} (\cos\theta + 1)\boldsymbol{d\varphi}, \tag{9.45}$$

suggests that we look at the following "unitary operator":

$$\tilde{U} = e^{i \frac{q}{\hbar} \chi} = e^{-i \frac{\kappa}{\hbar} \varphi}. \tag{9.76}$$

This "unitary operator" is not well-defined unless it is a periodic function of φ; i.e., $\frac{\kappa}{\hbar}$ is an integer! When $\frac{\kappa}{\hbar}$ is an integer, we have previously seen that both choices of the gauge are admissible, and therefore we have shown that the Schwinger formalism and the Dirac formalism are unitarily equivalent in this case.

When $\frac{\kappa}{\hbar}$ is a half-integer, we have previously seen that only the Dirac formalism is admissible. The unitary operator (9.76) that formally transforms it into the Schwinger formalism breaks down. Observe, however, that for instance, $U = e^{2i \frac{\kappa}{\hbar} \varphi}$ is well-defined in that case. It corresponds to the "gauge" transformation

$$\mathbf{A} \curvearrowright \mathbf{A} + 2\frac{g}{4\pi} \boldsymbol{d\varphi}; \qquad \text{i.e.,} \quad \frac{-g}{4\pi} (\cos\theta + 1)\boldsymbol{d\varphi} \curvearrowright \frac{-g}{4\pi} (\cos\theta - 1)\boldsymbol{d\varphi},$$

where we have moved the semi-infinite string from the positive z-axis to the negative z-axis. Thus, although we cannot convert the semi-infinite string into an infinite string, we can still move the string around, using unitary transformations.

Finally, we should observe that although the unitary operator (9.76) is periodic for suitable values of κ, it is still singular on the z-axis. This is inescapable, because the gauge potentials themselves are singular on the z-axis.

9.7 Quantization of the Magnetic Charge

We conclude this chapter by presenting two other arguments for the quantization of the magnetic charge:

1. Argument: On the classical level, the position of the string is chosen completely arbitrarily. Consider two different strings γ_1 and γ_2. They give rise to two different gauge potentials

$$\mathbf{A}_1 : (-\phi_1, \mathbf{A}_1) \quad \text{and} \quad \mathbf{A}_2 : (-\phi_2, \mathbf{A}_2),$$

which are singular along their respective strings. Each of these gauge potentials generates a Hamiltonian, which in the nonrelativistic limit is represented by

$$\tilde{H}_1 = \frac{1}{2m}(-i\hbar\nabla - q\mathbf{A}_1) + q\phi_1,$$

$$\text{respectively} \quad \tilde{H}_2 = \frac{1}{2m}(-i\hbar\nabla - q\mathbf{A}_2) + q\phi_2.$$

Thus we get two quantum-mechanical descriptions of the same system:

$$i\hbar\frac{\partial\psi}{\partial t} = \tilde{H}_1\psi, \quad \text{respectively} \quad i\hbar\frac{\partial\psi}{\partial t} = \tilde{H}_2\psi.$$

But since the position of the string is of *no* physical importance, the two situations must be indistinguishable; i.e., they must be *unitarily equivalent*. The two gauge potentials are connected through a singular gauge transformation, $\mathbf{A}_2 - \mathbf{A}_1 = \boldsymbol{d}\chi$, where χ is singular along the strings. Thus the two descriptions are only unitarily equivalent if $\exp\left[i\frac{q}{\hbar}\chi(\mathbf{r}, t)\right]$ is a single-valued function outside the strings. The jump of χ must therefore be quantized; i.e.,

$$\frac{q}{\hbar}\oint \boldsymbol{d}\chi = n2\pi. \tag{9.77}$$

This jump can now be computed using the following trick: Consider a sphere S^2, where we split into two hemispheres S^2_+ and S^2_- by the equator, γ; cf. Figure 9.13. Then we get

$$\oint_\gamma \boldsymbol{d}\chi = \oint_\gamma \mathbf{A}_2 - \oint_\gamma \mathbf{A}_1 = \oint_\gamma \mathbf{A}_2 + \oint_{-\gamma} \mathbf{A}_1.$$

Using Stokes's theorem we therefore get

$$\oint_\gamma \boldsymbol{d}\chi = \int_{S^2_+} \boldsymbol{d}\mathbf{A}_2 + \int_{S^2_-} \boldsymbol{d}\mathbf{A}_1 = \int_{S^2} \mathbf{F} = g.$$

(Notice that $\mathbf{A}_2$ is smooth on the hemisphere S^2_+ and similarly for $\mathbf{A}_1$.) Thus the jump of χ is simply given by the magnetic charge! The quantization rule (9.77) can therefore be rearranged as

$$\frac{qg}{\hbar} = n2\pi,$$

which is precisely equivalent to (9.34)!

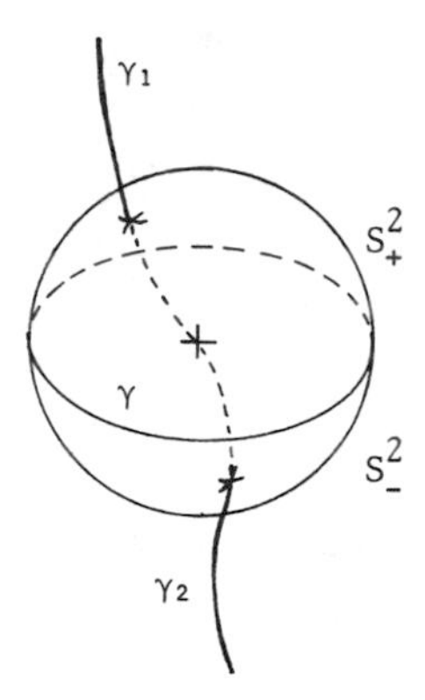

Figure 9.13.

2. Argument: While the first argument was very formal, the second argument will be more physical. It is based on path-integral techniques and is closely related to the Bohm–Aharonov effect.

   Consider two paths Γ_1 and Γ_2 that connect the initial point A with the final point B in our space diagram (see Figure 9.14). Suppose, furthermore, that the paths pass on each side of the string as shown in the figure. Then Γ_1 and Γ_2 form the boundary of a surface Ω that is hit by the string. (Strictly speaking, we should make a space–*time* diagram, but it is not easy to "visualize" this situation in four dimensions.)

   The amplitude $K(B \mid A)$ for a charged particle to propagate from A to B is the sum of all the amplitudes $\phi_\Gamma(B \mid A)$ for the particle to propagate along some specific path Γ (cf. (2.17)):

$$K(B \mid A) = \int \phi_\Gamma(B \mid A) D[\Gamma] = \int e^{\frac{i}{\hbar} S_\Gamma(B;A)} D[\Gamma].$$

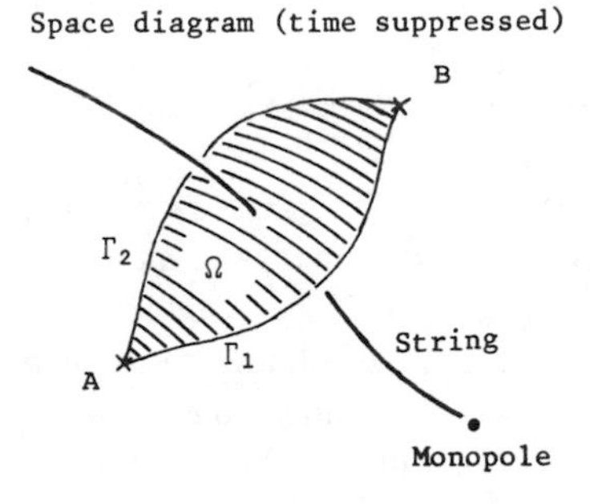

Figure 9.14.

When the charged particle moves in an external electromagnetic field, it picks up a change in phase given by

$$e^{\frac{iq}{\hbar}\int_\Gamma A_\alpha dx^\alpha}. \tag{2.33}$$

The contribution from Γ_1 and Γ_2 will interfere, and this interference is given by

$$e^{\frac{iq}{\hbar}\int_{\Gamma_2} A_\alpha dx^\alpha} e^{-\frac{iq}{\hbar}\int_{\Gamma_1} A_\alpha dx^\alpha} = e^{i\frac{q}{\hbar}\oint_\Gamma A_\alpha dx^\alpha}. \tag{9.78}$$

Using the notation of Section 9.2, the integral in the exponent can be rearranged as follows:

$$\oint_\Gamma \mathbf{A} = \oint_{\partial\Omega} \mathbf{A} = \int_\Omega d\mathbf{A} = \int_\Omega \mathbf{F} + \mathbf{S} = \int_\Omega \mathbf{F} + \int_\Omega \mathbf{S}.$$

Consequently, the change in phase due to the external field is given by

$$e^{\frac{iq}{\hbar}\int_\Omega \mathbf{F}} \cdot e^{\frac{iq}{\hbar}\int_\Omega S}. \tag{9.79}$$

The first term is okay. The charged particle ought to be influenced by the monopole field, but the second term must *not* contribute, since otherwise we could experimentally determine the position of the string using the Bohm–Aharonov effect! We are therefore forced to put

$$e^{\frac{iq}{\hbar}\int_\Omega \mathbf{S}} = 1. \tag{9.80}$$

But $\int_\Omega \mathbf{S}$ is the magnetic flux concentrated in the string that passes through Ω; i.e., it is equal to g. Consequently, equation (13.42) implies (9.34)

$$\frac{qg}{4\pi} = \frac{n}{2}\hbar.$$

It is instructive to compare the quantization of the monopole charge with the flux quantization of a magnetic vortex in a superconductor (Section 2.12). In both cases it is the magnetic flux that is limited to a discrete set of values. Furthermore, the basic mechanism behind the quantization is also very similar. In both cases the magnetic field interacts with electrically charged particles represented by a Schrödinger wave function $\psi(\mathbf{r}, t)$. In the case of a monopole, the phase of this wave function can make a quantized jump when we go once around the Dirac string, while in the case of the vortex, the phase can make a quantized jump when we go once around the vortex.

There is, however, one essential difference between the two cases: In the case of a monopole, we have been giving very general arguments to show that the jump of the phase is identical to the magnetic flux. Since we believe that a consistent description of this world must necessarily include quantum mechanics, we therefore conclude that if we should ever succeed in finding a magnetic monopole, its charge will be quantized. In the case of a magnetic flux string, this is no longer so! We can easily construct flux strings where the flux is *not* quantized. (Observe too that otherwise, the Bohm–Aharonov effect would disappear.) It is, of course, still true that the phase of a wave function can only make a quantized jump when we go once around the flux string, but this jump need not in general be related to the magnetic flux in the string. When a magnetic flux string happens to lie in a superconductor, this will have two very special consequences:

a. The length of ψ will be constant and nonzero inside the superconductor. (The length represents the density of the Cooper pairs.)
b. The gauge covariant derivative of ψ vanishes. (It represents essentially the Cooper current.) This implies the following relation between the phase φ and the gauge potential $\mathbf{A}$:

$$\nabla\varphi = \frac{q}{\hbar}\,\mathbf{A}. \tag{9.81}$$

Conditions (a) and (b) are characteristic of the so-called *ordered media*, and it is only through the interaction with an ordered medium that the magnetic flux in a flux string becomes quantized!

Solutions to Worked Exercises

Solution to 9.1.2

$$\begin{aligned}
&(\partial_\beta F^\alpha_{\ \gamma})F^{\gamma\beta} + \frac{1}{4}\,\eta^{\alpha\beta}\partial_\beta(F^{\gamma\delta}F_{\gamma\delta})\\
&= \frac{1}{2}\,(\partial_\beta F^\alpha_{\ \gamma})F^{\gamma\beta} + \frac{1}{2}\,(\partial_\beta F^\alpha_{\ \gamma})F^{\gamma\beta} + \frac{1}{2}\,\eta^{\alpha\beta}F^{\gamma\delta}(\partial_\beta F_{\gamma\delta})\\
&\qquad\downarrow \beta\to\delta \qquad\qquad\qquad \downarrow \beta\to\gamma;\ \gamma\to\delta\\
&= \frac{1}{2}\,(\partial_\delta F^\alpha_{\ \gamma})F^{\gamma\delta} + \frac{1}{2}\,(\partial_\gamma F^\alpha_{\ \delta})F^{\delta\gamma} + \frac{1}{2}\,\eta^{\alpha\beta}F^{\gamma\delta}(\partial_\beta F_{\gamma\delta})\\
&= \frac{1}{2}\,\eta_{\alpha\beta}[(\partial_\delta F_{\beta\gamma})F^{\gamma\delta} + (\partial_\gamma F_{\beta\delta})F^{\delta\gamma} + (\partial_\beta F_{\gamma\delta})F^{\gamma\delta}]\\
&= \frac{1}{2}\,\eta^{\alpha\beta}[\partial_\delta F_{\beta\gamma} + \partial_\gamma F_{\delta\beta} + \partial_\beta F_{\gamma\delta}]F^{\gamma\delta}.
\end{aligned}$$

But according to (9.2) we can exchange $(\partial_\delta F_{\beta\gamma} + F_{\delta\beta} + \partial_\beta F_{\gamma\delta})$ with $(\epsilon_{\beta\gamma\delta\epsilon}k^\epsilon)$

$$= \frac{1}{2}\,\eta^{\alpha\beta}[\epsilon_{\beta\gamma\delta\epsilon}k^\epsilon]F^{\gamma\delta} = \eta^{\alpha\beta}\overset{*}{F}_{\beta\epsilon}k^\epsilon = \overset{*}{F}{}^\alpha_{\ \beta}k^\beta.$$

Solution to 9.2.1

a. A trivial calculation gives

$$\frac{\partial R}{\partial x} = \frac{x}{R}, \quad \frac{\partial R}{\partial y} = \frac{y}{R}, \quad \frac{\partial R}{\partial z} = \frac{z}{R}.$$

Using this, we obtain

$$\frac{\partial}{\partial x}\left[\frac{1}{R(R-z)}\right] = -\frac{x(2R-z)}{R^3(R-z)^2};$$
$$\frac{\partial}{\partial y}\left[\frac{1}{R(R-z)}\right] = -\frac{y(2R-z)}{R^3(R-z)^2};$$
$$\frac{\partial}{\partial z}\left[\frac{1}{R(R-z)}\right] = \frac{1}{R^3}.$$

We can now calculate the curl! The x- and y-components offer no difficulties, but the z-component is a rather long story. If you split off the term z/R^3 and use the relation $x^2 + y^2 = R^2 - z^2 - \epsilon^2$, you should, however, obtain the result listed.

b. In the limit $\epsilon \to 0$ the first term reproduces the monopole field, so it is the remaining term that produces the string.

It is preferable to decompose the second term in the following way:

$$\epsilon^2(2R - z) = 2\epsilon^2(R - z) + \epsilon^2 z,$$

$$\mathbf{B}_\epsilon^{\text{STRING}} = -\frac{g_M}{4\pi}\frac{2\epsilon^2}{R^3(R-z)}\hat{\mathbf{k}} - \frac{g_M}{4\pi}\frac{\epsilon^2 z}{R^3(R-z)^2}\hat{\mathbf{k}}.$$

Now choose a point on the positive z-axis: $(0, 0, z_0)$. In the limit where $\epsilon \to 0$ we can replace $R - z$ by $\frac{\epsilon^2}{2z_0}$. Thus on the positive z-axis the two terms behave like

$$\text{(1)} \qquad -\frac{g_M}{\pi}\cdot\frac{1}{z_0^2} \qquad (\epsilon \ll 1),$$

$$\text{(2)} \qquad -\frac{g_M}{\pi}\cdot\frac{1}{\epsilon^2} \qquad (\epsilon \ll 1).$$

Thus the first term is approximately constant on the string, while the second term diverges. Consider the two terms as a function of the cylindrical radius ρ. They have a behavior somewhat like the behavior indicated in Figure 9.15.

Clearly, the first term vanishes in the limit as $\epsilon \to 0$, while the second may very well produce a δ-function. To investigate this, consider it as a function on the plane perpendicular to the string. We want to calculate the integral

$$\lim_{\epsilon\to 0} -\frac{g}{4\pi}\int_{\mathbf{R}^2}\frac{\epsilon^2 z}{R^3(R-z)^2}\,dx\,dy.$$

In the limit $\epsilon \to 0$ the main contribution comes from a region close to the string. But then we can safely perform the substitutions

$$R^3 \to z_0^3; \qquad R - z \to \frac{x^2 + y^2 + \epsilon^2}{2z_0}.$$

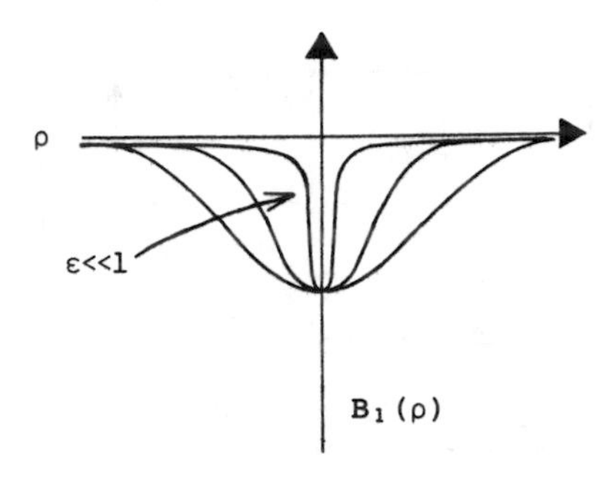

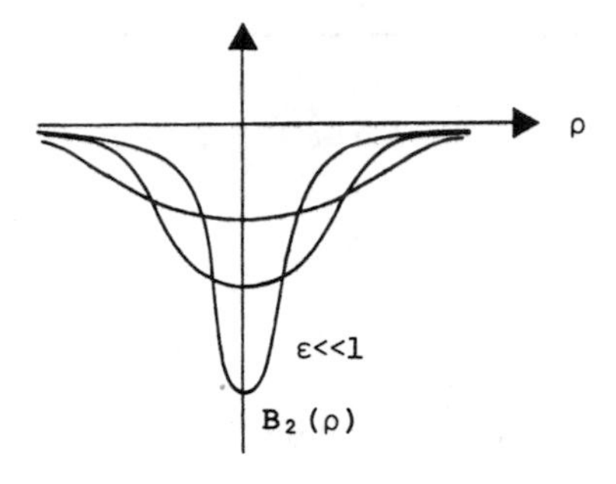

Figure 9.15.

Using polar coordinates, we therefore get

$$\lim_{\epsilon\to 0} -\frac{g_M}{4\pi}\int_{\mathbf{R}^2}\frac{\epsilon^2}{R^3(R-z)^2}\,dx\,dy = \lim_{\epsilon\to 0} -2g_M\int_{\rho=0}^{\infty}\frac{\epsilon^2\rho}{(\rho^2+\epsilon^2)^2}\,d\rho.$$

The latter integral is actually independent of ϵ, as we can easily see by performing the substitution $\rho = \epsilon t$. The integral then becomes a standard integral with the value $\frac{1}{2}$, and we finally get

$$\lim_{\epsilon\to 0} -\frac{g_M}{4\pi}\int_{\mathbf{R}^2}\frac{\epsilon^2 z}{R^3(R-z)^2}\,dx\,dy = -g_M.$$

But then the limit is a two-dimensional δ-function! For a point on the *negative* z-axis we know that $R - z \geq 2|z|$, and therefore the string terms vanish in the limit $\epsilon \to 0$.

Solution to 9.2.2

To calculate the integral explicitly, we first find a coordinate system in Minkowski space where the coordinate expression for $S_{\alpha\beta}(x)$ is especially simple. This is done in the following manner: Consider a space slice. Then we use the string itself as the positive x^3-axis (Figure 9.16). In these special coordinates the position of the

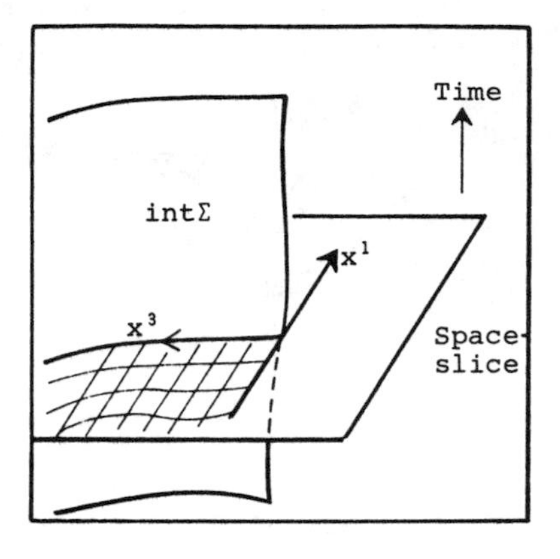

Figure 9.16.

monopole is always

$$(x^1, x^2, x^3) = (0, 0, 0),$$

and the string always occupies the positions

$$x^1 = x^2 = 0, \qquad x^3 \geq 0.$$

We say that these coordinates are *adapted* to the string. We can now evaluate the components of **S**,

$$S_{\alpha\beta}(x) = g_m \int_U \delta^4(x - \tilde{x}(\lambda^1, \lambda^2))\epsilon_{\alpha\beta\gamma\delta} \frac{\partial \tilde{x}^\gamma}{\partial \lambda^1} \frac{\partial \tilde{x}^\delta}{\partial \lambda^2} \, d\lambda^1 d\lambda^2.$$

As a parametrization of the sheet we simply choose

$$\tilde{x}^0 = \lambda^2, \qquad \tilde{x}^1 = 0, \qquad \tilde{x}^2 = 0, \qquad \tilde{x}^3 = -\lambda^1.$$

Then the expression for $S_{\alpha\beta}$ reduces to

$$S_{\alpha\beta}(x) = -\epsilon_{\alpha\beta 30} g_m \int_U \delta^4(x - \tilde{x}(\lambda^1, \lambda^2)) d\lambda^1 d\lambda^2.$$

If x^3 is negative, the sheet can never cross the point $x = (x^0, x^1, x^2, x^3)$, and the integral automatically vanishes. If x^3 is positive, we get

$$\begin{aligned}\int_U \delta^4(x - \tilde{x}(\lambda^1, \lambda^2)) d\lambda^1 d\lambda^2 &= \int_U \delta(x^0 - \tilde{x}^0(\lambda^1, \lambda^2))\delta(x^1)\delta(x^2) \\ &\quad \times \delta(x^3 - \tilde{x}^3(\lambda^1, \lambda^2)) d\lambda^1 d\lambda^2 \\ &= \delta(x^1)\delta(x^2).\end{aligned}$$

Consequently, we have shown that

$$S_{\alpha\beta}(x) = -g_m \epsilon_{\alpha\beta 30} \delta(x^1)\delta(x^2)\theta(x^3).$$

Now it is trivial to compute the magnetic flux associated with **S** through any surface. If the surface is hit by the string, we can use (x^1, x^2) to parametrize the surface in a neighborhood of the string:

$$\int_\Omega \mathbf{S} = \int -S_{12} dx^1 dx^2 = -g_M \int \delta(x^1)\delta(x^2) dx^1 dx^2 = -g_M.$$

Solution to 9.4.1

$$\mathbf{J} = \epsilon_0 \int \mathbf{r} \times (\mathbf{E} \times \mathbf{B}) d^3x = \epsilon_0 \int [(\mathbf{r} \cdot \mathbf{B})\mathbf{E} - (\mathbf{r} \cdot \mathbf{E})\mathbf{B}] d^3x$$
$$= \frac{g_M}{4\pi} \epsilon_0 \int \left[\frac{1}{r} \mathbf{E} - (\mathbf{r} \cdot \mathbf{E}) \frac{\mathbf{r}}{r^3} \right] d^3x.$$

The angular momentum in the direction of $\hat{\mathbf{r}}_0$ (where $\hat{\mathbf{r}}_0$ denotes a unit vector) is then given by

$$\hat{\mathbf{r}}_0 \cdot \mathbf{J} = \frac{g_M}{4\pi} \epsilon_0 \int \left[\frac{\hat{\mathbf{r}}_0 \cdot \mathbf{E}}{r} - \frac{(\mathbf{r} \cdot \mathbf{E})(\hat{\mathbf{r}}_0 \cdot \mathbf{r})}{r^3} \right] d^3x.$$

To evaluate this integral, we observe that

$$\nabla \cdot \left[\frac{(\hat{\mathbf{r}}_0 \cdot \mathbf{r})}{r} \mathbf{E} \right] = \left(\frac{\hat{\mathbf{r}}_0 \cdot \mathbf{r}}{r} \right) (\nabla \cdot \mathbf{E}) + \frac{\hat{\mathbf{r}}_0 \cdot \mathbf{E}}{r} - \frac{(\hat{\mathbf{r}}_0 \cdot \mathbf{r})(\mathbf{r} \cdot \mathbf{E})}{r^3}.$$

Inserting this relation we get

$$\hat{\mathbf{r}}_0 \cdot \mathbf{J} = -\frac{g_M}{4\pi} \epsilon_0 \int \frac{(\hat{\mathbf{r}}_0 \cdot \mathbf{r})}{r} (\nabla \cdot \mathbf{E}) d^3x + \frac{g_M}{4\pi} \epsilon_0 \int \nabla \cdot \left[\frac{(\hat{\mathbf{r}}_0 \cdot \mathbf{r})}{r} \mathbf{E} \right] d^3x$$

In the first integral we use that $\nabla \cdot \mathbf{E} = \frac{q}{\epsilon_0} \delta^3(\mathbf{r} - \mathbf{r}')$ where $\mathbf{r}'$ is the position of the charged particle. In the second integral we use Gauss's theorem to convert it to a "distorted" flux integral:

$$\int \nabla \cdot \left[\frac{(\hat{\mathbf{r}}_0 \cdot \mathbf{r})}{r} \mathbf{E} \right] d^3x = \int \frac{\hat{\mathbf{r}}_0 \cdot \mathbf{r}}{r} (\mathbf{E} \cdot \hat{\mathbf{n}}) dA$$
$$= \frac{q}{\epsilon_0} \int \cos\theta \sin\theta \, d\theta \, d\varphi = 0,$$

which thus vanishes automatically. Consequently,

$$\hat{\mathbf{r}}_0 \cdot \mathbf{J} = -\kappa (\hat{\mathbf{r}}_0 \cdot \hat{\mathbf{r}}').$$

This shows precisely that the component of **J** perpendicular to $\mathbf{r}'$ vanishes and that the component of **J** along the $\mathbf{r}'$ is given by $-\kappa$.

Solution to 9.4.2

Consider a beam of electrically charged particles carrying a specific energy E. A particle with impact parameter b will be scattered at a certain angle $\theta = \theta(E; b)$. Let us briefly recall the definition of the differential cross section. For a given scattering angle θ we consider the values of the impact parameter b corresponding to θ. (Remark: There may a priori be several different values of b producing the same θ!) Then we consider a small ring between θ and $\theta + \Delta\theta$ (Figure 9.17). On the unit sphere it represents the area

$$\Delta\Omega = 2\pi \sin\theta \Delta\theta.$$

The values of b corresponding to this ring will itself form a number of rings from b_i to $b_i + \Delta b_i$. The total area of these rings is given by

$$\Delta\sigma = \sum_i 2\pi b_i \Delta b_i.$$

The differential cross section $\frac{d\sigma}{d\Omega}$ is then defined as

$$\frac{d\sigma}{d\Omega}(\theta) = \frac{\text{beam area}}{\text{detector area}} = \frac{\Sigma 2\pi b \Delta b}{2\pi \sin\theta \Delta\theta} = \sum b \left| \frac{db}{d(\cos\theta)} \right|, \tag{9.82}$$

where we sum over all values of b corresponding to the given value of θ.

We know that a given particle moves on a cone. It is customary to characterize this cone by the angle χ, where $\pi - \chi$ is the angle spanned by the cone. This angle can be expressed in terms of ℓ and κ. Considering the triangle shown in Figure 9.18, we immediately get

$$\cot\frac{\chi}{2} = \frac{\ell}{|\kappa|} = \frac{mvb}{|\kappa|}. \tag{9.83}$$

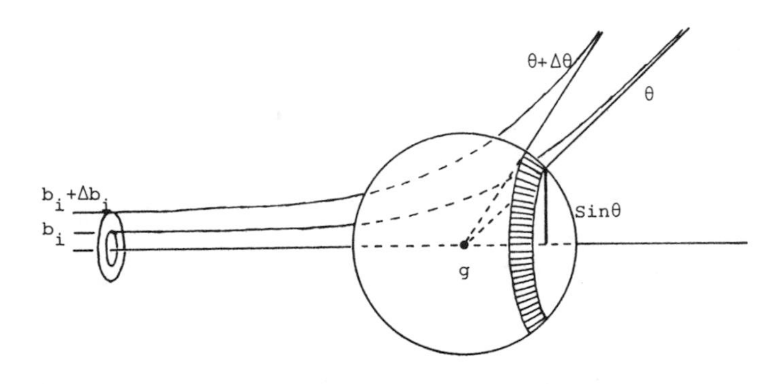

Figure 9.17.

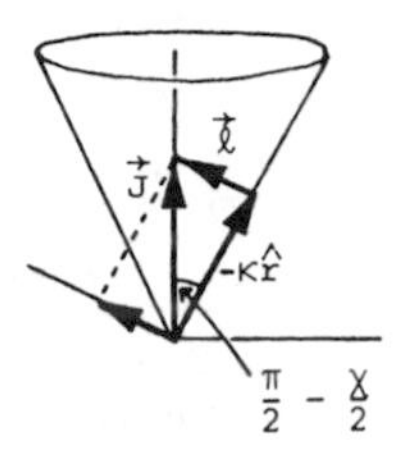

Figure 9.18.

As we shall see, the scattering angle θ depends only on χ. We can therefore rearrange equation (9.82) as

$$\frac{d\sigma}{d\Omega}(\theta) = \sum b \left| \frac{\frac{db}{d\chi}}{\frac{d(\cos\theta)}{d\chi}} \right| = \sum \frac{\kappa^2}{2m^2v^2} \frac{\cos\frac{\chi}{2}}{\sin^3\left(\frac{\chi}{2}\right)} \cdot \frac{1}{\left|\frac{d(\cos\theta)}{d\chi}\right|}. \tag{9.84}$$

It remains to determine θ as a function of χ. Here it is profitable to use an adapted coordinate system with the axis of symmetry as the third axis; i.e., $\mathbf{J}$ is pointing along the third axis. The position of the particle $\mathbf{r}(t)$ can now be expressed in terms of the radial distance r and the azimuthal angle ϕ. Using this, we get

$$\mathbf{v} = \frac{d\mathbf{r}}{dt} = \frac{dr}{dt}\hat{\mathbf{r}} + \frac{d\phi}{dt}\hat{\mathbf{k}} \times \mathbf{r}. \tag{9.85}$$

On the other hand, we obtain from equation (9.39) that

$$\mathbf{J} \times \mathbf{r} = (r \times m\mathbf{v}) \times \mathbf{r} = r^2 m\mathbf{v} - \mathbf{r}(m\mathbf{v} \cdot \mathbf{r}),$$

and consequently,

$$\mathbf{v} = (\mathbf{r} \cdot \mathbf{v})\hat{\mathbf{r}} + \frac{J}{mr^2}\hat{\mathbf{k}} \times \mathbf{r}. \tag{9.86}$$

By comparing expressions (9.85) and (9.86) we then deduce

$$\frac{d\phi}{dt} = \frac{J}{mr^2}. \tag{9.87}$$

Using equation (9.38), we can now integrate the above equation for ϕ (obviously we can choose $\phi(0) = 0$. Remember also that $b = d$.)

$$\phi(r_0) = \int_b^{r_0} \frac{J}{mr^2}\frac{dt}{dr}\,dr = \pm\frac{J}{mv}\left[\frac{\pi}{2} - \arcsin\frac{b}{r_0}\right],$$

where + refers to positive time and − refers to negative time. This leads to the following asymptotic values:

$$\lim_{t\to\pm\infty}(t) = \pm\frac{J}{mvb}\frac{\pi}{2}. \tag{9.88}$$

(Since ϕ has total increase

$$\Delta\phi = \frac{J}{mvb}\,\pi = \frac{J}{\ell}\,\pi,$$

we see that for $J/\ell = 2n$ the particle will move n times around the symmetry axis.)

Now we can finally express the scattering angle θ in terms of χ. From equation (9.86) we get

$$\mathbf{v} \approx (\hat{\mathbf{r}} \cdot \mathbf{v})\hat{\mathbf{r}} \qquad \text{for large } |t|,$$

and therefore the scattering angle θ is identical to the angle between $\hat{\mathbf{r}}(+\infty)$ and $-\hat{\mathbf{r}}(-\infty)$ (Figure 9.19): Thus we obtain

$$\begin{aligned}
\cos\theta &= -\hat{\mathbf{r}}(+\infty)\hat{\mathbf{r}}(-\infty) \\
&= -\cos^2\frac{\chi}{2}\,\left[\cos\phi(+\infty)\cos\phi(-\infty) + \sin\phi(+\infty)\sin\phi(-\infty)\right] \\
&\quad - \sin^2\frac{\chi}{2} \\
&= 2\cos^2\frac{\chi}{2}\cdot\sin^2\left[\frac{J}{mvb}\cdot\frac{\pi}{2}\right] - 1.
\end{aligned}$$

Using that

$$\frac{J}{mvb} = \frac{\sqrt{\ell^2+\kappa^2}}{\ell} = \sqrt{1+\tan^2\frac{\chi}{2}} = \frac{1}{\cos\frac{\chi}{2}},$$

we finally obtain the desired formula

$$\cos\theta = 2\cos^2\frac{\chi}{2}\,\sin^2\left[\frac{\frac{\pi}{2}}{\cos\frac{\chi}{2}}\right] - 1. \tag{9.89}$$

If we put

$$\zeta = \frac{\frac{\pi}{2}}{\cos\frac{\chi}{2}}, \tag{9.90}$$

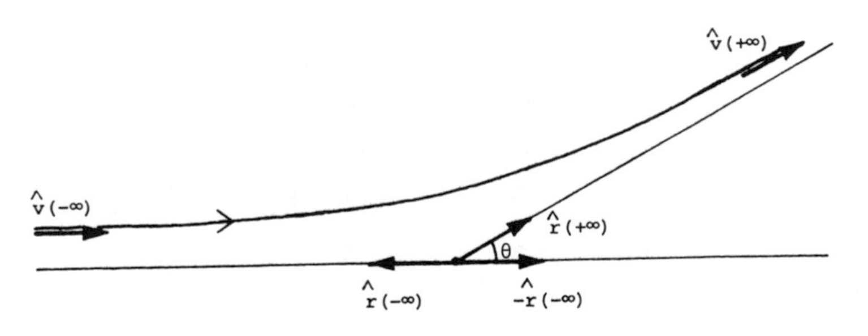

Figure 9.19.

this implies that

$$\frac{d(\cos\theta)}{d\chi} = 2\sin\frac{\chi}{2}\cos\frac{\chi}{2}\left[-\sin^2\zeta + \zeta\sin\zeta\cos\zeta\right].$$

Okay, substituting this into equation (9.84), we have derived the following marvelous formula for the differential cross section:

$$\frac{d\sigma}{d\Omega} = \frac{|\kappa|^2}{8mE}\sum\frac{1}{\sin^4\frac{\chi}{2}} \times \frac{1}{(-\sin^2\zeta + \zeta\sin\zeta\cos\zeta)}; \qquad \zeta = \frac{\frac{\pi}{2}}{\cos\frac{\chi}{2}}. \tag{9.91}$$

Having gone this far, let us make some comments. First, compare this with the Rutherford formula for Coulomb scattering,

$$\frac{d\sigma}{d\Omega} = \frac{1}{2}\left(\frac{e^2}{2E}\right)^2\frac{1}{\sin^4\frac{\theta}{2}}. \tag{9.92}$$

Observe that the energy dependence is different in the two cases:

$$\left(\frac{d\sigma}{d\Omega}\right)_{\text{monopole}} \propto \frac{1}{E}; \qquad \left(\frac{d\sigma}{d\Omega}\right)_{\text{Coulomb}} \propto \frac{1}{E^2}.$$

Then we consider forward scattering, where θ is very small. From equation (9.89) we see that to lowest order, $\theta \approx \chi$. Furthermore, we observe that

$$\sin\zeta \approx 1 \quad \text{and} \quad \cos\zeta \approx 0.$$

In the case of forward scattering we can therefore use the following approximation:

$$\theta \ll 1: \frac{d\sigma}{d\Omega} \approx \frac{|\kappa|^2}{8mE}\frac{1}{\sin^4\frac{\theta}{2}},$$

which has a structure very similar to the Rutherford formula.

For the case of backward scattering, things are more complicated. On the one hand, we now get contributions from several values of χ. On the other hand, $\frac{d\sigma}{d\Omega}$ becomes singular, i.e., infinite at a series of angles θ_n converging towards π. This is due to the term

$$\frac{1}{\sin^2\zeta - \zeta\sin\zeta\cos\zeta},$$

which can no longer be neglected. The denominator becomes zero when

$$\tan\zeta = \zeta \qquad (\zeta \neq 0).$$

For positive ζ this equation has an infinite number of solutions ζ_n. To each of these solutions corresponds a specific value of θ. The angles θ_n for which $\frac{d\sigma}{d\Omega} = \infty$ are called rainbow angles.

Solution to 9.5.2

a. $\hat{\mathbf{r}}\cdot\mathbf{L} = \hat{\mathbf{r}}\cdot[\mathbf{r}\times(-i\hbar\nabla - q\mathbf{A})] = -i\hbar\hat{\mathbf{r}}\cdot[\mathbf{r}\times\nabla]$.

b. $\mathbf{L} \cdot \hat{\mathbf{r}} = [\mathbf{r} \times (-i\hbar\nabla - q\mathbf{A})] \cdot \mathbf{r} = -i\hbar[\mathbf{r} \times \nabla] \cdot \hat{\mathbf{r}}$.

To facilitate the further computations, we switch to component expressions:

a. $\hat{\mathbf{r}} \cdot (\mathbf{r} \times \nabla) = \epsilon_{ijk} \frac{x^i}{r} (x^j \partial_k) = \frac{1}{r} \epsilon_{ijk} x^i x^j \partial_k = 0.$

b. $(\mathbf{r} \times \nabla) \cdot \hat{\mathbf{r}} = \epsilon_{ijk}(x^i \partial_j) \frac{x^k}{r} = \epsilon_{ijk} \frac{x^i x^k}{r} \partial_j + \epsilon_{ijk} x^i \partial_j \left(\frac{x^k}{r}\right)$
$= \epsilon_{ijk} x^i x^k \partial_j \left(\frac{1}{r}\right) + \epsilon_{ijk} \frac{x^i}{r} \delta^k_j = 0.$

Solution to 9.5.3

The analysis of the eigenvalue equation (9.65) is most easily performed using the raising and lowering operators $\tilde{J}_\pm = \tilde{J}_1 \pm i\tilde{J}_2$. Neglecting the normalization for a moment, we know that

$$\tilde{J}_\pm P_{jm} \propto P_{jm\pm 1}.$$

In the Schwinger formalism the raising and lowering operators are given by

$$\tilde{J}_\pm = -\hbar \left[-\frac{\partial}{\partial\theta} \pm \frac{m\cos\theta + \frac{\kappa}{\hbar}}{\sin\theta} \right] = -\hbar\sqrt{1-x^2} \left[\frac{\partial}{\partial x} \pm \frac{mx + \frac{\kappa}{\hbar}}{1-x^2} \right].$$

From this we obtain the following recurrence relations:

$$P_{jm\pm 1} \propto \sqrt{1-x^2} \left[\frac{\partial}{\partial x} \pm \frac{mx + \frac{\kappa}{\hbar}}{1-x^2} \right] P_{jm}.$$

These recurrence relations can be further rearranged using the identity

$$P' + fP = e^{\pm \int f(x)dx} \frac{d}{dx} \left[e^{\pm \int f(x)dx} P \right].$$

In this case,

$$f(x) = \frac{mx + \frac{\kappa}{\hbar}}{1-x^2} = \frac{1}{2} \frac{m + \frac{\kappa}{\hbar}}{1-x} - \frac{1}{2} \frac{m - \frac{\kappa}{\hbar}}{1+x},$$

and therefore,

$$\int f(x)dx = -\frac{1}{2}\left(m + \frac{\kappa}{\hbar}\right)\ln(1-x) - \frac{1}{2}\left(m - \frac{\kappa}{\hbar}\right)\ln(1+x).$$

The recurrence relations are then rearranged as

$$P_{jm\pm 1} \propto (1-x^2)^{1/2}(1-x)^{\frac{1}{2}\left(\pm m \pm \frac{\kappa}{\hbar}\right)}(1+x)^{\frac{1}{2}\left(\pm m \mp \frac{\kappa}{\hbar}\right)}$$
$$\times \frac{d}{dx}\left[(1-x)^{\mp\frac{1}{2}\left(m + \frac{\kappa}{\hbar}\right)}(1+x)^{\mp\frac{1}{2}\left(m - \frac{\kappa}{\hbar}\right)} P_{jm}\right].$$

By induction we then deduce

$$P_{jm+n}(x) \propto (1-x)^{\frac{1}{2}\left(n+m+\frac{\kappa}{\hbar}\right)}(1+x)^{\frac{1}{2}\left(n+m-\frac{\kappa}{\hbar}\right)} \cdot \frac{d^n}{dx^n}\left[(1-x)^{-\frac{1}{2}\left(m+\frac{\kappa}{\hbar}\right)}(1+x)^{-\frac{1}{2}\left(m-\frac{\kappa}{\hbar}\right)}P_{jm}(x)\right]. \qquad (*)$$

$$P_{jm-n}(x) \propto (1-x)^{\frac{1}{2}\left(n-m-\frac{\kappa}{\hbar}\right)}(1+x)^{\frac{1}{2}\left(n-m+\frac{\kappa}{\hbar}\right)} \cdot \frac{d^n}{dx^n}\left[(1-x)^{-\frac{1}{2}\left(m+\frac{\kappa}{\hbar}\right)}(1+x)^{\frac{1}{2}\left(m-\frac{\kappa}{\hbar}\right)}P_{jm}(x)\right]. \qquad (**)$$

This is our main result! From the general theory we know that P_{jm+n} and P_{jm-n} vanish for sufficiently large n. But this implies that

$$(1-x)^{-\frac{1}{2}\left(m+\frac{\kappa}{\hbar}\right)}(1+x)^{-\frac{1}{2}\left(m-\frac{\kappa}{\hbar}\right)}P_{jm}(x) = Q'_{jm}(x)$$

and

$$(1-x)^{\frac{1}{2}\left(m+\frac{\kappa}{\hbar}\right)}(1+x)^{\frac{1}{2}\left(m-\frac{\kappa}{\hbar}\right)}P_{jm}(x) = Q_{jm}(x)$$

are polynomials. Eliminating $P_{jm}(x)$, we get the following relation between Q'_{jm} and Q_{jm}:

$$Q'_{jm}(x) = (1-x)^{m+\frac{\kappa}{\hbar}}(1+x)^{m-\frac{\kappa}{\hbar}}Q_{jm}(x).$$

But as Q_{jm} and Q'_{jm} are polynomials this is consistent only if $m+\frac{\kappa}{\hbar}$ and $m-\frac{\kappa}{\hbar}$ are both integers! Using the relations

$$m = \frac{1}{2}\left(m+\frac{\kappa}{\hbar}\right) + \frac{1}{2}\left(m-\frac{\kappa}{\hbar}\right) \qquad \text{and}$$

$$\frac{\kappa}{\hbar} = \frac{1}{2}\left(m+\frac{\kappa}{\hbar}\right) - \frac{1}{2}\left(m-\frac{\kappa}{\hbar}\right),$$

we see that m and $\frac{\kappa}{\hbar}$ are both either half-integers or integers!

From the recurrence relations $(*)$, $(**)$ we can also obtain explicit expressions for the functions $P_{jm}(x)$. We know that $P_{jj}(x)$ is killed by the raising operator $\tilde{J}_+$. Therefore, we get

$$P_{jj}(x) = C\cdot(1-x)^{\frac{1}{2}\left(j+\frac{\kappa}{\hbar}\right)}(1+x)^{\frac{1}{2}\left(j-\frac{\kappa}{\hbar}\right)}.$$

This is a regular normalizable solution provided that $j \geq \frac{|\kappa|}{\hbar}$. Using the lowering operator, we then obtain the rest of the functions:

$$P_{jm}(x) = N_{jm}\cdot(1-x)^{-\frac{1}{2}\left(m+\frac{\kappa}{\hbar}\right)}(1+x)^{-\frac{1}{2}\left(m-\frac{\kappa}{\hbar}\right)} \cdot \frac{d^{j-m}}{dx^{j-m}}\left[(1-x)^{\left(j+\frac{\kappa}{\hbar}\right)}(1+x)^{\left(j-\frac{\kappa}{\hbar}\right)}\right].$$

CHAPTER 10

Smooth Maps—Winding Numbers

10.1 Local Properties of Smooth Maps

In this chapter we are going to examine maps from one manifold to another, and unfortunately, our work is going to consist of some rather technical investigations, but there are at least two main reasons for doing this anyhow:

First of all, maps actually play an important role in physics. When, for example, we discuss space–time symmetries, we consider certain maps from space–time into itself, such as rotations or translations, and we then investigate what happens to field configurations during such symmetry transformations. (This will be explored carefully in Sections 11.4–11.6.) It is also worth noticing that field configurations themselves are maps from space–time into a field space. E.g., a complex scalar field, say a charged Klein–Gordon field, is simply a smooth map from space–time into the complex numbers, and properties of such maps may contain nontrivial information about the system we are considering. (Examples of such nontrivial properties will be given in Sections 10.8–10.9.)

The second reason is that it turns out that we can use maps to transfer differential forms from one manifold to another in such a way that the various operations of the exterior calculus, i.e., the exterior derivative, the wedge product, etc., are preserved. This gives a much greater flexibility in the computation of various quantities.

Okay, as we cannot talk ourselves out of it, let M^m and N^n be differentiable manifolds. We want to consider what happens when we construct a map from one manifold to another, $f : M^m \curvearrowright N^n$. (See Figure 10.1.) We want to explain what we mean by f being smooth at a point P. Let f be a continuous map. Consider a point P_0 in M and the corresponding point $Q_0 = f(P_0)$ in N. If we introduce coordinate systems covering P_0 and Q_0, then f is represented by an ordinary Euclidean function:

$$(y^1, \ldots, y^n) = \bar{f}(x^1, \ldots, x^m).$$

Let U be an open neighborhood around Q_0. Then $f^{-1}(U)$ is an open neighborhood around P_0 because f is continuous. But this shows that the Euclidean representative f is well-defined in a neighborhood of $(x_0^1, \ldots, x_0^m)$, where $(x_0^1, \ldots, x_0^m)$ are the coordinates of P_0.

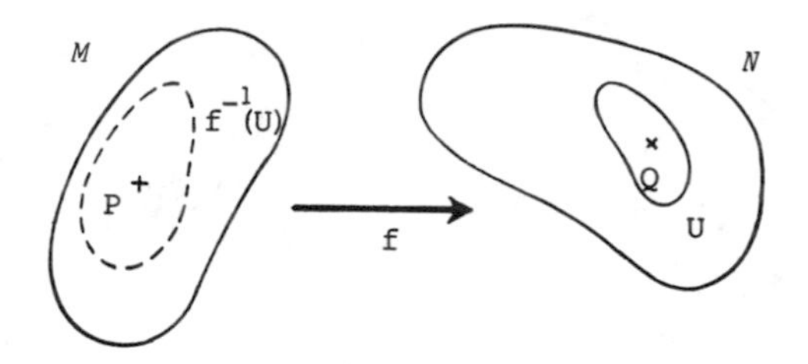

Figure 10.1.

Thus it has meaning to ask if the Euclidean representative $\bar{f}$ is smooth at $(x_0^1, \ldots, x_0^m)$. (If f is not required to be continuous, then the Euclidean representative $\bar{f}$ need not be defined at points close to $(x_0^1, \ldots, x_0^m)$, and we can no longer work out partial derivatives!) This motivates the following definition:

Definition 10.1. A continuous map $f : M^m \curvearrowright N^n$ is said to be *smooth* at a point P in M if the coordinate representative $\bar{f}$ is an ordinary smooth Euclidean map at the corresponding coordinate point $(x_0^1, \ldots, x_0^m)$.

Observe that if just one coordinate representative is smooth, then they are all smooth, because different coordinate systems depend smoothly upon each other.

Consider now a smooth map f. Then a coordinate representative will have partial derivatives of arbitrarily high order. The Jacobian matrix $[\partial y^i / \partial x^j]$ is of particular interest to us, as it controls the local properties of f. In a neighborhood of a point $P(x_0^1, \ldots, x_0^m)$ in M we have the Taylor expansion

$$\bar{f}(x_0^j + \Delta x^j) = \bar{f}(x_0^j) + \frac{\partial y^i}{\partial x^j}\Big|_{x_0} \Delta x^j + \text{higher-order terms.}$$

Thus when we approximate f in a neighborhood of P with a linear map, this linear map will precisely be generated by the Jacobian matrix. We can make these vague ideas more precise in the following way:

Consider the tangent spaces $\mathbf{T}_P(M)$ and $\mathbf{T}_{f(P)}(N)$. We will now show that f generates, in a canonical fashion, a linear map

$$f_* : \mathbf{T}_P(M) \curvearrowright \mathbf{T}_{f(P)}(N)$$

that controls the local properties of f and that is represented by the Jacobian matrix!

To see how this comes about, we proceed in the following way: Let $\mathbf{v}_P$ be a tangent vector at P. Then $\mathbf{v}_P$ is generated by a smooth curve $\lambda(t)$. (See Figure 10.2.) But we can use the smooth map f to transport the curve $\lambda(t)$ into a smooth curve $\lambda'(t) = f(\lambda(t))$ on N. The transferred curve λ' generates a tangent vector $\mathbf{u}_Q$ at $Q = f(P)$, which we define to be the transport of $\mathbf{v}_P$. However, this makes sense only if we can show that $\mathbf{u}_Q$ is independent of the particular curve

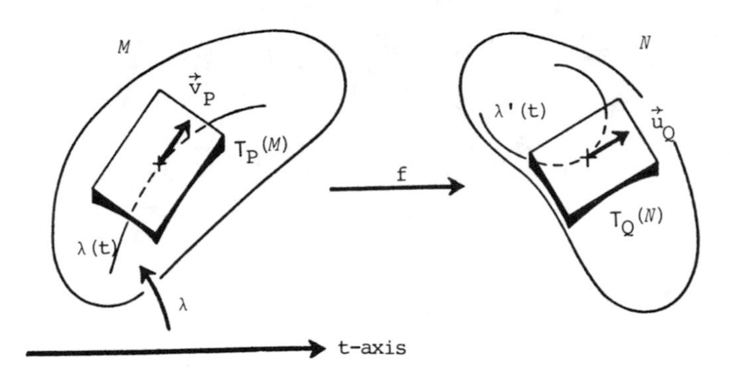

Figure 10.2.

$\lambda(t)$ we chose to generate $\mathbf{v}_P$. To see that this is actually the case, we work out the coordinates of $\mathbf{u}_Q$: Let λ be given by the parametrization $x^i = x^i(t)$, with P corresponding to $x^i(0)$. Then $\mathbf{v}_P$ is characterized by the coordinates

$$a^i = \frac{dx^i}{dt}\Big|_{t=0}.$$

The curve $\lambda' = f(\lambda)$ is characterized by the parametrization $y^i = \bar{f}^i(x^i(t))$. Using the chain rule, we therefore find the following coordinates of $\mathbf{u}_Q$:

$$b^i = \frac{dy^i}{dt}\Big|_{t=0} = \frac{\partial y^i}{\partial x^j}\frac{dx^j}{dt}\Big|_{t=0} = \left[\frac{\partial y^i}{\partial x^j}\right]_{|P} \cdot a^i.$$

But this clearly shows that the coordinates of $\mathbf{u}_Q$ depend only on the coordinates of $\mathbf{v}_P$, not on the particular curve λ. Furthermore, we immediately get that the induced map

$$f_* : \mathbf{T}_P(M) \curvearrowright \mathbf{T}_Q(N)$$

is represented by the Jacobian matrix:

$$\bar{\bar{D}}_f(P) = \left[\frac{\partial y^i}{\partial x^j}\right]_{|P}. \tag{10.1}$$

In particular, it is a linear map! The induced map satisfies a natural composition rule:

Theorem 10.1. *Let there be given three differentiable manifolds M, N, and L and two smooth maps $L \xrightarrow{f} M \xrightarrow{g} N$. Then*

$$(g \circ f)_* = g_* \circ f_*. \tag{10.2}$$

PROOF. This is an immediate consequence of the chain rule. Let us introduce three coordinate systems:

$$\begin{aligned} (x^1, \ldots, x^1) &\quad \text{covering } P \text{ in } L, \\ (y^1, \ldots, y^m) &\quad \text{covering } Q = f(P) \text{ in } M, \\ (z^1, \ldots, z^n) &\quad \text{covering } R = g(Q) \text{ in } N. \end{aligned}$$

Then $(g \circ f)_*$ is the linear map $\mathbf{T}_P(L) \to \mathbf{T}_R(N)$ generated by the Jacobian matrix,

$$\frac{\partial z^i}{\partial x^j} = \frac{\partial z^i}{\partial y^k} \frac{\partial y^k}{\partial x^j},$$

or equivalently,

$$\bar{\bar{D}}_{g \circ f}(P) = \bar{\bar{D}}_g(Q) \cdot \bar{\bar{D}}_f(P).$$

□

We conclude this section with a discussion of the behavior of vector fields. Let there be given a smooth vector field $\mathbf{V}(x)$ on M. We know that f generates a pointwise map

$$f_* : \mathbf{T}_P(M) \curvearrowright \mathbf{T}_{f(P)}(N).$$

Does this mean that f_* actually maps the vector field $\mathbf{V}$ onto another vector field $f_*(\mathbf{V})$? No, not in general! Let us explain where the trouble may come from:

a. If f is not surjective, there are points on M at which we do not attach any tangent vector.
b. If f is not injective, there exist two points P_1 and P_2 that are mapped into the same point Q. Consequently, the two tangent vectors $\mathbf{V}(P_1)$ and $\mathbf{V}(P_2)$ may very well be mapped into *two different* tangent vectors at $Q = f(P_1) = f(P_2)$.

So when we transport a vector field from M to N, the resulting "vector field" will in general be a "multivalued" vector field that is not "globally defined on N." (Consider, for instance, a constant map f that maps the whole manifold M into a single point.)

But even if f is bijective, we may be in trouble. Consider the coordinate expression for the transported vector field:

$$b^i(y^1, \ldots, y^n) = \frac{\partial y^i}{\partial x^j} \cdot a^j(x^1, \ldots, x^n).$$

We must demand that it depend smoothly on the coordinates $(y^1, \ldots, y^n)$. If the inverse function f^{-1} is smooth too, then everything works out fine, and we get

$$b^i(y) = \frac{\partial y^i}{\partial x^j}\Big|_{f^{-1}(y)} \cdot a^j(f^{-1}(y)), \tag{10.3}$$

where the right-hand side is smooth. But if f^{-1} is not smooth, we are in trouble! We must therefore demand that f be not only bijective, but that it be a diffeomorphism too!

Conclusion: *Only a diffeomorphism generates a well-defined transport of vector fields from one manifold to another.*

We proceed to study the local properties of f in terms of the induced map f. Let us define the rank of f at a point P to be the dimension of the vector space $f_*[\mathbf{T}_P(M)]$. (Using coordinate systems around P and $f(P)$, it is well known that the rank of f is equal to the rank of the Jacobian matrix $\overset{=}{D}_f(P)$.) As in the Euclidean case, we can then make the following definition.

Definition 10.2. A smooth map f is *regular* at a point P provided that the induced map f has maximal rank.

We can also introduce another useful concept. Let $f : M \curvearrowright N$ be a smooth map and let Q be a point in N. Then we consider the preimage $f^{-1}(Q)$ (which can be empty if $f(M)$ does not cover Q):

Definition 10.3[1]**.** We say that Q is a *regular value* if f is regular at all points in the preimage of Q. If Q is not regular, then it is called a *critical value*. (Remark: Q is also counted as a regular value if the preimage is empty.)

Remember that the tangent spaces have the same dimension as the underlying manifold. We can then distinguish between three cases:

$$\dim M = \dim N, \qquad \dim M < \dim N, \qquad \dim M > \dim N.$$

We will now discuss the three cases in more detail:

(a) $\dim M = \dim N$: As the first case we consider the especially important one, where M and N have the same dimension n. Then $\mathbf{T}_P(M)$ and $\mathbf{T}_Q(N)$ are vector spaces of the same dimension n, and f is regular at a point P exactly when $\underset{=}{f_*}$ maps $\mathbf{T}_P(M)$ isomorphically onto $\mathbf{T}_Q(N)$. This means that the Jacobian matrix $\overset{=}{D}_f(P)$ is a regular square matrix; i.e., it has nonzero determinant. But then the inverse function theorem from analysis tells us that f actually maps an open neighborhood U of P onto an open neighborhood V of Q and furthermore, that f restricts to a diffeomorphism, $f : U \curvearrowright V$. Thus f behaves extremely nicely locally! (See Figure 10.3.)

The inverse map $f^{-1}_{|V}$ then generates the reciprocal map:

$$(f_*)^{-1} : \mathbf{T}_Q(M) \curvearrowright \mathbf{T}_P(N),$$

where

$$(f_*)^{-1} = (f^{-1})_*. \tag{10.4}$$

[1] The above definition differs slightly from the definition used by mathematicians in the case where $\dim M < \dim N$. See e.g., Guillemin/Pollack (1974).

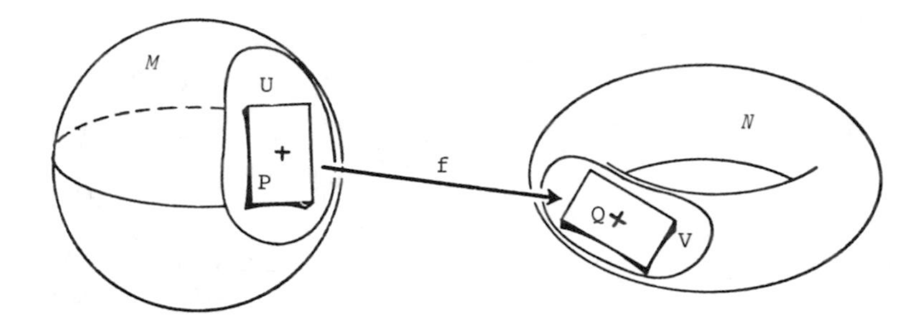

Figure 10.3.

This is a trivial consequence of the composition theorem (10.2), since the composite map of $f_{|U}$ and $f^{-1}_{|U}$ reduces to the identity map.

We can also consider a regular value Q_0 in N. If $f^{-1}(Q_0)$ is nonempty, we can find P_0 such that $f(P_0) = Q_0$, and by definition f is regular at P_0. It therefore maps an open neighborhood U of P_0 onto an open neighborhood V of Q_0. This argument can be strengthened: Suppose $f^{-1}(Q_0)$ is *finite*, and put

$$f^{-1}(Q_0) = \{P_1, \dots, P_k\}.$$

Then there exist disjoint open neighborhoods U_i of P_i and open neighborhoods V_i of Q such that f restricts to diffeomorphisms $f : U_i \curvearrowright V_i$. But here we can safely replace $V_1, \dots, V_k$ by their intersections $V = \bigcap_{i=1}^{k} V_i$, which will be an open neighborhood of Q_0. We can then shrink U_i to $U_i' = U_i \cap f^{-1}(V)$, and f will now restrict to diffeomorphisms,

$$f : U_i' \curvearrowright V.$$

Thus we have shown the following useful Lemma:

Lemma 10.1. *Let $f : M^n \curvearrowright N^n$ be a smooth map and let Q be a regular value such that $f^{-1}(Q)$ is finite, say*

$$f^{-1}(Q) = \{P_1, \dots, P_k\}.$$

(See Figure 10.4.*) Then there exists a single open neighborhood V of Q such that $f^{-1}(V)$ decomposes into disjoint neighborhoods $U_1, \dots, U_k$ of $P_1, \dots, P_k$ and such that the map f restricts to diffeomorphisms*

$$f : U_i \curvearrowright V, \qquad i = 1, \dots, k.$$

Remark. In the above lemma it is essential that $f^{-1}(Q)$ is finite. It is, in general, false when $f^{-1}(Q)$ is infinite.

(b) dim M < dim N: A smooth map $f : M^m \curvearrowright N^n$ ($m < n$) that is everywhere regular is called an *immersion*. Let f be such an immersion. In analogy

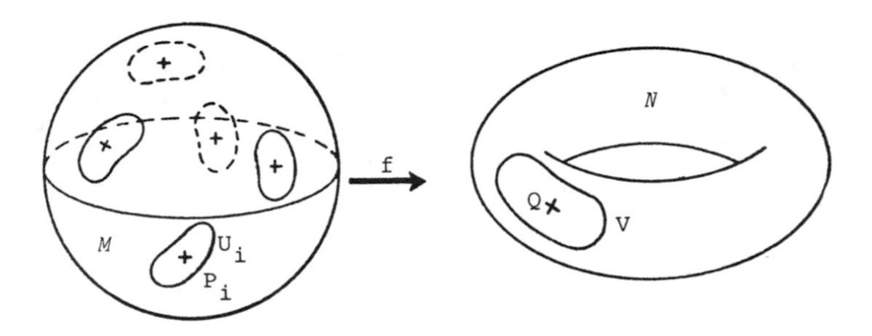

Figure 10.4.

with the discussion of Euclidean manifolds, we expect $f(M)$ to be a submanifold of N. Locally, everything works out all right, and we can transfer coordinate systems from M to $f(M)$. But globally, we may be in trouble, because f need not be injective, so that $f(M)$ can have self-intersections. (See Figure 10.5.) Consequently, we must demand that f be injective. But that is not enough. We still have to worry about the topology, since the inverse map $f^{-1} : f(M) \curvearrowright M$ need not be continuous. We eliminate this by demanding that f be a homeomorphism; i.e., f is injective, and both f and f^{-1} are continuous. Then it should come as no great surprise that the following theorem holds:

Theorem 10.2. *Let $f : M^m \curvearrowright N^n$ $(m < n)$ be a smooth map such that*
1. *f is an immersion;*
2. *f is a homeomorphism.*

Then $f(M)$ is a submanifold of N, and $f : M \curvearrowright f(M)$ a diffeomorphism.

A smooth map that satisfies both the above properties is called an *embedding*, and we say that M is embedded in N.

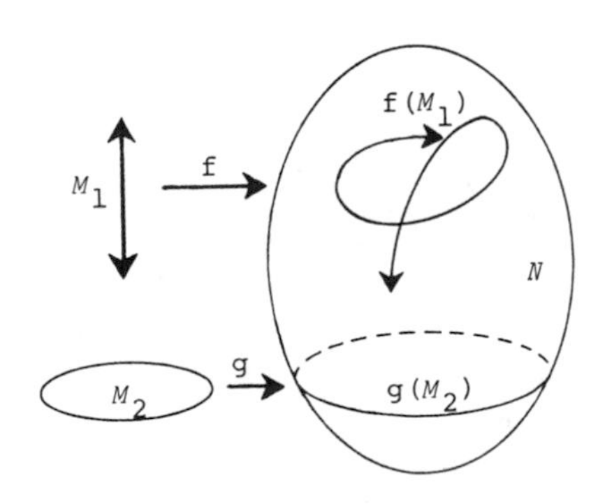

Figure 10.5.

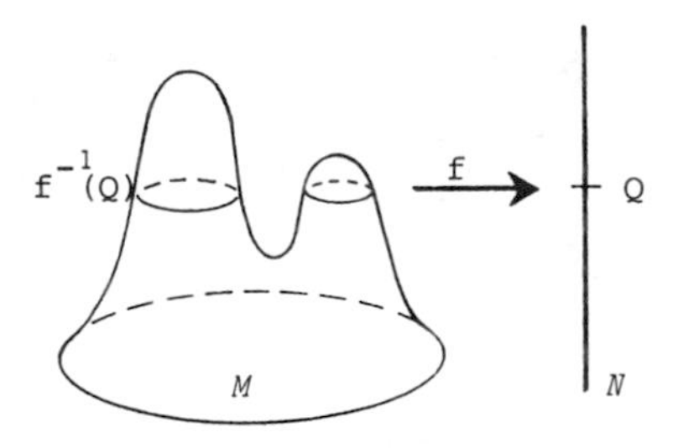

Figure 10.6.

(c) dim $M >$ dim N: A smooth map $f : M^m \curvearrowright N^n$ $(m > n)$ that is everywhere regular is called a *submersion*. Now, let f be a smooth map and let Q_0 be a point in N. We consider the preimage $f^{-1}(Q_0)$; cf. Figure 10.6, which consists of all solutions to the equation

$$f(P) = Q_0 \qquad \text{i.e.,} \qquad \begin{aligned} f^1(x^1, \ldots, x^m) &= y_0, \\ \vdots \qquad & \quad \vdots \\ f^n(x^1, \ldots, x^m) &= y_0^n. \end{aligned}$$

We have previously discussed how to construct Euclidean manifolds using equations of constraints (see Section 6.3). Using those techniques we can generalize Theorem 6.2 of Section 6.3 to the following theorem:

Theorem 10.3. *Let $f : M^m \curvearrowright N^n$ $(m > n)$ be a smooth map. If Q is a regular value in N, then either*
1. *$f^{-1}(Q)$ is empty, or*
2. *$f^{-1}(Q)$ is an $(n - m)$-dimensional submanifold of M.*

We recapitulate the preceding discussion in the following scheme. (See also Figures 10.5 and 10.6.)

$f : M^m \curvearrowright N^n$	If f is a regular map then it is called	Such a map can be used to construct submanifolds in
$m < n$	an immersion	N of dimension m
$m > n$	a submersion	M of dimension $m - n$

In the preceding discussion we introduced the notion of regular points and regular values. They were important in the characterization of the behavior of smooth maps. But to use the theorems just obtained, it will be necessary to control the existence of regular points and values. For a specific map we can, of course, compute the Jacobian matrix and check its rank and thereby determine the actual

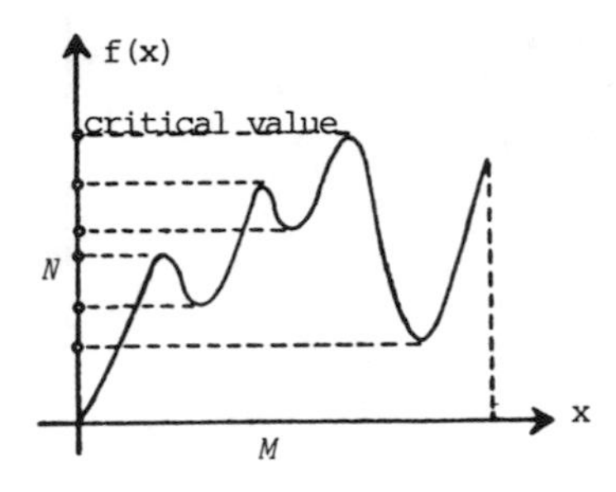

Figure 10.7.

positions of the regular points and the regular values. But interestingly enough, it turns out that it is possible to make general statements that hold for any smooth map.

Consider for simplicity ordinary functions $f : \mathbb{R} \curvearrowright \mathbb{R}$. Then a regular point is a point where $f'(x) \neq 0$. It is easy to construct smooth functions that are everywhere regular, e.g., $f(x) \equiv x$. It is equally easy to construct smooth functions that are nowhere regular, e.g., the constant function $f(x) = 0$. Thus we cannot hope for general statements about regular points. But consider now regular values. Even in the worst cases of a constant function, there is only a single critical value, so the critical value seems to be very exceptional.

This turns out to be a general feature of a smooth map. Consider a smooth map $f : M^m \curvearrowright N^n$. The points in M can be divided into two types:

a. Critical points,
b. Regular points.

Similarly, the points in N can be divided into 3 types:

a. Image of a critical point,
b. Images of only regular points,
c. Not an image of any point.

A famous theorem of Sard now states that *the set of critical points is mapped into a zero set of N; i.e., the set of critical values is a zero set in N*. Equivalently, we can say that almost every point in N is a regular value. (For a complete discussion of zero sets on manifolds one may consult, e.g., Spivac (1970) or Guillemin and Pollack (1974).)

Remark. By abuse of notation we will call a smooth map regular even if it does possess critical points, as long as the critical points are all isolated. E.g., the function indicated in Figure 10.7 will thus, slightly incorrectly, be referred to as a regular function.

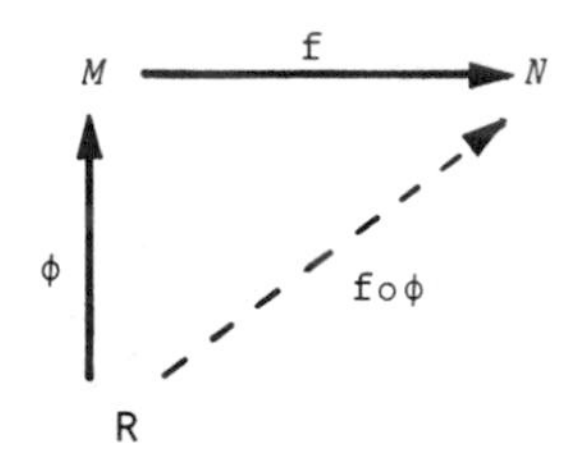

Figure 10.8.

10.2 Pullbacks of Cotensors

Let $f : M^m \curvearrowright N^n$ be a smooth map. We have seen that we run into difficulties when we try to transport vector fields from one manifold to another. We will now investigate what happens when we try to transfer other objects. It is instructive to consider maps first:

Suppose $\phi : \mathbb{R} \curvearrowright M$ is a smooth map into M (Figure 10.8). Then we can clearly push it forward to a smooth map $f \circ \phi : \mathbb{R} \curvearrowright N$ into N. This is connected with the transport of vectors. A smooth map $\phi : \mathbb{R} \curvearrowright M$ represents a curve on M. But smooth curves generate tangent vectors, and that is the basic mechanism behind the transport of tangent vectors as explained in Section 10.1

Suppose, then, that $\phi : N \curvearrowright \mathbb{R}$ is a smooth function on N (see Figure 10.9). This can be pulled back to a smooth function $\phi \circ f : M \curvearrowright \mathbb{R}$ on M. This can now be used to transport covectors. Consider a vector ω on N. Then we can find a smooth function $\phi : N \curvearrowright \mathbb{R}$ that generates ω; i.e., $\omega = \boldsymbol{d}\phi$, and this smooth function is pulled back to the smooth function $\phi \circ f : N \curvearrowright \mathbb{R}$. (See Figure 10.9.) If we can show that $\boldsymbol{d}(\phi \circ f)$ is independent of the particular function ϕ chosen, we can therefore pull back ω to $\boldsymbol{d}(\phi \circ f)$. This also allows us to determine the coordinates of the pulled-back covector, which will be denoted by $f^*\omega$.

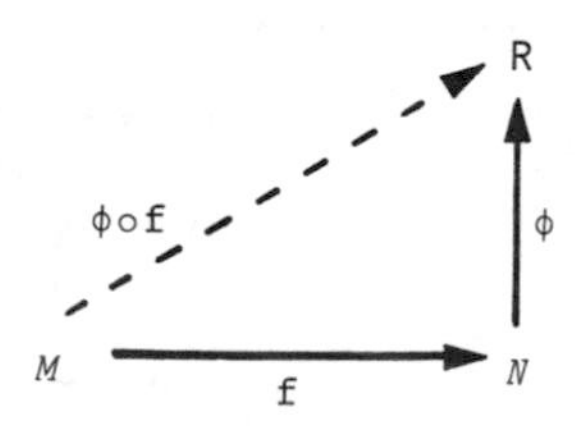

Figure 10.9.

The above considerations suggest that it is natural to push forward tangent vectors and to pull back covectors, but it even turns out that pullbacks have nicer global properties than vector transports.

This we will now investigate in some detail.

We start by reinvestigating the pullback of covectors: Let ω be a covector field on N. We want to construct an associated covector field on M that will be denoted by $f^*\omega$. Consider a point P on M. Then a tangent vector in $\mathbf{T}_P(M)$ is transported to a unique tangent vector in $\mathbf{T}_{f(P)}(N)$. (See Figure 10.10.) This motivates the following definition:

Definition 10.4.

$$\langle f^*\omega \mid \mathbf{v}_P\rangle \stackrel{\text{def}}{=} \langle \omega \mid f_*\mathbf{v}_P\rangle. \tag{10.5}$$

Clearly, $f^* : \mathbf{T}_P(M) \curvearrowright \mathbb{R}$ is a linear map, since $f_* : \mathbf{T}_P(M) \curvearrowright \mathbf{T}_{f(P)}(N)$ is linearly; but then we have constructed a globally defined covector field $f^*\omega$ on M. It remains to be shown that it is *smooth*. To do that we introduce coordinates. Then $f^*\omega$ is characterized by the component functions

$$\begin{aligned} b_i(x) &= \langle f^*\omega \mid \mathbf{e}_i\rangle = \langle \omega \mid f_*\mathbf{e}_i\rangle \\ &= a_j(y(x))\frac{\partial y^j}{\partial x^k}\,\delta_i^k = a_j(y(x))\frac{\partial y^j}{\partial x^i}\,. \end{aligned} \tag{10.6}$$

But the y-coordinates depend smoothly on the x-coordinates, since f is a smooth map; and the coefficients $a_j(y)$ are smooth functions, since ω is a smooth covector field on N. This shows that the right-hand side depends smoothly on x.

So the miracle has happened. We never get in trouble when we try to pull back covector fields! Clearly, the above analysis may be extended to arbitrary cotensor fields:

Theorem 10.4. *Let $f : M^m \curvearrowright N^n$ be a smooth map. Then f generates a map*

$$f^* : \mathbf{T}^{(0,k)}(N) \curvearrowright \mathbf{T}^{(0,k)}(M);$$

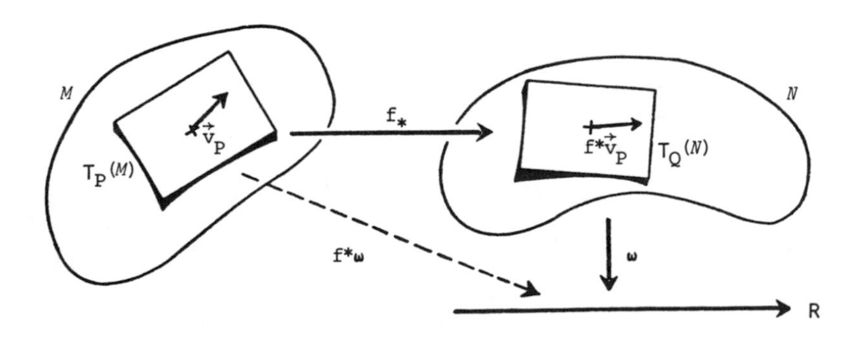

Figure 10.10.

i.e., f^* pulls back a smooth cotensor field on N to a smooth cotensor field on M. If $\mathbf{T}$ is a smooth cotensor field on N, then the pullback is given by

$$f^*\mathbf{T}(\mathbf{v}_1, \ldots, \mathbf{v}_k) = \mathbf{T}(f_*\mathbf{v}_1, \ldots, f_*\mathbf{v}_k), \tag{10.7}$$

which can be written out in components as follows:

$$(f^*\mathbf{T})_{i_1\ldots i_k}(x) = T_{j_1\ldots j_k}(y(x))\,\frac{\partial y^{j_1}}{\partial x^{i_1}} \cdots \frac{\partial y^{j_k}}{\partial x^{i_k}}\,. \tag{10.8}$$

Observe that if $M = N$ and f is the identical map, then we recover the usual transformation formula for the exchange of coordinates on a manifold!

The pullback has several simple properties:

Theorem 10.5.

a. *f is linear; i.e.,*

$$f^*(\mathbf{T} + \mathbf{S}) = f^*(\mathbf{T}) + f^*(\mathbf{S}); \qquad f^*(\lambda\mathbf{T}) = \lambda f^*(\mathbf{T}). \tag{10.9}$$

b. *f^* commutes with the tensor product; i.e.,*

$$f^*(\mathbf{T} \otimes \mathbf{S}) = f^*(\mathbf{T}) \otimes f^*(\mathbf{S}). \tag{10.10}$$

c. *If we have three manifolds L, M, N and two smooth maps $L \xrightarrow{f} M \xrightarrow{g} N$, then*

$$(g \circ f)^* = f^* \circ g^*. \tag{10.11}$$

PROOF. Only the composition rule is worth consideration. It can be derived directly from the composition rule for vector transports (10.2):

$$\begin{aligned}(g \circ f)^*\mathbf{T}(\mathbf{v}_1, \ldots, \mathbf{v}_k) &\stackrel{\text{def}}{=} \mathbf{T}[(g \circ f)_*\mathbf{v}_1, \ldots, (g \circ f)_*\mathbf{v}_k] \\ &= \mathbf{T}[(g_*(f_*\mathbf{v}_1), \ldots, g_*(f_*\mathbf{v}_k)] \\ &= (g^*\mathbf{T})[f_*\mathbf{v}_1, \ldots, f_*\mathbf{v}_k] \\ &= f^*(g^*\mathbf{T})[\mathbf{v}_1, \ldots, \mathbf{v}_k].\end{aligned}$$

If we prefer, it can of course also be derived from the component expression for the pullback using the chain rule. □

We shall especially use the pullback in two cases:

(a) *Metrics*: If $\boldsymbol{g}$ is a metric on N, then $\boldsymbol{g}$ is a symmetric, nondegenerate cotensor field of rank 2. Consequently, we can pull it back to a cotensor field $f^*\boldsymbol{g}$ of rank 2 on M.

It will obviously be symmetric, but it need *not* be *non-degenerate*! (If, e.g., f is constant, then $f^*\boldsymbol{g}$ vanishes everywhere.) Thus $f^*\boldsymbol{g}$ need not be a metric on M.

Exercise 10.2.1
Problem: Let N be a Riemannian manifold with the positive definite metric $\boldsymbol{g}$. Show that $f^*\boldsymbol{g}$ is a metric on M if and only if f_* is everywhere injective.

Exercise 10.2.2
Problem: (a) Let M and N be manifolds of the *same* dimension. Suppose N is equipped with a Minkowski metric, and let $f : M \curvearrowright N$ be a smooth regular map. Show that $f^*\mathbf{g}$ is a Minkowski metric on M.

(b) Let N be the Minkowski space $\mathbb{R}^4$ with the usual metric and let M be the real line $\mathbb{R}$. As the smooth map we consider

$$f(t) = (t; t; 0; 0).$$

Show that f_* is everywhere injective, but that $f^*\mathbf{g}$ vanishes identically.

(b) *Differential forms:* If $\mathbf{T}$ is a differential form on N, then $\mathbf{T}$ is by definition a skew-symmetric cotensor field. Consequently, we can pull it back to a cotensor field $f^*\mathbf{T}$, which will be skew-symmetric; i.e., $f^*\mathbf{T}$ is a differential form too! The pullback of a differential form is particularly simple. Let us introduce coordinates on M and N. Then we may decompose $\mathbf{T}$ as

$$\mathbf{T} = \frac{1}{k!} T_{i_1 \dots i_k}(y) \mathbf{d}y^{i_1} \wedge \dots \wedge \mathbf{d}y^{i_k}. \tag{7.8}$$

From the transformation rule (Theorem 10.4) we get that $f^*\mathbf{T}$ is characterized by the components

$$T_{j_1 \dots j_k}(y(x)) \frac{\partial y^{j_1}}{\partial x^{i_1}} \cdots \frac{\partial y^{j_k}}{\partial x^{i_k}};$$

i.e., it can be decomposed as

$$f^*\mathbf{T} = \frac{1}{k!} T_{j_1 \dots j_k}(y(x)) \frac{\partial y^{j_1}}{\partial x^{i_1}} \cdots \frac{\partial y^{i_k}}{\partial x^{i_k}} \mathbf{d}x^{i_1} \wedge \dots \wedge \mathbf{d}x^{i_k}. \tag{10.12}$$

But then we see that $f^*\mathbf{T}$ is obtained from $\mathbf{T}$ by performing the substitution

$$\mathbf{d}y^i \to \frac{\partial y^i}{\partial x^j} \mathbf{d}x^j, \tag{10.13}$$

so as usual the formalism produces the simplest possible answer.

Before we proceed with the investigation of differential forms, we will briefly discuss how one can extend the transport of tangent vectors to a transport of arbitrary tensors. One must then restrict to diffeomorphisms.

Definition 10.5. A diffeomorphism, $f : M^n \curvearrowright N^n$, generates a pullback of tangent vectors from N to M,

$$f_* : \mathbf{T}_Q(N) \to \mathbf{T}_{f^{-1}(Q)}(M).$$

If $\mathbf{v}_Q$ is a tangent vector at Q in N, then $f_*\mathbf{v}_Q$ is given by

$$f_*\mathbf{v}_Q = (f^{-1})^*\mathbf{v}_Q, \tag{10.14}$$

and in coordinates it is characterized by the components

$$[f_*\mathbf{v}_Q]^i = \frac{\partial x^i}{\partial y^j} [\mathbf{v}_Q]^j. \tag{10.15}$$

But once we can pullback both tangent vectors and covectors, we can clearly push forward tensors of arbitrary type.

Definition 10.6. A diffeomorphism, $f : M^n \curvearrowright N^n$, generates a transport of tensors of type (p, q) from M to N:

$$f_* : \mathbf{T}_P^{(p,q)}(M) \curvearrowright \mathbf{T}_{f(P)}^{(p,q)}(N).$$

If $\mathbf{T}$ is a tensor of type (p, q) at the point P, then $f_*\mathbf{T}$ is given by

$$(f_*\mathbf{T})(\omega_1, \ldots, \omega_p; \mathbf{v}_1, \ldots, \mathbf{v}_q) = \mathbf{T}(f^*\omega_1, \ldots, f^*\omega_p; f^*\mathbf{v}_1, \ldots, f^*\mathbf{v}_q), \tag{10.16}$$

and in coordinates it is characterized by the components

$$(f_*\mathbf{T})^{i_1\ldots i_p}{}_{j_1\ldots j_q}(y) = \frac{\partial y^{i_1}}{\partial x^{k_1}} \cdots \frac{\partial y^{i_p}}{\partial x^{k_p}} T^{k_1\ldots k_p}{}_{\ell_1\ldots\ell_q}(f^{-1}(y)) \frac{\partial x^{\ell_1}}{\partial y^{j_1}} \cdots \frac{\partial y^{\ell_q}}{\partial y^{j_q}}. \tag{10.17}$$

Observe that in this way smooth tensor fields are transported into smooth tensor fields. Of course, we could equally well pull them back using the inverse map. Theorem 10.5 can now be generalized as follows:

Theorem 10.6. *Suppose $f : M^n \curvearrowright N^n$ is a diffeomorphism inducing the tensor transport $f_* : \mathbf{T}^{(p,q)}(M) \curvearrowright \mathbf{T}^{(p,q)}(N)$. Then*

a. *f_* is linear; i.e.,*

$$f_*(\mathbf{T} + \mathbf{S}) = f_*(\mathbf{T}) + f_*(\mathbf{S}); \qquad f_*(\lambda\mathbf{T}) = \lambda f_*(\mathbf{T}). \tag{10.18}$$

b. *f_* commutes with the tensor product; i.e.,*

$$f_*(\mathbf{T} \otimes \mathbf{S}) = f_*(\mathbf{T}) \otimes f_*(\mathbf{S}). \tag{10.19}$$

c. *f_* commutes with contractions; i.e.,*

$$f_*(\mathbf{T}^C) = [f_*\mathbf{T}]^C. \tag{10.20}$$

d. *If we have 3 manifolds L^n, M^n, N^n of the same dimension and 2 diffeomorphisms $L \xrightarrow{f} M \xrightarrow{g} N$, then*

$$(g \circ f)_* = g_* \circ f_*. \tag{10.21}$$

PROOF. There is no need to go through all the details. It suffices to observe that when $f : M \curvearrowright N$ is a diffeomorphism, we can use f to transport a coordinate system from M to N; i.e., if $\phi : U \curvearrowright M$ is a coordinate system on M, then $f \circ \phi : U \curvearrowright N$ is a coordinate system on N. By the use of such associated coordinate systems, the coordinate expression for f reduces to the identity map $y^i = x^i$, and $\mathbf{T}$ and $f_*\mathbf{T}$ get identical components. □

In the rest of this section we restrict our consideration to differential forms. Because of the exterior calculus, the pullback of differential forms is a very powerful tool. Consider a smooth map $f : M \curvearrowright N$. It generates pullbacks

$$f^* : \Lambda^k(N) \curvearrowright \Lambda^k(M), \qquad k = 0, 1, 2 \ldots,$$

which we know are linear (Theorem 10.5). We will investigate that part of the exterior calculus that depends only on the manifold structure and not, e.g., on a metric. Thus, for the moment, we are concerned only with the wedge product, the exterior derivative, and the integral of differential forms.

Theorem 10.7. *The pullback commutes with the wedge product and the exterior derivative; i.e.,*

$$f^*(\mathbf{T} \wedge \mathbf{S}) = (f^*\mathbf{T}) \wedge (f^*\mathbf{S}). \tag{10.22}$$

$$f^*(\boldsymbol{d}\mathbf{T}) = \boldsymbol{d}(f^*\mathbf{T}). \tag{10.23}$$

PROOF. To avoid drowning in indices, we work it only out in the case of one-forms.

(a)
$$\begin{aligned} f^*(\mathbf{A} \wedge \mathbf{B}) &= f^*(\mathbf{A} \otimes \mathbf{B} - \mathbf{B} \otimes \mathbf{A}) \\ &= f^*(\mathbf{A}) \otimes f^*(\mathbf{B}) - f^*(\mathbf{B}) \otimes f^*(\mathbf{A}) \\ &= f^*(\mathbf{A}) \wedge f^*(\mathbf{B}), \end{aligned}$$

so the result is a trivial consequence of Theorem 10.6b.

(b)
$$\begin{aligned} [f^*(\boldsymbol{d}(\mathbf{A})]_{k\ell} &= [\boldsymbol{d}\mathbf{A}]_{ij} \frac{\partial y^i}{\partial x^k} \frac{\partial y^j}{\partial x^\ell} \\ &= \left(\frac{\partial A_j}{\partial y^i} - \frac{\partial A_i}{\partial y^j} \right) \frac{\partial y^i}{\partial x^k} \frac{\partial y^j}{\partial x^\ell} \\ &= \frac{\partial A_j}{\partial x^k} \frac{\partial y^j}{\partial x^\ell} - \frac{\partial A_i}{\partial x^\ell} \frac{\partial y^i}{\partial x^k} \end{aligned}$$

$$\begin{aligned} \text{(NB!)} &= \frac{\partial}{\partial x^k} \left(A_j \frac{\partial y^j}{\partial x^\ell} \right) - \frac{\partial}{\partial x^\ell} \left(A_i \frac{\partial y^i}{\partial x^k} \right) \\ &= \frac{\partial}{\partial x^k} [f^*\mathbf{A}]_\ell - \frac{\partial}{\partial x^\ell} [f^*\mathbf{A}]_k = [\boldsymbol{d}(f^*\mathbf{A})]_{k\ell}. \end{aligned}$$

Observe that there is a subtle point involved in these rearrangements: If we actually carry out the differentiations involved, we will pick up two extra terms involving the mixed partial derivatives $\frac{\partial^2 y^i}{\partial x^k \partial x^1}$ and $\frac{\partial^2 y^i}{\partial x^1 \partial x^k}$, but since partial derivatives commute, they cancel automatically. □

Mathematicians have a very efficient way of representing computation rules like those of Theorem 10.7. Consider especially the rule concerning the exterior derivative

$$f^* \circ \boldsymbol{d} = \boldsymbol{d} \circ f^*. \tag{10.23}$$

When we pull back a differential form and take the exterior derivative, it does not matter in which order we perform the operations. This is clearly an example of a *commutative* law as we know it from elementary number theory. Now consider Figure 10.11, where we have represented the various maps involved in (10.23) as arrows. If we take an element in the upper right corner, we can map it along the arrows, and evidently there are two possible ways of mapping it down to the lower left corner, where it can arrive either as $\boldsymbol{d}(f^*\mathbf{T})$ or as $f^*(\boldsymbol{d}\mathbf{T})$. But according to (10.23), they are identical, so the end result is independent of which route we actually follow in the diagram. This is expressed by saying that *the diagram is commutative.* As another example, consider the composition rule expressed in Theorem 10.6d. Here we have 3 maps involved corresponding to Figure 10.12, and the computation rule in Theorem 10.6d simply states that this diagram is commutative.

When we come to integration of differential forms, we must be somewhat more careful. We want to find out what happens to an expression like $\int_\Omega \mathbf{T}$ when we pull it back. Here $\mathbf{T}$ is a differential form of rank k, and Ω is a k-dimensional orientable regular domain in N. We must therefore also investigate what happens to Ω when we pull that back! In general, $f^{-1}(\Omega)$ will not be a k-dimensional orientable regular domain of M for the following reasons: It need not be compact,

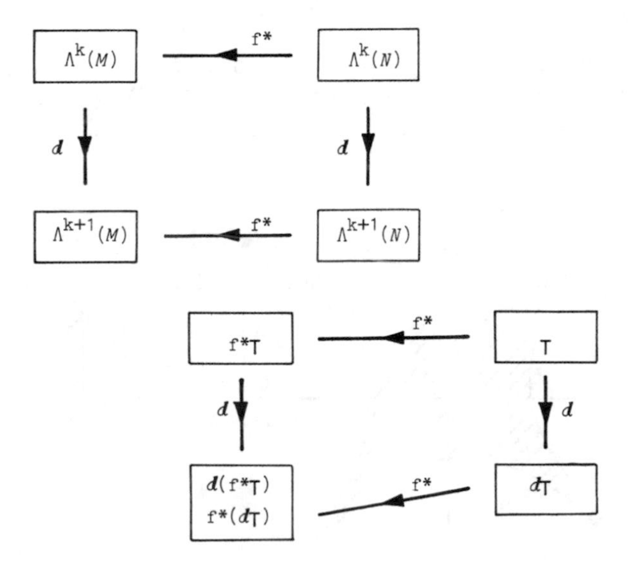

Figure 10.11.

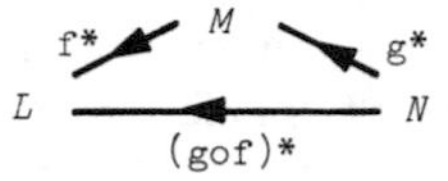

M
f* g*
L N
(gof)*

Figure 10.12.

it need not be a submanifold, and it need not be k-dimensional! To see this, consider the following example:

EXAMPLE 10.1. Let $M = \mathbb{R}^2$, $N = \mathbb{R}$, and put

$$f(t, x) = t^2 - x^2.$$

Then $f : M \curvearrowright N$ is a smooth map, although it is not regular at $(0, 0)$. The set $\Omega = [-1, 1]$ is a 1-dimensional regular domain in N, and $f^{-1}(\Omega)$ is the "strip" between the hyperbolas, as shown in Figure 10.13. It is not compact, and it is not one-dimensional.

Consequently, we must proceed a little differently. We must try to find a regular domain Ω' in M such that f maps Ω' diffeomorphically onto the regular domain Ω. Consider the following example:

EXAMPLE 10.2. Let $M = N = \mathbb{R}$, and consider the smooth map

$$f(x) = \frac{x}{1 + x^2},$$

which as graph depicted in Figure 10.14. Then $\Omega = [-1/2, 1/2]$ is a 1-dimensional regular domain in N. The preimage $f^{-1}(\Omega) = \mathbb{R}$ is not a regular domain, but $\Omega' = [-1, 1] \subseteq M$ is a 1-dimensional regular domain in M that is

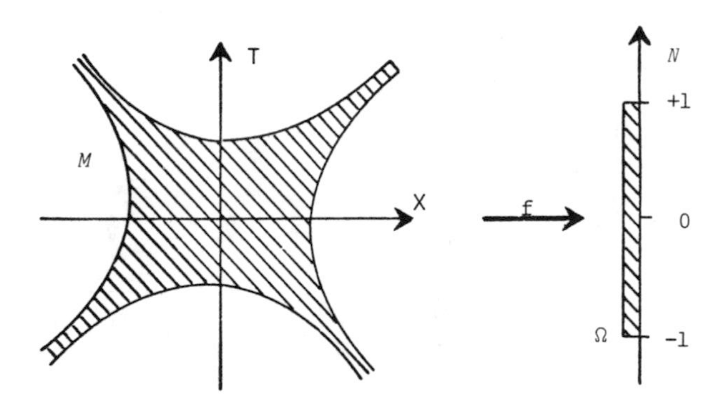

Figure 10.13.

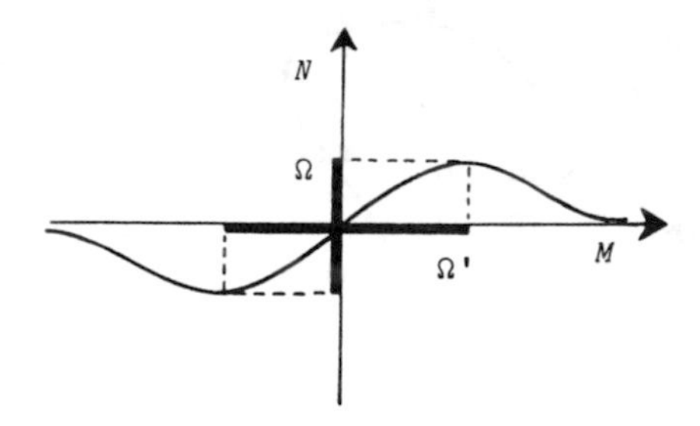

Figure 10.14.

mapped diffeomorphically onto Ω by f. (In one dimension this essentially means that f is monotonic when we restrict f to Ω'.)

Exercise 10.2.3
Problem: Reexamine Example 10.1 and try to find a suitable 1-dimensional regular domain Ω' that is mapped diffeomorphically onto $\Omega = [-1; 1]$.

Okay, let Ω be a given regular domain such that we can find Ω' that is mapped diffeomorphically onto Ω. Then we can compare the integrals

$$\int_{\Omega'} f^*\mathbf{T} \quad \text{and} \quad \int_{\Omega=f(\Omega')} \mathbf{T}.$$

Theorem 10.8. *Let* $\mathbf{T}$ *be a differential form on* N *of rank* k. *Let* Ω *be a* k*-dimensional orientable regular domain in* M *such that* f *maps* Ω *diffeomorphically onto a* k*-dimensional orientable regular domain in* N. *Then*

$$\int_{\Omega} f^*\mathbf{T} = \begin{cases} +\int_{f(\Omega)} \mathbf{T} & \text{if } f \text{ preserves the orientation,} \\ -\int_{f(\Omega)} \mathbf{T} & \text{if } f \text{ reserves the orientation.} \end{cases} \tag{10.24}$$

PROOF. (Outline): The proof will proceed in two steps.

Step 1: Here we assume that int Ω is a simple submanifold. Then we may cover it with a single coordinate system (ϕ, U) that generates the positive orientation (Figure 10.15):

$$\phi : U \xrightarrow{\text{bij.}} \text{int } \Omega \subset M; \qquad \text{i.e., } x^i = \phi^i(\lambda^1, \ldots, \lambda^k).$$

As f maps int Ω diffeomorphically onto int $f(\Omega)$, we can use f to generate a coordinate system on int $f(\Omega)$:

$$f \circ \phi : U \xrightarrow{\text{bij.}} \text{int } f(\Omega) \subset N.$$

In these coordinates the restriction of f to int Ω is represented by the Euclidean function

$$\mu^i = \lambda^i; \qquad i = 1, \ldots, k,$$

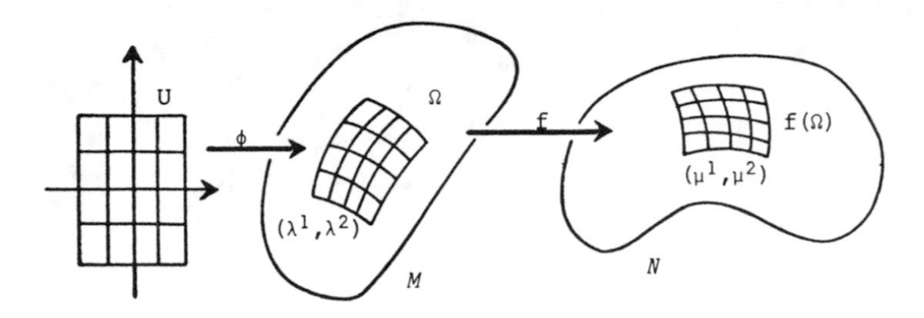

Figure 10.15.

i.e., the identity map. If f preserves the orientation, then $(f \circ \phi, U)$; i.e., the $(\mu^1, \ldots, \mu^k)$-coordinates generate a positive orientation on $f(\Omega)$. If f reserves the orientation, then $(f \circ \phi, U)$ generates a negative orientation, and this costs a sign when we evaluate the integral using these coordinates.

Okay, working out the two integrals in these especially adapted coordinate systems, we get for the case where f actually preserves the orientation

$$\begin{aligned}
\int_\Omega f^*\mathbf{T} &= \int_U (f^*\mathbf{T})_{i_1\ldots i_k}(x(\lambda)) \frac{\partial x^{i_1}}{\partial \lambda^1} \cdots \frac{\partial x^{i_j}}{\partial \lambda^k} d\lambda^1 \ldots d\lambda^k \\
&= \int_\Omega T_{j_1\ldots j_k}(y(x)) \frac{\partial y^{j_1}}{\partial x^{i_1}} \cdots \frac{\partial y^{j_k}}{\partial x^{i_k}} \frac{\partial x^{i_1}}{\partial \lambda^1} \cdots \frac{\partial x^{j_k}}{\partial \lambda^k} d\lambda^1 \ldots d\lambda^k \\
&= \int_U T_{j_1\ldots j_k}(y(x(\lambda))) \frac{\partial y^{j_1}}{\partial \lambda^1} \cdots \frac{\partial y^{j_k}}{\partial \lambda^k} d\lambda^1 \ldots d\lambda^k \\
&= \int_U T_{j_1\ldots j_k}(y(\mu)) \frac{\partial y^{j_1}}{\partial \mu^1} \cdots \frac{\partial y^{j_k}}{\partial \mu^k} d\mu^1 \ldots d\mu^k = \int_{f(\Omega)} \mathbf{T}.
\end{aligned}$$

Step 2: Here int Ω is an arbitrary manifold. We can then cover it with a positively oriented atlas

$$\text{int}\,\Omega \subset \bigcup_{i \in I} \phi_i(U_i).$$

Using a partition of unity, we now cut up $\mathbf{T}$ into small pieces so that each little piece is effectively concentrated in one of the simple manifolds $\phi(U_i)$. From step 1, the linearity of f^*, and the linearity of the integral we now deduce the desired formula for the general case. □

If we restrict ourselves to diffeomorphisms, we can, in particular, push differential forms forward. The statements of Theorems 10.7 and 10.8 now carry over trivially for the transport of differential forms using diffeomorphisms; i.e., the transport commutes with the wedge product and the exterior derivative, and it preserves integrals.

10.3 Isometries and Conformal Maps

In the preceding section we considered pullbacks of differential forms. This time we will concentrate on metrical aspects of smooth maps. Suppose $(M^n, \mathbf{g}_1)$ and $(N^n, \mathbf{g}_2)$ are differentiable manifolds of the same dimension and with metrics $\mathbf{g}_1$ and $\mathbf{g}_2$. If $f : M \to N$ is a smooth regular map, we have seen that we can pull back $\mathbf{g}_2$ to a new metric $f^*\mathbf{g}_2$ on M. We have now two metrics on the same initial manifold M, and we can then investigate various metrical aspects of f by comparing $\mathbf{g}_1$ and $f^*\mathbf{g}_2$.

Consider, e.g., a smooth curve Γ on M (Figure 10.16). It is transferred to a smooth curve $f(\Gamma)$ on N. Now suppose we want to compare the lengths of corresponding arcs. Then we get

$$\left\{\text{Arc-length of } \widehat{P_1P_2} \text{ on } M\right\} = \int_{\lambda_1}^{\lambda_2} \sqrt{g_{1_{\alpha\beta}} \frac{dx^\alpha}{d\lambda} \frac{dx^\beta}{d\lambda}}\, d\lambda$$

$$= \int_{\lambda_1}^{\lambda_2} \sqrt{\mathbf{g}_1\left(\frac{dP}{d\lambda}, \frac{dP}{d\lambda}\right)} d\lambda,$$

and using that $\frac{df(P)}{d\lambda} = f_* \frac{dP}{d\lambda}$, we furthermore get

$$\left\{\text{Arc-length of } \widehat{f(P_1)f(P_2)} \text{ on } N\right\} = \int_{\lambda_1}^{\lambda_2} \sqrt{\mathbf{g}_2\left(\frac{df(P)}{d\lambda}, \frac{df(P)}{d\lambda}\right)} d\lambda$$

$$= \int_{\lambda_1}^{\lambda_2} \sqrt{\mathbf{g}_2\left(f_*\frac{dP}{d\lambda}, f_*\frac{dP}{d\lambda}\right)} d\lambda = \int_{\lambda_1}^{\lambda_2} \sqrt{f^*\mathbf{g}_2\left(\frac{dP}{d\lambda}, \frac{dP}{d\lambda}\right)} d\lambda$$

$$= \left\{\text{Arc-length of } \widehat{P_1P_2} \text{ on } M \text{ relative to the metric } f^*\mathbf{g}_2.\right\}$$

But then we see that we have only to consider the initial manifold M, where we can compute the arc-lengths of $\widehat{P_1P_2}$ relative to the two metrics $\mathbf{g}_1$ and $f^*\mathbf{g}_2$.

The simplest possible behavior of a smooth map f is when $\mathbf{g}_1$ and $f^*\mathbf{g}_2$ coincide.

Definition 10.7. A smooth regular map $f : M^n \curvearrowright N^n$ is called a *local isometry* if it preserves the metric; i.e.,

$$f^*\mathbf{g}_2 = \mathbf{g}_1.$$

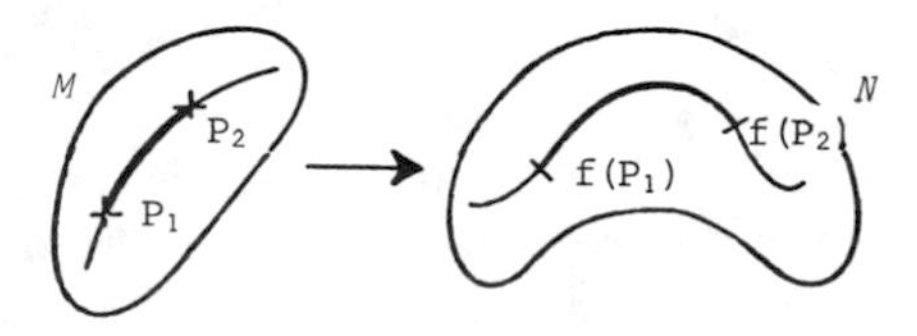

Figure 10.16.

It is called a *global isometry* if it is a diffeomorphism too.

Observe that isometries preserve arc-lengths. Isometries are, however, very special maps, so it is useful to discuss a broader class of maps where we still have some control over what is going on.

Definition 10.8. A smooth regular map $f : M^n \curvearrowright N^n$ is called a *conformal map* if it rescales the metric, i.e., there is a strictly positive scalar field $\Omega^2(x)$ on M such that

$$f^*g_2 = \Omega^2(x)g_1.$$

Remember that by abuse of notation we call a smooth map regular even in the presence of critical points, as long as they are all isolated.

Let us first give some general remarks about isometries and conformal maps to acquaint the reader with these new concepts. Suppose P is a point in M. As we have seen, a smooth map f generates a linear map f_* from the tangent space $\mathbf{T}_P(M)$ to the tangent space $\mathbf{T}_{f(P)}(N)$.

Consider first Riemannian manifolds, where the tangent spaces are Euclidean. When f is an isometry, then

$$g_1(\mathbf{u}, \mathbf{v}) = f^*g_2(\mathbf{u}, \mathbf{v}) = g_2(f_*\mathbf{u}, f_*\mathbf{v}),$$

so that f_* preserves the inner product between two tangent vectors; i.e., an isometry preserves the length of a vector and the angle between vectors.

When f is a conformal map, then

$$\Omega^2(p)g_1(\mathbf{u}, \mathbf{v}) = f^*g_2(\mathbf{u}, \mathbf{v}) = g_2(f_*\mathbf{u}, f_*\mathbf{v}),$$

so this time the inner product between two tangent vectors is rescaled. Evidently, the lengths are no longer preserved, but angles are, since

$$\cos(\mathbf{u}, \mathbf{v}) = -\frac{g(\mathbf{u}, \mathbf{v})}{\sqrt{g(\mathbf{u}, \mathbf{u})}\sqrt{g(\mathbf{v}, \mathbf{v})}},$$

and this is obviously independent of a rescaling. Thus a conformal map preserves the angles between arbitrary tangent vectors. But the converse also holds; i.e., any angle-preserving map is a conformal map. In fact, we can even show

Lemma 10.2. *A regular map $f : M \curvearrowright N$ is conformal if and only if it preserves right angles.*

PROOF. Clearly, it suffices to show that if two Euclidean metrics determine the same right angles, then they are proportional. Let $\mathbf{a}$ be a fixed vector. Any vector $\mathbf{b}$ now has a unique decomposition in a composant parallel to $\mathbf{a}$ and a composant orthogonal to $\mathbf{a}$. (See Figure 10.17.) Observe that this decomposition is independent of the choice of metric because they determine the same right angles. It is well-

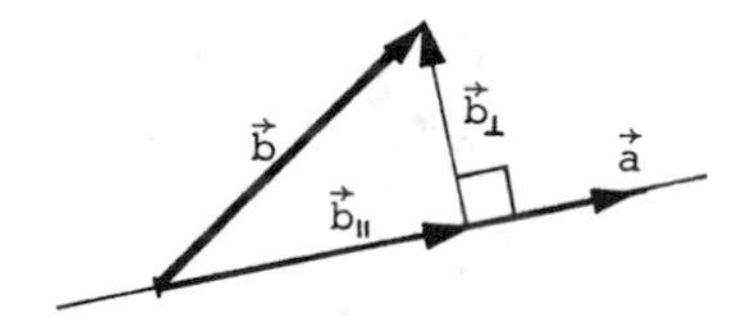

Figure 10.17.

known that the parallel composant is given

$$\mathbf{b}_{\|} = g(\mathbf{b}, \mathbf{a}) \frac{\mathbf{a}}{g(\mathbf{a}, \mathbf{a})} .$$

We thus get

$$\frac{g'(\mathbf{b}, \mathbf{a})}{g'(\mathbf{a}, \mathbf{a})} = \frac{g(\mathbf{b}, \mathbf{a})}{g(\mathbf{a}, \mathbf{a})} ; \qquad \text{i.e.,} \; \frac{g'(\mathbf{b}, \mathbf{a})}{g(\mathbf{b}, \mathbf{a})} = \frac{g'(\mathbf{a}, \mathbf{a})}{g(\mathbf{a}, \mathbf{a})} .$$

Thus the ratio

$$\frac{g'(\mathbf{b}, \mathbf{a})}{g(\mathbf{b}, \mathbf{a})}$$

is independent of **b**. Since the metrics g and g' are symmetric cotensors, it must also be independent of **a**; i.e., the metrics g and g' are proportional. □

Next we consider manifolds with Minkowski metrics. Now angles are no longer well-defined objects, but observe that the null-cone structure is preserved by a conformal map; i.e., time-like vectors are mapped into time-like vectors, null vectors into null vectors, and space-like vectors into space-like vectors.

This time we can then show

Lemma 10.3. *A regular map $f : M \curvearrowright N$ is conformal if and only if it preserves the light-cone structure.*

PROOF. Clearly, it suffices to show that if two Minkowski metrics generate the same light-cone structure, they are proportional. Let **a** be a time-like vector **b** a space-like vector. Then the line $\mathbf{a} + \lambda\mathbf{b}$ intersects the light cone in two different points corresponding to the two values λ_1 and λ_2. (See Figure 10.18.) On the other hand, λ_1 and λ_2 solve the equation

$$0 = g(\mathbf{a} + \lambda\mathbf{b}, \mathbf{a} + \lambda\mathbf{b}) = g(\mathbf{a}, \mathbf{a}) + 2\lambda g(\mathbf{a}, \mathbf{b}) + \lambda^2 g(\mathbf{b}, \mathbf{b}).$$

(Observe that this actually has two roots, since $g(\mathbf{a}, \mathbf{a})$ is strictly positive and $g(\mathbf{b}, \mathbf{b})$ is strictly negative.) We then get

$$\lambda_1 \cdot \lambda_2 = \frac{g(\mathbf{a}, \mathbf{a})}{g(\mathbf{b}, \mathbf{b})} .$$

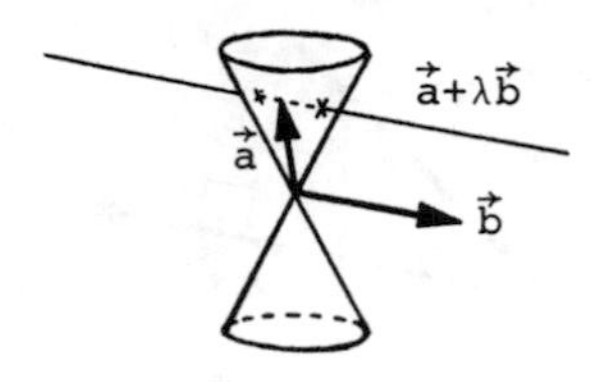

Figure 10.18.

But here λ_1 and λ_2 are independent of the choice of metric, so we conclude that

$$\frac{g'(\mathbf{a}, \mathbf{a})}{g'(\mathbf{b}, \mathbf{b})} = \frac{g(\mathbf{a}, \mathbf{a})}{g(\mathbf{b}, \mathbf{b})} ;$$

i.e., $g'(\mathbf{c}, \mathbf{c})/g(\mathbf{c}, \mathbf{c})$ is independent of $\mathbf{c}$. Let us put it equal to a constant k. Then we finally get

$$\begin{aligned} g'(\mathbf{u}, \mathbf{v}) &= \frac{1}{2}\left\{g'(\mathbf{u}+\mathbf{v}, \mathbf{u}+\mathbf{v}) - g'(\mathbf{u}, \mathbf{u}) - g'(\mathbf{v}, \mathbf{v})\right\} \\ &= \frac{k}{2}\left\{g(\mathbf{u}+\mathbf{v}, \mathbf{u}+\mathbf{v}) - g(\mathbf{u}, \mathbf{u}) - g(\mathbf{v}, \mathbf{v})\right\} \\ &= kg(\mathbf{u}, \mathbf{v}). \end{aligned}$$

□

Exercise 10.3.1
Introduction: Let $\mathbb{R}^{p+q}$ be the pseudo-Cartesian space equipped with the metric

$$\langle a \mid b\rangle = a^1b^1 + \cdots + a^pb^p - a^{p+1}b^{p+1} - \cdots - a^{p+q}b^{p+q}.$$

A *dilatation* is a map represented by $D : y^i = \lambda x^i$. An *inversion* is a map represented by $I : y^i = \frac{x^i}{\langle x|x\rangle}$. A *translation* is a map represented by $T : y^i = x^i + a^i$.

Problem: (a) Show that a dilatation is a conformal map with the conformal factor

$$\Omega^2(x) = \lambda^2.$$

(b) Show that an inversion is a conformal map with the conformal factor

$$\Omega^2(x) = \frac{1}{\langle x \mid x\rangle^2} .$$

(Strictly speaking, we must restrict ourselves to the manifold $M = \mathbb{R}^{p+q} \setminus \{x \mid \langle x \mid x\rangle \neq 0\}$, since the inversion breaks down on the "cone" $\langle x \mid x\rangle = 0$.)

(c) Show that the transformation $C = ITI$ is given by

$$y^i = \frac{x^i + a^i\langle x \mid x\rangle}{1 + 2\langle a \mid x\rangle + \langle a \mid a\rangle\langle x \mid x\rangle}$$

(which is, strictly speaking, well-defined only on the manifold $\mathbb{R}^{p+q} \setminus \{x \mid 1 + 2\langle a \mid x\rangle + \langle a \mid a\rangle\langle x \mid x\rangle \neq 0\}$). Show, furthermore, that it is a conformal map with the conformal factor

$$\Omega^2(x) = \frac{1}{(1 + 2\langle a \mid x\rangle + \langle a \mid a\rangle\langle x \mid x\rangle)^2} .$$

The transformation C is called a *special conformal transformation*. (Hint: Show that the composite of two conformal maps f and g is again conformal, with the conformal factor given by

$$\Omega^2_{g\circ f}(x) = \Omega^2_g(f(x)) \cdot \Omega^2_f(x).)$$

Worked Exercise 10.3.2
Problem: Show that the stereographic projection from the sphere onto the plane $\pi : S^2 \setminus \{N\} \curvearrowright \mathbb{R}^2$ is an orientation-reversing conformal map with the conformal factor

$$\Omega^2(\theta, \varphi) = \frac{1}{\sin^2 \frac{\theta}{2}} \, ,$$

where θ is the polar angle.

Okay, by now you should feel comfortable about isometries and conformal maps. We proceed to investigate various concepts that can be derived from the metric.

Let M^n be a manifold with metric $\boldsymbol{g}$. Then we have previously introduced an equivalence relation between tensors of various types. (See Section 6.9.) In coordinates, this corresponds to the raising and lowering of indices using the components of the metric tensor. Thus the cotensor T_{ij} is equivalent to the mixed tensor with components $T^i{}_j = g^{ik}T_{kj}$, etc. Now, when we use a diffeomorphism to transport tensors from one manifold to another, we should be careful. Suppose $f : (M^n, \boldsymbol{g}_1) \curvearrowright (N^n, \boldsymbol{g}_2)$ is a diffeomorphism and that $\mathbf{T}$ and $\mathbf{T}'$ are equivalent tensors on M. Then there is no reason why $f_*\mathbf{T}$ and $f_*\mathbf{T}'$ should be equivalent tensors on N (with respect to $\boldsymbol{g}_2$). But we know that f_* commutes with tensor products and contractions. If, for instance,

$$(\mathbf{T}')^i{}_j = (\boldsymbol{g}_1)^{ik}(\mathbf{T})_{kj} \, ,$$

we therefore get

$$(f_*\mathbf{T})^i{}_j = (f_*\boldsymbol{g}_1)^{ik}(f_*\mathbf{T})_{kj} \, .$$

Consequently, we conclude the following:

Lemma 10.4. *Suppose $f : (M^n, \boldsymbol{g}_1) \to (N^n, \boldsymbol{g}_2)$ is a diffeomorphism. If $\mathbf{T}$ and $\mathbf{T}'$ are equivalent tensors on M with respect to $\boldsymbol{g}_1$, then $f_*\mathbf{T}$ and $f_*\mathbf{T}'$ are equivalent on N with respect to $f_*\boldsymbol{g}_1$.*

Thus we see that unless f is an isometry, it will not respect the equivalence relations induced by the initial metrics $\boldsymbol{g}_1$ and $\boldsymbol{g}_2$ on M and N.

This observation is of vital importance in physics. Consider a scalar field ϕ, and let $(M, \boldsymbol{g})$ be the Euclidean space $\mathbb{R}^3$ (with the standard metric) representing physical space. The energy density is then given by

$$H = \frac{1}{2}\, \partial_i\phi\partial^i\phi.$$

But here we have used the equivalence relation induced by the metric! From the field itself we can only construct the covector $\boldsymbol{d\phi}$ with the components $\partial_i\phi$. So when we use the contravariant components $\partial^i\phi$, it is implicitly understood in this expression that we have used the metric components to raise the index. Thus it would be more correct to write out the energy density as

$$H = \frac{1}{2} g^{ij}\partial_i\phi\partial_j\phi.$$

This is a very common situation in physics: Indices are contracted using the metric. Now observe that the metric is a fixed geometrical quantity. It is physically measurable; e.g., the arc-length of a curve in a physical space can be measured with great accuracy in the laboratory. We are not free to exchange this metric if we want to compare the predictions of the theory with experimental results. Suppose now that we have been given a diffeomorphism of space into itself,

$$f : \mathbb{R}^3 \curvearrowright \mathbb{R}^3.$$

Then we can investigate the transformed field configuration $\phi' = f_*\phi$. E.g., we can compare the energy densities of the original and the transformed field configurations at corresponding points

$$\frac{1}{2} g^{ij}\partial_i\phi\partial_j\phi|_P \quad \text{and} \quad \frac{1}{2} g^{ij}\partial_i\phi'\partial_j\phi'|_{f(P)}.$$

But they are only identical if f is an isometry, since the metric is fixed. This distinguishes the isometries from a physical point of view: They leave various physical quantities invariant; i.e., they act as *symmetry transformations*.

We have previously discussed time-like geodesics on a manifold with Minkowski metric. (See Section 6.7.) We will now extend the concept of a geodesic. Motivated by the discussion in Section 6.7, we make the following definition.

Definition 10.9. A *geodesic* on a manifold M with a Minkowski metric $\mathbf{g}$ is a curve parametrized by $x^\alpha = x^\alpha(\lambda)$ that satisfies the *geodesic equation*

$$\frac{d^2x^\mu}{d\lambda^2} + \Gamma^\mu{}_{\alpha\beta}\frac{dx^\alpha}{d\lambda}\frac{dx^\beta}{d\lambda} = 0. \tag{6.49}$$

The parameter λ involved in the geodesic equation is called an *affine parameter.*

Observe that the affine parameter is determined only up to an affine parameter shift

$$\lambda = as + b \qquad (a \neq 0);$$

i.e., a geodesic satisfies the same equation (6.49) with respect to the new parameter s.

Consider a point P in M and a nontrivial tangent vector $\mathbf{v}_P$ at P (Figure 10.19). *Then there is exactly one geodesic that passes through P and has tangent vector $\mathbf{v}_P$.* To see this, we introduce coordinates around P. In these coordinates P is represented by coordinates x_0^i, and $\mathbf{v}_P$ by a^i. A geodesic through P with tangent

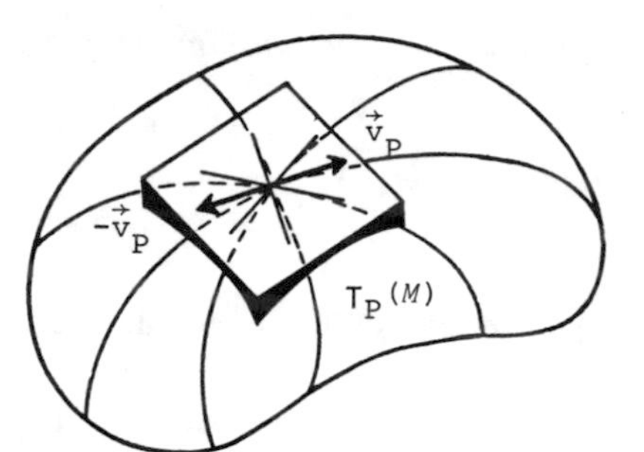

Figure 10.19.

vector $\mathbf{v}_P$ then has to satisfy the geodesic equation

$$\frac{d^2x^i}{d\lambda^2} + \Gamma^i{}_{jk}\frac{dx^j}{d\lambda}\frac{dx^k}{d\lambda} = 0$$

with the boundary conditions

$$x^i(0) = x_0^i \quad \text{and} \quad \frac{dx^i}{d\lambda}\Big|_{\lambda=0} = a^i.$$

But this second-order differential equation has a unique solution by the well-known uniqueness and existence theorem for ordinary differential equations. If we perform an affine parameter shift

$$\lambda = as + b \qquad (a \neq 0),$$

the tangent vector $\mathbf{v}_P$ is replaced by $a\mathbf{v}_P$, so we have actually shown that to each point P and each direction at P there corresponds exactly one geodesic.

Worked Exercise 10.3.3
Problem: (a) Let Γ be a curve on M parametrized by $x^\alpha = x^\alpha(s)$. Show that Γ is a geodesic if and only if it satisfies an equation of the form

$$\frac{d^2x^\mu}{ds} + \Gamma^\mu{}_{\alpha\beta}\frac{dx^\alpha}{ds}\frac{dx^\beta}{ds} = A(s)\frac{dx^\mu}{ds}. \tag{10.25}$$

(b) Show that the corresponding affine parameter λ is given by

$$\lambda = \int_0^s \exp\left[\int_0^{s_2} A(s_1)ds_1\right] ds_2 \tag{10.26}$$

(up to a linear change of the parameter).

The affine parameter is not only characterized by the simplest possible form of the geodesic equation, it is also distinguished by the following property:

Theorem 10.9. *Let λ be an affine parameter on a geodesic. Then the tangent vector*

$$\mathbf{v} = \frac{dP}{d\lambda}$$

has constant length; i.e., $\mathbf{g}(\mathbf{v}(\lambda), \mathbf{v}(\lambda))$ *is constant along the curve.*

PROOF.

$$\frac{d}{d\lambda}\left[g_{\alpha\beta}\frac{dx^\alpha}{d\lambda}\frac{dx^\beta}{d\lambda}\right] = 2g_{\alpha\beta}\frac{dx^\alpha}{d\lambda}\left[\frac{d^2x^\beta}{d\lambda^2} + \frac{1}{2}g^{\beta\gamma}\partial_\mu g_{\nu\gamma}\frac{dx^\mu}{d\lambda}\frac{dx^\nu}{d\lambda}\right].$$

Here we can exchange $\frac{1}{2}g^{\beta\gamma}\partial_\mu g_{\nu\gamma}$ by $\Gamma^\beta_{\mu\nu}$ due to the identity

$$g_{\alpha\beta}\frac{dx^\alpha}{d\lambda}\left[\frac{1}{2}g^{\beta\gamma}\partial_\mu g_{\nu\gamma}\right]\frac{dx^\nu}{d\lambda} = g_{\alpha\beta}\frac{dx^\alpha}{d\lambda}\left[\Gamma^\beta{}_{\mu\nu}\right]\frac{dx^\nu}{d\lambda}.$$

Thus we get

$$\frac{d}{d\lambda}\left[g_{\alpha\beta}\frac{dx^\alpha}{d\lambda}\frac{dx^\beta}{d\lambda}\right] = 2g_{\alpha\beta}\frac{dx^\alpha}{d\lambda}\left[\frac{d^2x^\beta}{d\lambda^2} + \Gamma^\beta{}_{\mu\nu}\frac{dx^\mu}{d\lambda}\frac{dx^\nu}{d\lambda}\right],$$

and this last expression vanishes automatically for a geodesic. □

So the affine parameter is a natural parameter on a geodesic! We can now divide the geodesics on M into three classes:

a. *Time-like geodesics*, where all the tangent vectors are time-like. (The affine parameter can then be normalized so that $g_{\alpha\beta}\frac{dx^\alpha}{d\lambda}\frac{dx^\beta}{d\lambda} = -1$.)
b. *Null geodesics*, where all the tangent vectors are null vectors.
c. *Space-like geodesics*, where all the tangent vectors are space-like. (The affine parameter can then be normalized so that $g_{\alpha\beta}\frac{dx^\alpha}{d\lambda}\frac{dx^\beta}{d\lambda} = +1$.)

We can then show

Theorem 10.10.
1. *Isometries map geodesics into geodesics and preserve the affine parameter. (This is valid for Riemannian manifolds too.)*
2. *Conformal maps preserve null geodesics. If* λ *is an affine parameter on the null geodesic* Γ, *then* $\int \Omega^2(\lambda)d\lambda$ *is an affine parameter on* $f(\Gamma)$, *where* Ω^2 *is the conformal factor.*

PROOF. The first proposition is almost trivial, since isometries preserve arc-lengths and since geodesics are in general characterized as extremizing the arc-length. Special care should, however, be paid to null geodesics, but here the result will follow from the second proposition.

To prove (2) it suffices to consider a single manifold M and to consider a rescaling of the metric $\mathbf{g} \to \Omega^2\mathbf{g} = \tilde{\mathbf{g}}$. Now, if Γ is a null geodesic, then in particular, it satisfies the geodesic equation

$$\frac{d^2x^\mu}{d\lambda^2} + \Gamma^\mu{}_{\alpha\beta}\frac{dx^\alpha}{d\lambda}\frac{dx^\beta}{d\lambda} = 0. \tag{6.49}$$

Under a rescaling

$$g_{\alpha\beta} \to \tilde{g}_{\alpha\beta} = \Omega^2(x)g_{\alpha\beta},$$

the Christoffel field $\Gamma^{\mu}_{\ \alpha\beta}$ given by (6.47) is changed into

$$\tilde{\Gamma}^{\mu}_{\ \alpha\beta} = \Gamma^{\mu}_{\ \alpha\beta} + \delta^{\mu}_{\alpha}\partial_{\beta}\ln\Omega + \delta^{\mu}_{\beta}\partial_{\alpha}\ln\Omega - g_{\alpha\beta}\partial^{\mu}\ln\Omega.$$

Thus in the rescaled metric we get

$$\frac{d^2x^{\mu}}{d\lambda^2} + \tilde{\Gamma}^{\mu}_{\ \alpha\beta}\frac{dx^{\alpha}}{d\lambda}\frac{dx^{\beta}}{d\lambda} = \left[\frac{d^2x^{\mu}}{d\lambda^2} + \Gamma^{\mu}_{\ \alpha\beta}\frac{dx^{\alpha}}{d\lambda}\frac{dx^{\beta}}{d\lambda}\right]$$
$$+ 2\frac{dx^{\mu}}{d\lambda}\frac{d}{d\lambda}\ln\Omega - [\partial^{\mu}\ln\Omega]g_{\alpha\beta}\frac{dx^{\alpha}}{d\lambda}\frac{dx^{\beta}}{d\lambda}.$$

Here the first term vanishes because Γ is geodesic, and the third term vanishes because the tangent vectors are null vectors. Thus we get

$$\frac{d^2x^{\mu}}{d\lambda^2} + \tilde{\Gamma}^{\mu}_{\ \alpha\beta}\frac{dx^{\alpha}}{d\lambda}\frac{dx^{\beta}}{d\lambda} = \frac{d\ln\Omega^2}{d\lambda}\frac{dx^{\mu}}{d\lambda}.$$

But according to Exercise 10.3.3, this shows that Γ is a null geodesic with respect to the rescaled metric. However, it gets the new affine parameter given by

$$\mu(\lambda) = \int_0^{\lambda}\exp\left[\int_0^{s_2}\frac{d}{ds}(\ln\Omega^2)ds_1\right]ds_2 = \frac{1}{\Omega^2(0)}\int_0^{\lambda}\Omega^2 d\lambda.$$

□

In many applications it is preferable to control the set of all possible global isometries: Consider a manifold M with metric $\mathbf{g}$, and suppose ϕ is a global isometry. Then so is ϕ^{-1}. This follows immediately from the composition rule (Theorem 10.5):

$$(\phi^{-1})^*\mathbf{g} = (\phi^{-1})^*\phi^*\mathbf{g} = [\phi\circ\phi^{-1}]^*\mathbf{g} = \mathrm{id}^*\mathbf{g} = \mathbf{g}.$$

By the same argument, it follows that if ϕ_1 and ϕ_2 are global isometries, then so is the composite map $\phi_2\circ\phi_1$. Thus we conclude that the set of global isometries form a group, which we call the *isometry group of the manifold M*, and which we denote by Isom$[M, \mathbf{g}]$. Let us now turn our attention to space–time, i.e., Minkowski space. Here we can determine the isometry group explicitly. It will be preferable to restrict to an inertial frame. Then the following holds:

Theorem 10.11. *A diffeomorphism $\phi : M \curvearrowright M$ is a global isometry if and only if it is of the form*

$$y^{\alpha} = A^{\alpha}_{\beta}x^{\beta} + b^{\alpha}, \tag{10.27}$$

where $\overline{\overline{A}} = (A^{\alpha}_{\beta})$ is a Lorentz matrix.

PROOF. The proposition is intuitively clear, since an isometry preserves geodesics, i.e., it maps straight lines into straight lines, and thus it must be an affine map of the form

$$y^{\alpha} = A^{\alpha}_{\beta}x^{\beta} + b^{\alpha}$$

with respect to an inertial set of coordinates. From this, the rest follows easily: The pulled back metric $\phi^*\mathbf{g}$ is characterized by the components

$$g_{\alpha\beta}(x) = \eta_{\rho\sigma}\frac{\partial y^\rho}{\partial x^\alpha}\frac{\partial y^\sigma}{\partial x^\beta} = \eta_{\rho\sigma}A^\rho_\alpha A^\sigma_\beta;$$

i.e.,

$$\overline{\overline{G}}(x) = \overline{\overline{A}}^+\overline{\overline{\eta}}\overline{\overline{A}}.$$

Consequently, the metric is preserved if and only if

$$\overline{\overline{\eta}} = \overline{\overline{A}}^+\overline{\overline{\eta}}\overline{\overline{A}},$$

i.e., if and only if $\overline{\overline{A}}$ is a Lorentz matrix. (Compare the discussion in Section 6.6.) □

The transformation (10.27) should be compared with the Poincaré transformation (Theorem 6.3), which has exactly the same form although a different meaning! The Poincaré transformation (Theorem 6.3) was a coordinate transformation: The points are fixed, while their inertial coordinates were exchanged. The new transformation (10.27), on the contrary, moves the points while keeping their inertial coordinates fixed. It is called an *active Poincaré transformation*. We have thus shown that the isometry group of Minkowski space–time is the Poincaré group.

If we start out with the Euclidean space $\mathbb{R}^n$ equipped with the usual Cartesian metric, then the isometry group is called the *Euclidean group of motions*. In Cartesian coordinates an isometry will then be represented by a linear map,

$$\overline{\overline{y}}_| = \overline{\overline{A}}\overline{\overline{x}}_| + \overline{\overline{b}}_|, \tag{10.28}$$

where $\overline{\overline{A}}$ is an orthogonal matrix; i.e., $\overline{\overline{A}}^+\overline{\overline{A}} = \overline{\overline{A}}\overline{\overline{A}}^+ = \overline{\overline{I}}$.

Exercise 10.3.4
Problem: Consider the unit sphere S^2 in $\mathbb{R}^3$.

a. Show that an orthogonal transformation in $\mathbb{R}^3$

$$\overline{\overline{y}}_| = \overline{\overline{A}}\overline{\overline{x}}_|; \qquad \text{where } \overline{\overline{A}} \in O(3),$$

generates a global isometry on S^2.
b. Show that the geodesics on S^2 are the great circles.
c. Let Γ be a geodesic and $\phi(\Gamma)$ its image under a global isometry ϕ. Show that the common points of Γ and $\phi(\Gamma)$ must be fixed points of ϕ.
d. Show that a global isometry on S^2 belongs to one of the following 3 types:
   1. The identity map.
   2. A reflection in a plane through the origin.
   3. A rotation around a line through the origin.

10.4 The Conformal Group

We will now study conformal transformations in Minkowski space in more detail. The basic "defect" of conformal transformations in Minkowski space, as compared with isometries, is that they do not act as linear transformations. Nevertheless, one can generate them from linear transformations in a higher-dimensional space. The linear representation of the group of conformal transformations will be the main topic of this section.

It turns out to be instructive to consider the general case of a pseudo-Cartesian space $\mathbb{R}^p \times \mathbb{R}^q$ equipped with the standard inner product:

$$\langle a \mid b \rangle = -a^1b^1 - \cdots - a^pb^p + a^{p+1}b^{p+1} + \cdots + a^{p+q}b^{p+q}.$$

Naively, we should consider only conformal transformations, which are smooth everywhere. This turns out to be too restrictive. In that case, the only nontrivial conformal transformations are the dilations. We will therefore admit conformal maps that break down at a suitable subset. The standard example of such a conformal transformation is the *inversion*:

$$y^\alpha = \frac{x^\alpha}{\langle x \mid x \rangle} \tag{10.29}$$

(cf. Exercise 10.3.1). It breaks down at the null cone through the origin: $\langle x \mid x \rangle = 0$. From the relation $y^2 = x^{-2}$ it furthermore follows that the closer a point is at the null cone through the origin, the farther away the image is. In some sense the inversion therefore maps the entire null cone through the origin to infinity. We would therefore like to enlarge the pseudo-Cartesian space by adding a "null-cone at infinity." In the enlarged space, the inversion will then be a diffeomorphism that exchanges the null cone through the origin with the null cone at infinity.

To get an idea of what this actually means, let us take a quick look at the Euclidean space $\mathbb{R}^n$. In this case, the "null-cone" is just the origin itself. We must therefore add a single point to $\mathbb{R}^n$ so that the inversion can exchange this point with the origin. For this purpose we consider the unit sphere S^n in $\mathbb{R}^{n+1}$. Using a stereographic projection from the north pole, we can then map $\mathbb{R}^n$ onto $S^n - \{N\}$.

Thus we can replace $\mathbb{R}^n$ by $S^n - \{N\}$. The north pole now represents the point at infinity. Furthermore, the stereographic projection is a conformal map (cf. Exercise 10.3.2); i.e., the metric on the sphere and the transferred metric from

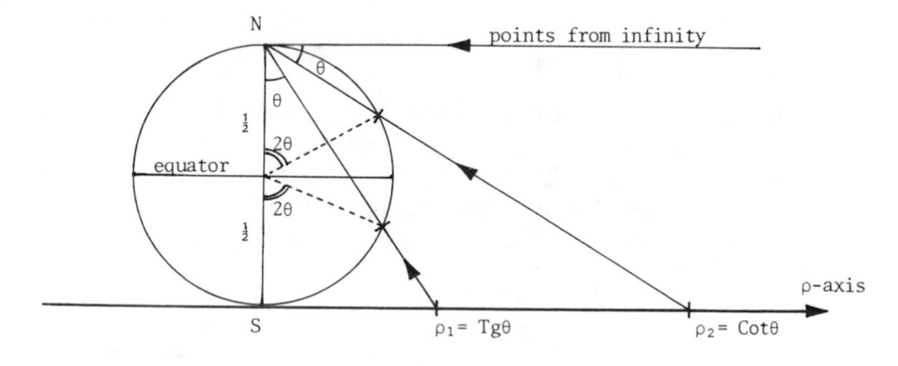

Figure 10.20.

$\mathbb{R}^n$ are conformally related. From the point of view of investigating conformal transformations, we can therefore equally as well work on S^n.

Now suppose we identify the Euclidean space $\mathbb{R}^n$ with $S^n - \{N\}$. What does the inversion look like on the sphere? As indicated in Figure 10.20, it is easy to see that the inversion corresponds to a reflection in the equation. On the sphere the inversion is thus a nice diffeomorphism that exchanges the south pole (i.e., the origin) with the north pole (i.e., the point at infinity).

Remark. The sphere S^n is a *compact* manifold. By adding a point at infinity we have therefore *compactified* the plane $\mathbb{R}^n$. In the mathematically oriented literature the sphere is therefore called the *one-point compactification* of Euclidean space.

Motivated by this example we now return to the pseudo-Cartesian space $\mathbb{R}^p \times \mathbb{R}^q$. First we enlarge the pseudo-Cartesian space by adding two extra dimensions: One time-like, labeled u, and one space-like, labeled v. In this way we obtain the pseudo-Cartesian space $\mathbb{R}^{1+p} \times \mathbb{R}^{q+1}$, in which a typical point will be denoted by

$$w = (u; x^1, \ldots, x^p; y^1, \ldots, y^q; v) = (u; z^\alpha; v).$$

The goal is to embed the pseudo-Cartesian space $\mathbb{R}^p \times \mathbb{R}^q$ as a suitable subset of $\mathbb{R}^{1+p} \times \mathbb{R}^{q+1}$ (similar to the embedding of the Euclidean space $\mathbb{R}^n$ as a sphere S^n inside $\mathbb{R}^{n+1}$). This will be done in a tricky manner! In the enlarged space $\mathbb{R}^{1+p} \times \mathbb{R}^{q+1}$ we introduce the null cone through the origin K. In the first step we then embed $\mathbb{R}^p \times \mathbb{R}^q$ isometrically as a subset of K. Define M to be the intersection of the null cone K and the hyperplane $u - v = 1$, and consider the bijective map $\phi : \mathbb{R}^p \times \mathbb{R}^p \curvearrowright M$ constructed in the following way:

$$\phi(z) = \left(\frac{\langle z \mid z \rangle + 1}{2} \,; z \,; \frac{\langle z \mid z \rangle - 1}{2} \right).$$

By construction, ϕ is an isometry. To see this, we observe that ϕ actually generates a global coordinate system on M. The basic frame vectors are given by

$$\tilde{\mathbf{e}}_a = [(\pm)z^a; \mathbf{e}_a; (\pm)z^a].$$

Consequently, the metric coefficients of the induced metric on M reduce to

$$g_{ab} = \langle \tilde{\mathbf{e}}_a \mid \tilde{\mathbf{e}}_b \rangle = (\mp)z^a z^b + \langle \mathbf{e}_a \mid \mathbf{e}_b \rangle (\pm) z^a z^b = \langle \mathbf{e}_a \mid \mathbf{e}_b \rangle.$$

Since the embedding is an isometry, we can simply identify $\mathbb{R}^p \times \mathbb{R}^q$ with this particular section M, which henceforth will be denoted by $M(\mathbb{R}^p \times \mathbb{R}^q)$.

The section $M(\mathbb{R}^p \times \mathbb{R}^q)$ will ultimately be replaced by another section of the mull cone K, but before we proceed with the construction, we must take a closer look at the conformal structure of the null cone K. A generator of K is a null vector

$$\mathbf{w} = (u, \mathbf{z}, v); \qquad \mathbf{w}^2 = 0.$$

It generates a line $\ell_\mathbf{w}$ on K,

$$\ell_\mathbf{w} = \lambda(u, \mathbf{z}, v); \qquad -\infty < \lambda < +\infty,$$

called a characteristic line. A given characteristic line $\ell_{\mathbf{w}}$ will intersect the section $M(\mathbb{R}^p \times \mathbb{R}^q)$ at most once, and there are characteristic lines that do not intersect $M(\mathbb{R}^p \times \mathbb{R}^q)$ at all. They correspond to the lines that are parallel to the hyperplane $u - v = 1$; i.e., they are generated by null vectors $\mathbf{w} = (u, \mathbf{z}, v)$, where $u = v$. But these are precisely the null vectors where $\mathbf{z}^2 = 0$. Thus there is a one-to-one correspondence between characteristic lines missing $M(\mathbb{R}^p \times \mathbb{R}^q)$ and points on the null cone through the origin in the original pseudo-Cartesian space $\mathbb{R}^p \times \mathbb{R}^q$ (cf. Figure 10.21a). Consequently, the exceptional lines represent points on the null cone at infinity!

Consider now two local sections N_1 and N_2 on the null cone K. Suppose furthermore that the characteristic lines intersect N_1 and N_2 at most once. Then we have a natural map

$$\pi : N_1 \curvearrowright N_2$$

obtained by projection along the characteristic lines (Figure 10.21b). The basic observation is the following:

Lemma 10.5. *The projection along characteristic lines, $\pi : N_1 \curvearrowright N_2$, is a conformal map.*

PROOF. It is preferable to introduce new coordinates in $\mathbb{R}^{1+p} \times \mathbb{R}^{q+1}$. We need two radial variables r_1, r_2 and $p + q$ homogeneous variables $\theta^1, \ldots, \theta^p, \phi^1, \ldots, \phi^p$:

$$r_1^2 = u^2 + (x^1)^2 + \ldots + (x^p)^2; \qquad \theta^1 = x^1/u, \ldots, \theta^p = x^p/u;$$
$$r_2^2 = v^2 + (y^1)^2 + \ldots + (y^q)^2; \qquad \phi^1 = y^1/v, \ldots, \phi^q = y^q/v.$$

(As usual, we have trouble with the homogeneous coordinates, which break down when $u = 0$ or $v = 0$. Since we are only interested in a local result, we will simply assume that $u = 0$ and $v = 0$ do not intersect N_1 or N_2.) The null cone K is then characterized by the equation

$$r_1 = r_2.$$

If we put the common value equal to r, we can introduce the following intrinsic coordinates:

$$(r, \theta^1, \ldots, \theta^p, \phi^1, \ldots, \phi^q)$$

on the null cone K. A characteristic line is then given by a fixed set of the homogeneous coordinates. Furthermore, the local sections can be parametrized as follows:

$$N_1 : r = f(\theta^1, \ldots, \theta^p, \phi^1, \ldots, \phi^q),$$
$$N_2 : r = g(\theta^1, \ldots, \theta^p, \phi^1, \ldots, \phi^q).$$

We can therefore use $(\theta^1, \ldots, \theta^p, \phi^1, \ldots, \phi^q)$ as intrinsic coordinates on N_1 and N_2. With this choice of coordinates, the projection map π is simply represented by the identity map!

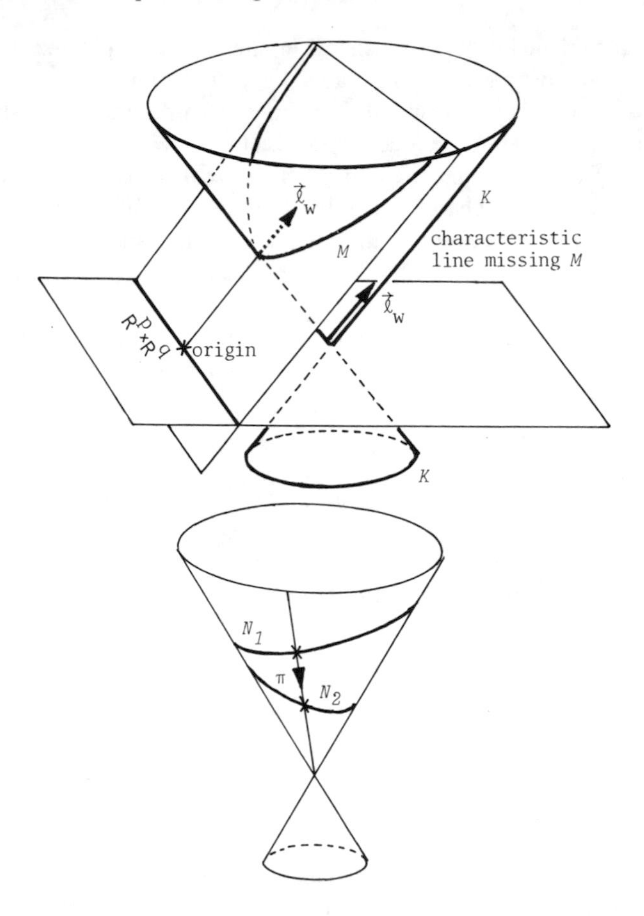

Figure 10.21.

We must now determine the various metrics involved in the game. Consider first the complete pseudo-Cartesian space $\mathbb{R}^{1+p} \times \mathbb{R}^{q+1}$ with the canonical metric

$$ds^2 = -du^2 - (dx^1)^2 - \cdots - (dx^p)^2 + (dy^1)^2 + \cdots + (dy^q)^2 + dv^2.$$

In the new coordinates this is reexpressed as

$$ds^2 = -dr_1^2 - r_1^2 c_{ij}(\theta) d\theta^i d\theta^j + r_2^2 d_{kl}(\phi) d\phi^k d\phi^l + dr_2^2.$$

The null cone K, where $r_1 = r_2 = r$, now gets the induced metric

$$ds^2_{1|K} = r^2[-c_{ij}(\theta) d\theta^i d\theta^j + d_{k1}(\phi) d\phi^k d\phi^l].$$

Finally, the two sections N_1 and N_2 are equipped with the induced metrics

$$N_1 : ds_1^2 = f^2(\theta, \phi)[-c_{ij}(\theta) d\theta^i d\theta^j + d_{kl}(\phi) d\phi^k d\phi^l]$$

$$N_2 : ds_2^2 = g^2(\theta, \theta)[-c_{ij}(\theta) d\theta^i d\theta^j + d_{kl}(\phi) d\phi^k d\phi^l].$$

Consequently,

$$ds_2^2 = \frac{g^2(\theta, \phi)}{f^2(\theta, \phi)}\, ds_1^2;$$

i.e.,

$$\pi_* \boldsymbol{g}_2 = \Omega^2(\theta, \phi)\boldsymbol{g}_1,$$

where

$$\Omega^2(\theta, \phi) = \frac{g^2(\theta, \phi)}{f^2(\theta, \phi)}.$$

□

Exercise 10.4.1
Introduction: In the above discussion we have embedded $\mathbb{R}^p \times \mathbb{R}^q$ as the intersection M_1 of the null cone K and the hyperplane $u - v = 1$ using the bijective map

$$\phi_1(z_1) = \left(\frac{\langle z_1 \mid z_1 \rangle + 1}{2}\, ; z_1; \frac{\langle z_1 \mid z_1 \rangle - 1}{2} \right).$$

We could equally well have embedded $\mathbb{R}^p \times \mathbb{R}^q$ as the intersection M_2 of the null cone K and the hyperplane $u + v = 1$ using the bijective map

$$\phi_2(z_2) = \left(\frac{1 + \langle z_2 \mid z_2 \rangle}{2}\, ; z_2; \frac{1 - \langle z_2 \mid z_2 \rangle}{2} \right).$$

Problem: Show that the projection along the characteristic lines

$$\pi : M_1 \curvearrowright M_2$$

corresponds to an inversion in $\mathbb{R}^p \times \mathbb{R}^q$.

Having obtained this lemma, we can now apply it to project $M(\mathbb{R}^p \times \mathbb{R}^q)$ into a suitable section of K (analogous to the sphere in the Euclidean case). This new section should then include the "null cone at infinity"; i.e., each characteristic line should intersect it exactly once. Unfortunately, we run into a slight technical problem: In general, it is not possible to find a single section that is intersected exactly once by each characteristic line. We shall therefore adopt the following strategy:

Let r_1 and r_2 denote the radial variables in the complete pseudo-Cartesian space $\mathbb{R}^{1+p} \times \mathbb{R}^{q+1}$:

$$r_1^2 = u^2 + (x^1)^2 + \cdots + (x^p)^2; \qquad r_2^2 = (y^1)^2 + \cdots + (y^q)^2 + v^2.$$

Denote by N the intersection to the null cone $K(r_1 = r_2)$ with the hypersphere

$$r_1^2 + r_2^2 = 2.$$

Clearly, N is a submanifold defined by the equations of constraint

$$r_1 = r_2 = 1.$$

Topologically, N is therefore a product of the two unit spheres S^p in $\mathbb{R}^{1+p}$ and S^q in $\mathbb{R}^{q+1}$. Consequently, N is a hypertorus:

$$N = S^p \times S^q.$$

The hypertorus $S^p \times S^q$ is a nice section on K, but each characteristic line will actually intersect it twice in antipodal points (see Figure 10.22). Consequently, *each point in the origin pseudo-Cartesian space $\mathbb{R}^p \times \mathbb{R}^q$ is represented by a pair of antipodal points on $S^p \times S^q$.*

It follows from Lemma 10.5 that the projection map $\pi : M(\mathbb{R}^p \times \mathbb{R}^q) \to S^p \times S^q$ is a conformal map. From the point of view of investigating conformal transformations, we can therefore equally well work on $S^p \times S^q$, except that we must restrict ourselves to transformations that map pairs of antipodal points into pairs of antipodal points.

In coordinates, the projection from $S^p \times S^q$ to $\mathbb{R}^p \times \mathbb{R}^q$ is given by

$$\pi(u, z^\mu, v) = \frac{1}{u - v}\, z^\mu.$$

Consequently, the points where u and v coincide are "sent to infinity." But according to our previous analysis, these points are in one-to-one correspondence with the null cone in $\mathbb{R}^p \times \mathbb{R}^q$. Thus $S^p \times S^q$ is obtained from $\mathbb{R}^p \times \mathbb{R}^q$ by adding a "cone at infinity." Because $S^p \times S^q$ is a *compact* subset of $\mathbb{R}^{1+p} \times \mathbb{R}^{q+1}$, it is often referred to as the *conformal compactification* of $\mathbb{R}^p \times \mathbb{R}^q$.

Consider once more the Euclidean space $\mathbb{R}^q$. In this case the conformal compactification reduces to $S^0 \times S^q$, *but the zero-dimensional sphere S^0 consists only of two points*, $u = \pm 1$. In this case (and only in this case!) the conformal

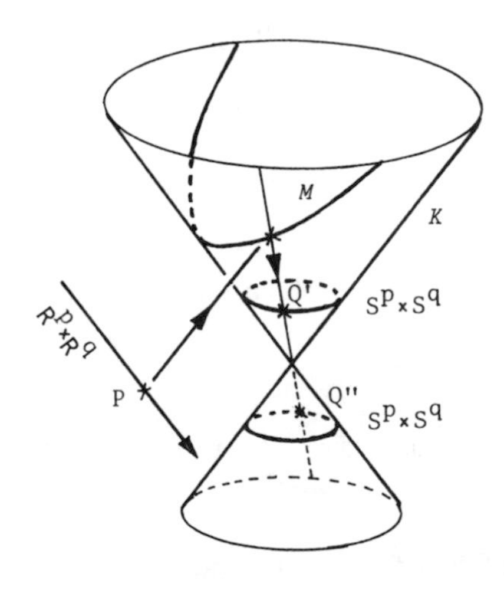

Figure 10.22.

compactification therefore breaks up into two disconnected components,

$$\{+1\} \times S^q \quad \text{and} \quad \{-1\} \times S^q.$$

Thus we need not double-count the points in the conformal compactification because we can simply throw away the component $\{-1\} \times S^q$! But then it is superfluous to enlarge the space with a time-like coordinate; i.e., we simply enlarge $\mathbb{R}^q$ to the Euclidean space $\mathbb{R}^{q+1}$ and use the unit sphere S^q in this enlarged space as the conformal compactification. It is now easy to show that the point at infinity corresponds to the north pole and that the projection along the characteristic lines corresponds to the stereographic projection (cf. Figure 10.23). Thus we have a nice, simplified picture in the Euclidean case.

Using the conformal compactification of the pseudo-Cartesian space $\mathbb{R}^p \times \mathbb{R}^q$, it is easy to construct conformal transformations on $\mathbb{R}^p \times \mathbb{R}^q$. Consider the matrix group $O(1+p; q+1)$ consisting of pseudo-orthogonal matrices operating on $\mathbb{R}^{1+p} \times \mathbb{R}^{q+1}$. Each matrix in $O(1+p; q+1)$ generates a conformal transformation on $S^p \times S^q$ in the following way:

Notice first that a pseudo-orthogonal matrix $\overline{\overline{S}}$ preserves the inner product in $\mathbb{R}^{1+p} \times \mathbb{R}^{q+1}$. In particular, it maps the null cone K into itself. Consequently, it maps the hypertorus $S^p \times S^q$ *isometrically* onto a new subset $\overline{\overline{S}}[S^p \times S^q]$, of K. To get back to the hypertorus, we then project along the characteristic lines! In this way we have constructed a mapping of the hypertorus into itself, which we denote by $\pi[\overline{\overline{S}}]$ (Figure 10.24). Alternatively, we can describe $\pi[\overline{\overline{S}}]$ in the following way: Each pair of antipodal points $\{P, -P\}$ on $S^p \times S^q$ generates a unique characteristic line ℓ_P on the null cone K. The pseudo-orthogonal transformation $\overline{\overline{S}}$ maps this characteristic line into another characteristic line $\ell_{\overline{\overline{S}}(P)}$. The image of P is then the intersection of $\ell_{\overline{\overline{S}}(P)}$ and $S^p \times S^q$. According to Lemma 10.5, the combined

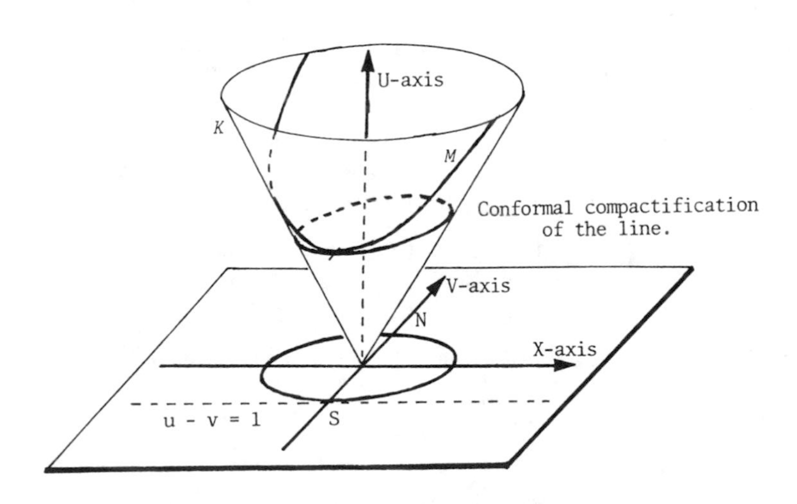

Figure 10.23.

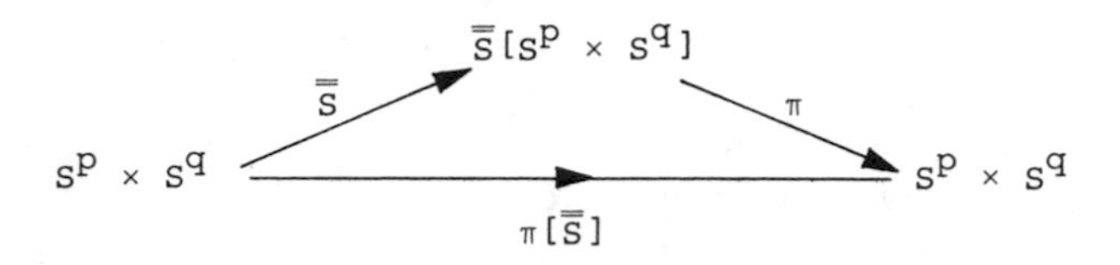

Figure 10.24.

transformation $\pi[\bar{\bar{S}}]$ is now a conformal transformation of the hypertorus into itself.

From the construction follows immediately some basic properties of $\pi[\bar{\bar{S}}]$. First, it maps a pair of antipodal points into a pair of antipodal points; i.e., it can also be considered a conformal transformation on $\mathbb{R}^p \times \mathbb{R}^q$. Next, the assignment of a conformal transformation to each pseudo-orthogonal transformation constitutes a *representation* of the pseudo-orthogonal group; i.e.,

$$\pi[\bar{\bar{S}}_2\bar{\bar{S}}_1] = \pi[\bar{\bar{S}}_2] \cdot \pi[\bar{\bar{S}}_1].$$

Notice, however, that the correspondence between pseudo-orthogonal transformations in $O(1 + p; q + 1)$ and conformal transformations in $\mathbb{R}^p \times \mathbb{R}^q$ is not one-to-one. This is because a point in $\mathbb{R}^p \times \mathbb{R}^q$ corresponds to a pair of antipodal points on $S^p \times S^q$. A pseudo-orthogonal matrix that interchanges antipodal points will therefore generate the identity. There is precisely one such pseudo-orthogonal matrix: $-\bar{\bar{I}}$. It follows that each conformal transformation is in fact generated by a pair of pseudo-orthogonal matrices $\{\bar{\bar{S}}, -\bar{\bar{S}}\}$. This is the price we have to pay when we want to represent conformal transformations by matrices!

The conformal transformations generated from $O(1 + p, q + 1)$ evidently constitute a group, known as the *conformal group* and denoted by $C(p, q)$. *In analogy with the conformal compactification of* $\mathbb{R}^p \times \mathbb{R}^q$, *we can now represent each conformal transformation in* $C(p, q)$ *by a pair of "antipodal" matrices in* $O(1 + p, q + 1)$.

Let us investigate the structure of the conformal group a little more closely.

Suppose that the pseudo-orthogonal transformation $\bar{\bar{S}}$ actually preserves the additional coordinates (u, v), i.e., that it is of the form

$$\bar{\bar{S}} = \begin{bmatrix} 1 & 0 & 0 \\ 0 & \bar{\bar{s}} & 0 \\ 0 & 0 & 1 \end{bmatrix}, \qquad \bar{\bar{s}} \in O(p, q). \tag{10.30a}$$

Then the corresponding transformation on $\mathbb{R}^p \times \mathbb{R}^q$ reduces to

$$\begin{array}{ccccccc} z^\alpha & \curvearrowright & (u; (u-v)z^\alpha; v) & \overset{\bar{\bar{S}}}{\curvearrowright} & (u; (u-v)s^\alpha_\beta z^\beta; v) & \curvearrowright & s^\alpha_\beta z^\beta \\ \mathbb{R}^p \times \mathbb{R}^q & \curvearrowright & S^p \times S^q & \curvearrowright & S^p \times S^q & \curvearrowright & \mathbb{R}^p \times \mathbb{R}^q. \end{array}$$

Consequently, *the conformal group* $C(p, q)$ *contains the group of origin-preserving isometries* $O(p, q)$.

In the remaining investigation it suffices to consider pseudo-orthogonal matrices of the form

$$\bar{\bar{S}} = \begin{bmatrix} * & * & * \\ * & \bar{\bar{I}} & * \\ * & * & * \end{bmatrix}.$$

They will be divided into four general types:

$$S(I) = \begin{bmatrix} 1 & 0 & 0 \\ 0 & \bar{\bar{I}} & 0 \\ 0 & 0 & -1 \end{bmatrix};$$

$$\bar{\bar{T}}(c^\mu) = \begin{bmatrix} 1 + \frac{\langle c|c\rangle}{2} & c_\nu & -\frac{\langle c|c\rangle}{2} \\ c^\mu & \bar{\bar{I}} & -c^\mu \\ \frac{\langle c|c\rangle}{2} & c_\nu & 1 - \frac{\langle c|c\rangle}{2} \end{bmatrix}; \tag{10.30b}$$

$$\bar{\bar{D}}(\lambda) = \begin{bmatrix} \cosh\lambda & 0 & -\sinh\lambda \\ 0 & \bar{\bar{I}} & 0 \\ -\sinh\lambda & 0 & \cosh\lambda \end{bmatrix};$$

$$\bar{\bar{C}}(c^\mu) = \begin{bmatrix} 1 + \frac{\langle c|c\rangle}{2} & c_\nu & \frac{\langle c|c\rangle}{2} \\ c^\mu & \bar{\bar{I}} & c^\mu \\ -\frac{\langle c|c\rangle}{2} & -c_\nu & 1 - \frac{\langle c|c\rangle}{2} \end{bmatrix}. \tag{10.30c}$$

Each of these types constitutes a representation of a particular subgroup of $C(p, q)$, i.e.,

$$\bar{\bar{S}}^2(I) = \bar{\bar{I}}; \qquad \bar{\bar{T}}(b^\mu)\bar{\bar{T}}(a^\mu) = \bar{\bar{T}}(b^\mu + a^\mu);$$
$$\bar{\bar{D}}(\lambda_2)\bar{\bar{D}}(\lambda_1) = \bar{\bar{D}}(\lambda_2 + \lambda_2); \qquad \bar{\bar{C}}(b^\mu)\bar{\bar{C}}(a^\mu) = \bar{\bar{C}}(b^\mu + a^\mu).$$

The first type, $S(I)$, represents the *inversion* (10.29)

$$z^\mu \curvearrowright \frac{z^\mu}{\langle z \mid z\rangle}.$$

The second type, $\bar{\bar{T}}(c^\mu)$, represents the *group of translations*

$$z^\mu \curvearrowright z^\mu + c^\mu. \tag{10.31}$$

The third type, $\bar{\bar{D}}(\lambda)$, represents the *group of dilatations*

$$z^\mu \curvearrowright e^\lambda z^\mu. \tag{10.32}$$

Finally, the fourth type, $\bar{\bar{C}}(c^\mu)$, represents the *group of special conformal translations*

$$z^\mu \curvearrowright \frac{z^\mu + c^\mu\langle z \mid z\rangle}{1 + 2\langle c \mid z\rangle + \langle c \mid c\rangle\langle z \mid z\rangle}. \tag{10.33}$$

Let us check the last of these statements just to illustrate the principle: A point z^μ in $\mathbb{R}^p \times \mathbb{R}^q$ is represented by the point $[u;\, (u - v)z^\mu;\, v]$ on the hypertorus

$S^p \times S^q$. By the pseudo-orthogonal transformation $\overline{\overline{C}}(c^\mu)$ this is mapped into the point $[u'; z'^\mu; v']$, where

$$u' = u + \langle c \mid z\rangle(u - v) + \frac{\langle c \mid c\rangle}{2}(u + v),$$
$$z'^\mu = (u - v)z^\mu + c^\mu(u + v),$$
$$v' = v - \langle c \mid z\rangle(u - v) - \frac{\langle c \mid c\rangle}{2}(u + v).$$

In $\mathbb{R}^p \times \mathbb{R}^q$ this corresponds to the point

$$\frac{z'^\mu}{u' - v'} = \frac{(u - v)z^\mu + c^\mu(u + v)}{(u - v) + 2\langle c \mid z\rangle(u - v) + \langle c \mid c\rangle(u + v)}$$
$$= \frac{z^\mu + c^\mu \frac{u+v}{u-v}}{1 + 2\langle c \mid z\rangle + \langle c \mid c\rangle \frac{u+v}{u-v}} .$$

It remains to determine the factor $(u + v)/(u - v)$. Since $[u; (u - v)z^\mu; z]$ is a point on the hypertorus, it follows that

$$u^2 + (u - v)^2 x^2 = v^2 + (u - v)^2 y^2 = 1.$$

From this relation we get in particular that

$$u^2 - v^2 = (u - v)^2(y^2 - x^2) = (u - v)^2\langle z \mid z\rangle;$$

i.e.,

$$\langle z \mid z\rangle = \frac{u^2 - v^2}{(u - v)^2} = \frac{u + v}{u - v} .$$

Inserting this, the induced mapping from $\mathbb{R}^p \times \mathbb{R}^q$ into itself finally reduces to

$$z^\mu \curvearrowright \frac{z^\mu + c^\mu\langle z \mid z\rangle}{1 + 2\langle c \mid z\rangle + \langle c \mid c\rangle\langle z \mid z\rangle} .$$

We also summarize the above findings in Table 10.1

Table 10.1.

Pseudo-orthogonal transformations in $\mathbb{R}^{1+p} \times \mathbb{R}^{q+1}$:	Conformal transformations in $\mathbb{R}^p \times \mathbb{R}^q$
Transformations preserving both u and v	Pseudo-orthogonal transformations: $z^\alpha \curvearrowright S^\alpha_\beta z^\alpha$, $\overline{\overline{S}} \in 0(p, q)$
Transformations preserving $u - v$	Translations: $z^\alpha \curvearrowright z^\alpha + c^\alpha$
Transformations preserving $u + v$	Special conformal transformations: $z^\alpha \curvearrowright \dfrac{z^\alpha + c^\alpha\langle z\vert z\rangle}{1 + 2\langle c\vert z\rangle + \langle c\vert c\rangle\langle z\vert z\rangle}$
Transformations in the (u, v)-plane	Dilatations: $z^\alpha \curvearrowright e^\lambda z^\alpha$
Reflection in the (u, z)-hyperplane	Inversion $z^\alpha \curvearrowright \frac{z^\alpha}{\langle z\vert z\rangle}$

By now the reader should feel comfortable about the general structure of the conformal group $C(p, q)$. It has the following simple characterization:

Lemma 10.6. *On a pseudo-Cartesian space $\mathbb{R}^p \times \mathbb{R}^q$ the conformal group $C(p, q)$ is the smallest group containing the isometry group and the inversion.*

PROOF. First we make the trivial observation that we can generate special conformal transformations using only translations and the inversion. This is due to the identity

$$\bar{\bar{C}}(a^\mu) = \bar{\bar{S}}(I)\bar{\bar{T}}(a^\mu)\bar{\bar{S}}(I)$$

(cf. Exercise 10.3.1). It remains to show that we can also generate dilatations. But that follows from the identity

$$\bar{\bar{D}}\left(\frac{1}{1 + \langle c \mid c\rangle}\right) = \bar{\bar{T}}\left(\frac{-c^\alpha}{1 + \langle c \mid c\rangle}\right)\bar{\bar{C}}(c^\alpha)\bar{\bar{T}}(c^\alpha)\bar{\bar{C}}\left(\frac{-c^\alpha}{1 + \langle c \mid c\rangle}\right).$$

(The verification is rather tedious but straightforward. Perhaps the quickest way to verify it is to show that

$$\left[T\left(\frac{-c^\alpha}{1 + \langle c \mid c\rangle}\right)C(c^\alpha)T(c^\alpha)\right](x) =$$

$$\left[D\left(\frac{1}{1 + \langle c \mid c\rangle}\right)C\left(\frac{-c^\alpha}{1 + \langle c \mid c\rangle}\right)\right](x),$$

where we work directly with the conformal transformations on $\mathbb{R}^p \times \mathbb{R}^q$.) □

We conclude this section with a few remarks about whether or not $C(p, q)$ in fact contains all possible conformal transformations. The answer is affirmative if the dimension of $\mathbb{R}^p \times \mathbb{R}^q$ is greater than two, though we will not prove it here. The two-dimensional case is an exceptional case because any holomorphic map of the complex plane into itself is, in fact, conformal. This will be discussed in Section 10.6. Nevertheless, $C(2)$ still denotes the conformal group constructed above. Notice that $C(2)$ is essentially the Lorentz group $O(1, 3)$!

Exercise 10.4.2
Problem (a) Show that the conformal group $C(2)$ consists of Möbius transformations of the complex plane into itself:

$$w = \frac{\alpha z + \beta}{\gamma z + \delta} \quad \text{and} \quad w = \frac{\alpha \bar{z} + \beta}{\gamma \bar{z} + \delta}, \qquad (\alpha\delta - \beta\gamma = 1).$$

(b) In the following we represent the Möbius transformation $w(z) = (\alpha z + \beta)/(\gamma z + \delta)$ by the 2×2-complex matrix

$$\begin{bmatrix} \alpha & \beta \\ \gamma & \delta \end{bmatrix}.$$

Show that the matrix group $SL(2, \mathbb{C})$ (consisting of 2×2 complex matrices with determinant 1) constitutes a prepresentation of $C_0(2)$ (consisting of orientation-preserving Möbius transformations).

10.5 The Dual Map

It still remains to investigate that part of the exterior calculus that also depends on a metric (and orientation), i.e., the dual map and its related concepts, like the inner product between differential forms, the codifferential δ and the Laplace–Beltrami operator $\boldsymbol{\Delta}$.

The dual map plays a special role, because it is connected to the dimension of the manifold. Let M be a manifold of dimension m with a dual map $*$. Then $*$ converts a differential form of rank k into a differential form of rank $(m-k)$. Okay, suppose M and N are manifolds with dual maps $*_1$ and $*_2$, and let $f : M^m \to N^n$ be a smooth map. Then we would like to compare $*_1(f^*\mathbf{T})$ and $f^*(*_2\mathbf{T})$, but they can have the same rank only if M and N are of the same dimension! For this reason we will assume in what follows that M and N have the same dimension n.

To begin with, we let N be an n-dimensional orientable manifold with a metric $\boldsymbol{g}$. We will, of course, assume that M is an n-dimensional orientable manifold, but a priori we do not assign any metric to M. We will let $f : M \curvearrowright N$ be a smooth regular map. Then we can pull back the metric $\boldsymbol{g}$ to a metric $f^*\boldsymbol{g}$ on M (cf. Exercise 10.2.2). Equipped with this metric we have a Levi-Civita form ϵ_f and a dual map $*_f$ on M. We can then show the following theorem:

Theorem 10.12. *Let the regular map $f : M^n \curvearrowright N^n$ be orientation-preserving (-reversing). Then*

$$f^*\epsilon = \epsilon_f \qquad (f^*\epsilon = -\epsilon_f), \tag{10.34}$$

and f commutes (anticommutes) with the dual map; i.e.,

$$f^*[*\mathbf{T}] = *_f[f^*\mathbf{T}] \qquad (f^*[*\mathbf{T}] = -*_f[f^*\mathbf{T}]). \tag{10.35}$$

PROOF. It is sufficient to check the case where f preserves the orientation. Let P be a point in M. As f is regular, we know that it maps an open neighborhood U of P diffeomorphically onto an open neighborhood V of Q in N. (See Figure 10.25.) We may safely assume that U is in the range of some positively oriented coordinate system (ψ, O). But then we can use $[f \circ \psi, O]$ as a positively oriented coordinate system covering V. If we express f in these coordinates, then f reduces

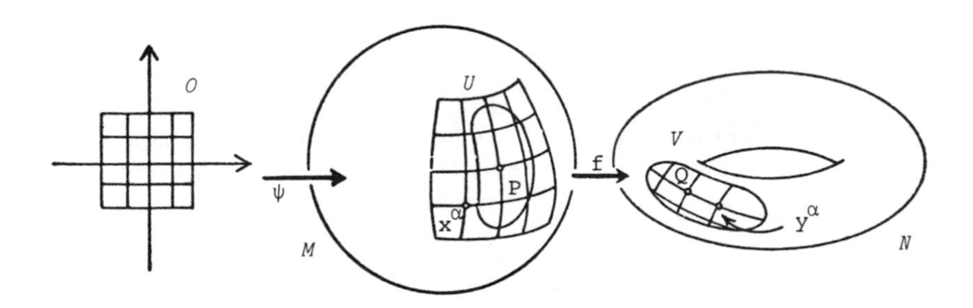

Figure 10.25.

to the identity map: $y^i = x^i$. But that is not all! When we pull back the metric, then the metric components are unchanged:

$$g_{k\ell}(x) = g_{ij}(y)\frac{\partial y^i}{\partial x^k}\frac{\partial y^j}{\partial x^\ell} = g_{ij}(y)\delta^i_k\delta^i_\ell = g_{k\ell}(y(x)),$$

as are the components of any differential form $\mathbf{T}$ we might like to pull back. Thus locally, the two manifolds are indistinguishable if we identify them using the map f. □

We now address ourselves to the more realistic situation where we have been given two orientable manifolds with metrics

$$(M^n; \mathbf{g}_1) \quad \text{and} \quad (N^n; \mathbf{g}_2).$$

If $f : M \curvearrowright N$ is a smooth regular map, then we want to compare the dual maps $*_1$ and $*_2$ on M and N. But observe that $\mathbf{g}_2$ is pulled back to a metric $f^*\mathbf{g}_2$ on M, and the relationship between the dual maps generated by $f^*\mathbf{g}_2$ and $\mathbf{g}_2$ is completely controlled by Theorem 10.12. We are thus really left with the problem of comparing the two metrics $\mathbf{g}_1$ and $f^*\mathbf{g}_2$. This reduces the problem to the following:

We consider just a single manifold M and see what happens with the dual map (the codifferential, the inner product, etc.) when we exchange one metric with another. Of course, everything turns out to be simplest when we do not exchange the metric at all! When we work with isometries we can therefore immediately take over the results obtained in Theorem 10.12. Combining this with Theorems 10.7 and 10.8, we then get

Theorem 10.13. *Let $f : M \curvearrowright N$ be an orientation-preserving local isometry. Then*

a. *f commutes with the dual map; i.e.,*

$$*_1[f^*\mathbf{T}] = f^*[*_2\mathbf{T}]. \tag{10.36}$$

b. *f commutes with the codifferential and the Laplace–Beltrami operator; i.e.,*

$$\delta f^*\mathbf{T} = f^*\delta\mathbf{T}; \qquad \mathbf{\Delta} f^*\mathbf{T} = f^*\mathbf{\Delta}\mathbf{T}. \tag{10.37}$$

c. *If f is a global isometry, it preserves inner products:*

$$\langle f^*\mathbf{T} \mid f^*\mathbf{S}\rangle = \langle \mathbf{T} \mid \mathbf{S}\rangle.$$

Almost equally simple is the case when f is a conformal map. It will be preferable to consider at first a single manifold M and see what happens if we perform a rescaling of the metric.

Theorem 10.14. *Let $\mathbf{g}$ and $\mathbf{g}' = \Omega^2(x)\mathbf{g}$ be conformally related metrics on the manifold M^n. Then*

a. *ϵ and ϵ' are conformally related:*

$$\epsilon' = \Omega^n(x)\epsilon. \tag{10.38}$$

b. *Let* **T** *be a differential form of rank k. Then* $^{*\prime}\mathbf{T}$ *and* $^{*}\mathbf{T}$ *are conformally related:*

$$^{*\prime}\mathbf{T} = \Omega^{n-2k}(x)^{*}\mathbf{T}. \tag{10.39}$$

PROOF.

a. $\overline{\overline{G}} = (g_{ij})$ is an $n \times n$ matrix. Thus we have

$$g'(x) = \det \overline{\overline{G}}' = \det[\Omega^2(x)\overline{\overline{G}}] = \Omega^{2n}(x) \det \overline{\overline{G}} = \Omega^{2n} g(x),$$

from which we immediately get

$$\sqrt{g'(x)}\, \epsilon_{a_1 \ldots a_n} = \Omega^n(x)\sqrt{g(x)}\, \epsilon_{a_1 \ldots a_n}.$$

b. When we compute the components of $^{*}\mathbf{T}$, we must raise the index. This involves the contravariant components of the metric, g^{ij}. But they are the components of the reciprocal matrix $\overline{\overline{G}}^{-1}$ Therefore, they transform with the reciprocal factor

$$g'^{ij} = \Omega^{-2}(x) g^{ij}.$$

Using this, we immediately get

$$\frac{1}{k}\sqrt{g'(x)}\, \epsilon_{a_1 \ldots a_{n-k} b_1 \ldots b_k} g'^{b_1 c_1} \ldots g'^{b_k c_k} T_{c_1 \ldots c_k}$$

$$= \Omega^{n-2k}(x) \frac{1}{k!} \sqrt{g(x)}\, \epsilon_{a_1 \ldots a_{n-k} b_1 \ldots b_k} g^{b_1 c_1} \ldots g^{b_k c_k} T_{c_1 \ldots c_k}.$$

□

Combining the results of Theorems 10.12 and 10.14, we then finally obtain:

Theorem 10.15. *Suppose* $(M^n, \mathbf{g}_1)$ *and* $(N^n, \mathbf{g}_2)$ *are orientable manifolds. Let* f *be an orientation-preserving (-reversing) conformal map, so that* $f^*\mathbf{g}_2 = \Omega^2(x)\mathbf{g}_1$. *Then*

a.

$$f^*[\epsilon_2] = \Omega^n(x)\epsilon_1 \qquad (f^*\epsilon_2 = -\Omega^n(x)\epsilon_1) \tag{10.40}$$

b.

$$f^*[*_2\mathbf{T}] = \Omega^{n-2k}(x) *_1 [f^*\mathbf{T}] \qquad (f^*[*_2\mathbf{T}] = -\Omega^{n-2k}(x) *_1 [f^*\mathbf{T}]) \tag{10.41}$$

Observe that if n is *even* and **T** has rank $k = \frac{n}{2}$ then f^* actually commutes (anti-commutes) with the dual map

$$f^*[*_2\mathbf{T}] = *_1[f^*\mathbf{T}] \qquad (f^*[*_2\mathbf{T}] = - *_1 [f^*\mathbf{T}]) \tag{10.42}$$

10.6 The Self-Duality Equation

We have previously introduced the concept of self-dual and anti-self-dual forms (see Exercises 7.5.3 and 7.6.9). They can only be constructed in spaces of even dimension, and they satisfy the equation

$$^{*}\mathbf{T} = \lambda \mathbf{T}; \qquad \text{rank } \mathbf{T} = \frac{1}{2} \dim M, \tag{10.43}$$

but due to the identity (7.43), λ is constrained to very special values:

$$\pm\mathbf{T} = {}^{**}\mathbf{T} = {}^{*}(\lambda\mathbf{T}) = \lambda^2\mathbf{T}; \qquad \text{i.e., } \lambda = \begin{cases} \pm 1 & \text{if } {}^{**}\mathbf{T} = \mathbf{T}, \\ \pm i & \text{if } {}^{**}\mathbf{T} = -\mathbf{T}. \end{cases}$$

Observe that in the latter case the self-duality equation works only for *complex-valued* differential forms. For future reference we collect the various possibilities in the following scheme

		The self-duality equations		
Dimension	T	Euclidean metric	Minkowski Metric	
2	1-form	$^{*}T = \pm iT$	$^{*}T = \pm T$	(10.44)
4	2-form	$^{*}T = \pm T$	$^{*}T = \pm iT$	

Exercise 10.6.1
Problem: (a) Let $n = \dim M$ be even and let rank $\mathbf{T} = \frac{n}{2}$. Show that $\mathbf{T}$ can be decomposed uniquely into a self-dual and an anti-self-dual part:

$$\mathbf{T} = \mathbf{T}^{+} + \mathbf{T}^{-}.$$

(b) Show that the above decomposition is an orthogonal decomposition; i.e., $(\mathbf{T}^{+} \mid \mathbf{T}^{-}) = 0$ in the case of a Riemannian metric.

Next, we observe that the self-duality equation has the following important invariance property:

Theorem 10.16. *The self-duality equation is conformally invariant.*

PROOF. This is an immediate consequence of (10.42). If $\mathbf{T}$ is self-dual or anti-self-dual, then $\dim M = n$ is even, and rank $\mathbf{T} = \frac{n}{2}$, so that a conformal map commutes (anti-commutes) with the dual map. □

Observe that an orientation-reversing map actually maps a solution to the self-duality equation $^{*}\mathbf{T} = \lambda\mathbf{T}$ into a solution of $^{*}\mathbf{T} = -\lambda\mathbf{T}$.

The self-duality equation is by far the most important example of a conformally invariant equation, and it comprises many well-known partial differential equations in disguise. To get familiar with it, we will study in some detail the historically most famous self-duality equation:

ILLUSTRATIVE EXAMPLE: THE CAUCHY-RIEMANN EQUATIONS.
Here we consider maps from $\mathbb{R}^2$ into itself:

$$w : \mathbb{R}^2 \curvearrowright \mathbb{R}^2; \qquad \text{i.e., } \left\{ \begin{array}{l} w_1 = w_2(x, y) \\ w_2 = w_1(x, y) \end{array} \right\}.$$

Such a map generates a complex-valued 1-form

$$\boldsymbol{d}w = \boldsymbol{d}w_1 + i\boldsymbol{d}w_2,$$

and we will demand that this be anti–self dual,

$$^*\boldsymbol{d}w = -i\boldsymbol{d}w.$$

If we write out the components, we will immediately find that this is equivalent to the following partial differential equations:

$$\frac{\partial w_1}{\partial x} = \frac{\partial w_2}{\partial y}\,; \qquad \frac{\partial w_1}{\partial y} = -\frac{\partial w_2}{\partial x}\,, \tag{10.45}$$

which are nothing but the famous Cauchy-Riemann equations!

Here it is preferable to introduce complex numbers, i.e., to identify $\mathbb{R}^2$ with $\mathbb{C}$ in the canonical fashion and consider complex functions

$$w : \mathbb{C} \curvearrowright \mathbb{C}, \qquad \text{i.e., } w = w(z).$$

Such a complex function generates the complex-valued 1-form

$$\boldsymbol{d}w = \frac{\partial w}{\partial z}\,\boldsymbol{d}z + \frac{\partial w}{\partial \bar{z}}\,\boldsymbol{d}\bar{z}.$$

(For the notation consult Exercises 7.6.9 and 7.4.3.) But here $\boldsymbol{d}z$ is anti–self dual and $\boldsymbol{d}\bar{z}$ is self dual, so this is precisely the decomposition of $\boldsymbol{d}w$ into a self-dual and an anti–self-dual part. (Compare with Exercise 10.6.1.) But then $\boldsymbol{d}w$ is anti–self-dual if and only if $\frac{\partial w}{\partial \bar{z}} = 0$, which is the complex form of the Cauchy–Riemann equations.

To summarize, we have thus shown (compare with Exercise 7.6.9):

Lemma 10.7. *A complex function $w = w(z)$ is holomorphic (antiholomorphic) if and only if it generates an anti–self-dual (self-dual) 1-form $\boldsymbol{d}w$.*

Let us now characterize the holomorphic functions geometrically. Here we have the following well-known characterization:

Theorem 10.17. *A nontrivial complex function $w = w(z)$ is holomorphic (antiholomorphic) if and only if it is an orientation-preserving (orientation-reversing) conformal map.*

PROOF. We know that the Cauchy–Riemann equations are conformally invariant; i.e., if w is any holomorphic function and $f : \mathbb{C} \curvearrowright \mathbb{C}$ is an orientation-preserving conformal map, then $w(f(z))$ is a holomorphic function too. Applying this to the identity function $w(z) = z$, which is trivially holomorphic, we see that f is holomorphic too.

To prove the remaining part of Theorem 10.17, we consider the standard metric on $\mathbb{C}$ given by

$$\boldsymbol{g} = \boldsymbol{d}x \otimes \boldsymbol{d}x + \boldsymbol{d}y \otimes \boldsymbol{d}y = \frac{1}{2}\,(\boldsymbol{d}z \otimes \boldsymbol{d}\bar{z} + \boldsymbol{d}\bar{z} \otimes \boldsymbol{d}z). \tag{10.46}$$

If $f : \mathbb{C} \curvearrowright \mathbb{C}$ is any smooth map represented by the complex function $w = f(z)$, we therefore obtain the following pulled-back metric:

$$f^*\mathbf{g} = \frac{1}{2}\,(\boldsymbol{d}w \otimes \boldsymbol{d}\bar{w} + \boldsymbol{d}\bar{w} \otimes \boldsymbol{d}w)$$
$$= \left(\left|\frac{\partial w}{\partial z}\right|^2 + \left|\frac{\partial w}{\partial \bar{z}}\right|^2\right)\mathbf{g} + \left(\frac{\partial w}{\partial z}\,\overline{\frac{\partial w}{\partial \bar{z}}}\right)\boldsymbol{d}z \times \boldsymbol{d}z$$
$$+ \left(\overline{\frac{\partial w}{\partial z}}\,\frac{\partial w}{\partial \bar{z}}\right)\boldsymbol{d}\bar{z} \times \boldsymbol{d}\bar{z}.$$

When f is holomorphic (antiholomorphic), this reduces to

$$f^*\mathbf{g} = \left|\frac{\partial w}{\partial z}\right|^2 \mathbf{g} \qquad \left(f^*\mathbf{g} = \left|\frac{\partial w}{\partial \bar{z}}\right|^2 \mathbf{g}\right),$$

which shows that in both cases f is a conformal map! (Strictly speaking, we should cut out the points where w is stationary, i.e., $\frac{\partial w}{\partial z} = 0$. They corresponds to the points where the map is not regular, and for a nontrivial holomorphic function they are isolated.) □

One of the reasons for the importance of the Cauchy–Riemann equations to physics lies in the fact that any solution to the Cauchy–Riemann equations is automatically a harmonic function, i.e., a solution to the Laplace equation in 2 dimensions:

$$\boldsymbol{d}^*\boldsymbol{d}\phi = 0; \tag{10.47}$$

i.e.,

$$\left(\frac{\partial^2}{\partial x^2} + \frac{\partial^2}{\partial y^2}\right)\phi = 0.$$

To see this well-known result in our new formalism, we let $w = w(z)$ be a holomorphic function. Then $\boldsymbol{d}w$ is anti–self dual, and we get

$$\boldsymbol{d}^*\boldsymbol{d}w = \boldsymbol{d}[-i\boldsymbol{d}w] = i\boldsymbol{d}^2 w = 0.$$

Thus the real and imaginary parts w_1 and w_2 are harmonic functions. (The same argument holds for antiholomorphic functions.) This is a common feature in many applications of the self-duality equations. They are brought into the game to simplify the search for a solution to a second-order equation. In this case it is the Laplace equation

$$\boldsymbol{d}^*\boldsymbol{d}\phi = 0, \tag{10.47}$$

which is trivially solved by solutions to the self-duality equations,

$$^*\boldsymbol{d}\phi = \pm i\boldsymbol{d}\phi. \tag{10.48}$$

But the self-duality equations are only first-order equations and are thus easier to handle!

Interestingly enough, it turns out in many applications that the "relevant" solutions to the second-order equations automatically solve the first-order equations! ("Relevant" means solutions that satisfy appropriate boundary conditions, integrability conditions, or other conditions imposed by the problem at hand.)

Remark. In the example concerning the Laplace equation this is precisely known to be the case; i.e., not only is the real part of a holomorphic function automatically a harmonic function, but all harmonic functions are generated this way. Let us also take a look at this well-known result in our new formalism:

Suppose ϕ is a real harmonic function and take a look at the *complex*-valued 1-form

$$2\,\frac{\partial\phi}{\partial z}\,\boldsymbol{d}z.$$

It is necessarily anti–self dual, and because ϕ is harmonic, it is a closed form

$$\boldsymbol{d}\left(2\,\frac{\partial\phi}{\partial z}\,\boldsymbol{d}z\right) = 2\,\frac{\partial^2\phi}{\partial\bar{z}\partial z}\,\boldsymbol{d}\bar{z}\wedge\boldsymbol{d}z = 0.$$

But then it is locally generated by a complex-valued function f,

$$\boldsymbol{d}f = 2\,\frac{\partial\phi}{\partial z}\,\boldsymbol{d}z, \tag{$*$}$$

and by Lemma 10.7 this function is holomorphic. It remains to be shown that ϕ is the real part of f. Now, observe that since ϕ is real, we have

$$\boldsymbol{d}\bar{f} = 2\,\frac{\partial\phi}{\partial\bar{z}}\,\boldsymbol{d}\bar{z}. \tag{$**$}$$

Combining $(*)$ and $(**)$, we now get

$$\frac{1}{2}\,\boldsymbol{d}(f+\bar{f}) = \frac{\partial\phi}{\partial z}\,\boldsymbol{d}z + \frac{\partial\phi}{\partial\bar{z}}\,\boldsymbol{d}\bar{z} = \boldsymbol{d}\phi;$$

i.e., $\phi = \frac{1}{2}\,(f+\bar{f})$ (up to an irrelevant constant).

Exercise 10.6.2
(The wave equation.) Problem: (a) Show, using Theorem 10.15 directly, that the Klein–Gordon equation for a massless scale field ϕ in $(1+1)$-dimensional space–time,

$$\boldsymbol{d}^*\boldsymbol{d}\phi = 0; \qquad \left[\left(\frac{\partial^2}{\partial t^2} - \frac{\partial^2}{\partial x^2}\right)\phi = 0\right], \tag{10.49}$$

is conformally invariant; i.e., the *wave equation* is conformally invariant.

(b) Suppose ϕ is a massless scalar field generating a self-dual (anti–self-dual) field strength

$$^*\boldsymbol{d}\phi = \boldsymbol{d}\phi; \qquad [^*\boldsymbol{d}\phi = -\boldsymbol{d}\phi].$$

Show then that it automatically solves the wave equation.

(c) Show that the self-duality equations reduce to

$$\frac{\partial \phi}{\partial t} = -\frac{\partial \phi}{\partial x}; \qquad \left[\frac{\partial \phi}{\partial t} = \frac{\partial \phi}{\partial x}\right],$$

and that the complete solution is of the form

$$\phi(x,t) = f(x-t); \qquad [\phi(x,t) = f(x+t)],$$

where f is an arbitrary smooth function.

(d) Let ϕ be an arbitrary solution to the wave equation. Show that it is of the form

$$\phi = \phi_1 + \phi_2,$$

where $\boldsymbol{d}\phi_1$ is self dual and $\boldsymbol{d}\phi_2$ is anti–self dual (This shows that the general solution to the wave equation is of the form

$$\phi(x,t) = f(x-t) + g(x+t).) \tag{10.50}$$

After this long digression on the Cauchy–Riemann equations, we conclude with the most important example of a conformally invariant field equation in classical physics:

Theorem 10.18. *Maxwell's equations for the electromagnetic field are conformally invariant in* $(3+1)$*-dimensional space–time.*

PROOF. This can be proved directly using Theorem 10.15 (compare with Exercises 10.6.2 and 10.6.3 below). It is, however, instructive to reduce the Maxwell equations directly to the self-duality equation, although it is a little bit tricky:

First, we complexify the field strength, cf. Exercise 9.1.3,

$$\boldsymbol{F} = \mathbf{F} + i{}^*\mathbf{F}. \tag{10.51}$$

In this way we produce an anti-self-dual 2-form

$${}^*\boldsymbol{F} = {}^*\mathbf{F} - i\mathbf{F} = -i\boldsymbol{F}.$$

On the other hand, if $\boldsymbol{F}$ is an anti-self-dual 2-form, then it is necessarily of the form (10.51).

Maxwell's equations can now be cast into the following complex form:

$${}^*\boldsymbol{F} = -i\boldsymbol{F}, \qquad \boldsymbol{dF} = 0. \tag{10.52}$$

Here the second equation is invariant under arbitrary smooth transformations, while the first one is invariant under orientation-preserving conformal transformations. (Under an orientation-reversing map, $\boldsymbol{F}$ is mapped into a self-dual form $\boldsymbol{F}'$; i.e.,

$${}^*\boldsymbol{F}' = i\boldsymbol{F}.$$

But then $\overline{\boldsymbol{F}'}$ is anti–self dual and has the same real part, so the real part of $\boldsymbol{F}'$ will still solve Maxwell's equations.) □

Exercise 10.6.3

Problem: Show, using Theorem 10.15 directly, that the Maxwell equation for the massless gauge potential $\mathbf{A}$ in $(3+1)$-dimensional space–time,

$$\boldsymbol{d}^*\boldsymbol{d}\mathbf{A} = 0; \qquad [(\partial_\mu \partial^\mu)A_\nu - \partial_\nu(\partial^\mu A_\mu) = 0],$$

is conformally invariant.

ILLUSTRATIVE EXAMPLE: CONFORMAL MAPS ON THE SPHERE.

In the previous example we have discussed conformal maps of the plane into itself and we have seen that they were given by the nontrivial holomorphic functions $f : \mathbb{C} \curvearrowright \mathbb{C}$. This time we will consider conformal maps from the sphere into itself, $f : S^2 \curvearrowright S^2$. It is worth observing that this example is closely related to the preceding example, since the stereographic projection from the sphere into the plane is itself an (orientation-reversing) conformal map. (See Exercise 10.3.2.)

You might think, then, that the results obtained in the previous example can be carried over trivially; i.e., the conformal maps from the plane into itself are transferred to the conformal maps from the sphere into itself using the conformal stereographic projection. But as we shall see, this is not quite true, so let us take a closer look.

Suppose we have been given a conformal map, $f : \mathbb{C} \curvearrowright \mathbb{C}$. We would like to lift it to a map $\tilde{f} : S^2 \curvearrowright S^2$. But here we can get into trouble, since the lifted map is a priori not well-defined at the north pole (which corresponds to infinity in $\mathbb{C}$). (See Figure 10.26.) Consequently, f can be lifted only if it has a well-defined limit at infinity, since only in this case can we extend $\tilde{f}$ by continuity at the north pole.

Consider then a map from the sphere into itself, $g : S^2 \curvearrowright S^2$. We want to project it down to a map $\tilde{g} : \mathbb{C} \curvearrowright \mathbb{C}$. But here we should be careful about the points that are mapped into the north pole. They give rise to singularities in the projected map $\tilde{g}$. In the case where g is conformal, i.e., $\tilde{g}$ holomorphic (antiholomorphic), this means that $\tilde{g}$ will get poles.

Combining these observations, you might now guess the answer:

Theorem 10.19. *An orientation-preserving (-reversing) map $g : S^2 \curvearrowright S^2$ is conformal if and only if it is projected down to an algebraic function*

$$w(z) = \frac{P(z)}{Q(z)}, \qquad \left(w(z) = \frac{P(\bar{z})}{Q(\bar{z})}\right), \tag{10.53}$$

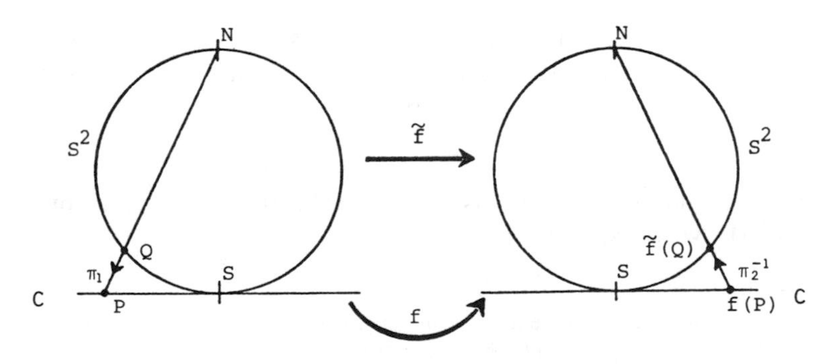

Figure 10.26.

where P, Q are arbitrary polynomials.

We now show this in detail:

The basic idea is to treat the sphere S^2 as a *complex manifold*. We already know what a real manifold is (cf. the discussion in Sections 6.1–6.3), and the corresponding definition of a complex manifold is a straightforward generalization:

We identify $\mathbb{R}^{2n}$ with $\mathbb{C}^n$. Let U be an open subset of $\mathbb{C}^n$ and $f : U \curvearrowright \mathbb{C}$ a smooth function of n complex variables: $f(z^1, \ldots, z^n)$. We say that f is *holomorphic* (i.e., complex-analytic) if either of the following two equivalent conditions are fulfilled:

1. $\partial f / \partial \bar{z}^i = 0, i = 1, \ldots, n$.
2. The function f can be expanded in a convergent power series around each point $(z_0^1, \ldots, z_0^n)$ in U:

$$f(z^1, \ldots, z^n) = \sum_{p_1, \ldots, p_n}^{\infty} a_{p_1, \ldots, p_n} (z^1 - z_0^1)^{p_1} \ldots (z^n - z_0^n)^{p_n}.$$

Consider, furthermore, a mapping ϕ from an open subset U of $\mathbb{C}^n$ into $\mathbb{C}^k$:

$$\phi(z^1, \ldots, z^k) = (\phi^1(z^1, \ldots, z^n); \ldots; \phi^k(z^1, \ldots, z^n)).$$

We say that the mapping ϕ is holomorphic provided that each of its components $\phi^1, \ldots, \phi^k$ is a holomorphic function.

Suppose now that M is a real differentiable manifold of dimension $2n$. A *complex coordinate system* on M is a smooth regular homeomorphism ϕ of an open subset U in $\mathbb{C}^n$ onto an open subset $\phi(U)$ in M. Furthermore, two complex coordinate systems are compatible provided that the transition functions are *holomorphic maps*. With these preparations we finally arrive at the main definition:

Definition 10.10. A *complex manifold* is a real differentiable manifold M equipped with an atlas of compatible complex coordinate systems.

Notice that in the complex case the concept of a smooth function has been replaced by the concept of a holomorphic function.

We can now show that the sphere S^2 is a complex manifold. As complex coordinate systems we use stereographic projections from $\mathbb{C}$ into S^2. It suffices to consider two such coordinate systems: one corresponding to a stereographic projection from the north pole and one corresponding to a stereographic projection from the south pole (cf. Figure 10.27).

The transition function is then given by

$$\phi_{21}(z) = \frac{1}{z},$$

which clearly is a holomorphic function! Equipped with these coordinate systems the sphere S^2 is thus a complex manifold known as the *Riemann sphere*. The

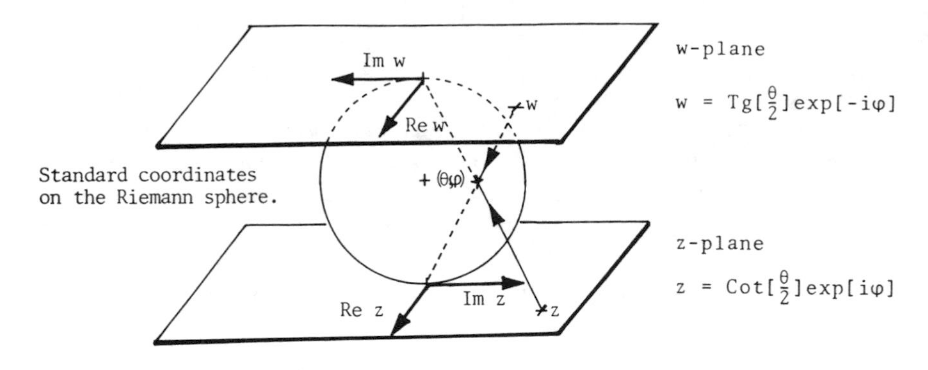

Figure 10.27.

complex coordinates generated by stereographic projections from the north pole and the south pole are referred to as *standard coordinates*.

As in the real case, we say that a continuous mapping, $f : M \curvearrowright N$ between two complex manifolds is *holomorphic* provided that it is represented by a holomorphic map when we introduce complex coordinates. We can now show, using the same arguments as in the complex plane, that *a smooth mapping from the Riemann sphere into itself is conformal if and only if it is holomorphic.* All we need is therefore a closer examination of the holomorphic maps from the Riemann sphere into itself.

Remarkably enough, we can now classify completely the holomorphic maps from the Riemann sphere into itself. This requires some preliminary knowledge about holomorphic functions, which can be found in any elementary textbook on complex analysis:

Suppose f is a holomorphic function with an isolated singularity at the point z_0 (i.e., f is discontinuous at z_0). Then f can be expanded in a Laurent series around z_0:

$$f(z) = \sum_{n=-\infty}^{\infty} a_n(z - z_0)^n.$$

Since f is discontinuous at z_0, at least one of the coefficients a_n corresponding to a negative integer n is nonvanishing. We distinguish between two cases: Either there exists an integer k such that $a_{-k} \neq 0$ which $a_n = 0$ when $n < -k$, or there exist an infinite number of negative integers n such that a_n is nonvanishing. In the first case we say that *f has a pole of degree k at z_0*. In the second case we say that *f has an isolated singular at* z_0.

Notice that if f has a pole of degree k at z_0, then we can rewrite it in the form

$$f(z) = \frac{g(z)}{(z - z_0)^n},$$

where $g(z)$ is holomorphic given at the point z_0 (i.e., the singularity can be removed by multiplying f by a polynomial).

If, on the other hand, f has an isolated singularity at z_0, then the famous *theorem of Weierstrass* states that f *maps each punctuated neighborhood of* z_0 *into a dense subset of* $\mathbb{C}$.

We will also need *Liouville's theorem*, which states that *a bounded holomorphic function defined on the whole complex plane is necessarily constant.*

The basic properties of holomorphic maps from the Riemann sphere into itself can now be summarized in the following way:

Lemma 10.8. *A nontrivial holomorphic map* $f : S^2 \curvearrowright S^2$ *is characterized by the following properties:*

a. *The preimage of a point is necessarily finite.*
b. *Using standard coordinates, it is represented by a holomorphic function* $w = w(z)$ *that has only a finite number of singularities, all of which are poles.*

PROOF.

a. In terms of local complex coordinates, the preimage of the point w_0 corresponds to the zero set of the holomorphic function

$$w(z) - w_0.$$

Consequently, the preimage consists of isolated points. But in a compact set, like S^2, a subset consisting of isolated points can only be finite.

b. Suppose f is represented by the holomorphic function $w = w(z)$, where we have introduced standard coordinates by means of a stereographic projection from the north pole. A singularity of $w(z)$ corresponds to a point z_0 where $w(z_0) = \infty$, i.e., to a point in the preimage of the north pole. By (a), there can be only a finite number of points in this preimage. Since f is continuous, it maps a small neighborhood of z_0 into a small neighborhood of the north pole. Due to Weierstrass's theorem, it consequently cannot correspond to an essential singularity of $w(z)$.

□

Using this lemma we can then easily prove the following powerful theorem:

Theorem 10.20. *Using standard coordinates, a given holomorphic map from the Riemann sphere into itself,* $f : S^2 \curvearrowright S^2$, *is represented by an algebraic function; i.e., it is of the form*

$$f(z) = \frac{P(z)}{Q(z)},$$

where P, Q *are polynomials.*

PROOF. We already know that f is represented by a holomorphic function $w(z)$ with a finite number of poles. If we denote the poles by z_i and the corresponding degrees by k_i, we can remove the singularities by multiplying $w(z)$ by the polynomial

$$Q(z) = \prod_i (z - z_i)^{k_i}.$$

Consider now the function

$$P(z) = Q(z)w(z).$$

By construction it is a holomorphic function without poles. Thus it can be expanded in a Taylor series

$$P(z) = \sum_{n=0}^{\infty} a_n z^n$$

that is everywhere convergent. If P is nontrivial, the corresponding function

$$P\left(\frac{1}{z}\right) = Q\left(\frac{1}{z}\right) w\left(\frac{1}{z}\right) = \sum a_n z^{-n}$$

will have a singularity at $z = 0$. According to Lemma 10.8, the function $w(1/z)$ cannot have an essential singularity at $z = 0$. It follows that $P(1/z)$ has a pole of some finite degree k; i.e., $a_n = 0$ when $n > k$. Consequently, $P(z)$ itself must be a polynomial,

$$P(z) = \sum_0^k a_n z^n.$$

□

10.7 Winding Numbers

As an application of the preceding machinery, we now specialize to the case where M and N are compact orientable manifolds of the same dimension n. Let $f : M^n \curvearrowright N^n$ be a smooth map. Then we know that almost all points in N are regular values (Sard's theorem). We can now show

Lemma 10.9. *Let Q be a regular value in N. Then either $f^{-1}(Q)$ is empty or it consists of a finite number of points.*

PROOF. The proof is somewhat technical and may be skipped without loss of continuity. If $f^{-1}(Q)$ is nonempty, we consider a point P in it. Then f_* maps $\mathbf{T}_P(M)$ isomorphically onto $\mathbf{T}_Q(N)$, and there exist open neighborhoods U, V around P and Q, such that f maps U diffeomorphically onto V (compare the discussion in Section 10.1). In particular, P is an *isolated* point in the preimage $f^{-1}(Q)$.

On the other hand, $f^{-1}(Q)$ is a closed subset of M. As M is compact, $f^{-1}(Q)$ must itself be compact. If $f^{-1}(Q)$ were infinite, it would then contain an accumulation point, which by definition is not isolated. (Each neighborhood of the accumulation point contains other points from the preimage.) Therefore, $f^{-1}(Q)$ can be at most finite. □

Let us now introduce coordinates that generate positive orientations on M and N. Let $Q : (y^1, \dots, y^n)$ and $P : (x^1, \dots, x^n)$ be chosen such that P lies in the preimage of the regular value Q; i.e., $Q = f(P)$. Since f_* maps $\mathbf{T}_P(M)$ isomorphically onto $\mathbf{T}_Q(M)$, we know that the Jacobian, $\det[\partial y^i/\partial x^j]_{|P}$, is not zero! Furthermore, f maps a neighborhood U of P diffeomorphically onto a neighborhood V of Q. If the Jacobian is positive, then $f : U \to V$ preserves the orientation, and if it is negative, then $f : U \to V$ reverses the orientation. Since the preimage $f^{-1}(Q)$ can contain at most a finite number of points, we can now define:

Definition 10.11. The *degree* of a smooth map $f : M^n \curvearrowright N^n$ at the regular value Q is the integer

$$\mathrm{Deg}(f; Q) = \sum_{P_i \in f^{-1}(Q)} \mathrm{Sgn}|\partial y^i/\partial x^j|_{|P_i} \tag{10.54}$$

Here $\mathrm{Sgn}[\partial y^i/\partial x^j] = \pm 1$ according to whether f_* preserves or reverses the orientation. Consequently, $\mathrm{Deg}(f; Q)$ simply counts the effective number of times Q is covered by the map f. If $f^{-1}(Q)$ is empty, it is understood that $\mathrm{Deg}(f; Q) = 0$.

If we let p denote the number of points in $f^{-1}(Q)$ with positive Jacobian and q the number of points in $f^{-1}(Q)$ with negative Jacobian, then p and q need not be constants as Q ranges through the compact manifold N. This is easily seen already from a 1-dimensional example.

EXAMPLE 10.1. Consider the map $f : S^1 \curvearrowright S^1$, where we have represented f by a periodic function (Figure 10.28):

But the following deep theorem holds, which we shall not attempt to prove:

Lemma 10.10 (Brouwer's lemma). *The degree of a smooth map $f : M \curvearrowright N$ is the same for all regular values in N.*

This common value is then simply denoted by $\mathrm{Deg}(f)$, and it measures the effective number of times N is covered by M by the map f. The integer $\mathrm{Deg}(f)$ is therefore referred to as the *winding number*, since it counts how many times M is wound around N. In the literature it is also referred to as the *Brouwer degree*.

Lemma 10.11. *If $f : M \curvearrowright N$ fails to be surjective, then the winding number is* 0.

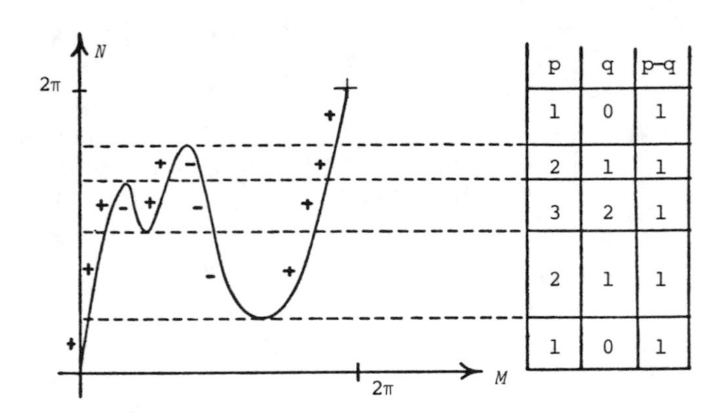

Figure 10.28.

PROOF. If f fails to be surjective, there exists a point Q such that $f^{-1}(Q)$ is empty. Such a point Q is therefore trivially a regular point with $\operatorname{Deg}(f; Q) = 0$. □

At this point we take a look at some illuminating examples:

EXAMPLE 10.2. As an example in dimension $n = 1$ we look at $M = N = S^1$. If we introduce polar coordinates, we may consider the map f given by

$$f(\varphi) = \begin{bmatrix} \cos n\varphi \\ \sin n\varphi \end{bmatrix}; \qquad \text{i.e., } \varphi' = n\varphi.$$

This is obviously a smooth map with winding number n.

EXAMPLE 10.3. In $n = 2$ dimensions we look at $M = N = S^2$. If we introduce polar coordinates, we may consider the map f given by

$$f(\theta, \varphi) = \begin{bmatrix} f^1(\theta, \varphi) \\ f^2(\theta, \varphi) \\ f^3(\theta, \varphi) \end{bmatrix} = \begin{bmatrix} \sin\theta & \cos n\varphi \\ \sin\theta & \sin n\varphi \\ \cos\theta & \end{bmatrix}; \qquad \text{i.e., } (\theta', \varphi') = (\theta, n\varphi).$$

This is clearly a smooth map outside the poles, and it maps the first sphere n times around the second sphere. In fact, it maps each little circle $\theta = \theta_0$ in M the number n times on the corresponding little circle in N. (See Figure 10.29.) If f were a smooth map, it would therefore have winding number n. Unfortunately, f is not necessarily smooth! In general, it is singular at the poles. This cannot be seen from the above coordinate expression, since the polar coordinates themselves break down at the north pole and south pole. This is a subtle point, so let us investigate it in some detail.

We may introduce smooth coordinates at the north pole, say the standard coordinates x and y. They are related to the polar coordinates in the following

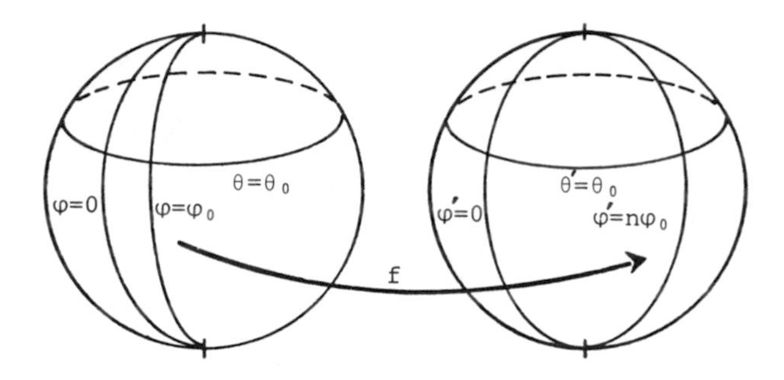

Figure 10.29.

way:

$$x = \sin\theta\cos\varphi \qquad \theta = \arcsin\sqrt{x^2+y^2}$$
$$y = \sin\theta\sin\varphi \qquad \varphi = \arctan\frac{y}{x}\,.$$

If f is smooth, then the partial derivatives

$$\mathbf{e}_x = \frac{\partial f}{\partial x} \quad \text{and} \quad \mathbf{e}_y = \frac{\partial f}{\partial y}$$

will depend smoothly upon (x, y). For simplicity, we compute the *extrinsic* coordinates $\mathbf{e}_x$ and $\mathbf{e}_y$ that are tangent vectors to the second sphere. Using that

$$\begin{bmatrix} \frac{\partial\theta}{\partial x} & \frac{\partial\theta}{\partial y} \\ \frac{\partial\varphi}{\partial x} & \frac{\partial\varphi}{\partial y} \end{bmatrix} = \begin{bmatrix} \frac{\partial x}{\partial\theta} & \frac{\partial x}{\partial\varphi} \\ \frac{\partial y}{\partial\theta} & \frac{\partial y}{\partial\varphi} \end{bmatrix}^{-1} = \begin{bmatrix} \frac{\cos\varphi}{\cos\theta} & \frac{\sin\varphi}{\cos\theta} \\ -\frac{\sin\varphi}{\sin\theta} & \frac{\cos\varphi}{\sin\theta} \end{bmatrix},$$

we easily find the following extrinsic components of $\mathbf{e}_x$ and $\mathbf{e}_y$:

$$\mathbf{e}_x = \begin{bmatrix} \cos\varphi\cos n\varphi + n\sin\varphi\sin n\varphi \\ \cos\varphi\sin n\varphi - n\sin\varphi\cos n\varphi \\ -\cos\varphi\tan\theta \end{bmatrix};$$

$$\mathbf{e}_y = \begin{bmatrix} \sin\varphi\cos n\varphi - n\cos\varphi\sin n\varphi \\ \sin\varphi\sin n\varphi + n\cos\varphi\cos n\varphi \\ -\sin\varphi\tan\theta \end{bmatrix}.$$

To be continuous at the north pole ($\theta = 0$) these vectors must be independent of φ when we put $\theta = 0$! But the two first components depend explicitly on φ unless they reduce to 0 or $\cos^2 n\varphi + \sin^2 n\varphi = 1$. This miracle only happens for $n = \pm 1$. Hence if $n \neq \pm 1$, then the above map is not smooth! For $n = 1$ it reduces to the identity map of a sphere onto itself. For $n = -1$ it reduces to the antipodal map of the sphere onto itself. They are obviously smooth.

Exercise 10.7.1
Introduction: Let $\chi : [0, \pi] \to [0, \pi]$ be a surjective increasing smooth function such that all the derivatives vanish when $\theta = 0$ or $\theta = \pi$. (You can, e.g., put

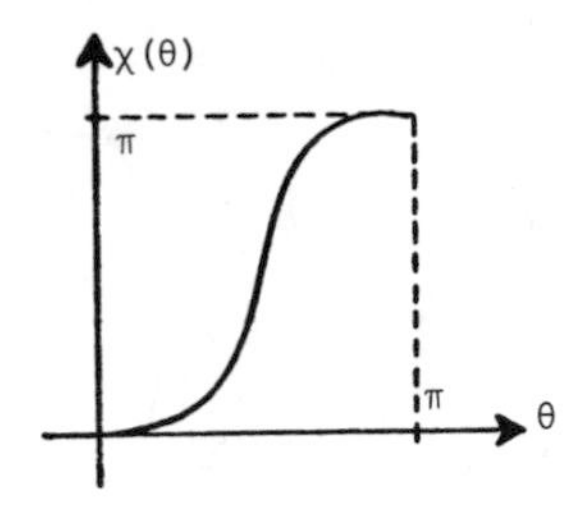

Figure 10.30.

(Figure 10.30)

$$\chi(\theta) = \frac{\pi}{A} \int_0^\theta e^{-\frac{1}{t(\pi-t)}}\, dt$$

with

$$A = \int_0^\pi e^{-\frac{1}{t(\pi-t)}}\, dt.)$$

Problem: (a) Show that the following map is smooth:

$$g(\theta, \varphi) = \begin{bmatrix} \sin \chi(\theta) \cdot \cos n\varphi \\ \sin \chi(\theta) \cdot \sin n\varphi \\ \cos \chi(\theta) \end{bmatrix}.$$

(b) Show that the only critical values are the north pole and the south pole.

The examples above can clearly be extended to produce smooth maps $S^n \curvearrowright S^n$ with arbitrary winding numbers.

Consider once more two arbitrary compact orientable manifolds M and N of the same dimension n, and let $f : M \curvearrowright N$ be a smooth map. If $\mathbf{T}$ is an n-form on N, then we can pull it back to an n-form $f^*\mathbf{T}$ on M. We want to compare the integrals

$$\int_M f^*\mathbf{T} \quad \text{and} \quad \int_N \mathbf{T}.$$

Since f need not be a diffeomorphism, they are not necessarily identical. However, f is characterized by an integer, the winding number $\text{Deg}(f)$, which tells how many times N is effectively covered by M. We can therefore generalize Theorem 10.8 of Section 10.2. The proof is omitted since it requires a greater machinery than we have developed yet. (See, e.g., Guillemin/Pollack (1974).)

Theorem 10.21 (Brouwer's theorem).

$$\int_M f^*\mathbf{T} = \text{Deg}(f) \int_N \mathbf{T}. \tag{10.55}$$

Remark. Let Q be a regular value in N with a nonempty preimage. Then the preimage is finite:

$$f^{-1}(Q) = \{P_1, \dots, P_k\}.$$

According to Lemma 10.1 of Section 10.1, we can find a single open neighborhood V of Q and disjoint open neighborhoods U_i of P_i such that f restricts to diffeomorphisms:

$$f : U_i \to V \qquad i = 1, \dots, k.$$

With these preparations, we consider a differential form $\mathbf{T}$ of rank n that vanishes outside V. Then $f^*\mathbf{T}$ vanishes outside $U_1 \cup \dots \cup U_n$. Therefore, an application of Theorem 10.8, of Section 10.2 gives

$$\int_M f^*\mathbf{T} = \sum_{i=1}^{n} \int_{U_i} f^*\mathbf{T} = \sum_{i=1}^{n} \operatorname{Sgn} \left| \frac{\partial y^i}{\partial x^j} \right|_{P_i} \int_V \mathbf{T}.$$

But $\int_V \mathbf{T}$ is independent of i. Consequently, we get

$$\int_M f^*\mathbf{T} = \operatorname{Deg}(f, Q) \cdot \int_N \mathbf{T} = \operatorname{Deg}(f) \cdot \int_N \mathbf{T},$$

where we have used Brouwer's lemma. This shows that the formula works for a differential form with a suitable small support.

Using Brouwer's theorem we can now construct an integral formula for the winding number. Let there be given a Riemannian metric on N. Then it induces a volume form, the Levi-Civita form:

$$\epsilon = \sqrt{g}\, \mathbf{dx}^1 \wedge \dots \wedge \mathbf{dx}^n.$$

Because N is compact, it can be shown to have a finite positive volume:

$$\operatorname{Vol}(N) = \int_N \epsilon. \tag{10.56}$$

If $f : M \curvearrowright N$ is a smooth map, we therefore get

$$\int_M f^*\epsilon = \operatorname{Deg}[f] \int_N \epsilon = \operatorname{Deg}[f] \cdot \operatorname{Vol}[N],$$

or

$$\operatorname{Deg}[f] = \frac{\int_N f^*\epsilon}{\operatorname{Vol}[N]}. \tag{10.57}$$

This is the formula we are after! To be able to apply it, we must construct the volume form first. Consider especially the sphere S^{n-1}.

Lemma 10.12. *Let $(x^1, \dots, x^n)$ be the extrinsic Cartesian coordinates on the unit sphere:*

$$S^{n-1} = \{(x^1, \dots, x^n) \mid (x^1)^2 + \dots + (x^n)^2 = 1\}.$$

Then the volume form with respect to the induced metric is given by

$$\boldsymbol{\Omega} = \frac{1}{(n-1)!}\,\epsilon_{a_1\ldots a_n} x^{a_1}\,\boldsymbol{d}x^{a_2} \wedge \cdots \wedge \boldsymbol{d}x^{a_n}. \tag{10.58}$$

PROOF. The induced metric on S^{n-1} is the metric inherited from the surrounding Euclidean space $\mathbb{R}^n$. Let $(\mathbf{v}_1, \ldots, \mathbf{v}_{n-1})$ be a set of tangent vectors at a point P on the sphere. What we must show then is that $\boldsymbol{\Omega}(\mathbf{v}_1, \ldots \mathbf{v}_{n-1})$ is the Euclidean volume of the parallelepiped spanned by $\mathbf{v}_1, \ldots, \mathbf{v}_{n-1}$ in $\mathbb{R}^n$ (compare Exercise 10.7.2 below). For simplicity, we prove this only in the case $n = 3$, i.e., for the two-sphere.

Let P be a point with coordinates (x^1, x^2, x^3). Then the radial vector $\mathbf{r} = \overrightarrow{OP}$ also has coordinates (x^1, x^2, x^3). Let $(\mathbf{u}, \mathbf{v})$ be two tangent vectors at P with coordinates (u^1, u^2, u^3) and (v^1, v^2, v^3). We then get

$$\boldsymbol{\Omega}(\mathbf{u}, \mathbf{v}) = \frac{1}{2}\epsilon_{abc} x^a \boldsymbol{d}x^b \wedge \boldsymbol{d}x^c(\mathbf{u}, \mathbf{v}) = \epsilon_{abc} x^a u^b v^c.$$

This is the three-dimensional volume of the parallelepiped spanned by $\mathbf{r}$, $\mathbf{u}$ and $\mathbf{v}$. (See Figure 10.31.) But $\mathbf{r}$ is a unit vector that is orthogonal to $\mathbf{u}$ and $\mathbf{v}$. Consequently, the three-dimensional volume of the parallelepiped is equal to the two-dimensional area of the parallelogram spanned by $\mathbf{u}$ and $\mathbf{v}$. We have thus shown that $\Omega(\mathbf{u}, \mathbf{v})$ is the area of parallelogram spanned by $\mathbf{u}$, $\mathbf{v}$, and we are through. □

Consider a compact orientable manifold M of dimension n, and let $\phi : M \curvearrowright S^n$ be a smooth map, which we parametrize as

$$\phi^a = \phi^a(x); \qquad \sum_{a=0}^{n} [\phi^a(x)]^2 = 1.$$

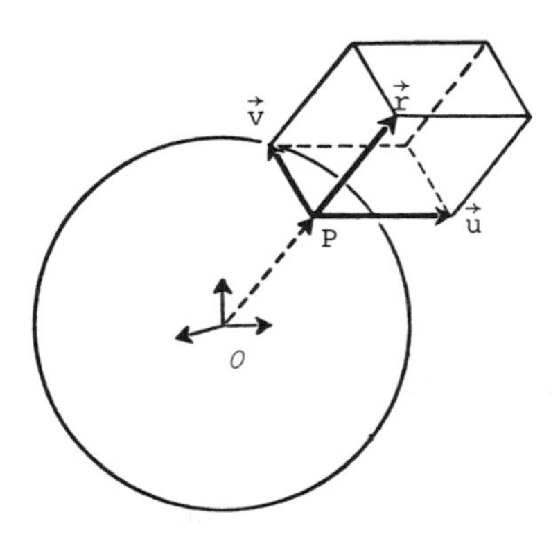

Figure 10.31.

Here we have used the *extrinsic* coordinates of the sphere. The winding number of ϕ is now given by the formula

$$\begin{aligned}\mathrm{Deg}[\phi] &= \frac{1}{\mathrm{Vol}(S^n)}\int_M \phi^*\boldsymbol{\Omega} \\ &= \frac{1}{n!\mathrm{Vol}(S^n)}\int_M \epsilon_{a_0\ldots a_n}\phi^{a_0}\boldsymbol{d}\phi^{a_1}\wedge\ldots\wedge\boldsymbol{d}\phi^{a_n}.\end{aligned} \tag{10.59}$$

This formula, which is based upon a particular choice of volume form on the sphere, has turned out to be very useful in various applications.

Remark. It can be shown that

$$\mathrm{Vol}[S^n] = \frac{2\pi^{\left(\frac{n+1}{2}\right)}}{\Gamma\left(\frac{n+1}{2}\right)}, \tag{10.60}$$

where Γ is the Γ-function, characterized by the recurrence relation

$$\Gamma\left(\frac{1}{2}\right) = \sqrt{\pi}, \qquad \Gamma(1) = 1, \qquad \Gamma(x+1) = x\Gamma(x). \tag{10.61}$$

We list a few particularly useful cases:

$$\mathrm{Vol}[S^1] = 2\pi; \qquad \mathrm{Vol}[S^2] = 4\pi; \qquad \mathrm{Vol}[S^3] = 2\pi^2; \qquad \mathrm{Vol}[S^4] = \frac{8}{3}\pi^2. \tag{10.62}$$

Exercise 10.7.2
Problem: Consider the sphere S^n and let us introduce *intrinsic* coordinates on the sphere $(\theta^1, \ldots, \theta^n)$. They induce canonical frames in the tangent spaces: $\mathbf{e}_1, \ldots, \mathbf{e}_n$. The induced metric is now characterized by the components $g_{ij} = \mathbf{e}_i \cdot \mathbf{e}_j$, and the volume form is given by

$$\epsilon = \sqrt{g}\,\boldsymbol{d}\theta^1 \wedge \cdots \wedge \boldsymbol{d}\theta^n.$$

1. Show that $\sqrt{g}$ = volume of the parallelepiped in $\mathbf{T}_P(S^n)$ spanned by $\mathbf{e}_1, \ldots, \mathbf{e}_n$.
2. Let $(\mathbf{v}_1, \ldots, \mathbf{v}_n)$ be any set of tangent vectors, and show that $\epsilon(\mathbf{v}_1, \ldots, \mathbf{v}_n)$ = volume of the parallelepiped in $\mathbf{T}_P(S^n)$ spanned out by $\mathbf{v}_1, \ldots, \mathbf{v}_n$.

(Hint: Use the *Legendre identity*:

$$\det(\mathbf{A}_i \cdot \mathbf{B}_j) = \det(A^{ij})\det(B^{ij}),$$

where A^{ij} are the Cartesian components of $\mathbf{A}_i$, and B^{ij} are the Cartesian components of $\mathbf{B}_i$.)

Exercise 10.7.3
Problem: (a) Consider the unit sphere S^1, and introduce the usual polar coordinate φ. Show by explicit calculation that the volume form (10.58) reduces to

$$\boldsymbol{\Omega} = \boldsymbol{d}\varphi.$$

(b) Consider the unit sphere S^2, and introduce the usual polar coordinates (θ, φ). Show by explicit computation that the volume form (10.58) reduces to

$$\boldsymbol{\Omega} = \sin\theta\,\boldsymbol{d}\theta \wedge \boldsymbol{d}\varphi.$$

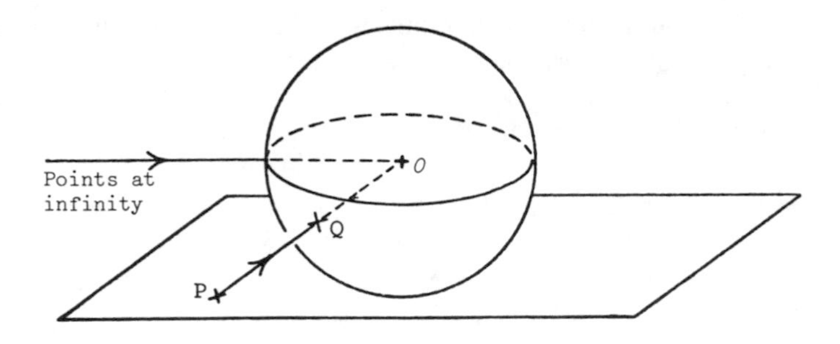

Figure 10.32.

Remark. It is important to observe that the concept of a winding number is well-defined only for smooth maps between *compact* manifolds. If M is not compact, then the preimage may consist of infinitely many points, so that the local Brouwer degree is not well-defined. But even if it is well-defined, it need not be constant on N. One might still be tempted to consider an integral expression like

$$\int_M \frac{1}{n!\mathrm{Vol}\,[N]}\,\epsilon_{a_0\ldots a_n}\phi^{a_0}\boldsymbol{d}\phi^{a_1}\wedge\ldots\wedge\boldsymbol{d}\phi^{a_n}. \tag{$*$}$$

But here the integral need not converge, and even if it converges, it need not reproduce an integer! Consider, for instance, the smooth map $\phi : \mathbb{R}^2 \curvearrowright S^2$, where ϕ denotes a stereographic projection from the center (cf. Figure 10.32).

It covers the lower hemisphere. Points on the upper hemisphere have Brouwer degree 0, while points on the lower hemisphere have Brouwer degree 1. Similarly, the integral $(*)$ takes the value $\frac{1}{2}$. All this is intuitively clear, since ϕ covers half of the sphere. So you may be tempted to attribute the winding number $\frac{1}{2}$ to this map, and that is okay with me, as long as you recognize that $\frac{1}{2}$ is not an integer.

10.8 The Heisenberg Ferromagnet

As an exemplification of the machinery built up in this chapter we will now look at a famous model from solid state physics. We have previously studied superconductivity (see Sections 2.12 and 8.7). This time we will concentrate on a ferromagnet. A single atom in the ferromagnet may be considered a small magnet with a magnetic moment proportional to the spin. At high temperature the interaction energy between the local magnets is very small compared with the thermal energy. As a consequence, the direction of the local magnets will be randomly distributed, due to thermal vibrations. For sufficiently low temperature, however, this interaction between the local magnets become dominant, and the local magnets tend to line up. We thus get an ordered state characterized by an order parameter, which we may choose to be the direction of the local spin vector.

Thus the order parameter in a ferromagnet is a unit vector. If we introduce Cartesian coordinates, it can be represented by a triple of scalar fields $[\hat{\phi}^1(x); \hat{\phi}^2(x); \hat{\phi}^3(x)]$ subject to the constraint $\hat{\phi}^a\hat{\phi}^a = 1$. (See Figure 10.33.)

We are going to study equilibrium configurations in a ferromagnet. Consider a static configuration $\hat{\phi}(\mathbf{x})$, where $\hat{\phi}(\mathbf{x})$ is a slowly varying spatial function. In the *Heisenberg model* for ferromagnetism one assumes that the static energy functional is given by

$$H[\hat{\phi}] = \frac{J}{2}\langle \boldsymbol{d}\hat{\phi}^a \mid \boldsymbol{d}\hat{\phi}^a\rangle = \frac{J}{2}\int \partial_i\hat{\phi}^a\partial^i\hat{\phi}^a\sqrt{g}\,d^Dx; \qquad \hat{\phi}^a\hat{\phi}^a = 1 \quad (10.63)$$

corresponding to a coupling "between nearest neighbors."

To obtain the field equations for the equilibriums states, we must vary the fields $\hat{\phi}^a$. But here we must pay due attention to the constraint $\hat{\phi}^a\hat{\phi}^a = 1$. The order parameter $\hat{\phi}$ is *not* a linear vector field; i.e., we are not allowed to form superpositions!

Theorem. *The field equations for the equilibrium configurations are given by*

$$\Delta\hat{\phi}^a + \hat{\phi}^a(\boldsymbol{d}\hat{\phi}^b \mid \boldsymbol{d}\hat{\phi}^b) = 0; \qquad \hat{\phi}^a\hat{\phi}^a = 1, \quad (10.64a)$$

or equivalently,

$$\boldsymbol{d}^*\boldsymbol{d}\hat{\phi}^a + \hat{\phi}^a({}^*\boldsymbol{d}\hat{\phi}^b \wedge \boldsymbol{d}\hat{\phi}^b) = 0; \qquad \hat{\phi}^a\hat{\phi}^a = 1. \quad (10.64b)$$

PROOF. We incorporate the constraint by the method of Lagrange multipliers. Consider, therefore, the modified energy functional

$$\tilde{H}[\phi^a(x); \lambda(x)] = \frac{J}{2}\langle \boldsymbol{d}\phi^a \mid \boldsymbol{d}\phi^a\rangle + \langle \lambda(x) \mid \phi^2(x) - 1\rangle.$$

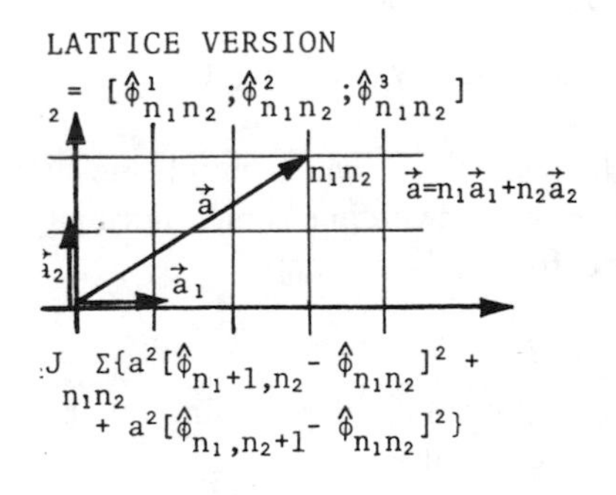

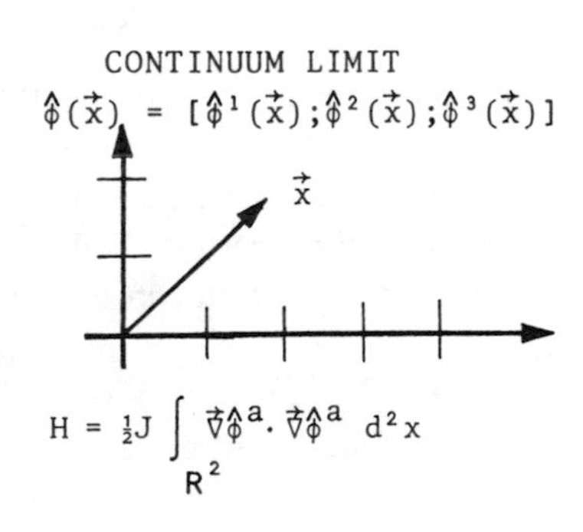

Figure 10.33.

Now, *all* the fields, $\phi^a(x)$ and $\lambda(x)$ must be varied independently of each other. Performing the variations

$$\lambda(x) \curvearrowright \lambda(x) + \epsilon\mu(x) \quad \text{and} \quad \phi^a(x) \curvearrowright \phi^a(x) + \epsilon\psi^a(x),$$

we obtain the "displaced" energy functional

$$\begin{aligned}\tilde{H}[\epsilon] = \tilde{H}[0] &+ \epsilon J \langle \boldsymbol{d}\psi^a \mid \boldsymbol{d}\phi^a \rangle + \epsilon^2 \frac{J}{2} \langle \boldsymbol{d}\psi^a \mid \boldsymbol{d}\psi^a \rangle \\ &+ 2\epsilon \langle \lambda(x) \mid \phi^a(x)\psi^a(x) \rangle + \epsilon^2 \langle \lambda(x) \mid \psi^2(x) \rangle \\ &+ \epsilon \langle \mu(x) \mid \phi^2(x) - 1 \rangle.\end{aligned}$$

As usual, we then demand that

$$\begin{aligned}0 = \frac{d\tilde{H}[\epsilon]}{d\epsilon}\Big|_{\epsilon=0} &= J\langle \boldsymbol{d}\psi^a \mid \boldsymbol{d}\phi^a \rangle + 2\langle \lambda(x) \mid \phi^a(x)\psi^a(x) \rangle \\ &+ \langle \mu(x) \mid \phi^2(x) - 1 \rangle \\ &= \langle \psi^a \mid J\boldsymbol{\delta d}\phi^a + 2\lambda(x)\phi^a(x) \rangle + \langle \mu(x) \mid \phi^2(x) - 1 \rangle.\end{aligned}$$

This leads to the Euler–Lagrange equations

$$J\boldsymbol{\Delta}\phi^a = 2\lambda(x)\phi^a(x); \qquad \phi^2(x) = 1.$$

As usual, the equation of motion for the Lagrange multipler degenerates to the equation of constraint. The constraint can now be used to eliminate $\lambda(x)$ from the equations of motion. Differentiating the constraint twice, we obtain

$$\phi^a \boldsymbol{\Delta}\phi^a = -(\boldsymbol{d}\phi^a \mid \boldsymbol{d}\phi^a). \tag{$*$}$$

Consequently,

$$2\lambda(x) = [2\lambda(x)\phi^a(x)]\phi^a(x) = J(\boldsymbol{\Delta}\phi^a)\phi^a = -J(\boldsymbol{d}\phi^a \mid \boldsymbol{d}\phi^a)$$

where we have used the equation of motion and the identity ($*$). Inserting this back in the Euler–Lagrange equations we finally obtain

$$\boldsymbol{\Delta}\phi^a = -(\boldsymbol{d}\phi^a \mid \boldsymbol{d}\phi^a)\phi^a; \qquad \phi^a\phi^a = 1.$$

(The equivalent version of this equation is obtained by dualizing it!) □

From now on we restrict ourselves to the case $D = 2$, which has the most interesting topological properties. Thus we consider a two-dimensional ferromagnet, where a spin configuration consequently corresponds to a map

$$\hat{\phi} : \mathbb{R}^2 \curvearrowright S^2.$$

This suggests that we introduce a topological quantity, the winding number Q, which tells us how many times the sphere is covered:

$$Q[\hat{\phi}] = \frac{1}{8\pi}\int_{\mathbb{R}^2} \epsilon_{abc}\hat{\phi}^a \boldsymbol{d}\hat{\phi}^b \wedge \boldsymbol{d}\hat{\phi}^c. \tag{10.65}$$

But here we encounter the usual problem, as the integral need not be well-defined because $\mathbb{R}^2$ is not a compact manifold. Furthermore, it need not be an integer even if it is well-defined (cf. the discussion in the preceding paragraph). We must therefore invoke a boundary condition, and fortunately we have in this model the natural choice: *The only spin configurations that are physically relevant are those with finite energy; i.e.,*

$$H[\hat{\phi}] = \frac{J}{2} \langle \boldsymbol{d}\hat{\phi}^a \mid \boldsymbol{d}\hat{\phi}^a \rangle < \infty.$$

This leads to the boundary condition

$$\lim_{\rho \to \infty} \boldsymbol{d}\hat{\phi}^a = 0; \tag{10.66}$$

i.e., $\hat{\phi}^a$ must be asymptotically constant:

$$\lim_{\rho \to \infty} \hat{\phi}^a(x) = \hat{\phi}^a_0. \tag{10.67}$$

Exactly which components the constant vector at infinity has we cannot say. All unit vectors are equally likely to be the one Nature chooses. But observe that once we have fixed the unit vector at infinity, we cannot change it without breaking the condition of finite energy. In what follows we choose the north pole as the asymptotic value of the order parameter:

$$\lim_{\rho \to \infty} \hat{\phi}^a(x) = [0; 0; 1]. \tag{10.68}$$

The boundary condition (10.68) has the important consequence that we can now compactify the base space. As usual, we perform a stereographic projection of the plane into the sphere, $\pi : S^2 \curvearrowright \mathbb{R}^2$. Notice that the north pole corresponds to the points at infinity. We can now lift a spin configuration in the plane, $\hat{\phi}^a(x)$, to a spin configuration on the sphere, $\tilde{\phi}^a = \pi^*\hat{\phi}^a$. We then extend the lifted field $\tilde{\phi}^a$ to the north pole by continuity:

$$\tilde{\phi}^a(N) = \begin{bmatrix} 0 \\ 0 \\ 1 \end{bmatrix};$$

cf. the discussion in Section 10.6. As a consequence of our boundary conditions we therefore see that *a spin configuration with finite energy can be lifted to a map $S^2 \curvearrowright S^2$ that maps the north pole into the north pole.* This shows that the winding number is well-defined for all the permitted configurations. The space E consisting of all the smooth finite energy configurations therefore breaks up into disconnected sectors E_n, where each sector is characterized by an integer n:

$$n = \frac{1}{8\pi} \int_{S^2} \epsilon_{abc}\hat{\phi}^a \boldsymbol{d}\hat{\phi}^b \wedge \boldsymbol{d}\hat{\phi}^c = \frac{1}{8\pi} \int_{\mathbb{R}^2} \epsilon_{abc}\hat{\phi}^a \boldsymbol{d}\hat{\phi}^b \wedge \boldsymbol{d}\hat{\phi}^c \tag{10.69}$$

(or equivalently,

$$n = \frac{1}{4\pi} \int \epsilon_{abc}\epsilon_{ij}\hat{\phi}^a \partial_i \hat{\phi}^b \partial_j \hat{\phi}^c dx^1 dx^2).$$

Remark. As a consequence of (10.42), the energy functional is conformally invariant:

$$\begin{aligned} H_{S^2}[\tilde{\phi}] &= \frac{1}{2} J \langle \boldsymbol{d}\tilde{\phi}^a \mid \boldsymbol{d}\tilde{\phi}^a \rangle_{S^2} = \frac{1}{2} J \int_{S^2} *\boldsymbol{d}\tilde{\phi}^a \wedge \boldsymbol{d}\tilde{\phi}^a \\ &= \frac{1}{2} J \int_{S^2} *\pi^*(\boldsymbol{d}\hat{\phi}^a) \wedge \pi^*(\boldsymbol{d}\hat{\phi}^a) = -\frac{1}{2} J \int_{S^2} \pi^*\{*\boldsymbol{d}\hat{\phi}^a \wedge \boldsymbol{d}\hat{\phi}^a\} \\ &= \frac{1}{2} J \int_{\mathbb{R}^2} *\boldsymbol{d}\hat{\phi}^a \wedge \boldsymbol{d}\hat{\phi}^a = \frac{1}{2} J \langle \boldsymbol{d}\hat{\phi}^a \mid \boldsymbol{d}\hat{\phi}^a \rangle_{\mathbb{R}^2} = H_{\mathbb{R}^2}[\hat{\phi}]. \end{aligned}$$

Consequently, we can solve the field equation (10.64) on the sphere whenever it is advantageous, since any solution to the equations of motion on the sphere automatically projects down to a solution of the equation of motion in the plane.

We proceed to investigate various equilibrium configurations. Let us first determine the *vacuum configuration* for the Heisenberg ferromagnet. The classical vacuum is characterized by having vanishing energy density. This implies that $\boldsymbol{d}\hat{\phi}^a$ vanishes; i.e., $\hat{\phi}^a$ is constant. Thus the classical vacuum corresponds to a spin configuration where all the local spin vectors point upwards. Clearly, it has winding number zero.

We then proceed to examine the nontrivial sectors. The fundamental problem is whether we can find ground states for the nontrivial sectors, i.e., whether we can determine a configuration in the sector E_n that has the lowest possible energy among all the configurations with winding number n. We call such a ground state a *spin wave*. (Notice that the energy functional (10.63) contains only a "kinetic" term. Although the model is based upon scalar fields, it thus corresponds precisely to the exceptional two-dimensional case, where Derrick's scalings argument does *not* apply; cf. the discussion in Section 4.7.) Observe that by definition, a spin wave is a local minimum for the energy functional (10.63), and consequently, it represents a solution to the second-order differential equation (10.64). It is a remarkable property of the Heisenberg ferromagnet that we can actually explicitly determine all the spin waves.

The first step in the analysis of spin waves consists in a reduction of the second-order differential equation for the spin wave to a first-order differential equation. This reduction is due to a *Bogomolny decomposition* of the energy functional (10.63). Consider the quantity $\boldsymbol{d}\phi^a$, i.e., $\partial_i\phi^a$. It carries two indices: A space index i, referring to the physical space, and a field index a, referring to the field space. Let us concentrate on the field index for a moment. Consider the vectors

$$\partial_1\hat{\phi}^a, \qquad \partial_2\hat{\phi}^a, \qquad \partial_3\hat{\phi}^a.$$

They are tangent vectors on the unit sphere in field space. We may now introduce a duality operation in this tangent space. As usual, for a two-dimensional vector space it corresponds to a rotation of $\frac{1}{2}\pi$. We denote it by # and observe that in terms of the Cartesian components for the tangent vector the duality operation is given by

$$\#\boldsymbol{d}\hat{\phi}^a = \epsilon_{abc}\hat{\phi}^b\boldsymbol{d}\hat{\phi}^c. \tag{10.72}$$

Observe especially that $\#^2 = -1$ and that $\#$ preserves the inner product in the tangent space.

Worked exercise 10.8.1
Problem: (a) Show that

$$\langle d\hat{\phi}^a \mid d\hat{\phi}^a\rangle = \langle \#d\hat{\phi}^a \mid \#d\hat{\phi}^a\rangle = \langle *d\hat{\phi}^a \mid *d\hat{\phi}^a\rangle \tag{10.73}$$

(b) Show that the topological charge, i.e., the winding number, is given by

$$n[\hat{\phi}] = \frac{1}{8\pi}\,\langle *d\hat{\phi}^a \mid \#d\hat{\phi}^a\rangle. \tag{10.74}$$

(c) Show that *any* field configuration automatically satisfies the differential equation

$$d\#d\hat{\phi}^a = \hat{\phi}^a(\#d\hat{\phi}^b \wedge d\hat{\phi}^b). \tag{10.75}$$

(Compare this with (10.64b)!)

From Exercise 10.8.1 we now easily obtain the desired Bogomolny decomposition:

$$H[\hat{\phi}] = \frac{J}{4}\,\| * d\hat{\phi}^a \mp \#d\hat{\phi}^a\|^2 \pm 4\pi J n[\hat{\phi}] \qquad (\text{with } \|\ \|^2 = \langle\ |\ \rangle). \tag{10.76}$$

But then we conclude that

a. *The energy in each section E_n is bounded below by*

$$H[\hat{\phi}] \geq 4\pi J|n[\hat{\phi}]|. \tag{10.77}$$

b. *A configuration with winding number n is a spin wave (i.e., a ground state for the sector E_n) if and only if it satisfies the first-order differential equation*

$$*d\hat{\phi}^a = \begin{cases} \#d\hat{\phi}^a & (\textit{when } n \textit{ is postive}), \\ -\#d\hat{\phi}^a & (\textit{when } n \textit{ is negative}). \end{cases} \tag{10.78}$$

(*This is known as a double self-duality equation*).

Exercise 10.8.2
Problem: Show, by explicit computation, that the first-order equation (10.78) implies the second-order equation (10.64).

The next step will be to show that every solution to the double self-duality equation (10.78) corresponds to a well-known geometrical object. In our case it turns out that we get the following nice characterization of spin waves:

A spin configuration $\hat{\phi} : \mathbb{R}^2 \curvearrowright S^2$ is a spin wave exactly when it generates a conformal map from the plane into the sphere.

PROOF. Let $\hat{e}_1, \hat{e}_2$ be the canonical frame vectors generated from the Cartesian coordinates in $\mathbb{R}^2$. Notice that they are lifted to the tangent vectors $\partial_1\hat{\phi}$, $\partial_2\hat{\phi}$. If $\hat{\phi}$ is conformal, this forces $\partial_1\hat{\phi}$, $\partial_2\hat{\phi}$ to be the orthogonal vectors of the same length. We leave it to the reader to verify that the converse holds, i.e., that the above property actually characterizes the conformal maps from the plane into the sphere.

Now suppose $\hat{\phi}$ is a configuration with winding number n that solves (10.78). This first-order differential equation can be rearranged as follows:

$$\partial_1\hat{\phi} = \#\partial_2\hat{\phi}, \qquad \partial_2\hat{\phi} = -\#\partial_1\hat{\phi}.$$

But from these equations it follows immediately that $\partial_1\hat{\phi}$, $\partial_2\hat{\phi}$ are orthogonal vectors of the same length. Consequently, $\hat{\phi}$ is a conformal map. On the other hand, it is not too difficult to show that any conformal map actually solves (10.78). We leave the details as an exercise:

Worked Exercise 10.8.3
Introduction: Let $\hat{\phi} : \mathbb{R}^2 \curvearrowright S^2$ be a smooth map, and introduce the following vectors in field space:

$$\begin{aligned} \mathbf{P}^a &= *\boldsymbol{d}\hat{\phi}^a - \#\boldsymbol{d}\hat{\phi}^a; \quad \text{i.e., } \mathbf{P}_i = \epsilon_{ij}\partial_j\hat{\phi} - \hat{\phi}\times\partial_i\hat{\phi} \\ \mathbf{Q}^a &= *\boldsymbol{d}\hat{\phi}^a + \#\boldsymbol{d}\hat{\phi}^a; \quad \text{i.e., } \mathbf{Q}_i = \epsilon_{ij}\partial_j\hat{\phi} + \hat{\phi}\times\partial_i\hat{\phi}. \end{aligned}$$

Problem: (a) Show that if $\hat{\phi}$ is a conformal map, then the following holds:

$$\mathbf{P}_1\cdot\mathbf{P}_2 = \mathbf{Q}_1\cdot\mathbf{Q}_2 = \mathbf{P}_1\cdot\mathbf{Q}_1 = \mathbf{P}_2\cdot\mathbf{Q}_2 = \mathbf{P}_1\cdot\mathbf{Q}_2 = \mathbf{P}_2\cdot\mathbf{Q}_1 = 0.$$

(b) Show furthermore that if $\hat{\phi}$ is a conformal map, then either

$$\mathbf{P}_1 = \mathbf{P}_2 = 0$$

or

$$\mathbf{Q}_1 = \mathbf{Q}_2 = 0.$$

This concludes the proof! □

As an example, we notice that the stereographic projection itself is a conformal map, and thus it represents a spin wave with winding number -1 (it reverses the orientation!). Observe also that the stereographic projection can be used to lift any spin configuration $\hat{\phi} : \mathbb{R}^2 \curvearrowright \mathbb{R}^2$ to a spin configuration $\tilde{\phi} : S^2 \curvearrowright S^2$. Consequently, there is also a one-to-one correspondence between spin waves and conformal maps from the unit sphere into itself that map the north pole into itself.

In the final step we introduce *complex analysis*. In the present case it is completely trivial. It is well known that the sphere is isomorphic to the extended complex plane

$$S^2 \approx \mathbb{C}^* = \mathbb{C}\cup\{\infty\},$$

and that the orientation-preserving conformal maps are holomorphic; cf. the discussion of the Riemann sphere in Section 10.6. But holomorphic maps are necessarily algebraic; i.e., they are of the form

$$w(z) = \frac{P(z)}{Q(z)} \qquad \text{with } P, Q \text{ polynomials.} \tag{10.79a}$$

Furthermore, $w(\infty) = \infty$ implies that Deg $P >$ Deg Q. Thus we have finally explicitly constructed all spin waves with a negative winding number. (See Belavin and Polyakov, 1975.)

Similarly, a spin configuration with a positive winding number corresponds to an antiholomorphic map on the sphere; i.e., it is of the form

$$w(z) = \frac{P(\bar{z})}{Q(\bar{z})} \qquad \text{with } P,\ Q \text{ polynomials, Deg } P > \text{Deg } Q. \tag{10.79b}$$

It would also be nice to find a simple formula relating the winding number to the polynomials P and Q. That is easy enough: The map $w = P/Q$ is a smooth map, and according to Sard's theorem, there are plenty of regular values. Let w_0 be a regular value. Then the preimage consists of all solutions to the equation

$$P(z) - w_0 Q(z) = 0.$$

As Deg $P >$ Deg Q it has Deg P distinct solutions. Furthermore, a holomorphic map preserves the orientation. Thus w has winding number Deg P and this means that $\hat{\phi}$ has winding number $-$Deg P.

Remark. Rather than relating spin waves to conformal maps, one can relate them directly to holomorphic (or antiholomorphic) functions. Using a stereographic projection π from the unit sphere in field space to the complex plane, we can project the order parameter down to a single complex field given by

$$w = \frac{\hat{\phi}^1}{1-\hat{\phi}^3} + i\,\frac{\hat{\phi}^2}{1-\hat{\phi}^3}\,.$$

In the same way, the tangent vector $\partial_i \hat{\phi}^a$ on the unit sphere is projected down into the tangent vector $\partial_i w$ in the complex plane; i.e.,

$$\boldsymbol{d}\hat{\phi}^a \overset{\pi^*}{\curvearrowright} \boldsymbol{d}w.$$

Since π is conformal, it preserves right angles. Furthermore, it reverses orientation, so that

$$\epsilon_{abc}\hat{\phi}^b \partial_i \hat{\phi}^c \overset{\pi^*}{\curvearrowright} -i\partial_i w; \qquad \text{i.e., } \#\boldsymbol{d}\hat{\phi}^a \overset{\pi^*}{\curvearrowright} -i\boldsymbol{d}w.$$

Similarly, it follows from the linearity of π^* that

$$\epsilon_{ij}\partial_j \hat{\phi}^a \overset{\pi^*}{\curvearrowright} \epsilon_{ij}\partial_j w; \qquad \text{i.e., } *\,\boldsymbol{d}\hat{\phi}^a \overset{\pi^*}{\curvearrowright} *\boldsymbol{d}w.$$

Putting all this together, we therefore see that the double self-duality equation (10.78) is projected down to the usual self-duality equation for a holomorphic function:

$$*\boldsymbol{d}w = -i\boldsymbol{d}w.$$

Okay, this concludes our discussion of spin waves, i.e., the ground states for the various sectors. We might still ask whether there are other finite-energy solutions to the full field equations

$$\boldsymbol{\Delta}\hat{\phi} = \hat{\phi}(\boldsymbol{d}\hat{\phi} \mid \boldsymbol{d}\hat{\phi}). \tag{10.64}$$

A priori there could be local minima lying somewhat above the ground state, or there could be "saddle points." But Woo (1977) has investigated this problem, and using complex analysis he has shown that the answer is negative:

Any finite energy solution to the second-order equation (10.46) *automatically solves the first-order equation* (10.78).

Worked Exercise 10.8.4

Problem: (a) Show that a spin configuration represented by the complex-valued function $w(z)$ has energy density

$$H = \frac{1}{(1+w\bar{w})^2}\left\{\frac{\partial w}{\partial z}\frac{\partial \bar{w}}{\partial \bar{z}} + \frac{\partial w}{\partial \bar{z}}\frac{\overline{\partial w}}{\partial z}\right\}. \tag{10.80}$$

(b) Show that the corresponding equations of motion are given by

$$(1+w\bar{w})\frac{\partial^2 w}{\partial \bar{z}\partial z} = 2\bar{w}\frac{\partial w}{\partial z}\frac{\partial w}{\partial \bar{z}} \quad \text{and} \quad (1+w\bar{w})\frac{\partial^2 w}{\partial \bar{z}\partial z} = 2w\frac{\partial \bar{w}}{\partial z}\frac{\partial \bar{w}}{\partial \bar{z}}. \tag{10.81}$$

(c) Consider the following two complex-valued functions:

$$f(z) = \frac{\frac{\partial w}{\partial z}\frac{\partial \bar{w}}{\partial z}}{(1+w\bar{w})^2} \quad \text{and} \quad g(z) = \frac{\frac{\partial w}{\partial \bar{z}}\frac{\overline{\partial w}}{\partial \bar{z}}}{(1+w\bar{w})^2}. \tag{10.82}$$

Show that $f(z)$ is holomorphic and that $g(z)$ is antiholomorphic provided that $w(z)$ solves the equation of motion.

(d) Show that if f has no poles, then w itself must be either holomorphic or antiholomorphic.

In the above exercise we have almost proven that a solution to the full field equation (10.64), which has finite energy, is automatically represented by a holomorphic or an antiholomorphic function. There is, however, one possible loophole: The quantity

$$f(z) = \frac{\frac{\partial w}{\partial z}\frac{\overline{\partial w}}{\partial z}}{(1+w\bar{w})^2}$$

might have a pole, or equivalently, the energy density might have a "pole." Consider, e.g., the inadmissible solution

$$w(z) = z^{1/2}.$$

(It is inadmissible because it is multivalued!) Neglecting the branch cut for a moment, we notice that it produces the energy density

$$H = \frac{1}{2|z|(1+|z|)^2}.$$

Consequently, the energy density has a "pole" at $z = 0$. But it is still integrable (as we can easily see if we introduce polar coordinates). Thus we cannot exclude the possibility that $f(z)$ might have a pole. One must then investigate the solution to the first-order equation (10.82) very carefully to exclude that possibility too; i.e., near a pole of $f(z)$ we must show that $w(z)$ is necessarily multivalued like

$w(z) = z^{1/2}$. For further details the reader should consult the original paper of Woo.

Exercise 10.8.5
Introduction: Let F be an ordinary function: $F : \mathbb{R} \curvearrowright \mathbb{R}$ and let $w(z)$ be a complex-valued field characterized by the energy density

$$H = F(w\bar{w}) \left\{ \frac{\partial w}{\partial z} \frac{\partial \bar{w}}{\partial \bar{z}} + \frac{\partial w}{\partial \bar{z}} \frac{\partial \bar{w}}{\partial z} \right\} .$$

In analogy with the Heisenberg ferromagnet, we will only be interested in static configurations with finite energy. Problem: (a) Show that the equations of motion are given by

$$F \frac{\partial^2 w}{\partial \bar{z} \partial z} = -F' \bar{w} \frac{\partial w}{\partial z} \frac{\partial w}{\partial \bar{z}} \quad \text{and} \quad F \frac{\partial^2 w}{\partial \bar{z} \partial z} = -F' w \overline{\frac{\partial w}{\partial z}} \, \overline{\frac{\partial w}{\partial \bar{z}}} .$$

(b) Show that

$$f(z) = F(w\bar{w}) \frac{\partial w}{\partial z} \frac{\partial \bar{w}}{\partial z}$$

is a holomorphic function provided that $w(z)$ solves the equations of motion.
(c) Specialize to a static massless complex Klein–Gordon field in (2 + 1) space–time dimensions and try to characterize the static solutions with finite energy. (There are none except for the trivial solutions $w(z) =$ constant!) Hint: Show that it corresponds to the case $F = 1$.

10.9 The Exceptional ϕ^4-Model

As another interesting example we will consider a model in two space dimensions that on the one hand is related to the Ginzburg–Landau model for superconductivity and on the other to the abelian Higgs model (cf. the discussion in Sections 8.7–8.8). It will be based on the static energy functional

$$H = \frac{1}{2} \langle \mathbf{B} \mid \mathbf{B} \rangle + \frac{1}{2} \langle \boldsymbol{D}\phi \mid \boldsymbol{D}\phi \rangle + \int_{\mathbb{R}^2} U[\phi]\epsilon; \qquad \boldsymbol{D} = \boldsymbol{d} - ie\mathbf{A}; \tag{10.83}$$

cf. (8.60) and (8.88). For the moment we leave the potential unspecified, except that we assume that it is gauge invariant; i.e., it is a function of $|\phi|^2$, and furthermore that it only vanishes at the nonzero value $|\phi| = \phi_0$.

The field equations for a static equilibrium configuration are given by

$$-\boldsymbol{D}^* \boldsymbol{D}\phi = 2 \frac{\partial U}{\partial |\phi|^2} \phi\epsilon , \tag{10.84a}$$

$$-\delta \mathbf{B} = i \frac{e}{2} [\bar{\phi} \boldsymbol{D}\phi - \phi \overline{\boldsymbol{D}\phi}]; \tag{10.84b}$$

cf. the Worked Exercise 8.7.3.

As usual, we consider only finite energy configurations. This leads to the boundary conditions:

$$\lim_{\rho \to \infty} \mathbf{B} = 0; \qquad \lim_{\rho \to \infty} \boldsymbol{D}\phi = 0; \qquad \lim_{\rho \to \infty} |\phi| = \phi_0. \tag{10.85}$$

As in the case of ordinary superconductivity, we have, furthermore, flux quantization; i.e.,

$$\Phi = \int_{\mathbb{R}^2} \mathbf{B} = n\,\frac{2\pi}{e}\,, \tag{10.86}$$

where n is related to the jump in the phase of the Higgs field when we go once around the flux tube.

So far, nothing is new. We will now try to see whether we can find the ground state configurations for the nontrivial sections (where $n \neq 0$). As usual, the investigation of the ground states will be based upon a Bogomolny decomposition of the static energy functional.

We start by guessing a reasonable set of first-order differential equations. Guided by the $(1+1)$-dimensional models and the Heisenberg ferromagnet, we try the ansatz

$$\mathbf{B} = \sqrt{2U[\phi]}\epsilon = *\sqrt{2U[\phi]} \tag{10.87a}$$

$$^*\boldsymbol{D}\phi = i\boldsymbol{D}\phi \qquad \text{(i.e., the self-duality equation).} \tag{10.87b}$$

They lead to the following pair of second-order differential equations:

$$-\boldsymbol{D}^*\boldsymbol{D}\phi = -i\boldsymbol{D}^2\phi = -e\mathbf{B}\phi = -e\sqrt{2U[\phi]} \tag{10.88a}$$

(where we have used Exercise 8.7.2).

$$\begin{aligned} -\boldsymbol{\delta}\mathbf{B} &= -\boldsymbol{\delta}^*\sqrt{2U[\phi]} = -^*\boldsymbol{d}\sqrt{2U[\phi]} = \frac{-\frac{\partial U}{\partial|\phi|^2}}{\sqrt{2U[\phi]}}\;^*\boldsymbol{d}|\phi|^2 \\ &= \frac{-\frac{\partial U}{\partial|\phi|^2}}{\sqrt{2U[\phi]}}\;[\bar{\phi}^*\boldsymbol{d}\phi + \phi^*\boldsymbol{d}\bar{\phi}]. \end{aligned} \tag{$*$}$$

But from the self-duality equation we get

$$^*\boldsymbol{d}\phi = i\boldsymbol{D}\phi + ie\phi^*\mathbf{A}; \qquad ^*\boldsymbol{d}\bar{\phi} = -i\overline{\boldsymbol{D}\phi} - ie\bar{\phi}^*\mathbf{A}.$$

Consequently, we get

$$\bar{\phi}^*\boldsymbol{d}\phi + \phi^*\boldsymbol{d}\bar{\phi} = i\bar{\phi}\boldsymbol{D}\phi - i\phi\overline{\boldsymbol{D}\phi}.$$

Inserting this into equation $(*)$, we can rearrange this as

$$-\boldsymbol{\delta}\mathbf{B} = -\frac{1}{\sqrt{2U[\phi]}}\,\frac{\partial U}{\partial|\phi|^2}\;i[\bar{\phi}\boldsymbol{D}\phi - \phi\overline{\boldsymbol{D}\phi}]. \tag{10.88b}$$

Comparing (10.88a–b) with (10.84a–b), we see that the first-order differential equations are compatible with the second-order differential equations, provided that the potential satisfies the identity

$$\frac{\partial U}{\partial|\phi|^2} = -\frac{e}{2}\sqrt{2U[\phi]}.$$

This restricts U to be a fourth-order polynomial of the form

$$U[\phi] = \frac{e^2}{8}\,(|\phi|^2 - \phi_0)^2 \qquad \text{with } |\phi| < \phi_0. \tag{10.88}$$

Notice that the above potential represents a special case of the potential energy density in the abelian Higgs model, where we have put

$$\lambda = \frac{e^2}{2}\,. \tag{10.89}$$

The abelian Higgs model based upon the potential energy density (10.88) is known as the *exceptional ϕ^4-model*. In superconductivity it corresponds to the case where

$$\kappa = \frac{1}{\sqrt{2}}\,. \tag{10.90}$$

(κ is the Ginzburg–Landau parameter (8.67). The identity (10.90) follows immediately from (10.89) when we perform the substitutions $\lambda \curvearrowright \beta$ and $e \curvearrowright q/h$.)

Okay, in the exceptional ϕ^4-model we thus have the possibility of reducing the field equations (10.84) to the first-order equations (10.87). This suggests the following Bogomolny decomposition:

$$\begin{aligned} H = {} & \frac{1}{2}\langle \mathbf{B} \mp \sqrt{2U[\phi]}\epsilon \mid \mathbf{B} \mp \sqrt{2U[\phi]}\epsilon\rangle \\ & + \frac{1}{4}\langle {}^*\boldsymbol{D}\phi \mp i\boldsymbol{D}\phi \mid {}^*\boldsymbol{D}\phi \mp i\boldsymbol{D}\phi\rangle + \text{a remainder term.} \end{aligned}$$

All we must check is that the remainder term is a topological term, i.e., that it depends only on the "winding number" n. We leave this as an exercise to the reader:

Worked Exercise 10.9.1
Problem: (a) Show that the winding number is given by

$$n = \frac{1}{2\pi\phi_0^2}\langle {}^*\boldsymbol{d}\phi \mid i\boldsymbol{d}\phi\rangle. \tag{10.91}$$

(This should be compared with (10.74)!)

(b) Show that the exceptional ϕ^4-model possesses the Bogomolny decomposition

$$H = \frac{1}{2}\,\|\mathbf{B} \mp \sqrt{2U[\phi]}\epsilon\|^2 + \frac{1}{4}\,\|{}^*\boldsymbol{D} \mp i\boldsymbol{D}\phi\|^2 \pm n\pi\phi_0^2 \qquad (\text{with } \|\ \|^2 = \langle\ |\ \rangle). \tag{10.92}$$

We therefore conclude as usual:

Theorem.

a. *The energy in the sector E_n is bounded below by*

$$H \geq |n|\pi\phi_0^2. \tag{10.93}$$

b. *A configuration with winding number n is a ground state if and only if it satisfies the first-order differential equation*

$$\mathbf{B} = \pm\frac{1}{2}\,e\left(\phi_0^2 - |\phi|^2\right)\epsilon; \qquad {}^*\boldsymbol{D}\phi = \pm i\boldsymbol{D}\phi. \tag{10.87}$$

In the above discussion you might feel that the first order equation (10.87a), which preceded the Bogomolny decomposition, was sort of "pulled out of the air." We did try to justify it by referring to our earlier experience with (1 + 1)-dimensional models, but you may not find that very convincing. We shall therefore present another argument, which allows one to make reasonable guesses in such a situation. The argument is closely related to the reasoning behind Derrick's scaling argument. We know that a pure scalar field theory in two dimensions cannot possess stable nontrivial static solutions. In the above model it is thus the presence of a gauge field which makes it possible to stabilize a static configuration. Let us look at this in some detail.

Consider the scaling transformation

$$D_\lambda(\mathbf{x}) = \lambda\mathbf{x}.$$

It is a conformal transformation with the conformal factor

$$\Omega^2(x) = \lambda^2.$$

Let us furthermore introduce the scaled configuration

$$\begin{aligned} \phi_\lambda &= D_\lambda^*\phi \qquad (\text{i.e., } \phi_\lambda(\mathbf{x}) = \phi(\lambda\mathbf{x})), \\ \mathbf{A}_\lambda &= D_\lambda^*\mathbf{A} \qquad (\text{i.e., } \mathbf{A}_\lambda(\mathbf{x}) = \lambda\mathbf{A}(\lambda\mathbf{x})). \end{aligned}$$

From (10.40)–(10.41) (or by working out the coordinate expression directly) we now get

$$\begin{aligned} \langle \boldsymbol{d}\mathbf{A}_\lambda \mid \boldsymbol{d}\mathbf{A}_\lambda\rangle &= \int_{\mathbb{R}^2} *D_\lambda^*\boldsymbol{d}\mathbf{A} \wedge D_\lambda^*\boldsymbol{d}\mathbf{A} = \lambda^2 \int_{\mathbb{R}^2} D_\lambda^*[*\boldsymbol{d}\mathbf{A} \wedge \boldsymbol{d}\mathbf{A}] \\ &= \lambda^2 \int_{D_\lambda(\mathbb{R}^2)} *\boldsymbol{d}\mathbf{A} \wedge \boldsymbol{d}\mathbf{A} = \lambda^2\langle \boldsymbol{d}\mathbf{A} \mid \boldsymbol{d}\mathbf{A}\rangle, \\ \langle \boldsymbol{D}_\lambda\phi_\lambda \mid \boldsymbol{D}_\lambda\phi_\lambda\rangle &= \int_{\mathbb{R}^2} *D_\lambda^*\boldsymbol{D}\phi \wedge D_\lambda^*\boldsymbol{D}\phi = \int_{\mathbb{R}^2} D_\lambda^*\{*\boldsymbol{D}\phi \wedge \boldsymbol{D}\phi\} \\ &= \int_{D_\lambda(\mathbb{R}^2)} *\boldsymbol{D}\phi \wedge \boldsymbol{D}\phi = \langle \boldsymbol{D}\phi \mid \boldsymbol{D}\phi\rangle, \\ \int_{\mathbb{R}^2} U[\phi_\lambda]\epsilon &= \lambda^{-2}\int_{\mathbb{R}^2} U[D_\lambda^*\phi]D_\lambda^*\epsilon = \lambda^{-2}\int_{\mathbb{R}^2} D_\lambda^*\{U[\phi]\epsilon\} \\ &= \lambda^{-2}\int_{D_\lambda(\mathbb{R}^2)} U[\phi]\epsilon = \lambda^{-2}\int_{\mathbb{R}^2} U[\phi]\epsilon. \end{aligned}$$

This leads to the following displaced static energy:

$$H[\phi_\lambda; \mathbf{A}_\lambda] = \frac{1}{2}\lambda^2\langle \mathbf{B} \mid \mathbf{B}\rangle + \frac{1}{2}\langle \boldsymbol{D}\phi \mid \boldsymbol{D}\phi\rangle + \lambda^{-2}\int_{\mathbb{R}^2} U[\phi]\epsilon.$$

A stable configuration, and especially a ground state, must now satisfy

$$0 = \frac{dH}{d\lambda}\Big|_{\lambda=0} = \langle \mathbf{B} \mid \mathbf{B}\rangle - \int_{\mathbb{R}^2} 2U[\phi]\epsilon.$$

Thus the condition of stability leads to the "virial theorem": A stable configuration satisfies the identity:

$$\langle \mathbf{B} \mid \mathbf{B}\rangle = \langle\sqrt{2U[\phi]}\epsilon \mid \sqrt{2U[\phi]}\epsilon\rangle. \tag{10.88}$$

This should be compared with (4.75). In analogy with the $(1+1)$-dimensional case we now suggest that a ground state not only satisfies this identity globally, but that it in fact satisfies it pointwise, i.e.,

$$\mathbf{B} = \pm\sqrt{2U[\phi]}\epsilon.$$

Thus we recover precisely the equation (10.87a).

So all we have got to do is to solve the first-order equations (10.87). Unfortunately, that is a very hard task, since one cannot write down the explicit solution. Using advanced analysis it can be shown that the most general solution with n flux quanta depends upon $2n$ arbitrary parameters corresponding to the center positions of n flux tubes each carrying a unit flux. (For details about the machinery required to analyze the general solution the reader should consult the book of Taubes and Jaffe (1981).)

If one specializes to spherical symmetric configurations, the analysis becomes very simple. If we make the assumption that ϕ and $\mathbf{A}$ can be expressed in the following form

$$\phi(\mathbf{x}) = \phi_0\tilde{\phi}(\rho)e^{in\varphi}; \qquad \mathbf{A}(x) = \frac{n}{e}A(\rho)\boldsymbol{d}\varphi, \tag{10.89}$$

then it is easy to see that this represents a finite-energy configuration with n flux quanta, provided that $\tilde{\phi}(\rho)$ and $A(\rho)$ satisfy the boundary conditions

$$\lim_{\rho\to 0}\tilde{\phi}(\rho) = 0; \qquad \lim_{\rho\to\infty}\tilde{\phi}(\rho) = 1; \qquad \lim_{\rho\to 0}A(\rho) = 0; \qquad \lim_{\rho\to\infty}A(\rho) = 1; \tag{10.90}$$

cf. the discussion in Section 8.7.

Worked Exercise 10.9.2
Problem: Show that the field equations (10.87) reduce to

$$\frac{n}{\rho}\frac{dA}{d\rho} = \frac{1}{2}e^2\phi_0^2(1-\tilde{\phi}^2); \qquad \rho\frac{d\tilde{\phi}}{d\rho} = n\tilde{\phi}(1-A) \tag{10.91}$$

when we specialize to the equation (10.89).

It follows from standard theory of first-order differential equations that there is precisely one solution $(\phi(\rho), A(\rho))$ that interpolates between $(0, 0)$ and the "equilibrium point" $(1, 1)$.

Solutions to Worked Exercises

Solution to 10.3.2

In polar coordinates the stereographic projection is given by $\rho = 2\cot\frac{\theta}{2}; \phi = \varphi$. (See Figure 10.34.) The Jacobian matrix is given by

$$\left(\frac{\partial y^i}{\partial x^j}\right) = \begin{bmatrix} -\frac{1}{\sin^2\frac{\theta}{2}} & 0 \\ 0 & 1 \end{bmatrix}.$$

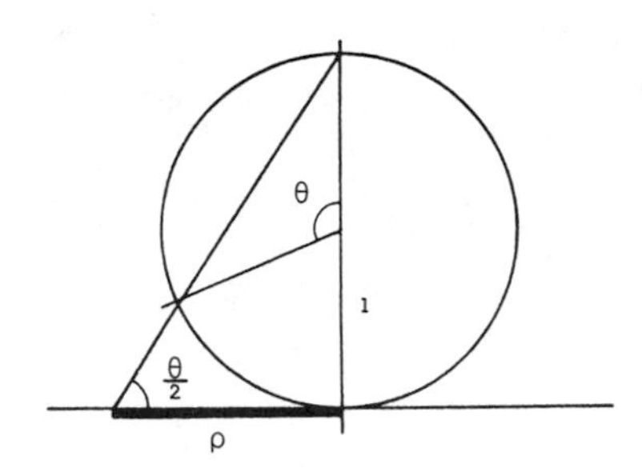

Figure 10.34.

(In particular, we observe that the Jacobian is negative; i.e., the stereographic projection is orientation-reversing.) The metrics $\boldsymbol{g}_1$ and $\boldsymbol{g}_2$ are characterized by the components

$$\overline{\overline{G}}_1 = \begin{bmatrix} 1 & 0 \\ 0 & \sin^2\theta \end{bmatrix}, \qquad \overline{\overline{G}}_2 = \begin{bmatrix} 1 & 0 \\ 0 & \rho^2 \end{bmatrix}.$$

The pull-backed metric is then characterized by the components

$$f^*\boldsymbol{g}_2 : \overline{\overline{D}}{}^{+}_{21}\overline{\overline{G}}_2\overline{\overline{D}}_{21} = \begin{bmatrix} \frac{1}{\sin^4\frac{\theta}{2}} & 0 \\ 0 & 4\cot^2\frac{\theta}{2} \end{bmatrix} = \frac{1}{\sin^4\frac{\theta}{2}}\begin{bmatrix} 1 & 0 \\ 0 & \sin^2\theta \end{bmatrix}.$$

Solution to 10.3.3

If we perform a general parameter shift $\lambda = \lambda(s)$, then

$$\frac{dx^\mu}{ds} = \frac{dx^\mu}{d\lambda}\frac{d\lambda}{ds} \quad \text{and} \quad \frac{d^2x^\mu}{ds^2} = \frac{d^2x^\mu}{d\lambda^2}\left(\frac{d\lambda}{ds}\right)^2 + \frac{dx^\mu}{d\lambda}\frac{d^2\lambda}{ds^2},$$

so the geodesic equation is transformed into

$$\frac{d^2x^\mu}{ds^2} + \Gamma^\mu{}_{\alpha\beta}\frac{dx^\alpha}{ds}\frac{dx^\beta}{ds} = \frac{dx^\mu}{ds}\frac{\lambda''(s)}{\lambda'(s)}.$$

Thus the most general covariant equation for a geodesic is of the form

$$\frac{d^2x^\mu}{ds^2} + \Gamma^\mu{}_{\alpha\beta}\frac{dx^\alpha}{ds}\frac{dx^\beta}{ds} = A(s)\frac{dx^\mu}{ds} \qquad \text{with } A(s) = \frac{\lambda''(s)}{\lambda'(s)}.$$

On the other hand, any smooth function $A(s)$ can be written in the form

$$A(s) = \frac{\lambda''(s)}{\lambda'(s)} = [\ln\lambda'(s)]'$$

with

$$\lambda = \int_0^s \exp\left[\int_0^{s_2} A(s_1)ds_1\right] ds_2.$$

Solution to 10.8.1

a. The first equality is a consequence of the fact that # preserves the inner product in field space, since it corresponds to a rotation. The second equality follows from Theorem 8.7

b. $\frac{1}{8\pi}\langle {}^*\boldsymbol{d}\hat{\phi}^a \mid \#\boldsymbol{d}\phi^a\rangle = \frac{1}{8\pi}\int * * [\boldsymbol{d}\hat{\phi}^a] \wedge \#\boldsymbol{d}\hat{\phi}^a = -\frac{1}{8\pi}\int \boldsymbol{d}\hat{\phi}^a \wedge \epsilon_{abc}\hat{\phi}^b\boldsymbol{d}\hat{\phi}^c = \frac{1}{8\pi}\int \epsilon_{abc}\hat{\phi}^a\boldsymbol{d}\hat{\phi}^b \wedge \boldsymbol{d}\hat{\phi}^c = n[\hat{\phi}]$.

c. $\boldsymbol{d}\#\boldsymbol{d}\hat{\phi}^a = \boldsymbol{d}[\epsilon_{abc}\hat{\phi}^b\boldsymbol{d}\hat{\phi}^c] = \epsilon_{abc}\boldsymbol{d}\hat{\phi}^b \wedge \boldsymbol{d}\hat{\phi}^c$.

Notice that $\mathbf{F}^a = \epsilon_{abc}\boldsymbol{d}\hat{\phi}^b \wedge \boldsymbol{d}\hat{\phi}^c$ is the vector in field space obtained by taking the cross product of the two tangent vectors $\boldsymbol{d}\hat{\phi}^b$ and $\boldsymbol{d}\hat{\phi}^c$. Consequently, it must be proportional to the radial vector $\hat{\phi}^a$, and since $\hat{\phi}^a$ is a unit vector, we get

$$\mathbf{F}^a = \hat{\phi}^a(\hat{\phi}^b\mathbf{F}^b) = \hat{\phi}^a\hat{\phi}^b\epsilon_{abc}\boldsymbol{d}\hat{\phi}^c \wedge \boldsymbol{d}\hat{\phi}^d = \hat{\phi}^a(\#\boldsymbol{d}\hat{\phi}^d \wedge \boldsymbol{d}\hat{\phi}^d).$$

Solution to 10.8.3

Let us write out $\mathbf{P}_1$, $\mathbf{P}_2$, $\mathbf{Q}_1$ and $\mathbf{Q}_2$ explicitly:

$$\mathbf{P}_1 = \partial_2\hat{\phi} - \hat{\phi}\times\partial_1\hat{\phi}, \qquad \mathbf{P}_2 = -\partial_1\hat{\phi} - \hat{\phi}\times\partial_2\hat{\phi},$$
$$\mathbf{Q}_1 = \partial_2\hat{\phi} + \hat{\phi}\times\partial_1\hat{\phi}, \qquad \mathbf{Q}_2 = -\partial_1\hat{\phi} + \hat{\phi}\times\partial_2\hat{\phi}.$$

Then we get

$$\mathbf{P}_1\cdot\mathbf{Q}_1 = \mathbf{P}_2\cdot\mathbf{Q}_2 = |\partial_1\hat{\phi}|^2 - |\partial_2\hat{\phi}|^2; \qquad \mathbf{P}_1\cdot\mathbf{P}_2 = \mathbf{Q}_1\cdot\mathbf{Q}_2 = 0$$

and

$$\mathbf{P}_1\cdot\mathbf{Q}_2 = \mathbf{P}_2\cdot\mathbf{Q}_1 = -2\partial_1\hat{\phi}\partial_2\hat{\phi}.$$

But for a conformal map $\hat{\phi} : \mathbb{R}^2 \curvearrowright S^2$ we know that $\partial_1\hat{\phi}$, $\partial_2\hat{\phi}$ are orthogonal tangent vectors of the same length, so this forces the right-hand side to vanish.

a. We start with the following important remark:

If $\partial_1\hat{\phi}$ (or $\partial_2\hat{\phi}$) vanishes, then all four tangent vectors $\mathbf{P}_1$, $\mathbf{P}_2$, $\mathbf{Q}_1$, and $\mathbf{Q}_1$ vanish.

To see this, let $\partial_1\hat{\phi}$ vanish. Then we get

$$\mathbf{P}_2 = -\hat{\phi}\times\partial_2\hat{\phi} \quad\text{and}\quad \mathbf{Q}_2 = \hat{\phi}\times\partial_2\hat{\phi}.$$

But they are orthogonal vectors, so this forces $\partial_2\hat{\phi}$ to vanish too. Clearly, $\mathbf{P}_1$, $\mathbf{P}_2$, $\mathbf{Q}_1$, and $\mathbf{Q}_2$ must now all vanish.

To prove (b) we now observe that $\mathbf{P}_1$, $\mathbf{P}_2$, $\mathbf{Q}_1$, and $\mathbf{Q}_2$ are all tangent vectors in the same tangent plane to the unit sphere. As they are mutually orthogonal at least two of them have to vanish. But apparently, there are more possibilities than those listed in (b).

Suppose $\mathbf{P}_1$ and $\mathbf{Q}_1$ vanish. Then $\partial_2\hat{\phi} = \frac{1}{2}(\mathbf{P}_1 + \mathbf{Q}_1)$ has to vanish too, and by the above remark they therefore all vanish. Similarly, $\mathbf{P}_2 = \mathbf{Q}_2 = 0$ makes then all vanish. Suppose then that $\mathbf{P}_1$ and $\mathbf{Q}_2$ vanish. From $\mathbf{Q}_2 = 0$ we get

$$\partial_1\hat{\phi} = \hat{\phi}\times\partial_2\hat{\phi};$$

and when we insert this in the expression for $\mathbf{P}_1$, we obtain

$$0 = \mathbf{P}_1 = \partial_2\hat{\phi} - \hat{\phi} \times (\hat{\phi} \times \partial_2\hat{\phi}) = 2\partial_2\hat{\phi}.$$

So again $\partial_2\hat{\phi}$ has to vanish, and therefore they all vanish. Similarly, $\mathbf{P}_2 = \mathbf{Q}_1 = 0$ makes them all vanish.

This leaves us with the two possibilities listed in (b).

Solution to 10.8.4

a. From Section 10.3 we know that the stereographic projection is a conformal map with the conformal factor

$$\Omega^2 = \sin^4\left(\frac{\theta}{2}\right) = \frac{1}{(1 + w\bar{w})^2}\,.$$

Consequently, we get

$$\frac{1}{2}\left(|\partial_1\hat{\phi}|^2 + |\partial_2\hat{\phi}|^2\right) = \frac{|\partial_1 w|^2 + |\partial_2 w|^2}{2(1 + w\bar{w})^2}\,.$$

Thus the energy density may be reexpressed in terms of complex variables as follows:

$$H = \frac{\partial_i w \partial_i \bar{w}}{2(1 + w\bar{w})^2} = \frac{2\left\{\frac{\partial w}{\partial z}\frac{\partial \bar{w}}{\partial \bar{z}} + \frac{\partial w}{\partial \bar{z}}\frac{\partial \bar{w}}{\partial z}\right\}}{(1 + w\bar{w})^2}\,.$$

b. The corresponding Euler–Lagrange equations can be written in complex form as

$$\frac{\partial H}{\partial w} = \frac{\partial}{\partial z}\left\{\frac{\partial H}{\partial \frac{\partial w}{\partial z}}\right\} + \frac{\partial}{\partial \bar{z}}\left\{\frac{\partial H}{\partial \frac{\partial w}{\partial \bar{z}}}\right\};$$

$$\frac{\partial H}{\partial \bar{w}} = \frac{\partial}{\partial z}\left\{\frac{\partial H}{\partial \frac{\partial \bar{w}}{\partial z}}\right\} + \frac{\partial}{\partial \bar{z}}\left\{\frac{\partial H}{\partial \frac{\partial \bar{w}}{\partial \bar{z}}}\right\}.$$

E.g., the first equation reduces to

$$\begin{aligned}
&-\frac{2(1 + w\bar{w})\bar{w}}{(1 + w\bar{w})^4}\frac{\partial w}{\partial z}\frac{\partial \bar{w}}{\partial \bar{z}} + \frac{\partial w}{\partial \bar{z}}\frac{\partial \bar{w}}{\partial z}\\
&= \frac{\partial}{\partial z}\frac{1}{(1 + w\bar{w})^2}\frac{\partial \bar{w}}{\partial \bar{z}} + \frac{\partial}{\partial \bar{z}}\frac{1}{(1 + w\bar{w})^2}\frac{\partial \bar{w}}{\partial z}\\
&= \frac{(1 + w\bar{w})^2\frac{\partial^2 \bar{w}}{\partial \bar{z}\partial z} - 2(1 + w\bar{w})\left[\frac{\partial w}{\partial z}\bar{w} + w\frac{\partial \bar{w}}{\partial z}\right]\frac{\partial \bar{w}}{\partial \bar{z}}}{(1 + w\bar{w})^4}\\
&\quad + \frac{(1 + w\bar{w})^2\frac{\partial^2 \bar{w}}{\partial \bar{z}\partial z} - 2(1 + w\bar{w})\left[\frac{\partial w}{\partial \bar{z}}\bar{w} + w\frac{\partial \bar{w}}{\partial \bar{z}}\right]\frac{\partial \bar{w}}{\partial z}}{(1 + w\bar{w})^4},
\end{aligned}$$

which after a long, but trivial, calculation reduces to

$$(1 + w\bar{w}) \frac{\partial^2 \bar{w}}{\partial \bar{z} \partial z} = 2w \frac{\partial \bar{w}}{\partial z} \frac{\partial \bar{w}}{\partial \bar{z}} .$$

c. We must show that f satisfies the Cauchy–Riemann equation $\frac{\partial f}{\partial \bar{z}} = 0$.

$$\begin{aligned}
\frac{\partial f}{\partial \bar{z}} &= \frac{(1 + w\bar{w})^2 \left\{ \frac{\partial^2 w}{\partial \bar{z} \partial z} \frac{\partial \bar{w}}{\partial z} + \frac{\partial w}{\partial z} \frac{\partial^2 \bar{w}}{\partial \bar{z} \partial z} \right\}}{(1 + w\bar{w})^4} \\
&\quad - \frac{\frac{\partial w}{\partial z} \frac{\partial \bar{w}}{\partial z} 2(1 + w\bar{w}) \left\{ \frac{\partial w}{\partial \bar{z}} \bar{w} + w \frac{\partial \bar{w}}{\partial \bar{z}} \right\}}{(1 + w\bar{w})^4} \\
&= \frac{1}{(1 + w\bar{w})^3} \left\{ (1 + w\bar{w}) \frac{\partial^2 w}{\partial \bar{z} \partial z} - 2\bar{w} \frac{\partial w}{\partial z} \frac{\partial w}{\partial \bar{z}} \right. \\
&\quad \left. + (1 + w\bar{w}) \frac{\partial^2 \bar{w}}{\partial \bar{z} \partial z} - 2w \frac{\partial \bar{w} \partial \bar{w}}{\partial z \partial \bar{z}} \right\}.
\end{aligned}$$

Inserting the equations of motion, we see that the right-hand side vanishes automatically. In the same way we can show that $g(z)$ is antiholomorphic.

d. If $f(z)$ has no poles, it is an *entire* function, i.e., a global holomorphic function. As the energy density vanishes at infinity, we conclude that

$$\lim_{\rho \to \infty} = \frac{\left| \frac{\partial w}{\partial z} \right|}{(1 + w\bar{w})} = 0; \qquad \lim_{\rho \to \infty} = \frac{\left| \frac{\partial \bar{w}}{\partial z} \right|}{(1 + w\bar{w})} = 0.$$

Consequently, $|f(z)| \to 0$ at infinity, and therefore, $f(z)$ must be bounded. But a bounded entire function is necessarily constant (Liouville's theorem). Therefore, $f(z)$ must, in fact, vanish identically. But then the equations of motion (10.81) reduce to $\partial w / \partial \bar{z} = 0$ or $\partial w / \partial z = 0$.

Solution to 10.9.1

a. $\langle {}^* \boldsymbol{d}\phi \mid i\boldsymbol{d}\phi \rangle = -i \int_{\mathbb{R}^2} \boldsymbol{d}\bar{\phi} \wedge \boldsymbol{d}\phi = -i \int_{\mathbb{R}^2} \boldsymbol{d}[\bar{\phi}\boldsymbol{d}\phi] = \lim_{\rho_0 \to \infty} -i \int_{\rho = \rho_0} \bar{\phi}\boldsymbol{d}\phi = -i\phi_0^2 \lim_{\rho_0 \to \infty} \int_{\rho = \rho_0} i\boldsymbol{d}\varphi = 2n\pi\phi_0^2$.

b. Expanding the norms, we immediately get

$$\begin{aligned}
&\frac{1}{2} \|\mathbf{B} - \sqrt{2U[\phi]}\epsilon\|^2 + \frac{1}{4} \|{}^*\boldsymbol{D}\phi - i\boldsymbol{D}\phi\|^2 \\
&= \frac{1}{2} \langle \mathbf{B} \mid \mathbf{B} \rangle + \langle \sqrt{U[\phi]}\epsilon \mid \sqrt{U[\phi]}\epsilon \rangle - \langle \mathbf{B} \mid \sqrt{2U[\phi]}\epsilon \rangle \\
&\quad + \frac{1}{2} \langle \boldsymbol{D}\phi \mid \boldsymbol{D}\phi \rangle - \frac{1}{2} \langle {}^*\boldsymbol{D}\phi \mid i\boldsymbol{D}\phi \rangle.
\end{aligned}$$

Consequently, the remainder term is given by

$$A + B = \langle \mathbf{B} \mid \sqrt{2U[\phi]}\epsilon\rangle + \frac{1}{2}\langle {}^*\boldsymbol{D}\phi \mid i\boldsymbol{D}\phi\rangle.$$

We will now determine when this is a topological term. Using that $\boldsymbol{D} = \boldsymbol{d} - ie\mathbf{A}$, we can expand the B-term as

$$\frac{1}{2}\langle {}^*\boldsymbol{d}\phi \mid i\boldsymbol{d}\phi\rangle + \frac{1}{2}\langle e\phi^*\mathbf{A} \mid \boldsymbol{d}\phi\rangle + \frac{1}{2}\langle {}^*\boldsymbol{d}\phi \mid e\phi\mathbf{A}\rangle$$
$$+ \frac{1}{2}ie^2\langle \phi^*\mathbf{A} \mid \phi\mathbf{A}\rangle = B_1 + B_2 + B_3 + B_4.$$

Here B_1 is the topological term we are after:

$$B_1 = \frac{1}{2}\langle {}^*\boldsymbol{d}\phi \mid i\boldsymbol{d}\phi\rangle = n\pi\phi_0^2.$$

Furthermore, B_4 vanishes automatically, since

$$B_4 = \frac{1}{2}ie^2\langle \phi^*\mathbf{A} \mid \phi\mathbf{A}\rangle = -\frac{1}{2}ie^2\int |\phi|^2\mathbf{A}\wedge\mathbf{A} \qquad \text{but } \mathbf{A}\wedge\mathbf{A} = 0.$$

Thus we are left with B_2 and B_3, which we can rearrange as

$$B_2 + B_3 = -\frac{1}{2}e\int(\bar{\phi}\boldsymbol{d}\phi + \phi\boldsymbol{d}\bar{\phi})\wedge\mathbf{A} = -\frac{1}{2}e\int \boldsymbol{d}|\phi|^2\wedge\mathbf{A}.$$

Here we can safely replace $|\phi|^2$ with $|\phi|^2 - \phi_0^2$, since the constant drops out anyway when we take the exterior derivative. Thus we get

$$B_2 + B_3 = -\frac{1}{2}e\int \boldsymbol{d}(|\phi|^2 - \phi_0^2)\wedge\mathbf{A}$$
$$= -\frac{1}{2}e\int \boldsymbol{d}[\mathbf{A}(|\phi|^2 - \phi_0^2)] + \frac{1}{2}e\int \mathbf{B}(|\phi|^2 - \phi_0^2).$$

The first term is converted to a line integral at infinity, and it now vanishes, since $|\phi|^2 - \phi_0^2$ vanishes at infinity. To summarize, the remainder term has thus been broken down to

$$\int_{\mathbb{R}} \mathbf{B}\sqrt{2U[\phi]} + \int_{\mathbb{R}} \frac{1}{2}\mathbf{B}e(|\phi|^2 - \phi_0^2) + n\pi\phi_0^2.$$

But here the first two terms obviously drop out, provided that

$$\sqrt{2U[\phi]} = \frac{1}{2}e(\phi_0^2 - |\phi|^2); \qquad \text{i.e., } U[\phi] = \frac{1}{8}e^2(|\phi|^2 - \phi_0^2)^2.$$

(In this way we have actually recovered the ϕ^4-model!)

Solution to 10.9.2

Let us first look at equation (10.84a):

$$\mathbf{B} = \frac{1}{2}e(\phi_0^2 - |\phi|^2)\epsilon.$$

On the other hand, we get

$$\mathbf{B} = \boldsymbol{dA} = \frac{n}{e}\frac{dA}{d\rho}\,\boldsymbol{d\rho} \wedge \boldsymbol{d\varphi} = \frac{n}{e\rho}\frac{dA}{d\rho}\,\epsilon.$$

By comparison, we thus obtain the first field equation.

The other one is a bit tricky. First we observe that (cf. Exercise 7.5.5)

$$^*\boldsymbol{d\rho} = -\rho\boldsymbol{d\varphi} \quad \text{and} \quad ^*\boldsymbol{d\varphi} = \frac{1}{\rho}\,\boldsymbol{d\rho}.$$

Then we get by a trivial calculation

$$\boldsymbol{D}\phi = \phi_0 e^{in\varphi}\frac{d\tilde{\phi}}{d\rho}\,\boldsymbol{d\rho} + \phi_0\tilde{\phi}e^{in\varphi}in(1-A)\boldsymbol{d\varphi}.$$

The two sides of the self-duality equation therefore reduce to

$$^*\boldsymbol{D}\phi = \phi_0 e^{in\varphi}\frac{d\tilde{\phi}}{d\rho}\,(-\rho\boldsymbol{d\varphi}) + \phi_0\tilde{\phi}e^{in\varphi}in(1-A)\frac{1}{\rho}\,\boldsymbol{d\rho},$$

$$i\boldsymbol{D}\phi = i\phi_0 e^{in\varphi}\frac{d\tilde{\phi}}{d\rho}\,\boldsymbol{d\rho} - \phi_0\tilde{\phi}e^{in\varphi}n(1-A)\boldsymbol{d\varphi}.$$

By equating the coefficients of $\boldsymbol{d\rho}$ or $\boldsymbol{d\varphi}$ we now finally obtain the second field equation.

CHAPTER 11

Symmetries and Conservation Laws

§ 11.1 Conservation Laws

We start by investigating the general structure of conservation laws in physics. They are concerned with a vector field **J** satisfying the equation of continuity (see (7.88))

$$\delta \mathbf{J} = 0 \quad ; \quad \left[\frac{1}{\sqrt{-g}} \partial_\mu \left(\sqrt{-g} J^\mu \right) = 0 \right]. \tag{11.1}$$

The vector field **J** is interpreted as a "current," and if Ω is a domain in a space slice, then the flux of **J** through Ω, given by the integral

$$Q = -\int {}^*\mathbf{J}, \tag{11.2}$$

is interpreted as a "charge." (The minus sign is conventional. Compare with (8.53).)

Consider now a tube as shown in Figure 11.1, and let us assume that the current **J** vanishes outside the tube. (The width of the tube can be very large, and the fact that

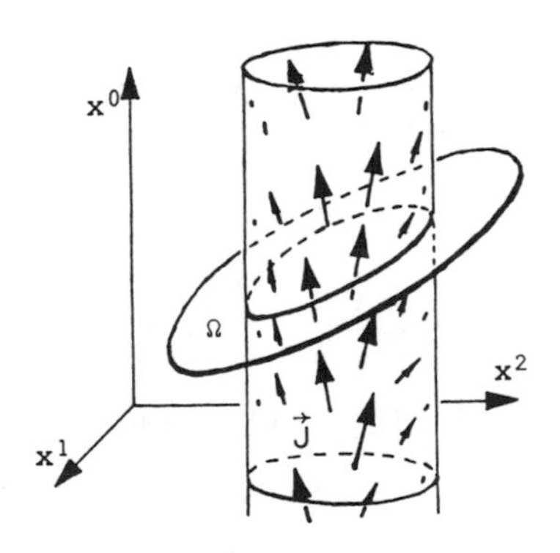

Figure 11.1.

the current vanishes outside the tube means in practice that "the current vanishes sufficiently fast at spatial infinity.") We can then show the following lemma:

Lemma 11.1 (The tube lemma). *A smooth "current"* $\mathbf{J}$ *that satisfies the continuity equation (11.1) will produce the same flux* $-\int_\Omega {}^*\mathbf{J}$ *through any closed hypersurface* Ω *intersected by the tube, or equivalently, the "charge" contained in a hypersurface will be the same for all hypersurfaces intersected by the tube.*

PROOF. Let Ω_1 and Ω_2 be two hypersurfaces intersected by the tube (Figure 11.2). We will assume that they are far apart. Together with the wall of the tube they then constitute the boundary of a 4-dimensional regular domain W. Be careful about the orientations relative to W! From Stokes's theorem we now obtain

$$0 = \int_W {}^*\delta\mathbf{J} = \int_W d^*\mathbf{J} = \int_{\partial W} {}^*\mathbf{J} = \int_{\Omega_2} {}^*\mathbf{J} - \int_{\Omega_1} {}^*\mathbf{J}.$$

(If Ω_1 and Ω_2 are not far apart, we introduce a third hypersurface Ω_3 that is far apart from Ω_1 and Ω_2 and use transitivity.) □

Okay, having constructed the machinery, we can now apply it to some important cases:

EXAMPLE 11.1: CONSERVATION OF ELECTRIC CHARGE.

The electric current is represented by a coclosed 1-form. Let us prove that this leads to the conservation of electric charge. To avoid convergence problems, we assume that the current is confined to a tube in space–time. Suppose S is an inertial frame of reference, and let Ω_1 and Ω_2 be three-dimensional regular domains contained in space slices relative to S such that Ω_1 surrounds the charges and currents at time t_1 and similarly for Ω_2 at time t_2 (see Figure 11.3). Then the total charge at time t_1, respectively t_2, is given by (cf. (8.53))

$$Q(t_1) = -\int_{\Omega_1} {}^*\mathbf{J}; \quad \text{respectively } Q(t_2) = -\int_{\Omega_2} {}^*\mathbf{J}.$$

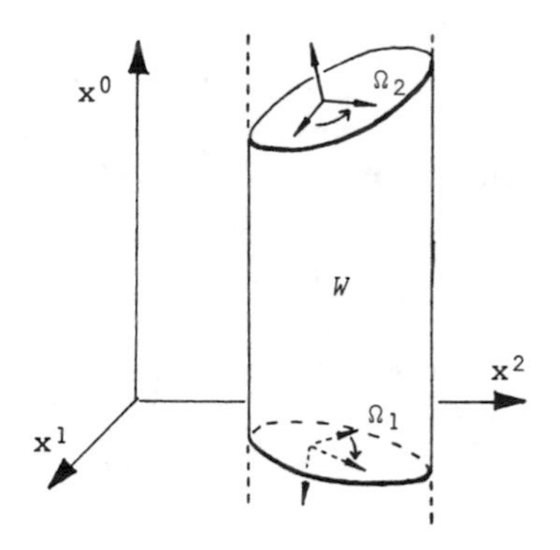

Figure 11.2.

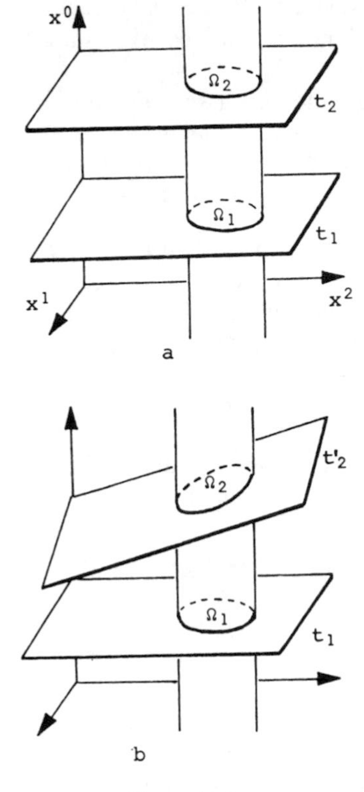

Figure 11.3.

But according to the tube lemma, they are identical; i.e., the charge measured by an observer in S is independent of time.

Actually, we can prove something more. Suppose S_1 and S_2 are two different inertial frames of reference. Then we can let Ω_1 be a regular domain contained in a space slice relative to S_1 and Ω_2 a regular domain contained in a space slice relative to S_2 (see Figure 11.3). As we have seen, the electric charge measured in S_1, respectively S_2, is given by the integral

$$Q_1 = -\int_{\Omega_1} {}^*\mathbf{J} \quad \text{respectively} \quad Q_2 = -\int_{\Omega_2} {}^*\mathbf{J}.$$

But by the tube lemma they are identical. Thus the electric charge is independent of the observer too.

We have now given the promised proof of the first half of Abraham's theorem (Section 1.6):

Theorem 11.1. *If* **J** *is a coclosed* 1*-form that vanishes sufficiently fast at spatial infinity, then the "charge"*

$$Q = -\int_{\Omega} {}^*\mathbf{J} = \int_{x^0=t^0} J^0 dx^1 dx^2 dx^3$$

is independent of time and the observer.

EXAMPLE 11.2: CONSERVATION OF ENERGY AND MOMENTUM.

We start by observing that the energy and momentum densities of a field configuration are represented by a *symmetric* tensor of rank 2 (cf. the discussion in Section 1.6). At first thought we may therefore think that the analysis of energy and momentum falls beyond the scope of exterior calculus, which deals only with *skew-symmetric* cotensors. On the other hand, we want to define the total energy and the total momentum of a field configuration. This involves integrals of the form

$$\int_{x^0=t^0} T^{\alpha\circ} d^3x,$$

and it would be nice to have an *invariant*, i.e., *geometrical*, characterization of these integrals.

To do that, we "mimic" the discussion of electric charge. We select an inertial frame of reference S. This inertial frame of reference is characterized by a set of inertial coordinates (x^0, x^1, x^2, x^3) generating the basic tangent vectors

$$\mathbf{e}_0, \mathbf{e}_1, \mathbf{e}_2, \mathbf{e}_3.$$

They constitute four vector fields, which, loosely speaking, specifies the direction of the Cartesian axes and the direction of time. By contracting the cotensor **T** with these four vector fields, we obtain four new 1-forms

$$\mathbf{P}_0 = \mathbf{T}(\mathbf{e}_0;\cdot); \quad \mathbf{P}_1 = \mathbf{T}(\mathbf{e}_1;\cdot); \quad \mathbf{P}_2 = \mathbf{T}(\mathbf{e}_2;\cdot); \quad \mathbf{P}_3 = \mathbf{T}(\mathbf{e}_3;\cdot), \qquad (11.3)$$

which in our inertial frame are characterized by the components

$$\mathbf{P}_0 : T_{\alpha\beta}\delta^\alpha_0 = T_{0\beta}; \quad \mathbf{P}_i : T_{\alpha\beta}\delta^\alpha_i = T_{i\beta}.$$

It follows immediately that $\mathbf{P}_0$ represents the energy current, while $\mathbf{P}_i$ represents the momentum current along the x^i-axis. Observe that although the currents $\mathbf{P}_0$, $\mathbf{P}_1$, $\mathbf{P}_2$, $\mathbf{P}_3$ are purely geometrical quantities, they depend strongly on the choice of the inertial frame of reference in question.

In terms of the inertial coordinates associated with S, we know that the energy momentum tensor **T** satisfies the continuity equation (1.38)

$$\partial_\beta T^{\alpha\beta} = 0,$$

and we therefore conclude that

$$\partial_\beta(\mathbf{P}_\alpha)^\beta = 0,$$

where $(\mathbf{P}_\alpha)^\beta$ are the contravariant components of $\mathbf{P}_\alpha$. Consequently, the current $\mathbf{P}_\alpha$ is conserved, which can be expressed in the geometrical form $\delta\mathbf{P}_\alpha = 0$. Thus $\mathbf{P}_0$, $\mathbf{P}_1$, $\mathbf{P}_2$, $\mathbf{P}_3$, all give rise to conserved quantities. If Ω is a space slice characterized by the equation $x^0 = t^0$, then the integrals

$$P_0 = -\int_\Omega {}^*\mathbf{P}_0 \quad , \quad P_i = -\int_\Omega {}^*\mathbf{P}_i \tag{11.4}$$

are all time-independent, provided that the energy and momentum currents vanish sufficiently fast at spatial infinity. Let us interpret these integrals. Using that $dx^0 = 0$ along Ω, we get

$$P_\alpha = \int_{x^0=t^0} (\mathbf{P}_\alpha)^0 dx^1 dx^2 dx^3 = \int_{x^0=t^0} T^0_\alpha dx^1 dx^2 dx^3.$$

Observing that in an inertial frame $-T^{00} = T^0_0, T^{io} = T^0_i$, we thus finally obtain that

$$-P_o = \int_{x^o=t^o} T^{oo} dx^1 dx^2 dx^3 = \text{the total energy,} \tag{11.5}$$

$$P_i = \int_{x^o=t^o} T^{io} dx^1 dx^2 dx^3 = \text{the total momentum along the } x^1\text{-axis.} \tag{11.6}$$

Consequently, the total energy and momentum are conserved. (Do not be confused by the signs. If we "raise" the index α, we get

$$[P^0, P^i] = [\text{total energy, total momentum along } x^i\text{-axis}]$$

with the correct signs as usual!)

We will next investigate what happens when we exchange the inertial frame of reference. Suppose S_2 is another inerrtial frame of rerference connected to S_1 through a Poincaré transformation

$$y^\alpha = A^\alpha{}_\beta x^\beta + b^\alpha.$$

It follows that the new basic tangent vectors $\underset{(2)^\alpha}{\mathbf{e}}$ are connected to the old $\underset{(1)^\alpha}{\mathbf{e}}$ through the formula (cf. (6.12))

$$\underset{(2)^\alpha}{\mathbf{e}} = \underset{(1)^\beta}{\mathbf{e}} \frac{\partial x^\beta}{\partial y^\alpha} = \underset{(1)^\beta}{\mathbf{e}} \breve{A}^\beta_\alpha,$$

where $\breve{A}^\beta_\alpha$ is the reciprocal Lorentz matrix. The new basic tangent vectors give rise to new energy and momentum currents

$$\underset{(2)^\alpha}{P} = \mathbf{T}\left(\underset{(2)^\alpha}{\mathbf{e}}; \cdot\right) = \mathbf{T}\left(\underset{(1)^\beta}{\mathbf{e}}; \cdot\right) \breve{A}^\beta{}_\alpha = \underset{(1)^\beta}{P} \breve{A}^\beta_\alpha. \tag{11.7}$$

Consequently, the new currents depend linearly upon the old currents. Let Ω_1 and Ω_2 be two three-dimensional volumes such that Ω_1 is contained in a space slice relative to S_1 and Ω_2 in a space slice relative to S_2 (See Figure 11.3). From the tube lemma we then get

$$\underset{(2)^\alpha}{P} = -\int_{\Omega_2} * \underset{(2)^\alpha}{\mathbf{P}} = -\int_{\Omega_1} * \underset{(2)^\alpha}{\mathbf{P}}.$$

Using (11.7), this is rearranged as

$$\underset{(2)^\alpha}{P} = -\int_{\Omega_1} * \underset{(1)^\alpha}{\mathbf{P}} - \int_{\Omega_1} * \left[\underset{(1)^\beta}{\mathbf{P}} \breve{A}^\beta{}_\alpha \right] = \left[\int_{\Omega_1} * \underset{(1)^\beta}{\mathbf{P}} \right] \breve{A}^\beta{}_\alpha = \underset{(1)^\beta}{P} \breve{A}^\beta{}_\alpha. \qquad (11.8)$$

Formally, this shows that the components (P_0, P_1, P_2, P_3) of the total energy–momentum vector transform like a Lorentz covector under a Poincaré transformation. However, we *cannot* interpret (P_0, P_1, P_2, P_3) as the components of some covector for the following reason: It has no base point, as it is not attached to any specific event in space–time!

To summarize, we have now given the promised proof of the second half of Abraham's theorem (Section 1.6):

Theorem 11.2. *Suppose* $\mathbf{T}$ *is a symmetric cotensor of rank 2 in Minkowski space that is conserved (i.e.,* $\partial_\beta T^{\alpha\beta} = 0$ *in an inertial frame). Let S be an inertial frame of reference generating the basic vector fields* $\mathbf{e}_0, \ldots, \mathbf{e}_3$*. Then the following hold:*

1. *The 1-forms* $\mathbf{P}_\alpha = \mathbf{T}(\mathbf{e}_\alpha, \cdot)$ *are coclosed.*
2. *The "charges"*

$$P_\alpha = -\int_\Omega {}^*\mathbf{P}_\alpha = \int_{x^0=t^0} T^0_\alpha \mathrm{d}x^1 \mathrm{d}x^2 \mathrm{d}x^3$$

are conserved, i.e., independent of time.

3. *If two inertial frames of references* S_1 *and* S_2 *are connected through a Poincaré transformation*

$$y^\alpha = A^\alpha{}_\beta x^\beta + b^\alpha,$$

then the "charge" transforms as a Lorentz-covariant quantity; i.e.,

$$\underset{(2)^\alpha}{P} = \underset{(1)^\beta}{P} \breve{A}^\beta{}_\alpha. \qquad (11.8)$$

Remark. In the theory of general relativity, space–time is curved; and therefore we cannot choose a global inertial frame. The energy–momentum currents then no longer exist, and this leads to difficulties in the interpretation of various quantities like "the total energy of a closed system."

§ 11.2 Symmetries and Conservation Laws in Quantum Mechanics

Now that we understand the formal aspects of conservation laws, we turn our attention to symmetries. To get a feeling for the machinery, we will investigate symmetry transformations in ordinary quantum mechanics. We will also indicate how they lead to conservation laws in quantum mechanics, but these are of a very trivial type involving no geometry, so this aspect is included only for physical reasons.

We start by considering a single particle moving in a potential $V(\mathbf{x})$. The system is completely characterized by its Schröinger wave function $\psi(t, \mathbf{x})$. Let us neglect dynamics for a moment and just consider the wave function as an ordinary complex-valued scalar field defined on the Euclidean space $\mathbf{R}^3$. Suppose now that $(f_\lambda)_{\lambda \in \mathbf{R}}$ is a family of diffeomorphisms. Then we can push forward the wave function

$$\psi_\lambda(\mathbf{x}) = (f_\lambda)_* \psi(\mathbf{x}). \tag{11.9}$$

But as this operation is linear, we may equivalently say that we have constructed a family of linear operators, $(\hat{U}_\lambda)_{\lambda \in \mathbf{R}}$, where

$$\hat{U}_\lambda \psi = (f_\lambda)_* \psi. \tag{11.10}$$

If furthermore, f_λ is a family of *isometrics*, then $\hat{U}_\lambda$ preserves the inner product (see Theorem 10.13 in Section 10.5):

$$\langle \hat{U}_\lambda \psi_2 \mid \hat{U}_\lambda \psi_1 \rangle = \langle (f_\lambda)_* \psi_2 \mid (f_\lambda)_* \psi_1 \rangle = \langle \psi_2 \mid \psi_1 \rangle_1.$$

Consequently, $(\hat{U}_\lambda)_{\lambda \in \mathbf{R}}$ is a family of *unitary* operators. Observe that $\hat{U}_\lambda \psi$ is normalized too, i.e., $\int |\hat{U}_\lambda \psi|^2 dV = 1$, so we can interpret $\hat{U}_\lambda \psi$ as a wave function; i.e., $\hat{U}_\lambda \psi$ represents another possible state of the system. Such families of unitary operators have been studied intensively both from a physical and a mathematical point of view.

Before we go into a further discussion, let us recapitulate a few basic facts from elementary quantum mechanics. Suppose $\hat{T}$ is a Hermitian operator representing a physical quantity T (e.g., the energy represented by $\hat{H} = -\frac{\hbar^2}{2m}\Delta + V(\mathbf{x})$ or the momentum represented by $\hat{p} = -i\hbar\nabla$). When we measure this quantity, the outcome of the experiment is a number, and the possible numbers we can measure are precisely the eigenvalues of the operator $\hat{T}$. Let ψ_n be a complete set of normalized eigenfunctions with the associated eigenvalues λ_n,

$$\hat{T}\psi_n = \lambda_n \psi_n.$$

If the state of the systems is charaterized by the wave function ψ, we can decompose it using the above eigenfunctions:

$$\psi = \Sigma a_n \psi_n \quad \text{with} \quad a_k = \langle \psi \mid \psi_k \rangle.$$

Here $|a_n|^2$ is the probability of measuring the number λ_n, and the *mean value* of the physical quantity $\hat{T}$ is given by

$$\langle T \rangle = \int \bar{\psi} \hat{T} \psi dV = \Sigma_n |a_n|^2 \lambda_n. \tag{11.11}$$

Now observe that a Hermitian operator $\hat{T}$ generates a family of unitary operators given by

$$\hat{U}_\lambda = \exp\left[-\frac{i}{\hbar}\lambda \hat{T}\right]. \tag{11.12}$$

This family is actually a one-parameter group; ie, it satisfies

$$\hat{U}_{\lambda_2+\lambda_1} = \hat{U}_{\lambda_2}\hat{U}_{\lambda_1},$$

and the Hermitian operator $\hat{T}$ is called the *infinitesimal generator* of this one-parameter group. On the other hand, if we have been given a one-parameter group of unitary transformations $\hat{U}_\lambda$, then a famous theorem of Stone guarantees that it is actually of the form

$$\hat{U}_\lambda = \exp\left[-\frac{i}{\hbar}\lambda\hat{T}\right] \tag{11.12}$$

for some Hermitian operator $\hat{T}$. (Of course, $\hbar$ has been introduced only for convenience. It will not occur in mathematical references!) Thus there is a bijective correspondence between one-parameter groups of unitary operators and physical quantities represented by Hermitian operators.

In the preceding discussion we have seen that a family of isometrics generates a family of unitary operators. If we assume that the family of isometries is a one-parameter group, i.e., it satisfies

$$f_{\lambda_2+\lambda_1} = f_{\lambda_2} \circ f_{\lambda_1},$$

then it will actually generate a one-parameter group of uitary transformations. In the end we will therefore find that isometries in space are linked up with various physical quantities!

When $(f_\lambda)_{\lambda\in\mathbf{R}}$ is a one-parameter group of isometries, we can associate a vector field $\mathbf{a}$ to this group. This vector field is the geometrical equivalent of the infinitesimal generator in physics. Geometrically, it is generated by the curves $\lambda \to f_\lambda(P)$ at $\lambda = 0$. (Observe that $f_\lambda(P)$ passes through P at $\lambda = 0$, thus generating a vector at P.) If we introduce coordinates, f_λ will be represented by $y^i = f^i_\lambda(x^j)$, and the components of the vector field $\mathbf{a}$ are thus given by

$$a^i = \frac{dy^i}{d\lambda}\Big|_{\lambda=0}. \tag{11.13}$$

We call this vector field *the characteristic vector field associated with the one-parameter group of isometries.*

NB! In what follows we will have to do many differentiations with respect to λ at $\lambda = 0$. For typographical reasons we will drop the suffix $|_{\lambda=0}$, so *in the rest of this chapter it will always be understood that differentiations are to be carried out at $\lambda = 0$.*

Using the characteristic vector field, we can now construct the infinitesimal generator $\hat{T}$:

$$\begin{aligned}
-\frac{i}{\hbar}\hat{T}\psi &= \frac{d}{d\lambda}\left(e^{-\frac{i}{\hbar}\lambda\hat{T}}\psi\right) = \frac{d}{d\lambda}(\hat{U}_\lambda\psi) \\
&= \frac{d}{d\lambda}(f_\lambda)_*\psi = \frac{d}{d\lambda}\psi(f_{-\lambda}(x)) \\
&= -\frac{\partial\psi}{\partial x^i}\frac{\partial y^i}{d\lambda} = -\langle \boldsymbol{d}\psi \mid \mathbf{a}\rangle.
\end{aligned}$$

Consequently, we have shown that

The infinitesimal generator $\hat{T}$ is the first-order differential operator given by

$$\hat{T}\psi = -i\hbar\langle d\psi \mid \mathbf{a}\rangle; \qquad \left[\hat{T} = -i\hbar a^j \frac{\partial}{\partial x^j}\right]. \tag{11.14}$$

We can also consider *vector particles*. Quantum-mechanically, they are characterized by a complex-valued vector field $\boldsymbol{\psi}(\mathbf{x})$ (rather than a complex-valued scalar field). But $\boldsymbol{\psi}(\mathbf{x})$ still has the usual interpretation. Its absolute square measures the probability density of finding the vector particle at the point $\mathbf{x}$:

$$P(\mathbf{x}) = \bar{\psi}_i(\mathbf{x})\psi^i(\mathbf{x}). \tag{11.15}$$

Again we can use a family of isometries $(f_\lambda)_{\lambda\in\mathbb{R}}$ to generate a group of unitary transformations $\hat{U}_\lambda\boldsymbol{\psi} = (f_\lambda)_*\boldsymbol{\psi}$, where the transformed wave function is characterized by the components $\psi^i_\lambda(\mathbf{x})$. We proceed to determine the infinitesimal generator T:

$$\begin{aligned} -\frac{i}{\hbar}\hat{T}\psi^i &= \frac{d}{d\lambda}(\hat{U}_\lambda\boldsymbol{\psi})^i = \frac{d}{d\lambda}[(f_\lambda)_*\boldsymbol{\psi}]^i \\ &= \frac{d}{d\lambda}\left[\frac{\partial y^i}{\partial x^j}\psi^j_\lambda(f_{-\lambda}(\mathbf{x}))\right] = \frac{\partial a^i}{\partial x^j}\psi^j - \frac{\partial \psi^i}{\partial x^j}a^j; \end{aligned}$$

i.e., in components we get

$$\hat{T}\psi^i = -i\hbar\left[(\partial_j\psi^i)a^j - (\partial_j a^i)\psi^j\right]. \tag{11.16}$$

This is a bit surprising! By construction we know that the quantity on the right-hand side is a covariant quantity, but the separate terms are not covariant quantities. However, we have a skew-symmetrization involved (in analogy with the exterior derivative!). Thus we have incidently discovered how to construct a new vector field out of two given vector fields, the *commutator* or the *bracket*.

Exercise 11.2.1
Problem: Let $\mathbf{u}$, $\mathbf{v}$ be two given vector fields on a manifold M characterized by the components $a^i(\mathbf{x})$ and $b^i(\mathbf{x})$. Show that

$$(\partial_j b^i)a^j - (\partial_j a^i)b^j \tag{11.17}$$

are the components of another vector field, which we denote by $[\mathbf{u}; \mathbf{v}]$.

Using the notation of exercise 11.2.1 we have thus shown:

The infinitesimal generator $\hat{T}$ is the first-order differential operator given by

$$\hat{T}\boldsymbol{\psi} = -i\hbar[\boldsymbol{\psi}; \mathbf{a}]. \tag{11.18}$$

Exercise 11.2.2
Problem: Show that the bracket satisfies the following rules:

1. Skew-symmetry: $[\mathbf{u}; \mathbf{v}] = -[\mathbf{v}, \mathbf{u}]$.
2. Bilinearity: $[\mathbf{u}; \mathbf{v}_1 + \mathbf{v}_2] = [\mathbf{u}; \mathbf{v}_1] + [\mathbf{u}; \mathbf{v}_2]$; $\quad [\mathbf{u}; \lambda\mathbf{v}] = \lambda[\mathbf{u}; \mathbf{v}]$.
3. Jacobi identity: $[\mathbf{u}; [\mathbf{v}; \mathbf{w}]] + [\mathbf{v}; [\mathbf{w}; \mathbf{u}]] + [\mathbf{w}; [\mathbf{u}; \mathbf{v}]] = 0$.

Exercise 11.2.3
Problem: Let $\hat{T}_1, \hat{T}_2$ be first-order differential operators, i.e., of the form

a. $\hat{T}\psi = -i\hbar\langle d\psi \mid \mathbf{a}\rangle$ (for a scalar particle),
b. $\hat{T}\psi = -i\hbar[\quad;\quad]$ (for a vector particle).

Show that the commutator $[\hat{T}_1, \hat{T}_2]$ is given by the first-order differential operators

$$\text{(a)}\quad [\hat{T}_1, \hat{T}_2]\psi = -\hbar^2\langle d\psi|[\mathbf{a}_1; \mathbf{a}_2]\rangle \quad \text{(for a scalar particle)}, \tag{11.19}$$

$$\text{(b)}\quad [\hat{T}_1, \hat{T}_2]\psi = -\hbar^2[\psi; [\mathbf{a}_1; \mathbf{a}_2]] \quad \text{(for a vector particle)}. \tag{11.20}$$

We are now in a position where we can investigate some symmetry transformations. In quantum mechanics, the dynamical evolution is controlled by the Hamiltonian $\hat{H}$, and a unitary transformation $\hat{U}$ is called a *symmetry transformation* if it commutes with the Hamiltonian; i.e., $[\hat{U}, \hat{H}] = 0$.

This has the following well-known consequence:

If ψ is a solution to the equations of motion, then so is the transformed state $\hat{U}\psi$.

PROOF. Putting $\psi' = \hat{U}\psi$, we get

$$i\hbar\frac{d}{dt}\psi' = i\hbar\frac{d}{dt}\hat{U}\psi = \hat{U}\left[i\hbar\frac{d}{dt}\psi\right] = \hat{U}\hat{H}\psi = \hat{H}\hat{U}\psi = \hat{H}\psi'. \qquad \square$$

But symmetry transformations also lead to conservation laws. If $\hat{T}$ is a Hermitian operator representing the physical quantity T, then $\int \bar{\psi}\hat{T}\psi dV$ is the mean value of this physical quantity, and we say that T is conserved if $\int \bar{\psi}\hat{T}\psi dV$ is constant in time. We can then easily show:

If $\hat{U}_\lambda$ is a one-parameter group of symmetry transformations, then the infinitesimal generator $\hat{T}$ is concerned.

PROOF. From $\hat{U}_\lambda\hat{H} = \hat{H}\hat{U}_\lambda$ we get by differentiation with respect to λ that $\hat{T}\hat{H} = \hat{H}\hat{T}$; i.e., the infinitesimal generator commutes with the Hamiltonian too. But then we obtain

$$\begin{aligned} ih\frac{d}{dt}\langle T\rangle &= i\hbar\frac{d}{dt}\int \bar{\psi}\hat{T}\psi dV \\ &= \int \overline{\left(-i\hbar\frac{d}{dt}\psi\right)}\hat{T}\psi dV + \int \bar{\psi}\hat{T}\left(i\hbar\frac{d}{dt}\psi\right)dV \\ &= -\int \overline{H\psi}\hat{T}\psi dV + \int \bar{\psi}\hat{T}\hat{H}\psi dV = \int \bar{\psi}[\hat{T}, \hat{H}]\psi dV = 0. \end{aligned}$$

$\square$

In our case, $\hat{U}_\lambda$ is generated by applying the space transformations f_λ. Consider the Hamiltonian

$$\hat{H} = -\frac{\hbar^2}{2m}\Delta + V(\bar{x}). \tag{11.21}$$

We know that f_λ commutes with the Laplacian (Theorem 10.13 in Section 10.5), which leaves us with

$$[\hat{U}_\lambda, \hat{H}] = \{(f_\lambda)_* V - V\} U_\lambda;$$

i.e.,

$\hat{U}_\lambda$ commutes with $\hat{H}$ if and only if the potential V is invariant under the transformation group f_λ; i.e., $(f_\lambda)_* V = V$.

§ 11.3 Conservation of Energy, Momentum and Angular Momentum in Quantum Mechanics

Okay, it is time to look at some simple applications. In $\mathbf{R}^3$ the isometry group consists of reflections (which are uninteresting for the present purpose), transformations, and rotations, (Cf. (10.28)).

Translations: In Cartesian coordinates a one-parameter group of translations is given by

$$y^i = x^i + \lambda a^i, \tag{11.22}$$

where a^i obviously are the components of the characteristic vector field. Since the components a^1 are constant, we get that the infinitesimal generator $\hat{T}$ reduces to the first-order differential operator

$$\hat{T} = -i\hbar a^j \frac{\partial}{\partial x^j}. \tag{11.23}$$

(This is valid for both scalar particles and vector particles!) For translations along the Cartesian axis, we get, in particular,

$$\hat{T}_1 = -i\hbar\frac{\partial}{\partial x}; \quad \hat{T}_2 = -i\hbar\frac{\partial}{\partial y}; \quad \hat{T}_3 = -i\hbar\frac{\partial}{\partial z};$$

i.e., the infinitesimal generator of translations represents the operator of momentum!

Furthermore, we see that the momentum is conserved if the potential is translation invariant, i.e., if V is constant. This is certainly reasonable enough: In the presence of a nontrivial potential, the particle will experience forces, and therefore its momentum will change.

Rotations: Consider first as an explicit example rotations around the z-axis. This one-parameter group is generated by the matrix

$$\bar{\bar{R}}_\lambda = \begin{bmatrix} \cos\lambda & -\sin\lambda & 0 \\ \sin\lambda & \cos\lambda & 0 \\ 0 & 0 & 1 \end{bmatrix}; \quad \text{i.e., } f_\lambda(\bar{\bar{x}}_|) = \bar{\bar{R}}_\lambda \bar{\bar{x}}_|.$$

By differentiation we get the characteristic vector field

$$\bar{a}_| = \frac{d}{d\lambda}\bar{\bar{R}}_\lambda \bar{x}_| = \begin{bmatrix} 0 & -1 & 0 \\ 1 & 0 & 0 \\ 0 & 0 & 0 \end{bmatrix} \begin{bmatrix} x \\ y \\ z \end{bmatrix} = \begin{bmatrix} -y \\ x \\ 0 \end{bmatrix},$$

which is circulating around the z-axis.

Exercise 11.3.1
Problem: Show that the rotation matrix $\bar{\bar{R}}_\lambda$ can be rearranged as

$$\bar{\bar{R}}_\lambda = \exp\left[\lambda \bar{\bar{S}}_z\right] \quad \text{with} \quad \bar{\bar{S}}_z = \begin{bmatrix} 0 & -1 & 0 \\ 1 & 0 & 0 \\ 0 & 0 & 0 \end{bmatrix}.$$

so that the characteristic vector field is given by,

$$\bar{a}_| = \bar{\bar{S}}_z \bar{x}_|.$$

Consider the case of a scalar particle. Then the corresponding infinitesimal generator (11.14) is the operator of orbital angular momentum around the z-axis

$$\hat{J}_z = -i\hbar\left[x\frac{\partial}{\partial y} - y\frac{\partial}{\partial x}\right] = \left[\mathbf{r} \times (-i\hbar\boldsymbol{\nabla})\right]_z \tag{11.24}$$

We get similar results for rotations around the z-axis and the y-axis, so we see that the infinitesimal generators for rotations are the operators of orbital angular momentum.

In the case of a vector particle things are slightly more complicated. Here the infinitesimal generator is given by the commutator, and we get an extra term

$$\hat{T}\psi^i = -i\hbar\left[a^j\partial_j\psi^i - \psi^j\partial_j a^i\right],$$

i.e.,

$$\hat{J}_z = -i\hbar\left[x\frac{\partial}{\partial y} - y\frac{\partial}{\partial x}\right] + i\hbar\bar{\bar{S}}_z = \hat{L}_z + \hat{S}_z \tag{11.25}$$

(compare with exercise 11.3.1). The first term is again the orbital angular momentum, but the second term is something new. It can be interpreted as the *spin-operator* for the vector particle. Observe that the operator $i\hbar\bar{\bar{S}}_z$ can have only the eigenvalues $0, \pm\hbar$, i.e., the spin projection along the z-axis can take only the values $0, \pm 1$ (in multiples of $\hbar$). So we see that the vector particle possesses an intrinsic spin of 1 unit. (Compare with the discussion in Section 3.6.) The operator

$$\hat{\mathbf{J}} = \hat{\mathbf{L}} + \hat{\mathbf{S}}, \quad \text{i.e.,} \quad \hat{J}_i = \hat{L}_i + \hat{S}_i \tag{11.26}$$

is then interpreted as the operator for the total angular momentum.

Observe that the angular momentum is conserved in both cases when the potential is rotationally invariant (i.e., when it depends only upon the radial distance r).

Exercise 11.3.2
Introduction: Let us introduce the skew-symmetric matrices $\bar{\bar{S}}_i$ characterized by the components

$$(\bar{\bar{S}}_i)^j{}_k = \epsilon_{ijk}. \tag{11.27}$$

Problem:

a. Show that

$$e^{\lambda \bar{\bar{S}}_i}$$

generates rotations around the x^i-axis.

b. Show that the matrices $\bar{\bar{S}}_i$, $i = 1, 2, 3$ satisfy the commutation rules

$$\left[\bar{\bar{S}}_i, \bar{\bar{S}}_j\right] = \epsilon_{ijk}\bar{\bar{S}}_k. \tag{11.28}$$

c. Let $\mathbf{a}_i$ be the corresponding characteristic vector fields. Show that they satisfy the commutation rules

$$\left[\mathbf{a}_i; \mathbf{a}_j\right] = \epsilon_{ijk}\mathbf{a}_k. \tag{11.29}$$

d. Show that the operators of angular momentum get the following commutation rules:

$$\left[\hat{J}_i, \hat{J}_j\right] = i\hbar\epsilon_{ijk}\hat{J}_k. \tag{11.30}$$

(It should be emnphasized that they hold both for scalar particles and for vector particles!)

Notice that it follows from exercise 11.3.2 that the operator of angular momentum for a scalar particle (11.24) and for a vector particle (11.26) both satisfies Dirac's commutation rule

$$\left[\hat{\mathbf{J}}_i, \hat{\mathbf{J}}_j\right] = i\hbar\epsilon_{ijk}\hat{\mathbf{J}}_k. \tag{9.50}$$

Exercise 11.3.3
Problem: Consider a vector particle. Show that the operator of angular momentum commutes with a spin-orbit coupling of the form

$$V_1(r)\hat{\mathbf{L}} \cdot \hat{\mathbf{S}},$$

where $\hat{\mathbf{L}}$ is the operator of orbital angular momentum and $\hat{\mathbf{S}}$ the spin operator.

It follows from exercise 11.3.3 that a Hamiltonian of the form

$$\hat{H} = -\frac{\hbar^2}{2m}\Delta + V(r) + V_1(r)\mathbf{L} \cdot \mathbf{S} \tag{11.31}$$

is rotationally invariant. (This is the Hamiltonian for an electron moving around a proton if we put $V(r) = -e^2/r$ corresponding to the Coulomb potential and $V_1(r) = \frac{e^2}{2m^2c^2r^3}$ corresponding to the spin-orbit coupling). We can easily show that the orbital angular momentum and the spin are not separately conserved, because they do not commute with the spin-orbit coupling. But according to the above exercise 11.3.3, the total angular momentum will be conserved!

The above discussion of the angular momentum can be generalized to more complicated systems. In the above cases the particle interacted with a scalar potential. We can also investigate what happens when the particle interacts with a gauge potential; i.e., it interacts with the electromagnetic field. For simplicity, we consider only the interaction of the electromagnetic field with a scalar particle.

An electrostatic field, like the Coulomb field, is generated from the electrostatic potential $\phi(\mathbf{r})$, which is completely analogous to the potential $V(\mathbf{r})$ in the above discussion! In that case the operator of angular momentum is therefore given by (11.24).

A monopole field, on the other hand, is generated from a (singular) vector potential $\mathbf{A}(\mathbf{r})$. In the case we have previously seen that the operator of angular momentum is given by

$$\hat{\mathbf{J}} = -i\hbar(\mathbf{r} \times \nabla) - (\mathbf{r} \times q\mathbf{A} + \kappa\hat{r}). \tag{9.47}$$

The correction term to the orbital angular momentum is determined by the similar correction term in the corresponding classical expression; cf. (9.39).

The correction term to the orbital angular momentum can be understood in the following way too: The vector potential $\mathbf{A}$ entering in the Hamiltonian (9.43) is not spherically symmetric. When (f_λ) is a one-parameter family of rotations, we therefore have

$$(f_\lambda)_*\mathbf{A} \neq \mathbf{A}.$$

But it generates the spherically symmetric monopole field, $\mathbf{B} = d\mathbf{A}$, and consequently, we have

$$\mathbf{B} = (f_\lambda)_*\mathbf{B} = (f_\lambda)_* d\mathbf{A} = d[(f_\lambda)_*\mathbf{A}],$$

i.e., $\mathbf{A}$ and $(f_\lambda)_*\mathbf{A}$ are gauge equivalent: $(f_\lambda)_*\mathbf{A} = \mathbf{A} + d\chi_\lambda$. Differentiating with respect to λ at $\lambda = 0$, we therefore conclude (cf. (11.18)) that

$$-q[\mathbf{a}; \mathbf{A}] = d\chi \quad \text{with} \quad \chi = q \cdot \frac{d\chi_\lambda}{d\lambda}_{|\lambda=0}. \tag{11.32}$$

We interpret χ as the infinitesimal generator of the gauge transformation necessary to compensate the rotation. It remains to identify χ. The characteristic vector field $\mathbf{a}$ associated to rotations around the unit vector $\hat{n}$ is given by

$$\mathbf{a} = \hat{n} \times \mathbf{r}.$$

The corresponding operator for the angular momentum around the $\hat{n}$-axis is according to (9.47) given by

$$\hat{\mathbf{n}} \cdot \hat{\mathbf{J}} = \hat{\mathbf{n}} \cdot \mathbf{l} + \hat{\mathbf{n}} \cdot [-\mathbf{r} \times q\mathbf{A} - \kappa\hat{r}].$$

We can now establish

Lemma 11.2. *The infinitesimal generator of the gauge transformation necessary to compensate for the rotation is identical to the correction term to the orbital angular momentum; i.e.,*

$$\chi = \hat{\mathbf{n}} \cdot [-\mathbf{r} \times q\mathbf{A} - \kappa\hat{r}]. \tag{11.33}$$

PROOF. We rewrite the expression for χ as

$$\chi = -(\hat{\mathbf{n}} \times \mathbf{r}) \cdot q\mathbf{A} - \kappa \hat{\mathbf{n}}\hat{r} = -q\left[\mathbf{a} \cdot \mathbf{A} + \frac{g_M}{4\pi}\hat{\mathbf{n}} \cdot \hat{r}\right].$$

From this we get

$$\partial_i \chi = -q\left[(\partial_i a^j)A^j + a^j(\partial_i A^j) + \frac{g_M}{4\pi} n^j \partial_i \left(\frac{x^j}{r}\right)\right].$$

Using that

$$\partial_i a^j + \partial_j a^i = 0,$$

while

$$\partial_i A^j - \partial_j A^i = \frac{g_M}{4\pi}\epsilon_{ijk}\frac{x^k}{r^3},$$

we can rearrange the above expression as

$$\partial_i \chi = -q\left[a^j(\partial_j A^i) - A^j(\partial_j a^i) + \frac{g_M}{4\pi}\left(\epsilon_{ijk} a^j \frac{x^k}{r^3} + n^j \partial_i\left(\frac{x^j}{r}\right)\right)\right].$$

Here the last two terms cancel automatically when we work them out, and we therefore end up with

$$\partial_i \chi = -q\left[a^j(\partial_j A^i) - A^j(\partial_j a^i)\right] = -q[\mathbf{a}; \mathbf{A}]^i. \qquad \square$$

§ 11.4 Symmetries and Conservation Laws in Classical Field Theory

Symmetries of a classical field theory have far-reaching consequences, as we have learned from the discussion in Sections 3.9 and 3.11. Let us now examine this topic once more in a slightly more general framework.

In what follows we will consider only *diffeomorphisms* of the manifold. Usually, we pull back differential forms, but when we restrict ourselves to diffeomorphisms, we can equally wel push them forward.

Consider a field theory consisting of a field configuration ϕ_α and the action functional

$$S_\Omega[\phi_\alpha] = \int_\Omega L[\phi_\alpha]\epsilon.$$

Here ϕ_α can stand for a collection of scalar fields or vector fields or something more complicated, say a cotensor field. Using a diffeomorphism f, we can push forward the space-time region Ω and the field configuration ϕ_α.

Definition 11.1. A diffeomorphism f is a symmetry transformation if

$$S_\Omega[\phi_\alpha] = S_{f(\Omega)}[f_*\phi_\alpha] \qquad (11.34)$$

for arbitrary regular domains Ω and arbitrary field configurations ϕ_α.

This implies, in particular, that $f_*\phi_\alpha$ is an extremum for the action; i.e., it is a solution to the field equation whenever ϕ_α is.

Following the arguments of the preceding section we now extend our considerations to families of diffeomorphisms:

Definition 11.2. A one-parameter family of diffeomorphisms $(f_\lambda)_{\lambda\in\mathbf{R}}$ is a collection of diffeomorphisms with the following properties:

1. $f_0 =$ the identity map.
2. For each point P the curve $\lambda \to f_\lambda(P)$ is smooth; i.e., f_λ depends smoothly upon λ.

Observe that the one-parameter family $(f_\lambda)_{\lambda\in\mathbf{R}}$ generates a tangent vector field on M in a canonical fashion. To each point P we have associated a smooth curve $f_\lambda(P)$ passing through P at $\lambda = 0$, thereby generating the tangent vector

$$\mathbf{a}_P = \frac{df_\lambda(P)}{d\lambda}. \tag{11.35}$$

(Here it is understood that differentiation is carried out with respect to $\lambda = 0$!). The corresponding vector field $\mathbf{a}$ is called the *characteristic vector field* associated with the one-parameter family of diffeomorphisms $(f_\lambda)_{\lambda\in\mathbf{R}}$.

Suppose ϕ_α is a field configuration in Minkowski space, and let us introduce coordinates (x^μ) on the region of space–time in question. (See Figure 11.4.)

Then f_λ is represented by the coordinate map $y^\mu = f_\lambda^\mu(x^\nu)$, which reduces to the identity map at $\lambda = 0$; i.e., $f_0{}^\mu(x^\nu) = x^\nu$. Observe that $dy^\mu/d\lambda$ are the components of the characteristic vector field $\mathbf{a}$. The space–time region Ω corresponds to the coordinate domain U and the displaced region to $f_\lambda(U)$. The field configuration ϕ_α is represented by ordinary Euclidean functions $\phi_\alpha(x^\mu)$, and so is the transport of ϕ_α; i.e., $(f_\lambda)_*\phi_{\alpha_\lambda}$ are represented by ordinary Euclidean functions too, which we denote by $\phi^\lambda{}_\alpha(x^\mu)$, reproducing $\phi_\alpha(x^\mu)$ for $\lambda = 0$. (The explicit form of $\phi^\lambda{}_\alpha$ will be calculated later on, when we need it.)

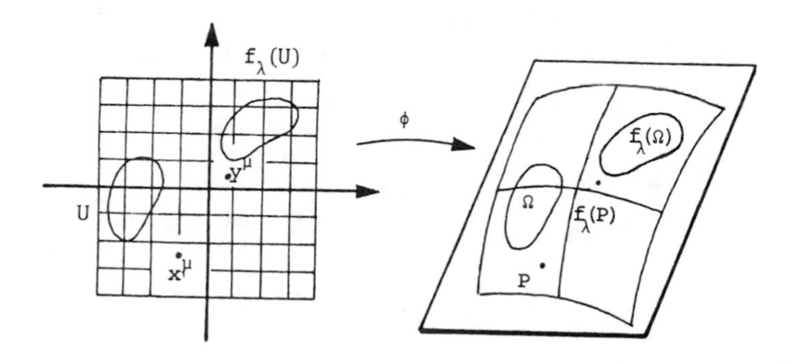

Figure 11.4.

Let us now specialize to a one-parameter family of symmetry transformations. Then we can finally state and prove the celebrated Noether's theorem:

Theorem 11.3 (Noether's theorem). *If a field theory with the canonical energy momentum tensor* $\mathring{T}^{\mu\nu}$ *(Section 3.2) admits a one-parameter family of symmetry transformations* $(f_\lambda)_{\lambda\in\mathbf{R}}$, *then the current*

$$J^\mu = a^\nu \mathring{T}_\nu{}^\mu + \frac{\partial L}{\partial(\partial_\mu \phi_\alpha)} \frac{d\phi_\alpha^\lambda(f_\lambda(x))}{d\lambda} \tag{11.36}$$

is conserved. Here a^ν *are the components of the characteristic vector field and* $\phi_\alpha{}^\lambda$ *the components of the displaced field configuration. (As usual, we calculate the derivative at* $\lambda = 0$!)

PROOF. The proof is quite technical, though interesting. It may be skipped on a first reading. First, we notice that the invariance implies

$$0 = \frac{d}{d\lambda} S_{f_\lambda(\Omega)}[(f_\lambda)_*\phi_\alpha] = \frac{d}{d\lambda}\int_{f_\lambda(\Omega)} L[(f_\lambda)_*\phi_\alpha]\epsilon.$$

The displaced action can then be evaluated as

$$S_{f_\lambda(\Omega)}[(f_\lambda)_*\phi_\alpha] = \int_{f_\lambda(U)} L\left[\phi_\alpha{}^\lambda(y);\ \frac{\partial}{\partial y^\mu}\phi_\alpha{}^\lambda(y);\ g_{\alpha\beta}(y)\right]\sqrt{-g(y)}d^n y.$$

Here we run into the technical difficulty that not only the integrand, but also the integration domain, depend on λ. However, when we calculate the integral using coordinates, we are free to use the same *coordinate domain* U in $\mathbf{R}^n$ for all the integrals involved. So let us transform the integral back to the λ-independent coordinate domain U, performing the substitution $y = f_\lambda(x)$ in the integral. This is legal, because f_λ is a diffeomorphism. We then get

$$S_{f_\lambda(\Omega)}[(f_\lambda)_*\phi_\alpha]$$
$$= \int_U L\left[\phi_\alpha{}^\lambda(y(x));\ \frac{\partial x^\nu}{\partial y^\mu}\frac{\partial}{\partial x^\nu}\phi_\alpha{}^\lambda(y(x));\ g_{\alpha\beta}(y)\right]\sqrt{-g(y(x))}\left|\frac{\partial y}{\partial x}\right| d^n x.$$

(It is important to notice that y still depends on λ!)

We can then finally perform the differentiation and obtain

$$\begin{aligned} 0 = \frac{dS}{d\lambda_{|\lambda=0}} = \int_U \Bigg\{\Bigg[& \frac{\partial L}{\partial \phi_\alpha}\frac{\partial\phi_\alpha{}^\lambda}{d\lambda} + \frac{\partial L}{\partial(\partial_\mu\phi_\alpha)}\frac{d}{d\lambda}\left(\frac{\partial x^\nu}{\partial y^\mu}\frac{\partial}{\partial x^\nu}\phi_\alpha{}^\lambda(y(x))\right) \\ & + \frac{\partial L}{\partial g_{\alpha\beta}}\frac{d}{d\lambda}g_{\alpha\beta}\Bigg]\sqrt{-g(x)} \\ & + L\left[\frac{d}{d\lambda}\sqrt{-g(y(x))} + \sqrt{-g(x)}\,\frac{d}{d\lambda}\left|\frac{\partial y}{\partial x}\right|\right]\Bigg\}d^n x. \end{aligned} \tag{11.37}$$

Here we need some identities, which we leave as an exercise: □

Worked Exercise 11.4.1
Problem: Recall that a^ν denotes the components of the characteristic vector field.
(a) Let $|\frac{\partial y}{\partial x}|$ denote the Jacobian. Show that

$$\frac{d}{d\lambda}\left|\frac{\partial y}{\partial x}\right| = \frac{d}{d\lambda}\left(\frac{\partial y^\mu}{\partial x^\mu}\right) = \partial_\mu a^\mu, \tag{11.38}$$

i.e., it suffices to differentiate the trace of the Jacobian matrix, which in turn gives the divergence of the characteristic vector field.
(b) Show that

$$\frac{d}{d\lambda_{|\lambda=0}}\left(\frac{\partial x^\nu}{\partial y^\mu}\right) = -\partial_\mu a^\nu. \tag{11.39}$$

Inserting this, (11.37) is reduced to

$$\begin{aligned} 0 = \int_U \Bigg\{ \Bigg[& \frac{\partial L}{\partial \phi_\alpha}\frac{d\phi_\alpha{}^\lambda}{d\lambda} + \frac{\partial L}{\partial(\partial_\mu \phi_\alpha)}\partial_\mu \frac{d\phi_\alpha{}^\lambda}{d\lambda} - \frac{\partial L}{\partial(\partial_\mu\phi_\alpha)}\partial_\nu\phi_\alpha\partial_\mu a^\nu \\ & + \frac{\partial L}{\partial g_{\alpha\beta}}\frac{d}{d\lambda}g_{\alpha\beta}\Bigg] \times \sqrt{-g(x)} \\ & + L\left[\partial_\mu\sqrt{-g(x)}a^\mu + \sqrt{-g(x)}\partial_\mu a^\mu\right]\Bigg\} d^n x. \end{aligned} \tag{11.40}$$

Here it is instructive to divide the terms into two groups. Those (A) containing $\frac{d\phi_\alpha{}^\lambda}{d\lambda}$ and those (B) containing a^ν:

$$0 = \int_U \{A + B\}\sqrt{-g}d^n x.$$

We want to rewrite A and B separately as divergences. Consider first A:

$$A = \frac{\partial L}{\partial \phi_\alpha}\frac{d\phi^\lambda_\alpha}{d\lambda} + \frac{\partial L}{\partial(\partial_\mu\phi_\alpha)}\partial_\mu\frac{d\phi^\lambda_\alpha}{d\lambda}.$$

Using the covariant equations of motion (8.35), A is immediately reduced to

$$A = \frac{1}{\sqrt{-g}}\partial_\mu\left(\sqrt{-g}\frac{\partial L}{\partial(\partial_\mu\phi_\alpha)}\cdot\frac{d\phi_\alpha{}^\lambda}{d\lambda}\right). \tag{11.46}$$

Next, we turn our attention to the second group:

$$\begin{aligned} B &= -\frac{\partial L}{\partial(\partial_\mu\phi_\alpha)}\partial_\nu\phi_\alpha\partial_\mu a^\nu + L\partial_\mu a^\mu + L\frac{1}{\sqrt{-g}}\partial_\mu(\sqrt{-g})a^\mu + \frac{\partial L}{\partial g_{\alpha\beta}}(\partial_\nu g_{\alpha\beta})a^\nu \\ &= \left[\frac{-\partial L}{\partial(\partial_\mu\phi_\alpha)}\partial_\nu\phi_\alpha + \delta^\mu{}_\nu L\right]\partial_\mu a^\nu + \left\{L\frac{1}{\sqrt{-g}}\partial_\nu(\sqrt{-g}) + \frac{\partial L}{\partial g_{\alpha\beta}}\partial_\nu g_{\alpha\beta}\right\}a^\nu. \end{aligned}$$

But here we recognize an old fellow in the first set of parentheses. It is nothing but the canonical energy momentum tensor $\mathring{T}_\nu{}^\mu$ (cf. (3.17)). The second term is due to the fact that we work in a covariant formalism where $\sqrt{-g}$ need not degenerate to a constant. We still need a useful identity:

Worked Exercise 11.4.2
Problem: Show that the following identity holds, provided that ϕ_α solves the equations of motion:

$$\frac{1}{\sqrt{-g}}\partial_\mu(\sqrt{-g}\mathring{T}_\nu{}^\mu) = \frac{1}{\sqrt{-g}}(\partial_\nu\sqrt{-g})L + \frac{\partial L}{\partial g_{\alpha\beta}}\partial_\nu g_{\alpha\beta}. \tag{11.42}$$

Using (11.42) we can finally rearrange the B-term as

$$B = \frac{1}{\sqrt{-g}}\partial_\mu(\sqrt{-g}\mathring{T}_\nu{}^\mu a^\nu). \tag{11.43}$$

Okay, it was a long tour de force, but now A and B has been rearranged as divergences, and we obtain the identity that we have looked forward to finding:

$$0 = \frac{dS}{d\lambda} = \int_U \left[\frac{1}{\sqrt{-g}}\partial_\mu\left\{\sqrt{-g}\left(\frac{\partial L}{\partial(\partial_\mu\phi_\alpha)}\frac{d\phi_\alpha{}^\lambda}{d\lambda} + a^\nu\mathring{T}_\nu{}^\mu\right)\right\}\right]\sqrt{-g}d^n x. \tag{11.44}$$

As U is arbitrary, this is consistent only if the integrand vanishes.

This is the main theorem! To apply it we must gain some familiarity with the current, expecially the last part. Let us try to calculate $d\phi_\alpha{}^\lambda/d\lambda$ explicitly. We consider first the case where ϕ_α is a collection of scalar fields. Then by definition,

$$(f_\lambda)_*\phi(x) = \phi(y^{-1}(x)); \text{ i.e.,}$$
$$\phi_\alpha^\lambda(f_\lambda(x)) = (f_\lambda)_*\phi_\alpha(y(x)) = \phi_\alpha(y^{-1}(y(x))) = \phi_\alpha(x),$$

so $\phi_\alpha^\lambda(f_\lambda(x))$ does not at all depend on λ, and the last term vanishes automatically! We thus conclude:

Theorem 11.4 (Noether's theorem for scalar fields). *If a field theory involving only scalar fields ϕ_α admits a one-parameter family of symmetry transformations $(f_\lambda)_{\lambda\in\mathbf{R}}$, then the following current is conserved:*

$$J^\mu = a^\nu\mathring{T}_\nu{}^\mu; \text{ i.e., } \mathbf{J} = \mathring{T}(\mathbf{a}, \cdot), \tag{11.45}$$

where $\mathring{T}$ is the canonical energy momentum tensor and $\mathbf{a}$ is the characteristic vector field associated with $(f_\lambda)_{\lambda\in\mathbf{R}}$.

However, in the case of a vector field, i.e., $\phi_\alpha = A_\alpha$, the last term no longer vanishes. By definition, we get

$$f_* A_\mu(x) = \frac{\partial x^\nu}{\partial y^\mu}A_\nu(y^{-1}(x)); \text{ i.e.,}$$

$$\phi_\alpha^\lambda(f_\lambda(x)) = (f_\lambda)_* A_\alpha(y(x)) = \frac{\partial x^\beta}{\partial y^\alpha}A_\beta(y^{-1}(y(x)) = \frac{\partial x^\beta}{\partial y^\alpha}A_\beta(x).$$

So in this case we get an extra factor $\partial x^\beta/\partial y^\alpha$, which *does* depend on λ! If we perform the differentiation and use Exercise 11.4.1(b) once more, we get

$$\frac{d\phi_\alpha{}^\lambda}{d\lambda_{|\lambda=0}} = -A_\beta\partial_\alpha\frac{dy^\beta}{d\lambda_{|\lambda=0}} = -A_\beta\partial_\alpha a^\beta;$$

i.e., this time we have a nonvanishing contribution to the current, which now looks as follows:

$$J^\mu = a^\nu \overset{\circ}{T}{}_\nu{}^\mu - \frac{\partial L}{\partial(\partial_\mu A_\alpha)} A_\beta \partial_\alpha a^\beta . \tag{11.46}$$

We will return to this peculiar current in Section 11.6 and show that it can actually be rewritten on the form $J^\mu = T^\mu{}_\nu a^\nu$, where $T^\mu{}_\nu$ is the *true* energy momentum tensor (cf. the discussion in Section 3.2).

In the above discussion we have considered only symmetry transformations generated by space–time transformations. We can, however, easily extend our discussion to internal symmetries as well, where an internal transformation is characterized by affecting only the field components. Suppose we have a one-parameter family of internal transformations,

$$T_\lambda : \phi_\alpha(x) \curvearrowright \phi_\alpha^\lambda(x).$$

We say that T is a symmetry transformation, provided that it leaves the action invariant, i.e.,

$$S_\Omega[T_\lambda(\phi)] = S_\Omega[\phi] \tag{11.47}$$

for any space–time region Ω. Then we can immediately take over the conclusions obtained in Theorem 11.3:

Theorem 11.5 (Noether's theorem for internal symmetries). *If a field theory admits a one-parameter family of internal symmetry transformations* $(T_\lambda)_{\lambda \in \mathbf{R}}$, *then the current*

$$J = \frac{\partial L}{\partial(\partial_\mu \phi_\alpha)} \frac{d\phi_\alpha^\lambda(x)}{d\lambda} \tag{11.48}$$

is conserved. (As usual, we calculate the derivative at $\lambda = 0$.*)*

The standard example of an internal transformation is the phase transformation

$$\phi(x) \curvearrowright \exp[i\lambda]\phi(x); \quad \bar{\phi}(x) \curvearrowright \exp[-i\lambda]\bar{\phi}(x),$$

where ϕ is a complex scalar field. The Noether current (11.48) then reduces to

$$J^\mu = \frac{\partial L}{\partial(\partial_\mu \phi)} \frac{d\phi^\lambda}{d\lambda} + \frac{\partial L}{\partial(\partial_\mu \bar{\phi})} \frac{d\bar{\phi}^\lambda}{d\lambda} = i\left[\phi \frac{\partial L}{\partial(\partial_\mu \phi)} - \bar{\phi} \frac{\partial L}{\partial(\partial_\mu \bar{\phi})}\right],$$

which is nothing but our old friend (3.67)!

We can also generalize the concept of a symmetry transformation. In the preceding discussion we have assumed that the displaced action is identical with the original action. From a dynamical point of view, however, two actions will lead to the same dynamics, provided that they just differ by a boundary term (i.e., their Lagrangian densities differ by a divergence). We may therefore say that a space–time transformation is a symmetry transformation in the generalized sense provided that it satisfies

$$S_{f(\Omega)}[f_*(\phi_\alpha)] = S_\Omega[\phi_\alpha] + S^f_{\partial\Omega}[\phi_\alpha]. \tag{11.49}$$

If we have a one-parameter family of symmetry transformations in the generalized sense, $(f_\lambda)_{\lambda\in\mathbf{R}}$, we must now be a little careful in the derivation of the Noether current. This time the displaced action is not constant, so we must first subtract the surface term, which can be rearranged in the form

$$S^\lambda_{\partial\Omega}[\phi_\alpha] = \int_\Omega \partial_\mu \chi^\mu_\lambda[\phi_\alpha]\sqrt{-g}d^n x.$$

This produces the following additional term to (11.44):

$$-\frac{dS^\lambda_{\partial\Omega}}{d\lambda} = -\int_\Omega \partial_\mu \left\{\frac{d\chi^\mu_\lambda}{d\lambda}[\phi_\alpha]\right\}\sqrt{-g}d^n x.$$

Consequently, the Noether current must now include an additional term of the form

$$-\frac{d\chi^\mu_\lambda}{d\lambda}[\phi_\alpha].$$

To conclude, we have therefore shown:

The more general Noether current consists of three terms:

a. *A piece*

$$J^\mu_1 = a^\nu \mathring{T}_\nu{}^\mu, \tag{11.50}$$

which comes from the displacements of the space–time points.

b. *A piece*

$$J^\mu_2 = \frac{\partial L}{\partial(\partial_\mu\phi_\alpha)}\frac{d\phi^\lambda_\alpha}{d\lambda}, \tag{11.51}$$

which comes from the displacement of the field. (Notice that it is generated partly from the space–time symmetries and partly from the internal symmetries involved.)

c. *A piece*

$$J^\mu_3 = -\frac{d\chi^\mu_\lambda[\phi_\alpha]}{d\lambda}, \tag{11.52}$$

which comes from the boundary term in the displaced Lagrangian. (As usual, we calculate the derivative at $\lambda = 0$.)

§ 11.5 Isometries as Symmetry Transformations

We can now investigate the consequences of the transformations in the isometry group. Consider Minkowski space equipped with a field configuration ϕ_α. The dynamics of ϕ_α are controlled by an action principle, i.e., we have a Lagrangian density constructed out of the fields and their first-order derivatives,

$$L = L[\mathbf{g}, \phi_\alpha, \boldsymbol{d}\phi_\alpha].$$

The crucial assumption is that $L[\mathbf{g}, \phi_\alpha, \boldsymbol{d}\phi_\alpha]$ itself is a scalar field on M, so that the action integral $\int L\epsilon$ is covariant. This guarantees that the field theory becomes covariant; i.e., the equations of motion will be covariant, etc. Now, using a global isometry f we can pull back the field ϕ_α and obtain the transformed configuration $f^*\phi_\alpha$. The important thing to recognize then is that

$$f^*L[\mathbf{g}, \phi_\alpha, \boldsymbol{d}\phi_\alpha] = L[\mathbf{g}, f^*\phi_\alpha, \boldsymbol{d}f^*\phi_\alpha] \tag{11.53}$$

This is not true for an arbitrary diffeomorphism! It works only because f is an isometry (in general we get $L[f^*\mathbf{g}, \ldots]$ on the right-hand side). Consequently, we get

$$\begin{aligned} S_\Omega[f^*\phi_\alpha] &= \int_{\Omega^*} L[\mathbf{g}, f^*\phi_\alpha, \boldsymbol{d}f^*\phi_\alpha] = \int_{\Omega^*} f^*L[\mathbf{g}, \phi_\alpha, \boldsymbol{d}\phi_\alpha] \\ &= \int_{\Omega} f^*(*L[\mathbf{g}, \phi_\alpha, \boldsymbol{d}\phi_\alpha]) = \int_{f(\Omega)} *L[\mathbf{g}, \phi_\alpha, \boldsymbol{d}\phi_\alpha] \\ S_{f(\Omega)}&[\phi_\alpha] \end{aligned} \qquad 11.54$$

because f is an isometry and therefore commutes with the dual map. We have thus shown:

A Poincard transformation is a symmetry transformation for any field theory, where the Lagrangian density is a scalar field constructed entirely out of the fields and their first-order derivatives. If ϕ_α is a solution to the equations of motion, we can transform it into another solution by applying a Poincard transformation.

The converse is not true. A specific field theory may very well admit a larger symmetry group than the isometry group. See exercise 11.5.1 below.

Worked Exercise 11.5.1
(Compare with exercises 10.6.2 and 10.6.3) Problem:

a. Show that the theory of a massless scalar field in $(1+1)$-dimensional space–time is conformally invariant; i.e., the conformal transformations are symmetry transformations.
b. Show also that the theory of electronmagnetism in $(3+1)$-dimensional space–time is conformally invariant.

Now we are ready to apply the machinery from the preceding section. All we have to do is to extract some suitable one-parameter families out of the full Poincaré group. In what follows we restrict to inertial coordinates.

EXAMPLE 11.3. *The group of translations.* A one-parameter group of translations is given by

$$y^\mu = x^\mu + \lambda a^\mu, \tag{11.55}$$

where a^μ are the componnents of a fixed four-vector **a**. Clearly, this constant vector field is the characteristic vector field associated with the one-parameter group. Consequently, we get from Noether's theorem that the current

$$J^\mu = a^\nu \mathring{T}_\nu{}^\mu$$

is conserved; i.e.,

$$0 = \partial_\mu J^\mu = a^\nu [\partial_\mu \mathring{T}_\nu{}^\mu].$$

(The last term in (11.46) vanishes because a^μ is constant.) As a^ν is arbitrary, we conclude that the canonical energy–momentum tensor itself is conserved. As a result, we see that *invariance under the translation group implies conservation of the canonical energy and momentum.*

EXAMPLE 11.4. *The group of Lorentz transformations*

By definition, a Lorentz matrix is a 4×4-matrix $\bar{\bar{A}}$ satisfying matrix conjugator

$$\bar{\bar{A}}^\dagger \bar{\bar{\eta}} \bar{\bar{A}} = \bar{\bar{A}} \bar{\bar{\eta}} \bar{\bar{A}}^\dagger = \bar{\bar{\eta}}; \quad \text{i.e.,} \quad \bar{\bar{A}}^\dagger \bar{\bar{\eta}} = \bar{\bar{\eta}} \bar{\bar{A}}^{-1}. \tag{6.32}$$

This is not a useful characterization when we want to construct one-parameter groups of Lorentz transformations. In analogy with the one-parameter groups of unitary transformations (Section 11.3) we will try to construct one-parameter groups of Lorentz matrices of the form $\exp[\lambda\bar{\bar{\omega}}]$ with some matrix $\bar{\bar{\omega}}$. Here it is useful to observe that $\exp[\bar{\bar{\omega}}]$ is a Lorentz matrix, provided that $\bar{\bar{\omega}}$ satisfies

$$\bar{\bar{\omega}}^\dagger \bar{\bar{\eta}} = -\bar{\bar{\eta}} \bar{\bar{\omega}};$$

i.e., $\bar{\bar{\eta}}\bar{\bar{\omega}}$ is skew-symmetric. If we write the matrix elements of $\bar{\bar{\omega}}$ as $\omega^\mu{}_\nu$ and if we rause and lower indices using the metric components, then $\bar{\bar{\eta}}\bar{\bar{\omega}}$ is characterized by the "covariant" components $\omega_{\mu\nu}$. (Observe that $\omega^\mu{}_\nu$ are not the components of a tensor!)

Whenever $\bar{\bar{\eta}}\bar{\bar{\omega}}$ is a skew-symmetric matrix we can therefore construct a one-parameter group of Lorentz transformations as

$$\bar{\bar{y}}_| = \exp[\lambda\bar{\bar{\omega}}]\bar{\bar{x}}_|. \tag{11.56}$$

By differentiation of this we get

$$\bar{\bar{a}}_| = \frac{d\bar{\bar{y}}_|}{d\lambda} = \bar{\bar{\omega}}\bar{\bar{x}}_|, \tag{11.57}$$

i.e., the characteristic vector field **a** is characterized by the components

$$a^\nu = \omega^\nu{}_\rho x^\rho. \tag{11.58}$$

Exercise 11.5.2

Introduction: Let us introduce six matrices labeled $\bar{\bar{\Omega}}_{\rho\sigma}$, where $\bar{\bar{\Omega}}_{\rho\sigma}$ is skew-symmetric in $(\rho\sigma)$. These matrices are going to be characterized by the "covariant" components $\eta_{\mu\rho}\eta_{\nu\sigma} - \eta_{\mu\sigma}\eta_{\nu\rho}$; i.e., $\bar{\bar{\Omega}}_{\rho\sigma}$ itself is characterized by the components.

$$(\bar{\bar{\Omega}}_{\rho\sigma})^\mu{}_\nu = \delta^\mu{}_\rho \eta_{\nu\sigma} - \delta^\mu{}_\sigma \eta_{\mu\rho}.$$

Problem:

a. Let (ijk) be an even permutation of (123). Show that $\exp[\lambda\bar{\bar{\Omega}}_{ij}]$ generates rotations around the kth axis through an angle λ.
b. Show that $\exp[\lambda\bar{\bar{\Omega}}_{0k}]$, $k = 1, 2, 3$, generates a special Lorentz transformation along the kth axis with velocity v given by the relation

$$\cosh\lambda = \frac{1}{\sqrt{1-u^2}}.$$

The parameter λ is called the *rapidity* and unlike the velocity v, it is additive under composition of special Lorentz transformations in the same direction.

Let us consider scalar fields first. We then get from Theorem 11.4 that $J^\mu = a^\nu \mathring{T}_\nu{}^\mu$ is conserved, i.e.,

$$0 = \partial_\mu J^\mu = \omega_{\nu\rho}(\partial_\mu x^\rho \mathring{T}^{\nu\mu}).$$

As $\omega_{\nu\rho}$ is an arbitrary skew-symetric matrix, we conclude that

$$0 = \partial_\mu[x^\rho \mathring{T}^{\nu\mu} - x^\nu \mathring{T}^{\rho\mu}]. \tag{11.59}$$

Consequently, invariance under the group of Lorentz transformation implies the conservation of the angular momentum!

It is convenient to introduce

$$\mathring{M}^{\rho\nu\mu} = x^\rho \mathring{T}^{\nu\mu} - x^\nu \mathring{T}^{\rho\mu}. \tag{11.60}$$

Then $\mathring{M}^{\rho\nu\mu}$ comprises the currents of the angular momentum density, but observe that it is not a covariant expression, not even if we restrict ourselves to inertial frames.

From the conservation of the angular momentum we now deduce

$$0 = \partial_\mu M^{\rho\nu\mu} = \delta_\mu{}^\rho \mathring{T}^{\nu\mu} - \delta_\mu{}^\nu \mathring{T}^{\rho\mu} + x^\rho(\partial_\mu \mathring{T}^{\nu\mu} - x^\nu(\partial_\mu \mathring{T}^{\rho\mu}).$$

Here the last two terms vanish, since $\mathring{T}^{\nu\mu}$ is conserved, and we end up with

$$0 = \mathring{T}^{\nu\rho} - \mathring{T}^{\rho\nu}; \tag{11.61}$$

i.e., the canonical energy momentum tensor is born symmetric. There is consequently no need to repair it. In the case of a scalar field, the *canonical energy–momentum tensor is the true energy–momentum tensor*!

In geometrical language, we start out with the inertial frame of reference S. We then associate to S six vector fields $\mathbf{e}_{\rho\sigma}$, where $\mathbf{e}_{\rho\sigma}$ is skew-symmetric in $(\rho\sigma)$. Here $\mathbf{e}_{\rho\sigma}$ is characterized by the components

$$(\mathbf{e}_{\rho\sigma})^\alpha = x_\rho \delta^\alpha{}_\sigma - x_\sigma \delta^\alpha{}_\rho; \quad \text{i.e.,} \quad \mathbf{e}_{\rho\sigma} = x_\rho \mathbf{e}_\sigma - x_\sigma \mathbf{e}_\rho. \tag{11.62}$$

When we contract them with the energy–momentum tensor; we obtain six conserved currents:

$$\mathbf{J}_{\rho\sigma} = \mathring{\mathbb{T}}(\mathbf{e}_{\rho\sigma}, \cdot); \quad \text{i.e.,} \quad (J_{\rho\sigma})^\mu = x_\rho \mathring{T}_\sigma{}^\mu - x_\sigma \mathring{T}_\rho{}^\mu, \tag{11.63}$$

which are interpreted as the currents of angular momentum relative to the (x^ρ, x^σ)-plane. If Ω denotes a space slice relative to S, then the associated charges

$$J_{\rho\sigma} = -\int_\Omega {}^*\mathbf{J}_{\rho\sigma} \tag{11.64}$$

are interpreted as the total angular momentum relative to the (x^ρ, x^σ)-plane.

Worked Exercise 11.5.3
Introduction: Consider two inertial frames of references S^1 and S^2 and the associated total angular momenta $J^1_{\rho\sigma}$, $J^2_{\rho\sigma}$ given by the above formula (11.64). The inertial coordinates (x^α) and (y^α) associated with S^1 and S^2 are assumed related through the Poincaré transformation

$$y^\alpha = A^\alpha{}_\beta x^\beta + b^\alpha.$$

Problem:

a. Show that the basic vector fields $\mathbf{e}^1_{\rho\sigma}$ and $\mathbf{e}^2_{\rho\sigma}$ are related through the formula

$$\mathbf{e}^2_{\rho\sigma} = \mathbf{e}^1_{\mu\nu}\breve{A}^\mu{}_\rho \breve{A}^\nu{}_\sigma + [b_\rho \mathbf{e}^2_\sigma - b_\sigma \mathbf{e}^2_\rho]. \tag{11.65}$$

b. Show that the total angular momenta measured in S^1 and S^2 are related through

$$J^2_{\rho\sigma} = J^1_{\mu\nu}\breve{A}^\mu{}_\rho \breve{A}^\nu{}_\sigma + [b_\rho P^2_\sigma - b_\sigma P^2_\rho], \tag{11.66}$$

where P^2_μ is the total energy omentum measured in S^2.

Next, we turn our attention to vector fields. Here the conserved current is given by (see (11.46) and (11.58))

$$J^\mu = \omega_{\nu\rho}\left(x^\rho \mathring{T}^{\nu\mu} - \frac{\partial L}{\partial(\partial_\mu A_\rho)} A^\nu\right),$$

which leads to the conservation of the slightly more complicated angular momentum

$$\mathring{M}^{\rho\nu\mu} = \left[x^\rho \mathring{T}^{\nu\mu} - x^\nu \mathring{T}^{\rho\mu}\right] + \left[\frac{\partial L}{\partial(\partial_\mu A_\nu)} A^\rho - \frac{\partial L}{\partial(\partial_\mu A_\rho)} A^\nu\right]. \tag{11.67}$$

Here the first term represents the conventional orbitral angular momentum of the field, but what is the origin of the last term? From a physical viewpoint the main difference between a scalar field and a vector field is that a scalar field carries no spin, while a vector field carries spin 1 (compare the discussion in Section 3.6). This is reflected in the fact that the scalar field carries no space–time index, while the vector field carries one. This again leads to the different behaviors under symmetry transformations. It is the nontrivial transformation property of A_α that on the one hand allows us to construct eigenfunctions for the rotation operator with nontrivial eigenvalues, while on the other hand it generates the last term in (11.67).

Now, for a field carrying spin we expect the total angular momentum to be composed of a spin part and a contribution form the orbital angular momentum:

$$J = L + S,$$

and it is only the complete angular momentum that is conserved, not the separate terms. Okay, this suggests then that the last term should be interpreted as the spin density.

§ 11.6 The True Energy–Momentum Tensor for Vector Fields

We have just seen that the conserved current (11.46) for a vector field has a complicated structure, involving not only the canonical energy–momentum tensor, but also a spin contribution. We also know that the canonical energy–momentum tensor for a vector field fails to be symmetric and has to be repaired (cf. the discussion in Sections 3.5 and 3.7). We will now extract the true energy–momentum tensor from the expression

$$J^{\mu} = a^{\nu}\mathring{T}_{\nu}{}^{\mu} - \frac{\partial L}{\partial(\partial_{\mu}A_{\alpha})}A_{\beta}\partial_{\alpha}a^{\beta} \tag{11.46}$$

The main clue is the observation that the expression

$$\mathring{T}_{\nu}{}^{\mu} = \frac{-\partial L}{\partial(\partial_{\mu}A_{\alpha})}\partial_{\nu}A_{\alpha} + \delta_{\nu}{}^{\mu}L \tag{3.17}$$

for the canonical energy–momentum tensor is not covariant; i.e., $\mathring{T}_{\nu}{}^{\mu}$ are not the components of a tensor. (Remember, we are working in arbitrary coordinates on an arbitrary manifold!) The trouble lies in the factor $\partial_{\nu}A_{\alpha}$, where we have differentiated through a vector field, which immediately destroys the covariance (compare the discussion in Section 7.1. For a vector field it is only the field strength $F_{\alpha\beta} = \partial_{\alpha}A_{\beta} - \partial_{\beta}A_{\alpha}$ that is covariant.) However, by construction, the complete expression for the current J^{μ} is covariant, and it would be nice to have this covariance explicitly built into the formula (11.46). This is where the last term comes into the game. It can be rearranged as follows:

$$\begin{aligned}\frac{\partial L}{\partial(\partial_{\mu}A_{\alpha})}A_{\nu}\partial_{\alpha}a^{\nu} &= \frac{1}{\sqrt{-g}}\partial_{\alpha}\left[\sqrt{-g}\,\frac{\partial L}{\partial(\partial_{\mu}A_{\alpha})}A_{\nu}a^{\nu}\right]\\ &\quad - \frac{1}{\sqrt{-g}}\partial_{\alpha}\left[\sqrt{-g}\,\frac{\partial L}{\partial(\partial_{\mu}A_{\alpha})}\right]A_{\nu}a^{\nu} - \frac{\partial L}{\partial(\partial_{\mu}A_{\alpha})}(\partial_{\alpha}A_{\nu})a^{\nu}.\end{aligned}$$

Let us pause to make a useful observation:

Worked Exercise 11.6.1
Problem: Prove that

$$\frac{\partial L}{\partial(\partial_{\mu}A_{\alpha})}$$

are the components of a skew-symmetric tensor.

Using the covariant equations of motion (8.35), we now rearrange the last term as follows:

$$\frac{\partial L}{\partial(\partial_\mu A_\alpha)} A_\nu \partial_\alpha a^\nu = \frac{1}{\sqrt{-g}} \partial_\alpha \left[\sqrt{-g} \frac{\partial L}{\partial(\partial_\mu A_\alpha)} A_\nu a^\nu \right] + \frac{\partial L}{\partial A_\mu} A_\nu a^\nu - \frac{\partial L}{\partial(\partial_\mu A_\alpha)} (\partial_\alpha A_\nu) a^\nu .$$

Inserting this, we can finally decompose the current into two parts:

$$\begin{aligned} J^\mu = &\quad J^\mu{}_{(1)} \quad + \quad J^\mu{}_{(2)} \\ = a^\nu &\left[\frac{-\partial L}{\partial A_\mu} A_\nu - \frac{\partial L}{\partial(\partial_\mu A_\alpha)} F_{\nu\alpha} + \delta^\mu{}_\nu L \right] - \frac{1}{\sqrt{-g}} \partial_\alpha \left[\sqrt{-g} \frac{\partial L}{\partial(\partial_\mu A_\alpha)} A_\nu a^\nu \right], \end{aligned} \tag{11.69}$$

which are manifestly covariant. From exercise 11.6.1 we learn that

$$S^{\mu\alpha} = \frac{\partial L}{\partial(\partial_\mu A_\alpha)} (A_\nu a^\nu) \tag{11.70}$$

are the contravariant components of a 2-form S and the last part is therefore given by

$$\underset{(2)}{\mathbf{J}} = \delta \mathbf{S}, \tag{11.71}$$

from which it immediately follows that it is conserved. Furthermore, it does not contribute to the total charge, since

$$\underset{(2)}{Q} = -\int_\Omega {}^*\underset{(2)}{\mathbf{J}} = -\int_\Omega {}^*\delta\mathbf{S} = \int_{\partial\Omega} {}^*\mathbf{S},$$

where we have used Gauss's theorem. Here the last integral vanishes, provided that the vector field dies off sufficiently fast at infinity. But then we can simply split off the second part. If we furthermore introduce the abbreviation

$$T_\nu{}^\mu = \frac{-\partial L}{\partial A_\mu} A_\nu - \frac{\partial L}{\partial(\partial_\mu A_\alpha)} F_{\nu\alpha} + \delta^\mu{}_\nu L, \tag{11.72}$$

we can now express the first component as

$$\underset{(1)}{\mathbf{J}} = \mathbf{T}(\mathbf{a}; \cdot). \tag{11.73}$$

We want to show that $T^{\mu\nu}$ defined by (11.72) is the true energy–momentum tensor. Consequently, we must show that it is symmetric, conserved; and that it produces the same energy and momentum as does $\mathring{T}^{\mu\nu}$. (Cf. the discussion in Section 3.2).

Since $\mathbf{J}$ and $\underset{(2)}{\mathbf{J}}$ are conserved, it follows that $\underset{(1)}{\mathbf{J}}$ is conserved too. From translational invariance we therefore get that $T^{\mu\nu}$ is conserved. Furthermore, $\underset{(2)}{\mathbf{J}}$ does not contribute to the total charge, so $T^{\mu\nu}$ and $\mathring{T}^{\mu\nu}$ produce the same total energy and momentum. Finally, it follows from Lorentz invariance that $T^{\mu\nu}$ produces the following conserved angular momentum:

$$M^{\rho\nu\mu} = x^{\rho}T^{\nu\mu} - x^{\nu}T^{\rho\mu};$$

i.e.,

$$0 = \partial_{\mu}M^{\rho\nu\mu} = T^{\nu\rho} - T^{\rho\nu}.$$

Consequently, $T^{\mu\nu}$ is a symmetric tensor.

Exercise 11.6.2
Problem:

a. Consider the theory of electromagnetism based on the gauge potential A_{μ}. Show that (11.72) reproduces the true energy momentum tensor (1.41).
b. Consider the theory of a massve vector field W_{μ}. Show that (11.72) reproduces the energy–momentum tensor given by exercise 3.7.1.

Using the true energy–momentum tensor for vector fields, we can now reformulate Noether's theorem in the case of vector fields:

Theorem 11.6 (Noether's theorem for vector fields). *If a field theory based upon a vector field A_{α} admits a one-parameter family of symmetry transformations $(f_{\lambda})_{\lambda\in\mathbf{R}}$, then the following current is conserved:*

$$\underset{(1)}{J^{\mu}} = a^{\nu}T_{\nu}{}^{\mu}; \quad \text{i.e.,} \quad \underset{(1)}{\mathbf{J}} = \mathbf{T}(\mathbf{a};\cdot), \tag{11.73}$$

where $\mathbf{T}$ *given by (11.72) is the true energy–momentum tensor and where* $\mathbf{a}$ *is the characteristic vector field associated with the one-parameter family.*

Exercise 11.6.3
Problem: Consider a massless scalar field in (1 + 1)-dimensional space–time, respectively the electromagnetic field in (3 + 1)-dimensional space–time.

a. Show that the conserved current and charge associated with the invariance under dilations are given by

$$J^{\mu} = x^{\nu}T_{\nu}{}^{\mu} \quad Q = \int_{M} x^{\nu}T_{\nu} \circ d^{D}x \quad (D = 1 \text{ or } 3) \tag{11.74}$$

b. Show that the conserved currents and charges associated with the invariance under special conformal transformations are given by

$$(\mathbf{J}_{\rho})^{\mu} = x^{2}T_{\rho}{}^{\mu} - 2x_{\rho}x^{\nu}T_{\nu}{}^{\mu} \quad Q_{\rho} = \int_{M}\left[x^{2}T_{\rho}{}^{0} - 2x_{\rho}x^{\nu}T_{\nu}{}^{0}\right]d^{D}x. \tag{11.75}$$

c. Show that the dilatational charge tranforms as a scalar under Lorentz transformations and that the special conformal charges transform as the components of a Lorentz vector under Lorentz transformations. How do they transform under general Poincaré transformations?

Exercise 11.6.4
Problem:

a. Show that the energy–momentum tensor of a conformal invariant field theory is traceless, $T^{\mu}{}_{\mu} = 0$. (Hint: Consider the conservation law corresponding to dilatational invariance.)
b. Consider a field theory possessing an energy–momentum tensor that is traceless, $T^{\mu}{}_{\mu} = 0$. Show that the charges (11.74) and (11.75) are conserved.

§ 11.7 Energy–Momentum Conservation as a Consequence of Covariance

In this section we will finally take up the problem of how to construct the true energy–momentum tensor from a more general point of view. We saw in Section 3.11 how gauge invariance could be used to construct a general expression for the electromagnetic current. In that case we have a Lagrangian depending on a charged field ϕ coupled to an external gauge potential A_μ. Performing a variation of the external field, $A_\mu \curvearrowright A_\mu + \epsilon\delta A_\mu$, the corresponding variation in the action will be linear in δA_μ,

$$\delta S = \frac{dS(\epsilon)}{d\epsilon_{|\epsilon=0}} = \int \frac{\partial L}{\partial A_\mu}\delta A_\mu \sqrt{-g}d^4x,$$

and we then showed that the coefficient in front of δA_μ actually was identical to the current:

$$J^\mu = \frac{\partial L}{\partial A_\mu}. \tag{3.76}$$

Furthermore, the gauge invariance of the action allowed us to prove that this current would be conserved as a consequence of the dynamics of the charged field ϕ.

In the present case we are considering an arbitrary field configuration ϕ_α. To construct the Lagrangian, which has to be a scalar field, we furthermore need the metric $\mathbf{g}$. We then construct an action functional of the general form

$$S = \int L[\mathbf{g}, \phi_\alpha, \boldsymbol{d}\phi_\alpha]\sqrt{-g}\boldsymbol{d}x^1 \wedge \ldots \wedge \boldsymbol{d}x^n.$$

(Observe that the metric is involved in the construction of the volume form too!) The basic idea is to treat the metric as an external field and see what happens when we perform a variation in the metric, $g_{\mu\nu} \curvearrowright g_{\mu\nu} + \epsilon\delta g_{\mu\nu}$. The corresponding variation of the action will be linear in $\delta g_{\mu\nu}$ and can therefore be written in the form

$$\delta S = \frac{dS(\epsilon)}{d\epsilon_{|\epsilon=0}} = \tfrac{1}{2}\int T^{\mu\nu}(x)\delta g_{\mu\nu}(x)\sqrt{-g}d^n x. \tag{11.76}$$

The coefficient $T^{\mu\nu}(x)$ is a symmetric tensor, and as we shall see, it reduces to the true energy–momentum tensor in the now well-known cases of scalar fields and vector fields. To obtain an explicit expression for $T^{\mu\nu}$ in terms of the Lagrangian, we need some useful identities involving the derivatives of the metric.

Worked Exercise 11.7.1
Problem: Deduce the following identities.

a.

$$\partial_\mu[\ln \det \bar{\bar{M}}(x)] = \mathrm{tr}[\bar{\bar{M}}^{-1}(x)\partial_\mu \bar{\bar{M}}(x)], \tag{11.77}$$

where $\bar{\bar{M}}(x)$ is an arbitrary matrix function.

b.

$$\Gamma^\alpha{}_{\mu\alpha} = \frac{1}{2} g^{\alpha\beta}\partial_\mu g_{\alpha\beta} = \frac{1}{\sqrt{|g|}}\partial_\mu\sqrt{|g|}. \tag{11.78}$$

c.

$$\frac{\partial\sqrt{|g|}}{\partial g_{\alpha\beta}} = \frac{1}{2}\sqrt{|g|}g^{\alpha\beta}. \tag{11.79}$$

d.

$$\frac{\partial g^{\alpha\beta}}{\partial g_{\mu\nu}} = -g^{\alpha\mu}g^{\beta\nu}. \tag{11.80}$$

Using (11.79), we now get

$$\delta S = \frac{d}{d\epsilon_{|\epsilon=0}}\int L\sqrt{-g}d^n x = \int\left[\frac{\partial L}{\partial g_{\mu\nu}}\delta g_{\mu\nu}\sqrt{-g} + L\frac{\partial\sqrt{-g}}{\partial g_{\mu\nu}}\delta g_{\mu\nu}\right]d^n x$$

$$= \int\left[\frac{\partial L}{\partial g_{\mu\nu}} + \tfrac{1}{2}g^{\mu\nu}L\right]\delta g_{\mu\nu}\sqrt{-g}d^n x,$$

from which we read off

$$T^{\mu\nu} = 2\frac{\partial L}{\partial g_{\mu\nu}} + g^{\mu\nu}L. \tag{11.81}$$

As an example, we take a look at the electromagnetic field. Here the Lagrangian density is given by (cf. (3.36))

$$L = -\frac{1}{4}F_{\rho\sigma}F^{\rho\sigma} = -\frac{1}{4}F_{\rho\sigma}F_{\alpha\beta}g^{\rho\alpha}g^{\sigma\beta}.$$

Note that L is gauge invariant, so $T^{\mu\nu}$ defined by (11.81) will automatically become a symmetric gauge-invariant tensor. Using (11.80), we find

$$\frac{\partial L}{\partial g_{\mu\nu}} = -\frac{1}{4}F_{\rho\sigma}F_{\alpha\beta}\left[-g^{\mu\rho}g^{\nu\alpha}g^{\sigma\beta} - g^{\rho\alpha}g^{\mu\sigma}g^{\nu\beta}\right]$$

$$= -\frac{1}{2}F^\mu{}_\alpha F^{\alpha\nu}.$$

Inserting this into (11.81), we then end up with

$$T^{\mu\nu} = -F^\mu{}_\alpha F^{\alpha\nu} - \frac{1}{4}g^{\mu\nu}F_{\rho\sigma}F^{\rho\sigma},$$

and that is precisely the true energy–momentum tensor for the elctromagnetic field (cf. 1.41). Motivated by this example, we will call the tensor defined by equation (11.81) the *metric* energy–momentum tensor.

Exercise 11.7.2
Introduction: We consider a theory consisting of a single scalar field ϕ and assume that the Lagrangian density is constructed as a function of the two scalar fields

$$\phi \quad \text{and} \quad \psi = g^{\alpha\beta}\partial_\alpha\phi\partial_\beta\phi = (\boldsymbol{d}\phi \mid \boldsymbol{d}\phi);$$

i.e.

$$L = L(\phi;\ g^{\alpha\beta}\partial_\alpha\phi\partial_\beta\phi).$$

Problem: Show that the metric energy–momentum tensor (11.81) coincides with the canonical energy–momentum tensor (3.17). [Hint: Show that

$$2\frac{\partial L}{\partial g_{\mu\nu}} = -\frac{\partial L}{\partial(\partial_\nu\phi)}\partial^\mu\phi = -2\frac{\partial L}{\partial\psi}\partial^\alpha\phi\partial^\beta\phi.]$$

Exercise 11.7.3
Introduction: We consider a theory consisting of a single vectorfield A_μ and assume that the Lagrangian density is constructed as a function of the two scalar fields

$$\theta = g^{\alpha\beta}A_\alpha A_\beta = (\mathbf{A} \mid \mathbf{A}) \quad \text{and} \quad \psi = g^{\alpha\rho}g^{\beta\sigma}F_{\alpha\beta}F_{\rho\sigma} = (\mathbf{F} \mid \mathbf{F}),$$

where $F_{\alpha\beta}$ is the usual field strength,

$$F_{\alpha\beta} = \partial_\alpha A_\beta - \partial_\beta A_\alpha;$$

i.e.,

$$L = L(g^{\alpha\beta}A_\alpha A_\beta;\ g^{\alpha\rho}g^{\beta\sigma}F_{\alpha\beta}F_{\rho\sigma}).$$

Problem: Show that the metric energy–momentum tensor (11.81) coincides with the true energy–momentum tensor (11.72).

Worked Exercise 11.7.4
Introduction: We consider a system consisting of a single relativistic particle and assume that the Lagrangian density is constructed as follows:

$$L = -m\int\sqrt{-g_{\alpha\beta}(x)\frac{dx^\alpha}{d\lambda}\frac{dx^\beta}{d\lambda}}\,\frac{1}{\sqrt{-g}}\,\delta^4(x - x(\lambda))d\lambda, \tag{11.82}$$

where $x^\alpha(\lambda)$ is a parametrization of the particle's world line.
Problem:

a. Show that the Lagrangian density (11.82) leads to the conventional action

$$S = -m\int\sqrt{-g_{\alpha\beta}(x(\lambda))\frac{dx^\alpha}{d\lambda}\frac{dx^\beta}{d\lambda}}\,d\lambda.$$

b. Show that the metric energy–momentum tensor (11.81) coincides with the conventional energy–momentum tensor (cf. the discussion in Section 1.6):

$$\overset{\circ}{T}{}^{\mu\nu} = m\int\frac{dx^\alpha}{d\tau}\frac{dx^\beta}{d\tau}\frac{1}{\sqrt{-g}}\,\delta^4(x - x(\tau))d\tau.$$

As shown in exercises 11.7.2–4, the metric energy–momentum tensor coincides with the true energy–momentum tensor for a system consisting of particles, scalar fields, and vector fields. Interestingly enough, we can now give a general proof for the conservation of the metric energy–momentum tensor based upon the covariance of the action. The proof is somewhat similar to the proof of Noether's theorem in

Section11.4. In particular, it is somewhat technical; and it may be skipped on a first reading:

Theorem 11.7. *The metric energy–momentum tensor*

$$T^{\mu\nu} = 2\frac{\partial L}{\partial g_{\mu\nu}} + g^{\mu\nu}L \tag{11.81}$$

will be conserved for any field theory based upon a covariant Lagrangian (i.e., the Lagrangian is a scalar field constructed from the metric, the fields, and their first-order derivatives). Furthermore, the conservation of the energy–momentum tensor can be expressed in a covariant way as

$$\frac{1}{\sqrt{-g}}\partial_\nu(\sqrt{-g}T^{\mu\nu}) + \Gamma^\mu{}_{\alpha\beta}T^{\alpha\beta} = 0. \tag{11.83}$$

PROOF. From covariance we get

$$S_\Omega[\phi_\alpha] = \int_{f_\lambda(\Omega)} L[f_\lambda^*\mathbf{g};\, f_\lambda^*\phi_\alpha;\, \boldsymbol{d}f_\lambda^*\phi_\alpha]f_\lambda^*\epsilon,$$

where f_λ is an arbitrary one-parameter family of diffeomorphisms. For the present purpose we choose the diffeomorphisms f_λ such that they map the interior of Ω into itself and such that f_λ reduces to the identity on the boundary. Observe that this impies that the characteristic vector field **a** vanishes on the boundary. Introducing coordinates, we now get

$$S_\Omega[\phi_\alpha] = \int_U L\left[g^\lambda_{\mu\nu}(y);\, \phi^\lambda_\alpha(y);\, \frac{\partial}{\partial y\mu}\phi^\lambda_\alpha(y)\right]\sqrt{-g^\lambda(y)}d^n y.$$

(Observe that the coordinate domain this time is independent of λ from the beginning.) We assume now that $\phi_\alpha(x)$ is a solution to the equations of motion. This has the important consequence that ϕ^λ_α does not contribute to the variation of the action integral. (Observe that the field this time really is varied. In the case of a scalar field we get, for instance, $\phi^\lambda_\alpha(y) = \phi_\alpha(f_\lambda^{-1}(y))$.) When we differentiate, we therefore need only take into consideration the new variations coming from $g^\lambda_{\mu\lambda}$:

$$\begin{aligned}
0 &= \int_U \frac{d}{d\lambda_{|\lambda=0}}\left[L\left(g^\lambda_{\mu\nu}(y);\, \phi_\alpha(y);\, \frac{\partial}{\partial y^\mu}\phi_\alpha(y)\right)\sqrt{-g^\lambda(y)}\right]d^n y\\
&= \int_U \left\{\frac{\partial L}{\partial g_{\mu\nu}}\sqrt{-g(y)} + L\frac{\partial\sqrt{-g(y)}}{\partial g_{\mu\nu}}\right\}\frac{dg^\lambda_{\mu\nu}(y)}{d\lambda}d^n y\\
&= \frac{1}{2}\int_U T^{\mu\nu}(y)\frac{dg^\lambda_{\mu\nu}(y)}{d\lambda}\sqrt{-g(y)}d^n y.
\end{aligned}$$

We must then compute $\frac{dg^\lambda_{\mu\nu}(y)}{d\lambda}$ explicitly. The displaced metric is given by

$$g^\lambda_{\mu\nu}(y) = (f_\lambda)_* g_{\mu\nu}(y) = \frac{\partial x^\alpha}{\partial y^\mu}\frac{\partial x^\beta}{\partial y^\nu}g_{\alpha\beta}(x)$$

with $x = f_\lambda^{-1}(y)$. Differentiating this with respect to λ at $\lambda = 0$, we get (using exercise 11.4.1b)

$$\begin{aligned}\frac{dg_{\mu\nu}}{d\lambda} &= -(\partial_\mu a^\alpha)g_{\alpha\nu} - (\partial_\nu a^\beta)g_{\mu\beta} - (\partial_\alpha g_{\mu\nu})a^\alpha \\ &= a^\alpha \partial_\mu g_{\alpha\nu} - \partial_\mu a_\nu + a^\beta \partial_\nu g_{\mu\beta} - \partial_\nu a_\mu - a^\alpha(\partial_\alpha g_{\mu\nu}).\end{aligned}$$

Inserting this, we find

$$0 = \int_U \left[T^{\mu\nu}\sqrt{-g}\partial_\nu a_\mu + \Gamma^\beta{}_{\mu\nu}T^{\mu\nu}a_\beta\sqrt{-g}\right] d^n y.$$

Performing a partial integration on the first term, we now end up with

$$0 = \int_U \left[\frac{1}{\sqrt{-g}}\partial_\nu(\sqrt{-g}T^{\beta\nu}) + \Gamma^\beta{}_{\mu\nu}T^{\mu\nu}\right] a_\beta\sqrt{-g}d^n y.$$

As the variation a_β is arbitrary. This is consistent only if

$$\frac{1}{\sqrt{-g}}\partial_\nu(\sqrt{-g}T^{\beta\nu}) + \Gamma^\beta_{\mu\nu}T^{\mu\nu} = 0 \tag{11.83}$$

Finally, we specialize to an inertial frame. Then the identity (11.83) reduces to

$$\partial_\nu T^{\beta\nu} = 0,$$

which is the standard expression (1.38) for the conservation of the energy–momentum tensor. □

From Theorem 11.8 we learn, too, that the covariant expression for the divergence of a symmetric rank 2 tensor is given by

$$\frac{1}{\sqrt{-g}}\partial_\nu(\sqrt{-g}T^{\beta\nu}) + \Gamma^\beta{}_{\mu\nu}T^{\nu\mu}. \tag{*}$$

This should be compared to the covariant expression for the divergence of a skew-symmetric rank 2 tensor:

$$\frac{1}{\sqrt{-g}}\partial_\nu(\sqrt{-g}F^{\beta\nu}). \tag{**}$$

In fact, $(*)$ is valid of a skew-symmetric tensor too, since in that case the contribution from the last term vanishes due to the symmetry of the Christoffel field in the indices $(\mu\nu)$. Thus $(*)$ is valid for arbitrary rank 2 tensors. This illustrates a general principle: The covariant expression for the partial derivatives of a tensor will contain the Christoffel field, but once we restrict to skew-symmetric tensors, the contributions from the Christoffel field vanish automatically, and we end up with the usual expressions from the exterior algebra.

§ 11.8 Scale Invariance in Classical Field Theories

In the final sections we will show how conformal transformations can be combined in a natural way with an internal transformation. The combined transformation then acts as a symmetry transformation for a large and interesting class of classical field theories.

As preparation, we consider scale transformations (dilatations) in n-dimensional Minkowski space. Using inertial coordinates, such a scale transformation is given by

$$D(x^\alpha) = \lambda x^\alpha. \tag{11.84}$$

Under a scale transformation, a scalar field transforms as

$$[D_*\phi][x] = \phi\left(\frac{x}{\lambda}\right).$$

The displaced action is therefore given by

$$\begin{aligned} S_{D(\Omega)}[D_*\phi] &= \int_{y\in\lambda\Omega} L[D_*\phi(y); \quad \frac{\partial}{\partial y^\mu} D_*\phi(y)]d^n y \\ &= \int_{y\in\lambda\Omega} L\left[\phi\left(\frac{y}{\lambda}\right); \quad \frac{\partial}{\partial y^\mu}\phi\left(\frac{y}{\lambda}\right)\right] d^n y. \end{aligned}$$

Changing variable, $y^\alpha = \lambda x^\alpha$, this is rearranged as

$$S_{D(\Omega)}[D_*\phi] = \int_\Omega L\left[\phi(x); \ \frac{1}{\lambda}\frac{\partial}{\partial x^\mu}\phi(x)\right]\lambda^n d^n x.$$

Consequently, a scale transformation is a symmetry transformation, provided that

$$\lambda^n L\left[\phi(x); \ \frac{1}{\lambda}\frac{\partial}{\partial x_\mu}\phi(x)\right] = L\left[\phi(x); \ \frac{\partial}{\partial x_\mu}\phi(x)\right]. \tag{11.85}$$

The standard Lagrangian of a scalar field is of the form

$$L = -\frac{1}{2}(\partial_\mu\phi)(\partial^\mu\phi) - U(\phi).$$

Usually, the potential energy density $U(\phi)$ consists only of a mass term, $\frac{1}{2}m^2\phi^2$, but in general, it can be an arbitrary positive function. If the theory is scale invariant, it follows from the kinetic term that n must be 2. Furthermore, the potential energy density must vanish. Thus we are left with the trivial example of a massless free scalar field in $(1+1)$-dimensions!

Let us now combine the scale transformation with the internal transformation

$$\phi(x) \to \lambda^{-d}\phi(x).$$

Here d is a real scalar, called *scaling dimension* of the field. This transformation is very similar to the phase transformation in the theory of complex scalar fields.

The displaced action is then changed to

$$S_{D(\Omega)}\left[\lambda^{-d}D_*\phi\right] = \int_\Omega L\left[\lambda^{-d}\phi(x); \quad \lambda^{-d-1}\partial_\mu\phi(x)\right]\lambda^n d^n x.$$

Consequently, the combined transformation is a symmetry transformation, provided that

$$\lambda^n L[\lambda^{-d}\phi(x); \lambda^{-d-1}\partial_\mu\phi(x)] = L[\phi(x); \partial_\mu\phi(x)]. \tag{11.86}$$

If the theory is scale invariant, it follows from the kinetic term that

$$d = \frac{n-2}{2} \text{ (scaling dimension of a scalar field).} \tag{11.87}$$

This number is then referred to as the *scaling dimension of a scalar field.* The potential energy term must furthermore satisfy the homogeneity property

$$\lambda^n U\left[\lambda^{\frac{2-n}{2}}\phi\right] = U[\phi].$$

Substituting $\phi \equiv 1$ and $x = \lambda^{1-\frac{n}{2}}$, we therefore see that U must be of the form

$$U(x) = U(1)x^{\frac{2n}{n-2}}; \quad n \geq 3. \tag{11.88}$$

e.g., in four dimensions this means that a term like

$$U[\phi] = \frac{g}{4!}\phi^4$$

is admissible. Notice, however, that a mass term is never admissible!

The same analysis can be applied to a vector field A_μ. The displaced field is given by

$$[D_*A]_\mu(x) = \frac{1}{\lambda}A_\mu\left(\frac{x}{\lambda}\right).$$

This leads to the displaced action

$$S_{D(\Omega)}[\lambda^{-d}D_*A] = \int_\Omega L[\lambda^{-d-1}A_\mu(x); \quad \lambda^{-d-2}\partial_\mu A_\nu(x)]\lambda^n d^n x.$$

A scale transformation is thus a symmetry transformation, provided that

$$\lambda^n L(\lambda^{-d-1}A_\mu; \quad \lambda^{-d-2}\partial_\mu A_\nu] = L[A_\mu, \partial_\mu A_\nu]. \tag{11.89}$$

If the theory is scale invariant, it follows from the kinetic term $-\frac{1}{4}F_{\mu\nu}F^{\mu\nu}$ that

$$d = \frac{n-4}{2} \text{ (scaling dimension of a vector field).} \tag{11.90}$$

This number is then referred to as the *scaling dimension of the vector field.* Furthermore, a potential energy term must satisfy

$$\lambda^n U\left[\lambda^{\frac{2-n}{2}}A_\mu\right] = U[A_\mu].$$

This is the same condition as before; i.e., in four dimensions this means that a term like

$$U[A_\mu] = \frac{g}{4!}(A_\mu A^\mu)^2$$

is admissible, while a mass term, e.g., is never admissible.

Notice that with the above choices of scaling dimensions, conditions (11.86) and (11.89) actually coincide! If we in general let ϕ_α denote a collection of scalar fields and/or vector fields, the preceding arguments thus show that a theory is scale invariant provided the Lagrangian satisfies the homogeniety property

$$\lambda^n L\left[\lambda^{\frac{2-n}{2}}\phi_\alpha; \lambda^{-\frac{n}{2}}\boldsymbol{d}\phi_\alpha\right] = L[\phi_\alpha; \boldsymbol{d}\phi_\alpha]. \tag{11.91}$$

This leads to the following rule:

A theory in Minkowski space is scale invariant provided that each term in the Lagrangian scales like λ^{-n}. The scaling of a term is found by scaling each field like $\lambda^{(2-n)/2}$ and each derivative like λ^{-1}.

As an important application of this rule we consider the case of a complex scalar field interacting with the Maxwell field through the rule of minimal coupling. In this case the standard kinetic term is modified to

$$-\tfrac{1}{2}(\overline{D_\alpha\phi})(D^\alpha\phi) = -\tfrac{1}{2}(\partial_\alpha\bar{\phi})(\partial^\alpha\phi) + \frac{ie}{2}A^\alpha[\bar{\phi}\partial_\alpha\phi - \phi\partial_\alpha\bar{\phi}] - \tfrac{1}{2}e^2 A_\alpha A^\alpha\bar{\phi}\phi.$$

The first term presents no problem, but the two last terms scale like $\lambda^{(4-3n)/2}$ and λ^{4-2n}. Thus then have the correct scaling behavior only in $n = 4$ dimensions. We therefore conclude:

Minimal coupling is scale invariant only in four dimensions.

In standard physics literature the above criterion for scale invariance is often stated in a slightly different way using dimensional analysis of the individual terms in the Lagrangian. In units where $\hbar = c = 1$, the action S itself is dimensionless. (In general, S and $\hbar$ have the same dimension, so that $S/\hbar$ is dimensionless; cf. Feynman's rule (2.20).) Any physical quantity T will now have a dimension, denoted by $[T]$, corresponding to a power of a length. E.g., the volume element $d^n x$ has dimension n. Each time in the Lagrangian density consequently has dimension $-n$.

Consider first the kinetic term

$$-\tfrac{1}{2}\partial_\mu\phi\partial^\mu\phi \quad \text{or} \quad -\tfrac{1}{4}F_{\mu\nu}F^{\mu\nu}.$$

Since a derivative corresponds to a division by a length, it reduces the dimension by 1. Consequently,

$$-n = 2([\phi] - 1);$$

i.e., the dimension of ϕ (and similarly of A_μ) is given by

$$[\phi] = \frac{2-n}{2}; \quad [A_\mu] = \frac{2-n}{2}. \tag{11.92}$$

Notice that the dimension of the field coincides with the scaling behavior of the field (cf. (11.91)).

Except for the kinetic term, all other terms will now contain coupling constants, which may or may not carry a dimension. Consider, e.g., a mass term

$$\tfrac{1}{2}m^2\phi^2.$$

In this case we get

$$-n = 2([m] + [\phi]) = 2\left([m] + \frac{2-n}{2}\right); \quad \text{i.e., } [m] = -1.$$

Consequently, the mass parameter m has dimensions -1 in any number of space–time dimensions. Any coupling constant that carries a dimension therefore generates a "mass-scale"; i.e., a suitable power of the coupling constant has the dimension of mass.

As a famous example we consider Newton's gravitational constant. As we have not yet developed a field theory of gravity, we will use Newton's second law,

$$m\frac{d^2x}{dt^2} = G\frac{mM}{r^2},$$

as a starting point. Notice that in units where $c = 1$, time will have the same dimension as length. Furthermore, the mass has dimension -1, so that we get

$$[G] = \left[\frac{d^2x}{dt^2}\right] + 2[r] - [M] = -1 + 2 + 1 = 2.$$

The mass scale in gravity is therefore given by

$$G^{-\frac{1}{2}}\left(\sqrt{\frac{\hbar c}{G}} = 5.5\ 10^{-8}\ \text{kg}\right), \tag{11.93}$$

which is known as the *Planck mass*. On the contrary, a dimensionless coupling constant cannot in any way be related to a mass: It simply generates a pure number. Consider, e.g., the coupling constant in electromagnetism. In four dimensions, A_μ has dimension -1. It follows from consideration of the covariant derivative, $\partial_\mu - ieA_\mu$, that e is dimensionless. Reintroducing $\hbar$ and c, the pure number it represents is given by

$$\frac{e^2}{\hbar c} \approx \frac{1}{137}, \tag{11.94}$$

which is known as the *fine-structure constant*.

Returning to the criterion of scale invariance, we can now reformulate it in the following way:

A theory is scale invariant precisely when it has the following two properties:

1. *It contains no mass terms;*
2. *All coupling constants are dimensionless.*

By abuse of language, a theory with these two properties is often referred to as a massless theory.

Given a scale invariant theory, there exists a corresponding conservation law. This can be found from Noether's theorem, where the displaced configuration now includes the action of the internal transformation. Consequently, a general field configuration $\phi_\alpha(x)$ transforms as follows:

$$\phi_\alpha^\lambda(x) = e^{-\lambda d}[(D_\lambda)_*\phi_\alpha](e^\lambda \cdot x); \quad D_\lambda(x^\alpha) = e^\lambda \cdot x^\alpha,$$

where d is the scaling dimension. We thus get an additional contribution to the conserved current coming from the factor $e^{-\lambda d}$ in front. Since the characteristic vector field associated with the one-parameter group of dilatations is given by

$$a^\nu = \frac{dy^\nu}{d\lambda}_{|\lambda=0} = x^\nu,$$

the conserved current reduces to

$$J^\mu = x^\nu T_\nu{}^\mu - d\phi_\alpha \frac{\partial L}{\partial(\partial_\mu\phi_\alpha)}. \tag{11.95}$$

This is a nice, except for one little detail. Unlike the case of ordinary space–time symmetries, the current is no longer of the form $J^\mu = a^\nu T_\nu{}^\mu$. This is very analogous to what happened in the case of vector fields, here we also obtained an additional spin contribution and consequently had to repair the canonical energy–momentum tensor (cf. (11.46)). This suggests that even the true energy–momentum tensor needs to be repaired in order to absorb the additional term in (11.95).

Actually, it can be repaired only in the case of scalar fields. This is due to a different structure in the two cases. In the scalar case the additional term is simply a gradient:

$$-d\phi \frac{\partial L}{\partial(\partial_\mu\phi)} = d\phi\partial^\mu\phi = \partial^\mu\left[\frac{d}{2}\phi^2\right]. \tag{11.96}$$

In the vector case we similarly get

$$-dA_\alpha \frac{\partial L}{\partial(\partial_\mu A_\alpha)} = dA_\alpha F^{\mu\alpha} \tag{11.97}$$

(cf. exercise 11.6.1). But this is *not* a gradient!

To see this, let us temporarily assume it is a gradient,

$$A_\alpha F^{\mu\alpha} \stackrel{?}{=} \partial^\mu \chi[A_\alpha]. \tag{*}$$

In this case, the functional

$$T_\Omega[A_\alpha] = \int_\Omega a_\mu A_\alpha F^{\mu\alpha} d^n x,$$

where a_μ is an arbitrary constant vector, can actually be converted to a pure boundary term

$$T_\Omega[A_\alpha] = \int_{\partial\Omega} a_\mu \chi[A_\alpha] n^\mu d^{n-1}x.$$

Consequently, the functional $T_\Omega[A_\alpha]$ is independent of variations in A_α as long as they vanish on the boundary. If we perform such a variation,

$$A_\alpha \curvearrowright A_\alpha + \epsilon B_\alpha,$$

we get

$$T_\Omega(\epsilon) = a_\mu \int_\Omega [A_\alpha(\partial^\mu A^\alpha - \partial^\mu A^\alpha) + \epsilon\{B_\alpha(\partial^\mu A^\alpha - \partial^\alpha A^\mu) + A_\alpha(\partial^\mu B^\alpha - \partial^\alpha B^\mu)\} \\ + \epsilon^2 B_\alpha\{\partial^\mu B^\alpha - \partial^\alpha B^\mu\}]d^n x.$$

We therefore obtain

$$0 = \frac{dT}{d\epsilon_{|\epsilon=0}} = a_\mu \int_\Omega B_\alpha(\partial^\mu A^\alpha - \partial^\alpha A^\mu) + A_\alpha(\partial^\mu B^\alpha - \partial^\alpha B^\mu)d^n x.$$

Performing a partial integration, this is rearranged as

$$0 = a_\mu \int [-\partial^\alpha A^\mu + g^{\mu\alpha}\partial_\nu A^\nu]B_\alpha d^n x.$$

This is consistent only if A^α satisfies the identity

$$\partial^\alpha A^\mu = g^{\mu\alpha}\partial_\nu A^\nu.$$

Taking the trace, we immediately see that $\partial^\alpha A^\nu$ must vanish! This contradicts our assumption that (*) was supposed to hold for an arbitrary configuration!

As we shall see later, this difference between the scalar case and the vector case also has the consequence that scale-invariant scalar theories are actually conformally invariant, whereas scale-invariant vector theories are *not* (except in the trivial four-dimensional case). In the following we shall therefore confine ourselves to the scalar case.

As usual, we modify the energy–momentum tensor by adding a term satisfying the requirements (3.18). This time, however, the construction is a bit tricky. It is based upon the properties of a particular tensor constructed directly from the metric:

$$g_{\alpha\beta\gamma\delta} \stackrel{\text{def}}{=} g_{\alpha\gamma}g_{\beta\delta} - g_{\alpha\delta}g_{\beta\gamma}. \tag{11.98}$$

Notice the following important symmetry properties:

$$g_{\beta\alpha\gamma\delta} = g_{\alpha\beta\delta\gamma} = -g_{\alpha\beta\gamma\delta}, \quad \text{skew-symmetry in the first and last pair of indices;} \tag{11.99a}$$

$$g_{\alpha\beta\gamma\delta} = g_{\gamma\delta\alpha\beta}. \quad \text{symmetry between the first and the last pair of indices} \tag{11.99b}$$

Using this, we now construct the following correction term:

$$\theta^{\mu\nu} = \frac{(n-2)}{4(n-1)}\partial_\lambda\partial_\rho[g^{\lambda\mu\rho\nu}\phi^2] = \frac{(n-2)}{4(n-1)}[g^{\mu\nu}\Box - \partial^\mu\partial^\nu]\phi^2. \tag{11.100}$$

From (11.99b), it follows that $\theta^{\mu\nu}$ is symmetric in μ and ν. Furthermore, (11.99a) implies that

$$\partial_\mu \theta^{\mu\nu} = \frac{(n-2)}{4(n-1)} \partial_\rho [\partial_\lambda \partial_\mu g^{\lambda\mu\rho\nu} \phi^2] = 0.$$

Finally, we get by explicit computation that

$$\theta^{00} = -\frac{(n-2)}{4(n-1)} \partial_i \partial^i \phi^2; \quad \theta^{i0} = -\frac{(n-2)}{4(n-1)} \partial^i \partial^0 \phi^2,$$

which shows that $\theta^{\mu\nu}$ does *not* contribute to the total energy and momentum. Thus (11.100) is an admissible correction term!

Using the tensor $g^{\lambda\mu\rho\nu}$, we can also construct a conserved current:

$$J^\mu_\theta = \frac{(n-2)}{4(n-1)} \partial_\lambda \partial_\rho [x_\nu g^{\lambda\mu\rho\nu} \phi^2]. \tag{11.101}$$

(Notice that the bracket is skew-symmetric in λ and μ.) This expression can be rearranged as

$$\begin{aligned} J^\mu_\theta &= \frac{(n-2)}{4(n-1)} x_\nu [\partial_\lambda \partial_\rho g^{\lambda\mu\rho\nu} \phi^2] + \frac{(n-2)}{4(n-1)} [g_{\lambda\nu} \partial_\rho (g^{\lambda\mu\rho\nu} \phi^2)] \\ &= x_\nu \theta^{\mu\nu} - \frac{(n-2)}{4} \partial^\mu \phi^2. \end{aligned}$$

On the other hand, scale invariance led to the following conserved current (cf. (11.95–96)):

$$J^\mu = x_\nu T^{\mu\nu} + \frac{(n-2)}{4} \partial^\mu \phi^2.$$

By adding these two conserved currents, we thus precisely cancel the unwanted term:

$$\tilde{J}^\mu = J^\mu + J^\mu_\theta = x_\nu \tilde{T}^{\mu\nu}, \quad \text{with} \quad \tilde{T}^{\mu\nu} = T^{\mu\nu} + \theta^{\mu\nu}.$$

Notice that since $\tilde{J}^\mu$ is conserved, we get

$$0 = \partial_\mu \tilde{J}^\mu = \partial_\mu (x_\nu \tilde{T}^{\mu\nu}) = \tilde{T}_\mu{}^\mu + x_\nu \partial_\mu \tilde{T}^{\mu\nu}.$$

But here the last term vanishes trivially because $\tilde{T}^{\mu\nu}$ is conserved. Consequently, the modified energy–momentum tensor is traceless!

Let us summarize the main outcome of the investigation:

Any scale-invariant scalar theory in ordinary Minkowski space possesses a conserved, symmetric, traceless energy–momentum tensor

$$\tilde{T}^{\mu\nu} = T^{\mu\nu} + \frac{(n-2)}{4(n-1)} [g^{\mu\nu} \Box - \partial^\mu \partial^\nu] \phi^2 \tag{11.102}$$

known as the conformal energy–momentum tensor. The associated conserved current

$$\tilde{J}^\mu = x_\nu \tilde{T}^{\mu\nu} \tag{11.103}$$

is known as the scale current.

§ 11.9 Conformal Transformations as Symmetry Transformations

Now that the group of scale transformations is under control, we return to the problem of conformal invariance in a classical field theory. The main difficulty is that we have to combine the conformal transformation with an internal transformation. Let us first look at this internal transformation in some detail:

Suppose M is an arbitrary manifold equipped with a metric $\mathbf{g}$ and that $f : M \curvearrowright M$ is a suitable diffeomorphism. We would like to assign a scale to the transformation f. If $\mathbf{g}$ and $f^*\mathbf{g}$ happened to be proportional, the constant of proportionality would define the square of the scale, but in general, there need not be any simple connection between $\mathbf{g}$ and $f^*\mathbf{g}$. Therefore, we turn our attention to the volume form instead. Since ϵ and $f^k\epsilon$ both are n-forms, they are necessarily proportional:

$$f^*\epsilon = J_f(x)\epsilon. \tag{11.104}$$

Consequently, we can always attach a *volume expansion* $J_f(x)$ to the transformation f. Notice that if f preserves the orientation, then according to (10.34), $f^*\epsilon$ reduces to ϵ_f (where ϵ_f is the volume form generated from the pulled-back metric $f^*\mathbf{g}$. This justifies the name $J_f(x)$. The scale λ associated with a scale transformation can now in the general case be replaced by the scale

$$\lambda_f(x) = [J_f(x)]^{1/n}.$$

Let us look at a few illustrative examples. If f is an isometry, then $f^*\mathbf{g} = \mathbf{g}$, and consequently,

$$J_f(x) = \pm 1,$$

according to whether f preserves or reverse the orientation. Similarly, if f is a conformal transformation, then $f^*\mathbf{g} = \Omega^2(x)\mathbf{g}$, and consequently,

$$J_f(x) = \pm\Omega^n(x), \tag{11.105}$$

according to whether f preserves or reverses the orientation (cf. (10.40)).

We can also work out a coordinate expression for $J_f(x)$. From (10.8) we get

$$\begin{aligned}[f^*\epsilon](x)_{1\ldots n} &= \sqrt{g(f(x))}\epsilon_{j_1\cdots j_n}\frac{\partial y^{j_1}}{\partial x^1}\cdots\frac{\partial y^{j_n}}{\partial x^n}\\ &= \sqrt{g(f(x))}\left|\frac{\partial y}{\partial x}\right|,\end{aligned}$$

where $|\frac{\partial y}{\partial x}|$ is the Jacobian of the transformation f. Consequently, the volume expansion is given by

$$J_f(x) = \frac{\sqrt{|g(f(x))|}}{\sqrt{|g(x)|}}\left|\frac{\partial y}{\partial x}\right|. \tag{11.106}$$

Notice that in ordinary Minkowski space, $J_f(x)$ simpy reduces to the Jacobian when we use inertial coordinates. (That is why the volume expansion has been denoted by capital J!)

From the coordinate expression follows an important group property. Suppose f_1, f_2 are successive transformations

$$z = f_2(y); \quad y = f_1(x).$$

Then

$$\begin{aligned} J_{f_2 \circ f_1}(x) &= \frac{\sqrt{|g(f_2(f_1(x)))|}}{\sqrt{|g(x)|}} \left| \frac{\partial z}{\partial x} \right| \qquad (11.107) \\ &= \frac{\sqrt{|g(f_2(f_1(x)))|}}{\sqrt{|g(f_1(x))|}} \sqrt{\frac{|g(f_1(x))|}{\sqrt{|g(x)|}}} \left| \frac{\partial z}{\partial y} \right| \left| \frac{\partial y}{\partial x} \right| \\ &= J_{f_2}(f_1(x)) \cdot J_{f_1}(x). \end{aligned}$$

We are now in a position where we can generalize the internal transformation associated with a scale transformation. Let ϕ_α denote a general field configuration consisting either of a set of scalar fields or a vector field. Under a diffeomorphism f it will then not only be pushed forward but also multiplied by an additional scale. The combined operation will be denoted by T_f, and it is given by

$$[T_f \phi_\alpha](x) = J_{f^{-1}}^{d/n}(x) \cdot [f_* \phi_\alpha](x). \qquad (11.108)$$

As usual, d denotes the scaling dimension of the field ϕ_α. In the case of a scale transformation, we know that $J_f(x) = \Omega^n(x) = \lambda^n$. Consequently, (11.108) reproduces the conventional transformation in the case of ordinary scale transformations. Notice too that when f is an orientation-preserving isometry, it just reduces to an ordinary space–time transformation.

The operation T_f satisfies the important group property

$$T_{f_2 \circ f_1} = T_{f_2} \circ T_{f_1}. \qquad (11.109)$$

To check this, we notice first that

$$[(f_*)(g \cdot \phi_\alpha)](x) = g(f^{-1}(x))[f_* \phi_\alpha(x)]$$

(cf. (10.17)). Then we apply the right-hand side of (11.109) to a field configuration ϕ_α:

$$\begin{aligned} [T_{f_2}[T_{f_1} \phi_\alpha]](x) &= J_{f_2^{-1}}^{d/n}(x)[(f_2)_*(T_{f_1} \phi_\alpha)](x) \\ &= J_{f_2^{-1}}^{d/n}(x)[(f_2)_*(J_{f_1^{-1}}^{d/n}(x) \cdot [(f_1)_* \phi_\alpha])](x) \\ &= J_{f_2^{-1}}^{d/n}(x) \cdot J_{f_1^{-1}}^{d/n}(f_2^{-1}(x)) \cdot [(f_2)_*(f_1)_* \phi_\alpha](x). \end{aligned}$$

Using the group properties (11.107) and (10.21), this reduces to

$$= J_{(f_2 \circ f_1)^{-1}}^{d/n}(x) \cdot [(f_2 \circ f_1)_* \phi_\alpha](x) = [T_{f_2 \circ f_1} \phi_\alpha](x).$$

(If you think this proof is too abstract, you should try to work it out in coordinates!) The essential content of the group property (11.109) is that whenever f belongs to a group of transformations on M, then the operation T_f constitutes a *representation* of this group. Notice also that T_f acts linearly upon the field ϕ_α.

For a given field theory we can now investigate whether a space–time transformation f is a symmetry transformation. Using the now familiar techniques, we calculate the displaced action:

$$S_{f(\Omega)}[T_f\phi_\alpha] = \int_{f(U)} L\left[T_f\phi_\alpha(y);\quad \frac{\partial}{\partial y^\mu}T_f\phi_\alpha;\quad g_{\mu\nu}(y)\right]\sqrt{-g(y)}d^n y.$$

Substituting $y = f(x)$, this is rearranged as

$$= \int_U L\left[(T_f\phi_\alpha)(f(x));\ \frac{\partial x^\nu}{\partial y^\mu}\frac{\partial}{\partial x_\nu}(T_f(\phi_\alpha)(f(x));\ g_{\mu\nu}(f(x))\right]$$
$$\times\sqrt{-g(f(x))}\left|\frac{\partial y}{\partial x}\right|d^n x.$$

Using (11.106), this is finally rearranged as

$$S_{f(\Omega)}[T_f\phi_\alpha] = \int_U J_f(x)L[T_f\phi_\alpha)(f(x));$$
$$\frac{\partial x^\nu}{\partial y^\mu}\frac{\partial}{\partial x}\nu(T_f\phi_\alpha)(f(x)); \tag{10.110}$$
$$g_{\mu\nu}(f(x))]\sqrt{-g(x)}d^n x.$$

Notice that apart from a scaling of the Lagrangian, the following substitutions have been performed:

$$\phi_\alpha(x) \to J_{f^{-1}}^{d/n}(f(x))\cdot[(f_*\phi_\alpha)](f(x)); \tag{11.111}$$

$$\partial_\mu\phi_\alpha(x) \to J_{f^{-1}}^{d/n}(f(x))\frac{\partial x^\nu}{\partial y^\mu}\frac{\partial[f_*\phi_\alpha]}{\partial x^\nu}(f(x))$$
$$+\frac{\partial x^\nu}{\partial y^\mu}\cdot[f_*\phi_\alpha](f(x))\cdot\partial\frac{[J_{f^{-1}}^{d/n}]}{\partial x^\nu}(f(x)); \tag{11.112}$$

$$g_{\mu\nu}(x) \to g_{\mu\nu}(f(x)) \tag{11.113}$$

The first substitution, associated with the potential term $U[\phi_\alpha]$, is quite harmless. E.g., in four dimensions the potential term $U[\phi] = \frac{g}{4!}\phi^4$ will be invariant under arbitrary space–time transformations! The second one, associated with the kinetic term $-\frac{1}{2}\langle \boldsymbol{d}\phi_\alpha \mid \boldsymbol{d}\phi_\alpha\rangle$, will, however, cause us trouble. When we square the derivative, we get additional terms that are linear in $\partial_\mu\phi_\alpha$. These terms will prevent the displaced action from being identical to the original action. All we can hope therefore is that they differ by a surface term, i.e., that the displaced Lagrangian is *essentially equivalent* to the original Lagrangian. The condition for f being a symmetry transformation is consequently

$$J_f(x)\cdot L\left[(T_f\phi_\alpha)(f(x));\ \frac{\partial x^\nu}{\partial y^\mu}\frac{\partial[T_f\phi_\alpha]}{\partial x^\nu}(f(x));\ g_{\mu\nu}(f(x))\right]$$
$$\approx L[\phi_\alpha;\ \partial_\mu\phi_\alpha;\ g_{\mu\nu}]. \qquad 11.114$$

(Here $\approx$ denotes that they may differ by a pure divergence.)

We can now apply this to investigate conformal invariance. For simplicity, we restrict ourselves to consideration of scalar fields. Furthermore, we need only worry about the kinetic term. If f is an orientation-preserving conformal transformation, we have previously seen that

$$J_f(x) = \Omega^n(x).$$

From the group property (11.107) it furthermore follows that

$$J_{f^{-1}}(f(x)) = \Omega^{-n}(x).$$

Consequently, the substitutions (11.111) and (11.112) reduce to

$$\phi(x) \to \Omega^{-d}(x)\phi(x)$$

and

$$\partial_\mu \phi(x) \to \frac{\partial x^\nu}{\partial y^\mu}[\Omega^{-d}\partial_\nu\phi + \phi\partial_\nu\Omega^{-d}].$$

Consequently, the kinetic term is replaced by

$$\Omega^n\left\{-\tfrac{1}{2}g^{\mu\nu}(f(x))\frac{\partial x^\rho}{\partial y^\mu}\frac{\partial x^\sigma}{\partial y^\nu}[\Omega^{-d}\partial_\rho\phi + \phi\partial_\rho\Omega^{-d}][\Omega^{-d}\partial_\sigma\phi + \phi\partial_\sigma\Omega^{-d}]\right\}.$$

But since f is a conformal map, we know that

$$g^{\mu\nu}(f(x))\frac{\partial x^\rho}{\partial y^\mu}\frac{\partial x^\sigma}{\partial y^\nu} = \Omega^{-2}(x)g^{\rho\sigma}(x).$$

Thus the displaced kinetic term reduces to

$$-\tfrac{1}{2}\Omega^{n-2}g^{\rho\sigma}[\Omega^{-d}\partial_\rho\phi + \phi\partial_\rho\Omega^{-d}][\Omega^{-d}\partial_\sigma\phi + \phi\partial_\sigma\Omega^{-d}].$$

Inserting the scaling dimension (11.87), this finally boils down to

$$\tfrac{1}{2}\partial_\mu\phi\partial^\mu\phi + \frac{n-2}{4}\{(\partial_\mu\phi^2)\partial^\mu \ln\Omega - \frac{n-2}{2}\phi^2\partial_\mu \ln\Omega\partial^\mu \ln\Omega\}.$$

The first term is precisely what we are after, but the last term is not a manifestly pure divergence. We therefore rearrange the displaced kinetic energy density as follows:

$$\begin{aligned}&-\tfrac{1}{2}\partial_\mu\phi\partial^\mu\phi + \frac{n-2}{4}\cdot\frac{1}{\sqrt{-g}}\partial_\mu\{\sqrt{-g}\phi^2\partial^\mu\ln\Omega\}\\ &\quad-\frac{n-2}{4}\phi^2\left\{\frac{1}{\sqrt{-g}}\partial_\mu(\sqrt{-g}\partial^\mu\ln\Omega) + \frac{n-2}{2}\partial_\mu\ln\Omega\partial^\mu\ln\Omega\right\}.\end{aligned} \tag{11.115}$$

It follows that a conformal transformation with the conformal factor $\Omega^2(x)$ is a symmetry transformation precisely when

$$\frac{1}{\sqrt{-g}}\partial_\mu(\sqrt{-g}\partial^\mu\ln\Omega) = \frac{2-n}{2}\partial_\mu\ln\Omega\partial^\mu\ln\Omega \tag{11.116a}$$

i.e.,

$$\Box \ln \Omega = \frac{2-n}{2} (\boldsymbol{d} \ln \Omega \mid \boldsymbol{d} \ln \Omega). \qquad (11.116b)$$

The latter condition is far from automatically fulfilled, but it turns out to be fulfilled for all conformal transformations in the ordinary flat Minkowski space. Since the conformal group is generated from isometries and the inversion, it suffices to check that the inversion is a symmetry transformation. In that case, we know from exercise 10.3.1 that the inversion has the conformal factor

$$\Omega(x) = x^{-2}.$$

Using that $\partial^\mu \ln \Omega = -2x^{-2}x^\mu$ and $\sqrt{-g} = 1$, it is now trivial to verify that (11.116) is satisfied. Notice, however, that the above argument applies only in ordinary "flat" Minkowski space. In a curved space–time, the theory of a massless scalar field is *not* conformally invariant, at least not when we use the standard Lagrangian!

Worked Exercise 11.9.1
Problem:

a. Consider a conformal transformation with the conformal factor $\Omega^2(x)$. Show that the kinetic term of a vector field is displaced as follows:

$$-\frac{1}{4}F_{\mu\nu}F^{\mu\nu} \to -\frac{1}{4}F_{\mu\nu}F^{\mu\nu} - \frac{n-4}{8}$$
$$\times \{4F^{\mu\nu}A_\mu\partial_\nu \ln \Omega + (n-4)[A_\mu A^\mu g^{\alpha\beta} - A^\alpha A^\beta]\partial_\alpha \ln \Omega \partial_\beta \ln \Omega\}. \qquad (11.117)$$

b. Let $\Omega_\lambda^2(x)$ be the conformal factor associated with the one-parameter group of special conformal transformations

$$c_\lambda(x) = \frac{x^\mu + \lambda b^\mu \langle x \mid x\rangle}{1 + 2\lambda\langle b \mid x\rangle + \lambda^2\langle b \mid b\rangle\langle x \mid x\rangle}.$$

Show that

$$\frac{d}{d\lambda}_{|\lambda=0}\{\partial^\nu \ln \Omega_\lambda\} = -2b^\nu \quad \text{and} \quad (\partial^\nu \ln \Omega_\lambda)_{|\lambda=0} = 0. \qquad (11.118)$$

(Hint: Use exercise 10.3.1.)

c. Show that the kinetic term associated with a vector field is *not* conformally invariant when $n \neq 4$. (Hint: Show that the correction term cannot be a pure divergence when we apply an infinitesimal special conformal transformation.)

Exercise 11.9.2
Introduction: In this exercise we restrict to the four-dimensional ordinary Minkowski space.

Problem: Show that minimal coupling is conformally invariant; i.e., show that the gauge invariant kinetic term

$$-\tfrac{1}{2}(\partial_\alpha - ieA_\alpha)\bar{\phi}(\partial^\alpha + ieA^\alpha)\phi$$

is conformally invariant.

We can summarize the above findings (including the results obtained in exercises 11.9.1 and 11.9.2) in the following way: *In the four-dimensional Minkowski space, any scale-invariant theory based upon scalar fields and vector fields is conformally invariant. In an arbitrary dimension different from four, it is only the scale-invariant theories based upon scalar fields that are conformally invariant.*

This explains among other things why we could not construct a conformal energy–momentum tensor in the case of vector fields (except, of course, in the trivial four-dimensional case.) The reason is the following: Suppose a theory is scale invariant and allows the construction of a conserved symmetric energy–momentum tensor $\tilde{T}^{\mu\nu}$ such that the associated scale current is given by

$$S^{\nu} = x_{\mu}\tilde{T}^{\mu\nu}.$$

Notice especially that the conservation of the scale current implies that $\tilde{T}^{\mu\nu}$ is traceless. From the vanishing trace it now follows trivially that the four current $\mathbf{S}_{\rho}$ given by

$$[\mathbf{S}_{\rho}]^{\nu} = (\langle x \mid x\rangle g_{\mu\rho} - 2x_{\rho}x_{\mu})\tilde{T}^{\mu\nu}$$

are conserved as well (cf. exercise 11.6.4). But these are precisely the currents associated with the special conformal transformations! In a massless scalar theory this is as expected, since such a theory is not only scale invariant but actually conformally invariant. However, in the case of a massless vector theory we do not expect any conserved currents associated with the special conformal transformations, since such a theory is not conformally invariant!

It is instructive to derive the conserved current (11.119) directly from Noether's theorem. Then we have to be careful, since the displaced action differs from the initial action by a divergence term (cf. (11.115)):

$$\frac{n-2}{4}\int \frac{1}{\sqrt{-g}}\,\partial_{\mu}(\sqrt{-g}\phi^2\partial^{\mu}\ln\Omega_{\lambda})\sqrt{-g}d^n x.$$

This must consequently be subtracted from the displaced action. The Noether current associated with a one-parameter group of conformal transformations is therefore given by

$$J^{\mu} = a\mathring{T}_{\nu}{}^{\mu} + \frac{\partial L}{\partial(\partial_{\mu}\phi)}\frac{d\phi^{\lambda}}{d\lambda_{|\lambda=0}}(f_{\lambda}(x)) - \frac{n-2}{4}\phi^2\partial^{\mu}\frac{d}{d\lambda_{|\lambda=0}}\ln\Omega_{\lambda}.$$

The displaced field configuration is given by

$$\phi^{\lambda}(f_{\lambda}(x)) = J_{f^{-1}}^{d/n}(f_{\lambda}(x))[(f_{\lambda})_{*}\phi](f_{\lambda}(x)) = \Omega_{\lambda}^{-d}(x)\phi(x).$$

Using this, the Noether current reduces to

$$J^{\mu} = a^{\nu}\mathring{T}_{\nu}{}^{\mu} + \frac{n-2}{4}\left[\frac{d\Omega_{\lambda}}{d\lambda_{|\lambda=0}}\right]^2\partial^{\mu}\left[\phi^2\left(\frac{d\Omega_{\lambda}}{d\lambda_{|\lambda=0}}\right)^{-1}\right], \qquad (11.120)$$

which generalizes (11.95).

In the case of special conformal transformations, where the one-parameter group is given by

$$y^\nu = \frac{x^\nu + \lambda b^\nu \langle x \mid x \rangle}{1 + 2\lambda \langle b \mid x \rangle + \lambda^2 \langle b \mid b \rangle \langle x \mid x \rangle},$$

we get the characteristic vector field

$$a^\nu = \frac{dy^\nu}{d\lambda_{|\lambda=0}} = b^\mu \langle x \mid x \rangle - 2x^\mu \langle b \mid x \rangle.$$

Furthermore, we know form exercise 10.3.1 that Ω_λ is given by

$$\Omega_\lambda = (1 + 2\lambda \langle b \mid x \rangle + \lambda^2 \langle b \mid b \rangle \langle x \mid x \rangle)^{-1}.$$

Consequently,

$$\frac{d\Omega_\lambda}{d\lambda_{|\lambda=0}} = -2\langle b \mid x \rangle.$$

Inserting this, the Noether current (11.120) reduces to

$$J^\mu = (b_\nu \langle x \mid x \rangle - 2x_\nu \langle b \mid x \rangle)\mathring{T}^{\mu\nu} + \frac{2-n}{2} \langle b \mid x \rangle^2 \partial^\mu \left[\frac{\phi^2}{\langle b \mid x \rangle} \right].$$

As in the case of scale transformations, we thus get an additional term, which we must somehow absorb in the canonical energy–momentum tensor. In analogy with (11.101), we therefore consider the current

$$J^\mu_\theta = \frac{(n-2)}{4(n-1)} \partial_\lambda \partial_\rho \{[b_\nu \langle x \mid x \rangle - 2x_\nu \langle b \mid x \rangle] g^{\lambda\mu\rho\nu} \phi^2\}.$$

Since

$$\begin{aligned} g^{\lambda\mu\rho\nu}&[b_\nu \langle x \mid x \rangle - 2x_\nu \langle b \mid x \rangle] \\ &= \langle x \mid x \rangle [b^\mu g^{\lambda\rho} - b^\lambda g^{\mu\rho}] - 2\langle b \mid x \rangle [x^\mu g^{\lambda\rho} - x^\lambda g^{\mu\rho}], \end{aligned}$$

which is skew-symmetric in μ and λ, it follows that J^μ_θ is trivially conserved. Performing the differentiations involved, we can now rearrange the expression for J^μ_θ as follows:

$$\begin{aligned} J^\mu_\theta = [b_\nu \langle x \mid x \rangle - 2x_\nu \langle b \mid x \rangle \theta^{\mu\nu} + \frac{n-2}{2} \langle b \mid x \rangle^2 \partial^\mu \left[\frac{\phi^2}{\langle b \mid x \rangle} \right] \\ + \frac{3(n-2)}{2(n-1)} \partial\rho[(b^\mu x^\rho - x^\mu b^\rho)\phi^2]. \end{aligned}$$

By adding these two conserved currents, we almost cancel the unwanted term:

$$\tilde{J}^\mu = J^\mu + J^\mu_\theta = [b_\nu \langle x \mid x \rangle - 2x_\nu \langle b \mid x \rangle \tilde{T}^{\mu\nu} + \frac{3(n-2)}{2(n-1)} \partial_\rho[(b^\mu x^\rho - x^\mu b^\rho)\phi^2].$$

Furthermore, the additional term constitutes itself a current that is trivially conserved. We can therefore throw it away, whereby we finally recover (11.119)!

Solutions to Worked Exercises

Solution to 11.4.1

a. The determinant is given by

$$\left| \frac{\partial y^\alpha}{\partial x^\beta} \right| = \epsilon_{\alpha_1 \ldots \alpha_n} \frac{\partial y^{\alpha_1}}{\partial x^1} \cdots \frac{\partial y^{\alpha_n}}{\partial x^n}.$$

Using Leibnitz's rule, we therefore get

$$\frac{d}{d\lambda} \left| \frac{\partial y^\alpha}{\partial x^\beta} \right| = \epsilon_{\alpha_1 \ldots \alpha_n} \frac{d}{d\lambda} \left(\frac{\partial y^{\alpha}{}_1}{\partial x^1} \right) \frac{\partial y^{\alpha_2}}{\partial x^2} \cdots \frac{\partial y^{\alpha_n}}{\partial x^n} + \cdots.$$

If we put $\lambda = 0$, this reduces to

$$\begin{aligned}
\frac{d}{d\lambda_{|\lambda=0}} \left| \frac{\partial y^\alpha}{\partial x^\beta} \right| &= \epsilon_{\alpha_1 \ldots \alpha_n} \frac{d}{d\lambda_{|\lambda=0}} \left(\frac{\partial y^{\alpha_1}}{\partial x^1} \right) \delta_2^{\alpha_2} \cdots \delta_n^{\alpha_n} + \cdots \\
&= \epsilon_{\alpha_1 2 \ldots n} \frac{d}{d\lambda_{|\lambda=0}} \left(\frac{\partial y^{\alpha_1}}{\partial x^1} \right) + \cdots \\
&= \frac{d}{d\lambda_{|\lambda=0}} \left(\frac{\partial y^1}{\partial x^1} \right) + \cdots + \frac{d}{d\lambda_{|\lambda=0}} \left(\frac{\partial y^n}{\partial x^n} \right).
\end{aligned}$$

b. We start with the identity

$$\frac{\partial x^\nu}{\partial y^\alpha} \frac{\partial y^\alpha}{\partial x^\mu} = \delta^\nu_\mu.$$

Differentiating this with respect to λ, we get

$$\frac{d}{d\lambda} \left(\frac{\partial x^\nu}{\partial y^\alpha} \right) \frac{\partial y^\alpha}{\partial x^\mu} + \frac{\partial x^\nu}{\partial y^\alpha} \frac{d}{d\lambda} \left(\frac{\partial y^\alpha}{\partial x^\mu} \right) = 0.$$

For $\lambda = 0$ this reduces to

$$\frac{d}{d\lambda_{|\lambda=0}} \left(\frac{\partial x^\nu}{\partial y^\alpha} \right) \delta^\alpha{}_\mu + \delta^\nu_\alpha \frac{d}{d\lambda_{|\lambda=0}} \left(\frac{\partial y^\alpha}{\partial x^\mu} \right) = 0,$$

i.e.,

$$\frac{d}{d\lambda_{|\lambda=0}} \left(\frac{\partial x^\nu}{\partial y^\mu} \right) + \frac{d}{d\lambda_{|\lambda=0}} \left(\frac{\partial y^\nu}{\partial x^\mu} \right) = 0.$$

Remark. When λ is close to 0, the Jacobian matrix is close to the identity matrix:

$$\left[\frac{\partial y^\alpha}{\partial x^\beta} \right] = \begin{bmatrix} 1 + \epsilon^1{}_1 & \epsilon^1{}_2 & \cdots \\ \epsilon^2{}_1 & \ddots & \\ \vdots & & 1 + \epsilon^n{}_n \end{bmatrix}.$$

If we work out the determinant and drop all higher-order terms, the off-diagonal contributions automotically cancel, and we get

$$\left| \frac{\partial y^\alpha}{\partial x^\beta} \right| = 1 + \epsilon^1{}_1 + \cdots + \epsilon^n{}_n + \text{higher-order terms},$$

and this is where the trace in (a) comes from!

Solution to 11.4.2

$$\begin{aligned}
\frac{1}{\sqrt{-d}} \partial_\mu(\sqrt{-g}\mathring{T}_\nu{}^\mu) &= \frac{1}{\sqrt{-g}} \partial_\mu \left(\sqrt{-g} \frac{-\partial L}{\partial(\partial_\mu \phi_\alpha)} \partial_\nu \phi^\alpha + \delta^\mu{}_\nu \sqrt{-g} L \right) \\
&= \frac{-1}{\sqrt{-g}} \partial_\mu \left(\sqrt{-g} \frac{\partial L}{\partial(\partial_\mu \phi_\alpha)} \right) \partial_\nu \phi^\alpha - \frac{\partial L}{\partial(\partial_\mu \phi_\alpha)} \partial_\mu \partial_\nu \phi^\alpha \\
&\quad + \left[\frac{1}{\sqrt{-g}} \partial_\nu(\sqrt{-g}) L + \partial_\nu L \right].
\end{aligned}$$

In the first term we can use the covariant equations of motion (8.35). In the second term we can interchange ∂_μ and ∂_ν. The first two terms can then be reduced to $-\partial_\nu L + \frac{\partial L}{\partial g_{\alpha\beta}} \partial_\nu g_{\alpha\beta}$, from which the result follows.

Solution to 11.5.1

a. Let f be a conformal map, and put $\psi = f^*\phi$. According to Theorem 10.15, we have

$$f^*[*d\phi] = \Omega^{n-2}(x) * f^*[d\phi] = *d\psi,$$

where we have used that $n = 2$. We then get

$$\begin{aligned}
\langle d\psi \mid d\psi \rangle_\Omega &= \int_\Omega *d\psi \wedge d\psi = \int_\Omega f^*[*d\phi \wedge d\phi] \\
&= \int_{f(\Omega)} *d\phi \wedge d\phi = \langle d\phi \mid d\phi \rangle_{f(\Omega)}.
\end{aligned}$$

b. Let f be a conformal map, and put $\mathbf{A}' = f^*\mathbf{A}$. From Theorem 10.15, we now get

$$f^*[*d_{\mathbf{A}}] = \Omega^{n-4}(x) * f^*[d_{\mathbf{A}}] = *d_{\mathbf{A}'},$$

where we have used $n = 4$. We then get

$$\begin{aligned}
\langle d_{\mathbf{A}'} \mid d_{\mathbf{A}'} \rangle_\Omega &= \int_\Omega *d_{\mathbf{A}'} \wedge d_{\mathbf{A}'} = \int_\Omega f^*[*d_{\mathbf{A}} \wedge d\mathbf{A}] \\
&= \int_{f(\Omega)} *d_{\mathbf{A}} \wedge d_{\mathbf{A}} = \langle d_{\mathbf{A}} \mid d_{\mathbf{A}} \rangle_{f(\Omega)}.
\end{aligned}$$

□

Solution to 11.5.3

a. We know that

$$\mathbf{e}_{\rho\sigma} = x_\rho \mathbf{e}_\sigma - x_\sigma \mathbf{e}_\rho \quad y_\alpha = x_\beta \breve{A}^{\beta}{}_{\alpha} + b_\alpha.$$

(NB! Observe that we are using covariant indices, not the usual contravariant indices.) Consequently, we get

$$\begin{aligned}
\mathbf{e}^2_{\rho\sigma} &= y_\rho \mathbf{e}^2_\sigma - y_\sigma \mathbf{e}^2_\rho = y_\rho \mathbf{e}^1_\beta \breve{A}^\beta_\sigma - y_\sigma \mathbf{e}^1_\alpha \breve{A}^\alpha{}_\rho \\
&= (x_\alpha \breve{A}^\alpha_\rho + b_\rho)\mathbf{e}^1_\beta \breve{A}^\beta_\sigma - (x_\beta \breve{A}^\beta_\sigma + b_\sigma)\mathbf{e}^1_\alpha \breve{A}^\alpha_\rho \\
&= [x_\alpha \mathbf{e}^1_\beta - x_\beta \mathbf{e}^1_\alpha]\breve{A}^\alpha_\rho \breve{A}^\beta_\sigma + [b_\rho \mathbf{e}^1_\beta \breve{A}^\beta_\sigma - b_\sigma \mathbf{e}^1_\alpha \breve{A}^\alpha_\rho] \\
&= \mathbf{e}^1_{\alpha\beta} \breve{A}^\alpha_\rho \breve{A}^\beta_\sigma + [b_\rho \mathbf{e}^2_\sigma - b_\sigma \mathbf{e}^2_\rho].
\end{aligned}$$

b. By definition, we have

$$J_{\rho\sigma} = \mathbf{T}(\mathbf{e}_{\rho\sigma}, \cdot) \quad \text{and} \quad P_\mu = \mathbf{T}(\mathbf{e}_\mu, \cdot),$$

and (b) follows immediately from (a).

Solution to 11.6.1

The Lagrangian density for a vector field depends only on $\partial_\mu A_\nu$ through the field strength $F_{\mu\nu} = \partial_\mu A_\nu - \partial_\nu A_\mu$. We therefore get

$$\begin{aligned}
\frac{\partial L}{\partial(\partial_\mu A_\alpha)} &= \frac{\partial L}{\partial F_{\rho\sigma}} \frac{\partial F_{\rho\sigma}}{\partial(\partial_\mu A_\alpha)} = \frac{\partial L}{\partial F_{\rho\sigma}} [\delta^\rho_\mu \delta^\sigma_\alpha - \delta^\rho_\alpha \delta^\sigma_\mu] \\
&= \frac{\partial L}{\partial F_{\mu\alpha}} - \frac{\partial L}{\partial F_{\alpha\mu}} = 2\frac{\partial L}{\partial F_{\mu\alpha}}.
\end{aligned}$$

□

Solution to 11.7.1

a. To prove (a), we observe that

$$\begin{aligned}
\partial_\mu \ln \det \bar{\bar{M}}(x)_{|x=x_0} &= \frac{\partial_\mu \det \bar{\bar{M}}(x)}{\det \bar{\bar{M}}(x_0)}\Big|_{x=x_0} = \partial_\mu \left[\frac{\det \bar{\bar{M}}(x)}{\det \bar{\bar{M}}(x_0)}\right]_{|x=x_0} \\
&= \partial_\mu \det[\bar{\bar{M}}^{-1}(x_0)\bar{\bar{M}}(x)].
\end{aligned}$$

But $\bar{\bar{M}}^{-1}(x_0)\bar{\bar{M}}(x)$ is a matrix function that reduces to $\bar{\bar{I}}$ at $x = x_0$. Repeating the argument from exercise 11.4.1, we now immediately get

$$\begin{aligned}
\partial_\mu \ln \det \bar{\bar{M}}(x)_{|x=x_0} &= \partial_\mu \operatorname{tr}[\bar{\bar{M}}^{-1}(x_0)\bar{\bar{M}}(x)] \\
&= \operatorname{tr}[\bar{\bar{M}}^{-1}(x_0)\partial_\mu \bar{\bar{M}}(x)].
\end{aligned}$$

b. From (6.47) we get

$$\Gamma^{\alpha}{}_{\mu\alpha} = \tfrac{1}{2} g^{\alpha\rho} \frac{\partial g_{\alpha\rho}}{\partial x^{\mu}}.$$

Introducing the matrices $\bar{\bar{G}} = (g_{\alpha\beta})$ and $\bar{\bar{G}}^{-1} = (g^{\alpha\beta})$, we can reformulate this as

$$\Gamma^{\alpha}{}_{\mu\alpha} = \tfrac{1}{2} \operatorname{tr} \bar{\bar{G}}^{-1} \partial_{\mu} \bar{\bar{G}},$$

which by (a) can be converted to

$$\Gamma^{\alpha}{}_{\mu\alpha} = \tfrac{1}{2} \partial_{\mu} \ln |\det \bar{\bar{G}}| = \tfrac{1}{2} \partial_{\mu} \ln |g| = \partial_{\mu} \ln \sqrt{|g|} = \frac{1}{\sqrt{|g|}} \partial_{\mu} \sqrt{|g|}.$$

c. Let the matrix $g_{\alpha\beta}$ depend in an arbitrary fashion on the parameter λ. Then we get from (a)

$$\tfrac{1}{2} g^{\alpha\beta} \frac{dg_{\alpha\beta}}{d\lambda} = \tfrac{1}{2} \frac{d}{d\lambda} \ln |g| = \frac{1}{\sqrt{|g|}} \frac{d}{d\lambda} \sqrt{|g|} = \frac{1}{\sqrt{|g|}} \frac{\partial \sqrt{|g|}}{2 g_{\alpha\beta}} \frac{dg_{\alpha\beta}}{d\lambda}.$$

From this we conclude that the symmetric matrices $\frac{1}{2} g^{\alpha\beta} \sqrt{|g|}$ and $\frac{\partial \sqrt{|g|}}{\partial g_{\alpha\beta}}$ must be identical.

d. Differentiating the identity $g^{\alpha\mu} g_{\mu\gamma} = \delta^{\alpha}{}_{\gamma}$, we immediately get

$$0 = \frac{\partial g^{\alpha\mu}}{\partial g_{\rho\sigma}} g_{\mu\gamma} + g^{\alpha\mu} \delta^{\mu}_{\rho} \delta^{\gamma}_{\sigma} = \frac{\partial g^{\alpha\mu}}{\partial g_{\rho\sigma}} g_{\mu\gamma} + g^{\alpha\rho} \delta^{\gamma}_{\sigma},$$

and multiplying by $g^{\beta\gamma}$, this is converted to

$$0 = \frac{\partial g^{\alpha\beta}}{\partial g_{\rho\sigma}} + g^{\alpha\rho} g^{\beta\sigma}.$$

□

Solution to 11.7.4

a.

$$\begin{aligned}
S &= \int_{\Omega} L \sqrt{-g} d^4 x \\
&= \int_{\Omega} \left[-\int m \sqrt{-g_{\alpha\beta}(x) \frac{dx^{\alpha}}{d\lambda} \frac{dx^{\beta}}{d\lambda}} \frac{1}{\sqrt{-g}} \delta^4(x - x(\lambda)) d\lambda \right] \sqrt{-g} d^4 x \\
&= \int \left[-\int_{\Omega} m \sqrt{-g_{\alpha\beta}(x) \frac{dx^{\alpha}}{d\lambda} \frac{dx^{\beta}}{d\lambda}} \delta^4(x - x(\lambda)) d^4 x \right] d\lambda \\
&= -\int m \sqrt{-g_{\alpha\beta}(x(\lambda)) \frac{dx^{\alpha}}{d\lambda} \frac{dx^{\beta}}{d\lambda}} d\lambda.
\end{aligned}$$

b.

$$\frac{\partial L}{\partial g_{\mu\nu}} = -m \int \left[\frac{\partial \sqrt{-g_{\alpha\beta} \frac{dx^\alpha}{d\lambda} \frac{dx^\beta}{d\lambda}}}{\partial g_{\mu\nu}} \frac{1}{\sqrt{-g}} \right.$$

$$\left. + \sqrt{-g_{\alpha\beta} \frac{dx^\alpha}{d\lambda} \frac{dx^\beta}{d\lambda}} \frac{\partial \left(\frac{1}{\sqrt{-g}} \right)}{\partial g_{\mu\nu}} \right] \delta^4(x - x(\lambda)) d\lambda$$

$$= m \int \frac{\frac{dx^\mu}{d\lambda} \frac{dx^\nu}{d\lambda}}{2\sqrt{-g_{\alpha\beta} \frac{dx^\alpha}{d\lambda} \frac{dx^\beta}{d\lambda}}} \frac{1}{\sqrt{-g}} \delta^4(x - x(\lambda)) d\lambda$$

$$+ \frac{m}{2} g^{\mu\nu} \int \sqrt{g_{\alpha\beta} \frac{dx^\alpha}{d\lambda} \frac{dx^\beta}{d\lambda}} \frac{1}{\sqrt{-g}} \delta^4(x - x(\lambda)) d\lambda$$

$$= \tfrac{1}{2} m \int \frac{dx^\mu}{d\tau} \frac{dx^\nu}{d\tau} \frac{1}{\sqrt{-g}} \delta^4(x - x(\tau)) d\tau - \tfrac{1}{2} g^{\mu\nu} L,$$

from which we conclude that

$$2 \frac{\partial L}{\partial g_{\mu\nu}} + g^{\mu\nu} L = m \int \frac{dx^\mu}{d\tau} \frac{dx^\nu}{d\tau} \frac{1}{\sqrt{-g}} \delta^4(x - x(\tau)) d\tau. \qquad \square$$

Solution to 11.9.1

a. For a conformal transformation, (11.112) reduces to

$$\partial_\mu \phi_\alpha(x) \curvearrowright \frac{\partial x^\rho}{\partial y^\mu} \left\{ \Omega^{-d}(x) \frac{\partial [f_* \phi_\alpha](f(x))}{\partial x^\rho} + [f_* \phi_\alpha](f(x)) \frac{\partial \Omega^{-d}}{\partial x^\rho} \right\}.$$

In the case of a vector field, we have

$$[f_* A_\alpha](f(x)) = \frac{\partial x^\sigma}{\partial y^\alpha} A_\sigma(x).$$

Therefore, (*) leads to the following displacement rule:

$$\partial_\mu A_\alpha \curvearrowright \frac{\partial x^\rho}{\partial y^\mu} \left\{ \Omega^{-d} \partial_\rho \left[\frac{\partial x^\sigma}{\partial y^\alpha} A_\sigma \right] + \frac{\partial x^\sigma}{\partial y^\alpha} A_\sigma \partial_\rho \Omega^{-d} \right\}.$$

Using that

$$\delta^\kappa_\sigma = \frac{\partial y^\lambda}{\partial x^\sigma} \frac{\partial x^\kappa}{\partial y^\lambda} \quad \text{and} \quad 0 = \frac{\partial y^\lambda}{\partial x^\sigma} \partial_\rho \frac{\partial x^\kappa}{\partial y^\lambda} + \partial_\rho \frac{\partial y^\lambda}{\partial x^\sigma} \frac{\partial x^\kappa}{\partial y^\lambda},$$

this can be rearranged as

$$\partial_\mu A_\alpha \curvearrowright \frac{\partial x^\rho}{\partial y^\mu} \frac{\partial x^\sigma}{\partial y^\alpha} \left\{ \Omega^{-d} \partial_\rho A_\sigma + A_\sigma \partial_\rho \Omega^{-d} - \Omega^{-d} A_\kappa \frac{\partial x^\kappa}{\partial y^\lambda} \frac{\partial^2 y^\lambda}{\partial x^\rho \partial x^\sigma} \right\}.$$

A skew-symmetrization in $(\mu\alpha)$, which on the right-hand side corresponds to a skew-symmetrization in $(\rho\sigma)$, then leads to the following displacement of the field strength:

$$F_{\mu\alpha} \curvearrowright \frac{\partial x^\rho}{\partial y^\mu}\frac{\partial x^\sigma}{\partial y^\alpha}\{\Omega^{-d}F_{\rho\sigma} + (A_\sigma\partial_\rho\Omega^{-d} - A_\rho\partial_\sigma\Omega^{-d})\}.$$

The displaced kinetic energy term therefore looks as follows:

$$\begin{aligned}-\frac{1}{4}\Omega^n(x)g^{\mu\nu}(f(x))g^{\alpha\beta}(f(x))&\frac{\partial x^\rho}{\partial y^\mu}\frac{\partial x^\sigma}{\partial y^\alpha}\frac{\partial x^\lambda}{\partial y^\nu}\frac{\partial x^\tau}{\partial y^\beta}\\ &\times\{\Omega^{-d}F_{\rho\sigma} + A_\sigma\partial_\rho\Omega^{-d} - A_\rho\partial_\sigma\Omega^{-d}\}\\ &\times\{\Omega^{-d}F_{\lambda\tau} + A_\tau\partial_\lambda\Omega^{-d} - A_\lambda\partial_\tau\Omega^{-d}\}.\end{aligned}$$

Since f is a conformal map, we know that

$$g^{\mu\nu}(f(x))g^{\alpha\beta}(f(x))\frac{\partial x^\rho}{\partial y^\mu}\frac{\partial x^\sigma}{\partial y^\alpha}\frac{\partial x^\lambda}{\partial y^\nu}\frac{\partial x^\tau}{\partial y^\beta} = g^{\rho\lambda}(x)g^{\sigma\tau}(x).$$

Furthermore, $d = (n-4)/2$, and the displaced kinetic term now reduces to (11.117).

b. According to exercise 10.3.1, we have

$$\Omega_\lambda(x) = \frac{1}{(1 + 2\lambda\langle b \mid x\rangle + \lambda^2\langle b \mid b\rangle\langle x \mid x\rangle)}.$$

Thus

$$\partial_\nu \ln \Omega_\lambda = \frac{2\lambda b_\nu + 2\lambda^2\langle b \mid b\rangle x_\nu}{(1 + 2\lambda\langle b \mid x\rangle + \lambda^2\langle b \mid b\rangle\langle x \mid x\rangle)},$$

from which (11.118) follows immediately.

c. If the kinetic term is conformally invariant, the displaced kinetic term must differ by at most a divergence term. In terms of inertial coordinates, we must therefore have

$$\begin{aligned}&-\frac{n-4}{8}\{4F^{\mu\nu}A_\mu\partial_\nu\ln\Omega + (n-4)[A_\mu A^\mu\eta^{\alpha\beta} - A^\alpha A^\beta]\partial_\alpha\ln\Omega\partial_\beta\ln\Omega\}\\ &= \partial_\mu\chi_f^\mu[A_\alpha]. \qquad **\end{aligned}$$

If we let f_λ represent the group of special conformal transformations, we get from (b) that

$$\partial_\mu\left\{\frac{d\chi_\lambda^\mu[A_\alpha]}{d\lambda_{|\lambda=0}}\right\} = (n-4)F^{\mu\nu}A_\mu b_\nu.$$

Thus we see that $F^{\mu\nu}A_\mu b_\nu$ must be a divergence (for an arbitrary choice of b^ν), but that is impossible according to the general argument given below (11.97).

Index